P9-CLQ-162

Harnessing AutoCAD® Land Development Desktop

Harnessing AutoCAD® Land Development Desktop

PHILLIP J. ZIMMERMAN

autodesk® press

Australia • Canada • Mexico • Singapore • Spain • United Kingdom • United States

autodesk® press

Harnessing AutoCAD® Land Development Desktop

Phillip J. Zimmerman

Autodesk Press Staff

Business Unit Director:
Alar Elken

Executive Editor:
Sandy Clark

Acquisitions Editor:
James DeVoe

Developmental Editor:
John Fisher

Editorial Assistant:
Jasmine Hartman

Executive Marketing Manager:
Maura Theriault

Marketing Coordinator:
Karen Smith

Executive Production Manager:
Mary Ellen Black

Production Manager:
Larry Main

Production Editor:
Stacy Masucci

Art and Design Coordinator:
Mary Beth Vought

Cover Image:
Matt McElligott

AutoCAD images reprinted with permission from and under the copyright of AutoDesk, Inc.

COPYRIGHT © 2001 Thomson Learning™.

Printed in Canada
2 3 4 5 XXX 06 05 04 02 01

For more information, contact Autodesk Press, 3 Columbia Circle, PO Box 15015, Albany, New York, 12212-5015.

Or find us on the World Wide Web at www.autodeskpress.com

All rights reserved. No part of this work covered by the copyright hereon may be reproduced or used in any form or by any means—graphic, electronic, or mechanical, including photocopying, recording, taping, Web distribution or information storage and retrieval systems—without written permission of the publisher.

For permission to use material from this text contact us by
Tel: 1-800-730-2214
Fax: 1-800-730-2215
www.thomsonrights.com

Library of Congress Cataloging-in-Publication Data

Zimmerman, Phillip J.
Harnessing AutoCAD Land Developer Desktop / Phillip Zimmerman.
p. cm.
ISBN 0-7668-2806-9
1. Civil engineering–Computer programs. 2. Surveying–Computer programs. 3. AutoCAD.
I. Title.
TA345 .Z54 2000
624'.0285'5369—dc21

Notice To The Reader

The publisher does not warrant or guarantee any of the products described herein or perform any independent analysis in connection with any of the product information contained herein. The publisher does not assume and expressly disclaim any obligation to obtain and include information other than that provided to it by the manufacturer.

The reader is expressly warned to consider and adopt all safety precautions that might be indicated by the activities described herein and to avoid all potential hazards. By following the instructions contained herein, the reader willingly assumes all risks in connection with such instructions.

The publisher makes no representations or warranties of any kind, including but not limited to, the warranties of fitness for particular purpose or merchantibility, nor are any such representations implied with respect to the material set forth herein, and the publisher and author take no responsibility with respect to such material. The publisher shall not be liable for any special, consequential, or exemplary damages resulting, in whole or in part, from the readers' use of, or reliance upon, this material.

Trademarks

Autodesk, the Autodesk logo, and AutoCAD are registered trademarks of Autodesk, Inc., in the USA and other countries. Thomson Learning is a trademark used under license. Online Companion is a trademark and Autodesk Press is an imprint of Thomson Learning. Thomson Learning uses "Autodesk Press" with permission from Autodesk, Inc., for certain purposes. All other trademarks, and/or product names are used solely for identification and belong to their respective holders.

CONTENTS

INTRODUCTION

INTRODUCTION

The AutoCAD Land Development Desktop program is an extremely powerful designing and drafting tool. AutoCAD Land Development Desktop combines elements from the Softdesk Civil/ Survey products into an industry specific AutoCAD. For example, functions that used to be found in six different menus and two different modules of Softdesk are now in a single menu. A second benefit of combining Softdesk Civil/Survey tools into a single program is the focusing of Land Development Desktop to the daily drafting and design tasks of a civil/survey firm. These tasks include creating points, lines and arcs, parcels, alignments, labels, and surfaces. The Civil Design Add-in broadens Land Development Desktop with tools for site grading, water runoff analysis and design, roadway design, and ponds. To use these tools effectively, you need a fundamental knowledge of the Civil Design process. The Survey Add-in handles the communication with data collectors, the reduction of field data, and the drafting of line work from the field. To use the tools of Survey, you need a basic knowledge of Surveying methods and terms.

Harnessing AutoCAD Land Development and Desktop, R2 provides a basic understanding of the tools found in Land Development Desktop and its two add-ins. The book, however, is *not* a Surveying or Engineering textbook, nor is it a tips and tricks book. If you need additional help with LDD, you can find newsgroups about LDD on the Internet or go to the support area of Land Development Desktop's WEB page at www.autodesk.com. This book provides explanations and exercises that help develop a basic understanding of the surveying and civil design tools found within the routines of Land Developer Desktop. Many of the examples came from professionals who use the software every day, while other examples were specifically designed to contain difficulties that demonstrate the capabilities of the Desktop.

There are two types of tools within the desktop: tools for the development of fundamental data and tools for civil engineering and roadway and site design. Fundamental data is point data (X, Y and Z), line, curves and spirals, and Digital Terrain Models (DTMs). These tools are found in the Desktop and the Survey Add-in. It is upon this foundation data that the second tool set, found in Civil Design, develops roadway, pond, and hydrological plans and results. An additional result of the design process can be new fundamental information, i.e., point data, lines and curves, and surfaces, which are used by Land Development Desktop or Survey to communicate the design to the field.

The tools in the Desktop and its Add-ins have four types of functions: creating, evaluating, editing, and annotating. Whether you are working with points, lines or surfaces, the tools used on these objects fall into one of the four above mentioned types. Exercises in this book emphasize the tasks each of the different tool sets provide to the user.

The first six sections of the book cover the fundamental data groups and the Land Development Desktop. These sections cover points, lines and curves, parcels, labeling, surfaces, and volumes. Sections 7-14 focus on the Civil Design Add-in for the Desktop. The tools in these sections cover hydrological analysis, grading, storm and sanitary piping design, and the roadway design process. The roadway design process uses lines and arcs, profiles, cross sections, templates, and volume calculations. Civil Design provides tools at each step along the roadway design process that create, evaluate, edit, and annotate the elements of the design.

Sections 15-17 explore the Survey Add-in to Land Development Desktop. The tools discussed in these sections provide the capability to download and upload to data collectors, reduce traverses, create figures, and provide tools that analyze field observations.

Section 18 covers the extensions Autodesk releases that enhance LDD, Civil Design, and Survey and the new capabilities of Land Development Desktop 2i.

HOW TO USE THE CD

The accompanying CD contains all of the files you need to complete the exercises in this book. The data files and drawings for each section are in corresponding section folders on the CD. The drawings are supplied in two forms; a drawing file and a template file. You have the choice of using either to start an exercise. The Data used in each exercise is also in the section folder. You can copy the files onto your computer or you can browse to their location on the CD.

An LDDTR2 prototype folder and various roadway templates are also contained on the CD. To use the LDDTR2 prototype you need to copy the prototype folder and it's subfolders to the *Prototypes* folder of your Land Development Desktop installation. The prototype folder is a subfolder of the *data* folder of LDD.

The roadway templates need to be copied to the *data\tplates\adtpl_i* subfolder of your installation of Land Development Desktop.

ABOUT THE AUTHOR

This book was developed from the training materials I used while working for an Autodesk reseller in the Chicago area. The need for materials and a thorough explanation of the workings of the program became apparent as I prepared for my classes. My students and friends never hesitated in providing material for class, especially if the material proved difficult or "didn't work." Even when the material proved challenging, there were ways to solve the problems facing us when working with the data. I appreciate their efforts and patience.

Prior to working at the reseller, I worked at a survey firm and a pavement management company. These experiences gave me my foundation in the use of AutoCAD and insight into the methods and needs of surveyors and civil engineers.

My educational background includes two degrees in Geography, a BA from the University of New Mexico and a MA from Western Michigan University.

Currently, I am the Land Technical Marketing Manager for the GIS Division of Autodesk. I develop the technical documents and data sets for the Land Solutions Group. I am one of many who participate in directing the development of new and existing Land Products.

SUPPLEMENTS

There is an Instructors Guide published for this book. The guide includes questions and answers for reviews and tests. The guide also includes some tips and comments on procedures and potential problems (Instructor's Manual, ISBN 0-7668-2807-7).

Periodically, Autodesk releases extensions for Land Development Desktop. These extensions are focused tools from user or industry requests and add to the existing tool sets of LDD, Civil Design, and Survey. The extension strategy has already introduced the FME data translators, cut/fill balancing, and slope annotation. Information about these extensions will be posted in the Online Companion for this text. The Online Companion can be found at www.autodeskpress.com.

ACKNOWLEDGMENTS

The lot exercise in Sections 2, 3, and 4 is from a Plat of Survey developed by Intech Consultants, Inc. of Downers Grove, Illinois. Tom Fahrenbach, P.L.S., Scott, and Joe also of Intech, helped with the lot exercises, and with the hydrology exercises on Tr-55 and HEC2 cross sections. The Hydrological exercise found in Section 12 is from a design created by the Forest Preserve of DuPage County, Illinois. Additional help in developing the exercises and book came from Tammi Hascek, Darcy Berg, Ross Hill and other planners and designers of the Forest Preserve of DuPage County, Illinois

Other data for the Hydrological example came from the TR-55 publication of the Soil Conservation Service of the USDA, Technical Publication 55. The Civil Design Pipes exercise of Section 14 is from a subdivision design by Gary Kurek for the Abonmarche Group in Benton Harbor, Michigan. The Plat of Survey example for the exercise in Section 15 is from a lot survey done by John Traverso, P.L.S., of Tri-County Engineering, Inc. Gail and Gil Evans at Balsamo/Olson Group Inc supplied the manual survey in Section 15. The network loop traverse in Section 16 is from data compiled by Bill Laster.

Special thanks go to Gerry Tener for performing a technical edit on the manuscript and verifying its content.

And thanks to the staff at Delmar, Thomson Learning: Sandy Clark, John Fisher, Mary Beth Vought, Stacy Masucci, Jasmine Hartman, and Larry Main. I want to give special thanks to my mom, sisters, and friends for their support while working on this book.

The Land Development Desktop Project and Drafting Environment

After you complete this section, you will be able to:

- Initialize Land Development Desktop
- Start a new drawing
- Create a project
- Set up a drawing
- Evaluate the prototype settings
- Evaluate the drawing settings
- Understand the subdirectory structure of external data
- Create a new prototype
- Create a series of drawing setup profiles

SECTION OVERVIEW

Land Development Desktop (LDD) is the result of the merger of AutoCAD and Softdesk. The modules of Softdesk are now combined and reduced into the Desktop and its two companion programs, Civil Design and Survey.

The structure of LDD, Civil Design, and Survey consists of the distillation of the routines from eight modules each containing approximately five menus of the Softdesk product line. LDD emphasizes a common set of tools for each person working on a project. Because of this, LDD emphasizes points, labeling, parcels, centerlines, and surfaces. The Civil Design product emphasizes the design functions of a project. The tool set for Civil Design includes designing ponds, evaluating water runoff, and designing and plotting roadway designs and site grading. The Survey product emphasizes the transfer of data between the field and the office. Additional tools in Survey evaluate the data recorded in the field, draw linework, and evaluate the coordinate geometry present in the project. Not every project will use every routine found in LDD, Civil Design, and Survey.

The focus of the first part of this book is LDD. Civil Design and Survey are covered in later sections, but they will be tied to routines and processes in LDD.

Land Development Desktop (LDD) is AutoCAD for people who work daily with points, boundary development, roadways (lineations), and surfaces. These four elements are the foundations upon which a project is built. The menus within LDD reflect these elements and the processes to create them and provide you with specific tools needed to complete the job. Each menu within the Desktop—Projects, Points, Lines/Curves, Alignments, Parcels, Labels, Terrain, and Inquiry—provides you with tools to create, analyze, edit, and annotate project information.

UNIT 1

Before you use the tools specific to Land Development Desktop (LDD), the Desktop requires you to complete three tasks: naming the drawing, assigning a project, and completing a drawing setup. After you complete these tasks, your drawing is ready for use within the Desktop. Almost all of the settings you encounter while using LDD are a result of values found in the selected project prototype.

These three tasks are required when you start a new drawing or open an existing one created without an LDD project assignment. The initiation of these tasks occurs automatically when you enter the drafting environment. You may decline the assignment of a project. If you do decline the project assignment, selecting any command specific to LDD will start the requests for a project assignment and finishing the setup. If you never use any LDD commands during the drafting session, the drafting environment is an AutoCAD Map only session.

The process of initializing Desktop is the topic of the first unit of this section.

UNIT 2

A Desktop drawing does not contain all of the settings and computed values of a project. To require a drawing to carry potentially so much information would create a large drawing file. The other problem with the idea of each drawing containing the project data is: how do other drawings share this information? The Desktop philosophy of creating and maintaining external data to the drawing allows you to create smaller-sized drawings. These drawings now contain only the necessary information from the project's external data. Desktop considers the drawing a user-defined view of the point database, surfaces, and design calculations. As to sharing data between drawings of the same project, any drawing assigned to a project "sees" all of the project's external data.

Land Development Desktop project data is a series of external files linked to the drawing by the project name. The name of the project is the lead entry in this data structure. Under the *Project* folder, Desktop creates additional folders to store more specific data. Each new folder contains default and computed values. Some folders contain data files that are primary data for the project. LDD provides a default folder

(*DWG*) in the project structure to hold the drawing(s) of the project. By backing up the entire directory structure, you archive all of the project data.

The project directory structure and its files are the topic of the second unit of this section.

UNIT 3

The projects of Land Development Desktop depend on the prototype to establish initial data files, annotation settings, and drawing settings. It is important to set the values in the prototype even if you use the default prototype. The settings of the prototype affect each drawing created for the project and its default values. The settings establish an "office standard" for the initial values of the drawing. If you never adjust the prototype's values, each new project will copy the prototype's settings, and you will have to adjust each new drawing's settings to be correct by your standards.

The Projects menu contains routines that create and modify the settings within a prototype. These routines are important to the consistent use of the power within the tools of Land Development Desktop.

The Drawing Setup dialog box or the Setup wizard allows you to create a drawing setup profile. A drawing setup profile contains additional information not included in the project data. The drawing setup profile defines the scale, text styles, orientation, and zone used by the drawing. These values may vary from drawing to drawing within a project.

A Drawing Setup Profile file creates drawings with consistent settings. A single selection in either the Setup wizard or the Drawing Settings dialog box sets the entire range of values under their control. Drawing Setup Profile Settings are scale dependent, so you must establish profiles for the most commonly used scales in your office.

The Prototype Manager, the routines that set the defaults for the prototype, and the Drawing Setup Profiles are the subject of the third unit of this section.

UNIT 1: DRAWING SETUP

NEW DRAWING AND PROJECT INITIALIZATION OVERVIEW

For Desktop to initialize correctly, you need to complete three tasks: name the drawing, assign it to a project, and define the drawing setup profile.

The first two tasks, naming the drawing and assigning a project, are a part of the New Drawing dialog box. You identify the name, project, and template of the drawing in the dialog box.

You can assign a project either by selecting an existing project or defining a new one. If the project is new, Land Development Desktop creates a project as a clone of a preexisting prototype that contains default values. There are files in the prototype that set values, prefixes, toggles, data files, and names of layers for every new drawing in a project. If you want different settings, you need to create a new prototype and edit its settings to your standards. After you edit the prototype, each new project using this revised prototype presents drawings with the desired settings.

The Projects menu contains routines that change the settings for the current drawing, but the routines do not affect the settings for a new drawing. If you want a new drawing to have the same settings as the previous drawing, you need to update the prototype settings based upon the changes in the current drawing. After you update the prototype, you reload the prototype settings into the project. The updated settings will affect only new drawings created after reloading the updated prototype settings.

If the project is new, you must set up the point database.

The last task is defining the drawing's setup values. This involves setting the base units, scale, precision, text types, style, and borders. This task is a part of the Drawing Setup Wizard.

STEP 1—STARTING A NEW DRAWING

Land Development Desktop requires you to name a drawing before you can use any of its routines. If you are starting a new drawing and project, select the New button in the LDD Start Up dialog box (see Figure 1–1) to display the New Drawing: Project Based dialog box (see Figure 1–2). If you are already in an LDD session, select the New Drawing icon to display the same New Drawing dialog box. The location of the drawing is set in the project definition. The default location of the drawing file is a folder (*DWG*) within the project folder. You can place the drawing in a different folder from the project folder by toggling on Fixed Path and entering the new path to the folder.

Land Development Desktop Template Drawing File

LDD provides a prototype drawing file named *aec_I.dwg* (*aec_M.dwg* for metric drawings). You can use your own drawing template files. They must be in the drawing template folder of LDD.

STEP 2—PROJECT ASSIGNMENT

The second task is assigning a project.

Creating a New Project

If the drawing does not belong to any existing project, you need to create a new one. The middle portion of the New Drawing dialog box contains buttons to set the location of the project and create the project (see Figure 1–2). The Browse button allows you to identify a folder anywhere on the network as the project location.

After setting the project location, you need to create the project folder by selecting the Create Project button.

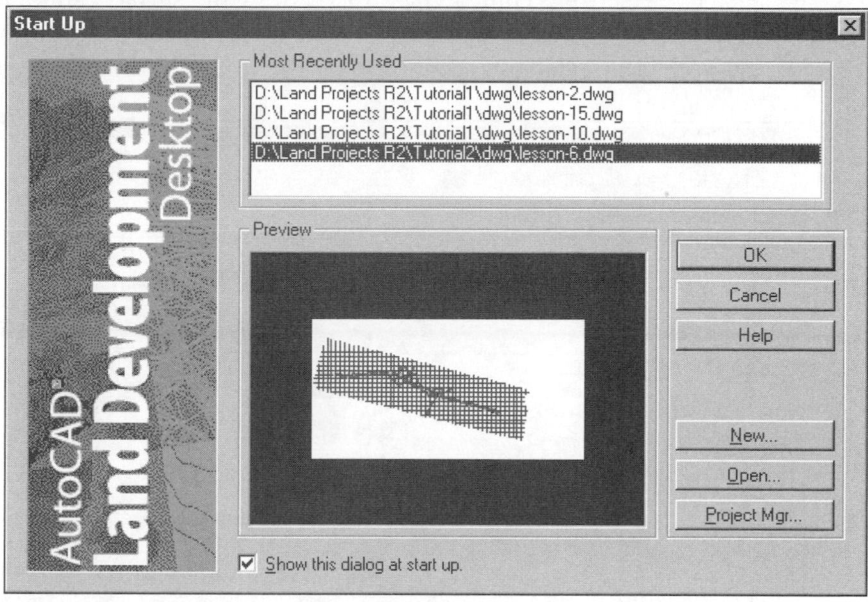

Figure 1–1

Figure 1–2

When creating a new project, select a prototype that contains the settings for your new project (see Figure 1–3). The prototype contains numerous default settings for LDD and data sets. These settings control the project, drawing, surface, annotation, and many other default values within Desktop. To make each project and drawing within a project show the same defaults, you must set the values in the prototype.

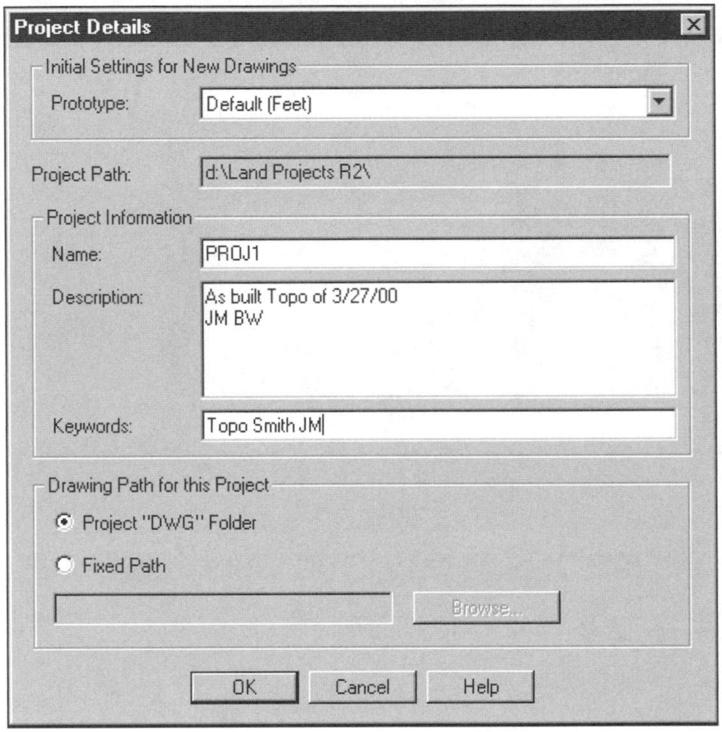

Figure 1–3

Each drawing's settings are read from its associated DFM file. If a project has a drawing named *ascot.dwg*, it will have an *ascot.dfm* file. The DFM file contains the default settings you see in the Drawing Settings dialog box. The settings, however, are a combination of values found in other project data files: *project.dfm*, *cgx.dfm*, and *default.dfm*. If you have installed the Civil Design and Survey plug-in, there will be a *cd.dfm* and a *sv.dfm* file. Each of these files exists in and is copied from the prototype. After you define a project and name the drawing, LDD creates a DFM from the settings found in the three files and sets them as the drawing defaults.

Because of the numerous settings and data values found within a prototype, it is a good starting point for office standards. The Project and Prototype Managers, Prototype Settings, Data Files, and Drawing Settings routines of the Projects menu allow you to create new prototypes and edit existing ones.

Land Development Desktop installs two prototypes, one for imperial and one for metric projects. You can create new prototypes from existing projects or define completely new prototypes by copying, editing, and setting values different from the original prototype.

The middle portion of the Create Project dialog box allows you to enter a description of the job and at the bottom enter a set of key words. The description will appear in some reports you print about the project, and the key words allow you to filter the display of projects to select from when searching for a project to assign to a drawing.

When you exit the Create Project dialog box, LDD creates a new project folder and copies the files (including subfolders) found within the prototype to the new project folder.

New Project Initial Point Database Settings

When you assign a drawing to a new project, LDD requires user settings for the point database (see Figure 1–4). The point database is an access database. The first setting determines the length of descriptions; the default is 32 characters. The second item is whether or not the project is using point names. Point names are alphanumeric point numbers, for example A1, Stn1, and so on. If you toggle on Use Point Names, the selection sets a default value so that the program will use the name column in the point database. By default a point name is 16 characters in length.

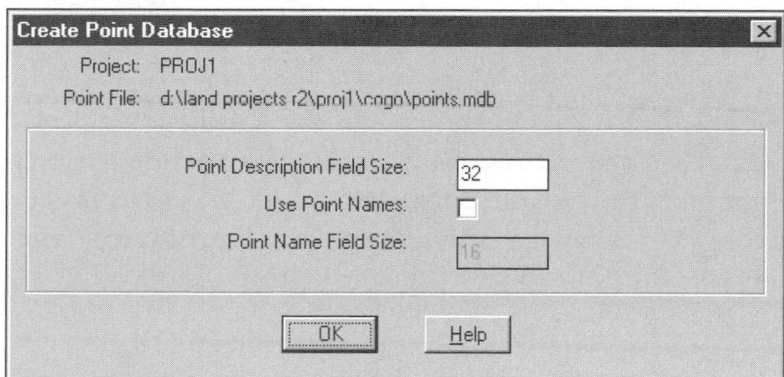

Figure 1–4

STEP 3—THE DRAWING SETUP

After you set the project name and the point database settings, the next task is setting up the drawing (see Figure 1–5). The Drawing Setup Wizard presents essential drawing values that you need to set. These values include base units (imperial or metric), measurement precision, coordinate zones, text styles, borders, and other drawing settings. To change the current values in the wizard, select the Next or Previous button at the bottom to change the information panels and edit the values in the associated panel. Each panel contains controls that modify the current drawing values.

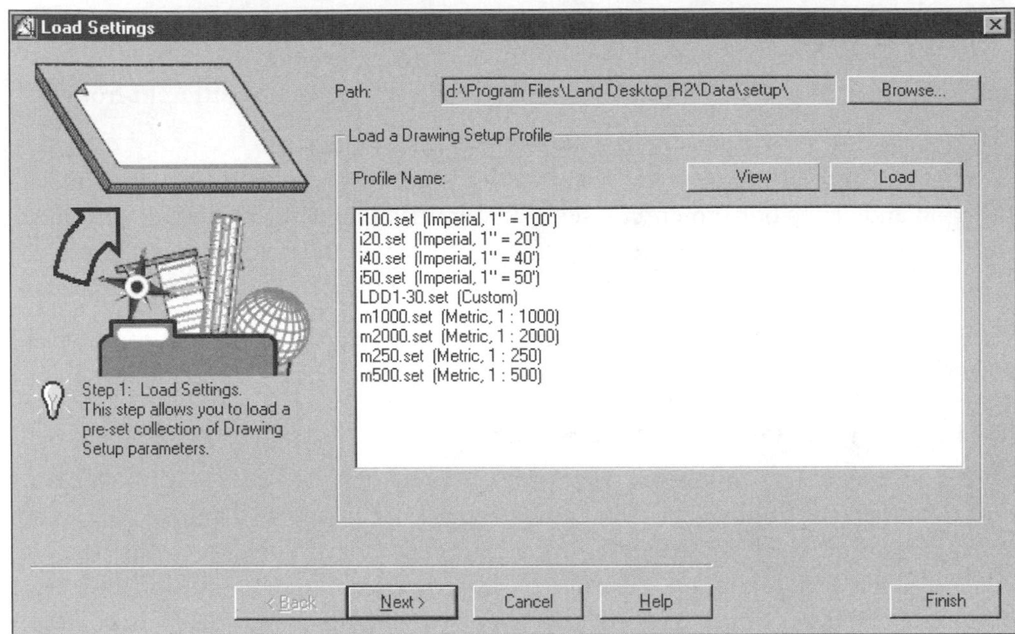

Figure 1–5

Load Settings

The first panel of the wizard is the Load Settings panel (see Figure 1–5). By selecting and loading an existing Drawing Setup Profile from the panel, you set all of the values in the wizard. If you are unfamiliar with the default settings of a setup, you should go through all of the panels and check the values before selecting the Finish button at the bottom of the wizard. Once you have set the values in the wizard, you can save the settings to a Drawing Setup Profile file. The next time you start a new drawing, all you need to do is to select this new profile and load it, and all of the values are set.

Units and Precision

The Units panel controls the measurement values in the drawing (see Figure 1–6). The panel sets the active measurement system (imperial or metric), angle units, angle display style, and precision, and illustrates how the units will display in the program.

The precision settings control the display of information only. Land Development Desktop holds all data to 15 decimal places. However, to make the precision more manageable, LDD controls the number of displayed decimal places for linear, angular, coordinate, and elevation measurements. Generally, the angular precision is 4 decimal places, because that setting displays a whole second in a reported angle.

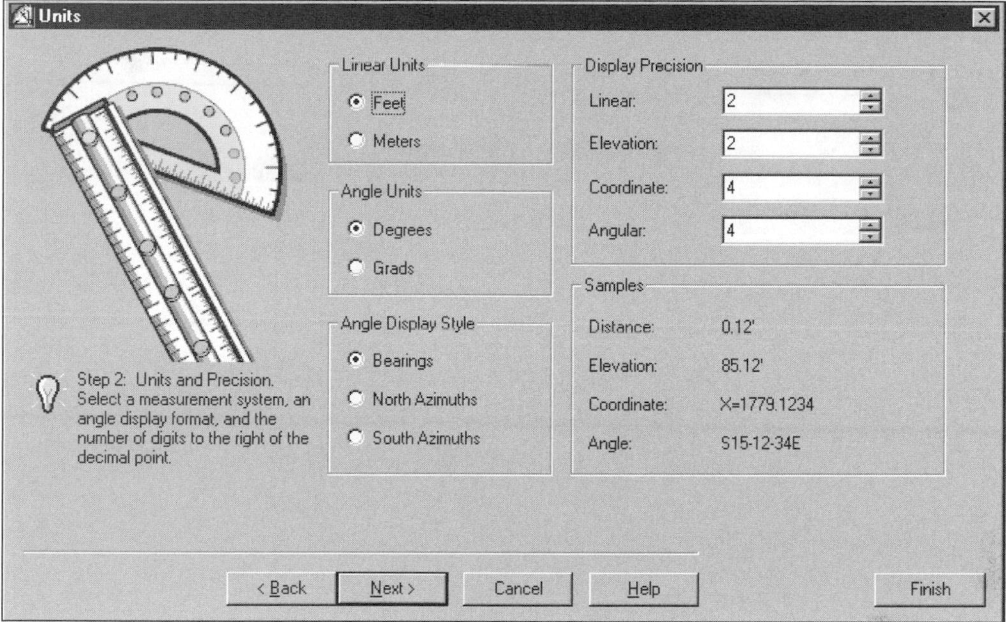

Figure 1–6

If you enter values with greater precision than the settings, the values remain correct in the external files. LDD displays their rounded values in reports according to the settings. When the precision settings change, the correct values will appear in the reports.

Scale

The Scale panel displays drawing scales and sheet sizes (see Figure 1–7). The horizontal scale applies to the planimetric drafting of lines and arcs in the drawing. The vertical scale applies to the 2D drafting of the vertical elements of design; for example, profiles and cross sections. The right side of the panel lists possible sheet sizes.

The panel displays traditional drafting scales and sheet sizes. To set nontraditional scales or sheet sizes, select Custom in each setting and edit the horizontal and vertical scales and the sheet size.

Coordinate Zone

The Zone panel sets the current coordinate system for the drawing (see Figure 1–8). By default the setting is No Datum, No Projection. This setting does not trigger any transformation of coordinate data when exporting or importing points.

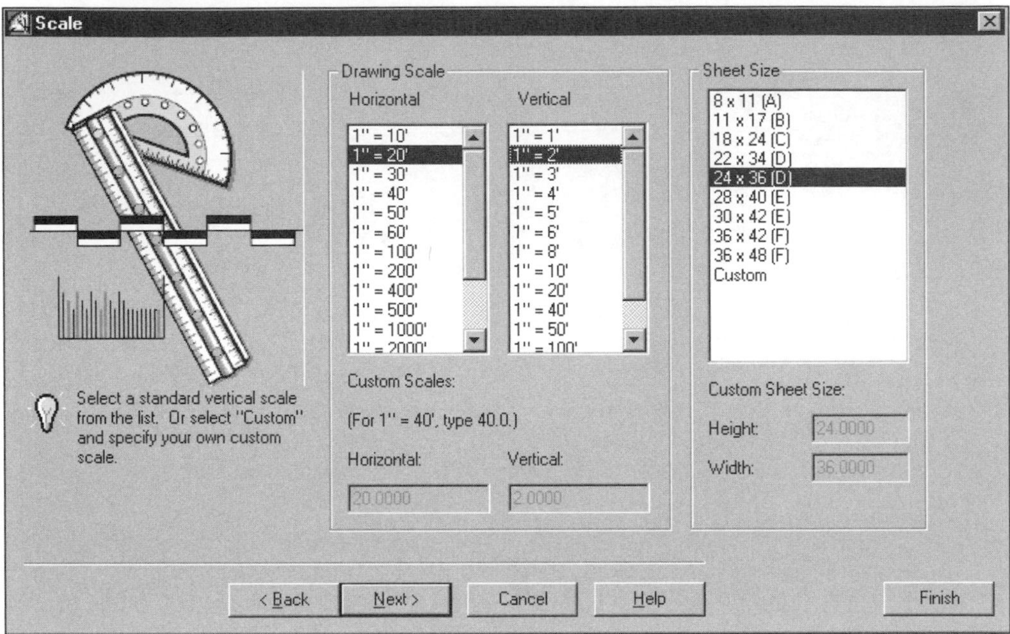

Figure 1–7

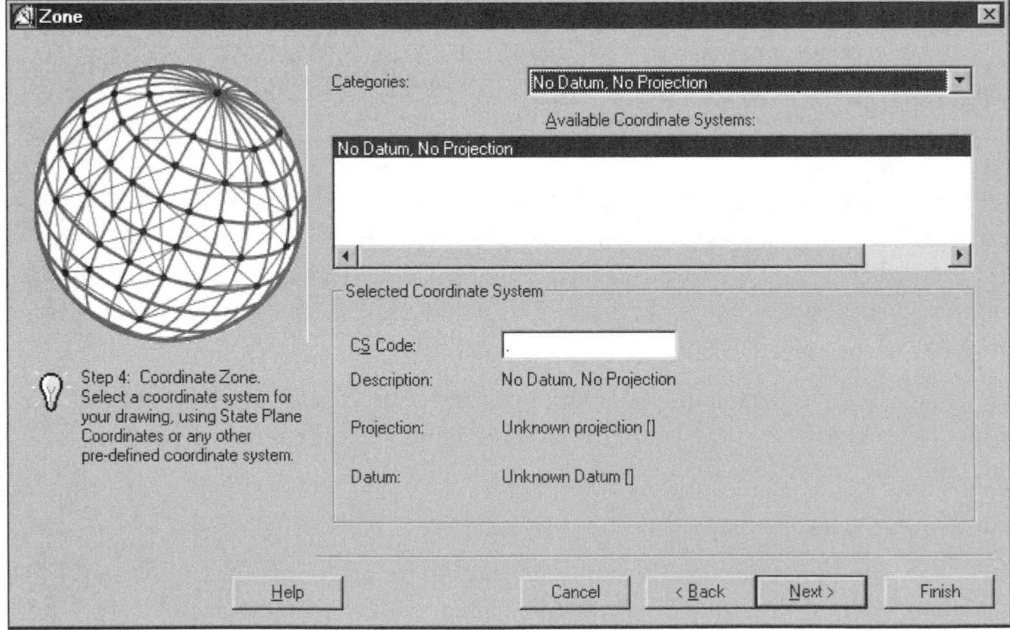

Figure 1–8

If the project is set in a specific coordinate system, the zone must be set correctly. Some projects require or are too large to be drafted on a flat surface coordinate system. The setting of a zone in the drawing does allow you to import points from different zones, for example, NAD83 US meter coordinates into a NAD27 US foot drawing.

Orientation

The Orientation panel displays entries allowing for the redefinition of the Northing/ Easting coordinate system (see Figure 1–9). You may need to reorientate a drawing for one of two situations. The first situation is when you are working in a drawing using state plane coordinates. When you work with coordinates in the millions, the precision AutoCAD had for the right side of the decimal point is shifted to the left to show the million values. Although this does not seem important, it is a problem when trying to use the Pedit Join routine of AutoCAD. Because the precision shifts to the whole number, AutoCAD fills the lower decimal places with less than accurate numbers. The Pedit routine requires all 15 decimal places to the right of the decimal to be the same before joining two segments. With the last decimal places being potentially different, the segments will not join even if you use the FILLET command to close the "gap." The impact of this shortcoming becomes apparent when you are defining a roadway centerline. The routine will define a centerline with some, but not all, of the segments. The Define Centerline by Objects routine uses Pedit to construct a temporary polyline to define the roadway. When the routine cannot add another segment because the coordinates do not match, it assumes the end of the centerline.

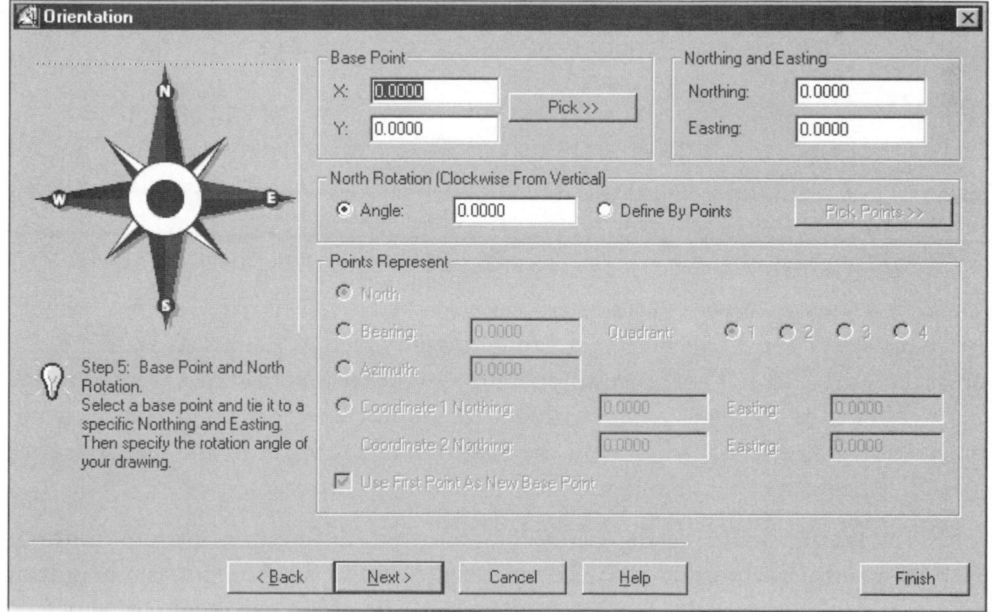

Figure 1–9

The remedy for this problem is to define a set of AutoCAD coordinates to have new LDD Northing/Easting values. This is the function of the Base Point section of the panel. An example of this is setting a base point of 0,0 (AutoCAD coordinates) and having it represent the Northing of 1,000,000 and an Easting of 1,000,000. Land Development Desktop will understand that the Northing/Easting values are in the millions, but AutoCAD will read the object coordinates as near 0 (zero). With the object coordinates near 0 (zero), the definition of a centerline will be successfully completed. The downside to this is that the Northings/Eastings do not match the coordinate system of AutoCAD.

The north rotation and base point routines remedy the problem when a drawing represents north as being any direction but up on the screen. These routines allow you to define a direction towards north and a new coordinate point. To use these routines in a drawing, draw a line in model space that represents the rotation of north and a set of known coordinates. The endpoint of the line representing the north rotation is a convenient location to indicate the known coordinates of a Northing/Easting system. After the routines have been run, LDD interprets the rotation angle not only for North but also for the Northing/Easting coordinate system.

You can define North by one of several methods. The first method is a rotation angle. The second is selecting two points in model space. After selecting the two points, you must identify the points as the North angle, a bearing with its quadrant, an azimuth, or two sets of coordinates. You can toggle on using the first selected point as the new base point.

LDD does not support User Coordinate System (UCS) definitions.

Text Style

The Text Style panel sets the types of text styles in the drawing (see Figure 1–10). The preferred text type is Leroy. After you select the Leroy text type, the adjacent window displays the styles in the text type. When you select the Load button, the text styles become a part of the drawing. The last window lets you select the current text style.

When you select the Load button, LDD reads a data file, *Leroy.stp*, to create the appropriate text styles. This process generates fixed text height styles, which means that the height of the style is proportional to the horizontal scale of the drawing. LDD reads the horizontal scale set in the Scale tab of the wizard. The text style files are ASCII files you can edit and add to, creating new text styles.

The text styles are similar to the Leroy template system found in manual drafting. Each Leroy template represents a different height of text, so changing the height of lettering means changing the Leroy template. In LDD this means that changing the text style changes the size of the lettering. The text height is in the style definition.

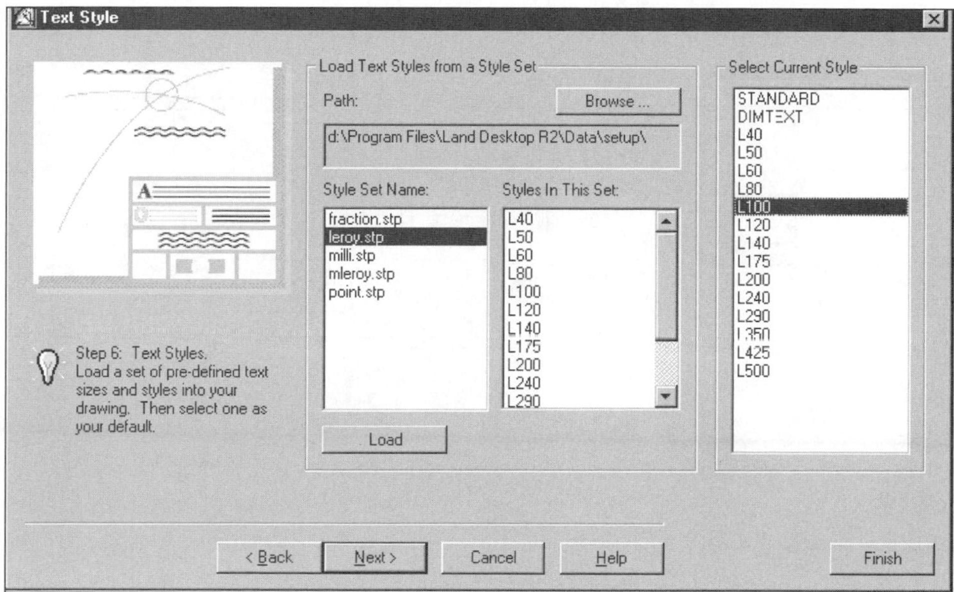

Figure 1–10

The formula for the height of a style is the height of the text on paper times the scale of the drawing. For example, if you want text to be 0.2" on the paper when plotting at a 1" = 50' scale, the text in the drawing must be 10 feet tall (0.2 × 50 = 10).

LDD has two additional text types: fraction and point. For metric drawings, LDD supplies Mleroy and Milleroy.

The *stp* file is a text file you can edit to add and/or modify the text style definitions. The text styles defined in a *stp* file can help create and maintain office standards through any series of drawings in a project.

A drawing may contain user-defined text styles, and using the Leroy style is not mandatory.

Border

The Border panel allows you to create or select a predefined border (see Figure 1–11). By toggling on any of the three options at the top of the panel, you create a border. The first border is a simple outline. The size of the frame is determined by the sheet size (Scale panel) and the values entered in the lower portion of the panel minus the margins. The border is a polyline inserted at coordinates 0,0.

The second method is to place an unscaled block (border) in the drawing. When you select this method, the lower portion of the panel reveals existing border files. When you exit the Setup wizard, LDD places an unscaled border into the drawing. If you drew the border 24 by 36 units, it will measure the same in the drawing. The border can then be moved to a layout for use.

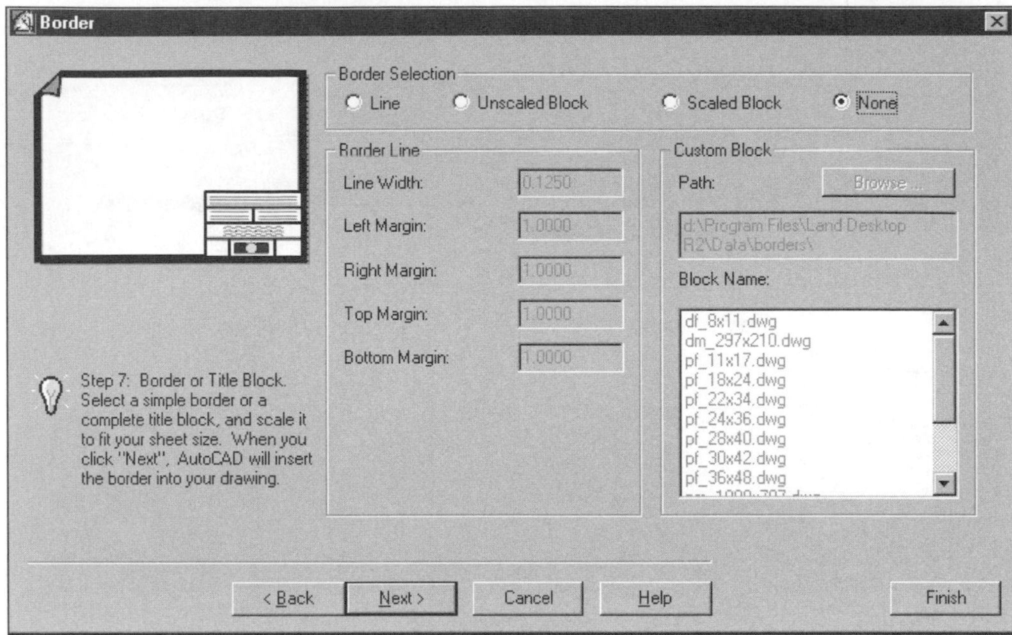

Figure 1–11

The last method is the same as the second; however, the frame is scaled (Scaled Block) by the drawing scale. The size of the frame is the scale times the size of the frame. For example, if you choose a 1"=20' scale and a D-size frame (24x36), the frame will measure 480 by 720 feet when inserted into the drawing.

LDD places several files into the border folder during the install process. You can place your files into the folder and select them from the list.

LDD inserts each border at the AutoCAD coordinates of 0,0.

Save Settings

The Save Settings panel allows you to save the current settings as a Drawing Setup Profile (see Figure 1–12). By the entering a name into the profile area at the center of the panel and selecting the Save button, you create a setup profile. If the next new drawing has the same settings as the current drawing, you can set up the drawing with a single selection. Each named setup you create reflects the standards in the office for precision, text styles, and borders.

After you select the Finish button in the Setup wizard, LDD displays a review dialog box. The box contains the current drawing settings. By selecting OK, you dismiss the dialog box.

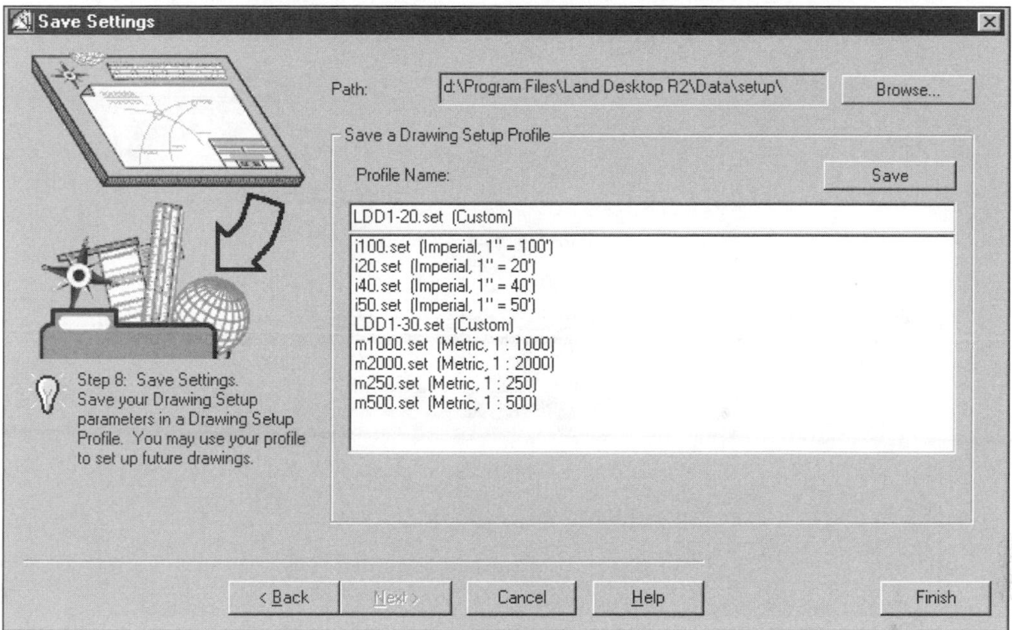

Figure 1–12

Existing Drawing Setup

There may be times when the drawing comes from another office with different software, or you don't have access to the project data. Where to locate the drawing depends upon how your company handles the location of drawing files and the project data. If the drawing is separate from the project data, you can place the drawing into the appropriate folder, open the drawing, create or select a project, and set the drawing setup profile. If drawings are located in the project structure, you first need to create a dummy drawing that creates the *DWG* directory in the project folder. Then copy the drawing into the *DWG* folder. After copying the file, you can open the file, assign it to the project, and set the drawing setup profile.

Not Assigning a Drawing to a Project

If you do not assign a project to a drawing, LDD will issue warning dialog boxes asking you to assign the drawing to a project (see Figure 1–13). After you answer Yes, the project presents a project assignment dialog box (see Figure 1–14). If you answer No, a dialog box appears reminding you that LDD routines will not work until you assign the drawing to a project (see Figure 1–15).

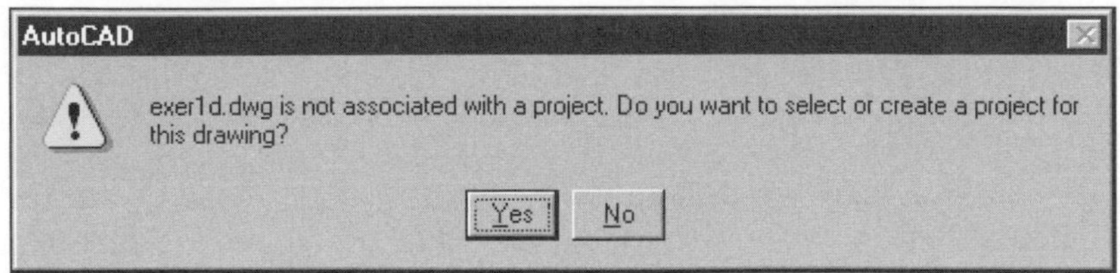

Figure 1–13

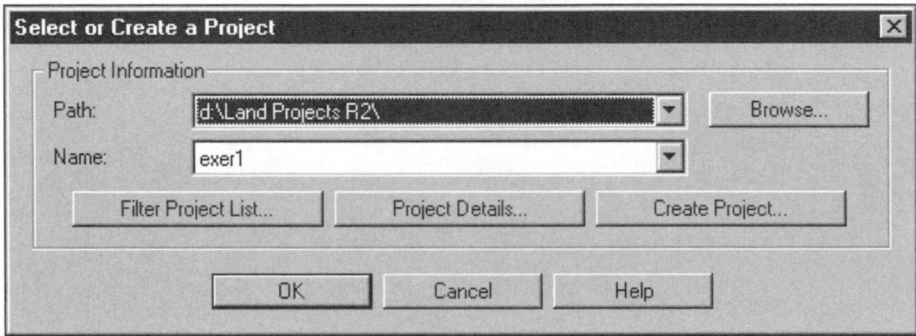

Figure 1–14

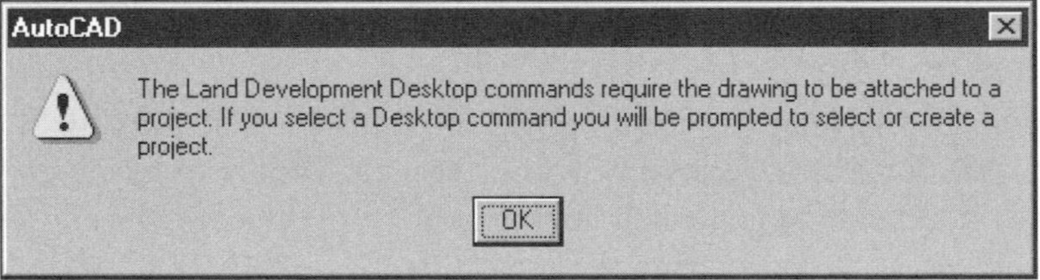

Figure 1–15

Exercise

The first exercise traces the steps to complete the three tasks of starting a Land Development Desktop project. These steps are naming a drawing, assigning a project, and establishing the drawing's Setup Profile. You must do these tasks each time you start a new drawing. The exercise assumes that you are entering a new session of LDD.

After you complete this exercise, you will be able to:

- Start a new drawing
- Initialize the Land Development Desktop environment
- Set the Point Database defaults
- Establish a drawing Setup Profile

CREATING A NEW DRAWING AND PROJECT

1. Select the New button in the LDD opening dialog box (see Figure 1–1).
2. In the Drawing Name area at the top of the New Drawing:Project Based dialog box, enter the name of the drawing (see Figure 1–2). Enter the name **LDDTR2**.
3. In the Project and Drawing Location area of the dialog box, select the Browse button to the right of the project path. The button allows you to view the folders you have access to on the network (see Figure 1–16). Select a directory different than the default (*C:\L* and Project *R2*) and select OK to exit the dialog box.

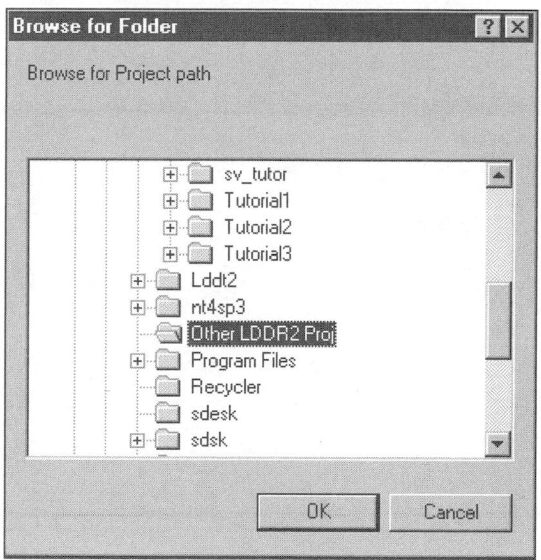

Figure 1–16

4. Select the Project Path drop list arrow and view the new path on the list. Set the Project Path to *C:\L* and Project *R2* (assuming the default install).
5. Select the Create Project button. The selection displays the Create Project dialog box. The top portion of the box identifies the prototype and the bottom portion identifies the project name, its description, and the key words associated with the project. Use Figure 1–3 as a guide for entering the information. When you have entered the information, select OK to exit the Create Project dialog box.

   Prototype: Default (Feet)

   Project Name: PROJ1

   Project Description: As built topo of 3/27/00 JM BW

   Project Key Words: Topo Smith JM

   Drawing Path for this Project: Project "DWG" folder

6. At the center of the New Drawing:Project Based dialog box, select the Project Details button to view the information you entered about the project (see Figure 1–17). Select OK to exit the dialog box.

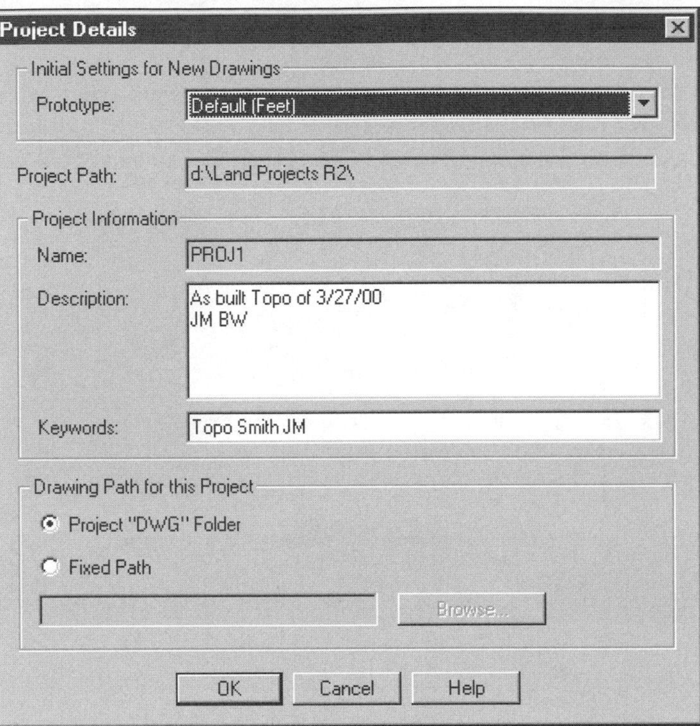

Figure 1–17

7. At the center of the New Drawing/Project Based dialog box, select the Filtered Project List button. The dialog box shows an unfiltered list of projects (see Figure 1–18). There may be no entries or only the Tutorial projects listed in the dialog box. Select either drop-list arrow to see whether any Key Words or Create By entries exist. If any entries exist, select one to view the change in the list of projects. Cancel out of the dialog box.

8. In the New Drawing:Project Based dialog box, select *aec_i.dwt* as the drawing template at the lower left of the New Drawing/ Project Based dialog box.

9. Review your entries in the New Drawing:Project Based dialog box to Figure 1–19. If the information is correct, select OK to exit the dialog box . Your drive letters may be different from the ones shown in the figure.

EXERCISE

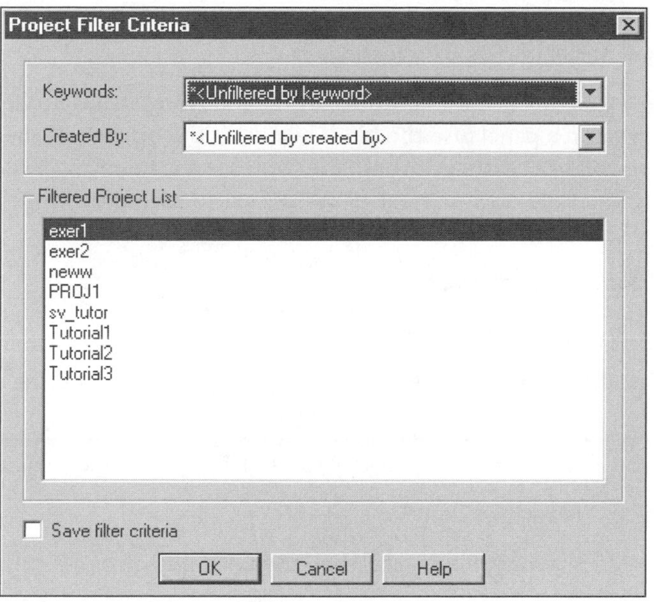

Figure 1–18

New Drawing: Project Based

Drawing Name

Name: LDDTR2.dwg

Project and Drawing Location

Project Path: d:\Land Projects R2\ Browse...

Project Name: PROJ1

Drawing Path: d:\Land Projects R2\PROJ1\dwg\

Filter Project List... Project Details... Create Project...

Select Drawing template

ACAD -Named Plot Styles.dwt

acad.dwt

ACADISO -Named Plot Styles.dwt

acadiso.dwt

aec_i.dwt

Show sub-folders Browse...

Preview

OK Cancel Help

Figure 1–19

10. The next dialog box displays the settings for the point database (see Figure 1–4). Accept the default values by selecting OK.

11. The Drawing Setup Wizard appears. Set the following values in the wizard. If no values are set in a panel, use the Next button at the bottom of the wizard to advance to the next settings panel. After you set the following values, select the Finish button at the bottom right of the wizard to save the changes and dismiss the dialog box.

    Load Settings – No values Set.

    Units and Precision – No values Set.

    Scale – Set the scale to 1"= 20' Horizontal, 1"= 2' Vertical and the sheet size to 24x36

    Coordinate Zone – Set the Category to No Datum, No Projection.

    Orientation – No values Set.

    Text Style – First select Leroy.stp, then the Load button and finally select L100 and the current text style.

    Border – Set the border selection to no border

    Save Settings – Enter LDD1-20 as the name for the drawing profile and select the save button in the upper right of the Save Settings panel to save the setup file.

12. The next dialog box is the Review dialog box. This dialog box contains all of the settings found in the previous wizard. Review the settings in the Finish dialog box. Dismiss the dialog box by selecting the OK button.

13. Select the Save icon, the floppy disc icon in the standard toolbar.

EXERCISE

This completes the first exercise. The three tasks you need to complete to use Land Development Desktop are complete: drawing name, project assignment, and drawing settings.

UNIT 2: PROTOTYPES AND PROJECTS

Land Development Desktop creates graphical data in the drawing and external data in a project folder structure. The initial values for the project folder come from a prototype. You can create and modify prototypes to promote standards. You can create any number of prototypes.

PROTOTYPE MANAGER AND PROJECT MANAGER

LDD has Prototype and Project Managers (see Figures 1–20 and 1–21). These managers create, modify, and report the status of prototypes and projects. There are routines that edit the settings of the prototypes. The User Preference routine allows you to change the initial behavior of LDD and set new directories for data files from their original location. The managers and their tools are found on the Projects menu of LDD.

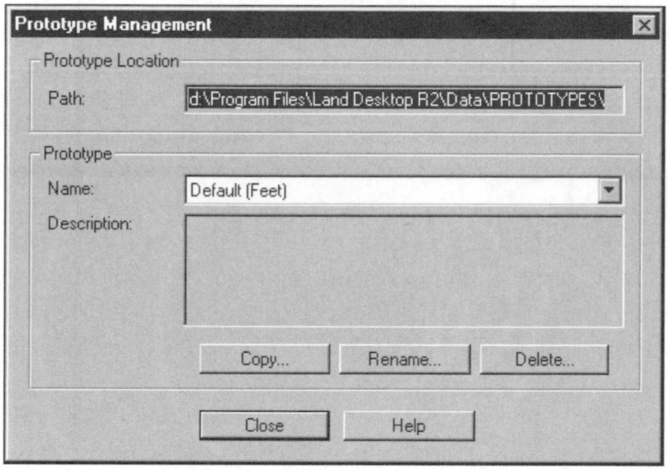

Figure 1–20

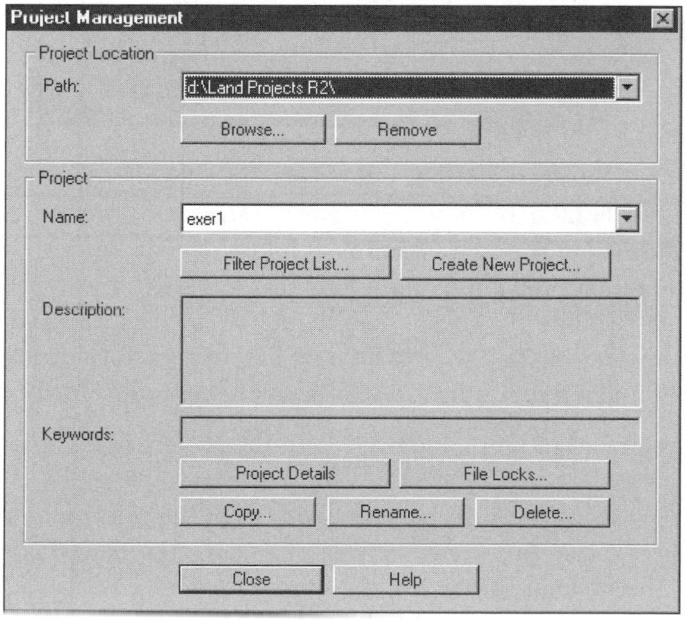

Figure 1–21

PROTOTYPE AND DRAWING SETTINGS

The routines that edit prototype settings are Prototype Settings and Drawing Settings. The Prototype Settings routine modifies only the settings in the prototype. The Drawing Settings routine modifies the settings of either the prototype or the drawing. You can edit, load, and save all or individual settings from the dialog box.

If you have an established project and you want to change the default values for new drawings you must edit and reload prototype settings. LDD looks at the project settings to establish the values for new drawings, and these settings are from files brought over from the prototype. To make new drawings follow a new set of project settings, you must edit, update, and reload the prototype settings files into the project. Then each new drawing will display the new settings. Because the Prototype Settings routine only edits the settings in the prototype, you must manually copy these files from the prototype to update the project. The files, *project.dfm* and *cgx.dfm*, are copied from the root of the prototype to the root of the project. The next file copied is *default.dfm*. Copy this file from the *DWG* folder of the prototype to the *DWG* folder of the project. If you have installed Survey and Civil Design, you need to modify their settings and copy them to the root of the project. When this is done, each new drawing will display the new settings.

PROJECT BASICS

As mentioned earlier, Land Development Desktop creates objects in the current drawing and external support files in the project folder structure. This methodology allows other drawings to share the common project data. Also, this allows an LDD drawing to reflect your decisions on which information needs to be included or excluded from a drawing. Many of the LDD routines create external data while at the same time creating objects in the drawing.

Land Development Desktop routines place data and information in a project folder structure (see Figure 1–22). Any drawing assigned to an existing project has unrestricted access to the information stored in the project folder structure. This means that you can import into any project's drawing a graphical representation of external data, such as alignment definitions, volume tables, points, surfaces, and so on. In other cases there are routines that allow you to view the numerical values of the external data, such as alignment definitions, volume calculations, and so forth.

In Land Development Desktop, you can locate the project folder anywhere on the network. If you have multiple locations, you must make sure the correct path is identified before creating the new project. If you create a project named LDDTR2, the Create Project dialog box creates a project folder named *LDDTR2* under the selected path. The skeletal source for the project structure is a prototype. In the prototype structure, LDD stores initial values for the project. This ensures that each project initially represents a set of standards.

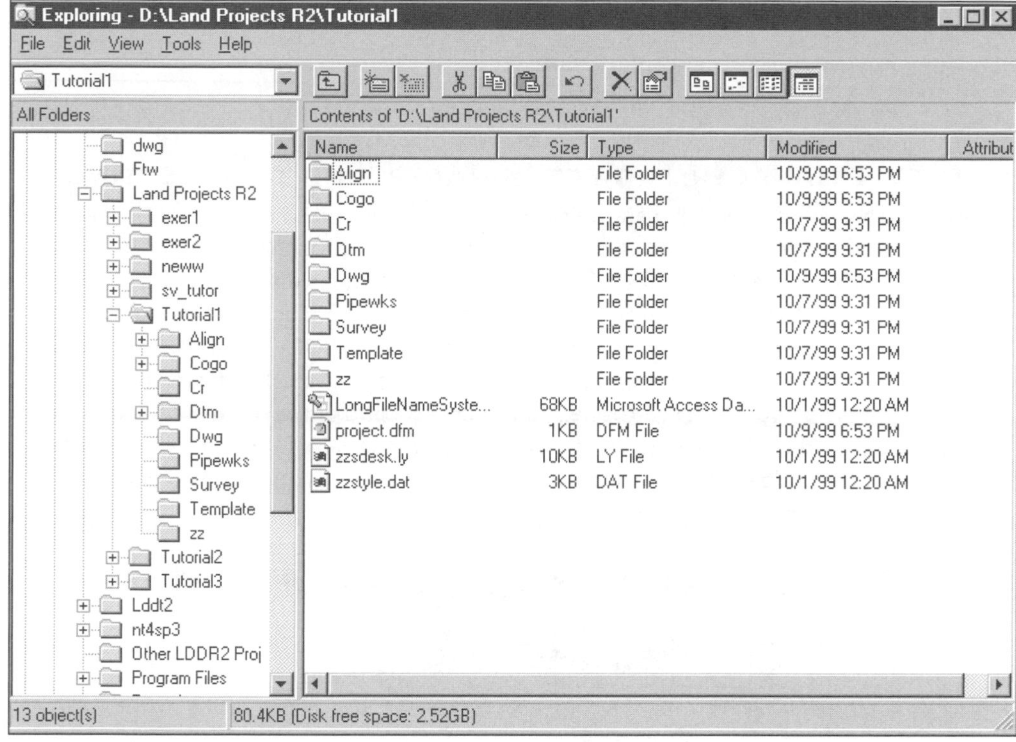

Figure 1–22

When creating a project, LDD by default provides a *DWG* directory in the project folder structure for the drawing files. When you are creating a project, you can specify a separate folder for the drawings of the project (see Figure 1–3).

LDD provides online help that contains an overview of the data files found in external data structure.

Exercise

The Projects menu contains the Project and Prototype Managers. These dialog boxes report and allow for some modification of projects and prototypes. The User Preferences, Data Files, Drawing Settings, and Setup routines set the initial behavior, the values of data files, and settings used by many of the routines found in LDD. This exercise reviews the functions and settings of each of these routines.

After you complete this exercise, you will be able to:

- Modify the start-up behavior of Land Development Desktop
- Work with the prototype and project managers

- Modify the Prototype Settings
- Review the current project status
- Modify the settings of the current drawing

PROTOTYPE MANAGER, PROJECT MANAGER, AND THEIR SETTINGS

1. Go to the Projects menu and select User Preferences (see Figure 1–23). The top portion of the dialog box, File Locations, controls the location of data files. All of the entries except temporary files are in the *Data* folder installed by LDD. You can change the location for each entry in the folder. The bottom left portion of the dialog box controls the initial startup behavior of LDD. These controls toggle on or off the Open, New, and LDD Start Up dialog boxes. The lower right area of the dialog box contains toggles you select to control the use of the Startup wizard or the tabbed Start Up dialog box and the default Drawing Setup Profile. Select the Help button to view the online help. Close Help and select Cancel to exit the dialog box.

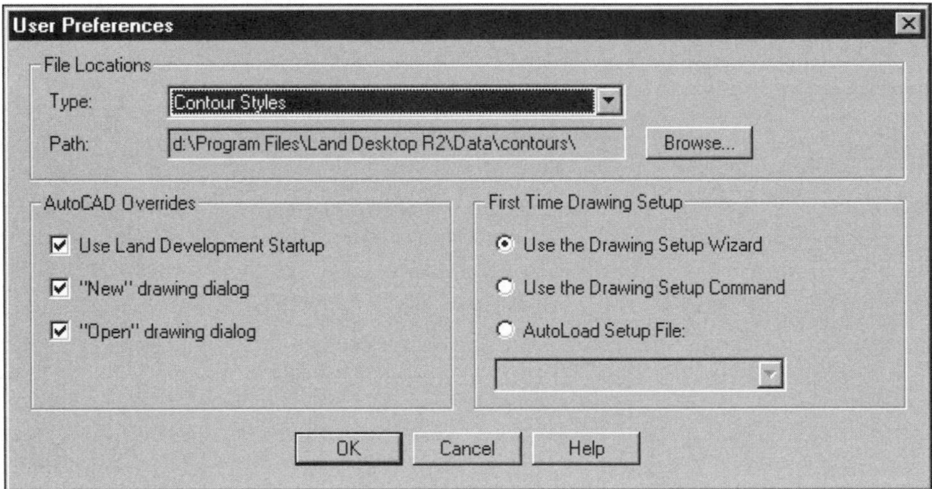

Figure 1–23

2. From the Projects menu, select Prototype Manager (see Figure 1–20). The manager allows you to set the current prototype. You can use other buttons in the dialog box to copy, rename, or delete existing prototypes. Select the Help button to view the online help. Close Help and select Close to exit the dialog box.
3. From the Projects menu select Prototype Settings. The first dialog box sets the current prototype. Select default (feet) and select OK to display the Edit Prototype Settings dialog box (see Figure 1–24). Select a few settings, view them, and cancel out of the settings. Select the Help button to view the online help. Close help and select Close to exit the dialog box.

EXERCISE

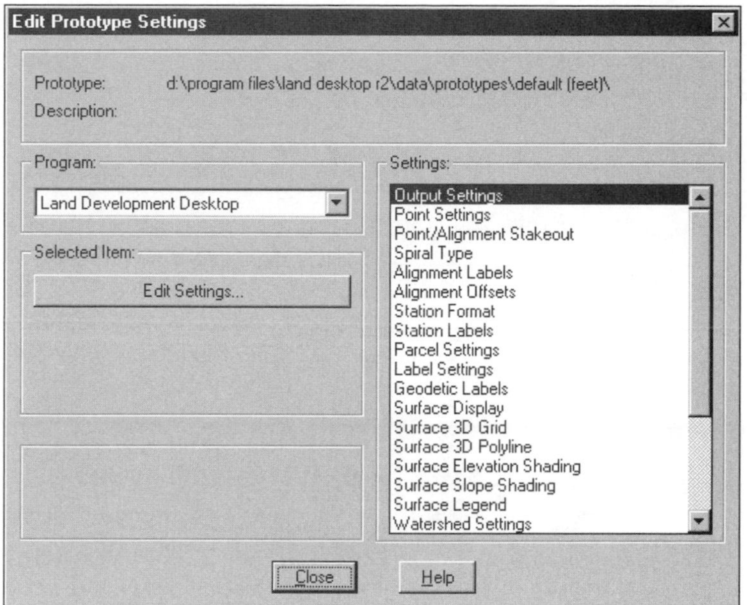

Figure 1–24

4. From the Projects menu, select Project Manager (see Figure 1–21). The top portion allows you to add (Browse) or remove project folder locations. The central portion displays the current project, its description, and key words. The Filter button will create a short list from which to select a project. The Key Words or Created By settings of a project allow the routine to filter the list of projects. The Create button does just that—creates a new project. The bottom portion contains several functions. The Copy, Rename, and Delete buttons allow you to prune out finished jobs, rename a misidentified project, or copy a project to a new project (phase one project is the seed for project phase two). The two remaining buttons show the details, Project Details, and the locks currently set, File Locks. If an LDD session crashes, there is a possibility that some file locks will not be removed. The Project File Lock dialog box allows you to view and remove locks (see Figure 1–25). Select the File Lock button to view the current locks. By default you see all locks. The By Owner button displays only your locks. Cancel out of the Locks dialog box. Select the Help button to view the online help. Close help and select Close to exit the dialog box.

5. From the Projects menu, select Data Files (see Figure 1–26). The Speed Tables file provides data for LDD routines that create curves for superelevation, Import/Export file formats define data file structures, and the remaining files are style definitions. Select one of the files and view the contents of the dialog box. After viewing the data files, select the Help button to view the online help. Close help and select Close to exit the dialog box.

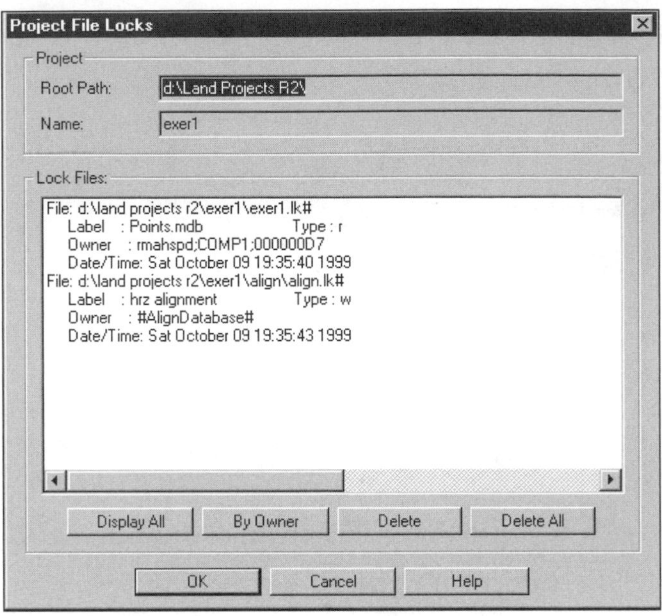

Figure 1–25

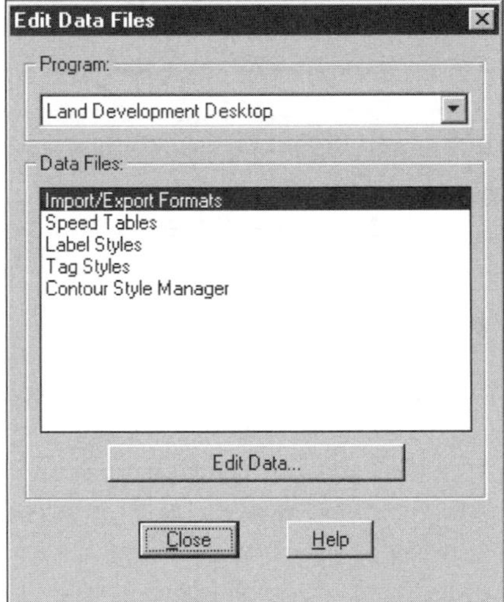

Figure 1–26

6. From the Projects menu, select Drawing Settings (see Figure 1–27). These settings are the project's settings for each new drawing. The dialog box allows you to load, edit, and save the drawing's settings. Editing these settings changes the settings only for this drawing or the selected prototype. Any changes you make to the prototype affect the next project you create using the modified settings. Select some of the settings to view their content. After viewing the settings, select the Help button to view the online help. Close Help and select Close to exit the dialog box.

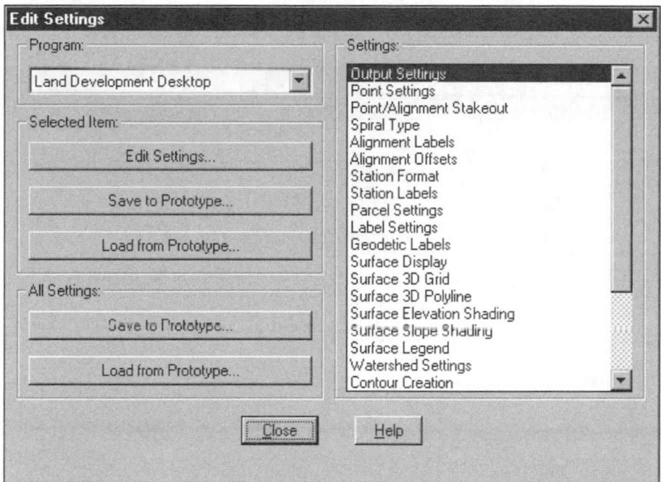

Figure 1–27

7. From the Projects menu (see Figure 1–28), select Drawing Setup. The Drawing setup dialog box is a condensed version of the Setup wizard. The dialog box should reflect the settings set in the New Drawing wizard. Select the Help button to view the online help. Close help and select Close to exit the dialog box.
8. Save the drawing by selecting the Save icon.

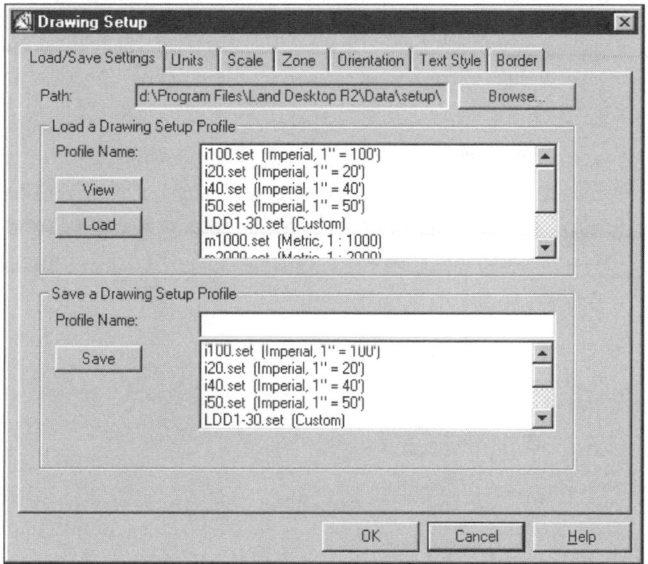

Figure 1–28

This completes the second exercise, viewing and becoming familiar with the settings found in a prototype and a Land Development Desktop drawing.

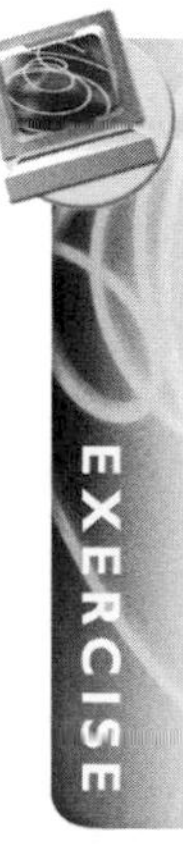

UNIT 3: CREATING AND MODIFYING A PROTOTYPE

The prototype is important to many aspects of a project. The settings found in the prototype affect each drawing of a project, establish consistent labeling standards, and can contain data files used by other routines in LDD. You must review and modify the settings in the prototype you select to be the seed for a project. If the prototype is never modified, each new drawing in the project will have to have its settings changed. Rather than editing the settings for each new drawing, it is better to fix the prototype once.

The easiest method to create a prototype is to copy an existing prototype with Prototype Manager. You can use either the Prototype Settings or Drawing Settings routine to modify the values of the new prototype. The difference between the routines is that Drawing Settings allows you to use the values established in a drawing to set the default values for the prototype.

The remaining task is to create a set of Drawing Setup Profile files (STP files). The Drawing Setup Profile contains the settings established in the Drawing Setup and/or the Setup wizard. By selecting and loading a Drawing Setup Profile, you set up the drawing with the information found in the seven tabbed sections of the Setup dialog box. The setup controls scale, borders, text styles, zone, and orientation settings. Since the setups are scale dependent, you need to create one setup for each used scale.

Exercise

This exercise will create a new prototype, LDDTR2. The settings for the prototype will be modified by the Prototype Settings routine. The second task of the exercise is establishing four new Drawing Setup Profiles. The Profiles are the saved settings found in the Drawing Setup dialog box. You have already created a new setup when you where setting up the current drawing in the New Drawing Wizard.

After you complete this exercise, you will be able to:

- Create a new prototype
- Modify the prototype settings
- Create drawing setup profiles

CREATING A NEW PROTOTYPE

1. Continuing in the current drawing (**LDDTR2)**, go to the Projects menu and select Prototype Manager. The routine displays the Prototype Manager dialog box (see Figure 1–20). From the drop list at the middle of the dialog box, select the prototype Default (Feet) to make it the current prototype.

2. Select the Copy button and the Copy Prototype dialog box appears (see Figure 1–29). Enter **LDDTR2** as the name of the new prototype. In the description box, enter the following description: **Working Prototype for exercises**. Select OK to close the dialog box to complete the copying process.

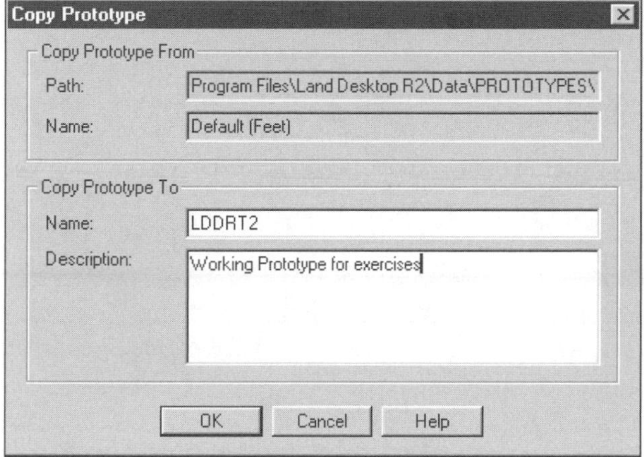

Figure 1–29

3. In Prototype Manager select the drop list arrow and view the new prototype on the list, LDDTR2. Select the Close button to finish the copying process.
4. From the Projects menu, select the Prototype Settings routine, select the LDDTR2 prototype, and edit and view its settings. Set the following prototype values:

Point Settings

Marker Tab – Custom Marker (see Figure 1–30)

Custom Marker Style a plus superimposed by a circle

Custom Marker Size in Absolute Units – 2 units

Text Tab (see Figure 1–31)

Style and Size – Size in Absolute Units – 4

Automatic Leaders – ON

Alignment Labels

Layer Prefix - *-

The prefix prevents the erasure of one alignment's labeling when erasing another alignment's labeling.

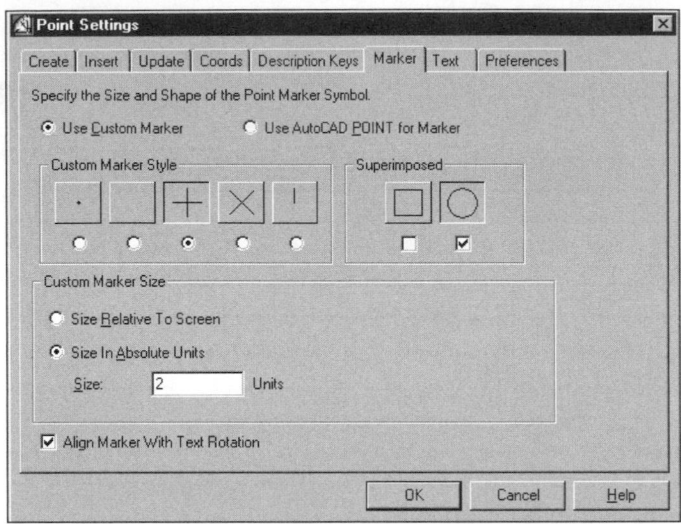

Figure 1–30

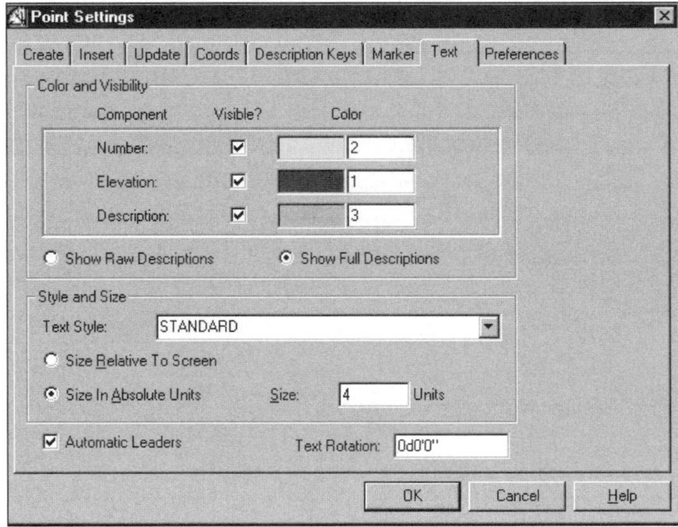

Figure 1–31

Parcel Settings

Parcel Numbering – Text Style - L200 and Prefix - Lot # (see Figure 1–32)

Square Feet/Meters Labeling – Text Style - L140

Acres/Hectares Labeling – Text Style - L100

Surface Display Settings

Layer Prefix - *-

The prefix prevents the erasure of one surface's objects when erasing another surface's objects.

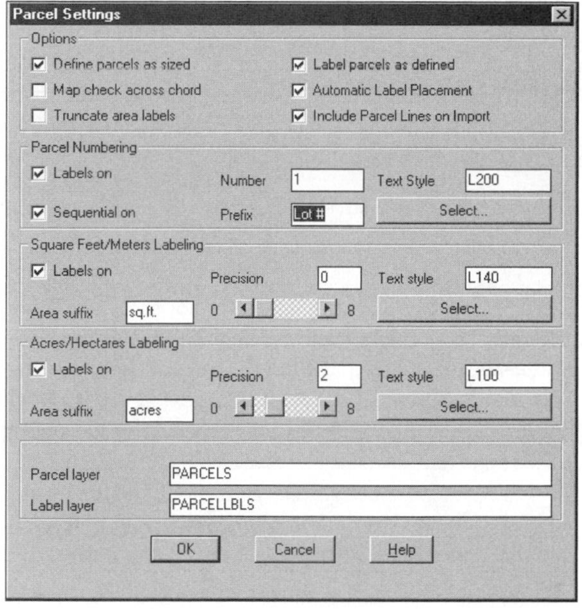

Figure 1–32

Surface Elevation Shading Settings

Layer Prefix - *-

The prefix prevents the erasure of one surface's objects when erasing another surface's objects.

Surface Slope Shading Settings

Layer Prefix - *-

The prefix prevents the erasure of one surface's objects when erasing another surface's objects.

Contour Creation

Minor Interval 1- Major Interval 5 (see Figure 1–33)

Contour Settings
Intervals
Both Minor and Major
Minor Only
Major Only
Minor Interval: 1
Layer: CONT-MNR
Major Interval: 5
Layer: CONT-MJR
Properties
Contour Objects
Contour Style: Standard
Polylines
Preview ...
Style Manager >>
OK
Cancel
Help

Figure 1–33

5. Select the Close button to close the Edit Prototype Settings dialog box.

CREATING DRAWING SETUP PROFILES

1. Continuing in the same drawing, **LDDTR2,** select Drawing Setup from the Projects menu to view the setup dialog box.
2. Select the Zone tab and make sure it is set to the Category of No Datum, No Projection.
3. Select the Border tab and make sure None is the border type.
4. Select the Scale tab. The scale should be set to 1"=20'. Set the horizontal scale to 1"=40' and the vertical scale to 1"=40'. Select the Load/Save Settings tab, enter the name LDD1-40, and save the current settings as a Drawing Setup Profile. This does not affect the current drawing. All this does is save the current settings as a new setup file.
5. Select the Scale tab. Change the horizontal scale to 1"=10' and the vertical scale to 1"=1'. Select the Load/Save Settings tab, enter the name LDD1-10 for the new profile, and save the settings as a Drawing Setup Profile. This does not affect the current drawing. All this does is save the current settings as a new setup file.
6. Select the Scale tab. Change the horizontal scale to 1"=30' and the vertical scale to 1"=3'. Select the Load/Save Settings tab, enter the name LDD1-30 for the new profile, and save the settings as a Drawing Setup Profile. This does not affect the current drawing. All this does is save the current settings as a new setup file.
7. Cancel out of the Drawing Setup dialog box to keep the current drawing scale.
8. The LDD1-20 Drawing Setup Profile does not need to be made because it was made in an earlier exercise. Save the drawing and exit LDD.

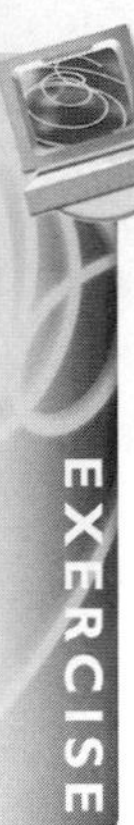

This completes the third exercise of this section. You now have a new prototype, LDDTR2, and a series of Drawing Setup Profiles. The new prototype has settings that are now considered "office standard" settings. Each new drawing created in a project using this prototype will reflect these default values. The series of Drawing Setup Profiles sets several defaults used by a drawing with a single selection. The main changes to the setups are the changing of the default zone to Lat Long, setting L100 as the default text style and inserting no border into the drawing.

Basic Data Group—Points

After you complete this section, you will be able to:

- Create point data
- Analyze, organize, and report point data
- Edit point data
- Annotate point data

SECTION OVERVIEW

Point data, as we mentioned in the last section, is one of Land Development Desktop's foundation data types. The Points menu of LDD contains tools that create, manipulate, analyze, and annotate coordinate data.

The manipulation, representation, and organization of points is complex, and LDD provides flexible and comprehensive point tools. The Points menu contains routines affecting points from creation to annotation (see Figure 2–1). The top portion of the menu contains routines to create, annotate, and organize points. The middle portion of the menu contains routines to edit and analyze points. The remainder of the menu contains routines to manipulate which points appear in the drawing, synchronize the point database and the drawing, plus point utilities and a point stakeout routine.

The number of tools to create point data reflects the multitude of situations that arise during work on a project. These tools allow you to create points from a list, an external file, AutoCAD entities, a surface, a roadway centerline, by interpolation between existing points, and from slopes and intersections. The organization of the point submenus by source of data helps you select which routine is appropriate for the job. (In this book, we will refer to the submenus as either flyouts or cascades.) The routines used depend upon the origins of the data and the tasks at hand. No one project will use all of the tools present in LDD.

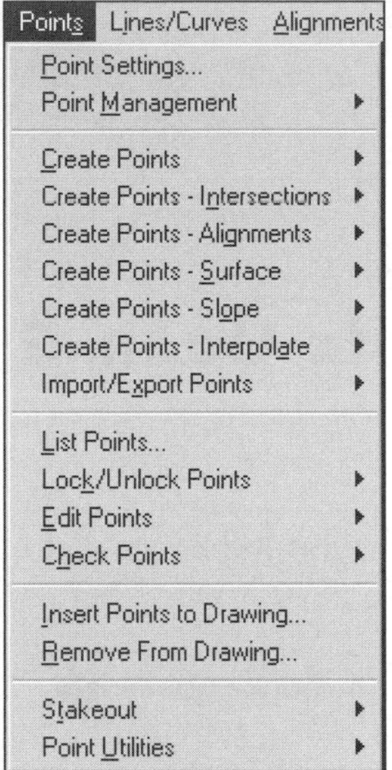

Figure 2–1

In Land Development Desktop, a point object represents coordinate data in the drawing. The object consists of a marker and three text labels: point number, elevation, and description. There is no layer assigned to the labels. This ability to separate the label from a layer allows you to display point number, elevation, and description independently of any layer settings. Other advantages of a point object are relative screen sizing, independent display options, rotation, and coordinate "locking." Coordinate locking prevents the AutoCAD move command from changing the location of the point object in the drawing.

Point objects on the screen represent an entry in an external data file, the point database. The point database is an access database. A drawing does not have to have points visible in the drawing to reference points in the project point database. All routines in LDD understand this fact. All drawings associated to a project have complete access to the point database. Each project has its own point database.

Here is a set of rules about the relationship between points, LDD, and the point database:

- All drawings associated to the same project share the same point database without restrictions.
- All LDD routines have access to the coordinates in the project point database.
- A drawing may have all, none, or some of the points from the point database.

- Generic AutoCAD routines edit only the AutoCAD properties of a point object.
- Land Development Desktop point editing routines update values both in the drawing and the external point database.

The effect of the fourth rule is that after editing a point object with a generic routine, there may be a difference between the new values in the drawing and the recorded values in the point database. You must address this difference manually to synchronize the values between the point database and the drawing. The fifth rule means that after you edit a point object with a LDD routine, there is no difference between the values in the drawing and the point database. LDD does the synchronization automatically.

In a Land Development Desktop project, there are three types of points. The first type is a point that does not require any action other than being placed into the drawing and point database. An example of this type of point is an elevation shot in topography. The second type of point requires a symbol in addition to the point object and marker. The symbol will be a part of the final product, but the point object and marker may not be present. Examples of this type of point are points representing manholes, signs, power poles, and so forth. The third type of point is one that is a part of a lineation or line. This type of point is found on a line or arc in the drawing. Examples of this type of point are centerlines, backs-of-curbs, sidewalks, and so forth.

UNIT 1

The point object and its creation and display on the screen are a result of user-defined settings. These settings are numerous and affect many aspects of the point's behavior. The routines that create points look at these settings to determine what point number is next, what symbols to use, what colors to display, and other settings. These settings are found in the top portion of the Points menu in the routine Point Settings. These settings are the focus of the first unit of this section.

UNIT 2

The second unit of this section reviews the point creation routines that deal with printed lists, external ASCII files, or calculated points (the Intersection flyout). Other point creation routines are in the Create Points and Import/Export Points flyouts of the Points menu and are discussed in their appropriate context in later sections of the book.

UNIT 3

The analysis and listing tools for points are the subject of the third unit of this section. The List Points routine and the ability to create named groupings of points enhance your ability to organize and list out points properly. The Point Utilities flyout of the Points menu contains routines to convert, view, and maintain point information. The ability of LDD to work with points that are a part of the point database, but not a part of a drawing means that you have tools showing the location of those absent points. Other tools aid in locating a point out of a list of several hundred points.

UNIT 4

The editing of points, including changing their number or location, is the subject of the fourth unit of this section. These routines are found in the Edit Points cascade of the Points menu.

UNIT 5

The annotation of points is accomplished through either description keys or labels. There is a need for two methods of point annotation. The first method, description keys, places a block at the coordinates of a point; for example, a symbol representing a manhole at the location of a manhole point. This symbol usually is a part of the final document while the point text may not be. Description keys have the power to translate field notation into office words. For example, IPF in the field may mean Iron Pipe Found in the office. A description key matrix is a table with entries indicating the field notation and the office translation. The description key also places the point marker, its text, and associated symbol on the user-defined layers. The field notations may contain numbers that define either or both a scaling factor or rotation angle used in conjunction with description keys.

The second method of point annotation is a point label style. A label style allows you to annotate a point with text and symbol values. The information for the label comes from the point database or an external data reference. Labeling and the description key file and its actions are the subject of the last unit of this section.

UNIT 1: POINT SETTINGS

When a point creation routine places points into a drawing, the values found in the Point Settings dialog box affect every aspect of the point's display. These settings control whether a point receives a symbol, what point number it is assigned, what colors it shows, whether it can be moved, and much more. All of the routines placing points into a drawing look to these settings for guidance. The Point Settings routine is found at the top of the Points menu (see Figure 2–1).

The Point Settings routine presents a multitabbed dialog box (see Figure 2–2). Each tab controls different aspects of point placement and display. Each tab displays a panel that contains several toggles, radio buttons, and initial data values. The initial defaults for this dialog box come from the prototype settings.

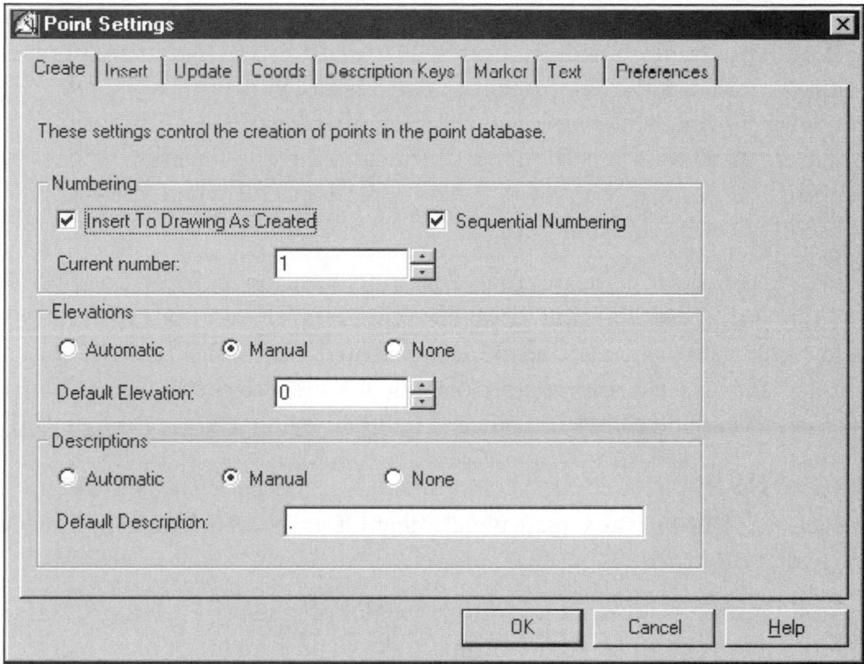

Figure 2–2

CREATE

NUMBERING

Point numbering is the first item in the Create panel (see Figure 2-2). The top portion of the panel contains two toggles. The first toggle, Insert to Drawing as Created, controls the placement of points into the drawing. When the toggle is on, a point creation routine inserts a point object into the drawing and makes an entry in the point database. When the toggle is off, the routine makes an entry only in the external point database and no point appears in the drawing.

The second toggle is Sequential Numbering. If the toggle is on, all of the point routines assume a sequential numbering system. If the toggle is off, the point routines prompt for a point number. By toggling off sequential point numbers, you indicate that each point number is a random (user supplied) number.

The last item sets the seed number for point numbers. The next point created that needs a point number will start with this number.

ELEVATION

The middle area of the panel sets the elevation parameters. The first part sets the method of assigning elevations. The remaining portion sets or reflects the current elevation assigned to a point.

The radio buttons in the elevation area of the panel set the elevation assignment method. The None radio button indicates no elevation is assigned to a point. The second radio button sets the assignment to Automatic. With this toggle on, the next point created is automatically assigned the Default Elevation. The routine creating the point enters the elevation without user verification. If the elevation value is wrong, you must edit the point to correct it. This toggle is useful when you are placing a series of points into the drawing with the same elevation.

The Manual radio button means that a routine creating a point prompts for an elevation. The initial default value for an elevation is a period (.) or no elevation. The default elevation value can also be the last assigned elevation. The Manual toggle allows you to verify the elevation before placing a point into the drawing. You accept the elevation by pressing ENTER or you can change it by entering a new elevation.

DESCRIPTIONS

The third set of defaults in the Create panel affect Descriptions. A description contains the identity of a point.

The first radio button is None. This button toggles off the use of descriptions. If this toggle is set to None, point creation routines will not prompt for or place a description on a point.

The second radio button sets the assignment to Automatic. With this toggle on, the next point created is automatically assigned the Default Description. The routine creating the point enters the description without user verification. If the description value is wrong, you must edit the point to correct it. This toggle is useful when you are placing a series of points into the drawing with the same description.

The Manual button means that a point creation routine prompts for a description. The initial default value is a period (.) or no description or the last used description. You accept the description by pressing ENTER or you can change it by entering a new description.

INSERT

The Insert panel controls the insertion of point objects, symbols, or labels (see Figure 2–3). The top portion of the panel sets the path to the symbol set used by the project. The default is a folder named *Cogo*. There are two additional installed folders, *Cogo_metric* and *APWA* (American Public Works Administration). These folders contain alternate symbol sets. If your office has a set of symbols and wants to use them, place them in a folder on a server and then set the Search Path for Symbol Block Drawing Files to that machine and folder.

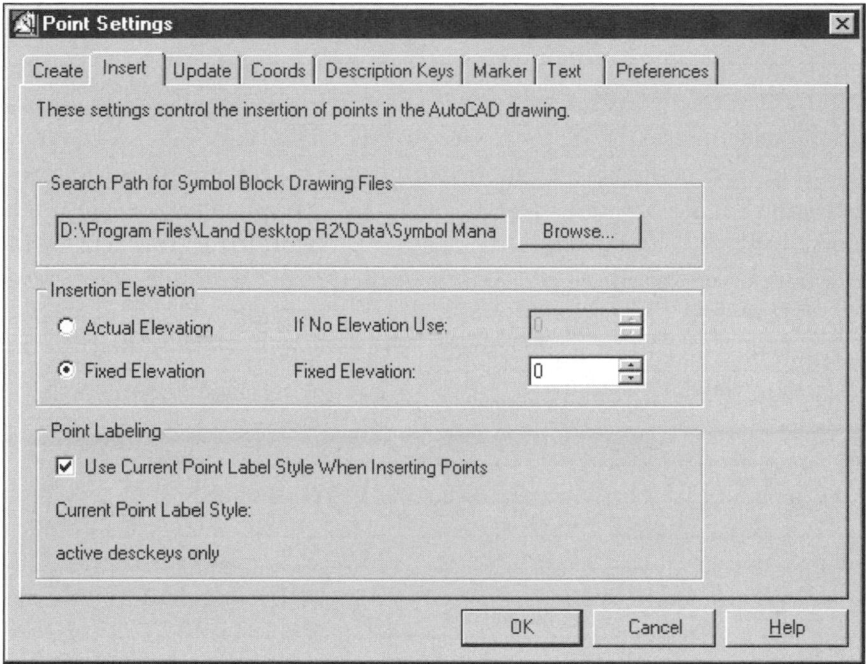

Figure 2–3

The middle portion of the panel determines the elevation of the inserted points. By default, all points in LDD have a 0 (zero) elevation. The reason for this is that if you were to draw a line from point node to point node, the resulting line would be the horizontal distance between the two points. The annotated lines of a survey are the horizontal distances between points, not the three-dimensional distance (slope distance). You can enter a value to the right of the toggle, but the elevation (Fixed) is applied to every point.

The Actual Elevation button places the points into the drawing at the elevation specified by the point. If the point has no elevation, it is inserted at the elevation found in the If No Elevation Use box.

The bottom portion of the panel toggles on or off labeling. This toggle must be on for description keys to work. Description keys are intimately tied to point label styles. As mentioned before, labels can be text, symbol, or both. Under the toggle, the panel lists the current labeling style. The style definition determines what actions take place when a point is inserted into the drawing.

UPDATE

The top portion of the Update panel toggles on point movement (see Figure 2–4). This toggle refers to the AutoCAD MOVE command and the behavior of all point objects in the selection set. If the toggle Allow Points To Be MOVE'd In Drawing is off, the marker of a point object cannot be moved. If you are attempting to move points with the toggle off, the moved point labels (point number, elevation, and description) display a leader back to their unmoved coordinate (marker) location. If the toggle is on, you can move the points. When you toggle on point moving, a second toggle box becomes active. If you select this second toggle, LDD will update the point database after moving the points. If the toggle is left off, you will need to update the point database manually.

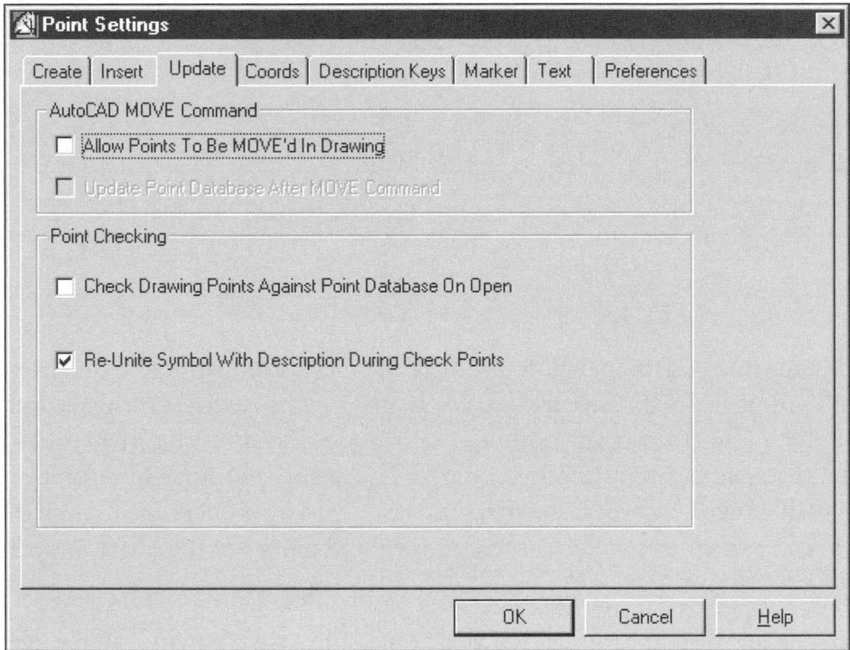

Figure 2–4

The Check Drawing Points Against Point Database On Open toggle forces LDD to compare the location of points in the drawing to their location in the point database. If the toggle is on, it will force LDD to modify the points present in the drawing when there is a difference between the points in the point database and drawing.

This toggle is useful when multiple users are working on a project. Each user may edit a point in a drawing and it is important that the edits are reflected in all other drawings referencing the point database. The routine updates an individual's drawing to show all of the changes created by the editing of other users.

The last toggle of this group is the Reunite Symbol With Description During Check Points. If any points in the drawing shift position during the point checking process, the Reunite toggle forces points with description key symbols to move the symbols to the new position of the point. When the toggle is off, the symbol does not move when the point changes it coordinate location.

COORDS

The Coords panel sets the default method LDD uses to report coordinates (see Figure 2–5). The default is Northing/Easting. Other methods are Easting/Northing, X-Y, or Y-X.

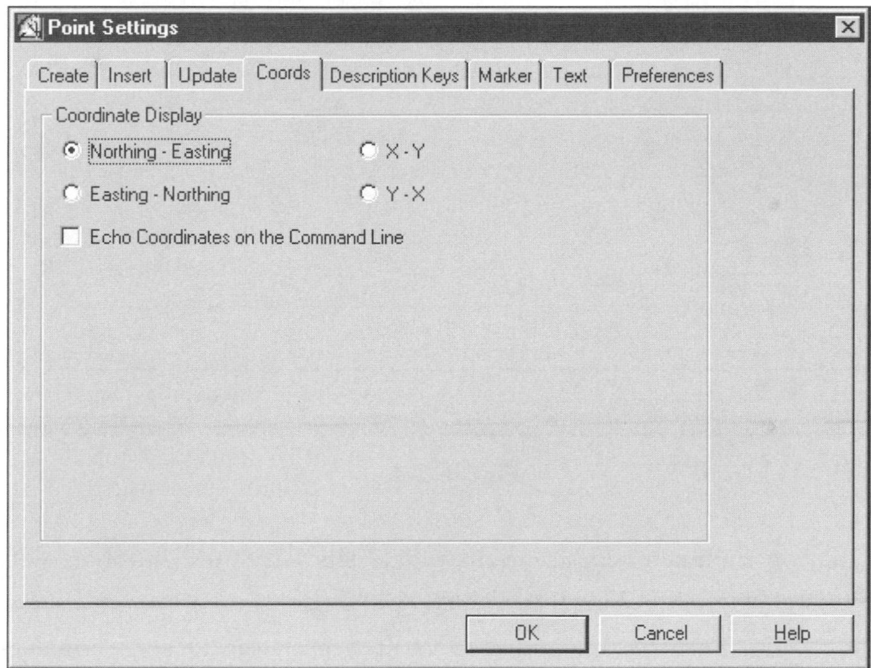

Figure 2–5

The Echo Coordinates On the Command Line toggle sets whether or not coordinates will appear on the Command line after placing a point.

DESCRIPTION KEYS

The Description Keys panel sets rules on reading the Description key file, parameter matching, and the search method to employ (see Figure 2–6). The radio buttons, Ascending and Descending, set how to read the Description key file. The only time this is significant is when you have two description keys starting with the same letters, for example ST* and STA*. Ascending uses the first key (ST*); the second is ignored. This is a top-down search through the description key list. Descending uses the second key (STA*) and if failing matches on the first (ST*). This is a bottom-to-top search of the description key list.

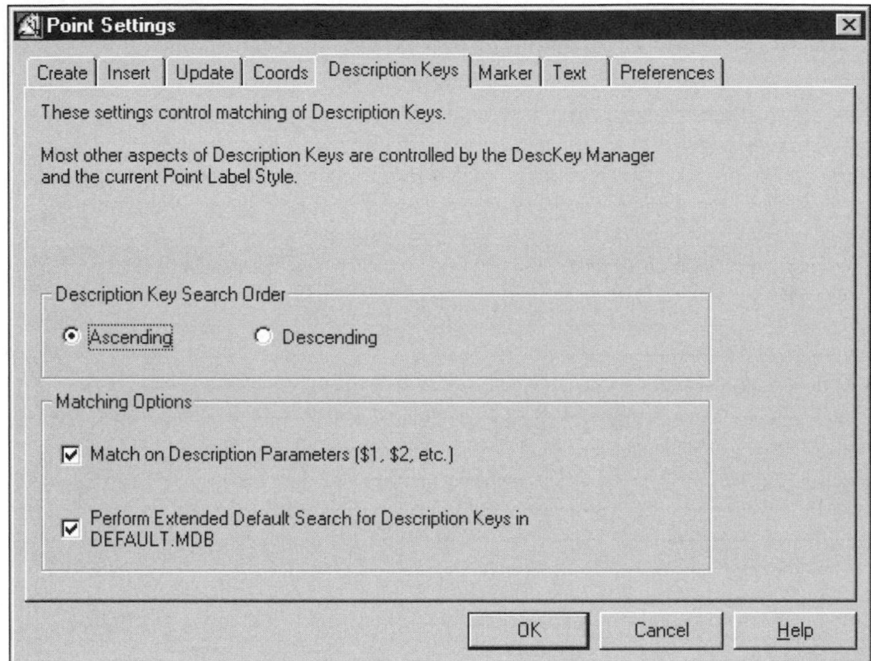

Figure 2–6

The Match on Parameter Descriptions toggle lets LDD use numbers within a description as scaling, rotating, and labeling values.

The last toggle, Perform Extended Default Search for Description Keys in DEFAULT.MDB, means that if searching a user Description key file (*mydesckey.mdb*) and no match is made, the routine is to continue the search into the *default.mdb* file.

MARKER

The Marker panel controls the way LDD identifies the coordinate location (see Figure 2–7). The marker can be an AutoCAD node or an LDD custom object. The custom object looks like a node, but you have more control over the object. The marker styles are similar to the node styles of generic AutoCAD. The Align Marker With Text Rotation toggle allows the marker to rotate to match a rotation angle set for the text of the point.

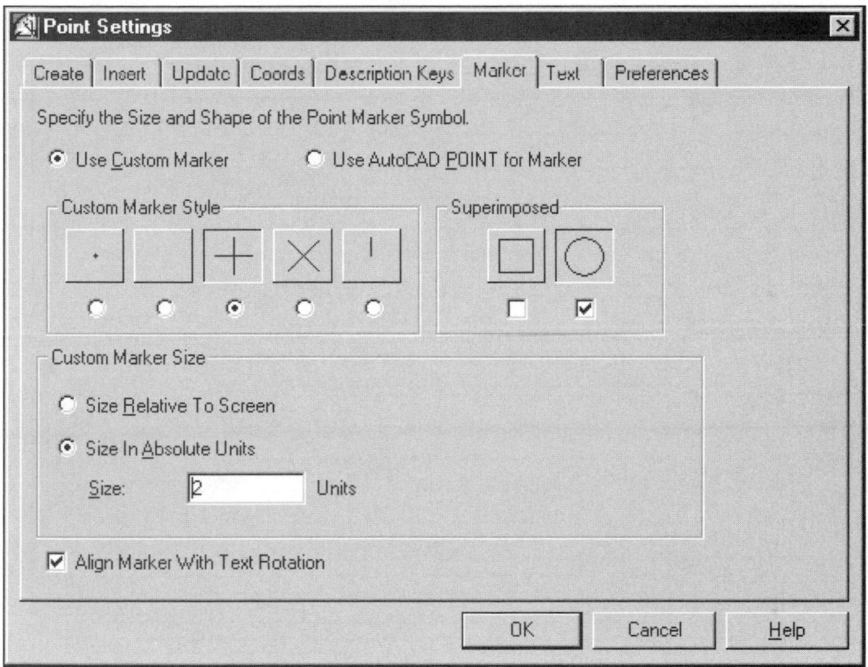

Figure 2–7

The difference between a custom marker and a generic node is the control over the sizing of the marker. The default setting is Size In Absolute Units. This toggle sets the marker to a fixed size of units. The size is set in the box adjacent to the toggle. The marker does not respond to changes in the size of the screen area. The Size Relative to Screen toggle sets the size of the marker relative to each new screen height set when you zoom. If you zoom in on a point (smaller screen height), the point resizes to a smaller size. If you zoom out (increase the screen height), it becomes larger. This sizing is related to the Always Regenerate Point After Zoom toggle on the Preferences panel. If the Regenerate toggle is off, the points will not resize until you cause a regeneration (regen) to occur. When on, this toggle regenerates only the points present in the drawing.

TEXT

The Text panel controls the traditional text around the point (see Figure 2–8). There are three components to the point text: point number, elevation, and description. You can toggle their visibility and color from the panel. In the middle portion of the panel is a toggle to have the point show the raw or full description. A raw description is a field-coded value and a full description is the translated description from Description keys (an office meaningful description).

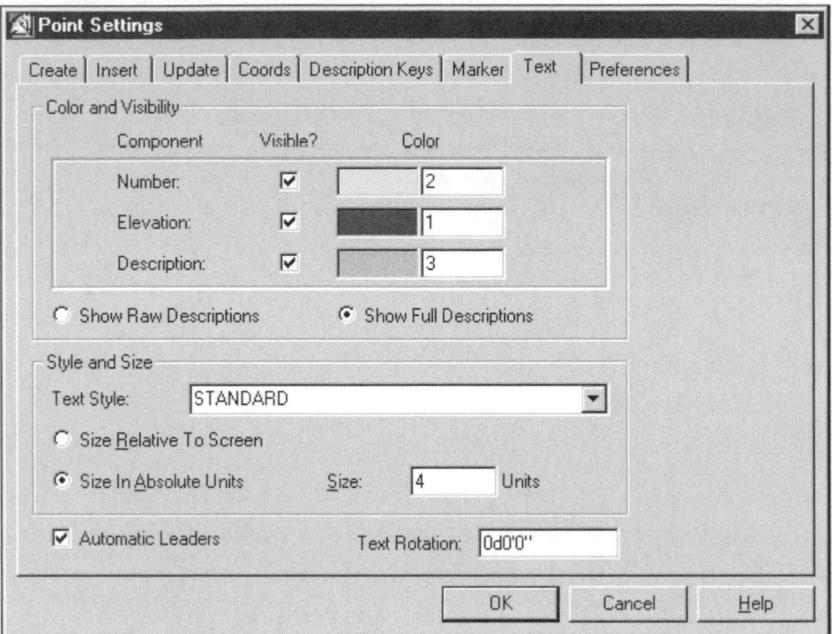

Figure 2–8

The bottom portion sets the text style, size, leadering, and rotation of the text. By default the text is a fixed (Absolute Units) size. When you zoom in, the text size remains constant. However, there is a toggle to make the text size relative to the screen height. If you zoom in and decrease the screen height, the points will resize as a percent of the current screen height. If you zoom out and increase the screen height, the text becomes larger. This sizing is related to the Always Regenerate Point After Zoom toggle on the Preferences panel. If the Regenerate toggle is off, the points will not resize until you cause a regeneration (regen) to occur. This toggle regenerates only the points present in the drawing.

PREFERENCES

The Preferences panel displays program performance settings (see Figure 2–9). The top toggles control the entry of point lists and the group names at the command prompt. The middle portion contains toggles affecting the resorting of point lists in the List Points dialog box. If these settings are toggled on, the list is resorted each time you remove or remove duplicates from the list.

The last toggle, Always Regenerate Point Display After Zoom, is important. As mentioned earlier, if marker and point text are set to size relative to screen, this setting has to be on for the resizing to work. If this toggle is off, the points will not resize until a regen occurs. When this toggle is on, the regen affects only points in the drawing.

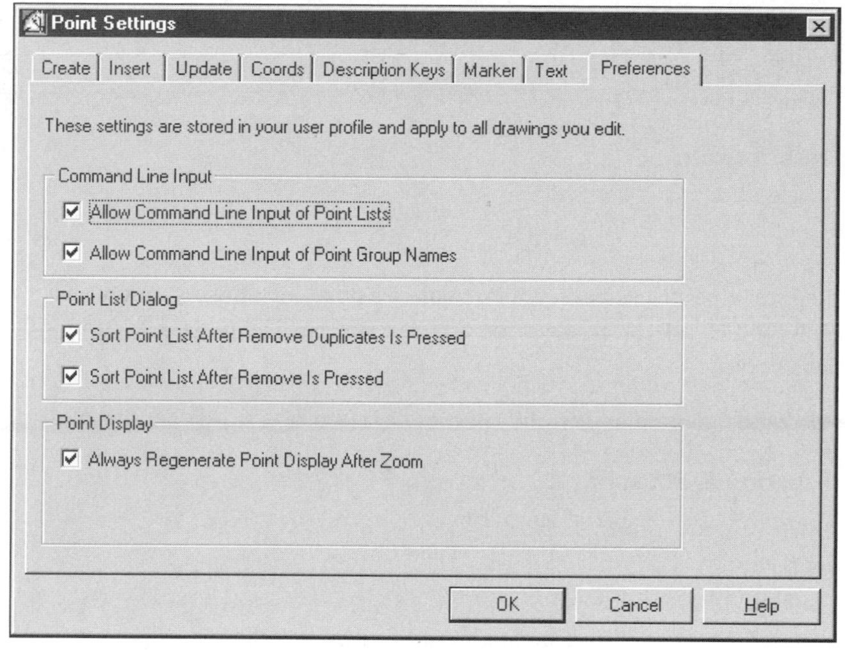

Figure 2–9

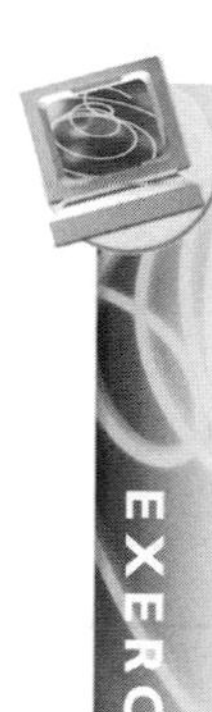

Exercise

Every routine that creates points uses the settings in the Point Settings dialog box. The settings you make in this dialog box control how a routine prompts for values, what symbols or labels to use, how to mark a coordinate, what color and text to use for a point object, and more.

After completing this exercise, you will be able to:

- Understand and set values in the Point Settings dialog box

POINT SETTINGS DIALOG BOX

1. If you are not in the drawing from the previous section, open the drawing *LDDTR2* that is in the proj1 project.
2. Select Point Settings from the Points menu.
3. Select the Create tab to view the Create panel settings (see Figure 2–2). Set the following values:

   Numbering:

   Toggle ON: Insert into Drawing as created

   Toggle ON: Sequential Numbering

   Current Number: 1

Elevations:

Toggle ON: Manual

Default Elevation: 0

Descriptions:

Toggle ON: Manual

Default: . (a period)

These settings tell a point creation routine to place a point into the current drawing, number it sequentially, and to ask for an elevation and description. The initial elevation is 0 and the description is none.

4. Select the Insert tab to view the Insert settings (see Figure 2–3). Set the following values:

Insertion Elevation:

Toggle to: Fixed Elevation

Set Fixed Elevation to: 0 (zero)

Toggle ON: Use Current Point Label Style When Inserting Points

The current Point Label style should be active desckeys only.

The symbol path should be set correctly when entering the Insert panel.

As mentioned earlier, by default all LDD points reside at elevation 0 (zero). The rationale is that the line distances between points have to be a horizontal distance. If you insert the points at their actual elevations, any line drawn between the point nodes will be a slope distance line. The distances placed on a plat of survey must be horizontal distances.

If the current point label style is not active desckeys only, then click on the OK button to exit the Point Settings dialog box. Next, go to the Labels menu and select the Settings routine. In the Settings dialog box select the Point Labels tab, click on the drop arrow to view the label list, and select active desckeys only to set it to the current label style. Select OK to exit the dialog box. Then rerun to the Point Settings routine from the Points menu.

5. Select the Update tab in the Point Settings dialog box (see Figure 2–4). Set the following values:

AutoCAD Move Command:

Toggle OFF: Allow Points To Be MOVE'd In Drawing

Point Checking:

Toggle OFF: Check Drawing Points Against Point Database On Open

Toggle ON: Re-unite Symbol with Description During Check Points

EXERCISE

These settings prevent points from being moved accidentally while in the generic AutoCAD Move routine. The Re-Unite Symbol toggle allows symbols to move to new locations when points change coordinates during the check points routine.

6. Select the Insert tab to view the Coords settings (see Figure 2–5). Set the following values:

   Coordinate Display:

   Toggle ON: Northing – Easting

   Toggle OFF: Echo Coordinates on the Command Line

These settings control how routines will prompt or list out coordinates for you.

7. Select the Description Keys tab to view the settings (see Figure 2–6). Set the following values:

   Description Key Search Order:

   Set Order to: Ascending

   Matching Options:

   Toggle ON: Match on Description Parameters

   Toggle ON: Perform Extended Default Search for Desc Keys in Default.mdb

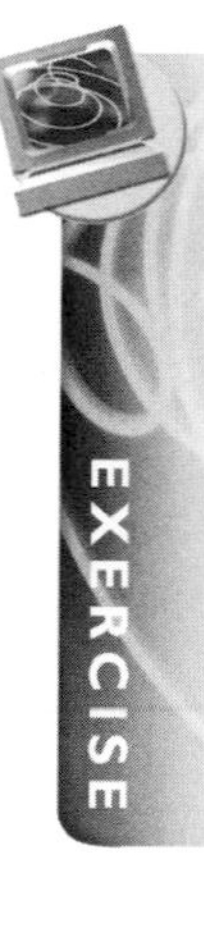

The Search order sets the matching process to go down the file (ascending the alphabet). The Descending toggle starts the search at the end of the file toward the beginning. Assume that the following entries are in a description key file:

MP Maple PNT-VEG CG_T33 SYM-VEG

MP Monument Post PNT-MON BM SYM-MON

The ascending search would find and match on MP (Maple) and not see MP (Monument Post). If the toggle is set to descending, the MP (Monument Post) is found and not MP (Maple).

A point label style can reference a user-defined description key file. If the key is not found in the user file, the search can be extended to the default key file.

8. Select the Marker tab to view the settings (see Figure 2–7). Set the following values:

   Custom Marker Style:

   Toggle on the + (plus)

   Superimposed: Toggle on the circle

EXERCISE

Custom Marker Size:

Toggle size to be Absolute

Set size to: 2 Units

Align Marker with Text Rotation: ON

This creates a custom coordinate marker, fixed at 2 units in size, showing a circle superimposed over a plus sign, and rotates the marker to the same angle as the text identifying the point.

9. Select the Text tab to view the current settings (see Figure 2–8). Set the following values:

Visibility and Color:

Number:	ON	Yellow
Elevation:	ON	Red
Description:	ON	Green

Description:

Show Full Description

Style and Size:

Style: Standard

Toggle size to Absolute

Set size to: 4 Units

Toggle ON: Automatic Leaders

Text Rotation: 0d0"0'

These settings create text adjacent to the marker that is fixed at 4 units, will create a leader to the marker if the text is moved, and will show point number, elevation, and description as yellow, red, and green text on the screen.

10. Select the Preferences tab to view the settings (see Figure 2–9). Set the following values:

    All toggles: ON

These settings allow for the command-line entry of point lists and group names, sorting of the point list in a point dialog box, and the regeneration of points in the drawing when they are assigned relative sizing.

11. Select OK to exit the dialog box.
12. Save the drawing. You may continue on to the next unit or exit the drawing session.

The Point Settings dialog box is extremely important when using description keys and point labels. You will be using this dialog box often when working in a project.

UNIT 2: CREATING POINTS

This unit reviews the point creation routines of the Create Points, Create Points–Intersection, and Import/Export Points flyouts of the Points menu.

CREATE POINTS

The Create Points flyout contains different types of methods to create points (see Figure 2–10). These methods include converting written data, mimicking surveying techniques, and using data from AutoCAD objects.

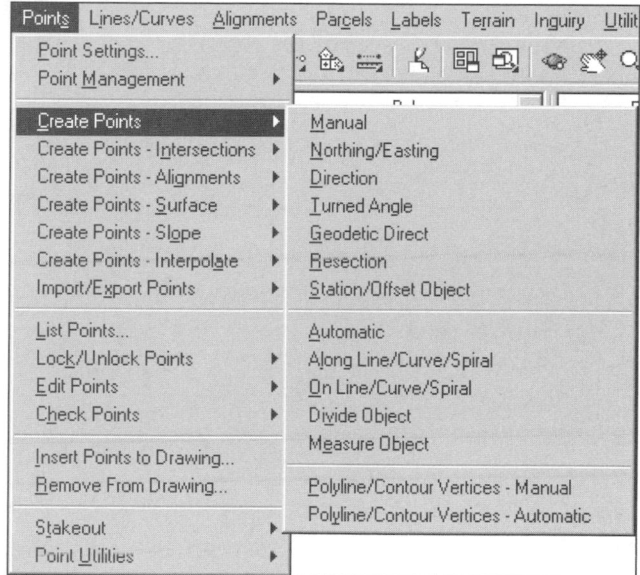

Figure 2–10

The Manual routine places points at entered AutoCAD coordinates, object snap selections, or LDD coordinates.

The surveying methods of placing points include setting points by northing and easting (Northing/Easting), direction (azimuths, bearing, and AutoCAD angles) and distance (Direction), by turning an angle and a distance (Turned Angle), Geodectic Direct, and Resection. The Northing/Easting routine uses direct coordinate input to create the points. The remaining routines require points or geometry (lines/curves/polylines) to be in the drawing.

Any LDD routine that prompts for a coordinate (point) location can use a coordinate toggle, that is, *.n*, *.p*, and *.g*. If a routine prompts for a starting point, the starting point can be an AutoCAD coordinate (*X-Y*), a northing/easting (*.n*), point number (*.p*), or a graphically (*.g*) selected point from the screen. To start the northing/easting mode, enter **.n** at the start point prompt. If you want to use a point number, enter **.p**, or to select a point graphically from the screen, enter **.g**. To toggle out of the mode, type the same toggle. For example, to toggle into northing/easting mode, type **.n.** and type **.n** to get out of northing/easting mode. The following routine snippets show how the Direction routine reacts to an AutoCAD object snap and the different . (dot) toggles.

```
Command:
COGO Set point Direction
Starting point: node of
Angular units: Degrees/Minutes/Seconds (DD.MMSS)
Quadrant (1-4) (Azimuth/POints):
```

- or -

```
Starting point: .p
Starting point:
 >>Point number: 1
Angular units: Degrees/Minutes/Seconds (DD.MMSS)
Quadrant (1-4) (Azimuth/POints):
```

- or -

```
Starting point: .n
Starting point:
 >>Northing: 5000
 >>Easting: 5000
Angular units: Degrees/Minutes/Seconds (DD.MMSS)
Quadrant (1-4) (Azimuth/POints):
```

- or -

```
Starting point: .g
Starting point:
 >>Select point object:
Angular units: Degrees/Minutes/Seconds (DD.MMSS)
Quadrant (1-4) (Azimuth/POints):
```

After you identify the starting point, the routine responds with a request for a direction. You can define a direction by one of four methods. The first is a bearing, the second is azimuth, the third is by typing **PO** and selecting coordinates using AutoCAD object snaps, and the fourth is by typing **.p** and identifying two point numbers from the point database defining a direction. This last method is not obvious. When the routine prompts for a Quadrant, enter **.p** and the prompting starts with the first point number. LDD identifies the current angular method as the first entry in the list of methods. If you enter an optional method, the next angular prompt will have the last used method first and a new list of optional methods. The following are examples of the different prompts issued by each angle entry method.

Bearing Method

```
COGO Set point by direction
Starting point:
 >>Point number: 1
Angular units: Degrees/Minutes/Seconds (DD.MMSS)
Quadrant (1-4) (Azimuth/POints): 1
Bearing (DD.MMSS): 88.0351
Distance: 495
COGO Set point by direction
Starting point:
```

Azimuth Method

```
COGO Set point by direction
Starting point:
 >>Point number: 1
Angular units: Degrees/Minutes/Seconds (DD.MMSS)
Quadrant (1-4) (Azimuth/POints): a
Azimuth (DDD.MMSS): 88.0351
Distance: 495.00
```

AutoCAD Object Snap Method

```
COGO Set point by direction
Starting point:
 >>Point number: 1
Angular units: Degrees/Minutes/Seconds (DD.MMSS)
Quadrant (1-4) (Azimuth/POints): po
 >First point: end of
 >Second point: end of
Distance: 495.00
```

Angle by Point Numbers (.*p*) Method

```
COGO Set point by direction
Starting point:
 >>Point number: 1
Angular units: Degrees/Minutes/Seconds (DD.MMSS)
Quadrant (1-4) (Azimuth/POints): .p
 >First point:
 >>Point number: 1
 >Second point:
 >>Point number: 2
Distance: 495
```

LDD uses a surveying convention for angular entries. This convention enters the angle of 34 degrees, 52 minutes, 18 seconds as 34.5218 (*dd.mmss*). Every LDD routine that uses an angle measurement uses this convention. Do not confuse this convention for generic AutoCAD's decimal degree entry. The convention in LDD can only go to 59 in the minute and second position.

The Azimuth method of angles assumes 0 degrees to be north and measures angles clockwise from 0 to 359.5959 (*dd.mmss*) (see Figure 2–11). Bearings divide a circle into four 90-degree segments. The top two quadrants, the northeast (quadrant 1) and northwest (quadrant 4), assume that a vector deflects from north to the east or west. So the angle of a vector varies from 0 degrees, a line traveling due north, to 90 degrees, a line traveling due east or west. The bottom two quadrants, the southeast (quadrant 2) and southwest (quadrant 3), assume that a vector deflects from south to east or west. So the angle of a vector varies from 0 degrees, a line traveling due south, to 90 degrees, a line traveling due east or west.

The Turned Angle routine simulates the station (pivot point), backsight and foresight process in Surveying. When you run this routine, you establish a station and a backsight by selecting a line or two points. When you are selecting a line, the endpoint nearest to the selection point is the pivot point and the farthest endpoint is the backsight point (the direction of the zero angle). When you are in point mode, you are selecting two points, the first is the instrument (pivot) point and the second is the backsight point (the direction of the zero angle). From either of these methods, angle and distance entries establish new points. The Turned Angle routine is static in that the pivot and backsight points remain the same until you change their location.

You toggle the Turned Angle routine into position mode by typing the letters **po** at the routine's prompt. In point mode the definition of a setup requires positions in coordinate space to establish the pivot and backsight points. In point mode, the first selected point becomes the instrument (pivot point) and the second point becomes the backsight point. Once in points mode, you indicate positions by

AutoCAD object snaps, point numbers (the *.p* toggle), graphically (the *.g* toggle), or by northings/eastings (the *.n* toggle). This lets any point object or coordinate be a pivot or a backsight point.

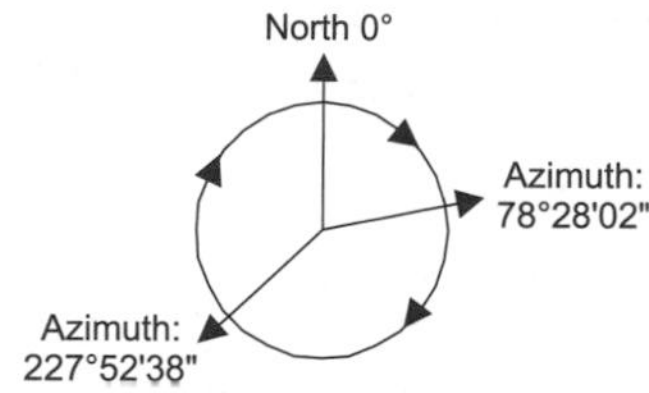

Azimuths:
Assumes North is 0°
Measures Angles Clockwise
Angles are between 0° and 360°

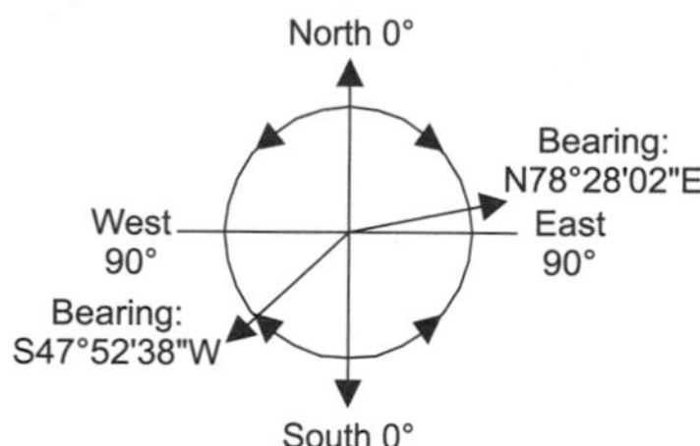

Bearing:
Assumes North or South is 0°
Angles are between 0° and 90°
East or West

Figure 2–11

There are many routines that have a POint option in the prompt. The By Turned Angle in the Create Points Flyout of the Points menu is an example of a routine using this option.

```
SELECT (POints/<Line>): (requires an AutoCAD line entity)
```

-or-

```
SELECT (POints/<Line>): PO  (toggles on Pointing mode)
SELECT (Line/<POint>):  (AutoCAD coordinates or object snap)
```

-or-

```
SELECT (POints/<Line>): PO   (toggles on Pointing mode)
SELECT (Line/<POint>):    (AutoCAD coordinates or object snap)
SELECT (Line/<POint>): .p  (toggles on Point Number mode)
SELECT (Line/<POint>): Point number:
```

-or-

```
SELECT (POints/<Line>): PO   (toggles on Pointing mode)
SELECT (Line/<POint>):  (AutoCAD coordinates or object snap)
SELECT (Line/<POint>): .n  (toggles on northing/easting mode)
SELECT (Line/<POint>): Northing: 4528.4562
Easting: 7524.6452
```

The reference methods place points in relation to an existing AutoCAD object. These routines include Automatic, Station Offset Object, Along a Line/Curve/Spiral, On Line/Curve/Spiral, Divide Object, and Measure Object. Each of these routines requires existing objects to reference or use.

The On Line/Curve/Spiral routine places point objects at the ends of lines, curves or spirals, radius points, and point of intersection (PI) points of arcs. The routine places points on a per-object basis, which does not check for points on adjacent endpoints. Because of this lack of checking, the routine will create duplicate points at adjacent endpoints.

The Automatic routine places points at the same locations as the On Line/Curve/Spiral routine, except for the PI point. The Auto routine works with a selection set of lines, curves and spirals. Before the routine places the points, it scans the selection set coordinates for duplicates and if it finds any, places only one point into the drawing at the coordinates.

Both the On Line/Curve/Spiral and Automatic routines do not scan the point database for preexisting points.

The Polyline/Contour Vertices Manual or Automatic routines create points whose elevation is the elevation found at all (Automatic) or selected (Manual) vertices of a polyline or contour. The elevation of the point will be 0 (zero), but the elevation represented by the point will be from the polyline's or contour's actual elevation.

CREATE POINTS–INTERSECTION

There are three types of intersection routines (see Figure 2–12). The first type of intersections is theoretical intersections, which include intersections between directions and distances from known points. The second type of intersections assumes the existence of line and arc entities in the drawing. The lines and arcs define directions and/or distances for the point computations. The last type of intersection works with defined alignments.

Many of the routines calculate a single solution. There are, however, routines that calculate two possible solutions. For example, the Distance/Distance routine creates two possible solutions. The routine identifies each solution with a red X in the drawing editor. You choose a solution by selecting nearest the red X with the mouse, or you can select All solutions to place two points.

Many of the routines allow for an offset to the defined direction. The convention of negative as left and positive as right towards the direction determines whether an offset is a negative or positive number.

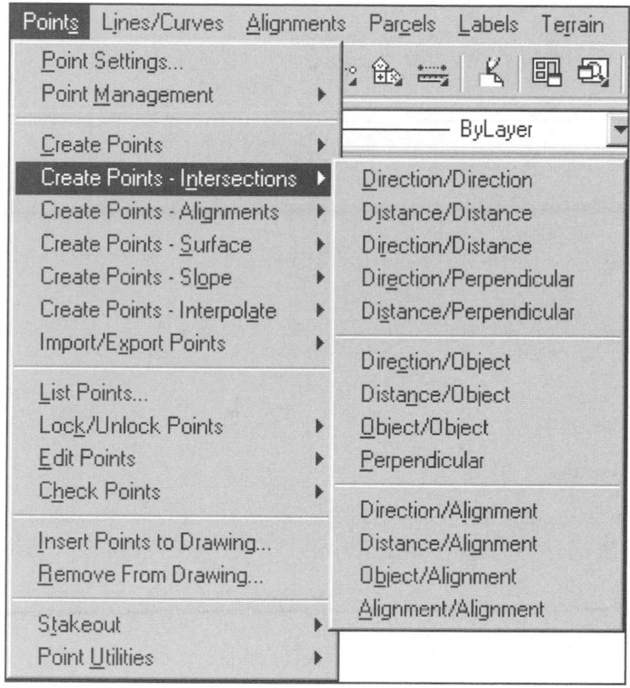

Figure 2–12

SURFACE/SLOPE/INTERPOLATE

The routines of these three flyout menus will be discussed in appropriate situations in later sections of this book.

IMPORT/EXPORT

At the middle of the Points menu are the Import/Export routines (see Figure 2–13). If you create or receive coordinate data in an ASCII file format, the Import routine converts the coordinates found in the file into point data for the point database. The Format Manager defines file formats that can accommodate almost any file format for importing or exporting point data.

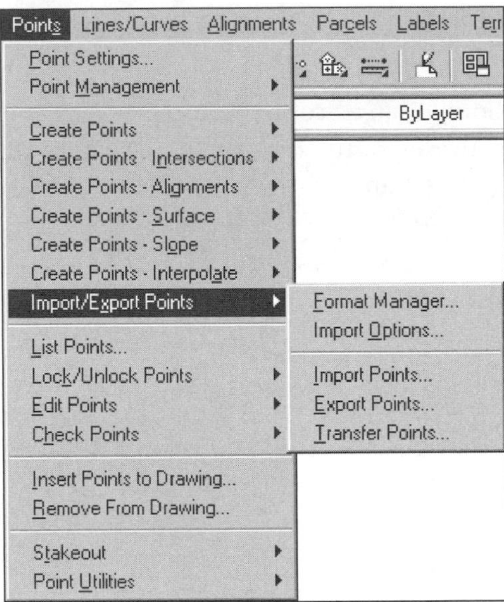

Figure 2–13

FORMAT MANAGER

The Import routine has several predefined file formats. These formats appear in Format Manager. If a particular format is not defined, Format Manager allows you to define new formats. As a result, the Import Points routine is able to read information from any external ASCII file format. The type of information imported depends upon what is in the file. If you define a coordinate system for the drawing, the Import routine can import longitude and latitude values into the drawing or points from a different coordinate system. When you import points from a different zone, the Import routine will calculate the new coordinates (transform) for the points.

The Import routine supports files that are comma-, space-, or column-delimited. The format of the file dictates which format the Import routine uses. You must exercise care when creating a column-delimited format for a file. In a column-delimited file, a decimal point is a part of the overall field width.

A coordinate file consists of columns and rows. The column portion of a file is the field, and a row is a record, or a unique occurrence of the fields in the collection of observations. In LDD, a field is any of the following values: point number, Northing, Easting, elevation, description, latitude, longitude, convergence, or scale factor. Each row in the imported file should have data for most or all the fields.

Before you can import a file, you must know its format. You can view the file in Notepad to determine the structure of the file. One very important piece of information about the file is which entry in the record represents the Northing and Easting values.

If you define the format incorrectly, the points will appear incorrectly in the drawing. Here is an excerpt from a coordinate file:

```
1,5530.83673423,4967.36847364,723.74,PP 60
2,5635.36849278,4952.84689347,722.84,IPF
```

The file contains the following: point number, northing, easting, elevation, and description. The file is comma-delimited. The Format Manager list includes a PNEZD comma-delimited format, which matches this file's structure.

IMPORT OPTIONS

The Import Options dialog box presents three decisions you must make when importing points into a drawing (see Figure 2–14). If there is more than one file to import, these options may change depending upon the information in the ASCII file.

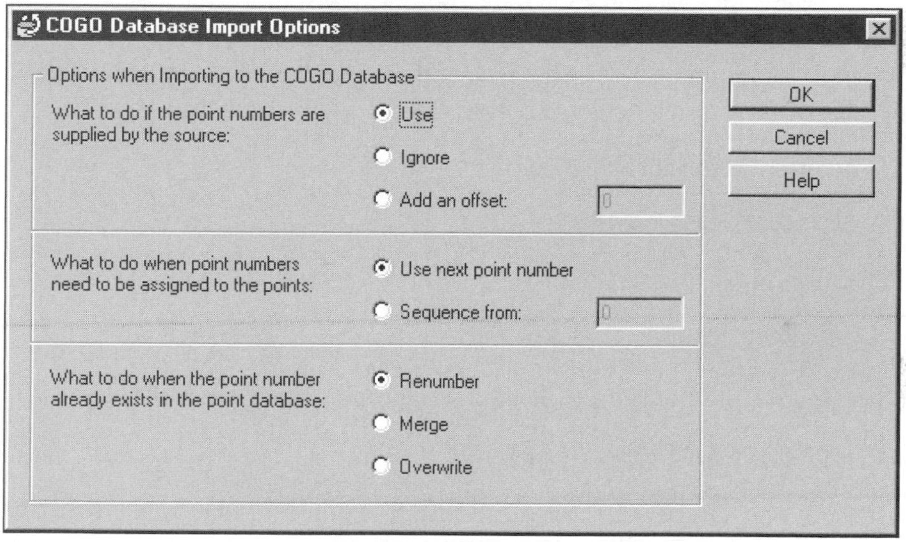

Figure 2–14

The upper portion of the panel concerns ASCII files that contain point numbers. The three choices are to Use, Ignore, or Add an offset to the point numbers. If Ignore is on, the middle portion of the dialog box determines how to assign the new point numbers. If Add an Offset is on, you need to set the offset value.

The middle of the panel determines which point numbers to assign. The first option is to use the next available point number. This means that if the ASCII file contains point numbers and you choose to ignore them, the Import routine will assign new point numbers by next available. For example, the ending point number of the point database is (109) and LDD assigns the next point number to the first record of the imported file, 110. The next available is a point number one higher than the last used point number.

If the Sequence From toggle is on, the Import routine assigns sequential numbers from this seed number. You must enter that number.

The bottom portion of the panel determines the method for handling point numbers from the ASCII file that have the same number as a point in the point database. There are three methods to handle this situation. The first method is to renumber the new point. How the point gets renumbered is set in the middle portion of the dialog box, Use Next Point Number or Sequence From.

The second option is to merge the point data. This method assumes the same point number in both the source file and project point database. The method takes the point data of the source file and applies it to points in the point database. The resulting entry is a combination of the source and point database information. For example, the source file has point number, northing and easting data. This source file data will update the values of the same point number in the point database.

Source File information for point 101:

```
101, 5000.0000, 6000.0000
```

Point Database entry for 101:

```
101, 100.0000, 100.0000, 723.84, IPF
```

With merging on and the source file read, the new entry for point 101 is:

```
101, 5000.0000, 6000.0000, 723.84, IPF
```

The last method is Overwrite. This means that the point number information in the import file will replace a duplicate point number's information in the point database. An example of this process is:

Source File information for point 101:

```
101, 5000.0000, 6000.0000
```

Point Database entry for 101:

```
101, 100.0000, 100.0000, 723.84, IPF
```

With overwrite on and the source file read, the new entry for point 101 is:

```
101, 5000.0000, 6000.0000, NULL, NULL
```

Exercise

You can create point data in many ways. The methods you use to develop points depend upon the initial data and job requirements. The point creation routines of LDD are for specific situations. You may never use all of the routines available in a single job. Some routines mimic survey field techniques, others assume input from a listing, and others transcribe notes into points.

After completing this exercise, you will be able to:

- Create points through manual methods
- Create points with intersection routines
- Create points by importing an ASCII file

MANUAL AND INTERSECTION METHODS

1. If you are not in the drawing from the previous unit, open the drawing *LDDTR2* that is in the proj1 project.

Figure 2–15 shows the starting point, bearings, and distances that will be used to complete this exercise. The first point is at the northing/easting of 10000,10000.

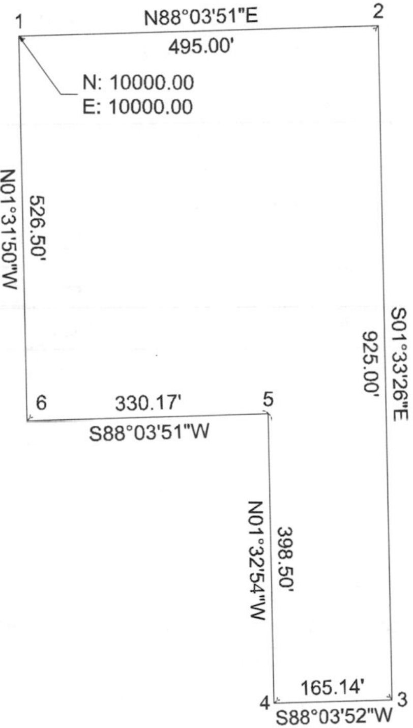

Figure 2–15

2. From the Create Points flyout of the Points menu, select the Northing/Easting routine. The routine prompts for each value separately. After entering the following values, press ENTER once to exit the routine.

   Northing: **10000**

   Easting: **10000**

   Description: **BCP**

   Elevation: **724.00**

   You will not see the point appear on the screen because the point coordinates are not within the current display.

   This is the only northing/easting to enter. If you have a long list of values, it would make sense to type the values into a file and then import the points through the Import/Export routine.

3. Use the Zoom To Point routine of the Point Utilities flyout of the Points menu. Enter **1** as the point number and **1375** as the zoom height. Use Pan to pan the point towards the top of the screen. If you have an IntelliMouse simply press down and hold the center wheel to pan the drawing.

4. From the Create points flyout of the Points menu, select the Direction routine. The routine prompts for a starting point, then a direction and distance. After you identify the starting point, the routine responds with a request for a direction and distance, in this case the direction and distance to point 2.

   Use the *.p* method to identify point 1 as the starting point and use the bearings and distances to travel around the boundary of the subdivision found in Figure 2–15. Remember that the bearing needs to have a quadrant specified. Quadrant 1 is traveling Northeast, 2 is Southeast, 3 Southwest and 4 is Northwest. Use the following elevations and descriptions for the new points.

Point Number	Elevation	Description
2	721.00	BCP
3	724.00	BCP
4	720.10	BCP
5	727.83	BCP
6	724.77	BCP

   Exit the Direction routine by pressing ENTER twice and save the drawing.

5. Use the Zoom Window command to zoom in on a point. Make sure you have a reasonable amount of space above and below the point.

6. Use the Zoom to Point routine in the Point Utilities flyout and zoom to point 1. The routine prompts you with the last screen height from step 5. Use a height of 300 to view the area around the point.

EXERCISE

7. From the Create Points flyout of the Points menu, select the Turned Angle routine. The initial prompt for the routine is line mode. Since there are no lines in the drawing, toggle the routine into point (POint) mode by typing **po**. Since we left the Direction routine in point number mode (*.p*) and did not exit the mode, the Turned Angle routine will start prompting for point numbers. The first point will be point number 1 and the second point will be point number 2. Use the following angles and distances to set points 7, 8, 9, and 10.

Angle Type/Angle	Distance	Description	Elevation
Turned 125.4523	35.7454	IPF	725.32
Turned 205.1634	45.6875	SMH	721.55
Turned 150.4642	55.5483	DMH	720.64
Turned 134.3422	65.5248	DMH	719.85

8. Press ENTER to exit the routine.
9. Click on the save icon to save the drawing.
10. Do a Zoom Extents to view all of the points.
11. Do a zoom with a .7x scale factor.

```
Command: z ZOOM
Specify corner of window, enter a scale factor (nX or nXP), or
[All/Center/Dynamic/Extents/Previous/Scale/Window] <real time>: .7x
```

12. From the Create Points–Intersections flyout of the Points menu, select the Distance/Distance intersection routine. Use the following values to place point 11. This is a situation where the routine creates only one solution. Select the red 'X' to place point 11.

Point 11

Point	Radius	Point	Radius	Desc	Elevation
1	165.32	2	329.68	CL-CL	721.22

13. From the Create Points–Intersection flyout of the Points menu, select the Direction/Direction intersection routine. The convention for offsets is to the right of the direction is a positive offset and to the left is a negative offset. For example, from point 1 to point 2 south of the line is positive and north is negative. Use the following data to create points 12 and 13. There is only one solution for points 12 and 13.

EXERCISE

Point 12

Start Point	Quadrant	Bearing	Offset	Elevation	Description
1	1	88.0351	33.00		
11	2	01.3150	33.00	723.44	LC

Point 13

Start Point	Quadrant	Bearing	Offset	Elevation	Description
1	1	88.0351	33.00		
11	2	01.3150	-33.00	724.14	LC

14. For the next three points, use the Distance/Distance intersection routine found in the Create Points–Intersections flyout of the Points menu. When prompted to select another solution, press ENTER to exit. Press ENTER to rerun the routine. Use the following data to create the points.

Point 14 – Select northern solution

Point	Distance	Point	Distance	Elevation	Desc
1	125.00	11	100.00	720.82	DMH

Point 15 – Select northern solution

Point	Distance	Point	Distance	Elevation	Desc
11	110.00	2	250.00	721.15	DMH

Point 16 – Select northern solution

Point	Distance	Point	Distance	Elevation	Desc
11	300.00	2	75.00	721.32	DMH

15. When finished entering the point data, save the drawing.

EXERCISE

IMPORTING COORDINATE DATA FROM ASCII FILES

The Import routine reads coordinate data that results from the processing of an external Coordinate Geometry program like PACSOFT, and data collector download packages like SDR Link or TFR Link. The routine reads the coordinate file and places LDD point objects in the drawing representing each record in the external file.

1. If you are not in the drawing from the previous exercise, open the drawing LDDTR2 that is in the proj1 project. Select the Format Manager routine from the Import/Export cascade of the Points menu. In the manager, select Add to add a new format (see Figures 2–13 and 2–16). When asked to select a format type, select User Point File.

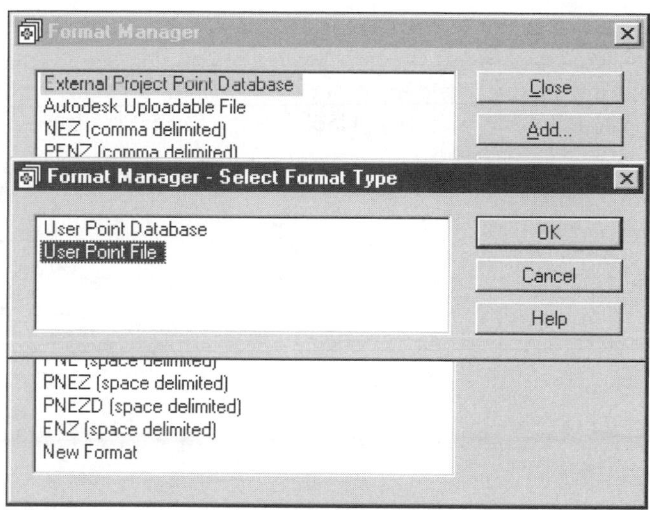

Figure 2–16

2. The first entry in format dialog box is the name of the format, STDNEZ.
3. Set the default extension to nez.
4. The next item defines the structure of the file. Is the file columnar or field with a delimiter? This will depend upon the structure of the ASCII file. The file we are working with has a field structure with a comma delimiter. Set the type to Delimited By and type a comma in the box to the right of the radio button.
5. If you want to support comments in the ASCII file, place the symbol that indicates that the following information is a comment. If the comment symbol is a # (pound sign), place a # (pound sign) in the Comment Tag box. When importing the ASCII file and the program encounters a number sign, the Import routine will ignore the information from the sign to the end of the line.
6. The Read No More Than setting limits the number of lines to read in from the file. If you want to import the first 100 lines, set the value to 100. The Sample Every value controls how often the Import routine reads a line of point data. Do not set a value for either Read No More Than or Sample Every.
7. The next setting, Coordinate Zone Transform, toggles on the translation of coordinates in the file. The entry below identifies the coordinate zone of the points in the import file. When the file is read, the routine will translate the coordinates into the coordinate system defined for the drawing. For this exercise, do not toggle on the transformation, and leave the entry blank.
8. Select the first header at the lower left of the dialog box. In the new dialog box, select the drop arrow and select number as the first entry in the file format. The format uses a . (period) to indicate a bad coordinate. Enter this in the lower portion of the current dialog box. Select OK to return to the format dialog box.

9. The format indicates the field order for each record by identifying the field topic from list of field types. Select the next header, and in the new dialog box select the drop arrow and choose a field topic. Set the topics to northing, easting, elevation, and description. Your format should look like the format in Figure 2–17.

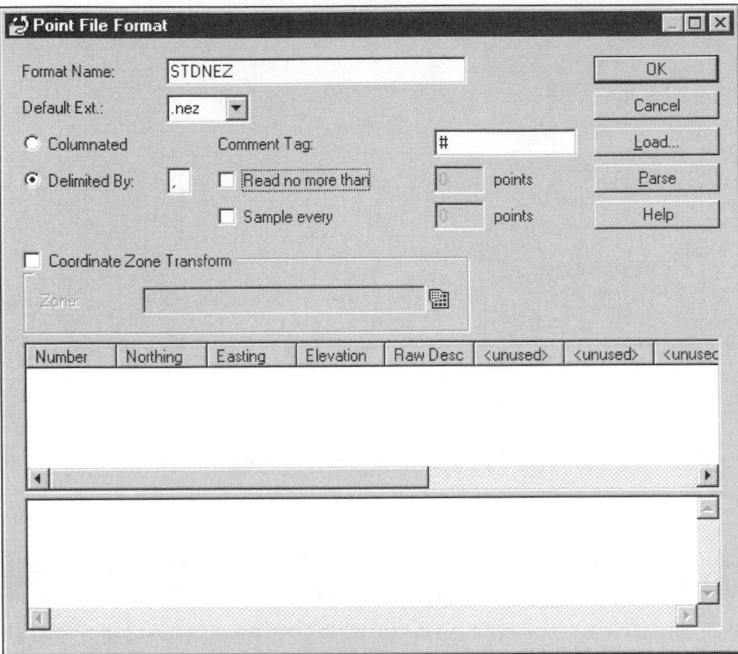

Figure 2–17

10. After you have defined the format, select the Load button and select the *napa.nez* file that is on the CD that comes with this book.

11. Select the Parse button to view the results of the format. If the format reads the file correctly, select OK to exit to the Format Manager. The new format should be on the list. Select the Close button to close the Format Manager.

12. Create a layer (neutral) and make it the current layer.

13. In the Import/Export Points flyout, select the Import Options routine. Set the method of adding points to the Cogo point database to Add an Offset. Set the offset to 10000 (see Figure 2–18).

14. Use the Import Points routine from the Import/Export cascade of the Points menu. In the dialog box, set the format to STDNEZ and select the *napa.nez* file. To see if any transformations are set, select the Advanced button to view the transformation settings. All of the toggles should be off. Select OK in the dialog box to import the points into the current drawing.

15. After importing the points, save the drawing and execute a zoom extents to view all of the new points in the drawing.

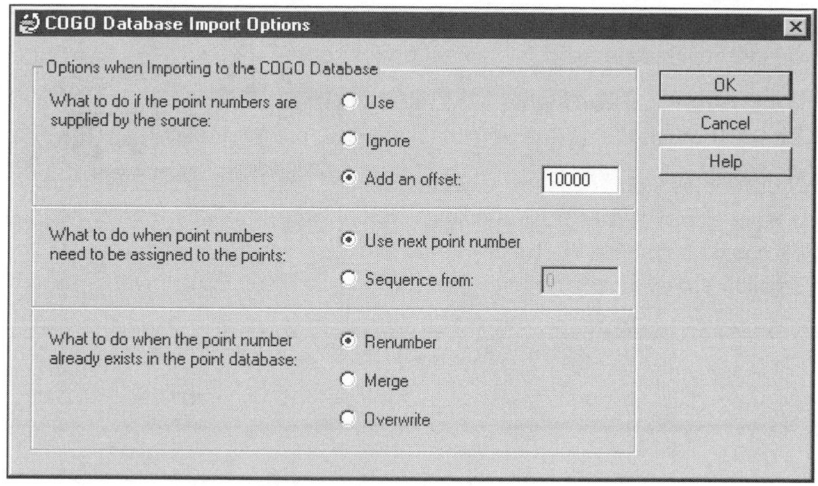

Figure 2–18

EXPORTING AN ASCII FILE

The Export routine of the Points menu exports some or all of the points in the project point database. The Export routine creates an ASCII file that you can distribute to colleagues needing the coordinate data.

1. Select Export Points from the Import/Export flyout, set the format to STDNEZ, select the file name icon and enter the name **napaall.nez**, and export all of the points.
2. Run the Windows Notepad program to view the new point ASCII file. The file should include all of the points 1-15001.

UNIT 3: ANALYZING AND REPORTING POINT DATA

LIST POINTS

The analysis of point data occurs in the List Points dialog box of the Points menu (see Figure 2–19). This dialog box allows you to create a list of points by simple or advanced filtering methods. The simple filtering methods are All Points, a Drawing Selection Set, or Point Group. When you select Drawing Selection Set, the Select button hides the dialog box and prompts you to create a selection set of points. The selected points appear in the bottom portion of the dialog box. The last method of selecting points is by group. Groups will be discussed later in this unit. A group is simply a collection of points known by a common name. Whichever method you use, the dialog box places the point numbers at the top and the point information at the bottom.

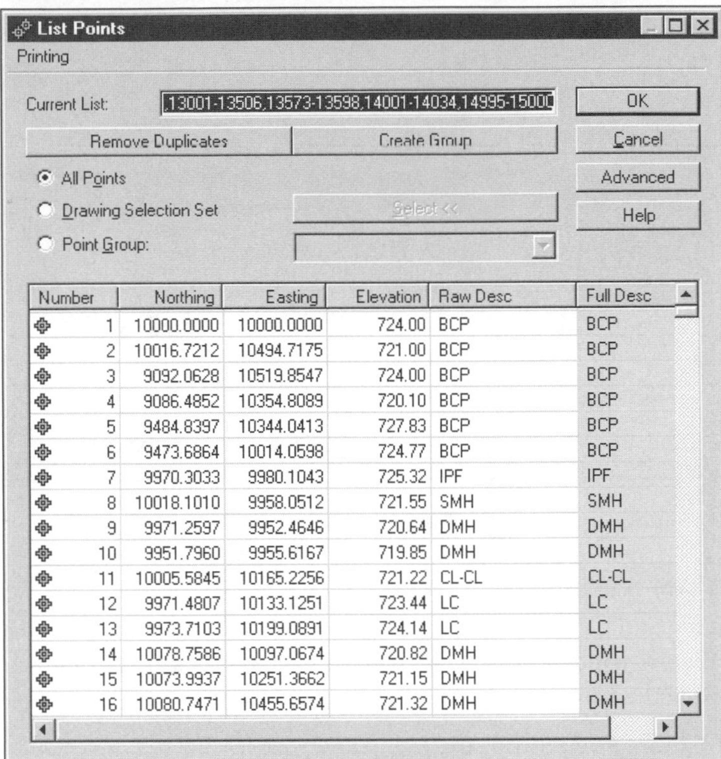

Figure 2–19

If it is necessary to sort the list, select the header of the column, and the list sorts. The first selection of the header sorts list in ascending (lowest first) order. If you select the same header again, the list sorts in descending (highest first) order. This is a result of the point database being an Access database. By selecting each header, you sort the list according to the values in the column.

The headers can be rearranged by clicking and holding down the left mouse button and moving the header to a new location. LDD retains the order of the headers from session to session.

Advanced filtering allows for a much more complex selection process (see Figure 2–20). The Advanced button displays a dialog box with three tabs: Source, Filter and List. The Source tab allows you to create a set of points for filtering. The source must be set before moving on to the filter tab. The Filter tab adds or removes points to the list from the source points (see Figure 2–21). The methods of filtering are point number or elevation range, point name (alphanumeric value), description (raw or full) and XDRef matching (point groups with overrides). A point number range can be sequential (1–10), out of sequence (1,4,7,2,18), or a combination of the two (1–5,10,12,20–25).

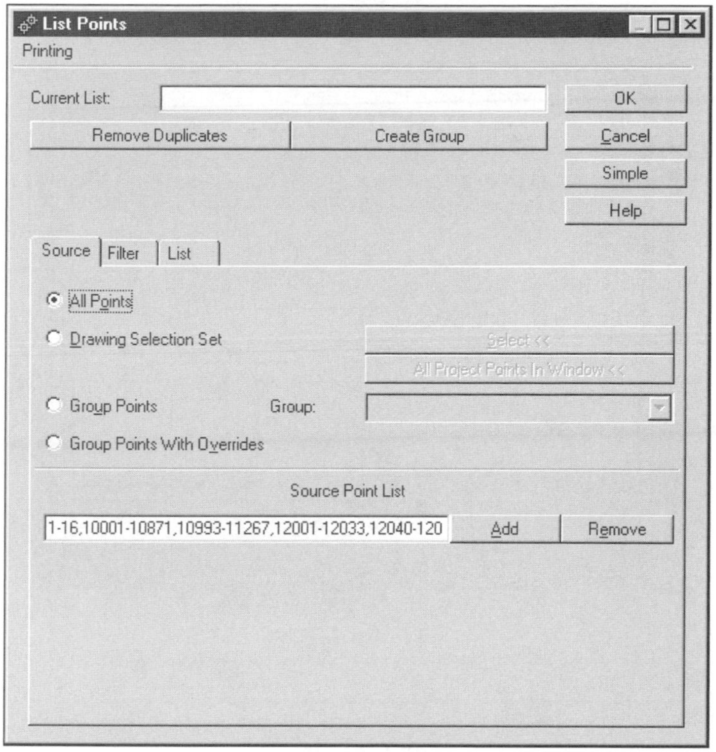

Figure 2–20

At the bottom of the dialog box is a toggle indicating whether to Include or Exclude points matching the values set in the middle of the pane. If you select the Filter button, the routine filters the points. The points matching the filter appear in the area to the right of the button. The last decision is to Add or Remove these selected points from the current list at the top of the dialog box. Once the list is complete, select the List tab and the panel shows the filtered points.

The list dialog box shows all of the pertinent information about the points including their raw and full description. The raw description is the original value for the point. The full description is a potentially converted description of the point. For example, the raw description OK 6 can be translated to 6 INCH OAK. This translation is done through description keys. Description keys are covered in Unit 5 of this section.

After you create a list in the dialog box, the next task is printing or creating a file. The Printing menu at the top of dialog box selects the method of output. If you select Print, the output goes directly to the system printer. The Print to File selection looks at the values in Output Settings to generate a report. In the Output Settings dialog box, you can name and define reports sent to a file (see Figure 2–22). Access to Output Settings is through the Drawing Settings routine of the Projects menu.

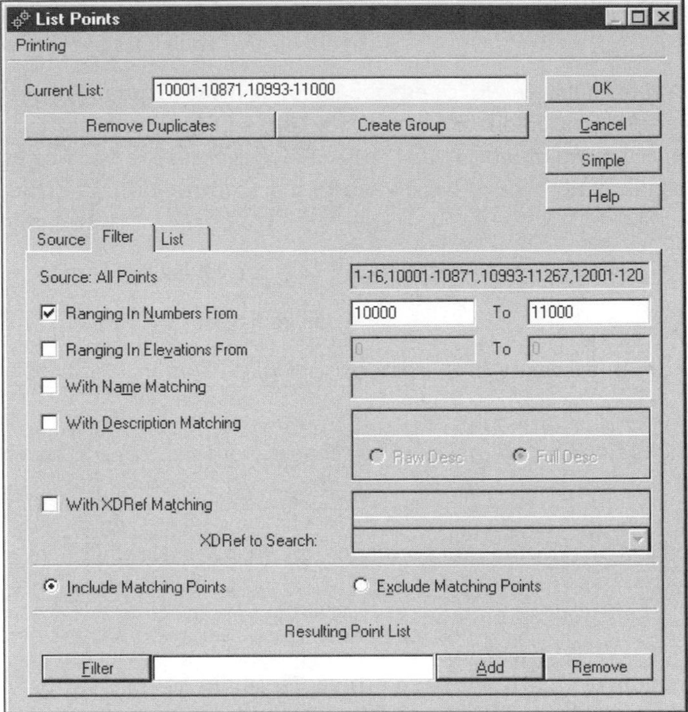

Figure 2–21

Output Settings

Output Options
File
Screen

Output Format
Date
Page numbers
Title
Sub headers
Page breaks
Overwrite file
Page length: 60
Page width: 80
Left margin: 0
Right margin: 0
Top margin: 0
Bottom margin: 0
Output File Name ...
output.prn
OK
Cancel
Help

Figure 2–22

When you activate the grips of a set of points and click the right mouse button, List Points is an available choice on the shortcut menu (see Figure 2–23). Rather than returning to the menus to execute a command, you can use the limited set of commands on the shortcut menu. The options in this menu change depending upon what type of object you select. If you select a group of points and a line, the options are different than if the selection set were all points. When creating a selection set of points you must be careful to select only points, otherwise the shortcut menu options will be different from what is in Figure 2–23.

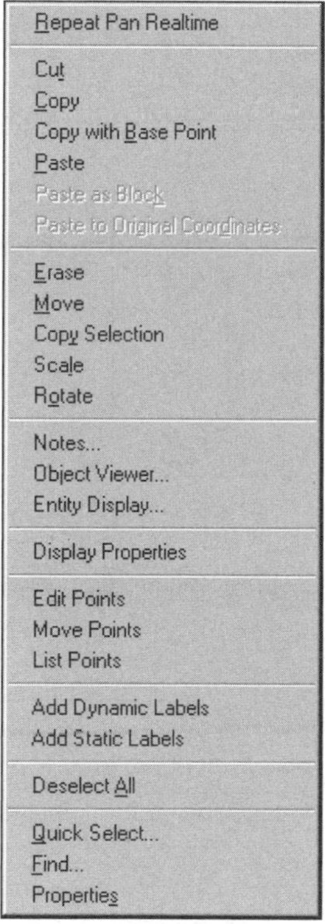

Figure 2–23

POINT GROUPS

Many times the reason for listing points is to view points that are similar either in description or location. There are times when you want the resulting list to be permanent and assigned a name. A point group allows you to assign a name to a list of points. By selecting the Create Group button at the top of the List Points dialog box, you create a named group of points (see Figure 2–19). There are routines in later sections of this book that require groups of points to be able to process their information.

The Point Group Manager is another place to create point groups. The routine is in the Point Management flyout of the Points menu (see Figure 2–24). The Group Manager dialog box lists the project's defined point groups (see Figure 2–25). The creation of groups is the same as creating a point list in the List Points routine except for saving the list as a named group.

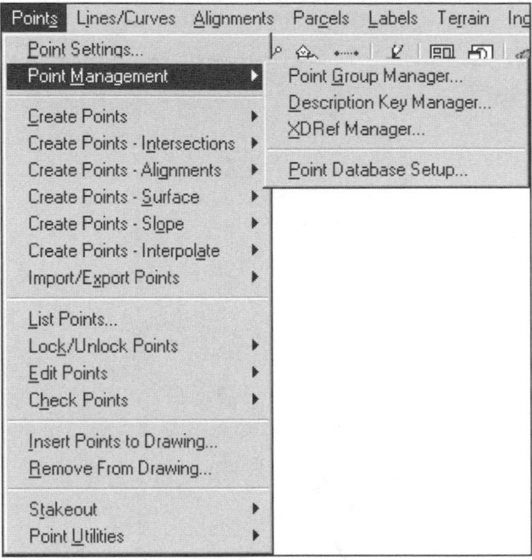

Figure 2–24

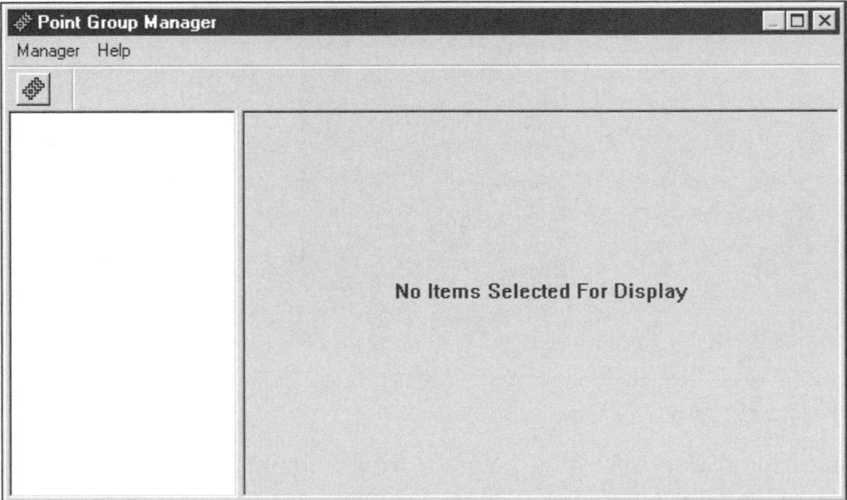

Figure 2–25

When you are in the Manager and select the Create Point Group icon, the Create Point Group dialog box displays (see Figure 2–26). You enter the name and the list of points. If you are going to create the list by selecting and/or filtering points, selecting the Build List button displays the same dialog boxes found in the List Points routine (see Figure 2–19). The two methods, Simple and Advanced, create the list of points later saved for the Group. The Simple method uses All points, a Selection Set of points, and a preexisting point Group. The second method uses the same sources of points as Simple, but allows for the filtering of points to add or subtract from the Group. See the earlier discussion about List Points.

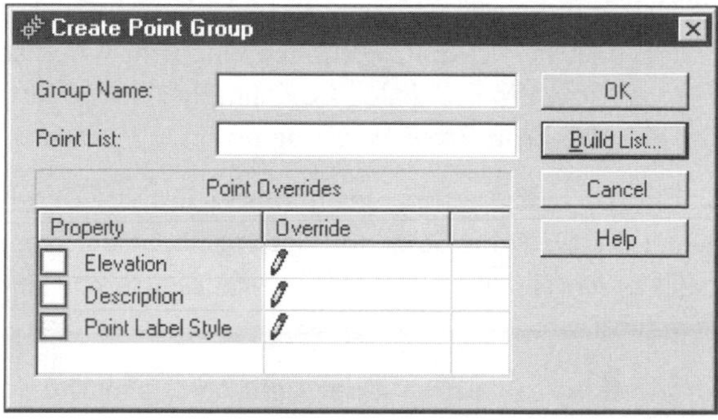

Figure 2–26

The Point Group Manager allows you to create overrides to the Elevation, Description, and Point Labeling Style. These overrides allow you to display vertical or over time measurements. For example, a series of material elevations come from a core log or daily reading of pollution amounts over time from a single sampling point. The point overrides allow you to show the same point with its different measurement and description values. The definition and use of point overrides will be a part of the Terrain Modeling section of this book.

Exercise

This exercise reviews uses the point Listing and Point Group Manager routines found on the Points menu.

After completing this exercise, you will be able to:

- Create a Simple point list
- Create an Advanced point list
- Create a point Group

POINT LISTING, ADVANCED SELECTION METHODS, AND POINT GROUPS

1. If you are not in the drawing from the previous unit, open the drawing LDDTR2 that is in the proj1 project.
2. Select the List Points routine from the Points menu, select the All button at the top of the dialog box, and view the points in the project.

3. To sort the point list, select a header. Try this with a couple of different headers. Remember, if you select a header twice, the sorting changes to descending instead of ascending.
4. If you want to rearrange the headers, select a header, hold down the left mouse button, and slide the header to the desired position.
5. The Print Preview button shows how the file will print when sent to the printer. To print the list as shown in the dialog box, simply select Print in the upper left of the dialog box. Exit the Print Preview dialog box.
6. Clear the point list by erasing the point numbers in the Current List area of the dialog box. Select the Drawing Selection Set toggle. Select the selection set button and build a new set of points. When you return to the dialog box, you can view the new point list.
7. Select the Advanced button at the upper right to display the advance filtering tabs. The points currently selected remain in the point list and are a source for filtering.

8. Clear the list at the top of the dialog box, select the Source tab, and set the All Points button on. The list of points at the top is not the final set of point numbers. It is the filtering of all points that produces the final list of point numbers you want to appear in this area.
9. Select the Filter tab. The Filter tab allows you to use several conditions to add and/or remove points from the final list. Currently our point source is all points in the project, but since there are no points in the list at the top portion of the dialog box, the points will not show when you select the List tab. After setting the filters for the points, you add or remove them to the list by clicking on the appropriate button at the bottom of the dialog box. To view the selected the points, you click on the List tab. After viewing the points, you can return to the Filter panel by clicking on the Filter tab. When back at the filter panel you can add, remove, or erase points from the list at the top. Use each of the following filters to create its point list. After creating the list, view the list, return to the filter panel, erase the current list of points, and create a new list with a new filter. When you use text to select points, the selection process is case sensitive.

Point Number Range:	1 – 1000
Point Number Range:	10000 – 20000
Raw Description:	IPF
Description:	BCP
Elevation Range:	700.00 – 800.00

10. If you have not done so, clear the points from the current point list at the top of the List Points dialog box.

11. Create two point groups using point number ranges. After selecting and viewing the point list, click the Create Point Group button to name the point group. Remember to clear the point list at the top before creating the next point list and group. Use the following point number ranges.

    Group Name: Exercise - Point Number Range 1 – 1000

    Group Name: Calculated – Point Number Range 10000 – 20000

12. Select OK to exit the List Points dialog box.
13. You can call the List point dialog box from the shortcut menu. Select a number of points to activate their grips. While the grips are active, click the right mouse button and select List Points from the shortcut menu.
14. Save the drawing.
15. From the Point Management flyout of the Points menu, select Point Group Manager. The manager will already have two point groups, Exercise and Calculated. Each of these groups can be inserted in the drawing by simply referring to their name in the Insert Point into Drawing routine.

The next portion of this exercise creates a new point group that is a subset of points from the Calculated group. The new group, subdivision, is a filtered group of points representing the Napa Subdivision.

1. To create the group, Napa Sub, select the Create Group icon. In the Create Group dialog box, enter the name of the group, **Napa Sub**. After entering the name, select the Build List button on the right side of the dialog box. When the simple filtering dialog box appears, select the Advanced button to display the Advanced filter button.
2. Select the Source tab and toggle the source to Group points, select the group Calculated, but do not click on the Add button. If the Current List at the top includes any points, erase the point numbers from the List.
3. Select the Filter tab, toggle on With Description Matching, set the description type to RAW, and enter **BP** (boundary point). Select the Filter button at the bottom of the dialog box, and when the points appear in the Resulting Point List area, add them to the Current List by clicking on the Add button. Select two more points by point number range, 10345 and 10754. You will need to enter each point number as the beginning and end of the search range. Your results should look like Figure 2–27. Select OK twice to exit the Filtering and Create Point Group dialog boxes.
4. Create one more group, Iron Pipes Found. The group is from the Calculated point group and is all points with the RAW description of IPF. After creating this group, exit to the Group Manager panel.

EXERCISE

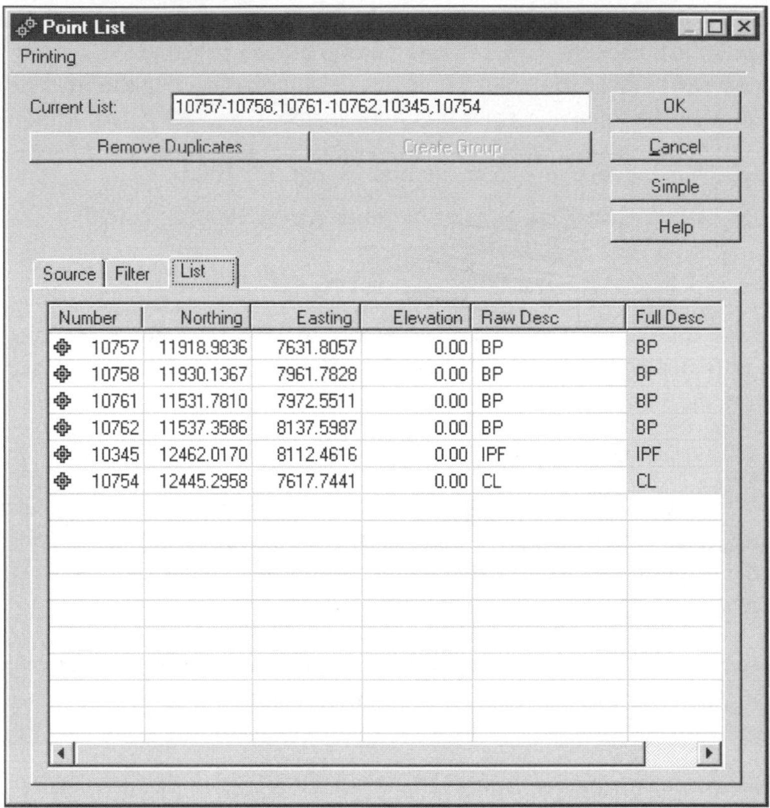

Figure 2–27

5. In Group Manager, select the + (plus sign) next to the group's name to view the point list. Select a single point in the list and the listing area will show only the one point. To view all of the points in the group, select the group name.
6. Exit the Point Group Manager by selecting Exit from the Manager menu or clicking the X at the upper right corner of the dialog box.
7. Click on the save icon to save the drawing.

UNIT 4: EDITING POINT DATA

EDIT POINTS MENU

You can edit point object values by using one of several editing tools found in the Edit Points cascade of the Points menu (see Figure 2-28). The routines of this menu edit the point's number, point name, elevation, description, and description key. The routines select points through the same methods found in the List Points dialog box. See the earlier discussion about the List Points dialog box.

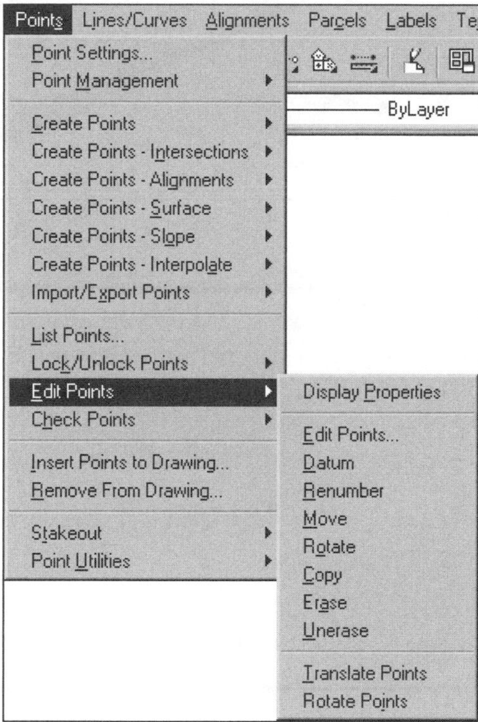

Figure 2–28

DISPLAY PROPERTIES

The Display Properties routine modifies the display settings for points currently in the drawing. After you select the points, the routine displays the Point Display Properties dialog box (see Figure 2–29). The three tabs, Marker, Text and Reset, control node size and symbol, text colors, relative or absolute sizing, and resetting the point. The routine also appears in the shortcut menu.

If you activate the grips of a set of points and then click the right mouse button, LDD displays a shortcut menu containing several editing routines and Display properties routine (see Figure 2–23). Rather than returning to the menus to execute some commands, you can use the commands that reside on the shortcut menu. The options in this menu change depending upon what type of objects are in the selection set. The routines that apply to editing and manipulating points only appear in the shortcut menu when the selection set contains only points.

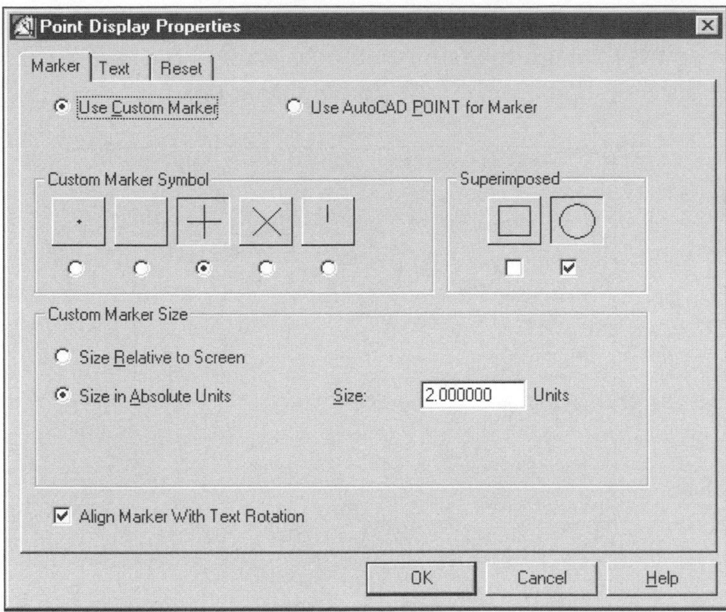

Figure 2–29

EDIT POINTS

The routine calls an editing dialog box. You first select the points and then edit any of the information associated with a point: number, point name, elevation, description, and raw description. This routine also appears in the shortcut menu. Editing the coordinates of a point will change the position of the point in the point database and the drawing.

DATUM

The Datum routine manipulates the elevation of the point. The change of elevation is either an absolute or a calculated value. For example, the field crew assumes a benchmark elevation of 100 feet for a topographic survey. After researching the benchmark, the surveyor determines the true elevation of the benchmark as 435.34. The Datum elevation routine prompts for the old and new bench elevations and then calculates the amount to add to each selected point, the difference between the original datum elevation and the new researched value. The change of elevation can be a positive or negative value. In the Absolute Change option, you supply a positive or negative value that the routine adds to the selected set of points.

Another situation where the Datum routine comes in handy is for a blown setup elevation of a survey. If the surveyor records a set of points with an incorrect setup elevation, all the point elevations recorded from the setup are incorrect. If the Surveyor discovers the error, he can use the routine to adjust all the affected points. In this case, you specify the wrong and the correct elevations or the difference between the two elevations and the datum routine adjusts all the selected points.

RENUMBER

The Renumber routine renumbers points by an additive factor. If the routine renumbers a point to a number already in use, the routine prompts for a solution. The prompt asks for one of two actions. First, the default choice gives the duplicate point the next available point number. The second overwrites the existing point in the point database with the newly numbered point.

MOVE

The Move routine physically moves the points in the coordinate system. This routine also appears in the shortcut menu. The options in the shortcut menu change depending upon what type of objects are in the selection set. The routines that apply to editing and manipulating points only appear in the shortcut menu when the selection set contains only points.

ROTATE

The Rotate routine rotates the points on the screen. After rotating the points, the routine updates the point database to reflect the current condition.

The Move and Rotate routines are different from the Translate Points and Rotate Points routines at the bottom of the Edit Points flyout. The Move and Rotate routines physically move and rotate the points. After moving and/or rotating the points, the routines update the project point database. The Translate Points and Rotate Points routines calculate rotation angles and offsets to determine where the points are in the northing/easting system. The routines do not physically rotate or translate the points in the drawing.

ERASE

The Erase routine erases the points from the screen and the point database.

COPY

The Copy routine copies the selected points, asks for an elevation, and assigns new point numbers.

UNERASE

The Unerase routine restores any points erased in the project. If you save the drawing, the routine can restore an erased point if the point number has not been reassigned.

ROTATE POINTS AND TRANSLATE POINTS

These two routines do not physically rotate or translate points in the drawing. The routines merely calculate rotation angles and offsets to report the new angles and coordinates. It would be better to use the Move and Rotate routine at the middle of the Edit Points flyout; the points will be in the correct position in the drawing after moving and rotating them.

INSERT POINTS TO DRAWING

To bring in a subset of points, use the Insert Points to Drawing routine of the Points menu (see Figure 2–1). The selection of points is by All Points, Point Number Range, Group, a Window (geographic area), or a point filtering dialog box. The dialog box option displays the same selection dialog box from List Points. After you generate a point list and exit the dialog box, the routine inserts the selected points into the drawing.

The Insert Points to Drawing routine recognizes and uses description keys. Description keys must be on (Insert panel of Point Settings) when you insert points into the current drawing.

REMOVE POINTS FROM DRAWING

There may be times when you want to remove points from the drawing but not from the project point database. One method is to use the generic ERASE command of AutoCAD. This erases the point object from the drawing, but not the point database. LDD provides a more powerful routine, Remove Points from Drawing. This routine is similar to Insert Points, but removes points from the current drawing. The selection of points is by All Points, Point Number Range, Group, a Window (geographic area), or point filtering dialog box. The dialog box option displays the same selection dialog box from List Points. After you generate a point list and exit the dialog box, the routine removes the selected points from the drawing.

When removing points from the drawing, the routine prompts for the removal of the symbols associated with the points. If you do not answer yes to the prompt, the next time you insert points with description keys active, you will have duplicate symbols at the point's location.

CHECK POINTS

The Land Development Desktop point database is separate from the drawing. There is a semiautomatic link between the drawing and the point database. Remember, generic AutoCAD tools manipulate only the graphical portion of a point. LDD-specific routines manipulate graphics of the object and the external point database entry. There will be times when the point database and the drawing do not contain the same information. For example, you may receive a drawing from another source that has LDD. However, all you have is the drawing and not the external data files. In this case the drawing contains points that are not a part of any current project. You need a routine to update the point database to reflect the points present in the drawing.

Or, you edit the point database in one drawing, and when you open a second drawing in the same project, its drawing points do not reflect the current state of the point database. This situation requires a tool that updates the drawing to reflect the current state of the point database. Whenever there is a difference between the drawing and the point database, the Check Points routines resolve these problems.

The Check Points routines recognize and use description keys, which must be on when bringing points into the current drawing.

MODIFY PROJECT

The Modify Project routine of the Check Points cascade updates, adds, or deletes project point data based upon the points present in a drawing (see Figure 2–30). As mentioned previously, the drawing may be from another firm and you are now adding the drawing and its points to a project. The drawing may contain point objects or Softdesk point blocks. Either way, the routine scans the drawing and updates the point database from the gleaned information.

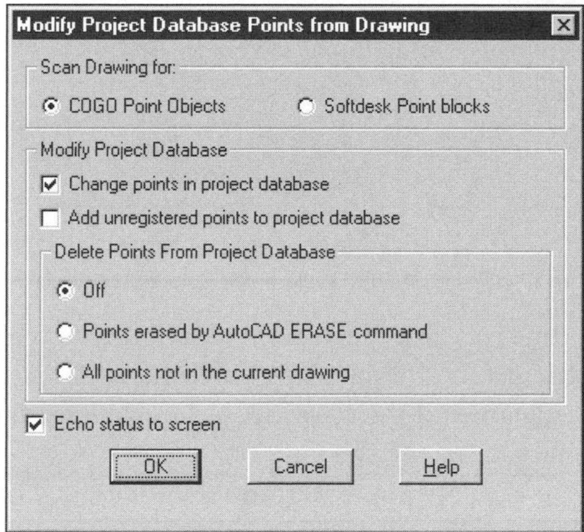

Figure 2–30

The routine performs only those functions you toggle on in the dialog box. Even though an option is on, the Check Points Modify Project routine acts only when there is a difference between what is in the drawing and the point database. If a point is missing in the project point database but is present in the drawing, the routine adds the point to the database. Therefore, the Modify Project routine updates the project point database from the state of the drawing. If there is a difference in the coordinates, elevation, or description of a point, the routine changes the point database to reflect the values in the drawing, if the Change option is on. This happens when you modify points with AutoCAD routines and want the changes to update the project point database. If points are absent in the drawing, but are present in the point database, the Modify Project routine deletes the points from the project point database, if the Delete Points option is on. The Check Points routine records the description key of a point as the routine sends the information of the point to the point database.

MODIFY DRAWING

Rather than having all the points present in the drawing, the opposite may be true. You may have a drawing where none or only a subset of the point database is present. The Modify Drawing routine of Check Points adds, changes, or deletes points in the current drawing to reflect the state of the point database (see Figure 2–31).

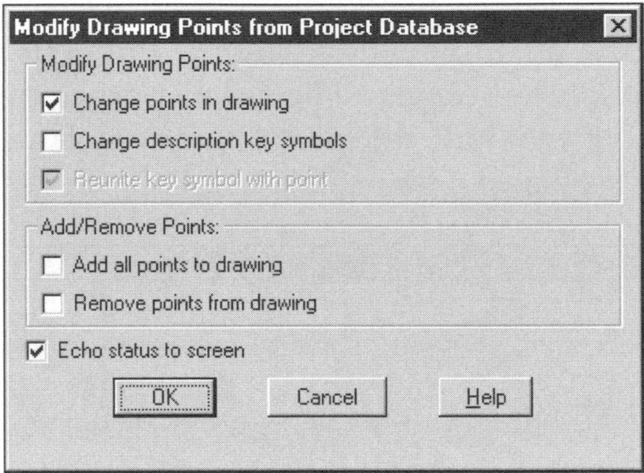

Figure 2–31

The Check Points routines perform only those functions you toggle on in the dialog box. Even if an option is on, Check Points acts only when there is a difference between what is in the point database and the drawing. When you use the Modify Drawing routine, if a point is missing in the drawing but present in the point database, the routine adds the point to the drawing, if the Add option is on. Therefore, the Modify Drawing routine brings points from the project point database into the current drawing. If there is a difference in coordinates, elevation, or description of a point, the routine updates the point in the drawing to reflect the values found in the point database, only if the change option is on. This happens when you modify point information in another drawing file and want the changes to update the points in the current drawing. If points are absent in the point database but are present in the drawing, the Modify Drawing routine deletes the points from the drawing, if the Delete option is on.

The only drawback to the Check Points Modify Drawing routine is that the routine is not selective about which points to add into a drawing. The Modify Drawing routine is an all-or-nothing proposition. If you want to be more selective about which points appear in the drawing, you should use the Insert Points routine. The Insert Points routine allows for the specification of which points to bring into the drawing.

You select points in the routine by All, a Point Number Range, a Window (geographical area), a Group, or through the filter dialog box.

The Modify Project and the Modify Drawing routines use the description key matrix, which sorts the points and symbols on to specific layers.

POINT UTILITIES CASCADE

The Point Utilities flyout provides many useful tools to locate, view, and convert points and objects in a drawing (see Figure 2–32). Many times you will be in a situation where the drawing does not contain any or a subset of the project points. The Quick View routine allows you to view the location of points not present in the drawing. Or, you may want to zoom to the location of a point. The Zoom to Point routine allows you to view the location of a specific point from the point database.

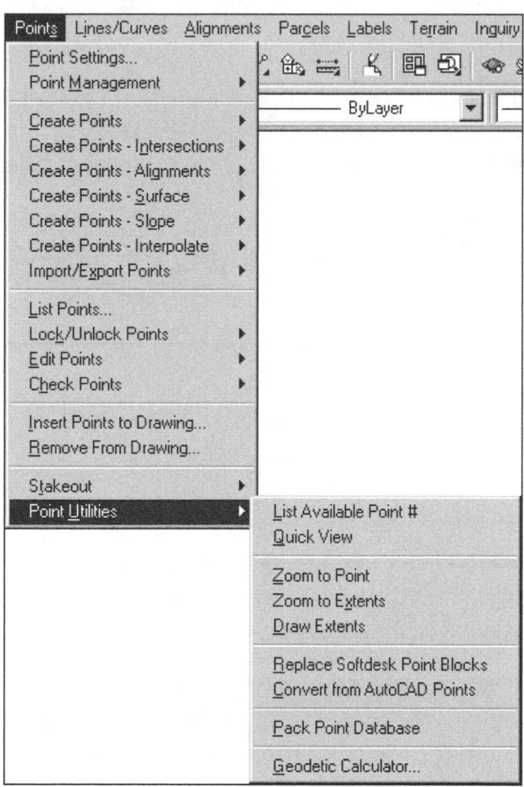

Figure 2–32

QUICK VIEW

The Quick View routine of the Point Utilities flyout displays all of the points in the point database as blips in the drawing. If you redraw the drawing, the blips disappear.

The blips allow you to view locations of points and then use them as a reference to defining a window in the Insert Point to Drawing routine.

DRAW EXTENTS

The Draw Extents routine draws a square representing the lowest and highest northing/easting in the point database.

LIST AVAILABLE

The List Available routine lists the currently used point numbers.

ZOOM TO A POINT

The Zoom to a Point routine executes a zoom center on a point from the point database. When you identify the point number, the routine prompts for a height of display.

STAKEOUT

The Stakeout flyout of the Points menu creates a printed list for staking out a selection set of points (see Figure 2–33). The menu contains routines for radial point, curve by direction or offsets, spiral by direction or offsets, and consecutive stakeout calculations.

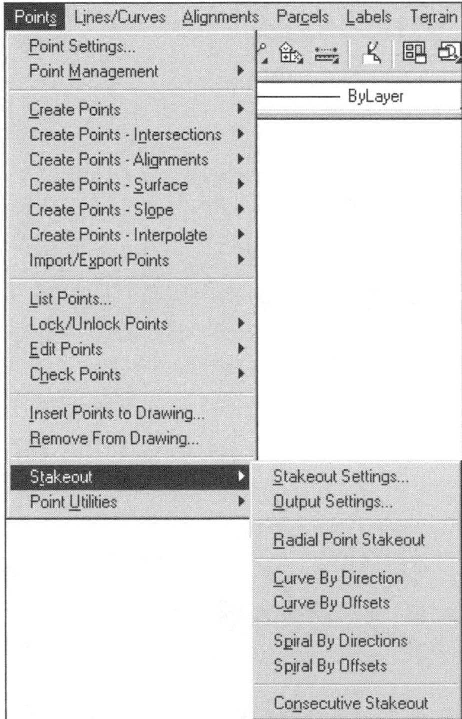

Figure 2–33

EXERCISE

Exercise

This exercise explores the editing tools of LDD. Each tool is for a situation that may arise during the course of a job. The ability to work with all or a subset of points allows you to work effectively and quickly within a drawing.

After completing this exercise, you will be able to:

- Use the point display routines
- Use the point editing routines
- Use the Check Points routine to update the drawing
- Use the Check Points routine to update the point database

EDIT POINTS

Display Properties

1. Open the LDDTR2 drawing of Proj1, perform a zoom extents, and remove all of the points from the drawing.
2. Insert the point group "exercise".
3. Zoom to Point 1 and set the height to 300.
4. Use the Display Properties routine of the Edit Points cascade from the Points menu to change some of the point values. Select only a few of the points. In the Display Properties dialog box, select the Text tab and turn off the display of point number and elevation. Select OK to exit the dialog box. Notice how the point objects rebuild themselves on the screen.
5. Select another set of points that have some elevation on and some off. When you enter the dialog box, select the Text tab and notice the check mark by the elevations toggle. The toggle is a gray check mark. A gray check mark means that some of the points in the selection set have the elevation off and others have it on. By selecting the toggle box, you can cycle between All Off, All On, and the original state at the time of selecting the points. Select OK to exit the dialog box.
6. Select all of the points, turn on point numbers and elevations, and select OK to exit the dialog box.
7. Use the AutoCAD MOVE command and select a couple of points. Select a new location for the points on the screen. The points show leaders back to their original coordinate location. Reselect the points and edit their display properties. Select the Text tab and turn off automatic leaders. Select OK to exit the dialog box. The leaders do not show. If you select the point text, LDD will highlight the marker as well as the point text.

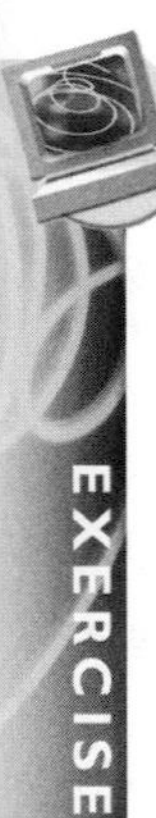

8. Reselect the points, click the right mouse button, and select Display Properties. In the dialog box, select the Text tab, and check on automatic leaders. To put the points back to their original locations near the coordinate marker, select the Reset tab and toggle on Move Marker Text Back to Marker Location. Select OK to exit the dialog box.
9. Select the Datum routine found in the Edit Points flyout of the Points menu. Set the change in elevation to 50 and select a few points.
10. Select the Renumber routine from the Edit Points flyout of the Points menu. Set the additive factor to 50 and select a few points.
11. Select the erase routine from the Edit Points flyout of the Points menu. Select the All option, and the routine returns the points just erased. Select two point from the screen. Go to the List routine to view all of the points in the point database. The two points erased are gone from the drawing and the point database. Select OK to exit the List Points dialog box.
12. Select the Unerase routine from the Edit Points flyout of the Points menu. Select the All option, and the routine returns the points just erased.
13. Select Group Manager from the Point Management flyout of the Points menu. Delete the Exercise point group.
14. Select the Create Group icon, enter **Exercise** as the point group's name, and select edit list.
15. Select the Advanced filter, set the source as all points, clear the point list at the top of the dialog box, and select the Filter tab. Toggle on Point Number Range and set it to 1–10000. Filter for the point numbers and add the points to the list. The group now has the renumbered points as a part of the group. Select OK until you exit the Group Manager dialog box.

EXERCISE

Unerase

The UNERASE command undoes the erasing of points from the drawing and the point database. The UNERASE command is good over editing sessions as long as the erased point number remains unused or the point data base is unpacked (a routine in the Point Utilities menu).

1. Save the drawing with the Save routine of the File menu.
2. Use the Remove Points from Drawing routine to remove some selected points from the drawing. After removing the points, use the List Points All toggle to view all of the points in the point database. The removed points remain in the project point database. Exit the List Points dialog box.

3. Use the AutoCAD ERASE command to erase two more points from the drawing. Use the List Points All option to view all of the points in the point database. Even though the points are not in the drawing, the points remain in the point database. Exit the List Points dialog box.
4. Use the Check Points Modify Drawing routine and toggle on Add All Points to the drawing. The missing points return to the drawing plus all of the points in the point database. The routine echoes back the addition of a number of points. The only way to reinsert the removed points is to use the Insert Points to Drawing routine.
5. Use the Remove Point from Drawing routine of the Points menu. Remove a set of points from the drawing. Use the point number range of 12000–12999.
6. Reopen the current drawing by using the Open File routine. Save the changes to the drawing.
7. When you re-enter the drawing, the points are still missing in the editor. Use the List Points routine, select the Advanced button, select the Source tab, and set the source to All. Select the Filter tab, set the point number range to 12000 to 12999, and filter and add the points to the point list. Select the List tab to view the points still in the point number database. Exit the List dialog box.
8. Select the New Drawing icon and start a new drawing. Name the drawing LDDTR2a, assign the drawing to the project PROJ1, and use the aec_i template file.
9. In the Setup wizard, assign the LDD1-20 setup to the drawing.
10. Display all of the points in the project database with the All toggle of the List routine. Even though the drawing does not have any points visible on the screen, it has access to the point database.
11. Use the Draw Extents routine in the Point Utilities cascade of the Points menu to draw the extents of the point database. Then use ZOOM EXTENTS to view the box drawn by the routine.
12. Run the Quick View routine of the Point Utilities cascade to view the points in the point database.
13. Run the Insert Points to Drawing routine and insert the group "exercise" into the drawing.
14. Use the Erase routine of the Edit Points menu and erase a few of the points. The points are erased from the drawing and the point database.
15. Reopen the LDDTR2 drawing and save the changes to LDDTR2a.

EXERCISE

EXERCISE

16. The points are no longer a part of the point database but are still present in the drawing. Use the Check Points Modify Drawing routine. Toggle on Delete all points not in the project. LDD presents an alert; select OK in the alert box, and continue. The routine deletes those points from the drawing that are no longer in the point database.
17. Use the Unerase routine in the Edit Points flyout to return the points to the point database and drawing.
18. Use the Remove Points from Drawing routine and remove all of the points.
19. Use the Insert Point into Drawing routine and insert the group exercise into the drawing.
20. Save the drawing.

UNIT 5: ANNOTATING POINT DATA

DESCRIPTION KEYS

Land Development Desktop uses description keys to annotate and sort point data in a drawing. The annotation of the point data occurs when you associate a symbol to a point with an entry in the description key matrix. When a routine places a point into the drawing, the description key matrix places a symbol at the coordinates of the point if the description you assign to the point has a corresponding entry in the matrix. Usually, the symbol is on a different layer than the point. For example, all centerline shots go to the layer CL_PNT, or curb shots to the layer CURB_PNT, and so forth, while the symbol may go to a different layer, such as TOPO_SYM. The sorting of points onto layers gives you the ability to control the visibility of points by turning off or freezing the layer of the point object. The Check Points and Insert Point routines follow description keys depending upon the current Point Label Style.

You turn on or off the actions of description keys when you change the status of the Use Point Label toggle of the Insert panel of the Point Settings dialog box. This assumes that the current point labeling style definition references a description key matrix. If the Use Point Label toggle is on and the current style does not reference a style using description keys, no symbols will appear. The style must reference a description key matrix to work. The default point style is Active Desckeys only (see Figure 2–3). The Insert panel of the Point Settings dialog box is the only place to toggle on Description keys.

A RAW description (the field entered description) is a shorthand notation of what a point represents. For example, MH in the field data represents a manhole in a drawing. The assignment of a description occurs in the field, data collectors, external coordinate files, or in any point creation routine in LDD. You need to translate the shorthand into office terms (FULL description). Generally, the office needs to have a more literal description of a point.

Description keys can be numerical, alphabetical, or a combination of both.

DESCRIPTION KEY MATRIX

You must keep three rules in mind when using description keys:

- The toggle Use Current Point Label Style When Inserting Points must be on in the Insert panel of the Point Settings dialog box.
- The current point label style definition must reference a valid description key file.
- Description keys are case sensitive.

Description keys accomplish three tasks based upon the information in the description key matrix. The first task is sorting point objects and symbols onto specific layers. No matter which routine creates a point, the description key matrix places similarly described points and symbols onto similar named layers. For example, placing curb shots on a layer CURB_PNT or tree and shrub shots on a VEG_PNT layer, manhole symbols on the TOPO_SYM layer, and tree symbols on the VEG_SYM layer, and so forth. Potentially the description key matrix specifies a layer for every point and symbol in the project. You control the visibility of the points and symbols by turning off or freezing and turning on or thawing layers in the drawing.

The second task is annotating points with a symbol indicated by an entry in the description key matrix. This entry sets the symbol name from a list of blocks. LDD installs three symbol libraries: Cogo, Cogo_metric, and APWA. The path to these symbol sets can be set in the User Preferences dialog box of the Projects menu. If you change the directory in User Preferences, you need to exit and restart LDD. The specific library of symbols is set in the Point Settings dialog box. The list of libraries reflects the User Preferences path setting. You are, however, allowed to change the path in the dialog box. This allows you to reference a block library irrespective of the User Preferences setting.

The third task is translating field note descriptions into office terms. In LDD terms, this is translating the RAW description into a FULL description. An example of this translation is the RAW description of OK 6 translating to 6 INCH OAK, the FULL description. The description matrix contains a format that it uses to create the FULL description for values within the RAW description.

When creating a point, the point routine references the Description Key matrix of the point style to compare the RAW description of the point to the code element of the key matrix. If they match, the point creation routine reads the remaining elements in the matrix to determine what to do. The match can be a simple character to character match, for example TOPO for the RAW description and TOPO for the code in the description key matrix.

The process of literal character matching does not work when you record additional values in the RAW description. Examples of this type of RAW description are MH1, MH2, OK 3, or OK 6. Each of these descriptions has common portions, manholes (MH) and oak trees (OK) and parts that change for each observation (manhole 1 and 2 and 3 and 6 inch diameter oak trees). The use of an * (asterisk) wildcard after the common elements in a code entry of the description key matrix solves the problem of trying to match RAW descriptions containing numbers or measurements from the field. For example, the description code of OK* in the description key matrix matches the RAW descriptions of OK 3 and OK 6. In the case of OK 3 and OK 6, only the OK of each description needs to match. If there is a match, the actions defined in the matrix occur.

The FULL description of a point is the result of a format found in the description key matrix. As before, the match of OK 6 with the code of OK* causes the description key matrix to generate a FULL description for the point. If the description format for OK* is OAK TREE, the resulting FULL description is OAK TREE. The resulting FULL description ignores the changing measurement found in the second part of each oak tree observation.

The use of parameters in the format entry of the description key matrix allows you to use numbers or measurements from the field to create FULL descriptions in the drawing reflecting those values. The RAW description of OK 6 has two parts: OK and 6. The OK represents the type of tree (oak) and the diameter of the tree is the number 6 (6 inch). In LDD each part of the description separated by a space is a parameter of the description. The description key matrix counts the parameters starting with 0 (zero). As a result, OK is parameter 0 (zero) and the 6 is parameter 1. To use the parameters of the RAW description in the format entry of the description key matrix, you reference them by using $0, $1, … $9 in the format entry. A raw description can have up to 10 parameters $0…$9. To create the FULL description of 6 INCH OAK from the RAW description of OK 6, you would use the format of $1 INCH OAK. This format uses the 6, the second parameter of the RAW description ($1) as the first part of the new FULL description.

If the FULL description is the same as the RAW description, the description key format entry is $* (dollar sign asterisk). An example of this would be the format entry of $* translating the RAW description of TOPO into the FULL description of TOPO. The format $* simply uses the RAW description as the FULL description.

After you install Land Development Desktop, the description key matrix is a file in the *Cogo* folder of the Prototype, *default.mdb*. The description key file of a project is a copy of the prototype's description key file. LDD allows for more than one description key file. The point style label definition references which description key file to use.

When you edit a description key file that resides in a project, the edits are specific to the project. When you add, change, or delete keys, the changes occur only with the project copy of the matrix. To make the matrix available to other projects, it needs to be copied to the *Cogo/DescKey* directory of the prototype and/or other projects. The Save Desckey File to Prototype routine of the Description Key manager copies the description key file to a selected prototype. The Load Desckey File from Prototype selection loads the selected file from a prototype to the project.

You create description keys in one of three ways. The first way is by calling up the Description Key Manager in the Point Management cascade of the Points menu (see Figure 2–34). Clicking on the Create Description Key icon displays another dialog box, Description Key Properties. This dialog box allows you to enter the necessary information to define the key (see Figure 2–35). When you have completed entering the information, select OK to save the data to the *default.mdb* file.

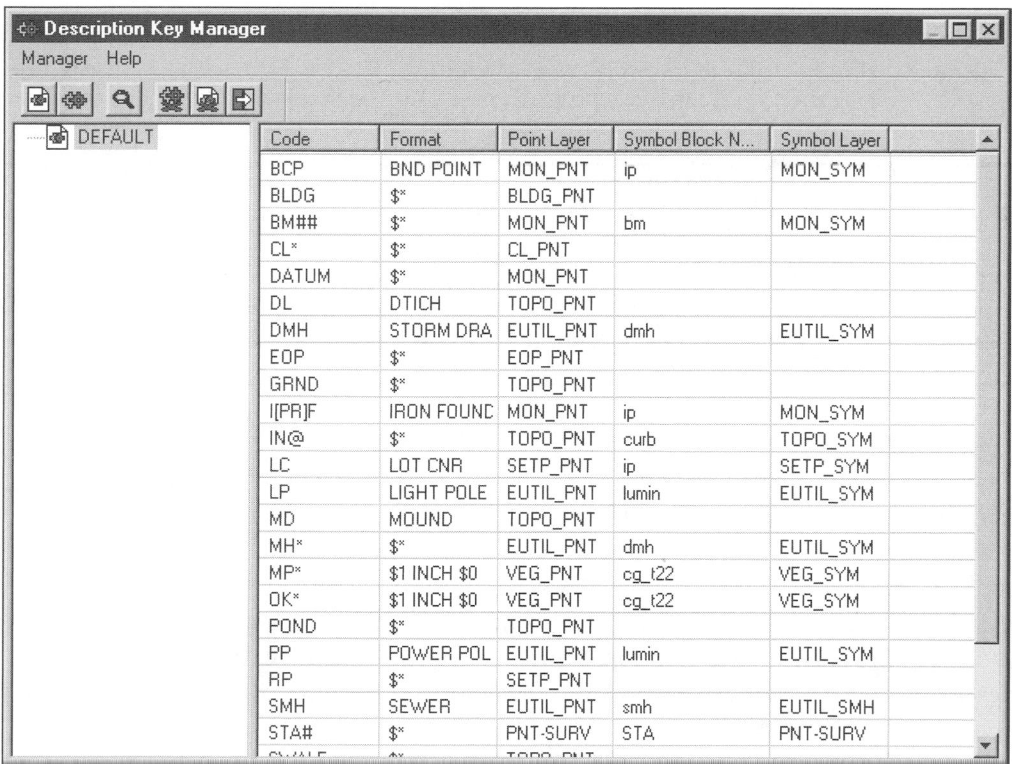

Code	Format	Point Layer	Symbol Block N...	Symbol Layer
BCP	BND POINT	MON_PNT	ip	MON_SYM
BLDG	$*	BLDG_PNT		
BM##	$*	MON_PNT	bm	MON_SYM
CL*	$*	CL_PNT		
DATUM	$*	MON_PNT		
DL	DTICH	TOPO_PNT		
DMH	STORM DRA	EUTIL_PNT	dmh	EUTIL_SYM
EOP	$*	EOP_PNT		
GRND	$*	TOPO_PNT		
I[PR]F	IRON FOUNC	MON_PNT	ip	MON_SYM
IN@	$*	TOPO_PNT	curb	TOPO_SYM
LC	LOT CNR	SETP_PNT	ip	SETP_SYM
LP	LIGHT POLE	EUTIL_PNT	lumin	EUTIL_SYM
MD	MOUND	TOPO_PNT		
MH*	$*	EUTIL_PNT	dmh	EUTIL_SYM
MP*	$1 INCH $0	VEG_PNT	cg_t22	VEG_SYM
OK*	$1 INCH $0	VEG_PNT	cg_t22	VEG_SYM
POND	$*	TOPO_PNT		
PP	POWER POL	EUTIL_PNT	lumin	EUTIL_SYM
RP	$*	SETP_PNT		
SMH	SEWER	EUTIL_PNT	smh	EUTIL_SMH
STA#	$*	PNT-SURV	STA	PNT-SURV

Figure 2–34

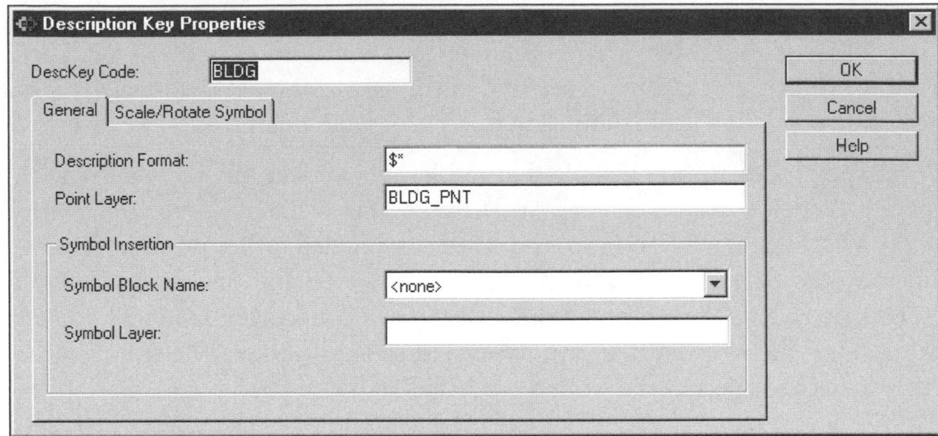

Figure 2–35

The second way to add keys to the *default.mdb* file is within Access. The *default.mdb* file is an Access database file located in the *Cogo\DescKey* folder.

The last way is to import a Softdesk *Project.dsc* or *dca.dsc* description key file. There is an Import DSC icon in the Description Key Manager dialog box.

The description key files from Softdesk are valid for LDD. LDD converts the earlier versions of description key file from Softdesk, the *dca.dsc* file (Version 12 and earlier) and *project.dsc* (S7 and S8) files. LDD will convert the files when opening the project for the first time. The Description Key manager can also import a previous version description key file with its Import DSC routine.

LDD defines seven parts to each description key. A key can use one or all of the parts of the matrix. A key may only sort points to layers in the drawing. In this case, you identify the key and the layer for the point and no symbol or symbol layer. The task(s) the key must do determines how many of the fields the key uses.

The Elements and Options of the Description Key Matrix

The seven parts of a description key matrix includes five elements and two options. The five elements are a description key code, description format, a layer for the point, what symbol to use, and a layer for the symbol. The description code can contain wildcards to group similar RAW descriptions together. For example, STA# matches STA1, STA2 throughSTA9. The description key code sees these points as being the same. A description format is the translator of the matched RAW description into a Full description. The two options for description keys are scaling and rotating of the symbol.

The following is a sample description key file listing.

Key	Description	Layer	Symbol	Symbol Layer	Scale	Rotate
BCP	BND POINT	MON_PNT	IP	MON_SYM	DWG	
BLDG	$*	BLDG_PNT				
BM###	$*	MON_PNT	BM	MON-SYM	DWG	
CL*	$*	CL_PNT				
DATUM	$*	MON_PNT				
DL	DITCH	TOPO_PNT				
DMH	STORM DRAIN	EUTIL_PNT	DMH	EUTIL_SYM	DWG	
EOP	$*	EOP_PNT				
IN@	$*	TOPO_PNT	CURB	TOPO_SYM	DWG	
I[PR]F	IRON FOUND	MON_PNT	IP	MON_SYM	DWG	
LC	LOT CNR	SETP_PNT	IP	SETP_SYM	DWG	
LP	LIGHT POLE	EUTIL_PNT	LUMIN	EUTIL_SYM	DWG	
MD	MOUND	TOPO_PNT				
MH*	MANHOLE	EUTIL_PNT	DMH	EUTIL_SYM	DWG	
MP*	$1 INCH $0	VEG_PNT	CG-T22	VEG_SYM	DWG	
OK*	$1 INCH $0	VEG_PNT	CG-T10	VEG_SYM	DWG	
POND	$*	TOPO_PNT				
PP*	POWER POLE	EUTIL_PNT	U_POLE	EUTIL_SYM	DWG	$1
RP	$*	SETP_PNT				
SMH	SEWER	EUTIL_PNT	SMH	EUTIL_SYM	DWG	
SWALE	$*	TOPO_PNT				
TOPO	$*	TOPO_PNT				
WV	W VALVE	EUTIL_PNT	WV	EUTIL_SYM	DWG	

Element 1—The Description Key Codes

Alphabetical Description Codes The first thing to remember about alphabetical codes is that they are case-sensitive; that is, IP is different from *Ip*, *iP*, or *ip*.

If the raw descriptions are alphabetical, the description codes may contain wildcards. These wildcards include: ? (alphanumerical character), # (number only), @ (alphabetical character only), . (any non-alphanumerical character), [] (a list), ~ (logical not), and * (anything). Wildcards cannot be at the beginning of a description code.

A description code wildcard is a literal match. If you set the code to STA#, it will match the raw descriptions of STA0 through STA9. If the range of station numbers is from 0 to 9999, the code needs to be STA####. With the code set to STA####, the raw description of STA0001 will match, but the raw description of STA1 will not match. The raw description does not match because it does not have enough numbers to match the code STA####. It is important to remember that the match is a literal per-character match.

Examples of this type of description keys are as follows:

STA#—matches only STA0 ... STA9

STA##—matches only STA01 ... STA99

STA###—matches only STA001 ... STA999

STA*—matches STA0 ... STA99999999 also STAKE, STAR12, or anything with STA first

LDD can represent the keys as a code list.

I[PR]F

The above description key code matches the following raw descriptions:

IPF

IRF

The list wildcard allows you to include and exclude letters from matches.

CP[A...M]—Control points with only the letters A through M match.

CP[~N...Z]—Control points not with the letters N through Z match.

The above description key codes match the following raw descriptions:

CPA

CPC

CPM

Numerical Raw Descriptions Raw descriptions can be numerical. Whatever the format is of the raw descriptions, the description key codes will have to reflect the raw descriptions in the description code entry to work properly.

How would a numerical raw description describe a 4" diameter tree? The numerical raw description ties the tree diameter to a single entry or an entry representing a range of diameters. For example, the key 1231 indicates a 1"–2.99" diameter deciduous tree and 1232 indicates a 3"–6" diameter deciduous tree. There will have to be two codes, 1231 and 1232, to handle the raw descriptions.

A RAW description can be a combination of alphabetic and numeric values, STA1, and so on.

Element 2—Description Format–Translation

The next part of the key matrix is the format of the translation of the RAW description to a FULL description in the drawing. There may be times when translation of a key is not needed and other times when it is necessary. The purpose of the translation of a description key is to make it more understandable to the people in the office. There are three basic types of translation. In the first method, the RAW description becomes the FULL description. The second method changes the order of and/or adds information to the FULL description from the RAW description. The last method completely replaces the RAW description to make the FULL description.

If a RAW description key does not need translation, enter $* (dollar sign asterisk) into the Description Format field. If the RAW description is TOPO and the FULL description is the same, the entry in the Description Format field is $*. The dollar sign asterisk ($*) means that no translation of the Raw Description occurs.

An example of a description format that adds to and changes the order of the FULL description from the RAW description is: $1 INCH OAK. If the RAW description is OK 6 and the description format is $1 INCH OAK, the resulting FULL description is 6 INCH OAK. This example uses the second ($1) parameter of the description code to create a tree diameter.

An example of replacing the RAW description to get a FULL description is an entry of IRON FOUND as the format entry. When the routine matches the description code of IPF, it translates the code to the FULL description of "IRON FOUND".

Element 3—Point Layer

Following the description format, the matrix contains the point layer name. If the entry specifies a layer not present in the drawing, the Description Key routine creates the layer. If there is no layer specified, the point goes to the current layer.

Element 4—Symbol Name

If appropriate, the next entry specifies the symbol name. If the symbol is not present in the drawing, the description key routine looks for a drawing file with the same name in the folder specified in the Search Path for Symbol Block Drawing Files. You can set this path in the Insert panel of the Point Settings dialog box (see Figure 2–3). If you want to use in-house symbols, you set this path to the folder containing the drawing files. If the symbol you reference is not found, the description key routine will issue an error and continue.

If you make your own symbols, there are some rules you should follow.

- All symbols are unit size.
- All symbols should be from entities on layer 0.
- You should draw all symbols in generic AutoCAD. Or if you draw symbols within a drawing already assigned a project, use the wblock routine to create the symbol drawing file.

The first rule states that you should draw symbols their plotted size. If a manhole symbol is to be 0.2 inch in diameter when plotted, its drawing size is a diameter 0.2 of an AutoCAD unit. When inserting the manhole symbol, the routine scales the symbol by the horizontal scale of the drawing. If the drawing scale is 1 inch = 20 feet, the scaling factor for the symbol is 20.

The second rule has to do with AutoCAD's handling of blocks. A symbol whose entities are on layer 0 will become a part of the insertion layer.

The third rule prevents possible project reassignment.

You can exaggerate the size of the symbols by setting a scaling factor in the scale option of the description key matrix. See the discussion on setting the scale factor (Option 1) for a key below.

Element 5—Symbol Layer

The final element in the description key matrix is the symbol layer. If the entry specifies a layer not present in the drawing, the Description Key routine creates the layer. If there is no layer specified, the point goes to the current layer.

Option 1–Symbol Scale

The first option sets the scaling of the symbol. By default the scaling is the same as the scale of the drawing (see Figure 2–36).

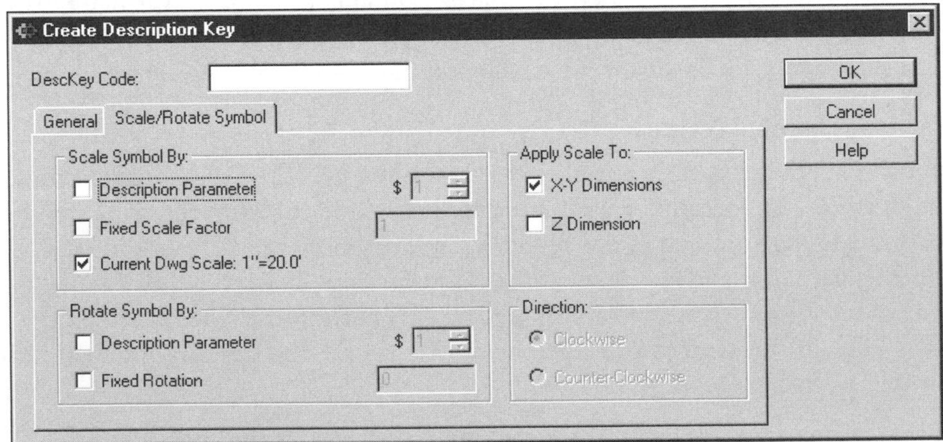

Figure 2–36

A second option within this group scales the symbol by a parameter found within the RAW description. For example, instead of measuring the diameter of each tree, you measure the drip line of the tree. What you want is a symbol representing the drip line diameter. The RAW descriptions MP 20, MP 15, and MP 7 indicate the diameter of each tree's drip line. In each description, the second parameter is the diameter of the drip line. If you toggle on only Description Parameter and set it to the correct parameter number ($1), the routine scales the block by the parameter.

The catch to parameters is that they count from 0 (zero). In the preceding example of MP 20, MP is parameter 0 (zero) and 20 is parameter 1. When toggling on the Description Parameter, the toggle requires you to set a parameter number. A RAW description can contain up to 10 parameters ($0–$9). A parameter is separated by a space in the RAW description.

The last option is a fixed scaling factor. You would toggle this on and set the factor. When a routine places a symbol into the drawing, it will have the same scale factor no matter what scale is set in the drawing.

All of the toggles in this group can be on or off in any combination. Be careful to understand the consequences of toggling on more than one scaling factor. If all of the toggles are on, the scaling of the symbol is the product of all of the options checked.

Option 2–Symbol Rotation

The Rotate Symbol By area allows you to change the default method of rotation symbols (see Figure 2–36). The default is no rotation. The first option is using a value in the RAW description as a rotation parameter. If this option is toggled on, the parameter number must be set. The last override is a fixed rotation angle. If this option is toggled on, the rotation angle needs setting.

If you are using a RAW description parameter to set the rotation angle, beware that parameters count from 0 zero. In the example LP 45, LP is parameter 0 (zero) and 45 is parameter 1 ($1). Then toggling on Description Parameter requires you to set the parameter number. A RAW description can contain up to 10 parameters ($0- $9). A parameter is separated by a space in the RAW description.

POINT LABEL STYLES

In previous releases, points were blocks with attributes. You could use the attribute editor to move and rotate portions of the point block to create a label of some value present in the point. There was no easy method of labeling coordinates, elevations, descriptions, and symbols as a single function. The introduction of point labeling styles offers an alternative method and gives greater flexibility to labeling point information. Like line labels, point labels can be dynamic or static. Dynamic means that they can be updated to display new values associated by the point without having to recreate the label.

LDD installs a number of point label styles. The default style is Active Description Keys. This style labels points with a symbol. You can extend or change this style to include additional, less, or different information about the point. A point label style can assign a symbol to a group of points with their own annotation different than their original values. When label styles are well used, they are powerful annotation tools.

The Edit Label Styles routine of the Labels menu displays the Edit Label Styles dialog box. If you select the Point Label Styles tab, the Point Label Styles panel appears (see Figure 2–37). The upper left portion of the Point Label Styles panel lists the name and the pieces of information (Data) available about a point. The name of the style may contain a path. When you highlight an entry in the Data area, you can place the selected data item in the Text (style definition) window by clicking on the >>Text button. The toggle Turn Off Marker Text means that the point number, elevation, and description of the point will not show when using this style to label the point.

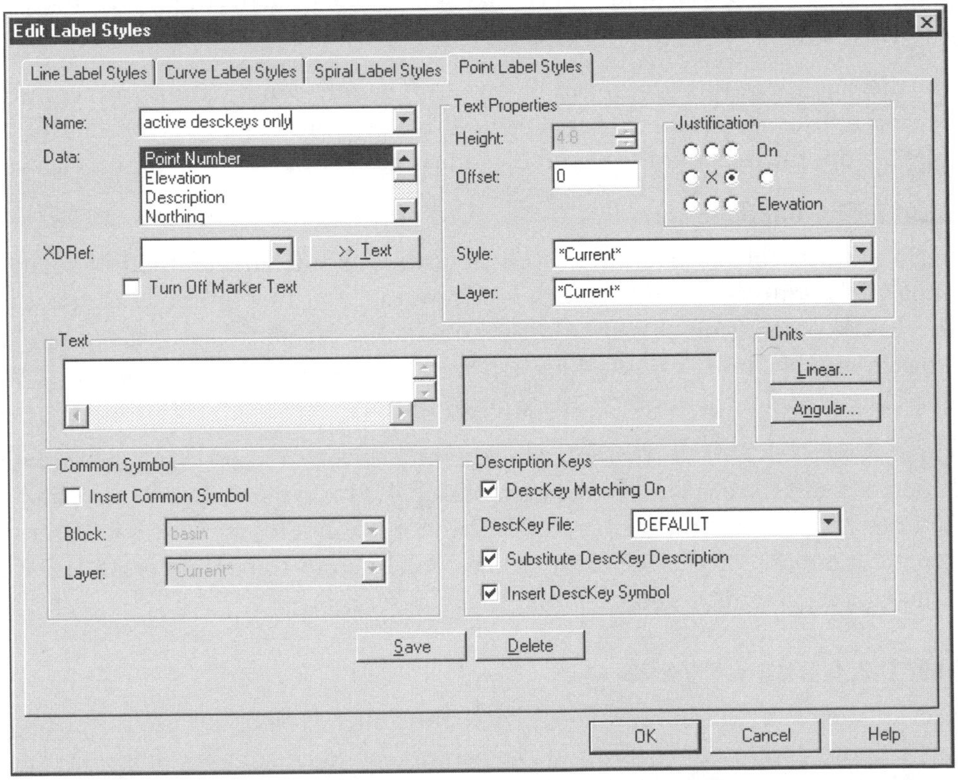

Figure 2–37

You can add text to the labeling definition of a point, for example, placing an N: and E: before the northing and easting coordinates of the point label definition. The process would be first typing in the N: in the text are of the style, then selecting

Northing from the Data list, and then clicking on the >>Text button to place Northing after the N: in the text area.

The upper right portion of the dialog box controls the text style, location, and layers for the label. Any text style can be used in a label. The size needs to be set relative to the scale of the drawing. You set the style by selecting it from a drop list. If the style has a height, the height box changes to the size; if not, you must make sure that the size is correct for the scale of the drawing. If you use a Leroy text style, the style already has a height definition relative to the scale of the drawing.

To the right of the text style is the location of the label relative to its point coordinates. You have the choice of eight positions to the side of the point or on the point's coordinates. If the label is too close to the point marker, you can add an offset to move the label from the marker. You set this value in the Offset box.

The last control in this area is the layer name. If you click the drop list arrow, you can select from a list of layers currently in the drawing. If the layer does not exist, you can type in the layer name and when a routine generates the label, the layer will be made.

The middle portion of the dialog box displays the style format, a common block identifier, and/or the use of description keys. The common block toggle places the same symbol on all of the labeled points when using the style, for example, labeling a group of survey benchmarks. Each benchmark will show the same text and symbol when labeled with this style. The style identifies the common symbol name and layer for the symbol.

The Description Keys toggle allows you to place symbols based upon a description key matrix in addition to text and data values you set at the top of the point style panel. The description key file can be different from the default key file. The two toggles below allow the translation of a RAW description to a FULL description and inserts the symbol called out in the Description Key matrix.

The Save button places a file containing the style definition in the data folder structure. With the file in the data folder, all projects can now use the point style. The Delete button deletes the style from the data folder structure.

Exercise

This exercise reviews and uses point creation routines found on the Points menu. In conjunction with the placing of points is their annotation, which is a part of the description key matrix of Softdesk.

After completing this exercise, you will be able to:

- Define and use the description key matrix
- Create a point labeling style

CREATING AND USING DESCRIPTION KEYS

1. If you are not in the LDDTR2 drawing, open the drawing.
2. Open the Points menu, cascade out Points Management, and select Description Key Manager. Select *default.mdb* to identify the current description key file and then select the Add Description Key icon. Add the keys listed in the heading "The Elements and Options of the Description Key Matrix" in the preceding discussion of description keys. If you have limited time, you can copy the default description key file (*default.mdb*) from the CD that comes with the book. Before you copy the file, you need to exit LDD and place the file in the *DescKey* folder of the current project (proj1). The only key that needs to be added to the copied description key file is the OK* key. See the discussion under the heading of "The Elements and Options of the Description Key Matrix" to see the values for the OK* description key.

Figure 2–38 shows the Description Key Properties dialog box for BCP.

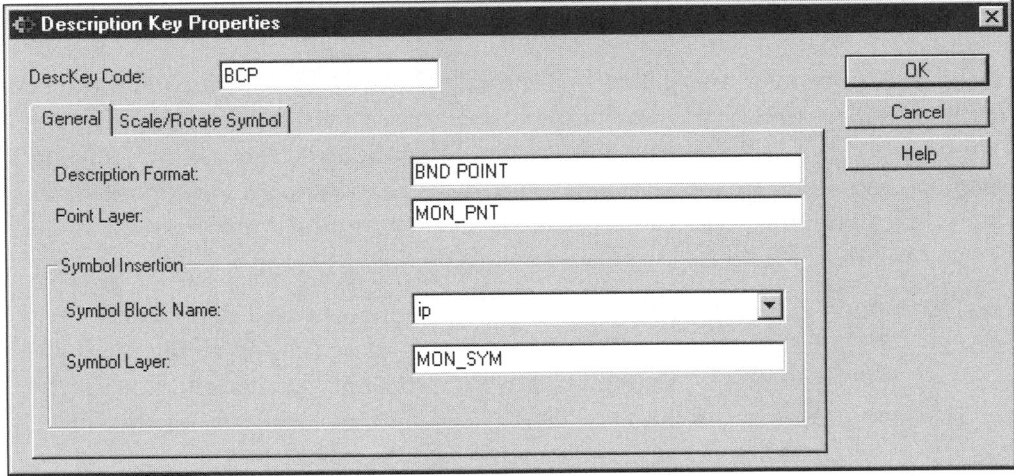

Figure 2–38

Figure 2–39 shows the Create Description Key dialog box for MP.

Figure 2–40 shows the Create Description Key dialog box rotation option for PP.

3. Remove all of the points from the drawing.
4. Use the Insert Points to Drawing routine to place the "exercise" group into the drawing.
5. Zoom to point 1 with a height of 100 units. Notice that the points in the area now have symbols and some new descriptions.

EXERCISE

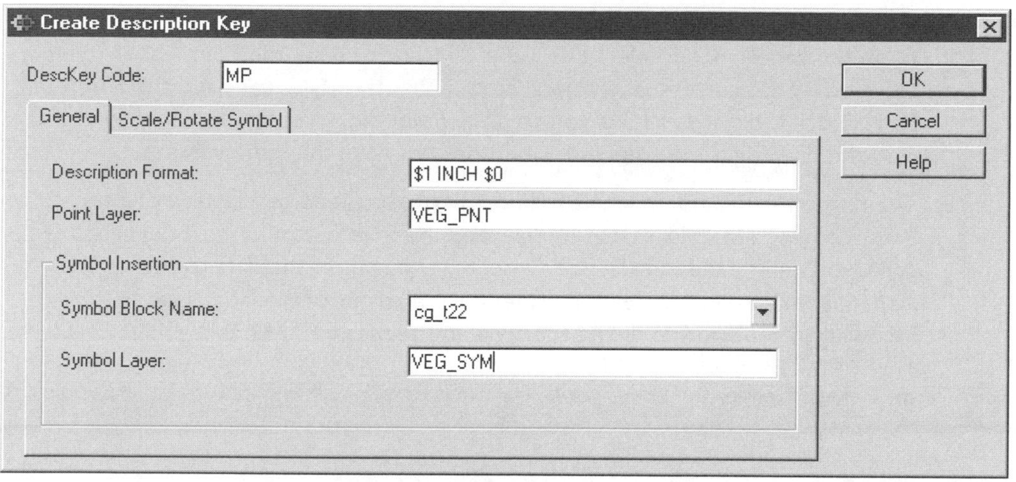

Figure 2–39

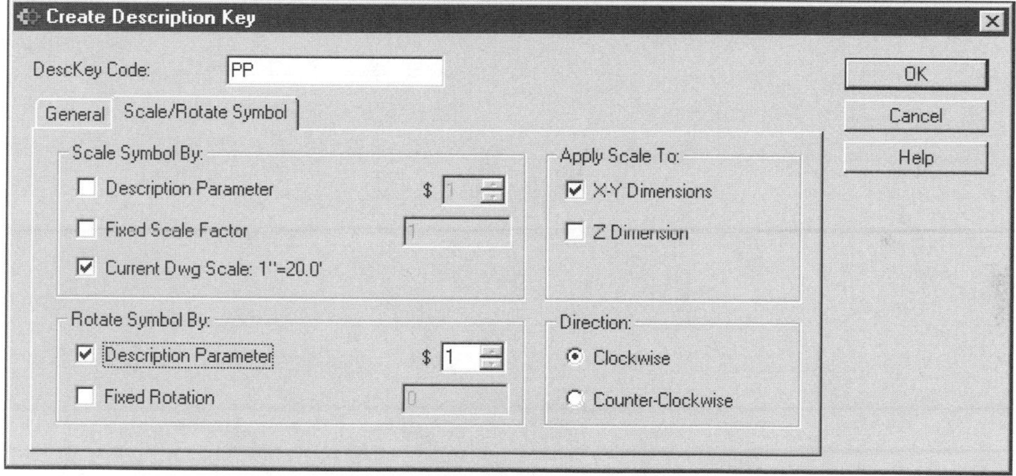

Figure 2–40

6. Use the AutoCAD List routine and select a point and its symbol. The selected items should be on the layers defined in the Description Key Matrix. Remember, a point that does not match a description key is put on the current layer.
7. Remove all of the points and symbols from the drawing.
8. Save the drawing file with the Save command in the File menu.

EXERCISE

CREATING AND USING POINT STYLES

9. From the Labels menu, select Edit Label Styles and select the Point Label Styles tab. Review the current label styles. Notice that the styles vary and the Softdesk Point Block Only style allows you to place point blocks in the drawing. So you can pass the drawing on to someone who does not have the same version.

10. The first style you create labels the northing/easting and description of a point. After defining the style, save it by selecting the Save button. When defining this style, you need to add N:, E:, and Desc: text in front of the data elements. You also must manually place a return after each element of the label. Use the following information to define the style and see Figure 2–41 as a guide.

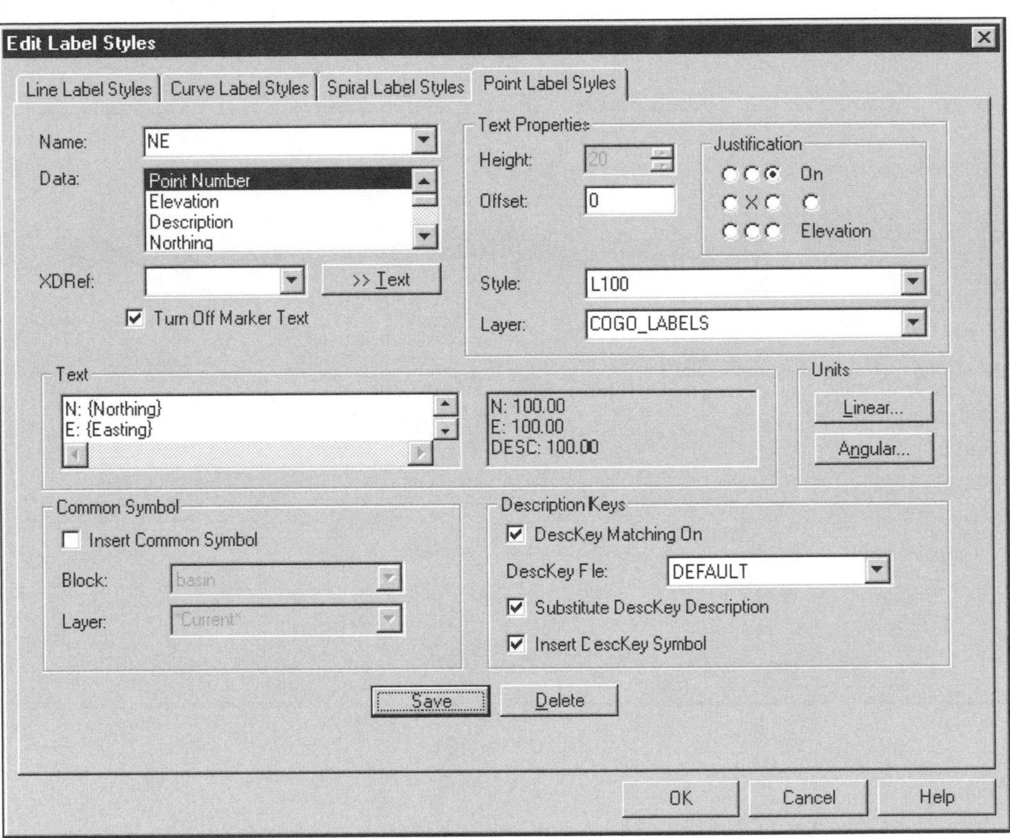

Figure 2–41

Name:	NorthEast
Toggle Turn Off Marker Text:	ON
Position:	Upper Right of the Coordinate point
Text Style:	L140

Layer:	POINT_LBL
Text:	N: {Northing} ENTER
	E: {Easting} ENTER
	DESC: {Description}
Toggle Common Block:	OFF
Toggle Description Keys:	ON
Toggle Substitute DescKey Description:	ON
Toggle Insert DescKey Symbol:	ON

11. The second style you create assigns a common block to points. After defining the style, save it by selecting the Save button. When defining this style, you need to add N:, E:, and Group Name: text in front of the data elements. You also must manually place a return after each element of the label. Use the following information to define the style, and see Figure 2–42 as a guide.

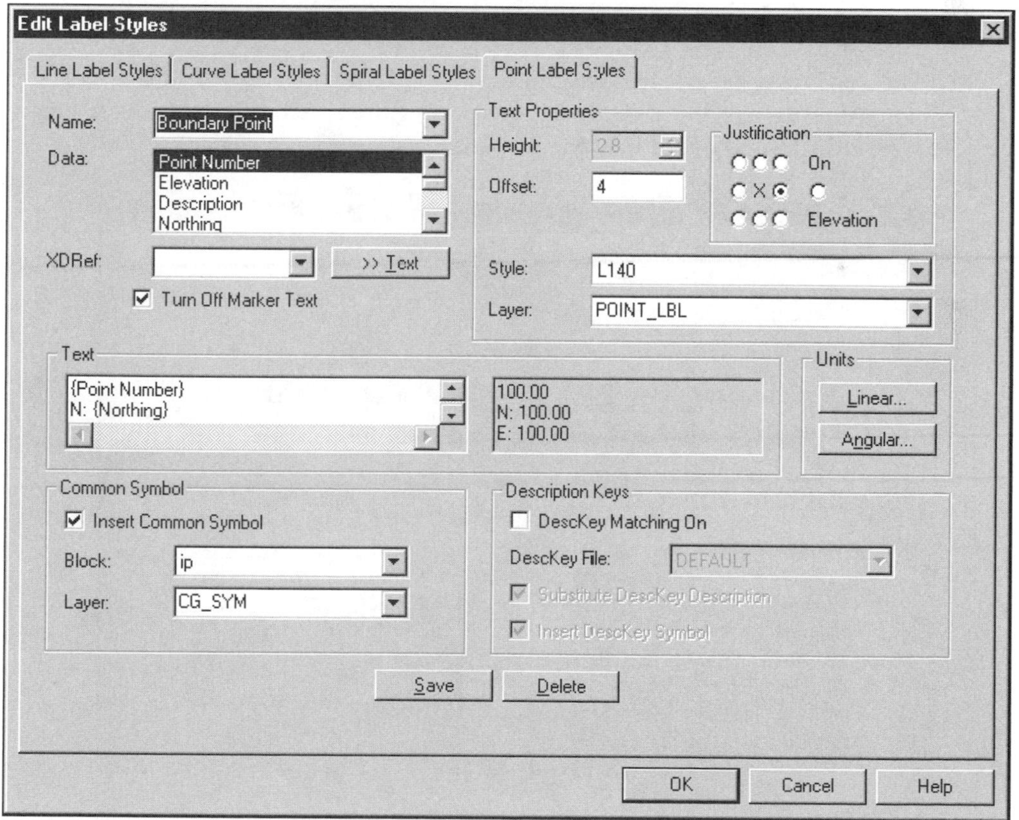

Figure 2–42

Name:	Boundary Point
Toggle Turn Off Marker Text:	ON
Position:	Right of the Coordinate point
Offset:	4
Text Style:	L140
Layer:	Point_LBL
Text:	{Point Number} ENTER
	N: {Northing} ENTER
	E: {Easting} ENTER
Group Name:	{Group Name}
Toggle Common Block:	ON
Common Block:	IP
Layer:	CG_SYM
Toggle Description Keys:	OFF

EXERCISE

12. The third style you create labels the elevation of a point at its coordinate location. After defining the style, save it by selecting the Save button. Use the following information to define the style, and see Figure 2–43 as a guide.

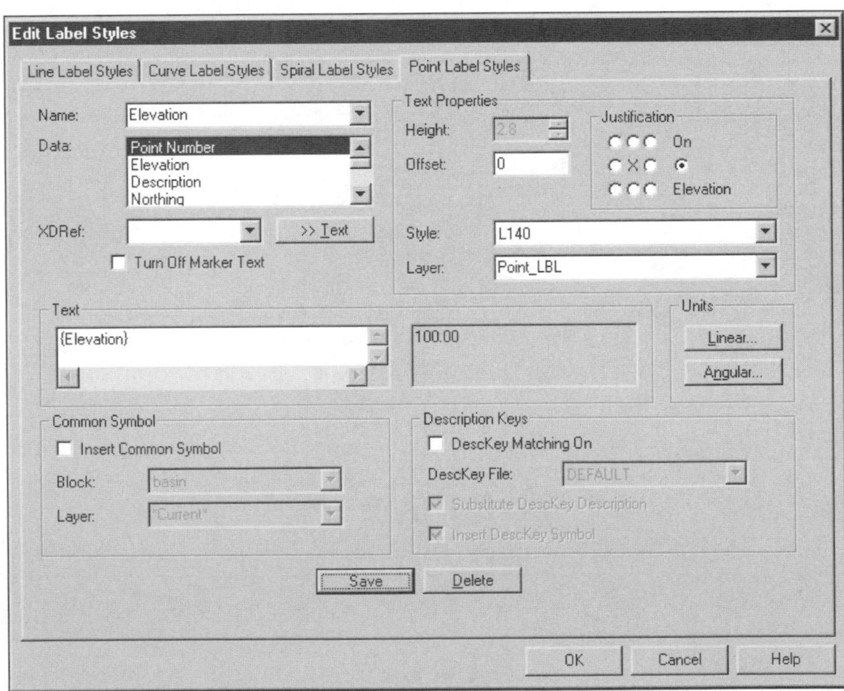

Figure 2–43

Name:	Elevation
Toggle Turn Off Marker Text:	ON
Position:	On Elevation
Offset:	0
Text Style:	L140
Layer:	Point_LBL
Text:	{Elevation}
Toggle Description Keys:	OFF

13. From the Labels menu, select Show Dialog Bar.

14. In the Labeling dialog bar, select the Point tab and set the current labeling style to NorthEast.

15. From the Points menu, run the Insert Point into Drawing routine and insert the group "Exercise". The points should have the labeling to the right of the point.

16. In the Labeling dialog bar, select the Point tab and set the current labeling style to Boundary Points.

17. From the Points menu, run the Insert Point into Drawing routine and insert the group "Exercise". In the warning dialog box, select the Replace All button to replace the current labeling with the new label style.

18. In the Labeling dialog bar, select the Point tab and set the current labeling style to Elevation.

19. From the Points menu, run the Insert Point into Drawing routine and insert the group "Exercise". In the warning dialog box, select the Replace All button to replace the current labeling with the new label style.

20. Rotate the elevation labels. Start by clicking on the Settings icon of the Style Properties dialog box to display the Label Settings dialog box. Next select the Point Labels tab and change the rotation angle to 45. Click on OK to exit the settings dialog box. The labels do not rotate.

21. Run the Update All Label Styles routine from the Labels menu to rotate the labels.

22. Remove all of the points from the drawing.

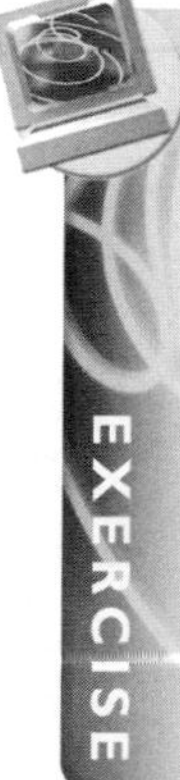

This concludes the section on the basic point tools of Land Development Desktop. The Points menu contains all of the tools to create, edit, analyze, and annotate project point data. There are many more point tools than what was covered in this section; however, more tools will be explained in later sections.

The point can represent the beginning or end of a line or curve, and many of the routines that are on the Points menu also appear on the Lines/Curves menu. The next foundation data type of a LDD project is lines and curves.

Line and Curve Data

After you complete this section, you will be able to:

- Use line and curve drafting routines
- List line and curve data
- Edit line and curve data
- Annotate lines and curves
- Create point data from lines and curves

SECTION OVERVIEW

The second data group of Land Development Desktop is the lines and curves of a drawing. These entities represent basic data and constructions; that is, boundaries, centerlines of roads, pipe runs, roadway transition control, and so forth.

UNIT 1

The Line and Arc routines of LDD (the Lines/Curves menu) represent a set of tools for the office to connect data points. The Line and By Point (#) Number Range routines create lines and arcs from points in the point database. Other routines use notes as a source for creating lines and curves. Although you can create a majority of lines and curves in the drawing from the Line, Arc, and Fillet routines of AutoCAD, the LDD routines represent a set of useful Civil/Survey drafting tools. The line drawing routines of LDD are the subject of the first unit of this section.

UNIT 2

LDD's listing routines report information about the lines and curves in the drawing beyond what is available in the generic AutoCAD LIST command. Routines on the Inquiry menu allow you to view all pertinent data about each entity type. Listing and analyzing lines and curves is the topic of the second unit of this section.

UNIT 3

The third unit covers the editing of lines and arcs beyond the editing tools of generic AutoCAD.

UNIT 4

The annotation of lines and curves is the subject of the fourth unit of this section. LDD has two methods of annotation, dynamic and static. The first method has labels that react to changes in the labeled object. This may be fine in some cases; however, in an office that trims the lines back from an intersection to clearly show a symbol, the reduction of the line now shows in the label of the line. The second method, static, uses label text that does not react to any changes in the line work.

UNIT 5

Unit 5 covers another method of labeling line work in a drawing. This alternate method uses line and curve tables. The table method requires you to create tags in the plat that relate to entries in a corresponding table.

UNIT 6

When you place points on line, arc, and spiral objects, their coordinate information becomes a part of the point database. The point database, however, does not understand that there is a relationship between the coordinates of the points and the lines or arcs in the drawing. It is up to you to manage the relationship between the points and the lines or arcs. The best way to manage this association is by the description value in the point object. The fifth unit of this section covers the routines that create points related to lines and curves.

UNIT 1: CREATING LINES AND CURVES

The AutoCAD LINE, CURVE, and FILLET commands can create most of the line and curve entities in a drawing. The Line and Arc routines found in LDD enhance standard tools or provide additional tools not found in AutoCAD; for example, best fit lines and curves, placing a tangential line or arc to an existing object, and spirals. Some of the LDD arc routines require more information than just a radius. These routines use tangent lengths, mid-ordinate distances, and other curve properties. You may not always know the needed additional item, but LDD has a Curve Solver routine (on the Utilities menu) that will calculate the needed values. The LDD routines are most useful when the task is transcribing computed line and arc values into a drawing.

The Lines/Curves menu contains the Line, Arc, and Spiral routines (see Figure 3–1) of LDD. This includes Special Lines or linear symbology; for example, lines that represent guard rails, tree lines, and so forth. These special lines are not lines with text or symbols as a part of the linetype definition, but a series of copied objects.

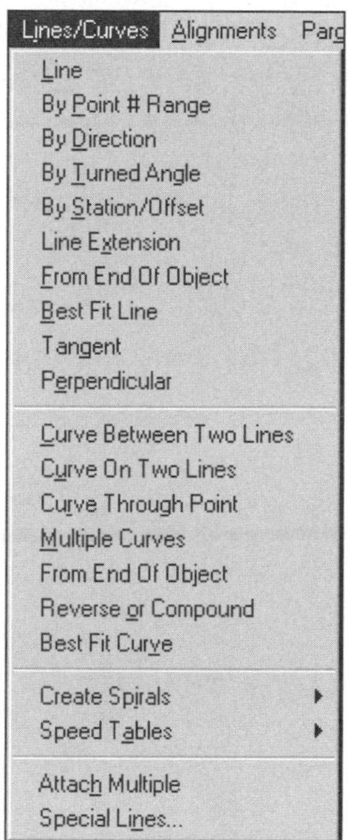

Figure 3–1

LINES

The Line routine of the Lines/Curves menu creates lines from user-selected points. The remaining line routines create lines by surveyor methods, lists of points, or intersections. The surveyor methods include Least Squares Fitting (Best Fit), By Direction, By Station and Offset, and By Turned Angle. The point list method is Line By Point # Range. The intersection routines include lines tangent to an arc and perpendicular to other lines. The From End of Object routine places a line tangent to the end of an arc.

THE LINE ROUTINE

The LDD Line routine allows you to create lines from point data in the point database, northings/eastings, and AutoCAD object snaps. To toggle an LDD routine into point number mode, use the dot p toggle (*.p*); to toggle into northing/easting mode, use the dot n toggle (*.n*); and to toggle into selecting a point from the screen, use the dot g toggle (*.g*).

The . (Dot) Toggles

When you are creating lines and arcs with the LDD routines, you can select coordinates from the screen in different ways including: AutoCAD positions, point numbers, northings/eastings, and graphically selecting points from the screen. The default is to select positions off the screen by using the usual AutoCAD object snaps or AutoCAD absolute coordinates. By using the dot toggles, *.p*, *.n*, or *.g*, in the routines, you can toggle the routine into point number mode, northing/easting, or graphical select mode. If you exit the routine and do not clear the dot toggle, any routine using the toggles will remember the last toggle set. The next routine starts by using the last used dot toggle and will not stop using the toggle until you toggle off its mode. You toggle off the dot toggle by typing the current dot toggle.

Here are some examples of toggling on point number, northing/easting, and graphical mode:

```
Starting point: .p
Starting point: Point number:
```

-or-

```
Starting point: .n
Starting point: Northing: 5214.3214
Easting: 5412.3121
```

-or-

```
Starting point: .g
Starting point: Select point block:
```

BY POINT # (NUMBER) RANGE

The line By Point # (Number) Range routine creates a line with vertices at each point in the range of point numbers. You can specify a point number range in three ways: sequential, individual, or mixed point number ranges.

In the first method, sequential, you denote a consecutive range of point numbers by placing a dash (–) between the first and last number in the sequence; that is, 100–200. This tells the routine to use number 100 as the starting point of the line with each successive number the next vertex of the line; that is, 100,101,102,103, and so forth.

The second method is for situations where successive points are not sequential. In this case, a comma separates the list of individual point numbers. For example, point numbers 1,5,10,15,1 create a line from point 1 to 5 to 10 to 15 and closing back to 1.

The third method combines sequential and individual point numbers to draw a line; for example 1,5,10,20–27,40,43,50–63,1.

BY DIRECTION

The By Direction routine creates lines that are the result of traveling a distance at a bearing, azimuth, or angle that you select through options in the graphics editor. The routine starts by prompting for a starting point. You can specify a starting point with an AutoCAD selection or by using the dot toggles *.p*, *.g*, or *.n* (point number, graphically, or northing/easting). The routine then prompts for an angle and distance until you exit. The next angle and distance are from the endpoint of the last drawn line segment. As you exit out of the routine, the routine prompts you about a starting point for a new line. Pressing ENTER at the prompts exits the routine. This routine is exactly like the By Direction routine found in the Create Points cascade of the Points menu.

When the routine prompts for a direction, you can indicate the direction by one of the following methods:

```
Bearing (Quadrant) Method
Start point: .p
Start point: Point Number: 1
Quadrant (1-4) (Azimuth/POints): 1
Bearing (DDD.MMSS): 85.5750
```

-or-

```
Azimuth Method
Start point: .p
Start point: Point Number: 100
Quadrant (1-4) (Azimuth/POints): a
Azimuth (DDD.MMSS): 350.2413
```

-or-

```
AutoCAD Point Method
Start point: .p
Start point: Point Number: 1
Quadrant (1-4) (Azimuth/POints): PO
First point: <select a point> Second point: <select a point>
```

-or-

```
Point Number Method
Start point: .p
Start point: Point Number: 1
Quadrant (1-4) (Azimuth/POints): .p
First point number: 1
Second point number: 2
```

The By Direction routine walks the perimeter of a boundary, lot, or similar type of object. When you exit the routine, the routine reports the distance traveled along the perimeter. The By Direction routine does not create arc segments. The best that can be done when coming across an arc is to travel the chord of the arc.

BY TURNED ANGLE

The By Turned Angle routine mimics a surveyor's setup. The routine defines a setup of an instrument and a backsight point by the two endpoints of the selected line. From this setup the routine turns an angle or deflection angle and travels a distance to draw the line. The routine has two modes for establishing the instrument point (the pivot point) and backsight point: line mode and select point mode. In line mode (the default), the routine assumes an existing line whose endpoint nearest to the selection point is the instrument's location and the farthest endpoint of the same line is backsight point (the direction of 0 degrees).

If no line is present, you can toggle the routine into selecting the AutoCAD positions to locate the pivot and backsight points by typing in **PO** at the routine's prompt. When in point mode, you can select positions from the screen by one of several methods: AutoCAD object snaps or absolute coordinates, point number mode (the dot p toggle *.p*), graphically (the dot g toggle *.g*), or northing/easting mode (the dot n toggle *.n*). The first point selects the instrument point and the second point the backsight. These actions mimic field activity when placing points from an instrument setup.

BY STATION OFFSET

The By Station Offset routine draws a line at a station and offset of an existing line or arc. When you select the line or arc, the routine places a blip 'X' at the endpoint nearest to where you selected the object. This 'X' represents the beginning of the stationing. After establishing the station, the routine then prompts for the offsets of the line segment. The offset is either a positive (the right) or negative (left side) value. The resulting line is perpendicular to a line or radial to an arc.

LINE EXTENSION

The Line extension routine is a line length editing routine. See the discussion on editing lines in unit 3 of this section.

BEST FIT LINE

The Best Fit Line routine draws a line from measurements around, but not necessarily on, the line. This occurs, for example, when a field crew surveys a back-of-curb with many shots, but you want a single line segment to represent the back-of-curb. The routine assigns a default error value (1.0) to each point along the line. You can assign a larger or smaller error value to a point to influence the drawing of the line. When you change the error values, you indicate to the Best Fit routine that a point describes a line better (lower value) or worse (higher value). If the error value is lowered, the line moves toward the point. If the error value is raised, the line moves away from the point. If a point is on the line, the error value of 0.0 forces the routine to place the line on the point.

TANGENT

The Tangent routine creates a line tangent to an existing arc. After you select an arc, the routine prompts for the starting point of the line on the arc. If the line attaches to the end of the arc, you select the endpoint to start the line at the end of the arc. The routine then prompts for a second point that defines the length of the line. If the second point selected is not tangential to the arc point, the routine uses the distance between the selected points as the length of line. If you select a starting point for the line that is somewhere along the arc, the routine creates a line tangent to that point on the arc and does not clean up after the intersection point of the arc and line.

PERPENDICULAR

The Perpendicular routine draws a perpendicular line from a selected point on an object. If the object is an arc, the routine draws a radial line. The routine then prompts for a second point that defines the length of the line.

SPECIAL LINES

LDD provides a series of special line types to denote linear features. These features include fences, trees, guard rails, lines with text and symbols, and so forth.

The line routines, Text and Symbol, generate unique lines in the drawing. If you indicate a water main or gas line, you can incorporate the letters W or G in the line. If you want to include a symbol from the symbol library in the line, identify the symbol by name and the routine includes the symbol in the line.

The Special Line routine controls the length of the dash segment of the line, the length of the gap between dash segments, the size of the symbol or text in the line, and the ability to change the symbol used in the line. The routine does not adjust the height of a fixed text height style.

The routine allows for the point number, graphic, or northing/easting dot toggles (*.p*, *.g*, or *.n*) to define endpoints of lines.

These lines are not the same as the lines with text and symbols found in the linetypes of AutoCAD. The special line routines, Text or Symbol, place blocks at appropriate points on the line and do not incorporate them into a "line" like their AutoCAD counterparts.

CURVES

The arc routines include creating arcs from existing entities or points (see Figure 3–1). The curve routines using existing entities are: From End of Object, Curve Between Two Lines, Curve On Two Lines, and Reverse or Compound Curves. Routines using points to create a curve are Best Fit Curve and Through Point. The Multiple Curves routine creates a series of arcs to solve a turning problem.

The From End of Object routine creates arcs from the end of an existing tangent line. The Curve Between Two Lines and Curve On Two Lines routines create an arc segment between two tangent lines. The Curve Between Two Lines routine trims the tangents, whereas the Curve On Two Lines routine does not. The Reverse or Compound Curves routine creates tangent arcs from an existing arc. Curves that turn the same way as the existing curve are compound and curves turning in the opposite direction are reverse.

The curve routines may prompt for curve information not easily available when working on a preliminary plan. The values the routine prompts for may be the result of mathematical calculations about a specific arc. To help give information about a curve, LDD provides a curve calculator. The Curve Solver calculator is on the Utilities menu. You must enter two of the first five parameters of the arc before the routine produces the values of the curve.

If you are working on a preliminary plan, it may be more useful to use the OFFSET and FILLET commands of AutoCAD than the heavily number-dependent routines of LDD.

CURVE BETWEEN OR ON TWO TANGENTS

The Curve Between Two Tangents routine defines a curve between two existing lines by any of the following methods: arc length, tangent length, external secant, degree of curvature, chord distance, middle ordinate distance, minimum distance, and radius. When you create a curve by degree of curvature, you use either a railroad (chord) or highway arc definition. The routine cleans off the tangent line segments like the AutoCAD Fillet routine. The Curve On Two Tangent routine places an arc on the two tangent lines, but does not remove the extra tangent line segments.

If the arc is a negative value, the routine produces a looping arc.

CURVE THROUGH POINT

The Curve Through Point routine creates a tangent arc from a line through a point. Take care when using this routine because the result may be a looping curve, not a tangential arc segment.

MULTIPLE CURVES

The Multiple Curve routine places multiple arc radii between two tangent lines. The routine prompts for curve data for all but one arc. The radius of the remaining arc floats to accommodate the data for the fixed arc segments. The routine can have up to 10 curve segments.

FROM END OF OBJECT

The From End of Object routine gives you the ability to create a curve from the end of an existing line segment. The routine starts the arc from the nearest endpoint of the line to the selection point on the line. Then you define a radius, enter one additional piece of curve data, and the routine draws a curve. A positive radius creates a right turning curve (clockwise) and a negative radius creates a left turning curve (counterclockwise). The additional value of the curve is one of the following: tangent length, chord length (highway definition only), include angle, or radius.

REVERSE OR COMPOUND CURVES

The Reverse or Compound Curves routine creates curves that are tangent to existing curves. A compound curve is tangent to and shares the radius line of the adjacent curve. A reverse curve is tangent to an adjacent curve but turns in the opposite direction from the adjacent curve.

BEST FIT CURVE

The Best Fit Curve routine draws a curve from measurements around, but not necessarily on, the curve. The routine assigns a default error value (1.0) to each point along the curve. You can assign a larger or smaller error value to a point to influence the drawing of the curve. When you change the error values, you indicate to the Best Fit routine that a point describes a curve better (lower value) or worse (higher value). If the error value is lowered, the curve moves toward the point. If the error value is raised, the curve moves away from the point. If the point is on the arc, the error value of 0.0 forces the routine to place the curve on the point.

ARCS BY POINT NUMBER RANGE

The By Point # (Number) Range command (near the top of the Lines/Curves menu) allows you to incorporate arcs into the line drawing process. The following letters control the type and direction of the circular arc segment.

- C – Defines an arc by S,E,C (start, endpoint, and center point). The C fixes the location of the radius point.
- F – Defines an arc with a start and endpoint and a variable radius point. The F floats the center point of the arc.
- R – Indicates that the curve turns to the right.
- L – Indicates that the curve turns to the left.

The following are valid curve entries for an arc in the By Point # (Number) Range routine.

Line segments with a curve having a starting, ending, and fixed radius point:

```
Point numbers: 5,10,C15R,24,58
```

This draws a line from 5 to 10, with 10 being the beginning of the arc. The arc turns to the right with point 15 being the radius point. The arc ends at point 24.

If the arc is a left turn, the entry would be:

```
Point numbers: 5,10,C15L,24,58
```

The preceding entry creates a fixed radius arc with a starting and ending point.

Line segments with a curve having a starting, ending, and floating radius point:

```
Point numbers: 5,10,F15R,24,58
```

This draws a floating radius point arc between points 10 and 24 with a radius point near 15. The arc turns to the right.

If the arc is a left turn, the entry would be:

```
Point numbers: 5,10,F15L,24,58
```

SPIRALS

The Spiral routines place spirals into existing geometry of the drawing (see Figure 3–2).

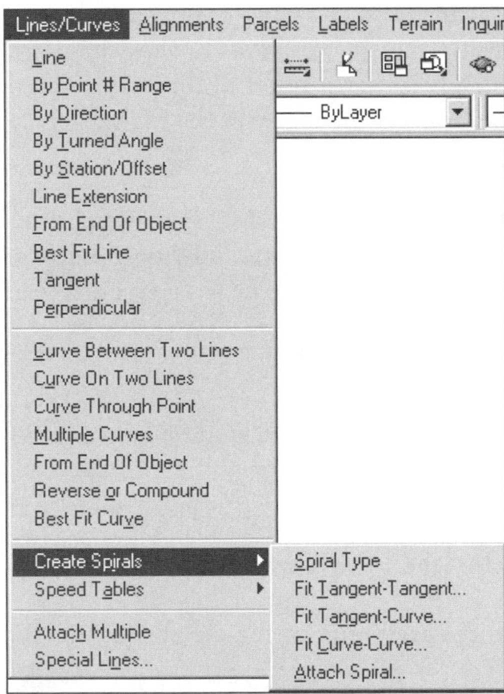

Figure 3–2

The Fit Tangent–Tangent routine places spirals between two tangents in a spiral-curve-spiral or spiral-spiral format. In the spiral-curve-spiral routine the beginning or ending spiral can have a 0 (zero) length. The remaining spiral routines place a spiral between a tangent and a curve (Fit Tangent–Curve) and between two curves (Fit Curve–Curve).

The Attach Spiral routine attaches a spiral to the end of a line, arc, or spiral. The routines include spiraling in or out, to a point, or creating a compound spiral.

SPEED TABLES

This section of the menu allows you to create and/or edit speed tables for superelevation roadway design (see Figure 3–3). The path routine sets the location of the table folders. The Edit Speed Tables routine allows you to select and edit the values in a speed table. The Create Curves routine creates curves from the definitions found in the speed tables.

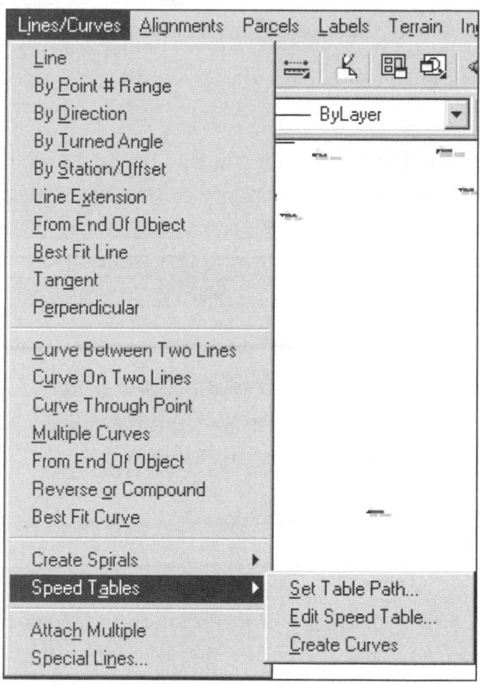

Figure 3–3

ATTACH MULTIPLE

The Attach Multiple routine attaches a line, arc, or spiral to the end of the last drawn entity. This routine frees you from having to use an object snap before adding the next segment.

EXERCISE

Exercise

After you complete this exercise, you will be able to:

- Use the line drafting routines from the Lines/Curves menu
- Use the *.n*, *.g*, and *.p* dot toggles
- Specify point number ranges
- Use the Line/Curve/Spiral routine to view design properties
- Use the Design Properties routine to view design properties

USING THE LINE DRAFTING ROUTINES

1. Open the drawing LDDTR2 of PROJ1. This drawing is from the last section's exercise.
2. Remove all of the description key symbols and points from the drawing.
3. Insert the exercise group of points. If you need to, do a zoom extents to view the points.
4. From the Lines/Curves menu, select the By Direction routine to draw the Subdivision boundary. Start the boundary at the Northing/Easting of 10000,10000 (the upper left of the boundary). Use the bearings and distances found in Figure 3–4 to the boundary lines. After drawing the line work, exit the routine by pressing ENTER twice.
5. Undo back to remove the line work.
6. Use the line By Point # (Number) Range routine to draw the boundary. Use the range of 1-6,1 to draw the line work. Exit the routine by pressing ENTER at the Point numbers: prompt.
7. Undo back to remove the line work.
8. From the Lines/Curves menu, select the Line routine. When the routine prompts for a starting point, enter the **.p** dot toggle and enter point number **1**. Enter points 2 through 6 to draw most of the boundary. When the last line segment needs to be drawn, toggle out of point number mode by typing **.p** and type the letter **C** to close the boundary. Exit the routine by pressing ENTER at the Starting Point: prompt.
9. Undo the line work and remove all the description key symbols and points from the drawing.
10. Insert the Napa Sub and Iron Pipes Found point groups and a range of point numbers. Use 10815-10850 as the point number range. If you receive a warning that a point is already in the drawing, select Skip All from the warning dialog box.
11. Do a zoom extents to view the new points.

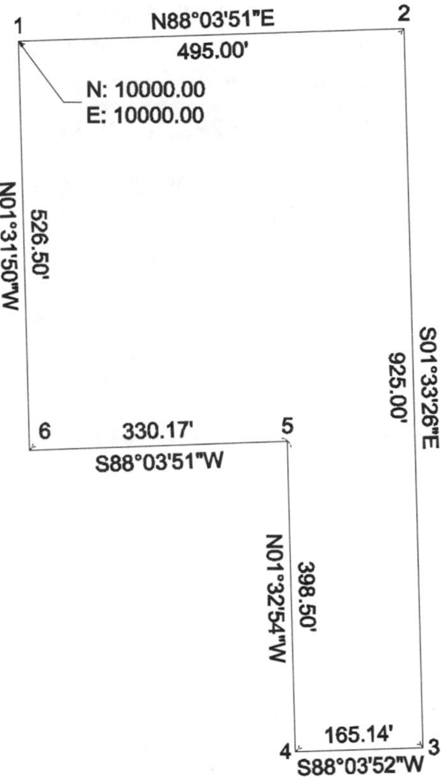

Figure 3–4

12. Use the Zoom to Point routine from the Point Utilities cascade of the Points menu. Zoom to point 10829 and use a height of 600.
13. Use the By Point # (Number) Range routine to draw the boundary line work. Use the point numbers 10754, 10345, 10762, 10761, 10758, 10757, 10754.

You will notice that point 10828 is to the south of point number 10761. This point needs to have its coordinates edited because they were incorrectly typed into the imported point file. Use the Edit Points routine to edit the coordinates of point 10828.

14. Select the Edit Points routine from the Edit Points flyout of the Points menu. Select point number 10828, change the northing from 11476.9135 to **11746.9135**, and select OK to exit the dialog box. The point should move to the north.
15. Use the Zoom to Point routine from the Point Utilities cascade of the Points menu to zoom in closer to point 10829. Use a screen height of 150.
16. Use the Curve Solver routine from the Utilities menu to calculate the following curves: Radius 400.00 feet, a chord of 55.86 feet, a second curve with a radius of 175.50 feet, and an arc length of 135.87 feet.

17. From the Lines/Curves menu, select the By Point # (Numbers) routine to draw two line segments and an arc between points 10828 and 11227. When the routine prompts for the point numbers, enter the following numbers in this format:

    Point numbers: **10828,10834,C10829R,10835,11227**

This entry draws a line segment between points 10828 and 10834 and 10835 and 11227. The arc is a right-handed arc starting at 10834, ending at 10835 and centered at 10829. Press ENTER to exit the routine. Your results should look like Figure 3–5.

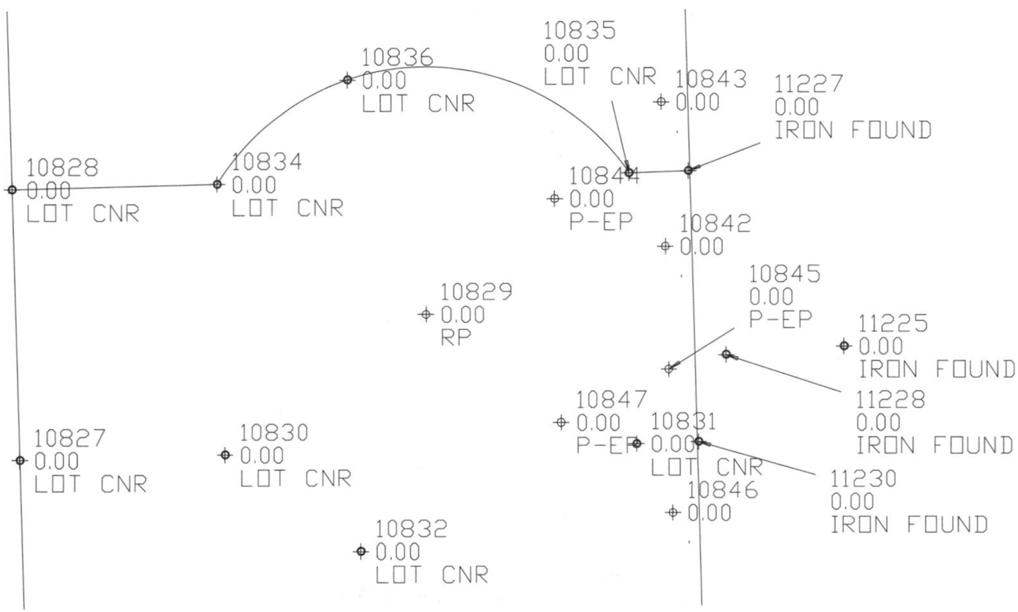

Figure 3–5

18. Use the Zoom to Point routine to zoom to point 10781. Use the height of 500.

19. Use the By Point # (Number) Range routine to draw the next line segments. The routine does not need the points in the list to be present in the drawing to be able to draw the line. Some of the lines created below will not show up on the screen until you zoom to a larger view of the drawing. Draw the following line segments.

```
Point numbers: 10768,10771,10776,10790
Point numbers: 13011,10768
Point numbers: 10775,10773
Point numbers: 10780,10823,10782,C10781L,10787
```

The last entry creates a line and arc segment. Your results should look like Figure 3–6.

Figure 3–6

20. Create a perpendicular line to the arc segment just drawn with the Perpendicular routine of the Lines/Curves menu. The routine asks you to identify the arc, then the perpendicular point, and the end of the arc. In this case the perpendicular and endpoint of the line is point number 10815. The routine creates the line from the radial perpendicular point of point number 10815 and ends the line at point 10815. The following is the prompting for this process. You may have to use the Zoom to Point routine to view point 10815. If you do use the routine set the height to 300.

```
Select entity (or POints): (select a point on the arc segment)
Locate start point of line: .p
Locate start point of line:
 >>Point number: 10815
Locate end point of line:
>>Point number: 10815
```

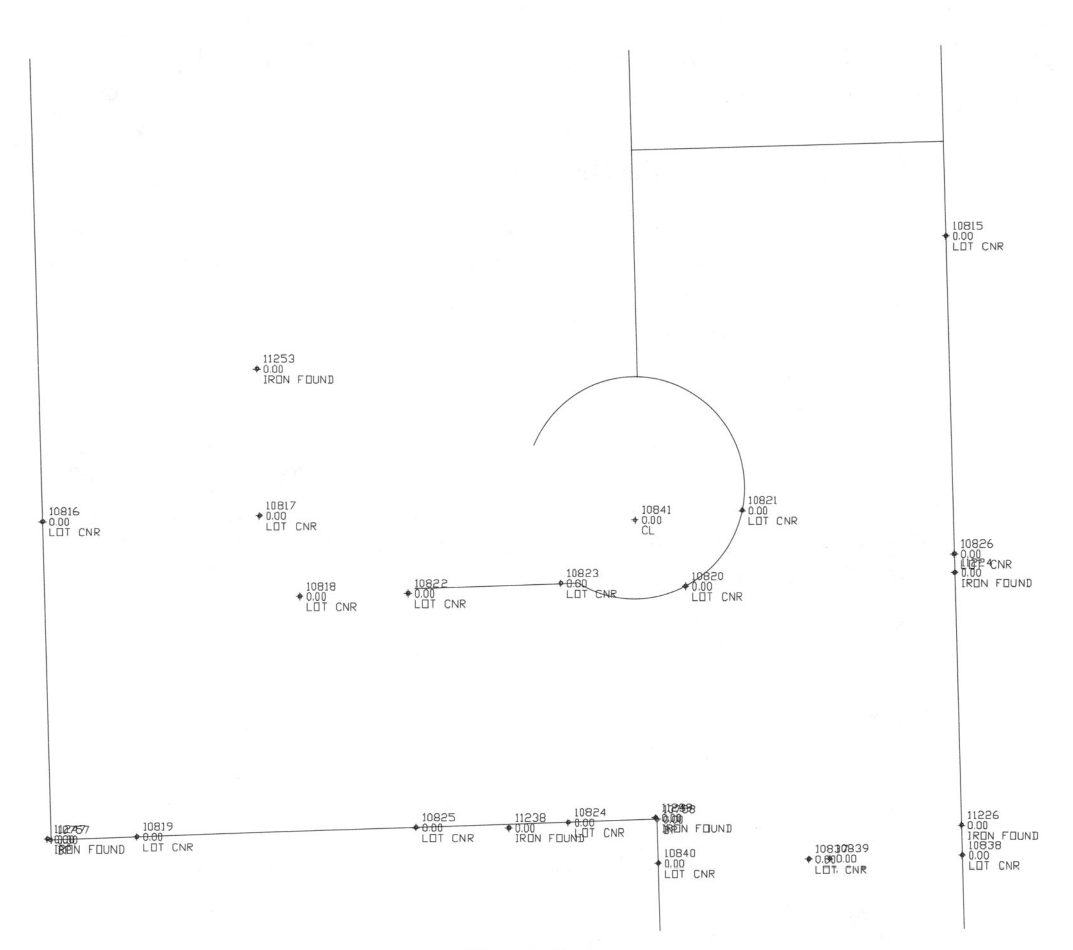

21. Use the Line By Point # (Number) routine to create the next line. Use the following values to create the line. After creating the line, exit the routine by pressing ENTER.

```
Point numbers: 11000,10778,10841,10781
```

22. At the Southwest intersection of the two new lines (point number 10778), use the Curve On Two Lines routine to create a 40-foot radius curve. Notice that the routine does not remove the tangent segments from the two lines. Undo to remove the curve segment.

23. At the same intersection use the Curve Between Two Lines routine to create an arc between the lines. After you select the line segments, the routine prompts for which value to create the arc from. The default value is a radius. Press ENTER to indicate that the next entry is a radius. After entering the radius, 40 (feet), press ENTER to create the arc. The routine creates the arc and cleans up the extra tangents of the two intersecting lines. Press ENTER one more time to exit the routine.

24. Save the drawing.

25. Use the Zoom to Point routine and zoom to point number 10787 with a height of 150.

26. Insert points 10787 and 10789 into the drawing.

27. Use the Line/Curve/Spiral routine from the Inquiry menu to measure the distance between points 10787 and 10789. You need to toggle the routine into POints mode (**po**), then into point number mode (**.p**). The routine prompts for three selections; however, since this is a distance, select only two points and press ENTER when prompted for the third point. When you press ENTER, the routine returns the line values. Finally, press ENTER one more time to exit the routine.

```
Select entity (or POints): po
First point: .p
First point:  >>Point number: 10787
Second point (end, center, or spi):
 >>Point number: 10789
Last point:
 >>Point number: .p (press ENTER)
- - - - - - - - - - - - - - - - - - - - - - - - - - - - - - - - - -
                              LINE DATA
- - - - - - - - - - - - - - - - - - - - - - - - - - - - - - - - - -
Begin . . . . .  North: 12132.7467           East:  7893.5313
End . . . . . .  North: 12123.4904           East:  7880.1776
                 Distance: 16.25             Course: S 55-16-18 W
Select entity (or POints):
```

EXERCISE

28. Use the Curve Solver routine to view the statistics on a curve with a chord distance you just measured and a radius of 15 feet. These measurements will be used for the reverse curve you will draw in the next step.

```
Included angle:
Radius: 15
Arc length:
Chord length: 16.25
Tangent length:

- - - - - - - - - - - - - - - - - - - - - - - - - - - - - - - - -
                    CURVE SOLVER RESULTS
- - - - - - - - - - - - - - - - - - - - - - - - - - - - - - - - -
 Included angle = 65-35-40
         Radius = 15.00'
 Tangent length = 9.67'
     Arc length = 17.17'
   Chord length = 16.25'
External secant = 2.84'
   Mid ordinate = 2.39'
Degree of curve = 21-58-19
```

29. Select the Reverse or Compound Curves routine from the Lines/Curves menu to create a reverse curve on the end of the existing curve. Select the existing curve at its northwestern end, press ENTER to select the default curve type (reverse), enter the radius of 15 feet, type the letter **C** for the chord option, and enter the chord length of **16.25** feet. The routine creates the curve and prompts for another arc to work with. Press ENTER to exit the routine.

```
COGO Reverse or compound curve
Select curve:
TYPE [Compound/<Reverse>]:  (press ENTER)
Proposed radius: 15
FACTOR [Tangent/Chord/Included/<Length>]: c
Chord length: 16.25
```

30. Next draw a tangent line from the end of the new arc whose length is from the end of the arc to point number 10777. Use the Tangent routine from the Lines/Curves menu. Select the end of the arc nearest to point 10789. If the routine is still in point number mode. toggle it off (.p), use the object snap of endpoint, and select the western end of the arc. When prompted for the end of the line, toggle on point number mode (.p), and enter point number **10777**. Your drawing should look like Figure 3–7.

EXERCISE

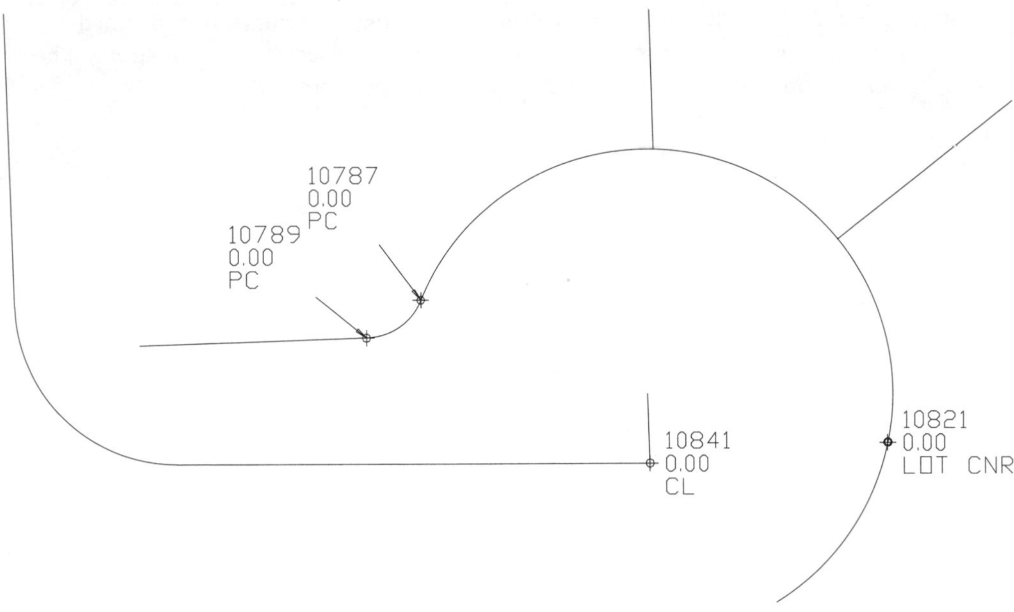

Figure 3–7

```
Select entity (or POints): (select near the west end of the arc)
Locate start point of line:
 >>Point number: .p
Locate start point of line: _endp of
Locate end point of line: .p
Locate end point of line:
 >>Point number: 10777
```

31. Save the drawing.

UNIT 2: ANALYZING LINE AND CURVE DATA

Land Development Desktop provides tools for listing out information about lines, arcs, and spirals. The most useful listing routines are Line/Curve/ Spiral from the Inquiry menu or Design Properties from the shortcut menu.

Both routines return a comprehensive report on line, curve, and spiral objects. The Line/Curve/Spiral routine lists the design information in the text window of AutoCAD. The Design Properties routine of the shortcut menu lists the information in a dialog box. Figures 3–8, 3–9, and 3–10 show the dialog box report for a line, an arc, and a spiral.

The following is an example of a line report from Line/Curve/Spiral:

LINE DATA

```
Begin    North: 5104.7717    East: 5126.3542
End      North: 5234.4171    East: 5086.8484
Distance:135.53      Course: N 16-56-50 W
SELECT POints/<entity>:
```

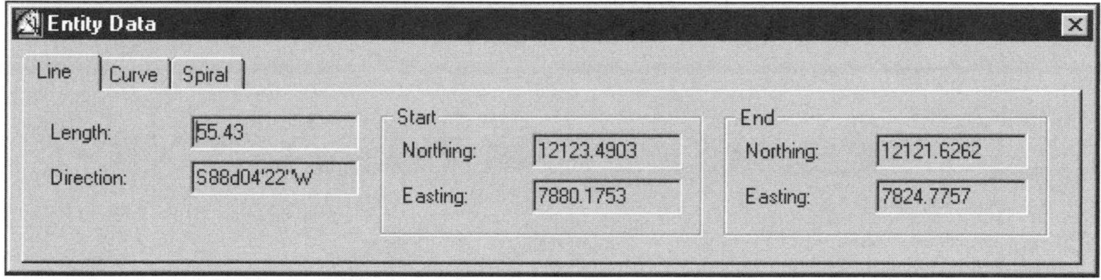

Figure 3–8

The following is an example of an arc segment report from Line/Curve/Spiral:

ARC DATA

```
Begin              North: 5104.9981    East: 4999.3658
Radial Point       North: 5104.8169    East: 4969.3663
End                North: 5105.1051    East: 4939.3677
PI                 North: 8939.4643    East: 4976.2035
Tangent:3834.54    Chord: 60.00        Course: N 89-54-52 W
Arc Length: 93.78  Radius: 30.00       Delta: 179-06-13
SELECT POint/<entity>:
```

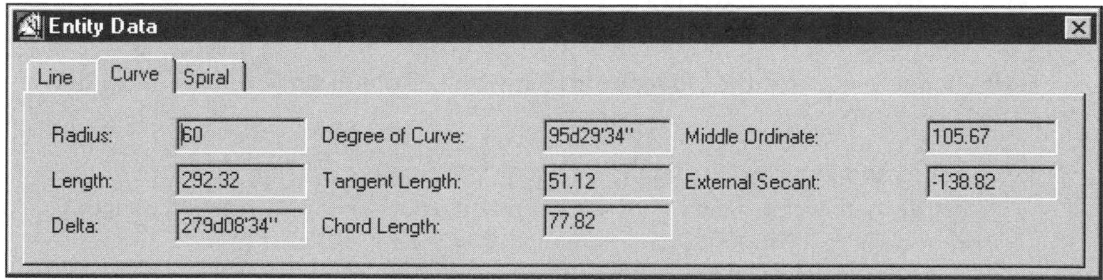

Figure 3–9

The following is an example of a spiral segment report from Line/Curve/Spiral:

SPIRAL DATA: Clothoid

```
TS . . . .  North: 5334.4397            East: 5631.3078
SPI. . . .  North: 5382.7558            East: 5677.3835
SC . . . .  North: 5402.7811            East: 5704.1417
                K : 49.9537               P : 1.3875
                A : 173.2051            Tau : 0.1667
                Xs: 99.7226               Ys: 5.5445
        Long Tan : 66.7639           Course: N 43-38-25 E
        Short Tan: 33.4218           Course: N 53-11-23 E
Length: 100.0000          Radius: 300.0000          Theta: 9-32-57
Select entity (or POints):
```

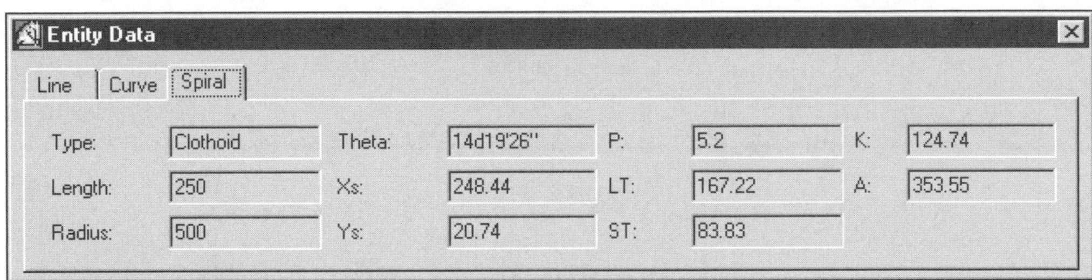

Figure 3–10

Exercise

After you complete this exercise, you will be able to:

- List line and curve data with the Line/Curve/Spiral routine
- List line and curve data from the shortcut menu

LISTING LINE AND CURVE DATA

1. If you are not in the LDDTR2 drawing, open the drawing and use the Zoom to a Point routine to set the current drawing view. Zoom in on point 10787 with a height of 150.
2. List lines and arcs in the current view first by selecting the line or arc to activate its grips, then clicking the right mouse button, and selecting Design Properties from the shortcut menu. The routine lists the data in a dialog box.
3. Use the Line/Curve/Spiral routine from the Inquiry menu. Select lines and arcs from the screen. Toggle the routines into point mode (PO) and while in pointing mode, select points on the screen with object snaps and point numbers (.p) to

get a report from the listing routine. The report is placed in the AutoCAD text screen. Exit the routine by pressing ENTER.

4. Save the drawing.

UNIT 3: EDITING LINE AND CURVE DATA

Most of the editing of line and arc data consists of erasing and redrawing the entity. The Line Extension routine changes a line by an overall length or by extending one of its endpoints. The only routines that change a line or arc before drawing it are the Best Fit routines found on the Lines/Curves menu.

LINE EXTENSION

The Line Extension routine extends a line from one endpoint. The amount of change at the end of the line is either a specified distance or an overall length. The routine extends the endpoint of the line nearest to the point of selection, placing an "X" at the active endpoint. The routine then prompts for a distance to add to the endpoint, which is a positive or negative value. The optional method of extending a line is to specify the total length of the line. The routine calculates the distance the line needs to change to attain the specified length and modifies the line accordingly.

Exercise

After you finish this exercise you will be able to:

- Extend a line from one end by overall length or extending one end.

EXTENDING A LINE

1. If you are not in the LDDTR2 drawing, open the drawing and use the Zoom to a Point routine to set the current drawing view. Zoom in on point 10787 with a height of 150.

2. Start the Line Extension routine from the Lines/Curves menu, then select the east-west line to the west of the reverse arc. An 'X' appears at the western end of the line and the routine reports the overall distance of the line. The default method adds a length to the line. At the Distance to Change prompt enter **10** to add 10 feet to the line.

```
COGO Line extension
Select the line:
End that is to change.
Length = 55.43
Distance to change (or Total): 10
```

3. While still in the routine reselect the line. This time change the line's total length by placing the routine into total mode. Change the routine into Total length mode by entering a **T** in at the routine's prompt and pressing ENTER. Enter the length of 55.43 and press ENTER twice to exit the routine.

```
Select the line:
End that is to change.
Length = 65.43.
Distance to change (or Total): t
New length: 55.43
Select the line:
```

4. Save the drawing.

UNIT 4: ANNOTATING LINES AND CURVES

The Labels menu presents the Land Development Desktop line, curve, and spiral labeling tools. There are two types of labeling methods, dynamic and static. A dynamic label reacts to changes in the characteristics of the line, arc, or spiral. A static label does not react to any changes of the object it labels. The Disassociate Labels routine creates static labels from dynamic labels. There are two types of labeling styles. The first is text labeling in the vicinity of the line, arc, or spiral. The second is a tag that contains information for the creation of a line, arc, or spiral table.

The Preferences routine of the Labels menu controls the resulting label on a line, arc, and spiral. The Edit Label and Tag Styles routines develop new and edit existing label styles. Many of the Labels menu routines appear on a shortcut menu, when a line, arc, or spiral displays grips.

ANNOTATION STYLES

The use of objects allows the lines, arcs, and spirals to understand that there is a relationship between their geometry and their annotation. For example, after you grip edit a line, the line knows that if there is annotation associated with it, it needs to change the text values displayed on the screen. The technology updates the text as soon as the change is made to the line, arc, or spiral.

The annotation can be sensitive to a change of scale. If you use one of the Leroy text styles and you change the scale of the drawing in the Drawing Setup dialog box, the Update All or Update Selected Labels routine changes the size of the annotation. The updating process not only changes the size of the text, but also adjusts the distance the text is from the line. The style defines the offset distance of the text from the line. The distance shown in the drawing changes as a function of the annotation style definition and scale of the drawing.

Land Development Desktop comes with three types of labeling styles. The first type is the current label style. This type uses the current text style to label lines, arcs, and spirals. The second type is Leroy. This type of label uses the predefined LDD Leroy text styles. The last style is Metric. This type of label uses the predefined LDD milli text styles. You set the path to these predefined styles in the General tab of the Label Settings dialog box of the Labels menu or in the Label Styles path in the User Preferences dialog box of the Projects menu.

ANATOMY OF A LABEL STYLE

The Edit Styles routine on the Labels menu calls a dialog box that displays the definition of a label style (see Figure 3–11). There are four types of labels in LDD, line, curve, spiral, and point. You can modify or create new styles with the same routine. An annotation style definition contains settings for values, text style, height, layer, offset, justification, arrows, crow's feet, and which values are above or below the line. A style may contain a formula to calculate a value that becomes a part of the annotation.

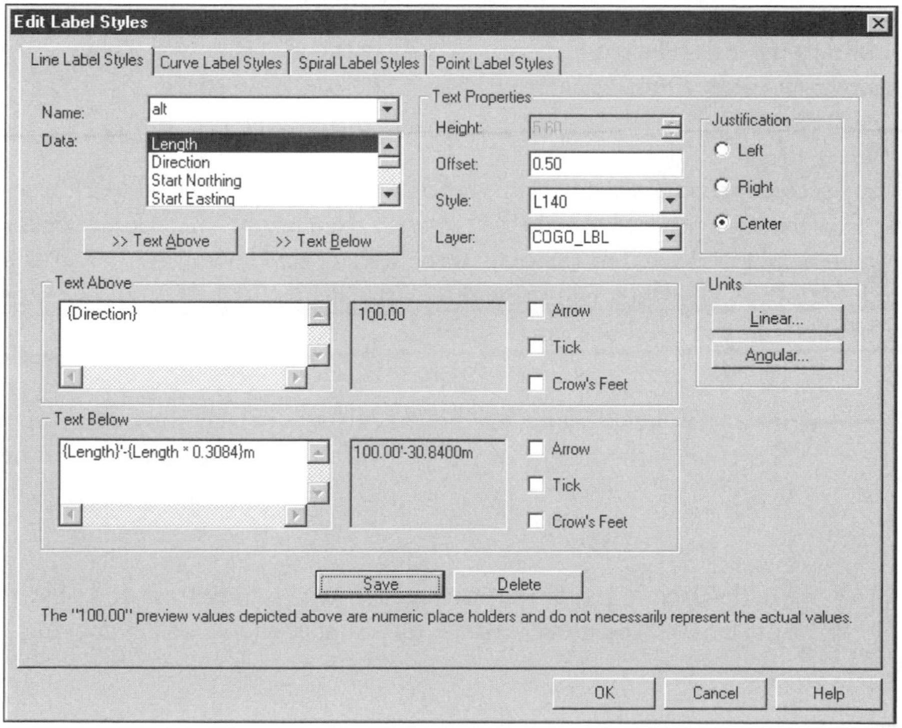

Figure 3–11

NAME AND DATA GROUP

The Name drop list area lists all currently defined labeling styles. The Data area lists all the information about each object that can be a part of a label. Each object, line, curve, and spiral has a different list of data elements to chose from.

The Text Above and Text Below buttons place a selected data item into the above or below area of the style definition.

Text Above and Text Below

When you view a style, the Text Above and Text Below areas of the dialog box display the elements that make up the style. In the two areas, the value to be labeled is shown in braces ({}). The text can contain a prefix or suffix outside of the braces that will appear with the values in the final annotation. The box to the right displays the resulting text.

The toggles to the right of the preview area control the use of arrows, ticks, and crow's feet in the annotation definition.

TEXT PROPERTIES GROUP

The text properties of the style control the text style, height, offset, layer, and justification.

Style

The annotation uses a selected text style to display the values on the screen. The text style must exist in the drawing before it is assigned to a label style.

Height

The style selected can be a fixed or variable height definition. The type of style affects how you need to set the height setting. If the text style is a fixed text height style, the dialog box displays the height as defined by the style. If the scale of the drawing is 1" = 50' and the style is L120, the height box will show 6.0 as the height (0.12*50). To change the height of the text, you need to change the style. If you choose L140, the text height will be 7.0 (0.14*50).

If the style is not a fixed height style, the height box will display the height last used for that style. To change the height of the text in the style, just change the height shown in the box.

Offset

The offset is the distance in plotted units that the text is away from the object. For example, if the offset is 0.5 and the scale is 50, the distance should be 2.5 drawing units.

Layer

This is the layer on which the text is placed into the drawing. By clicking the drop list arrow, you can select any existing layer in the drawing. If you want to use a layer not present in the drawing, type the name of the layer into the box. When labeling, the layer will be created if it is not already in the drawing.

Justify

There are three justifications for labeling text: left, center, and right. These justifications are in relation to the object being labeled.

UNITS GROUP

The Units group sets the precision values for linear and angular measurements (see Figure 3–11). This allows a label to have different precision settings relative to the drawing setup. These settings also control the precision of any calculated values.

Linear Units

The Linear Units dialog box controls three values: linear, coordinate, and formula (see Figure 3–12). The linear value controls the annotation of horizontal distances. The coordinate value controls the precision of northing/easting values. The formula precision applies to any formula that is entered as a part of the style definition. This capability allows annotation to display metric distances in an imperial drawing. An example of a formula in a style is:

{Length}'- {Length * 0.0348}m

This creates a style that annotates the length of the line in imperial and metric units.

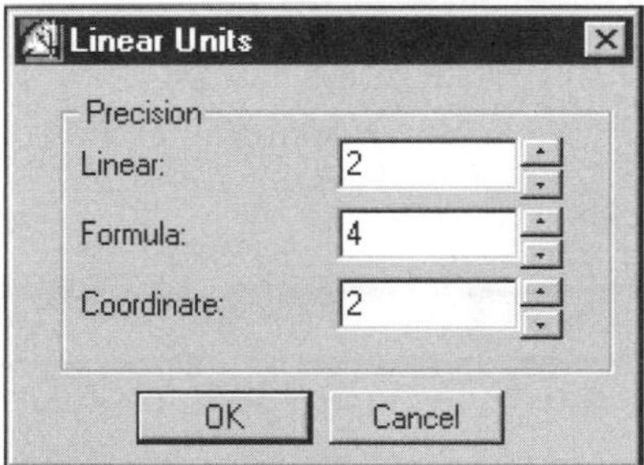

Figure 3–12

Angle Units

The Allow Text Spaces setting in the Angular Units dialog box places spaces between the text and the angle of a bearing.

The angle precision controls how the angle is displayed on the screen (see Figure 3–13). A precision of 0 shows only degrees. A precision setting of 4 displays degrees, minutes, and whole seconds of angle.

The formula setting controls the precision of any formula calculating a value for the annotation style. You may want to display the magnetic direction as well as the true bearing in the annotation. To be able to label this value requires the entry of a formula.

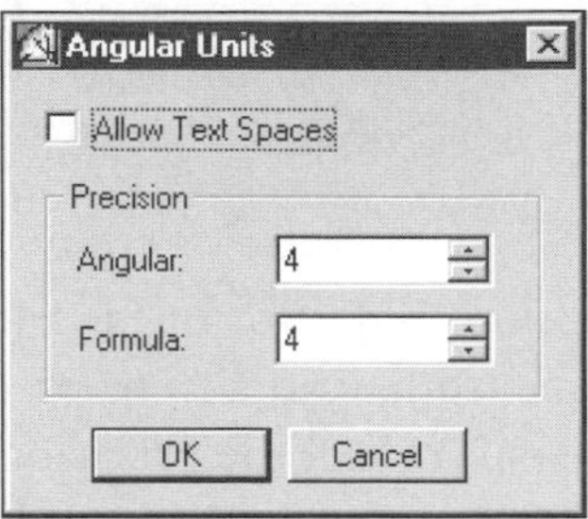

Figure 3–13

SAVE AND DELETE

The Save and Delete buttons at the bottom of the Line Label Styles dialog box save or delete the currently displayed style.

LABELS SETTINGS

The Labels Settings dialog box is a multitabbed dialog box that controls the initial behavior and location of the label styles. The initial tab, General, controls the default location of annotation styles and their two main behaviors: Update Labels When Style Changes and Update Labels When Objects Change (see Figure 3–14). If these two toggles are off and you either change a style definition or edit the object, the annotation will not change. These toggles can be on and off in any combination. However, if you run the Update All or Update Selected routine, the annotation will react according to the changes.

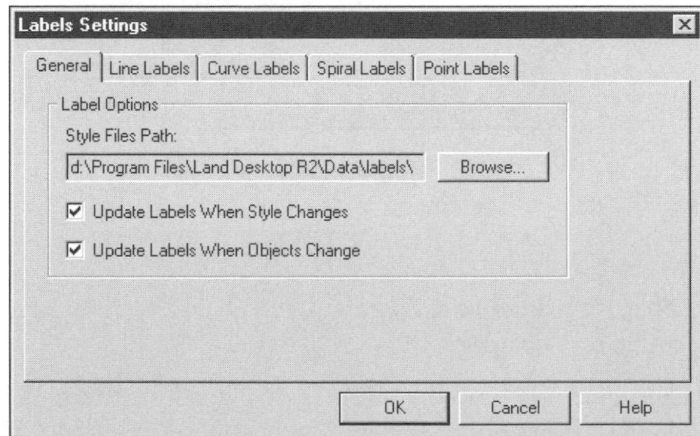

Figure 3–14

Land Development Desktop comes with three types of labeling styles. The first type is the current label style. This style uses the current text style to label lines, arcs, and spirals. The second type is Leroy. This style uses the predefined LDD Leroy text styles. The last style is metric. This style uses the predefined LDD milli text styles. You set the path to these predefined styles in the General tab of the Label Settings dialog box of the Labels menu or in the Label Styles path in the User Preferences dialog box of the Projects menu.

The remaining tabs display the default annotation settings for each type of object (see Figures 3–15, 3–16, and 3–17). Each panel sets the default labeling style, whether the label is aligned to the object, and if the annotation is not aligned to an object, its rotation angle. The Line panel has an additional toggle. This toggle controls whether the bearing is a result of how the line was drawn or forces the bearing to be either a northerly or southerly value.

The Tag Labels area controls the creation of data for a table displaying line, curve, or spiral information. This group sets the style and seed number for the tag.

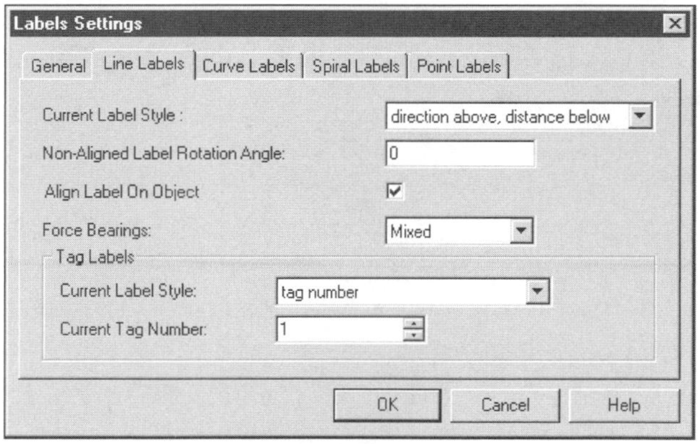

Figure 3–15

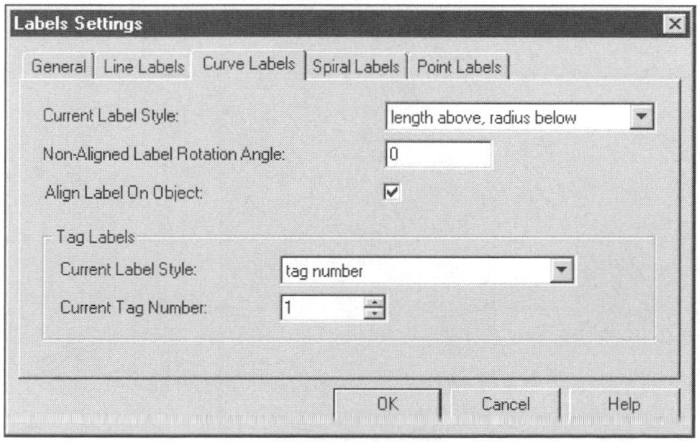

Figure 3–16

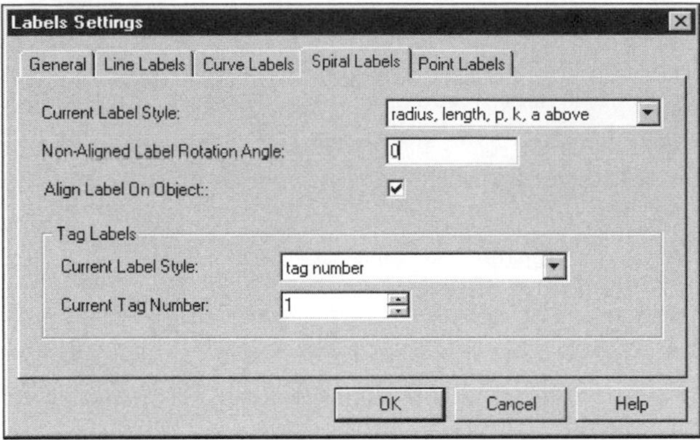

Figure 3–17

SHORTCUT MENU OPTIONS

If you click the right mouse button while some line, curve, or spiral objects are highlighted, LDD will display a shortcut menu displaying a number of routines found on the Labels menu (see Figure 3–18). The routines on the menu are Add Dynamic Label, Add Tag Label, Add Static Label, Update Labels, Flip Direction, Delete, and Disassociate Labels.

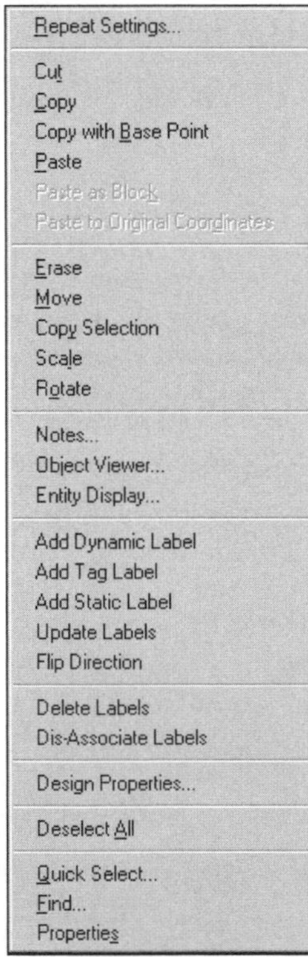

Figure 3–18

LABEL DIALOG BAR

The Label dialog bar is a labeling command center (see Figure 3–19). The bar sets the current line, curve, spiral, and point labeling style. The Align on Object toggle sets the alignment of the annotation with object. Clicking the Pencil icon calls the Edit Label Styles dialog box (see Figure 3–11). Clicking the Tag icon calls the Edit Tag Styles dialog box. The Labeling Mode toggle in the upper left of the dialog bar sets label or tag mode annotation.

This dialog bar eliminates a number of menu picks when you are labeling a plat.

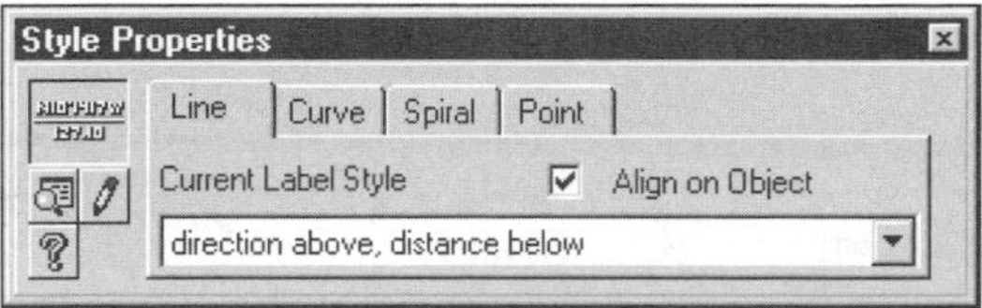

Figure 3–19

ADD STATIC LABELS

The Add Static Labels routine labels entire objects with non-reactive labels. The label does not respond to any change made to the object and it is your responsibility that the label shows the correct direction and/or distance. You can change the label from static to dynamic by selecting the label, pressing the right mouse button, selecting Edit Label Properties from the pop-up menu, and toggling on Dynamic Update Label Text. The Edit Label Properties dialog box also allows you to swap the location of the labeling text.

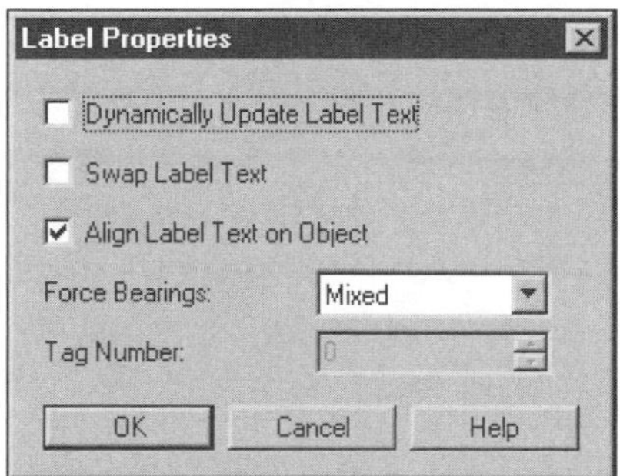

Figure 3–20

LABELING SEGMENTS

The Dynamic and Add Static labeling methods label the entire object. To label portions of lines and arcs, you have to use the Label Line By Points or Label Curve By Points routines. These routines have the option of selecting points to generate a label value. You can select the points by using AutoCAD coordinates, object snaps, point numbers, graphically selecting point objects, or northings/eastings. The . (dot) toggles call the last three selection methods.

The labels that result from the Labeling by Points routines are static labels. These labels do not react to changes, if any changes do occur, you must be sure to check the labeling to see if it needs to be changed. The Flip Direction, Swap Label Text, and Delete Label routines will not work with these labels because these routines work only with Dynamic or Static labels of entire objects.

This also means that if a style places the text on the wrong side of the line, you need to define a style that is the opposite of the one that does not label correctly. For example, if Direction Above and Distance Below labels incorrectly, you need to define the style Distance Above and Direction Below to be able to label the line correctly.

The drawing may contain both static and dynamic labels. This may not be a desirable mixture. You may want to make all of the labels static, bearing in mind that any changes means fixing the labeling displayed on the screen.

Exercise

After you complete this exercise, you will be able to:

- Define annotation styles
- Set and save annotation defaults
- Create dynamic labels
- Create static labels
- Annotate on or off an object
- Create annotation tags
- Create line and arc table

CREATING LABELS

The process of creating a label includes filling in the blanks and selecting the correct values within the Edit Label Styles dialog box. If you select the Save button, the style definition becomes a file in the *LABELS* subfolder of the *DATA* folder. The name of the style is in the Name area of the dialog box. Each label style has a unique name and different extensions for the style type, line (lns), curve (crs), or spiral (sps). The ltd, ctd, and std extensions indicate line, curve, and spiral table styles. These files are binary files and are not editable with any editor.

The procedure for creating a style is naming the style and defining what the style annotates, the text parameters, and the precision. After you set the values, the style becomes a permanent file when you select the Save button.

1. Open the LDDTR2 drawing and select the Label Settings routine from the Labels menu, set the following values, and select OK to exit the Settings dialog box. The path to the Leroy label styles assumes the default install (see Figure 3–21).

```
General Tab:
Style Files Path: C:\Program Files\Land Desktop
  R2\Data\labels\leroy
Update Labels When Style Changes: ON
Update Labels When Objects Change: ON
Line Labels Tab:
Current Style: direction above, distance below
Non-aligned Label Rotation Angle: 0
Align Label on Object: ON
Forced Bearings: Mixed
```

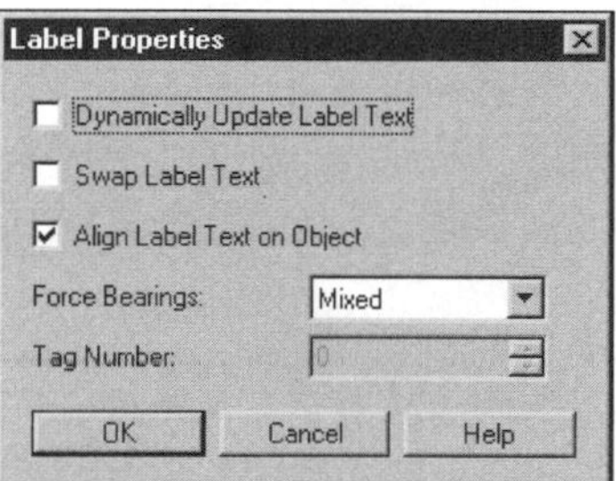

Figure 3–21

2. Select the Edit Label Styles routine from the Labels menu, select the Line Label Styles tab, then click in the Name box, and enter the new style's name, **Dist**. The Dist style annotates only the distance of a line or two selected points. The length is below the line and uses crow's feet. Use Figure 3–22 as a guide and enter the data below to define the line label style. Select the Save button to save the style after entering the values.

```
Name: Dist
Text Below: {Length}'  (add the foot mark after the length)
Arrow Toggle: OFF
Offset: .5
Text Style: L120
Layer: LINE LBL
Crow's Feet Toggle: ON
```

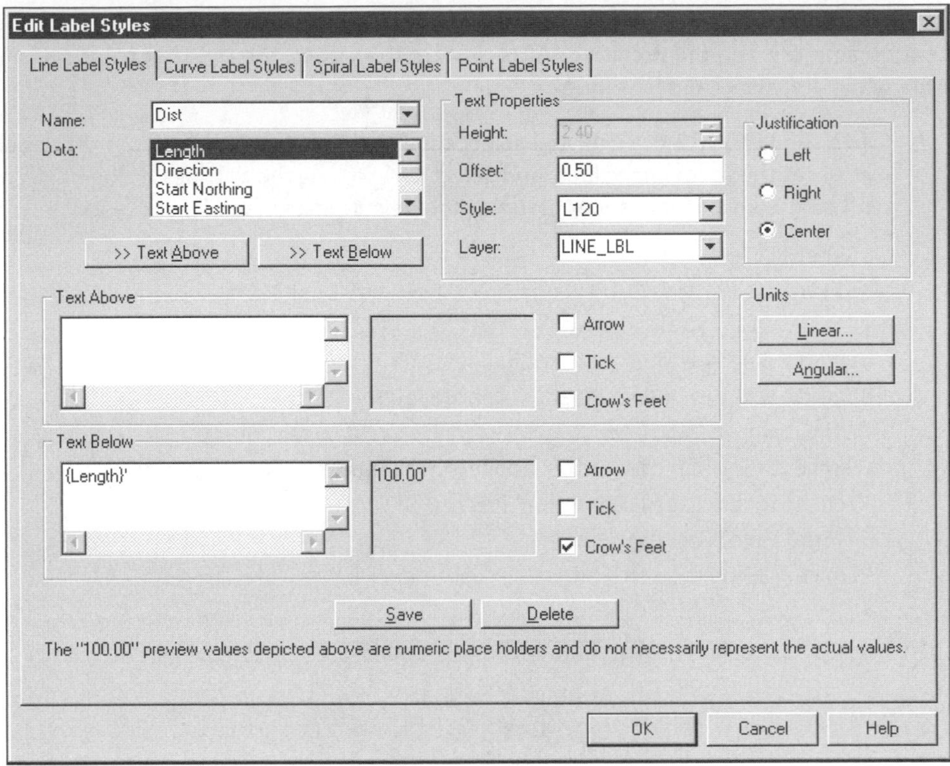

Figure 3–22

3. Click in the Name box, and enter the new style's name, **Alt**. The Alt style annotates Direction and the distance of a line or two selected points. The length is below the line and displays the distance as feet and meters. To make the style display metric, you add a formula in the text below box. In this case, you follow the length value with the formula **{Length * 0.0384}**. This formula returns the length in metric and used the text style to write the value for the label. Use Figure 3–23 as a guide and enter the following data to define the line label style. Select the Save button to save the style after entering the values. Select the OK button to exit the dialog box.

```
Name: Alt
Text Above: {Direction}
Text Below: {Length}'-{Length * 0.0384} (Add this to calculate Metric)
Offset: .5
Text Style: L120
Layer: LINE_LBL
Units: Linear - 2
Crow's Feet Toggle: OFF
```

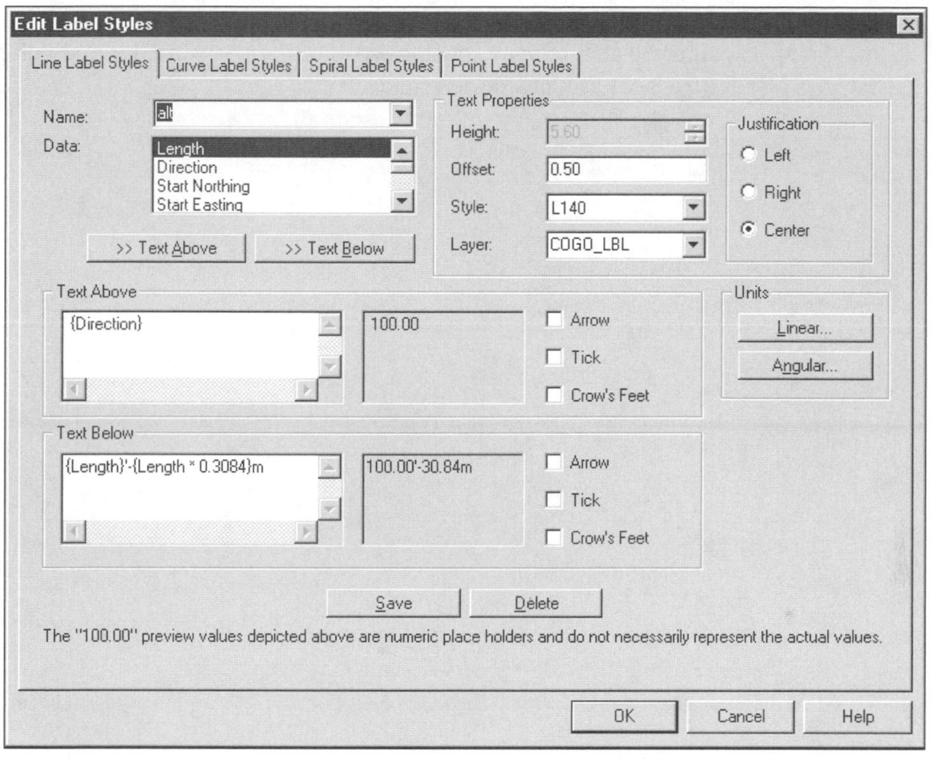

Figure 3–23

4. Select the Show Dialog Bar routine from the Labels menu to display the labeling command center (the Style Properties toolbar).
5. Click the down arrow of the Current Label Style list to view the available label styles, and select a new line label style as the default. By selecting each tab and then clicking the display list arrow, you can view all of the label styles for lines, curves, and spirals. Reselect the Line tab to make it the current panel.
6. Click the Pencil icon in the dialog box. This displays the Edit Styles routine. Select OK to exit the Edit Styles routine.
7. Click the Setting icon to the left of the pencil icon to display the Settings dialog box. Select OK to exit the Settings dialog box.
8. Click the Help (question mark) icon to view the associated help file. Exit the Help dialog box by selecting the upper right X in the dialog box.
9. Click the Labeling Mode icon that is above the Pencil icon. This switches labeling between text and tag labeling modes. Tags allow you to create line and arc tables for the display of line and arc information. Leave the Style Properties dialog box on the graphics screen.

10. Use the Zoom to Point routine. Zoom to point 10791 with a height of 300. You may need to pan to be able to view the line work of the triangular lot and cul-de-sac arc. Use Figure 3–24 as a positioning guide.

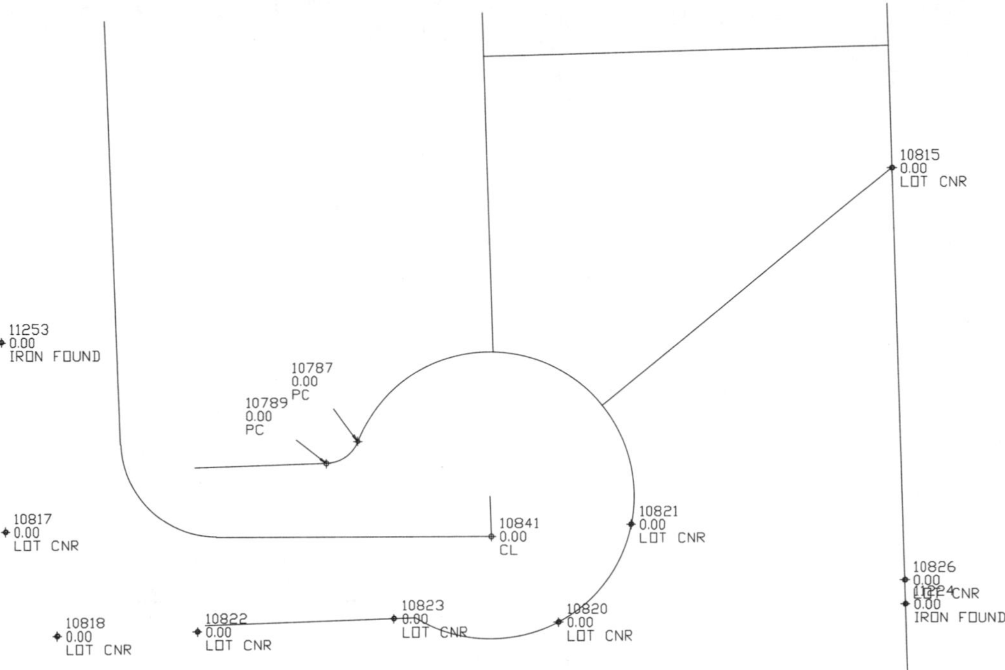

Figure 3–24

11. Make sure the default line annotation style is Direction Above and Distance Below. This should be the style named in the Label dialog bar. Select several lines, click the right mouse button, and select Add Dynamic Labels. Labels should appear on the line work.

12. Use the zoom window command to view the annotation.

13. Select the southerly boundary line of the triangular lot to activate its grips, select the southerly grip of the line, and move the end to a point in the northwest of the lot. Notice the annotation reacts to the changes. Undo the grip edit.

14. Again select the same line to activate it grips, click the right mouse button, and select Flip direction. This changes the line's annotation to the opposite direction.

15. Select the Swap Label Text routine from the Labels menu and select the same line. You select the line, not the annotation, because you are telling the line what to do with the label. This routine changes the sides of the line the text is on. After you use the routine, the bearing is outside the lot line and the distance is inside.

16. Undo back the changes to the southerly lot line so that it is back to its original position.
17. If you need to, relabel the lot lines of the triangular lot.
18. Use the Swap Label Text routine to change the other labels so that the distance is inside and the bearing is outside.

There is a problem with the west and east boundary of the lot. In the west, the lot line is two segments and the east line extends north and south of the lot segment. To label these segments correctly, you must label the segments by points. The labels that result from the Labeling by Points routine are static labels. Since static labels do not react to changes, if any changes do occur, you must be sure to check the labeling to see if it needs to be changed.

The Flip Direction, Swap Label Text, and Delete Label routines will not work with these static labels because the labels do not represent the entire object. If a labeling style places the text on the wrong side of the line, you need to define a style that is the opposite of the one that does not label correctly. For example, if Direction Above and Distance Below labels incorrectly, you need to define the style as Distance Above and Direction Below to be able to label the line correctly.

19. From the Labels menu, select Delete Label, select the westerly lot line, and delete the label. The label has to be selected from endpoint or point number to identify the entire length of the west lot line. Use the Label Line by Points routine to label the west lot line. The routine prompts for a Tag label; answer **No** to the prompt. The default mode is points, so activate object snaps (endpoint and intersection), and select the northerly end first then the southerly end of the line. The label is a southeast label and the bearing is in the interior of the lot. Exit the routine by pressing ENTER.
20. Use the Swap Label Text routine to change the text to different sides of the line. The routine does not work, because the label is a static label. The Flip Direction routine will not work for the same reason. The Delete Labels routine will not work because the labels are static labels.
21. Change the current line label style in the Properties Style dialog box to Distance Above and Direction Below and label the west lot line by points.
22. Change the line label style to Direction Above and Distance Below and label the east lot line by points.
23. Use the Label Arc by Points routine to label the arc segment at the front of the lot. Use the Length Above and Radius Below label style. Running object snaps do not work too well with this routine, so toggle off object snaps. With object snaps off, you will need to select the two endpoints of the arc (the PC and PT); however, the second selection is done without using an object snap. The second selection of the routine automatically uses the center object snap. The order in which you select the PC and PT does not matter. After labeling the front yard section of the arc, press ENTER to exit the routine.

24. In the Label Properties toolbar, select the Line tab, set the current style to Direction Above and Distance below, and click the Edit Style icon (the pencil). Change the text style to L200 and select OK to exit the dialog box. The labels change to have larger text, because the style definition has changed. Select Undo to undo the style change.

25. Click the Settings icon of the Style Properties toolbar, then the General tab, and toggle off Update Labels When Style Changes. Select OK to exit the dialog box.

26. Select the Edit Styles button, select the Line Labels tab, set the current style to Direction Above and Distance Below, and change the text style to L120. Select OK to exit the dialog box. The text on the screen does not change because of the toggle you turned off in the Settings dialog box.

27. You can force the change by using the Update Selected Labels routine. Use the Update Selected Labels routine and select the two lines to view the labels change.

28. Click the Settings icon on the Style Properties toolbar, then the General tab, and toggle off Update Labels When Objects Change. Select OK to exit the dialog box.

29. Select the southerly lot line to activate its grips, select the southern most grip, and select a new position for the endpoint. The labeling does not follow the line to its new location. You can reunite the label to the line by running the Update Selected routine from the Labels menu.

30. Use the Update Selected routine from the Labels menu to reunite the label to the line.

31. Undo the label update and the line change.

32. Click the Settings icon on the Style Properties toolbar, then the General tab, and toggle on both Update Labels When Objects Change and Update Labels When Style Changes. Select OK to exit the dialog box.

33. Delete the labeling.

34. Save the drawing.

35. Experiment with different settings and label styles. Try setting the forced bearing when labeling a line. The forced bearing toggle is in the Line Label tab of the Label Settings dialog box.

36. Erase all of the labels and save the drawing.

UNIT 5: LINE, ARC, AND SPIRAL TABLES

An annotation table comes from tags placed in a drawing. Tags are a labeling style applied to the selected object. The tags bind the data of the object to an entry in a table. With annotation tags in the drawing, the next step is to create the table. You can do this with one of the Add Table routines (see Figure 3–25). Each routine presents a dialog box displaying the current table definition. The definition can be modified or new styles can be created. After you exit the dialog box, the routine creates the table.

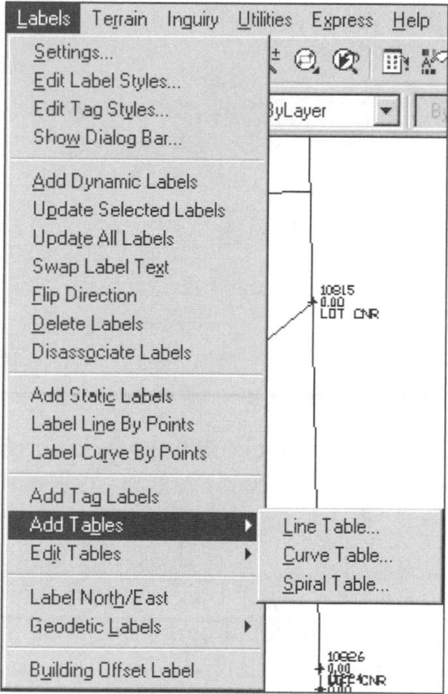

Figure 3–25

The Edit Tables flyout contains routines to edit, redraw (update), and delete tables in the current drawing (see Figure 3–26). The Edit Table Layout routine allows you to redefine an existing table in the drawing. The Redraw (update) Table routine rescans the drawing and updates the table entries in the drawing. The Delete Table routine does just that—deletes tables in the current drawing.

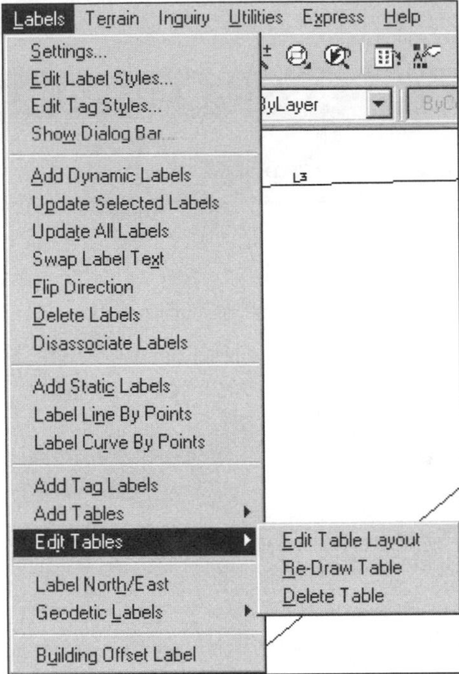

Figure 3–26

ANATOMY OF A TAG STYLE

The Edit Tag Styles routine of the Labels menu displays the settings for the current tag label style (see Figure 3–27). Although the Edit Tag Styles dialog box looks the same as the Edit Label Styles dialog box, there is only one style and one set of data. The only real control in this dialog box is the text height, text style, offset, and labeling layer. The reason for this is that a tag conveys the information of the line, arc, or spiral to the Add Table routine. The function of the tag is to identify the object and to assign it a number that becomes the numbered entry in the corresponding table.

If you edit a line labeled by a tag or delete a tag and/or labeled object, the tag-to-table relationship is not dynamic. You must use the Redraw Table routine from the Edit Tables cascade of the Labels menu (see Figure 3–26). The Redraw Table routine only works on tags that are dynamic. If you tag a line, arc, or spiral by point, the tag is static and will not be a part of any Redraw Table routine results.

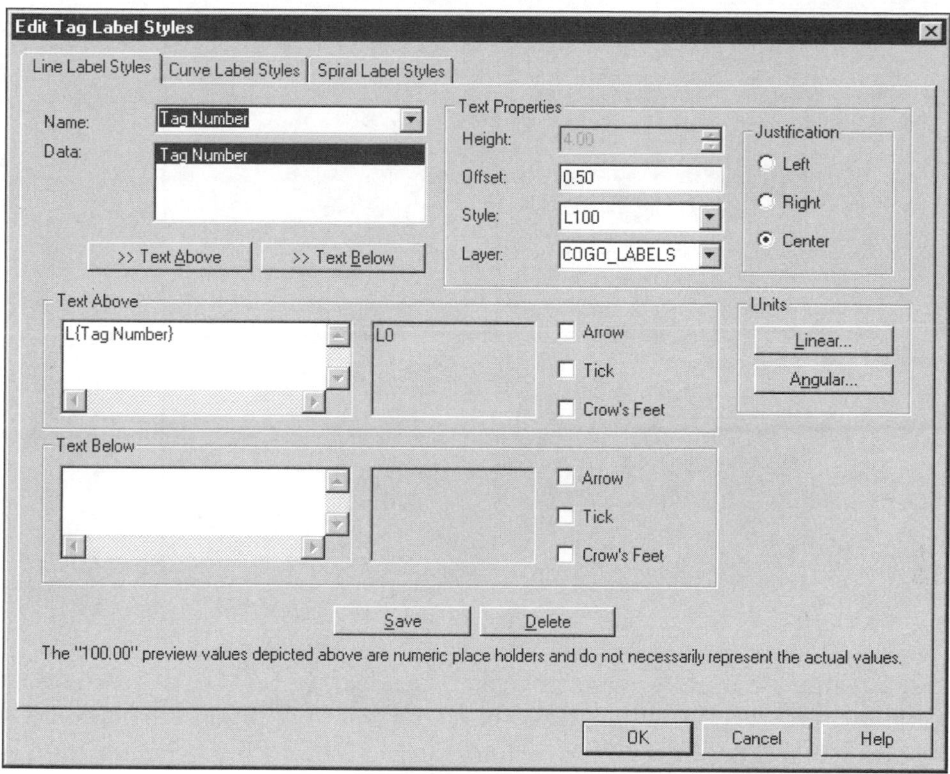

Figure 3–27

TABLE DIALOG BOX

The Table dialog box is broken into three sections; Table Title, Column Definition, and format toggles and values (see Figure 3–28). The Table Title portion contains settings for the title, the text style and size, and layers. The Column Definition portion lists what pieces of information are in the table. The definition of a table is similar to a spreadsheet. Each column can contain different object information: length, direction, starting northing or easting, ending northing or easting, tag number, or entity description. Below this area are buttons to edit the values listed in the table or to insert or delete new columns in the table. Between the Table Title and Column Definition areas are settings that control the maximum rows per page (zero means all rows), the layer of the table's line work, and toggles controlling table sorting, and drawing a border. At the bottom left of the dialog box are buttons to save and load the table definitions.

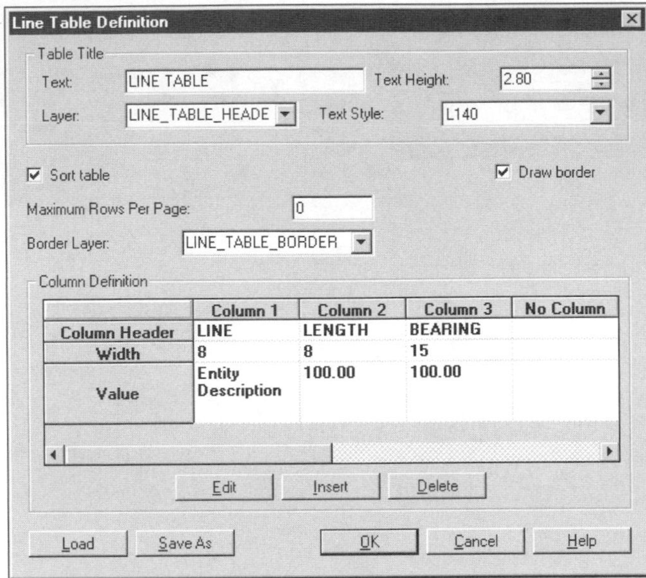

Figure 3–28

To edit a column, first select the column and then select the Edit button. This displays the Column #(number) Definition dialog box (see Figure 3–29). The top portion of the dialog box contains the name of the column, the width of the field, the text style, and justification. The middle portion allows you to select the data for the column in the table and it formatting properties. The lower portion previews the value and its precision.

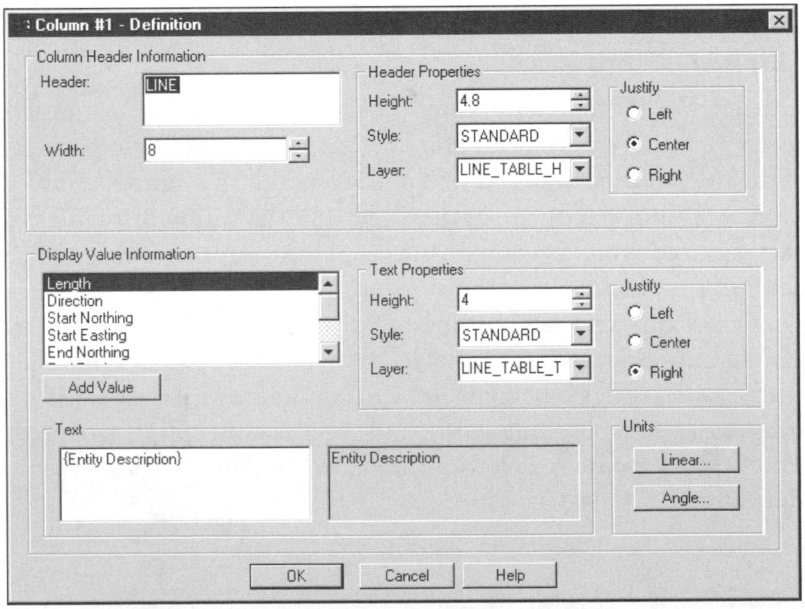

Figure 3–29

Exercise

After you complete this exercise, you will be able to:

- View and edit a table style
- Place tags on entities
- Create a table

WORKING WITH TABLE STYLES AND TAGS

1. If you have not removed the labeling from the entities on the screen, remove it now.
2. Click the Settings icon on the Style Properties dialog box, then select the General tab, and toggle on Update Labels When Style or Object Changes.
3. Select the Line Labels tab and make sure the tag number is set to 1. Select OK to exit the Settings dialog box.
4. Set the labeling mode to Tag by clicking the Toggle Label/Tag Mode icon above the Pencil icon in the Labeling Bar.
5. The northerly and southerly lot lines can be labeled with the shortcut menu, but the easterly, westerly, and the arc segment must be labeled as selected points. Create the two labels from the shortcut menu and label the remaining line and arc segments with the Labels By Points routines of the Labels menu.
6. Run the ZOOM command, use the Center option, select a point near the center of the lot, and set the height to 400. Use the PAN command to move the lot to the western part of the display. The line and arc tables will be placed just to the east of the boundary.
7. From the Add Table flyout of the Labels menu, select the Line Table routine. The routine displays the table form dialog box. Change the text style to L200, select OK to exit the dialog box, and select a point towards the top of the display and just east of the boundary line. The routine creates a line table to the right and down from the selected point.
8. From the Add Table flyout of the Labels menu, select the Curve Table routine. The routine displays the table form dialog box. Change the text style to L200, select OK to exit the dialog box, and select a point below the line table. The routine creates a curve table.
9. Select a line that was labeled by points, select the line's end grip, and change the location of the end of the line. The line changes location, but the tag does not. A By Points tag label is static.
10. From the Edit Tables flyout of the Labels menu, select and run the Re-Draw Table routine. The table does not update to the new data about the edited line.

The Re-Draw Table routine does not see the editing of the line's location if its label is not dynamic. The Re-Draw Table routine will only see the change to the line when you delete the tag, reset the number, and add a new tag to the line. Only after you edit the tag will the Re-Draw Table routine update its values.

11. Select a line that was labeled by the shortcut menu process to activate its grips.
12. Select one of the end grips and move the end to a new location. The tag follows the line and has the current line information.
13. From the Edit Tables flyout of the Labels menu, select the Re-Draw Table routine, and window the line table. The table now updates to the new data about the edited line.
14. Erase from the drawing the tables, tags, and the diagonal line in the triangular lot.
15. Save the drawing.

UNIT 6: POINT DATA FROM LINES AND ARCS

Now that you have lines and curves in the drawing, you can return to the Points menu of Land Development Desktop and use the routines found in the menu to place points on them. After you place the points, you have access to the coordinate data of the line, arc, or spiral they are on. Their descriptions should describe what the points represent. LDD does not dynamically link a point to an object. If you edit the line, the points on the line do not know anything has happened to the line. It is up to you to keep the relationship between points and objects correct.

The Create Points flyout of the Points menu contains the following routines to place points on AutoCAD objects (see Figure 3–30).

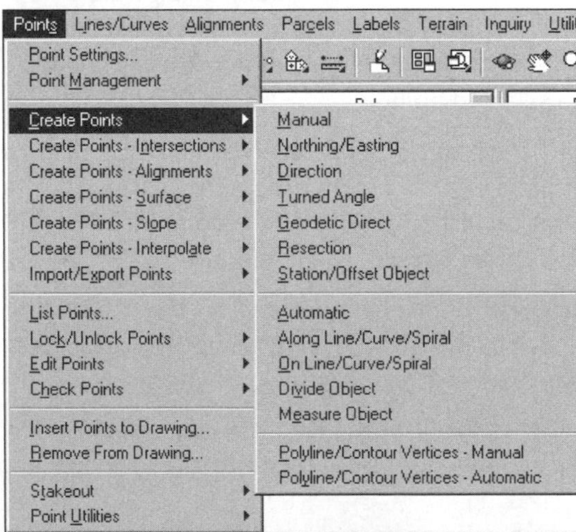

Figure 3–30

MANUAL

The Manual routine allows for the use of any reasonable object snap on an AutoCAD object.

AUTOMATIC AND ON LINE/CURVE/SPIRAL

The Automatic and On Line/Curve/Spiral routines place points on the endpoints of a selected object. If the object is an arc, both place a point at the radius point of the curve. The Only On Line/Curve/Spiral routine places a point at the PI (point of intersection) of a curve.

There is a difference, however, in how the Automatic and On Line/Curve/Spiral routines check for duplicate points at the end of the line or arc. The Automatic routine evaluates a selection set to see whether an endpoint of a line or an arc is the same as the endpoint of another line or arc. If there are duplicate endpoints, the routine places only one point.

The On Line/Curve/Spiral routine does not evaluate for duplicate endpoints or preexisting point objects at all because the routine works on an object-by-object basis, not on a selection set. If you select a line and then an adjacent arc, the routine will place two points at the common endpoint of the line and arc.

When you are working with arcs, the Automatic routine is different from the On Entity routine. The Automatic routine places points at the ends of the arc and at the radius point. The On Entity routine places points at the endpoints, radius point, and point of intersection (PI) of the arc. The order of placement is counterclockwise around the arc, endpoint, center, PI, and the other endpoint of the arc.

DIVIDE AND MEASURE

The Divide and Measure routines are similar to the routines in AutoCAD. The Divide routine marks where to divide an object into a user-specified number of segments with point objects. The Measure routine measures from an endpoint along a line, spiral, or arc segment. The routine places point objects on the segment at a user-specified spacing. Both routines ask for an offset distance and do not evaluate the point database for preexisting points for the same coordinate.

ALONG LINE/CURVE/SPIRAL

The Along Line/Curve/Spiral routine places points along an object at a distance from its endpoint. After you establish the endpoint, the routine prompts for a distance measured from the endpoint of the object having an "X." When you select the object, the routine finds the nearest endpoint to the selection point of the line.

Exercise

After you complete this exercise, you will be able to:

- Create point data from lines and curves
- Use Point routines from the Set Points cascade

CREATING POINT DATA FROM LINES AND CURVES

1. If you are not in the LDDTR2 drawing, open the drawing.
2. Use the Zoom to Point routine from the Point Utilities cascade of the Points menu, zoom to point 10841, and use a height of 250.
3. Use the Remove Points from Drawing routine of the Points menu, answer Yes to removing the symbols, and remove all of the points.
4. When you use the Manual, Automatic, On Line/Curve/Spiral, Along Line/Curve/Spiral, Divide, and Measure routines, they will look at the settings in the Point Setting dialog box to see what point number to start at and what options are set concerning elevations and descriptions. Select the Point Settings routine of the Points menu and set the following values in the dialog box (see Figure 3–31). Select OK to exit the dialog box.

    `Create Tab`
    `Current Number:` **5000**
    `Elevations and Descriptions: None`

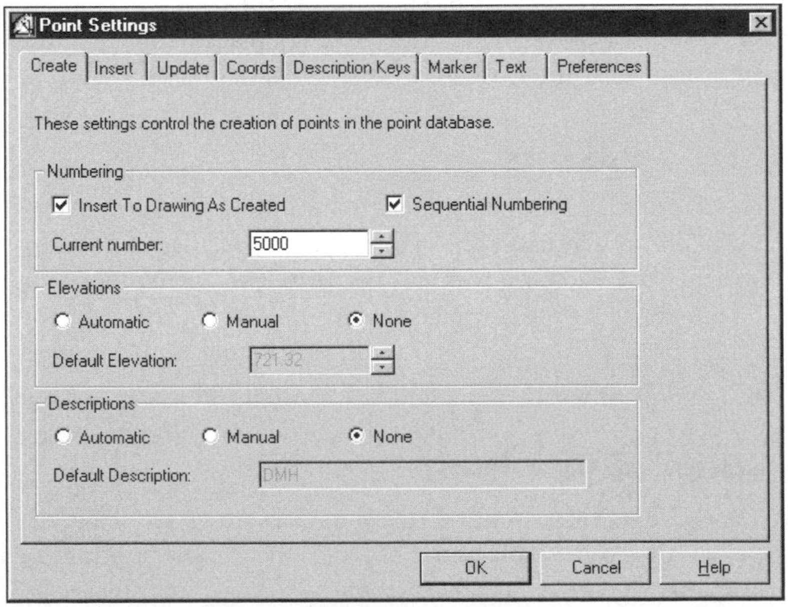

Figure 3–31

5. Select the Automatic routine from the Create Points cascade and select the ROW lines (north and south) and arcs to create points. The routine places points at each endpoint and the radius point of the arc.
6. Use the Zoom to Point routine, zoom to point 10787, and use the height of 50. Notice that the routine places only one point at the endpoints of the lines and arcs.
7. Use the ZOOM PREVIOUS command to return the last view of the line work.
8. Use the Remove Points from Drawing routine of the Points menu, answer Yes to removing the symbols, and remove all of the points.
9. Select the On Line/Curve/Spiral routine from the Create Points cascade and select the north and south ROW lines and arcs to create points. The routine places points at each endpoint, PI, and the radius point of the arc. You should select the small arc first before selecting the remaining lines and arcs.
10. Use the Zoom to Point routine, zoom to point 10787, and use the height of 50. Notice that the routine places duplicate points at the endpoints of the lines and arcs and a single point at the PI of the small arc.
11. Use the ZOOM PREVIOUS command to return the last view of the line work.
12. Use the Remove Points from Drawing routine of the Points menu, answer Yes to removing the symbols, and remove all of the points.
13. Use the Divide routine from the Create Points cascade to divide the northern ROW line into three equal segments. When the routine prompts for an offset distance, accept the default value of 0.00.
14. Use the Divide routine from the Create Points cascade to divide the large arc into five equal segments. The routine allows for points to be offset from the selected object. The side depends upon which endpoint of the arc the routine is starting at. Select the arc so the points offset to the outside of the arc. When the routine prompts for an offset distance, set the value to 15.00.
15. Use the Remove Points from Drawing routine of the Points menu, answer Yes to removing the symbols, and remove all of the points.
16. Use the Measure routine from the Create Points cascade to place points at a fixed spacing on the northern ROW line. Accept the starting station value of 0.00 and press ENTER to accept the beginning and ending station values. Set the spacing to 10 and when the routine prompts for an offset distance, accept the default value of 0.00.
17. Use the Measure routine from the Create Points cascade to place points on the large arc. Accept the starting station value of 0.00 and press ENTER to accept the beginning and ending station values. Set the spacing to 20 and when the routine prompts for an offset distance, set the value to 10.00. The offset of 10 will place the points to the side of the arc. The routine allows for points to be offset from the selected object. The side depends upon which endpoint of the arc the routine is starting at.

18. Use the Remove Points from Drawing routine of the Points menu, answer Yes to removing the symbols, and remove all of the points.
19. Select the Along Line/Curve/Spiral routine from the Create Points cascade of the Points menu to set points on the large arc. Select the arc near its southern endpoint. Place five points on the arc using the following distances to place the points:

    Distance: 20, 50, 100, 200, and 275.

20. Select the Along Line/Curve/Spiral routine from the Create Points cascade of the Points menu to set points on the northern ROW line. Select the line near its eastern endpoint. Place four points on the arc using the following distances to place the points.

    Distance: 10, 25, 40, and 50.

21. Use the Remove Points from Drawing routine of the Points menu, answer Yes to removing the symbols, and remove all of the points.
22. Use the Erase routine of the Edit Points cascade and erase the point number range 5000–6000 (the number option at the routines prompt).
23. Erase the southerly lot line of triangular lot, any tags, labels, or tables.
24. Save the drawing.

EXERCISE

Lines, curves, and spirals make up the second basic data group of Land Development Desktop. The origins of these objects can be AutoCAD commands or points from the point database. LDD contains tools to analyze their design values in terms understood by surveyors and civil engineers. Although the tool set for editing the objects is quite limited, LDD provides many tools to annotate and create points from lines, curves, and spirals.

The next section discusses the creation and organization of the lines and arcs into parcels.

SECTION 4

Parcels

After you complete this section, you will be able to:

- Set parcel creation defaults
- Use the Slide Bearing routine
- Use the Swing to Curve routine
- Use the Radial routine for parcel sizing
- Define Parcels by polylines
- Use Parcel Editor

SECTION OVERVIEW

From point, line, and arc data, you begin to create organizations of diverse data. These organizations are more than mere points, lines, and arcs. They create information and place data into coherent forms; for example, parcels, boundaries, and roadway centerlines. This section covers the organization of points, lines, and arcs as parcels. The Land Development Desktop parcel routines are found on the Parcels menu (see Figure 4–1). These horizontal shapes display themselves as AutoCAD objects and have an external database entry in a Land Development Desktop file. The link between the AutoCAD objects and the external data file is not dynamic. The external file is in essence a snapshot of the current conditions. If you modify the AutoCAD objects that define a parcel in the drawing, the external file no longer reflects the conditions in the drawing. If this occurs, you must update the external file to reflect the current state of the drawing.

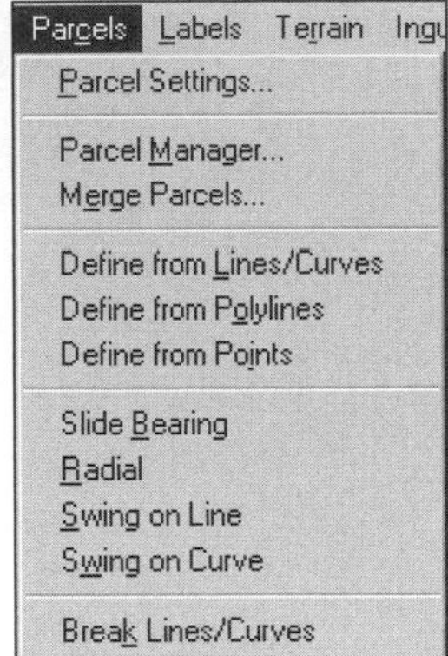

Figure 4–1

UNIT 1

The first unit of this section reviews the settings related to parcels in LDD. These settings control how the lot is annotated, reviewed, and whether its values are exported to an external database.

UNIT 2

Land Development Desktop contains several lot creation routines. These routines fall into two categories. The first category is routines that define a lot from preexisting geometry. The second category completes the geometry of the lot and defines the lot to a database if toggled on. These lot sizing routines are a part of the second unit.

UNIT 3

Editing parcels is a process of deleting and redefining them. Unit 3 reviews the limited set of tools for manipulating parcel data.

UNIT 4

Unit four covers the tools to review parcel definitions.

UNIT 5

The last unit covers the annotation of parcels. The annotation can occur when the lots are defined or after their definition.

UNIT 1: PARCEL SETTINGS

The values in the Parcel Settings dialog box affect the annotation and creation of external data of each lot you create or define (see Figure 4–2). The Parcel Settings dialog box sets automatic annotation, text styles, text size, methods, and toggles on the definition process for the lot defining or sizing routines. Whenever Land Development Desktop creates an external data file, there will be a Define routine.

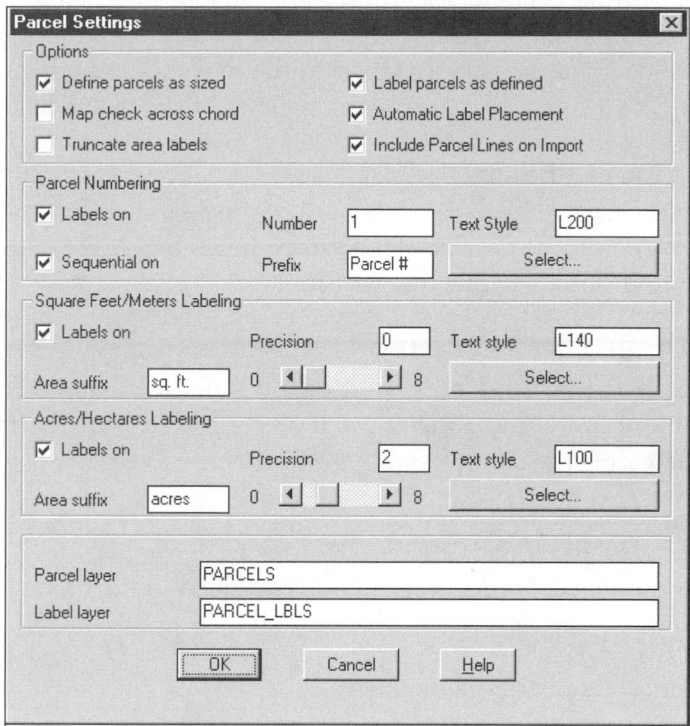

Figure 4–2

OPTIONS

The Options area of the Parcel Settings dialog box controls automatic annotation and data writing. These options are described in the following headings.

DEFINE PARCELS AS SIZED

When the Define Parcels as Sized toggle is on, a routine defining or creating a lot places the bearings and distances values of objects in the drawing to an external parcel database.

LABEL PARCELS AS DEFINED

When the Label Parcels as Defined toggle is on, a routine defining or creating a lot places the annotation of the parcel in the drawing automatically.

MAP CHECK ACROSS CHORD

When the Map Check Across Chord toggle is on, a routine defining or creating a lot creates evaluation data by the chord information of the lot. If this toggle is off, the mapcheck and inverse processes use the curve length.

AUTOMATIC LABEL PLACEMENT

When the Automatic Label Placement toggle is on, a routine defining or creating a lot places the label at the center of the parcel. If this toggle is off, a routine defining or creating a lot prompts you for the insertion point of the annotation. What the label consists of depends upon the settings found in the Parcel Numbering, Square Feet/Meters Labeling, and Acres/Hectares Labeling areas of the dialog box.

TRUNCATE AREA LABELS

When the Truncate Area Labels toggle is on, a routine defining or creating a lot will truncate to the nearest value the size of the parcel. If this toggle is off, the area will be rounded to the nearest digit set by the precision.

INCLUDE PARCEL LINES ON IMPORT

When the Include Parcel Lines on Import toggle is on, the Import Parcels routine will create both line work and annotation. If this toggle is off, the Import Parcels routine will create only the annotation.

PARCEL NUMBERING

The settings in the Parcel Numbering area are described in the following headings.

LABELS ON

Turns on the parcel number for annotation.

SEQUENTIAL ON

Turns on the automatic assignment of parcel numbers. The numbers would be sequential starting with the seed parcel number. If Sequential On is off, the assignment of lot numbers is random; that is, the routines will prompt for a lot number.

LOT PREFIX

The Lot Prefix entry sets a lot prefix for the lot name; for example, Parcel #, Lot #, and so forth. If you want a space between the prefix and the number, place a space in the prefix.

TEXT STYLE

You can enter a style name or select a style from a list of existing styles by selecting the Select button.

SQUARE FEET/METERS LABELING

The settings in the Square Feet/Meters Labeling area toggle the square footage portion of a lot label. If you toggle on Square Feet/Meters Labeling, the Define routine annotates the lot with the square unit values.

LABELS ON

Turns on the annotation.

AREA SUFFIX

Sets the annotation suffix for the square units measurements of a parcel.

PRECISION

The Precision setting sets the precision of the square unit annotation. You can enter the value in the reporting box or set the value by adjusting the slider.

TEXT STYLE

You can set the default text style by either entering the style name or selecting it from a list of existing styles when you select the Select button.

ACRES/HECTARES LABELING

The settings in the Acres/Hectares Numbering area are described in the following headings.

LABELS ON

Turns on the annotation.

AREA SUFFIX

Sets the annotation suffix for the square units measurements of a parcel.

PRECISION

The Precision setting sets the precision of the square unit annotation. You can enter the value in the reporting box or set the value by adjusting the slider.

TEXT STYLE

You can set the default text style by either entering the style name or selecting it from a list of existing styles when you select the Select button.

Exercise

After you complete this exercise, you will be able to:

- Understand and use the Parcel Settings dialog box

WORKING WITH PARCEL SETTINGS

1. Open the drawing LDDTR2from the PROJ1used in the previous section.
2. For this exercise, set the following settings for parcels by running the Parcel Settings routine from the Parcels menu (see Figure 4–2). Select the Select button to choose the text style from a list of styles.

```
Define to database: ON
Label Lots as Sized: ON
Automatic Label Placement: ON
Lot Numbering:
Sequential: OFF
Lot Description Label: 1
Lot Prefix: Parcel #
Labeling: ON
Style: L200
Square Unit Labeling:
Labeling: ON
Style: L140
Precision: 0
Area Unit Labeling:
Labeling: ON
Style: L100
Precision: 2
```

3. Select OK to exit the dialog box.

UNIT 2: CREATING PARCELS FROM ENTITIES

The parcel defining and sizing routines of the Parcels menu create lot definitions, which can be external to the drawing (see Figure 4–1). Any drawing that is a part of a project can view and import a lot definition.

The parcel defining and sizing tools create lots from preexisting entities in the drawing. Lines, arcs, polylines, and point objects form the basis for LDD lots. You need only indicate which objects constitute the lot.

LOT DEFINING AND SIZING ROUTINES

The Define Parcel from Line/Curves, Polylines, and Points routines read the existing objects in the drawing and create annotations and/or entries in the external parcel database. These routines scan the selected objects to create the lot information.

Land Development Desktop provides four lot sizing routines. Each routine is for a specific situation. The first tool is the Slide Bearing routine. This routine slides the last line of the lot along bearings represented by other vectors in the lot. The second routine, Radial, draws the last line of the lot by creating a line radial to an arc in the lot. The last two routines, Swing on Line or Swing on Curve, create the last line of the lot by holding one end of the line and "swinging" the other endpoint to either a line or arc.

The Swing On Curve routine does not calculate a radial line to create the last lot line. The line is a result of the area in the lot. If you want a radial line, you must use the Radial routine.

The four lot sizing routines create lots of an exact size. The precision of the routines may be more than is necessary for the preliminary state of the project. Because of the preliminary nature of the lot sizes, it may not be necessary to write the definitions to the lot database.

After the lots become final, you will want to define the lots. The quickest way to define a lot is by a polyline representing its boundary. The BPOLY command of AutoCAD quickly creates lot boundary polylines. When you use the Define by Polyline routine, the lots become a part of the external database.

Exercise

After you complete this exercise, you will be able to:

- Define parcels from line/curves/points
- Define parcels from polylines
- Define parcels from points
- Use the Slide Bearing routine
- Use the Radial routine for sizing
- Use the Swing on Line sizing routine
- Use the Swing on Curve sizing routine

CREATING PARCELS 1 AND 2–DEFINE PARCEL FROM POINTS

1. If you are not in the LDDTR2 drawing, open the drawing, use the Zoom to Point routine to zoom to point 10774, and use the height of 250. Use the PAN command to center the two lots on the screen. Use Figure 4–3 as a guide to setting the display.

2. If there is not a line dividing the rectangle into two segments, draw a line using point numbers 10772 and 10774. This should divide the rectangle into two parts (see Figure 4–3).

3. From the Parcels menu, run the Define Parcel from Points routine and define parcels 1 and 2 by points. Use the following point numbers to define the lots. You may want to insert the point numbers to view their location.

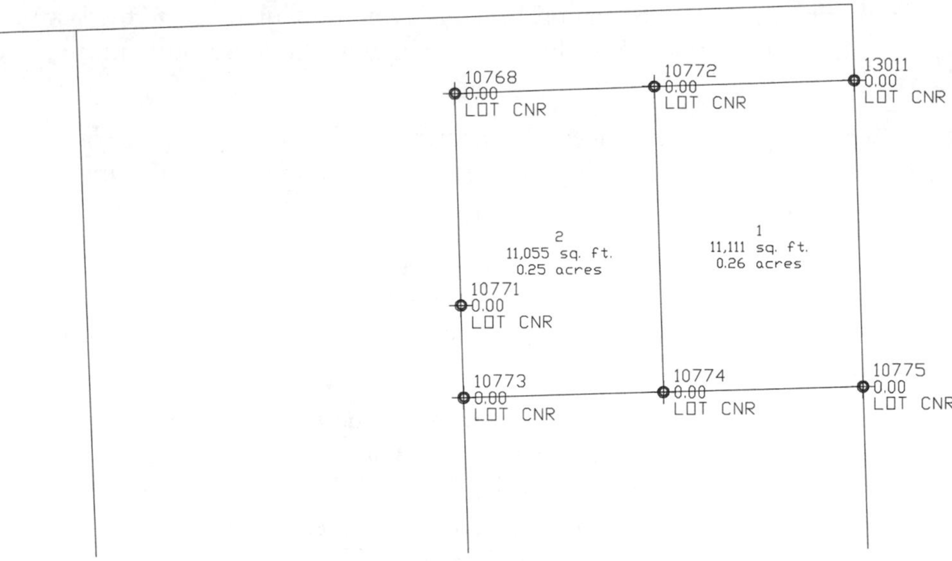

Figure 4–3

Lot	Points
1	13011, 10772, 10774, 10775
2	10772, 10768, 10771, 10773, 10774

When you get to the last point number, press ENTER twice to close the lot and to label the lot. The routine continues prompting for points defining the next lot. Enter the second lot's point numbers and press ENTER until you exit the routine. The Define from Points routine annotates each lot you close, and the annotation is centered in the lot.

CREATING PARCEL 3–DEFINE PARCEL FROM LINES/CURVES

1. Draw Lot 3 with the LINE command and use the following point numbers: 10768, 10767, 10769, and 10771. You may have to pan the display to view the new lot lines. You may want to insert the point numbers to view their location. Use Figure 4–4 as a guide.
2. From the Parcels menu, run the Define Parcel from Lines/Curves routine and define parcel 3 from the lines drawn in the last step. The routine prompts you to select the line/arc nearest the Point Of Beginning (POB). Select a point near the eastern end of the north property line of Lot 3 (point 10768). When you select the point, a red X appears at the line's eastern end. Select the remaining three lines to define the lot. The routine places the annotation at the center of the lot. Press ENTER twice to exit the routine.
3. Open the Layer dialog box, create a new layer, LOTS, assign it a color, make it the current layer, and exit the dialog box.
4. Save the drawing.

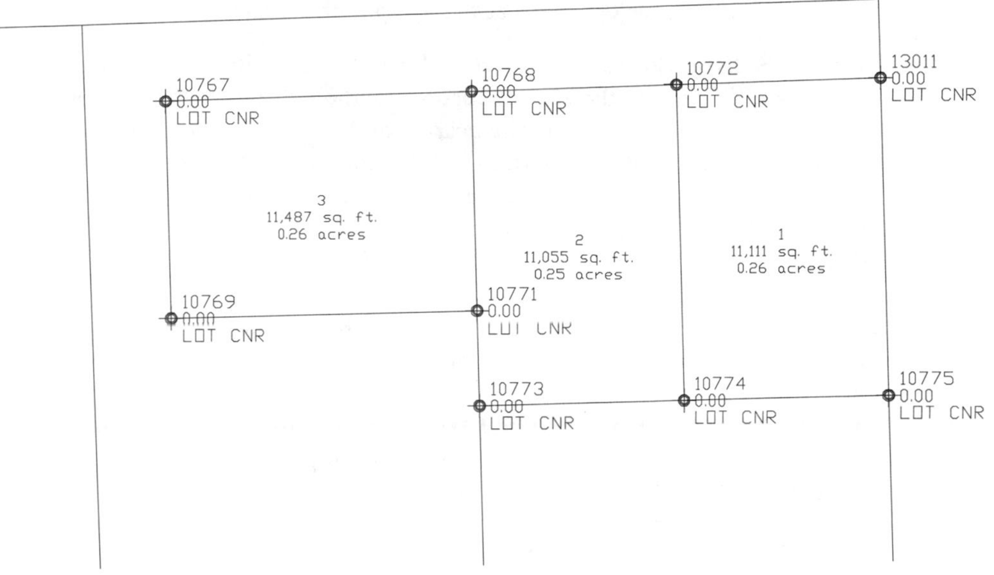

Figure 4–4

CREATING LOT 4–SLIDING BEARING ROUTINE

1. Draw a line from point number 10769 to 10777. You may want to insert the points to view their location.
2. Pan the display so that the southerly line of lot 3 is near the top of the screen. Use Figure 4–5 to set your display.

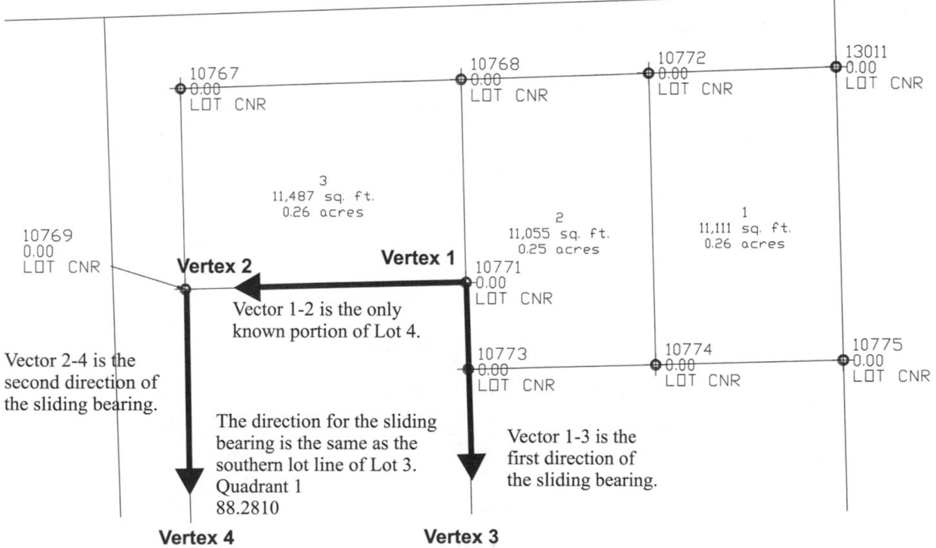

Figure 4–5

3. Toggle off Polar and Object Snaps at the bottom of the display.

4. Select the Slide Bearing routine to create lot 4. The Slide Bearing routine prompts you to indicate the known portions of the lot, the directions along which the last lot line slides, and the bearing of the sliding lot line. The first selection is the east endpoint of the south lot line of lot 3. The second selection is the west endpoint of the south lot line of lot 3. The southern line of lot 3, the two select points, is the only known portion of lot 4. The next two selections define the bearing for the east and west lines of the lot. Use the nearest object snap on the east and west lot line south of lot 3. Then you define the direction of the closing lot line. You can define the direction by bearing, azimuth, AutoCAD object snaps or coordinates, or point numbers. For this step, you will enter the bearing of the southerly line of lot 3 (88.2810). Finally, the routine prompts for the square footage of the lot. Enter **10160** as the needed area. The routine sizes and labels the lot. To return to the command prompt, press ENTER at the Parcel first point prompt. The prompting is as follows.

```
Parcel first point:
 >>Point number: .p
Parcel first point: endpoint
Next point (or Curve): endpoint
Next point (or Curve): (press ENTER)
First direction: _nea to
Second direction: _nea to
Choose new parcel line direction.
Angular units: Degrees/Minutes/Seconds (DD.MMSS)
Quadrant (1-4) (Azimuth/POints): 1
Bearing (DD.MMSS): 88.2810
The minimum possible area is approximately = 1 square units.
Desired area, in square units: 10160
Calculating, Please Wait!
Parcel first point:  (press ENTER)
Command:
```

5. Save the drawing.

CREATING LOT 5–DEFINE FROM POLYLINE

1. The next lot, lot 5, is immediately below lot 4. If you cannot see all of the line work for lot 5, use the PAN command to adjust the display. Use Figure 4–6 as a guide.

2. Select the AutoCAD Layer button, create a new layer, name the layer LOT, assign it a color, make it the current layer, and select OK to exit the Layer dialog box.

EXERCISE

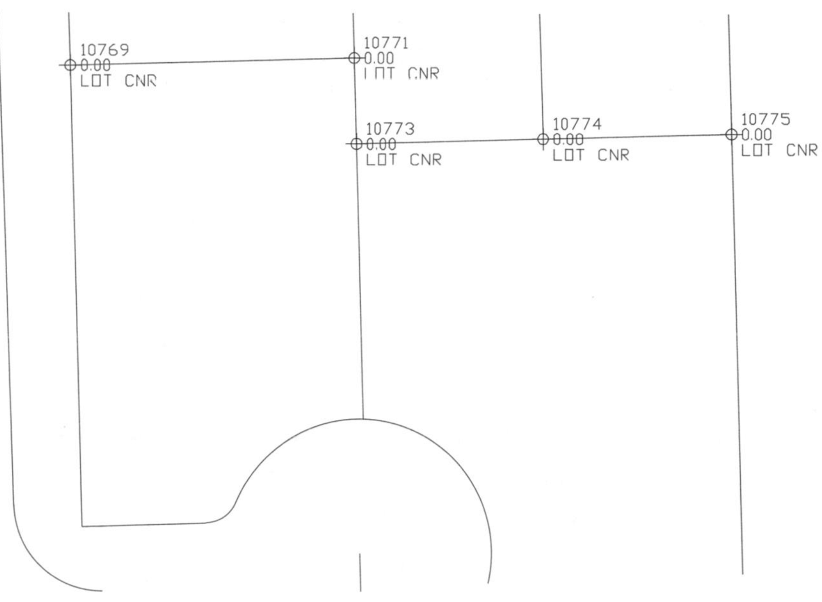

Figure 4–6

3. Use the BPOLY command to create a polyline covering the lot boundary. If the boundary created by BPOLY extends into lot 4, undo the BPOLY polyline, and retry the BPOLY command again. The second time you try the BPOLY command, click the Selection Set Button in the lower left of the dialog box. This allows you to create a new boundary set by selecting the only the lines and arcs that make up lot 5. After selecting the objects, press ENTER to return to the Bpoly dialog box. Then click the Pick Points icon to create the lot 5 polyline.

4. Use the Define from Polyline routine to define lot 5. When the routine prompts you for a polyline, type **last** to select the last object. The boundary should highlight. Accept the boundary and the routine labels the lot. Press ENTER to exit the routine.

5. Click the Layer icon and set the neutral layer to be the current layer, freeze the LOT layer, and select OK to exit the Layer dialog box.

6. Save the drawing.

CREATING LOT 6–SWING ON CURVE

The sizing of Lot 6 is merely a swinging line to the arc that creates the front yard of the lot. Currently the drawing contains four of the five sides of Lot 6. Remember that the routine does not calculate a radial line to create the new lot line. The line is a result of the lot area. What you supply to the routine is the swing point, the boundary including the arc to which to connect the last lot line. Point 10815 is the swing point for the lot and the beginning of the lot outline (see Figure 4–7). Since the last lot line swings to the arc and the arc is the last portion of the boundary, you do not include it as a curve in the defining of the known lot boundary. Also, you must use the intersection object snap to identify the intersection of the west lot line and the front yard curve (vertex 3). For some reason, if you use the end object snap, the routine fails to understand the relationship of the line and arc.

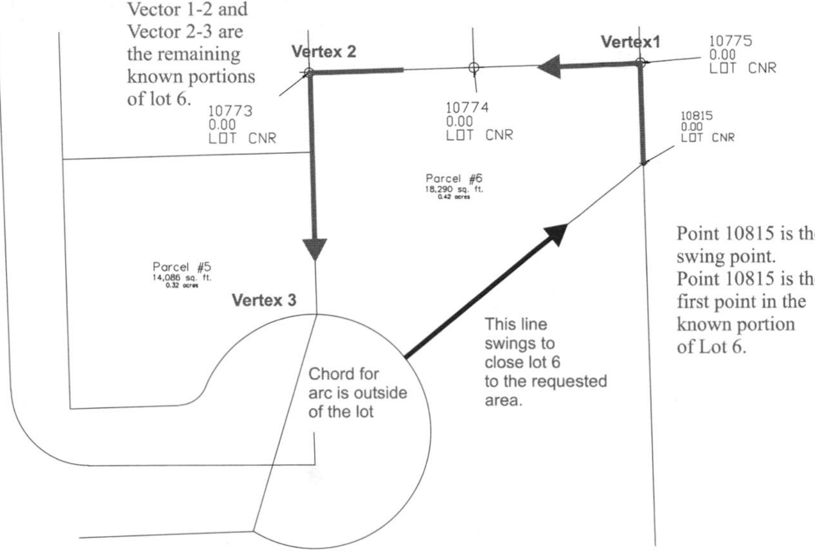

Figure 4–7

EXERCISE

1. Insert point number 10815.
2. You may have to pan and zoom out to be able to view all of the large cul-de-sac arc. Use Figure 4–7 to guide your adjustment.
3. Start the Swing On Curve routine from the Parcels menu.
4. The command runs as the following:

```
Point to swing from: .p
Point to swing from: Point number: 10815
Curve/<Next point>: Point number: .p
Curve/<Next point>: endp of (select vertex 1)
Curve/<Next point>: endp of (select vertex 2)
Curve/<Next point>: int of (select vertex 3)
Curve/<Next point>: (press ENTER because now is the time to iden-
  tify the arc to swing to)
Point on curve: (select any point on the arc segment)
Pick end of curve: (select near the southern end of the arc)
```

The routine draws a chord on the screen reflecting the two choices.

```
Position of chord to lot (Outside/Inside) <I>: o
The minimum possible area is approximately = 14,887 square units.
Desired area, in square units: 18290
Calculating, Please Wait!
```

The routine draws a southern lot line for Lot 6 and inserts the label.

```
Point to swing from: (press ENTER to exit)
```

5. Save the drawing.

CREATING LOTS 7 AND 8–RADIAL

The Radial routine creates the next two lots from the cul-de-sac arc. The only known part of Lot 7 is the northern lot line (southern line of Lot 6), the arc, and the east boundary line. The radial line starts on the arc and ends somewhere along the east property line. The Radial Lot routine asks about these known portions of the lot.

1. Draw the southern lot line of lot 8 by using the Line routine from the Lines/ Curves menu. Toggle the routine into point number mode (.p) and connect point 10838 to 10840. After connecting the point, press ENTER to exit the routine.
2. Use the Zoom to Point routine to zoom to point 10791 and use the height of 350.
3. Start the Radial routine from the Parcels menu. The routine prompts for the known portions of the lot. First identify the northern lot line by an endpoint object snap in the northeast corner and an intersection object snap where the lot line intersects the arc. Again do not include the arc, because it is the origin of the radial line. If there were lines after the arc, but before the arc next to which the radial line is drawn, you would identify this arc. Then identify the arc and its southerly end point. The routine draws the chord of the arc and asks if the chord is inside the lot. The chord in this case is outside the lot. Lastly, you identify the east boundary line by a nearest object snap. The routine uses this vector to search for a radial solution. Use Figure 4–8 as a guide.

The command sequence runs as follows:

```
LOT first point: endp of  (select vertex 1)
Curve/<Next point>: int of  (select vertex 2)
Curve/<Next point>: (press ENTER)
Point on curve to draw radial from: (select a point on the arc)
End of curve: endp of (select near the southern end of the arc)
Position of chord to lot (Outside/Inside) <I>: o (the chord is on
  the outside of the arc)
Direction point: nea to (select the east property line south of vertex 1)
The minimum possible area is approximately = 4 square units.
Desired area, in square units: 13330
Calculating, Please Wait!
```

EXERCISE

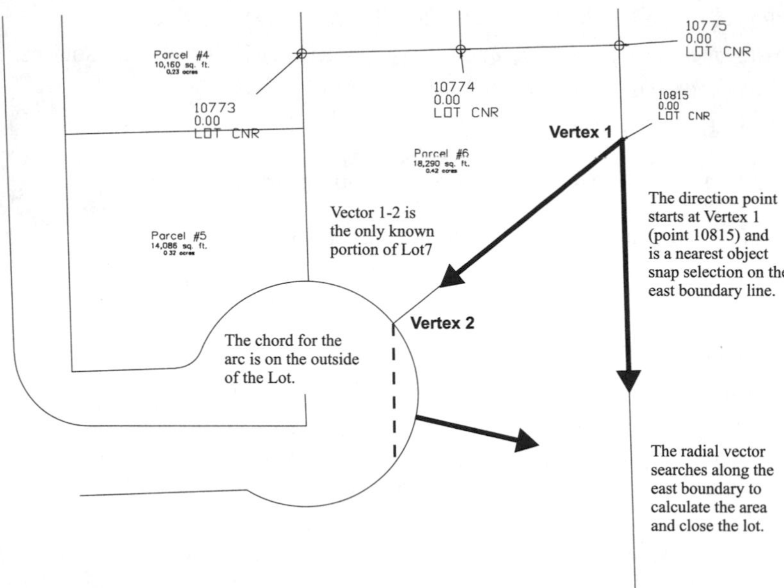

Figure 4–8

The routine draws a southern radial lot line for Lot 7 and labels the lot.

```
LOT first point:
```

After completing the construction of Lot 7, move on to Lot 8. You may need to pan the display up to view the southern lot line of Lot 8. Lot 8 is also a radial line lot (see Figure 4–9).

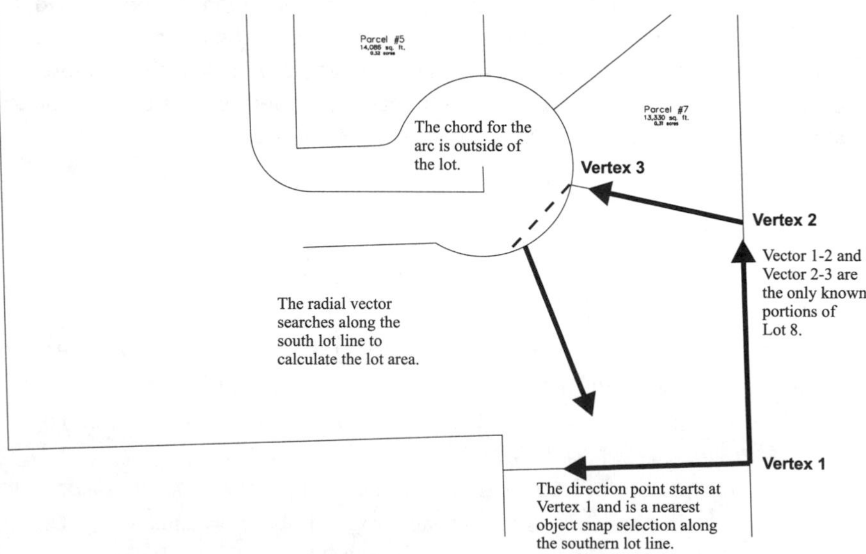

Figure 4–9

The only known part of Lot 8 is the eastern boundary line, a southern lot line, and the northern lot line. The northern lot line is the result of creating Lot 7. You identify the known portions of the lot by starting at Vertex 1 in the southeast corner, up to the southern line of Lot 7 (Vertex 2), and finally to the intersection of the lot line with the front yard arc (Vertex 3). You must use the intersection object snap when identifying the intersection of the side yard with the front yard arc (Vertex 3). Again do not include the arc, because it is the origin of the radial line. The radial line starts on the arc and ends somewhere along the southern lot line west of the eastern boundary. The southern lot line represents the direction for closure of the radial line. You use a nearest object snap on the southern lot line to define the direction point from Vertex 1.

```
LOT first point: endp of (the intersection of the east and south
  boundary lines, vertex 1)
Curve/<Next point>: int of (the east end of the southern line of
  Lot 7 and the east boundary line)
Curve/<Next point>: int of (the intersection of the arc and the
  north lot line)
Curve/<Next point>: (press ENTER)
Point on curve to draw radial from: (choose the arc)
End of the curve: endp of (select near the west end of the arc)
Position of the chord to lot (Outside/Inside)<I>: o
Direction Point:  nea to (a rubber band from the intersection of
  the boundary lines; choose on the south boundary line)
The minimum possible area is approximately = 12,967 square units.
Desired area, in square units: 19,850
Calculating, Please Wait!
```

The routine draws a westerly radial lot line for Lot 8 and labels the lot.

```
LOT first point: (press ENTER)
```

4. Save the drawing.

CREATING LOT 13–SWING ON LINE

The last routine, Swing to Line, closes a lot to an existing line segment. It is the same methodology as the Swing to Curve routine except for swinging to a line. You know more about the geometry of Lot 13 than you do about any of the other lots so far. The lot has several line segments, an eastern side yard, an arc, a back yard boundary line, and a front yard line. The question is where does the westerly side yard line intersect the front yard when you ask for a specific lot size?

The lot starts at point 10824 and travels easterly on the southern lot line. The next portion of the lot is the radial line that is the westerly line of lot 8. This time the arc is a known portion of the lot, so include it in the known portion of lot 13. The westerly end of the arc is the start of the direction for the lot to close on (Vertex 5). See Figure 4–10.

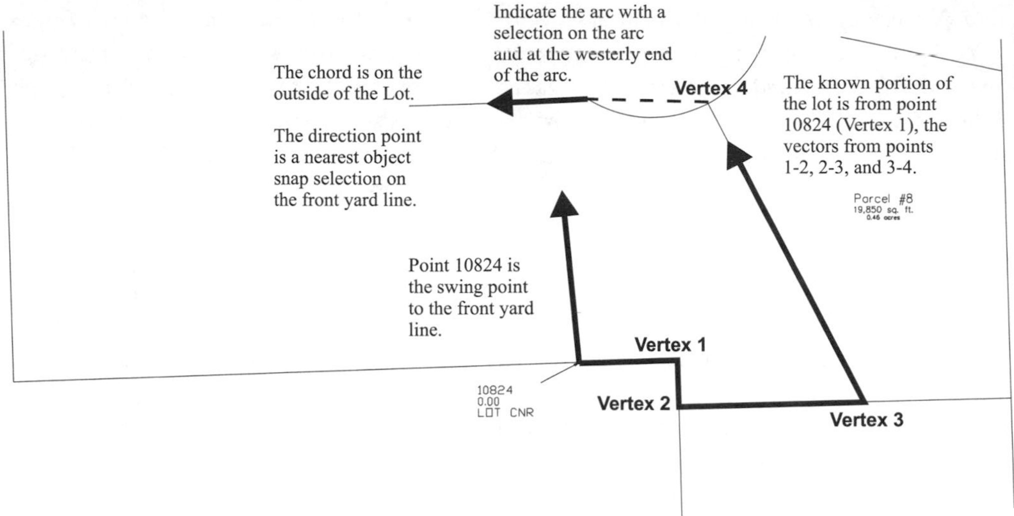

Figure 4–10

1. Insert the point number 10824 with the Insert Point into Drawing routine. Use the Number option to specify the point number.
2. Use the Zoom to Point routine and zoom to point 10824 with a height of 300. Use Figure 4–10 to set your display.
3. Since the next lot number is out of sequence, you need to reset the number for the Parcel Settings dialog box. Open the Parcel Settings dialog box and set the parcel number to 13. Select OK to exit the dialog box.
4. Create Lot 13 using the Swing to Line routine from the Parcels menu. Start by identifying point 10824 as the swing point. Toggle out of point number mode (.p) and use the intersection object snap to identify the lot boundary. This time the arc is a inner portion of the lot as opposed to the final item defining the lot perimeter. This allows you to use the C(urve) option while defining the known lot lines. The front yard line represents the direction of the vector to swing the line to.

```
Point to swing from: .p
Point to swing from: Point number: 10824
Curve/<Next point>: Point number: .p
Curve/<Next point>: int of  (select vertex 1)
Curve/<Next point>: int of  (select vertex 2)
Curve/<Next point>: int of  (select vertex 3)
```

```
Curve/<Next point>: int of  (select vertex 4)
Curve/<Next point>: c
Point on curve: (select a point on the curve)
End of curve in the lot: endp of  (select the endpoint of the arc
   at the west end)
Position of chord in the lot (Outside/Inside) <I>: o
Curve/<Next point>:  (press ENTER)
Direction point: nea to (select a point along the front yard line
   of Lot 13)
The minimum possible area is approximately = 13,868 square units.
Desired area, in square units: 15000
Calculating, Please Wait!
```

The routine draws a westerly radial lot line for Lot 8 and labels the lot.

```
Point to swing from: (press ENTER)
```

5. Save the drawing.

UNIT 3: EDITING LOTS

There are no specific lot editing tools. The lot routines allow only for the sizing and defining of lots to an external database. The external data files reflect the current state of the lot configuration. There is no dynamic connection between the entities in the drawing and the lot database definitions. If you modify the lots graphically in the AutoCAD editor with AutoCAD tools, you affect only the AutoCAD portion of the data. LDD does not transmit the new lot shapes or sizes to the external database. You must redefine the lot. When you redefine the lot, the routine overwrites the lot database to reflect the new lot definition.

UNIT 4: ANALYZING PARCEL DATA

LDD provides two methods of analyzing lots. The first is a series of listing routines that report only the area of the lot. These routines are in lower portion of the Inquiry menu (see Figure 4–11). The second method is analyzing the lot geometry from the entries in the external lot database through Parcel Manager (see Figure 4–12). The Parcel Manager generates Map Check and Inverse reports about each selected lot.

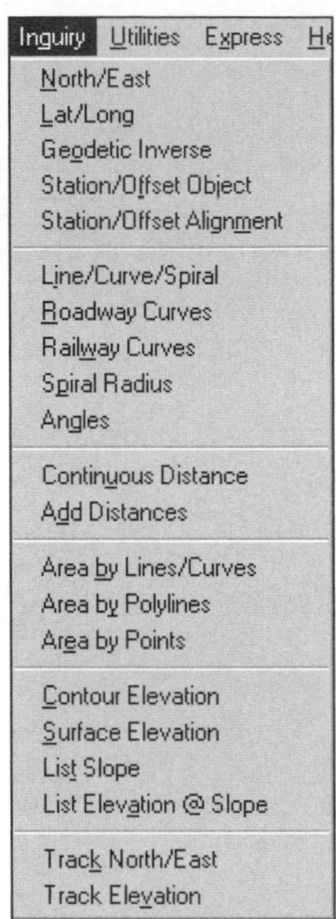

Figure 4–11

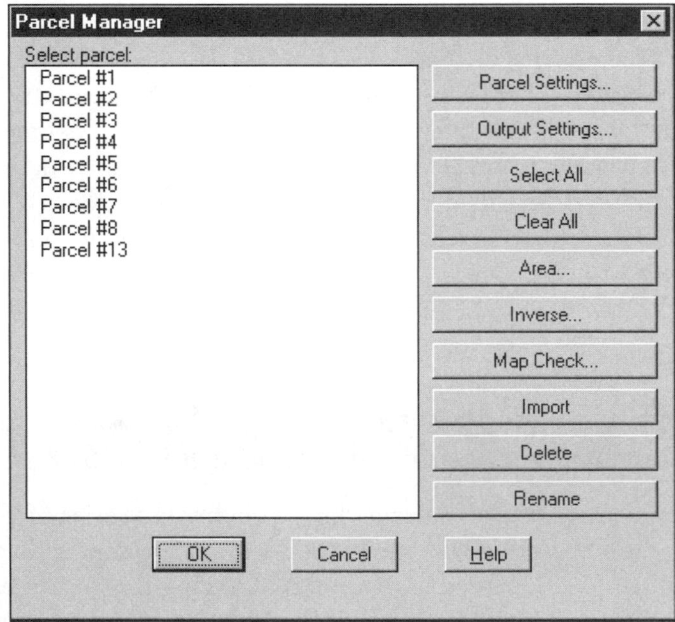

Figure 4–12

A Map Check report calculates the perimeter of the lot to the precision of the drawing. If the distance precision is set to 2, that is the quality of the Map Check report. If you adjust the linear precision to a higher number of decimal places, the report becomes more accurate. The Inverse report uses the precision of AutoCAD, 14 decimal places, to calculate the error in the lot definition.

If you place point objects on the lot lines, the coordinate information of the lot becomes a part of the point database. Again, you must remember that the point database does not store the coordinate data as representing lots. A description value associates the lot coordinate data with a point object. It rests with you to maintain that the points correctly represent a lot. With lot information as point database data, the coordinate information is available to any of the point database outputs.

Exercise

After you complete this exercise, you will be able to:

- List parcel areas from line/curves/points
- List parcel areas from points
- List parcel areas from polylines
- Create a Map Check report in Parcel Manager
- Create an Inverse report in Parcel Manager

EXERCISE

CREATING REPORTS ABOUT PARCELS

1. If you are not in the LDDTR2 drawing, open the drawing.
2. Open Parcel Manager by selecting Parcel Manager from the Parcels menu.
3. Select a parcel and then select the Map Check button to create a report about the parcel. Close the report window. With the parcel still selected click on the Inverse button to create an Inverse report. Close the Inverse report window. Select other parcels and view their statistics (see Figures 4–13 and 4–14).
4. If you would like, print the lot statistics to the printer.
5. Select the OK button until you exit Parcel Manager.

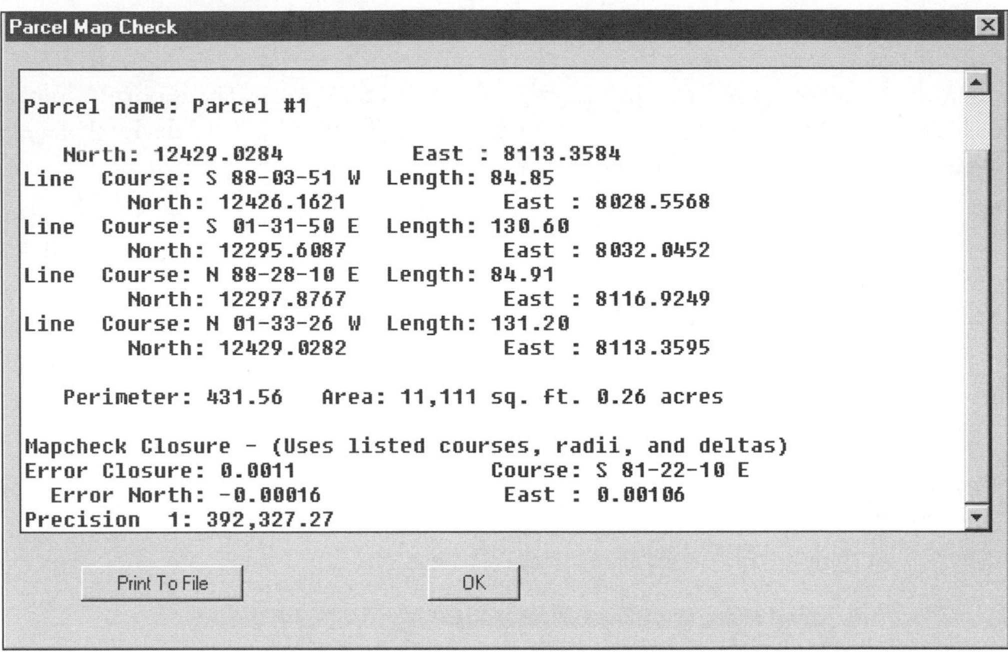
Parcel Map Check
Parcel name: Parcel #1
North: 12429.0284 East : 8113.3584
Line Course: S 88-03-51 W Length: 84.85
North: 12426.1621 East : 8028.5568
Line Course: S 01-31-50 E Length: 130.60
North: 12295.6087 East : 8032.0452
Line Course: N 88-28-10 E Length: 84.91
North: 12297.8767 East : 8116.9249
Line Course: N 01-33-26 W Length: 131.20
North: 12429.0282 East : 8113.3595
Perimeter: 431.56 Area: 11,111 sq. ft. 0.26 acres
Mapcheck Closure - (Uses listed courses, radii, and deltas)
Error Closure: 0.0011 Course: S 81-22-10 E
Error North: -0.00016 East : 0.00106
Precision 1: 392,327.27
Print To File
OK

Figure 4–13

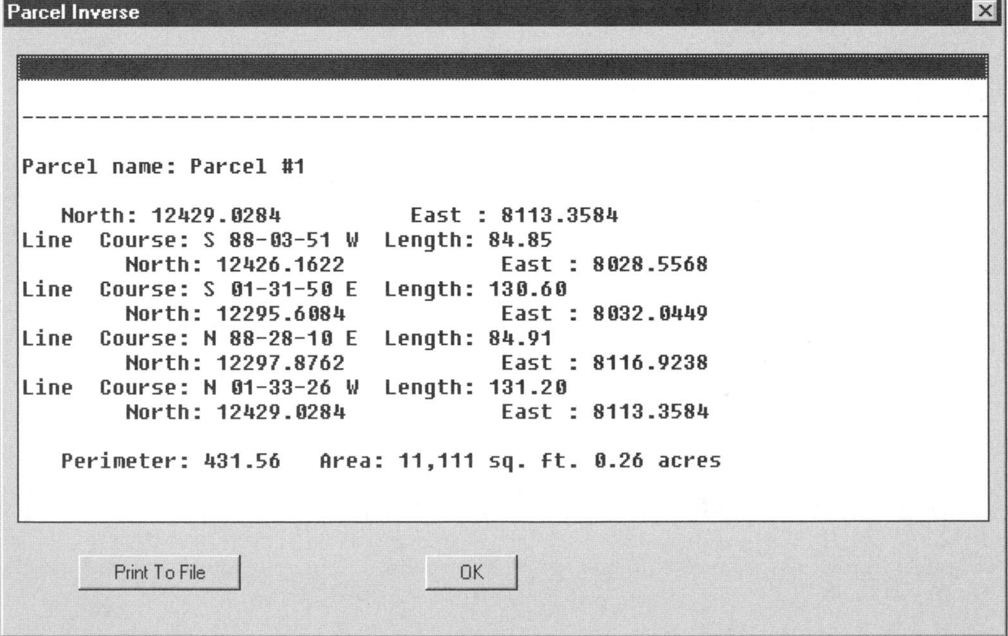
Parcel Inverse
Parcel name: Parcel #1
North: 12429.0284 East : 8113.3584
Line Course: S 88-03-51 W Length: 84.85
North: 12426.1622 East : 8028.5568
Line Course: S 01-31-50 E Length: 130.60
North: 12295.6084 East : 8032.0449
Line Course: N 88-28-10 E Length: 84.91
North: 12297.8762 East : 8116.9238
Line Course: N 01-33-26 W Length: 131.20
North: 12429.0284 East : 8113.3584
Perimeter: 431.56 Area: 11,111 sq. ft. 0.26 acres
Print To File
OK

Figure 4–14

EXERCISE

UNIT 5: ANNOTATING LOTS

In Land Development Desktop, there are two ways to annotate lots. The first is labeling the lots as they are defined. The second is labeling the lots after defining them. The method of annotating is set by toggling on Label as Lots Are Sized in the Parcel Settings. Both methods label according to the settings found in the Parcel Settings dialog box (see Figure 4–2). But it is only in the second method that you can select which lot to annotate. The annotation can contain the lot number and size of the lot. The size is either in square feet, acres, or both.

While creating lots with one of the lot sizing routines, the routine annotates the new lot.

Exercise

EXERCISE

After completing this exercise, you will be able to:

- Define a lot by polylines
- Review lot values in the Lot Editor

DEFINING LOTS

The second method is to draw or select entities that represent the lot boundary. After you select the line work, the Label routine annotates the lot. The lot definition routines create lots with AutoCAD lines, arcs, polylines, or choices.

While annotating lots, the routine may store the lot definitions in an external lot database. The lot database contains information to analyze each lot. The methods of analysis are map check, inversing, closure, and area.

1. Start a new drawing, name it LOTS, assign it to the proj1 project, and use the *aec_i.dwt* template file (see Figure 4–15). If you are prompted, save the changes to the LDDTR2 drawing.
2. In the Load Settings wizard, select the LDD1-40 setup, select the Load button in the upper right of the dialog box, and then select the Finish button (see Figure 4–16). When the routine closes the dialog box, a review dialog box appears. Select OK to close the dialog box.
3. Select the Parcel Manager routine from the Parcels menu.
4. In the Parcel Manager, select the Select All button and then select Import. This will place a copy of the lot boundary and annotation in the drawing.
5. Use the Zoom Extents routine to view the imported parcels.
6. Save the drawing.

New Drawing: Project Based
Drawing Name
Name: lots.dwg
Project and Drawing Location
Project Path: d:\Land Projects R2\ Browse...
Project Name: PROJ1
Drawing Path: d:\Land Projects R2\PROJ1\dwg\
Filter Project List... Project Details... Create Project...
Select Drawing template
ACAD -Named Plot Styles.dwt
acad.dwt
ACADISO -Named Plot Styles.dwt
acadiso.dwt
aec_i.dwt
Show sub-folders Browse...
Preview
OK Cancel Help

Figure 4–15

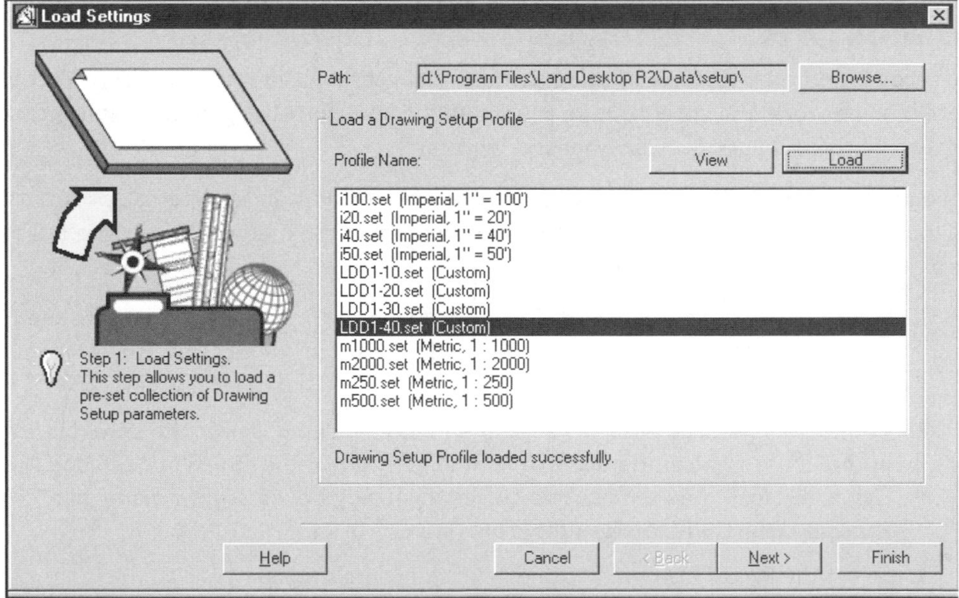

Figure 4–16

This completes the section on Parcels and the creation of lines and curves. Parcels are a method of organizing the subdivision of land within a boundary. Land Development Desktop gives you tools to size, analyze, and import parcels into a drawing.

Surfaces

After you complete this section, you will be able to:

- Use Terrain Model Explorer
- Use data gathering routines for Digital Terrain Model (DTM) surfaces
- Build a surface and evaluate its quality
- Control triangulation
- Use surface display routines
- View a surface in 3D
- Create and annotate contours

SECTION OVERVIEW

The last basic data group of Land Development Desktop is the terrain surface. Land Development Desktop represents the Point, Line, and Curve data groups as two-dimensional objects. The elevations represented by these two groups are only attributes of two-dimensional objects. The length of a line is its 2D length, not its slope distance, just as the distance between points is the horizontal distance. The responsibilities of plat measurements are two-dimensional in nature.

When the focus shifts to the three-dimensional representation of the site, LDD uses 3D objects and the actual elevations from the points. In Land Development Desktop, a surface is an electronic web representing elevations. Surfaces can show the current state of a site or its final design. In essence, a surface is the starting point and ending point of the site design process.

UNIT 1

The LDD routines for creation, editing, analysis, and annotation of surfaces are complex processes displayed on the Terrain menu. The Terrain routines read data files and create new data, layers, and objects in the drawing. The Edit Settings dialog box contains project defaults that the routines use when creating layers or assigning values. This dialog box and its settings are the first unit of this section.

UNIT 2

The Digital Terrain Model (DTM) or Triangular Irregular Network (TIN) surface is a construction from one or a combination of the following data types: point, breakline, and contour. Point data may come from external ASCII coordinate files, AutoCAD 3D objects, and/or the point database (point groups). Breaklines help control triangulation along linearly related points, for example, edges-of-pavements, breaks in slope, stream banks, swales, ditches, and other important surface slope break points. Breaklines can be drawn from the field or by the user. Breaklines can be either 2D or 3D. There is no difference between their effect upon a surface. The only condition when using 2D breaklines is that they must have corresponding point data. 3D breaklines do not have to correspond to any surface point data. Contour data is from 3D polylines or contours in the drawing. They are breaklines in that the triangulation is between them, not across them.

The Terrain Model Explorer is the command center for the creation of surfaces. The Build (Surface) routine of the Terrain Explorer reads the surface data and produces a triangulated representation of the data. The data types, their creation, evaluation, and their effect on the surface are the topic of the second unit of this section.

UNIT 3

The initial surface may need editing. There may be missing data, incorrect triangles, sliver triangles at the periphery—all indicative that the surface needs editing. The third unit covers the editing of a surface.

UNIT 4

The Terrain Model Explorer generates statistical reports about each surface. The surface display routines provide visualization and analysis tools to view the elevation and slope information of the surface in a more understandable way. Because the surface can represent the start and end of a project, it is critical that it represents the data correctly. Other analysis tools allow you to cut cross or long sections (profiles) from the surface. With AutoCAD entities representing the slopes, elevations, and relief of a surface, the drawing provides potential presentation materials. These entities can be passed to 3D Studio VIZ for animations and enhanced visualization techniques. The development of the initial and final surfaces can be time-consuming, yet these surfaces have a powerful visual impact on designers, developers, and potential consumers. The fourth unit of this section discusses the surface analysis and visualization tools of LDD.

UNIT 5

The annotation of a surface is by contours. Contours are objects (a special type of 3D polylines) that have a constant elevation. Contours connect points on the surface that have the same elevation. Contours are the subject of the fifth unit of this section.

UNIT 6

The last unit of this section covers routines that create new point data that is the result of having a surface.

UNIT 1: DTM SETTINGS

The DTM settings are available to you through the Drawing Settings routine on the Projects menu. LDD displays the Edit Settings dialog box from which you select the appropriate settings item (see Figure 5–1).

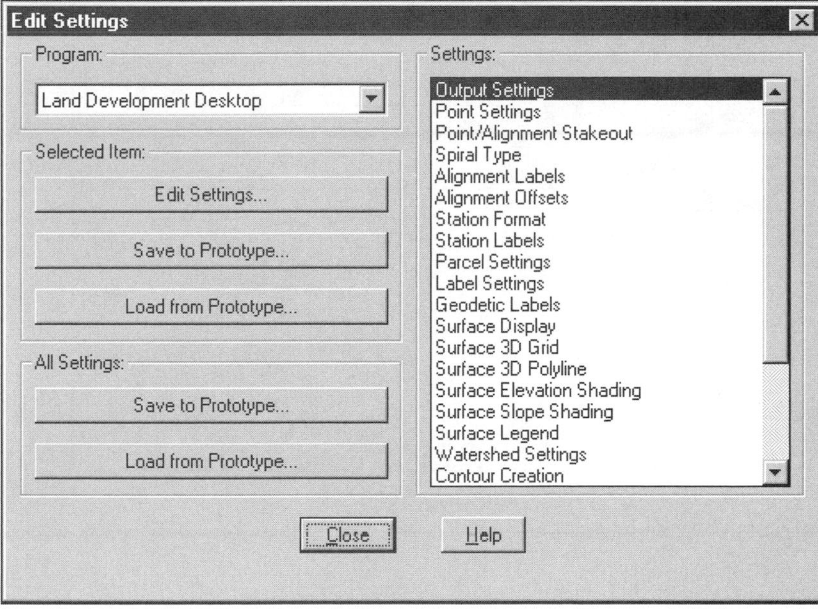

Figure 5–1

The settings that you edit in this dialog box apply to the current drawing and are a reflection of the project settings. Remember that the project settings come from the prototype you selected when the project was first created. The edited settings can be saved individually or en masse to the prototype. You can copy the settings from the prototype to change or update the settings in the project and current drawing.

The word "surface" prefixes all of the settings that affect the Terrain process in the Edit Settings dialog box. If a routine uses these settings, it will display the appropriate dialog box and ask you to verify its contents before completing its task.

SURFACE DISPLAY

The Surface Display Settings dialog box sets the common values for viewing surface triangulation, breaklines, and borders (see Figure 5–2). These elements can be 2D or 3D. If they are 3D, the routines will look at the two values in the lower right corner,

Base Elevation and Vertical Factor. If a routine creates a skirt (the Create Skirts toggle) around the surface, the Base Elevation sets the bottom elevation of the skirt. The vertical factor exaggerates the relief of the surface by exaggerating the highest points on the surface by the largest amount and the lowest points with the least amount of exaggeration. If you set these values in the dialog box, they are set for all dialog boxes that use these values.

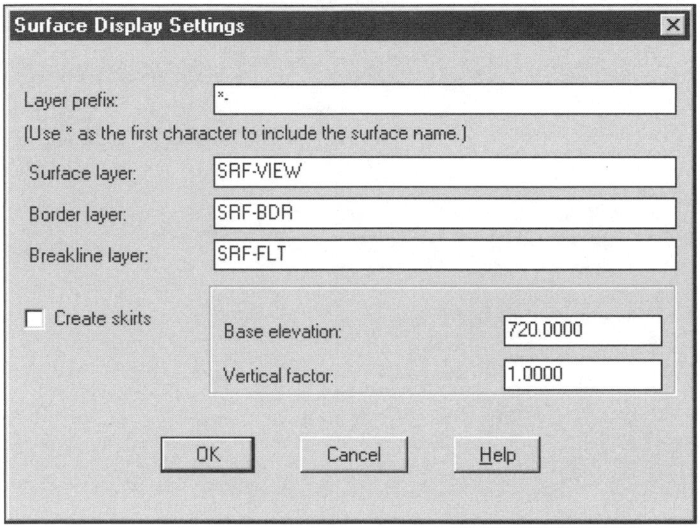

Figure 5–2

All of the surfaces built and displayed by the Terrain Model Explorer will use the same layer names. Each time you import the surface data, the routine will ask whether to delete any previous objects. If there are any previous objects and you answer Yes, all objects on the layers are erased. The layer prefix allows each surface to follow the naming convention, but to prefix the layers with its own name. The prefix is entered as an asterisk (*) in the top portion of the dialog box. However, the name of the surface would then be appended immediately to the layer names found in the middle of the dialog box. To make the layer naming clear, you need to enter a separator after the asterisk. The separator can be either a dash (-) or an underscore (_). If you set this value in the dialog box, it is set for all dialog boxes that use the prefix.

SURFACE 3D GRID

The Surface 3D Grid Generator routine creates a representation of the surface as user-sized squares (see Figure 5–3). The routine assigns an elevation to a corner where the corners of the squares intersect the surface. The size of the squares is set by one of two methods, absolute size or number. The first method uses an absolute size cell set by the user. The second method divides the grid area by the number of cells in the M (*x*) and N (*y*) directions. The reason for using M and N comes from the fact that the area defining the grid may be rotated. If the grid is rotated, the normals slicing the surface into squares are not the same direction as *X* or *Y*, hence their labels of M and N.

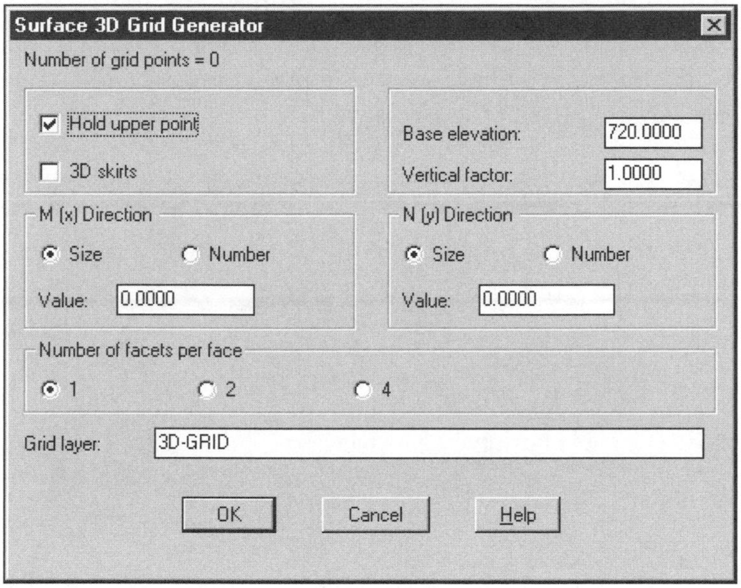

Figure 5–3

In the bottom portion of the dialog box, you can set the number of faces a square can have. Having more than one face allows the routine to show radical elevation changes over a square. A cell with one face averages the change of elevation over the entire square. Two faces allow two points of the cell to be independent while the two remaining points are tied together. This state is similar to the four points and diagonal of a triangulated surface. Four faces allow the cell to show radical changes in elevation within a cell.

The Surface 3D Grid Generator routine looks at the two additional values in the upper right corner, Base Elevation and Vertical Factor. If the routine creates a skirt (the 3D Skirts toggle) around the grid, the Base Elevation sets the bottom elevation of the skirt. The Vertical Factor exaggerates the relief of the surface by exaggerating the highest points on the surface by the largest amounts and the lowest points with the least amount of exaggeration. If you set these values in the dialog box, they are set for all dialog boxes that use the values. The Hold Upper Point toggle adjusts the cell size so there is an even number of cells in the M and N direction.

The last entry sets the layer on which the grid appears in the drawing. If you have set a layer prefix, the Surface 3D Grid Generator routine will append the surface name to the layer.

SURFACE 3D POLYLINE

The 3D Polyline Grid routine creates a polyline grid to represent a surface (see the Surface 3D Polyline Grid Settings dialog box in Figure 5–4). The settings are exactly the same as for Surface 3D Grid Generator, except the layers. If run, the 3D Polyline Grid routine uses or creates the two layers listed in the dialog box. The layers represent the M and N directions. If you have set a layer prefix, the 3D Polyline Grid routine will append the surface name to the layers.

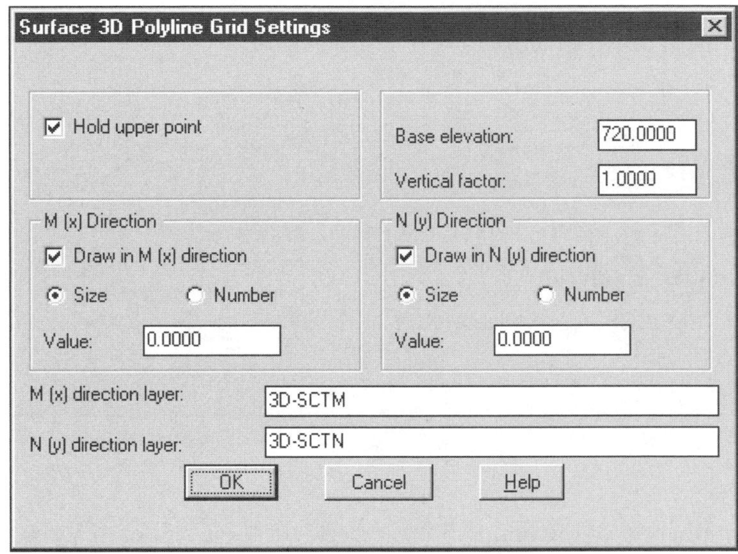

Figure 5–4

SURFACE ELEVATION SHADING SETTINGS

The analysis of a surface involves evaluating the distribution of elevations. The elevation routines create a ranged-based method of viewing the elevations on a surface. The values in the Surface Elevation Shading Settings dialog box control their color, number of ranges, and layers.

The upper portion of the dialog box contains settings for the Layer Prefix, Create Skirts, Base Elevation, and Vertical Factor (see Figure 5–5). These settings have been discussed earlier in this section. The middle portion of the dialog box sets the number of ranges the routine will use. After you set the number of ranges, the next step is how to define the range values. There are two methods of defining a range, automatic (Auto-Range) and user (User-Range). The Auto-Range method creates the range break points by dividing the overall range by the number of ranges. For example, if the range of slopes is 200 (0 to 200 percent) and the number of desired ranges is 8, Auto-Range sets the range breaks to every 25 percent. The automatic method is best for evenly or normally distributed populations. If the distribution of elevations is skewed (for example, most values are low and a few are scattered about at the high end), the first range may represent 90 percent of the population. This leaves 10 percent of the population to fill out seven more ranges. Land Development Desktop does not report the distribution of elevations on a surface. You may have to experiment with the settings to find the correct number of ranges and whether it should be an auto or manual setting process.

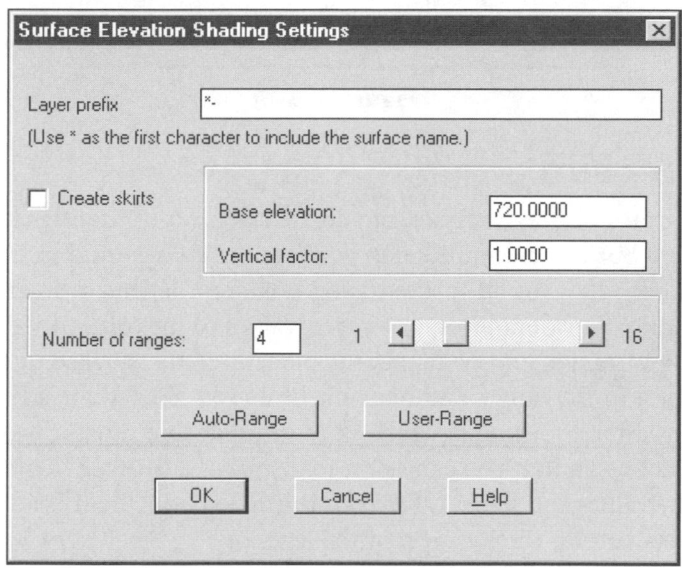

Figure 5–5

SURFACE SLOPE SHADING SETTINGS

The Surface Slope Shading Settings dialog box (see Figure 5–6) is the same as the Surface Elevation Shading Settings dialog box. The only difference is the surface property, slopes. LDD reports the distribution of slopes in the statistics area of Terrain Model Explorer. If the mean slope value is midway between the lowest and highest slope values, this indicates a fairly evenly distributed population of slopes. If the mean value is nearer the highest or lowest value, the distribution of slopes is skewed. See the discussion under the previous heading for setting ranges by either Auto-Range or User-Range methods.

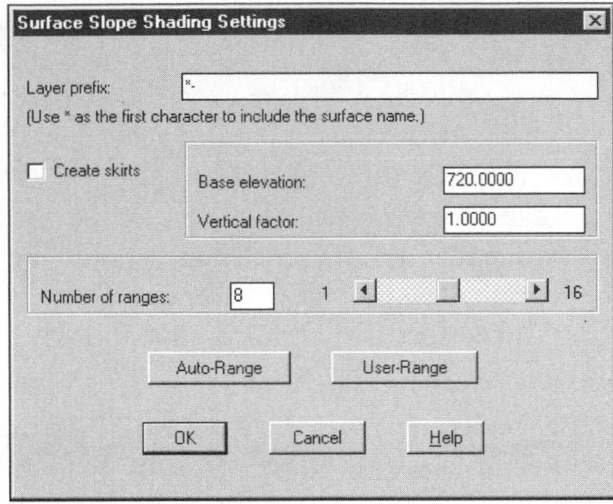

Figure 5–6

SURFACE LEGEND

Many of the surface display routines can create a legend to identify the ranges and their values. The Surface Legend dialog box sets the values used in the legend (see Figure 5–7). The values found in the dialog box are a legend title and six column settings. A column can denote one of seven values. The column values are color, layer, beginning of range, end of range, the percent of the surface in the range, the area of the surface in the range, and none of the above. You create a legend as a part of a routine, not after a routine has run. When the routine runs, it presents a dialog box containing a button that you can select to display the Surface Legend dialog box. After you set the values and select OK to exit the dialog box, the routine completes its task and prompts you for the location of the legend.

Surface Legend

Legend title: Elevations

Column 1	Column 2	Column 3	Column 4	Column 5	Column 6
Color	Color	Color	Color	Color	Color
Layer	Layer	Layer	Layer	Layer	Layer
Begin	Begin	Begin	Begin	Begin	Begin
End	End	End	End	End	End
%	%	%	%	%	%
Area	Area	Area	Area	Area	Area
None	None	None	None	None	None

OK Cancel Help

Figure 5–7

WATERSHED SETTINGS

The Watershed Settings dialog box sets the criteria for defining a watershed (see Figure 5–8). The main watershed criteria are area and depth. You set the size of both in the left portion of the dialog box and toggle on whether a watershed has to exceed both values to be a watershed. The lower left portion of the panel toggles on or off the filling in the watershed with AutoCAD solids, erasing any previous watershed objects, and labeling the watersheds with a number. The right side of the panel lists the layers that the Watershed routine will create and use. If you have set a layer prefix, the Watershed routine will append the surface name to the layers.

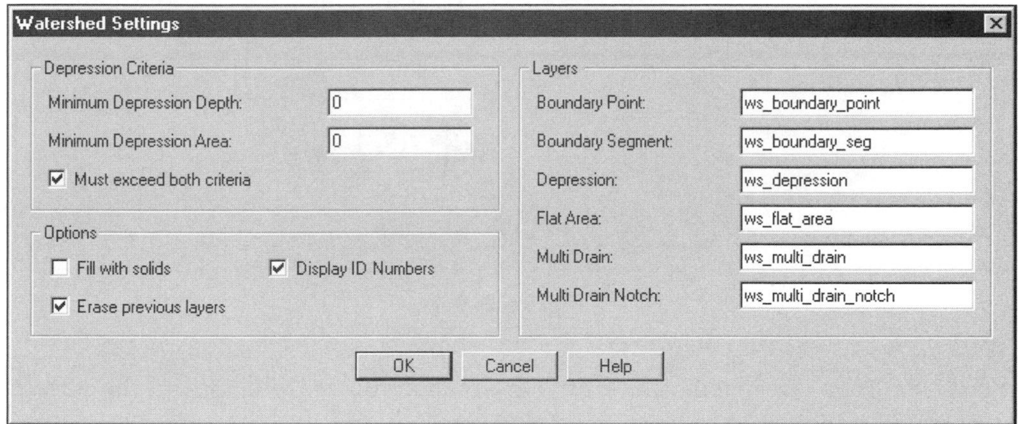

Figure 5–8

CONTOUR CREATION

The Contour Creation dialog box sets the contour interval, layers, types of objects, and style for creating contours (see Figure 5–9). The top portion of the dialog box sets the contour interval type as both, or only minor or major. Below the interval type, in the Intervals area, are the interval values. Generally, these values are 1 (for minor) and 5 (for major) feet. If you click the up or down arrows to the right of the minor value, the dialog box will adjust both minor and major values. For example, if the current settings are 1 and 5, you can click the up arrow to the right of the minor number to change the minor value to 2 and the major to 10. The geometric relation of the minor to major is preserved in both settings, that is, 1 to 5 and 2 to 10. If the initial numbers between minor and major are 1 and 10, you can click the up arrow to the right of the minor number to change the values to 2 and 20. If the minor value is 1 and you click the down arrow to the right of the minor number, the interval values become less than one.

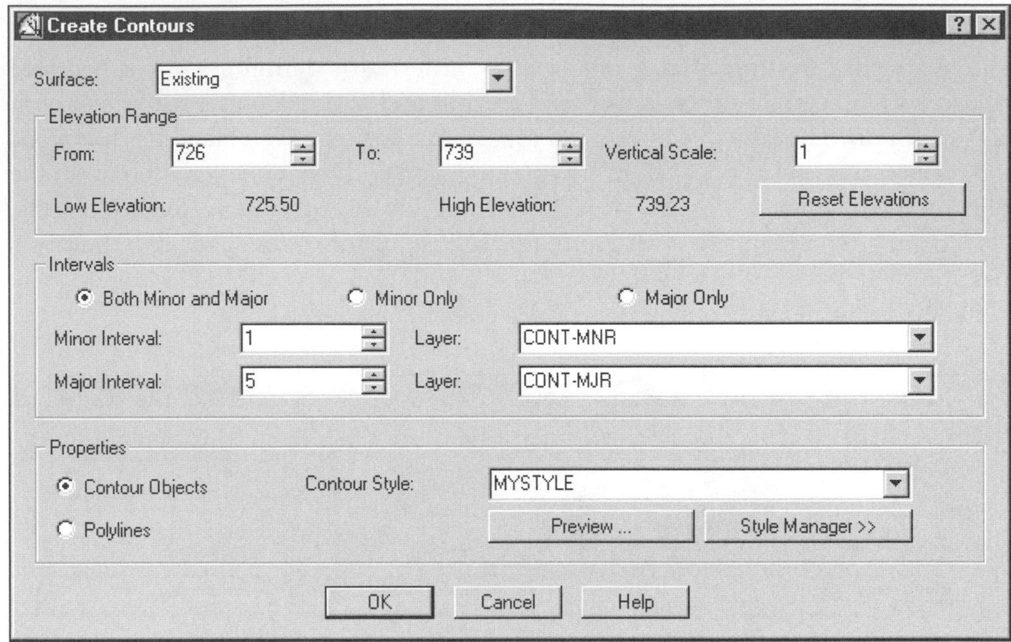

Figure 5–9

If you have set a layer prefix, the Contour Creation routine will append the surface name to the layer names listed in the dialog box.

The bottom portion of the dialog box sets the object type and contour style. The Contour Creation routine creates either 3D polylines or contour objects. If you need to share the contour data with someone who does not have LDD, it is best to provide them with 3D polylines. LDD does have an object enabler that will allow any vanilla AutoCAD to view and to a limited extent manipulate the objects. The enabler is a free download from the Autodesk Web page or you can copy it from the LDD CD (the enabler directory).

CONTOUR STYLES

A contour style controls several aspects of contour behavior. The aspects under a style's control are labeling, contour width, label position, smoothness, and grips. LDD stores the style definitions in the *Contours* folder of LDD's *Data* folder. The definitions can be shared if the *Data* folder is located in a directory. The Contour Style Manager dialog box contains several tabs to set the values under its control.

CONTOUR APPEARANCE

The Contour Appearance tab controls the contour and labeling display and smoothness settings (see Figure 5–10). The Contour Display settings allow a contour to have or not have grips and a preset width. A contour with grips can be moved on the screen. This however, creates a difference between their calculated and visual location. The Label Display values set whether a contour will display a label and grips, labels with no grips, or no labels at all. A contour label with grips allows you to relocate the label along the contour. The label is a part of the contour and will never separate from the contour even if you select a point far from the contour.

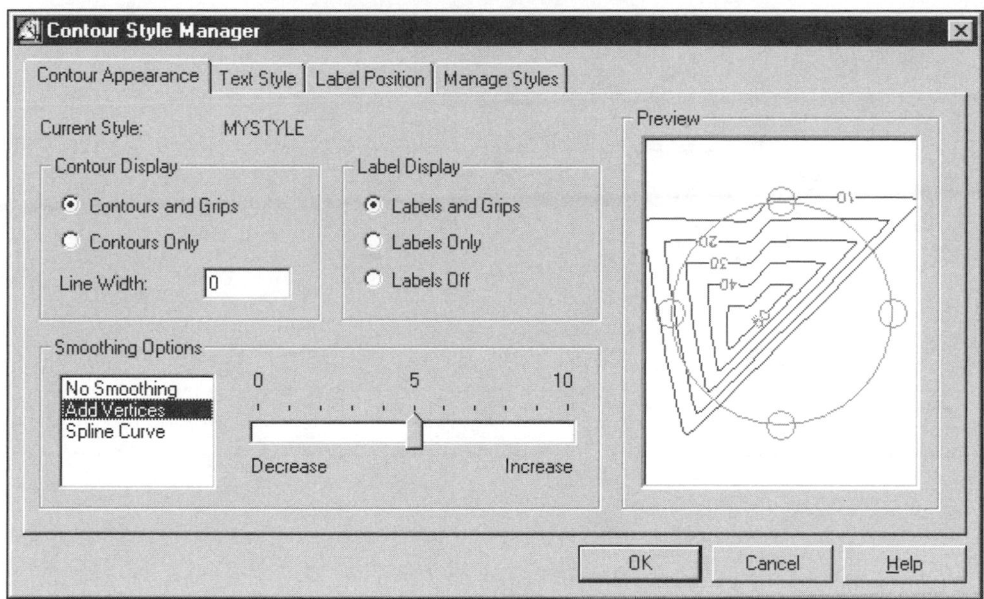

Figure 5–10

The settings for smoothness are none, vertex, or spline. The Add Vertices option will smooth the contours, but only a small amount even when set to the highest setting. The Add Vertices option adds vertices to the contour to smooth its appearance. Spline smoothing pulls contours away from their original location and may cause contours to cross over each other. Splining is extreme smoothing and in switchback situations, the Spline Curve option will cause the contours to cross over one another.

TEXT STYLE

The Text Style tab sets the text style, color, precision, and prefixes and suffixes to the elevation of the contour (see Figure 5–11). The Style drop list displays all currently defined text styles. If you select a text style with a fixed height (a Leroy style), the height is set in the definition of the style and the height setting at the bottom of the panel is grayed out. If you select a style that is a variable height style, the Height box allows you to set a height for the labeling text. The label of a contour is not on a layer in the drawing. To be able to differentiate the label from the contours, you use a color. You can plot this color with a different width by defining a thicker pen in an AutoCAD color plot table.

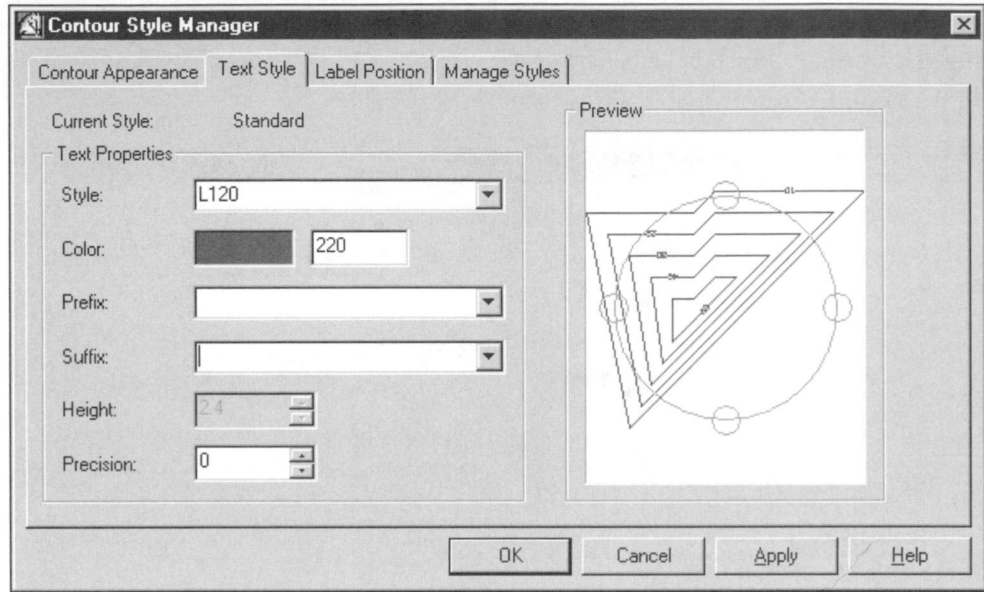

Figure 5–11

The drop arrows for Prefix and Suffix display a list of predefined labels for a contour. You can add to the list by typing either a prefix or suffix value.

The last setting controls the precision of the contour label. This setting is independent from all other precision settings in LDD.

LABEL POSITION

The Label Position panel contains settings that control the location of the contour label, its readability, and encapsulation (see Figure 5–12). The upper part of the panel sets the location of the label to be on, above, or below the contour. If the label is above or below the contour, the Offset setting offsets it from the contour. By default this offset distance is 1 text height from the contour. You can change that amount.

The Readability portion of the panel sets the behavior of contour labels when in layout viewports or when viewed up slope. If the Make Plan Readable option is on, contour labels will rotate themselves in the viewport to be horizontal in that viewport. The Label Positive Slope option orientates the label so that the top of the text is towards the next highest elevation.

The last setting, Border Around Label, controls the placement of a border around the contour label. There are three choices: no label, a rectangular label, and a circular label.

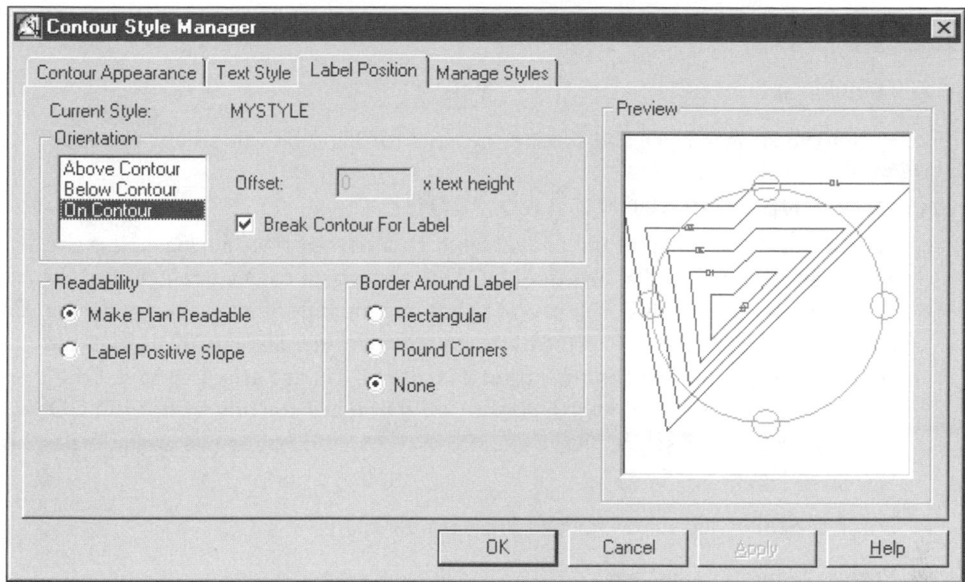

Figure 5–12

MANAGE STYLES

The Manage Styles panel saves, sets, and displays the current drawing contour style and any styles not loaded into the drawing (see Figure 5–13). This panel allows you to manage a library of styles and to create standards within the office environment.

Figure 5–13

Exercise

After you complete this exercise, you will be able to:

- Set the surface layer and settings defaults for Land Development Desktop

CREATING A NEW DRAWING AND PROJECT

1. Start a new drawing, Terrain, and assign it to a new project Surfaces, use the aec_I template file, and assign the LDDTR2 prototype to the project. Use Figures 5–14 and 5–15 as guides. The project path may be different on your machine. The pathing depends upon the installation at your machine. Select OK to exit and start the project/drawing initialization process. If you are asked to save the changes to the current drawing, select OK. If you are starting from the LDD opening dialog box, select New (see Figure 5–16) and you arrive at the dialog box shown in Figure 5–14.

   Enter the following values for the Project definition:

   Project Name: Surfaces

   Description: Section 5 Exercises

   Key words: surface terrain

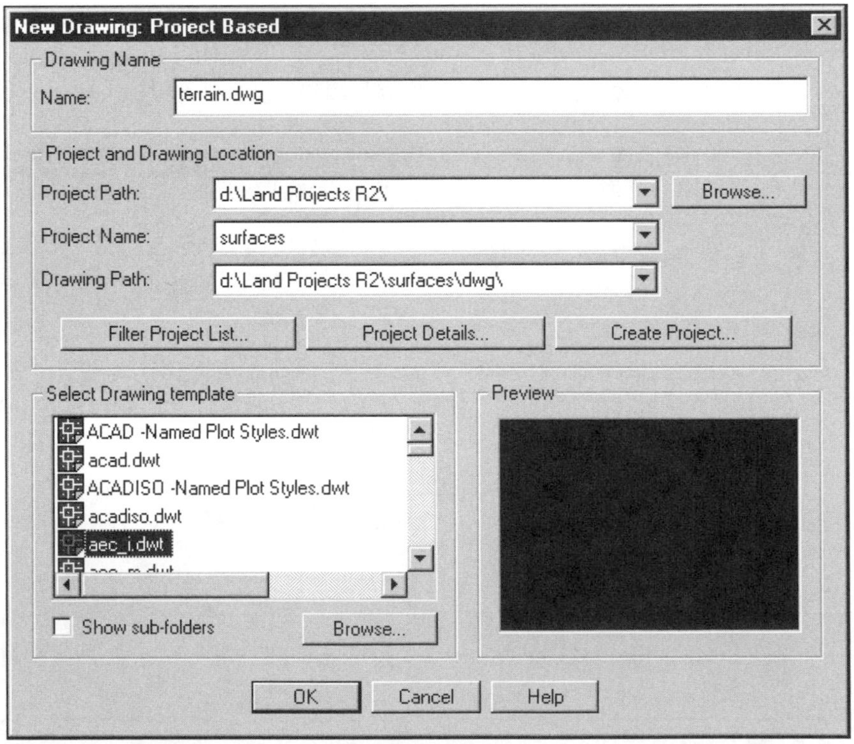

Figure 5–14

Project Details

Initial Settings for New Drawings

Prototype: LDDTR2

Project Path: d:\Land Projects R2\

Project Information

Name: surfaces

Description: Section 5 Exercises

Keywords: surface terrain

Drawing Path for this Project

Project "DWG" Folder

Fixed Path

Browse...

OK Cancel Help

Figure 5–15

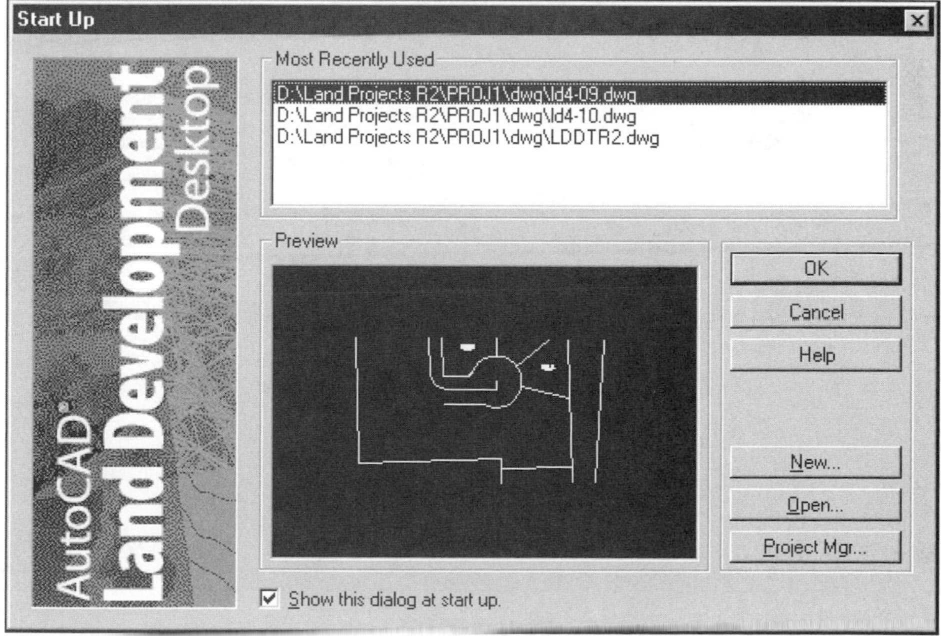

Figure 5–16

EXERCISE

2. The initialization process next prompts you about the creation of the point database (see Figure 5–17). Select OK to exit the dialog box and to continue initializing the project/drawing.

Create Point Database

Project: surfaces

Point File: d:\land projects r2\surfaces\cogo\points.mdb

Point Description Field Size: 32

Use Point Names:

Point Name Field Size: 16

OK

Help

Figure 5–17

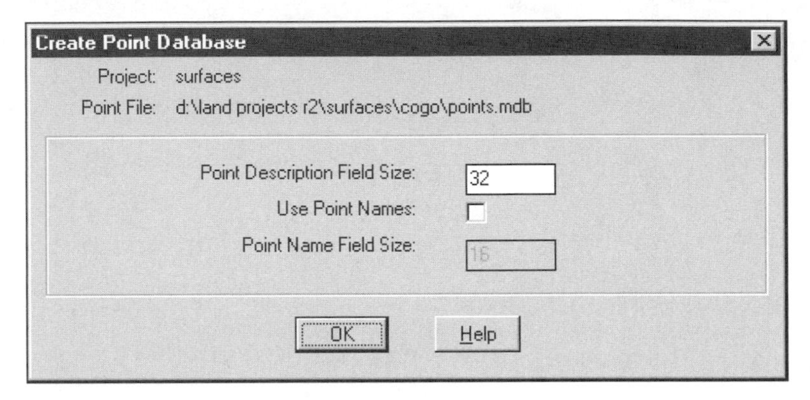

3. The next dialog box sets the drawing scale and other drawing values. First select the *LDD1-40.set* file to set the scale to 1 inch = 40 feet (see Figure 5–18) and then select the Load button in the upper right to create the settings and text styles for the drawing. Select the Finish button to exit the dialog box.

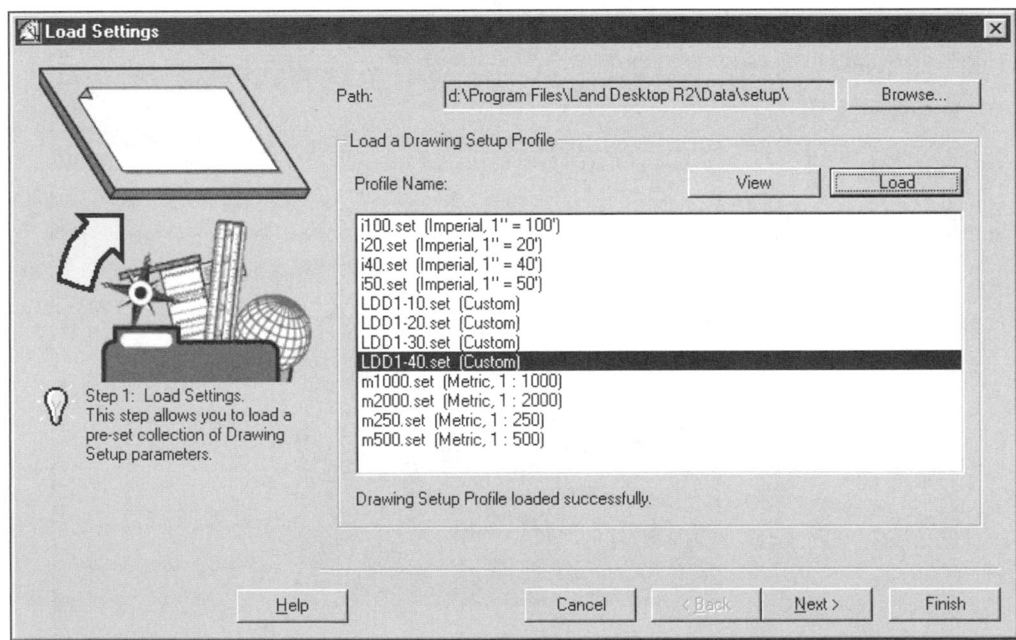

Figure 5–18

4. Lastly, review the settings in the Finish dialog box and select OK to exit the dialog box.

5. Click the Save Drawing icon to save the drawing.

DTM SETTINGS

1. From the Projects menu, select Drawing Settings. The routine displays the Edit Settings dialog box (see Figure 5–1).

2. Select Surface Display and then the Edit Settings button to edit the Display settings. The routine displays the Surface Display Settings dialog box (see Figure 5–2). Set the base elevation to 720 and enter an ***-** (asterisk dash) as the layer prefix. Instead of a dash you can use an underscore. The prefix appends the layers used by the routine with the name of the surface. The dash is a separator between the surface and layer name. The surface for this exercise does not have an elevation lower than 720 feet. After you set the values, select OK to exit the dialog box.

3. Select Surface 3D Grid and then the Edit Settings button to view the settings dialog box (see Figure 5–3). There is nothing to set in this dialog box at this time. Select OK to exit and return to the Edit Settings dialog box.

4. Select Surface 3D Polyline and then the Edit Settings button to view the settings dialog box (see Figure 5–4). There is nothing to set in this dialog box at this time. Select OK to exit and return to the Edit Settings dialog box.

5. Select Surface Elevation Shading and then the Edit Settings button to view the settings dialog box (see Figure 5–5). The range values and their colors will be set in a later exercise in this section. There is nothing to set in this dialog box at this time. Select OK to exit and return to the Edit Settings dialog box.

6. Select Surface Slope Shading and then the Edit Settings button to view the settings dialog box (see Figure 5–6). The range values and their colors will be set in a later exercise in this section. There is nothing to set in this dialog box at this time. Select OK to exit and return to the Edit Settings dialog box.

7. Select Surface Legend and then the Edit Settings button to view the settings dialog box. Use Figure 5–7 as a guide and enter the following values.

Title:	Elevation
Column 1:	Color
Column 2:	Begin (Range value)
Column 3:	End (Range value)
Column 4:	Percent (of site in this range)
Column 5:	Area (of site in this range)
Column 6:	None

EXERCISE

8. Select Watershed Settings and then the Edit Settings button to view the settings dialog box (see Figure 5–8). There is nothing to set in this dialog box at this time. Select OK to exit and return to the Edit Settings dialog box.
9. Select Contour Creation and then the Edit Settings button to view the settings dialog box. Use Figure 5–9 as a guide in setting the values. At this time, set the minor interval to 1 and the major interval to 5. Do not exit the dialog box.
10. Select the Style Manager button at the lower left of the Contour Settings dialog box. This displays the Contour Style Manager dialog box.
11. Select the Contour Appearance tab and set the values as shown in Figure 5–10.
12. Select the Text Style tab and set the values as shown in Figure 5–11. Select the black color swatch to display the color palette. You can select any color for the label text.
13. Select the Label Position tab and set the values as shown in Figure 5–12.
14. Select the Manage Styles tab, enter **Mystyle** as the name of the new contour style at the top right of the dialog box, select Add to add the style, and select OK to save the style in the drawing. With the Mystyle still highlighted, select the Save button in the center of the dialog box to save the style as an LDD contour style. Use Figure 5–13 as a guide.
15. Select OK to exit the Style Manager and the Contour Settings dialog box.
16. Select the Close button to save the settings and close the Edit Settings dialog box.

EXERCISE

UNIT 2: SURFACE DATA

A surface is the result of evaluating elevations. The surface building process creates triangles connecting known points of elevation. In Land Development Desktop there are three types of surface elevation data. The first data type is points. Points can come from a field survey, an ASCII coordinate file, AutoCAD 3D objects, and the point database. There is an inherent problem with point data for surfaces. Usually, the point data is not dense enough to preserve the integrity of linear features on a surface. Examples of linear features are ditches, edges-of-pavement, streams, and so on. LDD uses the second type of data, breaklines, to control and preserve linear features on a surface. The last data type is contours. This data uses the vertices found on 3D polylines or contours to create a triangulated surface.

The routines that create surface data and build the surfaces automatically save the data to external files. The external files are subfolders in the *DTM* folder of the project. The subfolder names are the same as the surfaces.

LDD collects data and builds surfaces in the Terrain Model Explorer (TME). The TME is the only place where you can create a surface. This dialog box displays all of the data types that contribute to a surface. By selecting the data type and pressing the right mouse button, the TME lists the different routines that collect and/or create the data. If any one of these data groups change, you must rebuild the surface. If the surface uses points, breaklines, and contours and you add new points to a point group, add a breakline, or change contours, you must rerun the Build (Surface) routine to make the changes a part of the surface. When you use a surface editing routine, the routine automatically updates the surface definition.

There are two types of surfaces in Land Development Desktop, terrain and volume. A terrain surface is the result of point, breakline, and contour data and represents an existing or proposed site. A volume surface is the result of a comparison between two terrain surfaces. The elevations of a volume surface are the differences in elevation between the two surfaces being compared. The terrain surface is the topic of this section and section 6 covers the volume surfaces.

POINTS

A TIN displays how the Build (Surface) routine processed surface data. Each data point is significant in defining the rise and fall of elevations over the surface. The Build (Surface) routine connects each elevation point as a vertex of a triangle. TINs from point data tend to have low triangle counts. The resulting triangles have two characteristics. The first characteristic is a known elevation at each corner of the triangle. The second characteristic is a constant slope along the legs and across the face of each triangle. A triangle face contains a slope whose value reflects the net effect of the slopes found on each leg of a triangle. This allows you to evaluate the slopes over an entire surface.

If there are unrecorded significant elevation points not in the data set, the integrity of the surface is in question. In this situation the current surface triangulation does not correctly define the surface. These unrecorded significant points need to be added to the data set or edited into the surface to correctly define the surface.

LDD creates triangles from the point data by using a nearest-neighbor analysis. The nearest-neighbor method may cause the surface building routine (Modeler) to misrepresent areas within the surface. The Modeler looks at four points, uses them to create a perimeter, and then places a diagonal line between the nearest two of the four points. The classic problem for the Modeler is a set of four points equally spaced: two points with the elevation of 721 and two points with the elevation of 719 (see Figure 5–19). As mentioned above, the TIN Modeler connects the outside four points with lines. The problem occurs when the Modeler creates the diagonal leg. There are two solutions to the situation. The first solution is connecting the 721 elevations with the diagonal, which means that the diagonal represents a ridge between the two lower elevations. The second solution is connecting the 719 elevations with the diagonal producing a valley between two higher points, a swale. The default solution is to create a diagonal that connects the two closest points.

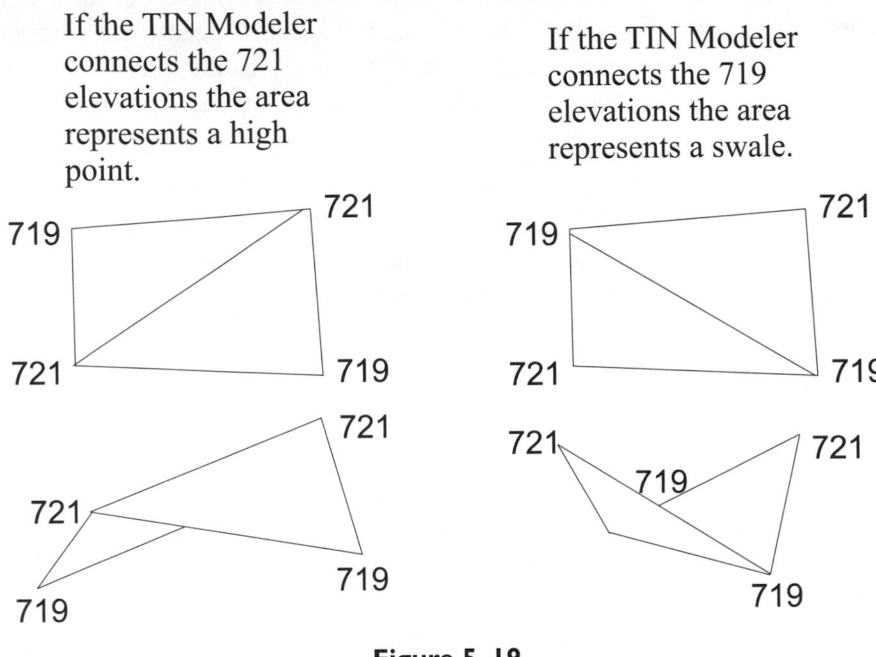

Figure 5–19

As a result of the nearest-neighbor analysis, the placement of a diagonal can radically change the interpretation of a group of points. When you are building a surface from point data only, the solution is solely a nearest-neighbor analysis. The Modeler does not understand that there may be significant linear features on the surface that the nearest-neighbor process destroys.

POINT DATABASE

You can select point database points by using point groups (see Figure 5–20). A point group can represent all or a selected portion of the points in the project point database. Point groups allow you to remember which points are in the surface and to create a group or groups excluding unwanted influences. For example, if the field crew includes fire hydrants in the data by observing the top of the hydrant, this point could be anywhere from 1 to 2 feet above the surrounding ground elevations. If these points are included in the surface, they will create a pile each time they are woven into the surface. Clearly, they should be removed from the surface data set, hence removed from the surface point group.

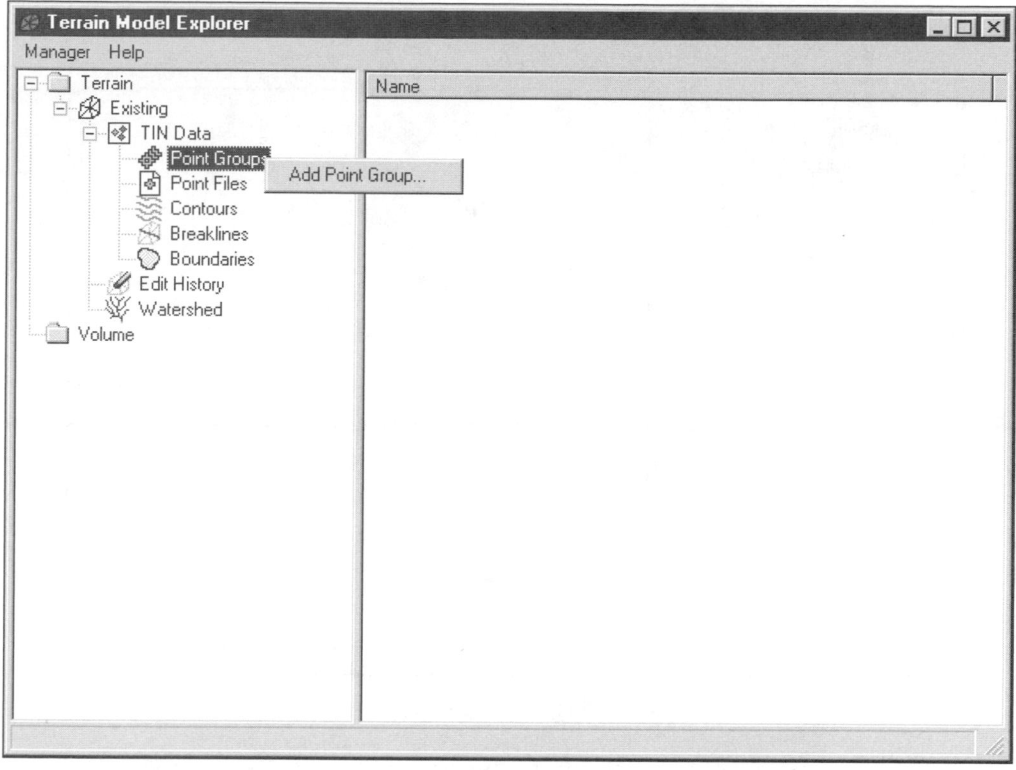

Figure 5–20

EXTERNAL POINT FILES

The second source of point data is an external coordinate file (see Figure 5–21). The Add Point File selection imports point data into the Surface point data area. The file can be an ASCII text file or an Access database file. The point data resulting from this process is not available to the project point database. If you want the coordinate data to be a part of the point database, import the points into the project point database through the Import/Export routine of the Points menu.

AUTOCAD ENTITIES

The last source for surface point data is AutoCAD objects (see Figure 5–22). If you select 3D objects in the drawing, these routines extract the *X*, *Y*, and *Z* coordinates of each vertex as point data for a surface. The routine works with AutoCAD nodes, lines, blocks, text, 3D faces, and polyface mesh objects. The data routine treats each vertex of an object as independent data points. The routine does not report or generate any connection between the endpoints and vertices of the objects. The visual links between the vertices are not saved with the point data.

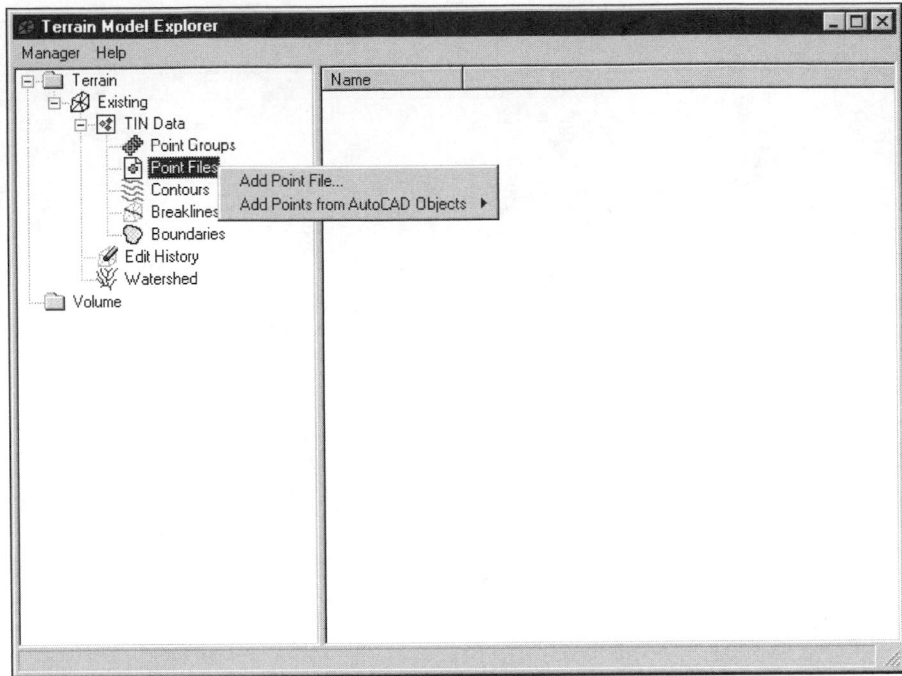

Figure 5–21

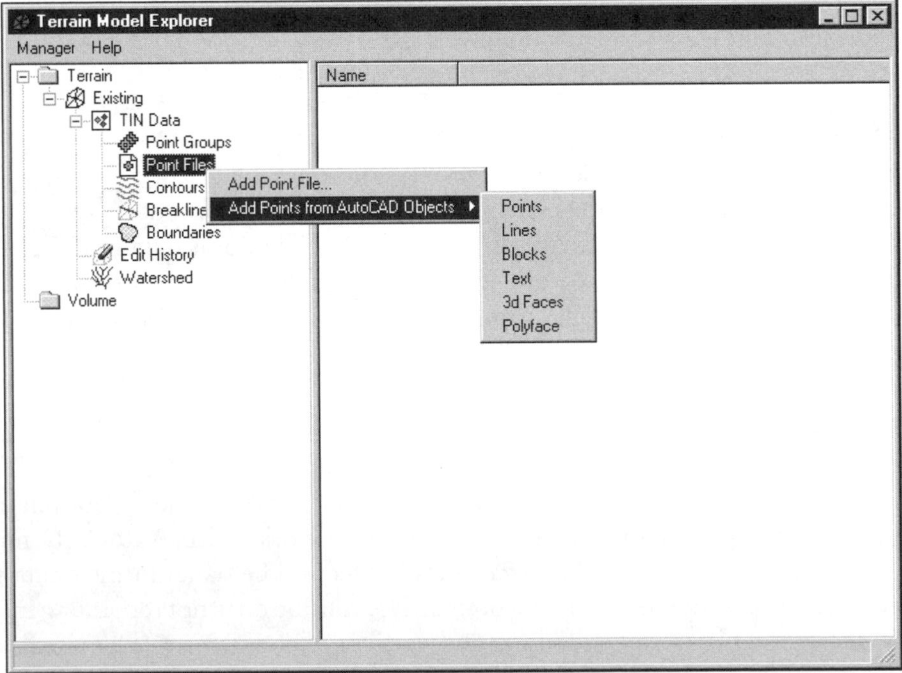

Figure 5–22

If you want to preserve the connection between the vertices, the objects must be identified as breakline data.

BREAKLINES

The misrepresentation of linear features occurs consistently in point data, because of the nearest-neighbor analysis. The field data must be dense enough along a linear feature to both describe and delineate it from all of the other intervening points of the survey. In many surveys, this is not the case. The most difficult task for the field crew is to view their survey data as triangulated point data. They see their data as points within the context of reality. For them, it is hard to understand why a point 100 feet away is not a part of the roadway edge in a triangulated surface. They view the point along the road as part of the roadway edge. But this point and a similar point separated by 100 feet may lose their linear connection when the Surface Modeler creates surface triangles from several surrounding points. The additional points may form triangles between the two edge-of-pavement shots, negating their linear and elevation relationship. This same problem occurs in situations where the data is too sparse to support a correct interpretation (see Figure 5–19). The situation shown in Figure 5–19 does not have enough data to correctly resolve the diagonal connection problem.

The solution to the above dilemmas is additional control over the triangulation process to preserve the linear and elevation relationships between points or to correctly control triangulation where the data is sparse. LDD uses breakline data to make up for this lack of information consistency. Breaklines imply a connection between points and prevent the Modeler from triangulating across the line connecting the points. The result is the correct interpretation of the linear situation.

LDD uses polylines to create breaklines. The breakline routines append extended entity data to the polyline containing the name and type of the breakline.

There are two types of breaklines, Standard and Proximity. The first type, Standard, is a 3D polyline. Since this polyline already contains elevations, it is not necessary to associate vertices on the polyline with LDD point objects. The second type, Proximity, is a 2D polyline. Since a proximity breakline is 2D, it *must* have a point object at each vertex of the polyline. The point objects provide the breakline with the necessary elevations, whereas the polyline provides the line across which no triangle can pass. Even though the breaklines are different types of objects, they produce the same effect on a surface.

The Surface Modeler will produce unpredictable results if the breakline data includes crossing breaklines. If this situation occurs, the Modeler does not understand your intentions. The result of crossing breaklines is the last one in the breakline data file controls the situation. However, it is best to avoid these situations at all cost.

STANDARD BREAKLINE

LDD represents a Standard breakline as a 3D polyline whose data comes from: existing point objects, 3D lines, an external file, or 3D polylines already present in the drawing (see Figure 5–23). A 3D polyline is a polyline with potentially different elevations at each vertex.

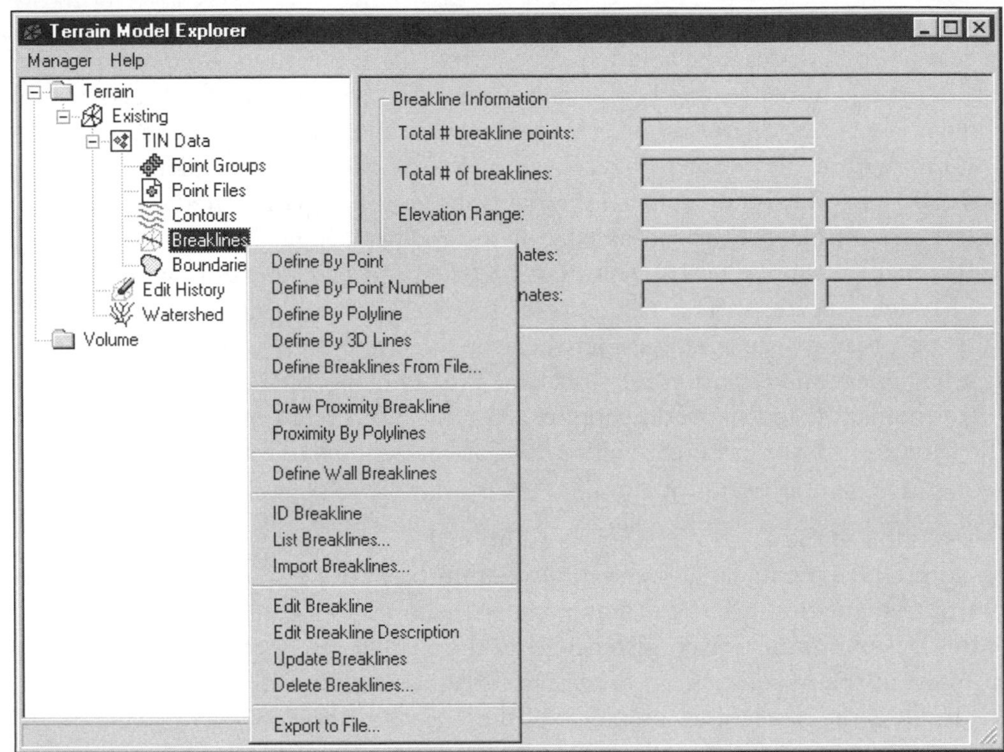

Figure 5–23

Define By Point and Define By Point Number

The Define By Point and Define By Point Number routines define 3D polyline breaklines (see Figure 5–23). The Define By Point routine creates the breakline from selected point objects along the linear path of the breakline. After you select the points, the routine prompts for a breakline name. The Define By Point Number routine defines a 3D polyline by a list of point numbers. You enter the point numbers as individual numbers separated by a comma (1,44,6,24,45), as a sequential number group separated by a dash (100–120), or as a combination of the two (1,34,62,30–33,73, 93–100). Both routines place the resulting 3D polyline on the SRF-FLT layer and append a number and breakline group name to each as extended entity data. If you have set a layer prefix, the routine will apply it to the SRF-FLT layer. The layer name for breaklines is set in the Surface Display settings dialog box.

Define By Polyline

The Define By Polyline breakline routine converts an existing 3D polyline into a breakline (see Figure 5–23). After you identify the 3D objects, the routine prompts for a breakline group name. The routine places the resulting 3D polylines on the SRF-FLT layer and appends each object with the number and name of the group as extended entity data. The routine offers the option of deleting the drawn objects before placing the breakline on the SRF_FLT layer. If you have set a layer prefix, the routine will apply it to the SRF-FLT layer. The layer name for breaklines is set in the Surface Display settings dialog box.

Define By 3D Line

The Define By 3D Line routine converts a 3D line into breakline data (see Figure 5–23). The routine uses the elevation at both endpoints and preserves the connection between them. The Define By 3D Line routine prompts for the selection of the 3D lines and a name for the breakline group. The routine does not convert the lines into polylines, nor append the fault data to them. You cannot edit 3D lines with the Edit Breakline routine.

To be able to edit the data represented by the 3D lines, you need to import the breakline data based upon the 3D lines. The Import Breakline routine creates 3D polylines representing the original 3D lines and appends the breakline name and number to them. The Edit Breakline routine can now edit the objects.

Define Breaklines From File

The Define Breaklines From File routine creates breaklines whose data is from an external ASCII file (see Figure 5–23). The routine generates both Standard and Proximity breaklines.

The file format identifies the type of breakline with a letter, with S indicating Standard, P indicating Proximity, L as a Left wall, and R as a Right wall breakline. The file has a specific format. The first line of the file contains the letter code of the breakline data type (S, P, L, or R), the easting (*X*), northing (*Y*), and elevation (*Z*), and the group name. Each line after the first line represents the next vertex on the polyline. This data starts at column 2 with spaces between *X Y Z* values. The file indicates a new breakline with the letter code of the breakline type, S, P, L, or R.

The following is an excerpt from the file.

```
# Warrenville Wholesale Project 5334
S1254.6458 2142.5155 718.25 Swale
 1257.5421 2146.5147 718.11
 1258.4522 2149.5426 717.93
P2418.6523 2748.6444 0.00 Berm
 2415.9856 2740.7824 0.00
 2436.1857 2732.8971 0.00
```

The preceding file excerpt contains two breaklines: Swale and Berm. The Swale breakline is 3D (S for Standard) and the Berm breakline is 2D (P for Proximity). The correct interpretation of a proximity breakline occurs when there are point objects at or near the coordinates of the Proximity fault vertices.

Define Wall Breaklines

The Define Wall Breaklines routine defines a sheer face data set (see Figure 5–23). The routine uses a preexisting polyline in the drawing. If the polyline is 2D, the routine prompts you for the elevations on the polyline. If the polyline is 3D, the routine prompts you to verify the elevations found on the polyline.

After you select the polyline, the routine prompts for an offset side, making the original polyline an edge of the wall. The routine creates blip arrows pointing away from the offset side to get or verify the wall elevation and towards the offset side to get the offset side's elevations.

The routine allows you to specify the offset elevation in one of two ways: as an absolute elevation or as an elevation difference. Elevation difference is a measure of the distance between the top and bottom of a wall. This measured distance allows the routine to calculate the needed elevations along the wall. A positive value for the offset side means that the offset side is the top of the wall. A negative difference means that the offset side is the bottom of the wall.

The routine places a 3D polyline on the SRF-FLT layer and appends it with the name of the fault group as extended entity data. If you have set a layer prefix, the routine will apply it to the SRF-FLT layer. The layer name for breaklines is set in the Surface Display settings dialog box.

PROXIMITY BREAKLINE

The second type of breakline is Proximity (see Figure 5–23). These breaklines are 2D polylines with a Z elevation of 0. To be able to create an elevation for a base or apex point of a triangle, there *must* be an LDD point object at or near each polyline vertex. The polyline defines a lineation that a triangle leg cannot cross. The point objects provide the correct elevations of the breakline to the Surface Modeler. The routine appends extended entity data to the 2D polylines that identifies the name and number of the breakline.

The TIN Modeler produces unpredictable results if there is no point block at the vertex of the polyline. If there are points along the polyline, but no vertex, the points are not a part of the breakline data. There *must* be a vertex at each point and there *must* be a point at each vertex of the breakline.

Draw Proximity Breakline

The Draw Proximity Breakline routine creates a 2D polyline (see Figure 5–23). When using this routine, you must select positions in the display near LDD point objects.

As the routine draws the fault, it enters the coordinates of the closest point object to the definition. It is from the point object that the routine extracts the *X*, *Y*, and *Z* values for the breakline. If there are several point objects in the vicinity of the vertex, it is the point nearest to the selected point that is the breakline entry. If you want to specify a point, use the dot toggles (.p, .n, or .g) or an object snap. As you draw the polyline, the routine does not snap the polyline to the associated point on the screen. You must import the breakline to view the actual polyline resulting from the routine. The routine places the resulting polyline on the SRF-FLT layer and appends the number and name of the breakline group to each object as extended entity data. If you have set a layer prefix, the routine will apply it to the layer name.

Proximity By Polylines

The Proximity By Polylines routine converts existing 2D polylines into breakline data (see Figure 5–23). The routine copies the polylines to the SRF-FLT layer. The routine asks whether or not to delete the original polyline. After making the copy of the polyline, the routine appends the entities with the breakline group name and number. If you have set a layer prefix, the routine will apply it to the layer name. Again, there must be an LDD point object at or near each vertex. The results are unpredictable if there is no point block at a vertex.

EDITING BREAKLINES

There are only two routines that edit breaklines (see Figure 5–23). The first routine, Edit Breakline edits all aspects of the breakline. The Edit Breakline routine is similar to the PEDIT command of AutoCAD. The second routine, Edit Breakline Description, edits the name of the breakline. Both routines do not distinguish between standard or proximity breaklines.

CONTOUR DATA

Contours provide a convenient way of developing a surface. Since the contours represent known elevations, you need only create a data file that represents the contours and then build the surface.

Surfaces from contour data usually do not have the same triangulation problems found in a TIN from points. Each contour line is in essence a breakline and forces the TIN Modeler to create triangles between each contour.

There are, however, two potential problems with contour surfaces. The first is the loss of high and low points on the surface. The second is a problem with contour peninsulas and bays (see Figure 5–24). In contour loops, the Surface Modeler constructs triangles that connect to the same contour because the data points nearest each other are from the same contour. The result is a flat spot on the surface. When you create a new set of contours from this data, the new contours will cut off the bays and peninsulas, because the contour generator follows the first triangle leg with the same elevation.

The net result of these errors is minimal, and they may even cancel each other out. Land Development Desktop attempts to optimize the diagonals to control this problem. In most situations, the optimization corrects the triangle leg problems; however, there are times when there is no solution to the problem (see Figure 5–25). To fix the problem, you may have to add breakline or point data to the surface data set.

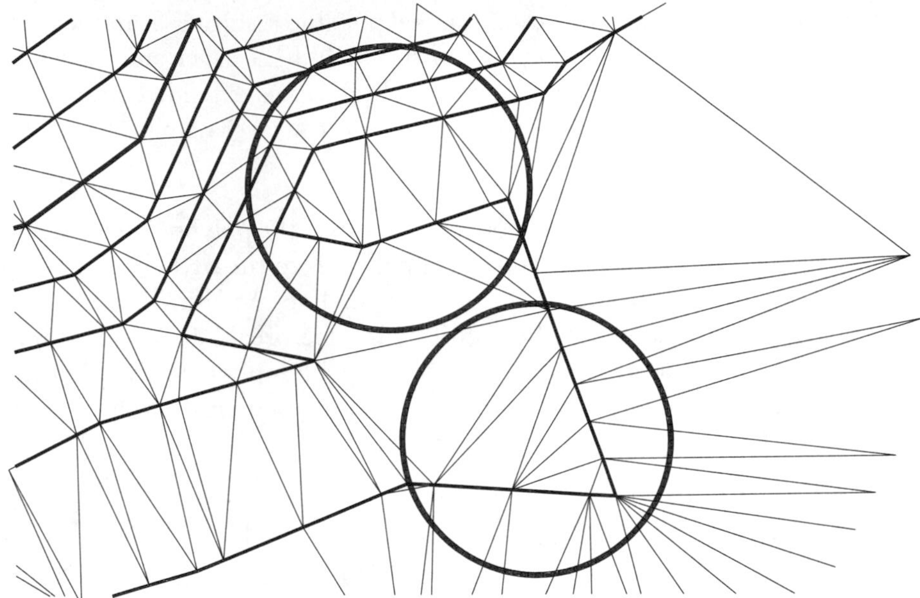

Figure 5–24

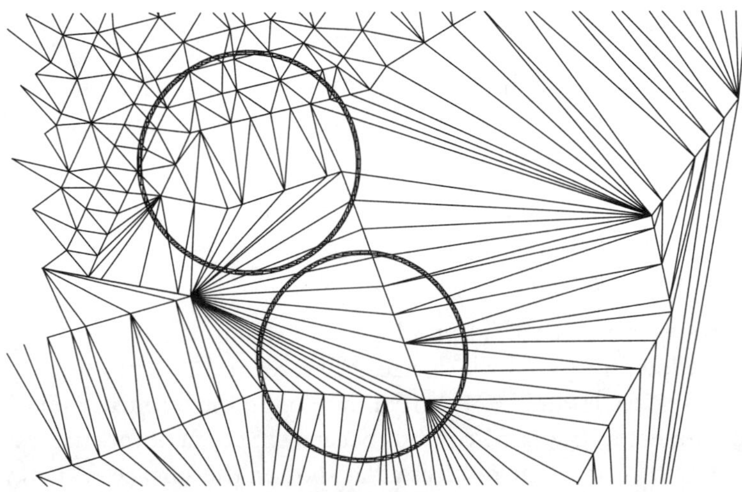

Figure 5–25

The Surface Modeler returns unpredictable results when contour lines cross over each other. Even though this occurs in reality, the TIN algorithm cannot cope with this situation. Sheer walls are a similar situation and without fault data, it is almost impossible to get the TIN Modeler to correctly interpret these features.

The Add Contour Data routine scans for polylines or contour objects in the drawing and creates an external data file for the TIN Modeler (see Figure 5–26). Before the scanning process starts, you must make sure that the Create as Contour Data toggle is on in the Contour Weeding dialog box (see Figure 5–27). The Create as Contour Data toggle makes the data from the contours breakline data. In addition to the data toggle, this dialog box sets weeding and supplementing factors affecting the resulting data. The weeding values influence the removal of vertices from the data set. The supplementing factors affect the adding of new vertices to the data set.

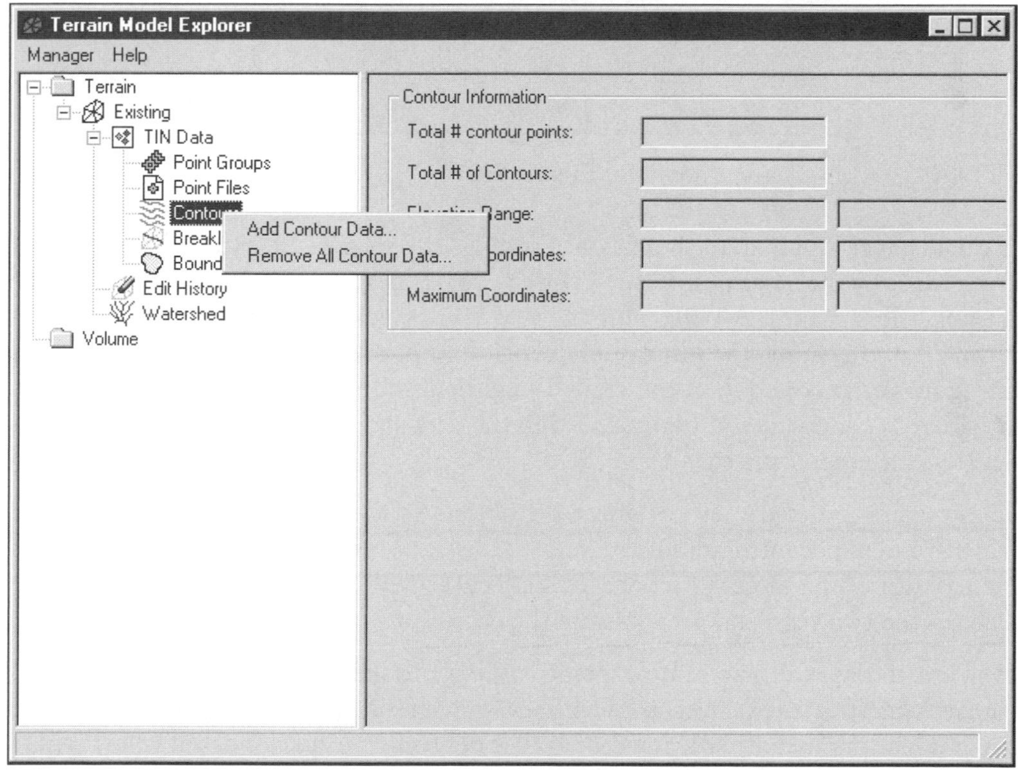

Figure 5–26

The values used by Weeding are always lower than those for Supplementing. Weeding removes potentially redundant polyline vertices. You should read an AND between the two weeding factors. When three adjacent vertices on the contour meet the weeding conditions, the routine removes the middle vertex from the data set (see Figure 5–28).

The first condition, Distance, applies to the length of a polyline segment containing three adjacent polyline vertices. If the overall segment length is less than the distance setting, the analysis can move on to the second condition. If the segment is longer than the distance setting, the routine moves on to find a set of three vertices with a length less than the distance setting.

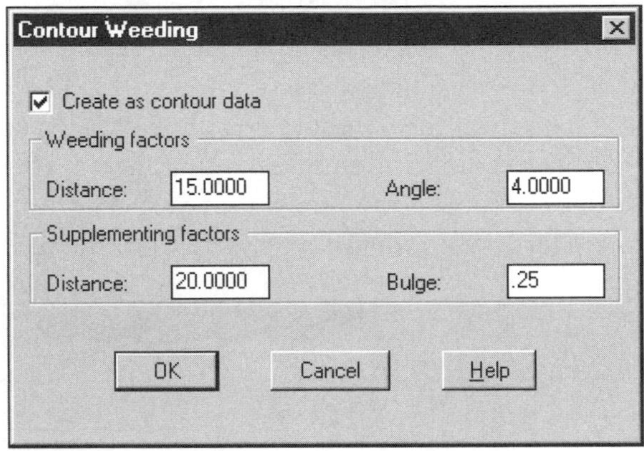

Figure 5–27

If the distance is less than the distance setting, the routine evaluates the deflection angle made by the three points (see Figure 5–28). A deflection angle is the amount of turning a line makes. A deflection angle of 90 degrees is a right turn. If the angle setting is 4, this limits the turning angle to 4 degrees or less to the right or left of the line from vertex 1 and 2. If the angle is less than 4.0 degrees, the second vertex is considered redundant and not included in the data set. If the angle is greater than 4 degrees, the second vertex is kept.

The weeding process does not affect the objects in the drawing, but it determines which vertices become data for the surface file. If two adjacent segments are longer than 10 feet or the three vertices have a deflection angle greater than 4.0 degrees, the data for the two segments is written to the data file.

How much data can you remove before having problems creating a surface? Some papers suggest that removing as much as 50 percent of the vertices produces a viable surface. The number of vertices weeded out is not really the question, but what vertices remain after the weeding process. The only way to evaluate the effects of weeding is to review the resulting triangulation of the surface.

The values used by Supplementing allow you to add new vertices to the data set. The first value relates to the length of a polyline or contour segment and the second to the bulge factor of an arc. The default values are 100.0 feet in length and a bulge factor of 1.0. If a polyline or contour segment is over 100 feet in length, the routine adds vertices to the data set. The routine adds the vertices to the data file, not to the polyline. If a polyline contains an arc segment with a bulge greater than 1.0, the routine adds a vertex to the data set. Again, the routine adds vertices not to the polyline segment but to the data set.

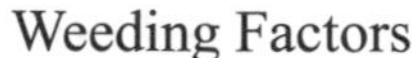

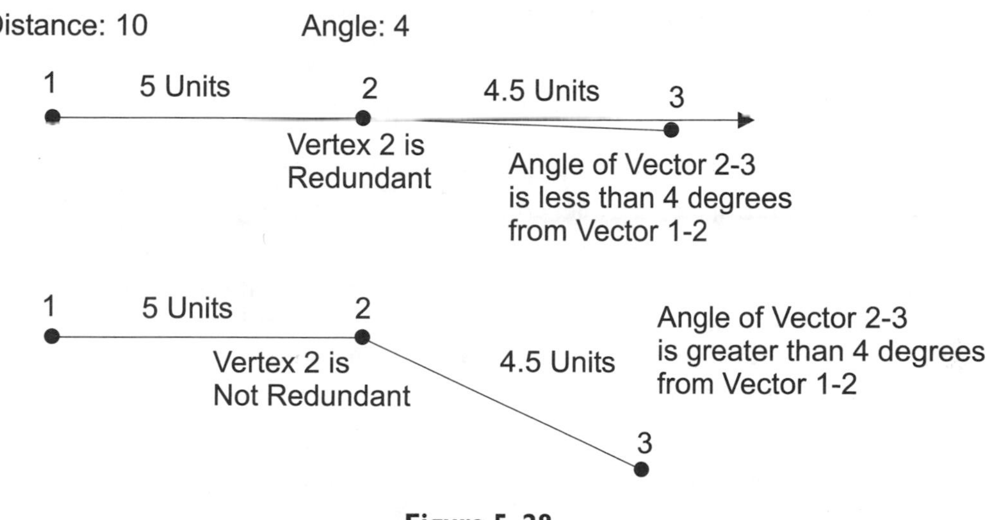

Figure 5–28

There are three methods of collecting contour data. The first method is by selecting a contour from the screen that represents a layer containing contours. The Create Contour Data routine selects all polylines and contours on the layer. The second method is by creating a selection set of contours from the screen. With the last method, you type the name(s) of the layer(s) and the routine selects all the polylines and contours on the layer. This last method allows you to select contours from layers that are either off or frozen.

Be aware that the Create Contour Data routine selects *all* polylines and contours on the identified layers. You must make sure that the polylines selected are truly contours. If not, the surface may contain erroneous data.

Caution: If you receive a contour drawing file that has been outside your control, you should thoroughly check the drawing and all the objects within it before using its data.

Problems with third-party contour data include spurious contours, zero elevation contours, contours with varying elevations (720 to 0 to 720), and crossing contours. You need to correct all such problems before using the data.

EXERCISE

Exercise

After you complete this section you will be able to:

- Use data gathering routines for surfaces
- Build an LDD surface
- Understand the types and functions of breaklines
- Edit breakline data
- Use Quick Cross Section to evaluate the quality of the surface

DEFINING A NEW DRAWING AND PROJECT

1. If you are not in the TERRAIN drawing of the surfaces project, open the file. Currently the drawing is empty.
2. Insert the drawing file, DTM, from the CD. The file is located in the *Terrain* folder. Insert the drawing at 0,0,0, with a scaling factor of 1 and a rotation angle of 0, and explode the drawing upon insertion. This drawing contains boundary lines for a proposed subdivision.
3. Save the drawing.
4. Use the Point Settings routine from the Points menu to set the following point settings:

Insert Tab:

Use Current Point Label Style When Inserting Points: ON

Current Labeling Style: Active Desckeys Only

Update Tab:

Allow Points to be MOVE'd in Drawing: OFF

Description Keys:

Matching Options: Match on Description Parameters: ON

Marker Tab:

Custom Marker Style	+ (plus)
	Size: Absolute 2.0 units
Superimpose	Circle

Text Tab:

Text Style Standard

Size: Absolute 4.0 units

Automatic Leaders: ON

Preferences:

Always Regenerate Point Display After Zoom: OFF

5. If the current Labeling Style is not Active Desckeys Only, select the Settings routine from the Labels menu, select the Point Labels tab, click the drop list arrow for a list of the available styles, and finally select Active Desckeys Only. Select OK to exit the dialog box. This sets the needed current point labeling style.

IMPORTING POINTS

1. Use the Description Key Manager to view the current list of Description Keys. The ASCII file that you will use for this exercise will use the following description keys. Compare this list to your current description key list. Add the necessary codes to the description key list so all of the points in the file will match a key.

Code	Format	Point Layer	Symbol	Symbol-Layer
BLDG	$*	BLDG_PNT		
CL	$*	CL_PNT		
DL	DITCH	TOPO_PNT		
EEFYD	EX FIRE HYD	EUTIL_PNT	E_HYD	EUTIL_SYM
EOP	$*	EOP_PNT		
HDWL	HEADWALL	ESTRT_PNT		
I[PR]F	IRON FOUND	MON_PNT	IP	MON_SYM
LC	LOT CORNER	SETP_PNT	IP	SETP_SYM
LP*	LIGHT POLE	EUTIL_PNT	LUMIN	EUTIL_SYM
MD	MOUND	TOPO_PNT		
MP*	$1 INCH MAPLE	VEG_PNT	CG_T10	VEG_SYM
OK*	$1 INCH OAK	VEG_PNT	CG_T15	VEG_SYM
POND	$*	TOPO_PNT		
PP*	POWER POLE	EUTIL_PNT	U_POLE	EUTIL_SYM
RP	$*	SETP_PNT		
SMH	SAN SEWER	EUTIL_PNT	SMH	TOPO-SYM
SWALE	$*	TOPO_PNT		
TOPO	$*	TOPO_PNT		

2. Run the Import Options routine of the Import/Export cascade of the Points menu and set the What to Do When Point Numbers Are Supplied toggle to USE. After setting the toggle click on the OK button to exit the dialog box.

3. Use the Import Points routine from the Points menu to import the coordinates from a file. The format of the import file is the STDNEZ format (PNEZD Comma Delimited format). The file contains point number, northing, easting, elevation, and description (RAW). Import the file *LDTERRAIN.nez* from the *terrain* folder of the CD.

4. After importing the points, save the drawing.
5. Use the Zoom to Point routine, zoom to point 84, and use the height of 200. Notice the translation of raw descriptions into full descriptions and the insertion of symbols.
6. Use the AutoCAD List routine, list point number 92, and the symbol associated with the point. They should be on different VEG layers.
7. Use the Zoom to Point routine, zoom to point 38, and use the height of 150. Notice the translation of raw descriptions into full descriptions and the insertion of symbols.
8. Use the Zoom Extents routine to view the entire drawing.
9. Save the drawing.

DEFINING A POINT GROUP

The points now in the drawing contain points that do not need to be in the surface. The way the fire hydrant points were surveyed makes them one to two feet higher than their surrounding ground shots. Also, the Terrain Modeler uses only point group points when using points from the point database. Use the Point Group routine to create a point group, Existing, that does not include fire hydrant shots.

1. Select the Point Group Manager routine from the Point Management cascade of the Points menu.
2. Click the Create Group icon in the upper left and enter the name of the group, **Existing**, in the top portion of the dialog box.
3. Select the Build List button to display the Point List dialog box.
4. Select the Advanced button to display the advanced filtering options.
5. In the Advanced filtering dialog box, set the source to All and Add the points to the list. The point numbers 1-228 should appear in the top of the list (see Figure 5–29).
6. Select the Filter tab to view the filtering options. First, toggle on With Description Matching, then set the description type to RAW, and enter **EFYD** as the raw description. The entry must be in uppercase for the match to occur (see Figure 5–30). At the bottom of the dialog box, select the Filter button to select the Fire Hydrant points. When the points appear in the Resulting Point List area, select the Remove button to remove the points from the point list. There will be some confirmation dialog boxes that appear during this process. Simply click on OK to dismiss these boxes.
7. Select the List tab and select the RAW description column header to sort the remaining list of points based on this field. Scroll down the list to the points with the raw description of EOP. The EFYD points are now absent from the list of points.

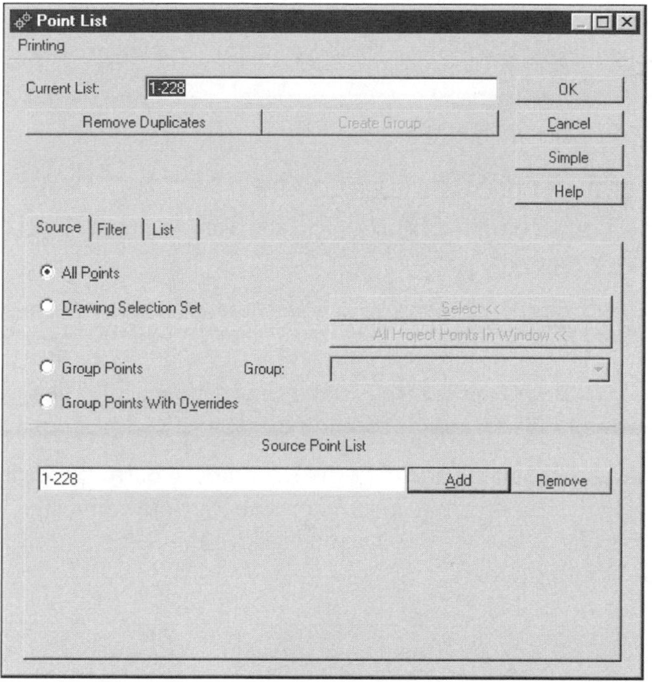

Figure 5–29

Point List
Printing
Current List: 1-228
OK
Remove Duplicates
Create Group
Cancel
Simple
Help
Source
Filter
List
Source: All Points
1-228
Ranging In Numbers From
1
To
1
Ranging In Elevations From
0
To
0
With Name Matching
With Description Matching
EFYD
Raw Desc
Full Desc
With XDRef Matching
XDRef to Search:
Include Matching Points
Exclude Matching Points
Resulting Point List
Filter
149,181,198,216
Add
Remove

Figure 5–30

8. Select OK twice, once to approve the new point list and once to exit the Create Point Group dialog box. There should be a point group Existing in the Point Group Manager.

9. Exit the Point Group Manager.

USING THE TERRAIN MODEL EXPLORER

1. Select the Terrain Model Explorer from the Terrain menu. The routine displays the Explorer window.

2. Select the word terrain, and once it is selected, click the right mouse button, and select Create New Surface.

3. Select the + (plus sign) under terrain and then under Surface1. The first selection exposes a list of surfaces and the second exposes a list of data types a surface can have.

4. Select the Surface1 name and click the right mouse button; this displays a short-cut menu with a list of commands. Select the Rename command and name the surface Existing. Select the OK button to close the Rename dialog box. Use Figure 5–31 as a guide.

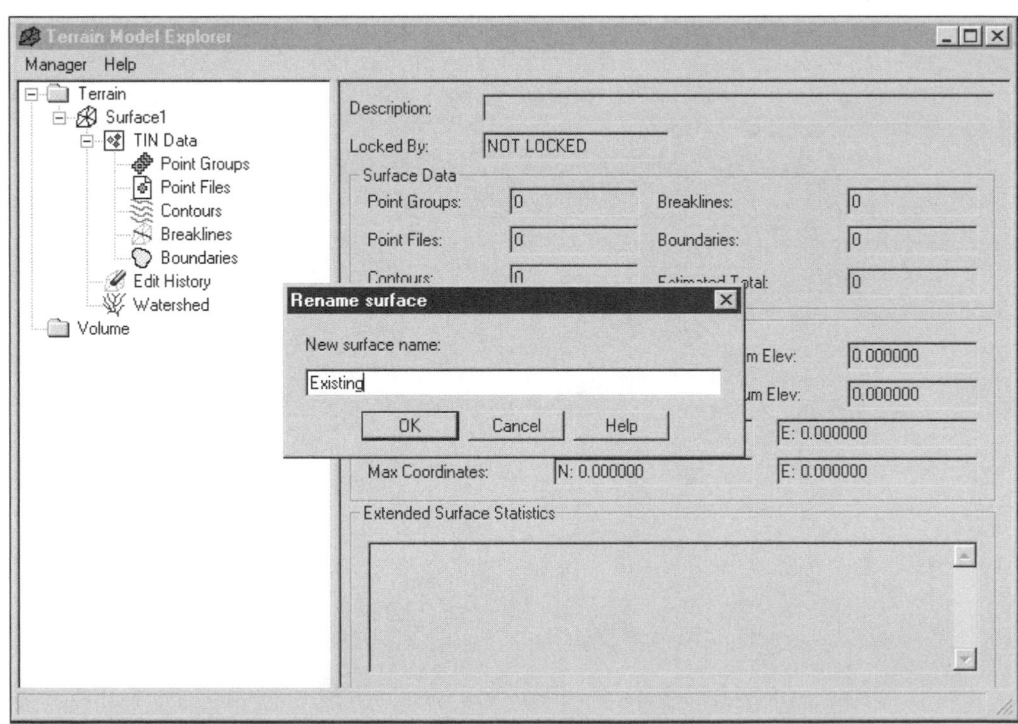

Figure 5–31

5. The surface tree rolls up when exiting the Rename command. Expand the surface data list by selecting the + (plus sign) under the Existing surface name.

6. To add the Existing point group to the surface data list, select Point Groups, click the right mouse button, and select Add Point Group. This displays a dialog box listing the only point group in the project. Select the point group, Existing, and then select the OK button to exit the dialog box.

7. Create the surface using only the point group data. Select the surface name (Existing), click the right mouse button, and select Build (surface). This displays the Build dialog box. Toggle on Compute Extended Statistics and then select OK to triangulate the point data. Use Figure 5–32 to set your build options. When the Modeler finishes triangulating the data, select OK to acknowledge the completion.

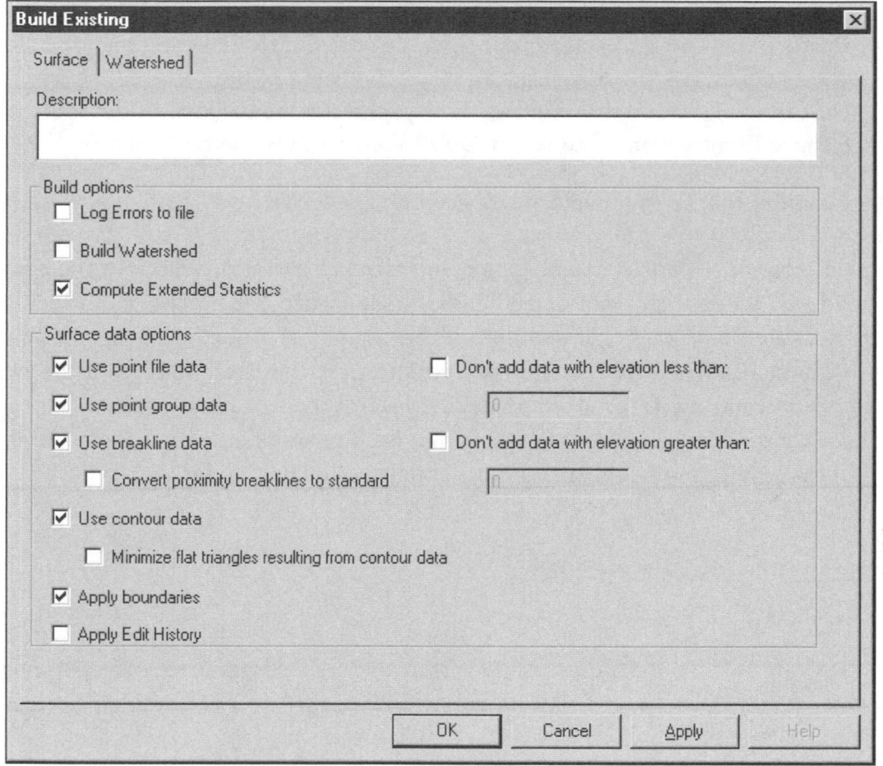

Figure 5–32

8. The results of the triangulation appear in the surface panel to the right of the Terrain tree. The information includes the number of points, minimum and maximum coordinates and elevations, and the surface statistics. Surface statistics includes information on area (2D and 3D), number of triangles, and slopes (minimum, maximum, and mean (average). Review the information in your panel.

9. To view the resulting triangulation, select the name of the surface, Existing, and click the right mouse button, this displays the surface commands. Cascade out the Surface Display menu and select Quick View to view the triangulation.

10. Minimize the Terrain Explorer to view the triangulation. Place the minimized TME in the upper right of the screen away from needed information.

11. Save the drawing.

Where the TIN triangles cross over the road in the southern portion of the drawing, you should be aware that the surface does not interpret the data correctly. At the intersection of the existing and proposed roads, you can see how the Modeler misinterprets the point data.

12. Use the Zoom to Point routine to view the interpretation problems. Zoom to point number 38 using the height of 175.

13. Restore the Terrain Explorer and again use the Quick View command to view the vectors. To view the triangulation, select the name of the surface, Existing, and click the right mouse button; this displays the surface commands. Cascade out the Surface Display menu, and select Quick View to view the triangulation.

14. Minimize the Terrain Explorer to view the triangulation.

This view displays the Modeler connecting points that are a part of a ditch to the centerline of the roadway; for example, points 180 and 38 (see Figure 5–33). The TIN Modeler also connects a swale point on the south side of the road to the northern edge-of-pavement (points 183 and 18). The TIN Modeler fails to interpret the linear nature of the edge-of-pavement, centerline, swale, or ditch points because of the lack of data clearly defining the lineations. Our everyday experiences tell us that each swale point connects to other swale points and that the edge-of-pavement points connect in a linear form.

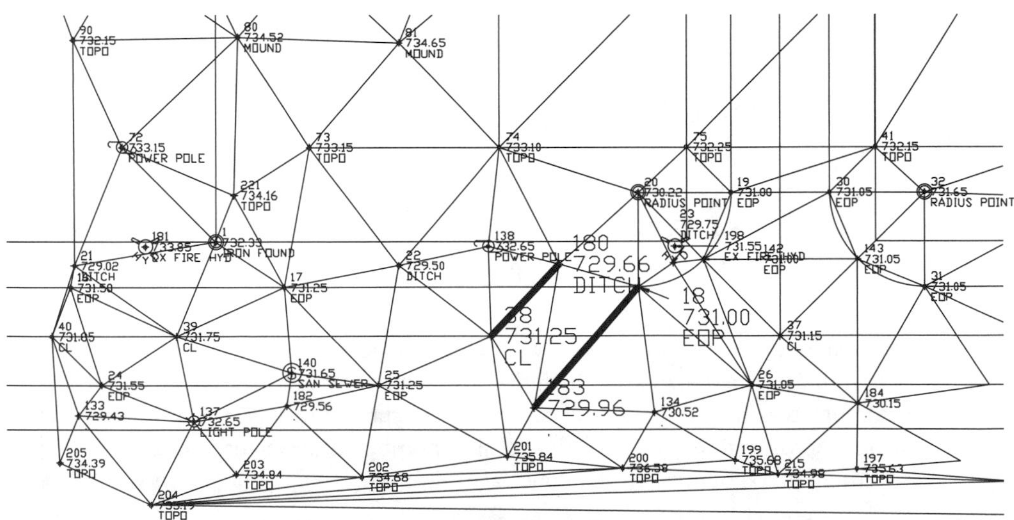

Figure 5–33

QUICK MULTIPLE CROSS SECTIONS

Errors in interpretation show best in sections across the ditch and roadway. A section will show the elevations of the ditch crossing the roadway to the southern edge-of-pavement (point 22 connecting to point 25) and the swale elevations south of the roadway extending to the ditch north of the road (point 183 connecting to point 180). In the following steps you will use the Create Section View routine from the Sections cascade of the Terrain menu to view the surface along and across the roadway.

1. Use the Create Section View routine in the Sections cascade of the Terrain menu to view the surface along and across the roadway. Toggle on ortho mode. Start the section line just to the east of point 205, to north of point 181, east past point 17, south between points 202 and 203, east to a point between 202 and 201, north past point 138, east pass point 180, and then south to a point between 200 and 201.
2. After placing the line into the drawing, the routine creates a long section window displaying the elevations along the line from the surface. The vertical lines are vertices along the section line. Our interest is in the first, third, fifth, and seventh segments along the line. These segments cross the roadway and do not display an appropriate road cross section.
3. Move the pointer into the section window. As you move the pointer, the distance along the line and the elevation from the surface is shown in the top portion of the window.
4. Select the Section menu from the section window. The Section menu allows you to save the section to the clipboard for use in another window application, save the section as a file (Windows Metafile), or to change the view properties of the section (color and grid settings).
5. Minimize the Section View dialog box. It should say Existing in the upper left corner.
6. Select OK to exit the View Properties dialog box.
7. Save the drawing.

ADDING BREAKLINE DATA

To control the misinterpretation of data, LDD uses breaklines. You need to use breaklines for the centerline and edges-of-pavement, the ditch on the north side of the road, and the swale along the southern roadway edge.

Standard–By Point Number

1. If you have panned or zoomed to view the triangulation in additional areas of the drawing, use the Zoom to Point routine to return to the triangulation problems along the roadway. Zoom to point number 38 using the height of 175.
2. If necessary, redraw the screen to remove the ghost surface triangles. The white line of the section will remain on the screen. Do not confuse this line with the triangulation lines.

EXERCISE

3. From the Points menu display the Edit Points cascade and select the Display Properties routine. Select all of the points, toggle off elevations in the Text tab, and select OK to exit the Properties dialog box. The points now display only point number and description. You may have to select some points and reposition them in the display to clearly see point numbers and descriptions.

4. Use the By Point routine to create the north ditch breakline. You can find the routine by restoring the Terrain Explorer, selecting breaklines, and selecting Define By Point (see Figure 5–23). Name the breakline ditch. The point numbers 21,22,180, and 23 define the breakline. Use Figure 5–34 as a guide. When the routine prompts you about deleting an object, select Yes. The routine draws the line on the current layer. If you answer No to the prompt, you will have the original object on the current layer and a copy on the SRF-FLT layer.

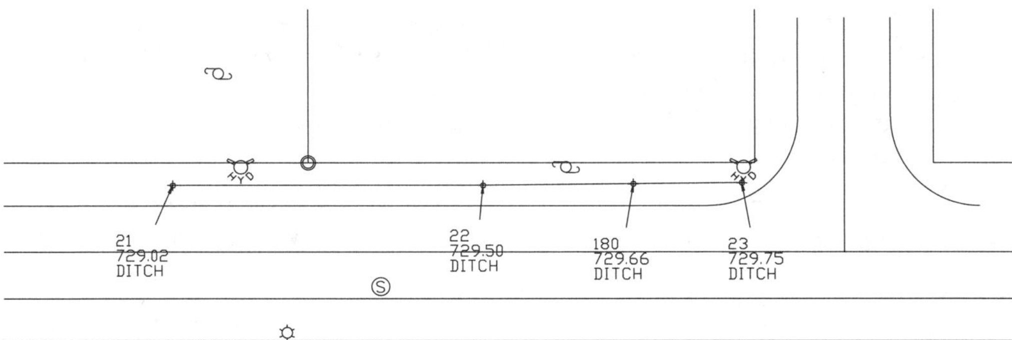

Figure 5–34

5. Use the same routine to create the north edge-of-pavement breakline. Name the breakline NEOP. The point numbers 16, 17, 18, 142, and 19 define the breakline. Use Figure 5–35 as a guide. Again, answer Yes to the deletion prompt.

6. The routine prompts for the starting point of the next breakline. Press ENTER to exit the routine and reenter the Terrain Explorer.

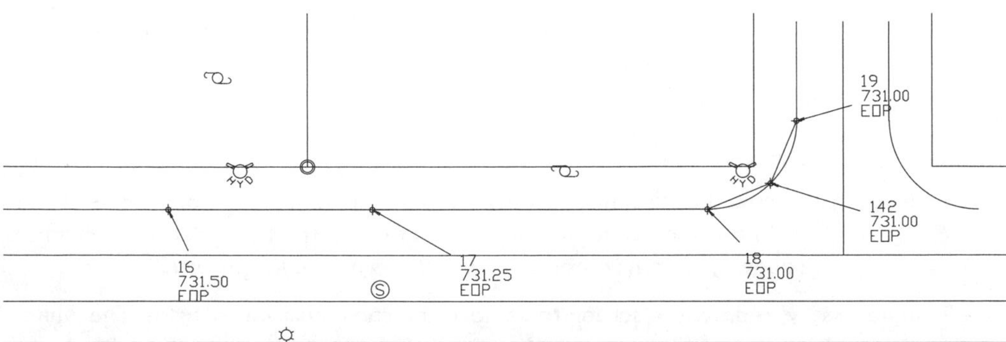

Figure 5–35

EXERCISE

7. Select Breaklines and click the right mouse button to display the breakline options (see Figure 5–23). Select List Breaklines to view the breakline list. The Terrain Explorer displays a list of defined breaklines. Select the ditch breakline and then select List. The List button displays the ditch data definition (see Figure 5–36). At the bottom of the dialog box is the number, name, and type of breakline being viewed.

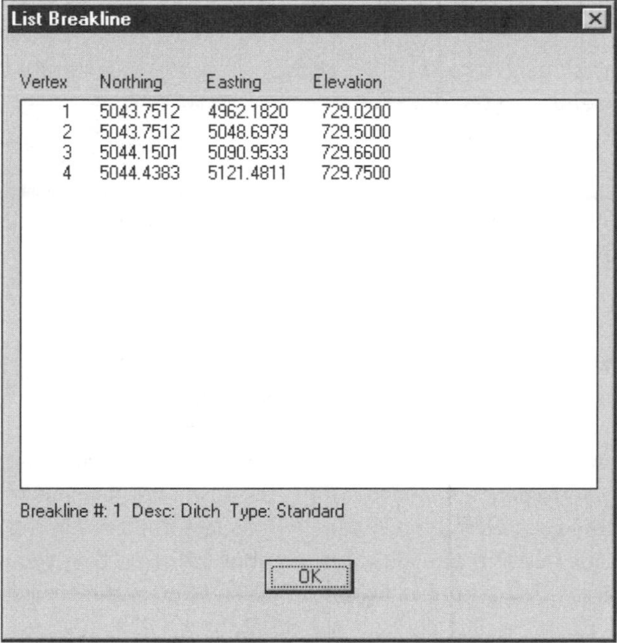

Figure 5–36

8. Select OK to exit the List dialog box. You return to the list of defined breaklines. In the List dialog box, select Cancel to return to the Terrain Explorer. The Surface data area now contains the number of data points the breaklines contribute to the surface.

9. Select surface name Existing, click the right mouse button, and select Build (surface) to rebuild the surface. When the Build Existing dialog box appears, select Use Breakline Data if it is not toggled on (see Figure 5–32). Select OK to build the surface and again to exit the Done building surface dialog boxes. The Surface now contains the new breaklines.

10. After rebuilding the surface, select the surface's name, Existing, click the right mouse button, cascade out the Surface Display menu, and select the Quick View routine to view the new triangulation. Minimize the Terrain Explorer to view the new triangulation. The new triangulation should use the DITCH and NEOP breaklines as points and legs of triangle. However, no triangles should cross over the breaklines.

11. Restore the Section View dialog box and use the Update Section Views routine to view the impact of the new triangulation on the sections. The changes to the triangulation show the ditch and edge-of-pavement more clearly in the first, third, fifth, and seventh segments along the section line. The Update Section Views routine is necessary because you are not changing the location of the section line. In this case the section line is not aware of the fact you made changes to the surface. If you changed the section line, grip edit it, it immediately rescans the surface to update the displayed profile.

12. Redraw the screen to remove the ghost triangulation, minimize the Section View dialog box, and save the drawing.

By Point Number

1. Use the Zoom to Point routine to set the next display. Zoom to point 37 with a height of 300.

2. Restore the Terrain Explorer to the screen.

3. Use the By Point Number routine to define the next breakline. Select Breaklines, click the right mouse button, and select Define By Point Number (see Figure 5–23). This routine allows you to enter a list of point numbers rather than selecting points from the screen. The point numbers can be individual, sequential, or a mix of the two. Individual numbers are separated by a comma and sequential numbers have a dash (-) separating the lowest and highest numbers in the sequence. The first breakline, CL, has one individual number followed by a sequential list. The numbers are 177, 35–40. See Figure 5–37. This starts the breakline at point 177, and continues it sequentially from points 35 to 40. After you press ENTER, the routine prompts for the breakline name. Enter **CL**, and again press ENTER. The routine creates the breakline in both the drawing and the external breakline data file. The routine returns to the point numbers prompt.

4. At the point numbers prompt. enter the next set of point numbers to create the south edge-of-pavement breakline. The point numbers for the breakline are: 24–29,178. Name the fault SEOP. See Figure 5–37.

5. At the point numbers prompt, enter the next set of point numbers to create the swale breakline south of the edge-of-pavement breakline. The point numbers for the breakline are: 190, 191, 185, 184, 134, 183, 182, and 133. Name the fault SWALE. See Figure 5–37.

6. Press ENTER to exit the Define By Point Number routine and reenter the Terrain Explorer.

EXERCISE

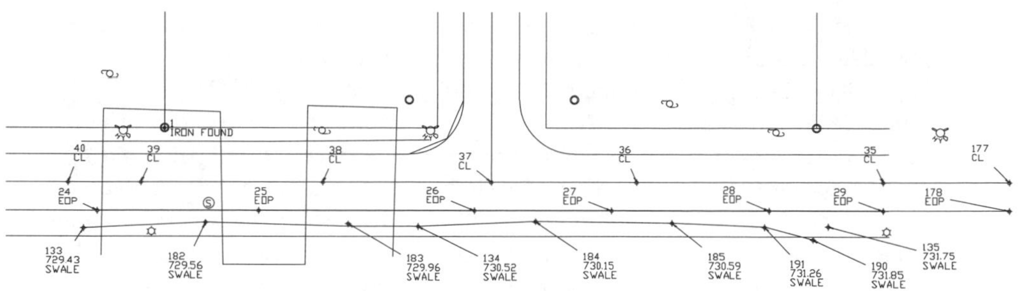

Figure 5–37

Proximity (2D) Breaklines

A Proximity breakline is a 2D polyline. The polyline has no elevation. The elevations for the breakline come from the point objects at each vertex in the polyline. The polyline defines the linear portion of the breakline and the point objects define the elevations.

1. Minimize the Terrain Explorer.
2. Use the Zoom to Point routine to zoom to point 2. Use the height of 150 to view the north edge-of-pavement for the east portion of the road. Use Figure 5–38 as a guide.

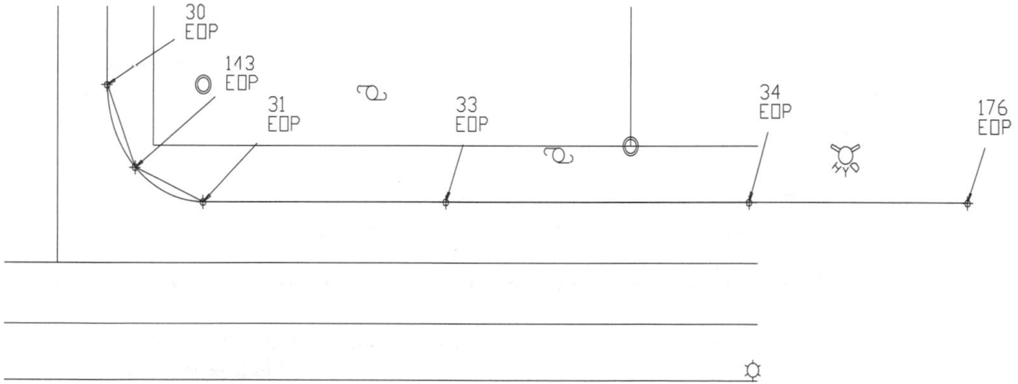

Figure 5–38

3. Restore the Terrain Explorer.

4. Use the Draw Proximity Breakline routine to define the next breakline. Select Breaklines, click the right mouse button, and select Draw Proximity Breakline (see Figure 5–23). The routine starts by prompting for a point. This is an AutoCAD coordinate point. You can use any appropriate object snap or you can use the dot toggles to select points by numbers, graphically, or by northing/easting values (.p, .g, and .n). If you toggle into point number mode (.p), you enter the point numbers. If you select graphically (.g), you select any part of the point object to select the point. Northing and easting are not really an option in this case. Whichever method you choose, you need to select points 30, 143, 31, 33, 34, and 176. After you enter the points, the routine prompts for a name. Name the breakline NEOPE. Press ENTER after entering the name and select Yes to delete the polyline the routine drew on the current layer. The routine then prompts for a new starting point defined by a point number. Press ENTER to exit the routine and reenter the Terrain Explorer.
5. Minimize the Terrain Explorer.
6. Use the AutoCAD List routine to view the object just drawn. The object is a 2D polyline.
7. Restore the Terrain Explorer, select breaklines, click the right mouse button, and select List Breaklines. In the List Breaklines dialog box, select NEOPE and then select the List button. The breakline has only vertices and their coordinates for data. The report at the bottom indicates that it is a proximity fault. First select OK and then the Cancel button to return to the Terrain Explorer.
8. Select surface name Existing, click the right mouse button, and select Build (surface) to rebuild the surface. When the Build Existing dialog box appears, make sure Use Breakline Data is on. The toggle below this selection converts the proximity breaklines to 3D faults. For this time, do not toggle on the conversion. Select OK to build the surface and again to exit the Done building surface dialog boxes. The Surface now contains the new breaklines.
9. Use the Zoom to Point routine to zoom to point 37 with a height of 300.
10. Next select the surface's name in the Terrain Explorer, Existing, click the right mouse button, cascade out the Surface Display menu, and select the Quick View routine to view the new triangulation. Minimize the Terrain Explorer to view the new triangulation. The new triangulation should use all of the breaklines as points and legs of triangles. However, no triangles should cross over the breaklines (see Figure 5–39).
11. Redraw the screen to remove the ghost triangles.
12. Use the Zoom to Point routine to zoom to point 53 with a height of 150.

EXERCISE

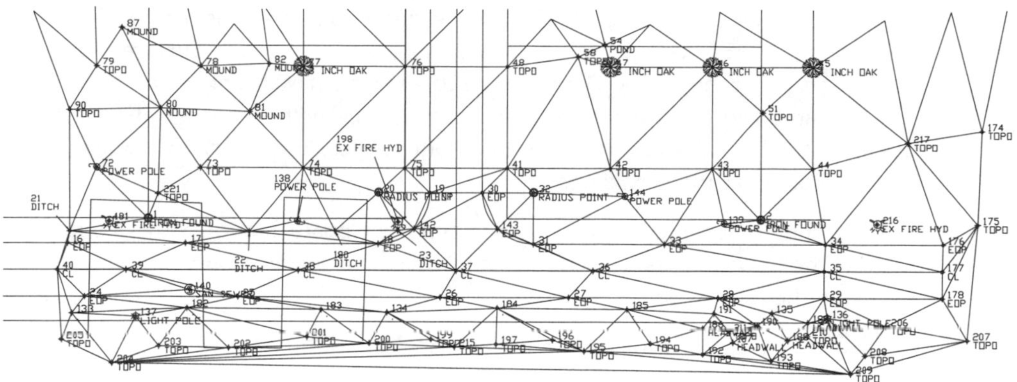

Figure 5–39

13. Create a Pond breakline with the Draw Proximity Breakline routine. Restore the TME, select Breaklines, click the right mouse button, and select Draw Proximity Breakline (see Figure 5–23). The routine starts by prompting for a point. This is an AutoCAD coordinate point. You can use any appropriate object snap or you can use the dot toggles to select point by numbers, graphically, or by northing/easting values (.p, .g, and .n). If you toggle into point number mode (.p), you enter the point numbers. If you select graphically (.g), you select any part of the point object to select the point. Northing and easting are not really an option in this case. Whichever method you choose, you need to select points 53, 54, 55, 56, 57, and Closing back to 53. You do not need to toggle out of the dot mode if you are closing. Just type the letter **c** at the point number prompt and the routine will close the breakline. After you enter the points, the routine prompts for a name. Name the breakline Pond. Press ENTER after entering the name and select Yes to delete the polyline the routine drew on the current layer. The routine then prompts for a new starting point defined by a point number. Press ENTER to exit the routine and reenter the Terrain Explorer.

14. In Terrain Model Explorer, select the surface name Existing, click the right mouse button, and select Build (surface) to rebuild the surface. When the Build Existing dialog box appears, select Use Breakline Data if it is not toggled on (see Figure 5–32). The toggle below this selection converts the proximity breaklines to 3D faults. Toggle on the conversion of 2D breaklines into standard. Select OK to build the surface and again to exit the Done building surface dialog boxes. The Surface now contains the new breaklines.

15. Minimize the Terrain Explorer.

16. Use the Zoom to Point routine and zoom to point 37 with a height of 400.

17. Restore the Terrain Explorer and select the surface's name, Existing, in the Terrain Explorer. Click the right mouse button, cascade out the Surface Display menu, and select the Quick View routine to view the new triangulation. Minimize the Terrain Explorer to view the new triangulation. The new triangulation should use all of the breaklines as points and legs of triangles. However, no triangles should cross over the breaklines.

18. Restore the Section View dialog box and run the Update Section Views routine to view the new cross section. The ditch, swale, and roadway should be clearly shown in the long section.

19. Redraw the screen to remove the ghost triangles.

20. Erase the section line. This will also erase the Section View dialog box.

21. Save the drawing.

EXERCISE

Creating a Wall Fault

Another problem in the TIN occurs with the headwall at the easterly end of the road (see Figure 5–40). The TIN Modeler does not interpret the headwall correctly. There are lines connecting points across the headwall (point 190 connecting to 193 and point 211 connecting to 192). The TIN lines crossing the headwall create the same problem found along the roadway edges and ditches. As with most lineations, you may (will) have to control the triangles that enter and leave the wall area. The Define Wall Breaklines routine creates a breakline that has both the bottom and top elevations of the wall. While running the Define Wall Breaklines routine in the next exercise, you define the top elevations of the wall as absolute elevations and the bottom of the wall as either absolute elevations or as differences in elevation.

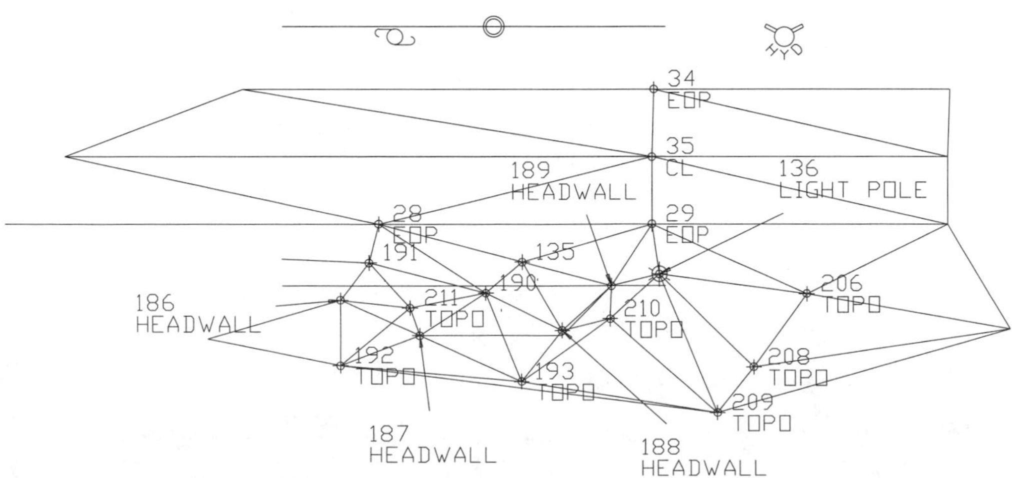

Figure 5–40

The Wall routine uses an existing polyline that connects the wall points, so if you draw a 2D polyline defining the top of the wall, the routine prompts you for the elevations of the top. If you draw a 3D polyline, the routine prompts you to verify the elevations found on the polyline.

After selecting the wall polyline, you specify an offset direction to the bottom of the wall. Then the routine prompts for the elevations at the top and bottom of the wall. The routine creates blip arrows pointing away from the offset side to get or verify the wall elevation and towards the offset side to get the offset side's elevations.

The Wall routine specifies the offset elevation in two ways: as an absolute elevation and as an elevation difference. Elevation difference is a measure of the distance between the top and bottom of a wall. This measured distance allows the routine to calculate the needed elevations along the wall. A positive value for the offset side means that the offset side is the top of the wall. A negative difference means that the offset side is the bottom of the wall.

If the polyline is a 2D polyline, enter the elevations for the polyline and the offset side. If the polyline is a 3D polyline, review the existing polyline elevation and enter the elevation for the offset side. If you want, you can change all of the elevations.

EXERCISE

1. Before creating the Wall breakline, restore the Terrain Model Explorer, select the surface Existing, and review the current slope values for the surface. The current slope values will be severely impacted by the wall slopes. Make a note of the current values; they should be similar to the following:

   Minimum Grade: 0.00%

   Maximum Slope: 205.66%

   Mean (Average) Slope: 6.40%

2. Use the Zoom to Point routine and zoom to point 135 with a height of 150.
3. In the TME select the Existing surface, press the right mouse button, cascade out the Surface Display menu, and select Quick View. Notice how the TIN lines cross the headwall.
4. Use the point label grip to rearrange the point labels to better view the headwall points, 186, 187, 188, and 189. You move the label by first activating the grip by selecting the point, next selecting the grip box, and then moving the point to the new location. If you move the point far enough away from the coordinate location a leader will appear between the coordinate point and the new location of the label.
5. Select points 186–189 to activate their grips. Click the right mouse button to display the shortcut menu. From the shortcut menu, select Display Properties, then the Text tab, and toggle on elevations. Select OK to exit the dialog box.

6. Use the Create By Elevation routine of the 3D Polylines cascade of the Terrain menu to create a 3D polyline representing the headwall. When the routine prompts for a From point, toggle into the graphic mode (.g). Select points 186–189 and accept their elevations, toggle out of graphic select by typing in **.g**, enter an **X** at the Select object prompt, and press ENTER to exit the routine.

7. Restore the Terrain Explorer, select Breaklines, click the right mouse button, and select Define Wall Breaklines. The first prompt is to name the breakline. Name the breakline Headwall and press ENTER. Next, select the 3D polyline. When the routine asks for the offset side, select a point near point 135 (see Figure 5–40). Follow the prompts. When the blip arrow points downward, verify the elevation, and when the blip points up, enter the offset side elevations as differences in elevation. When you have completed answering all the prompts, the routine asks for another wall line. Just press ENTER to exit the routine and to return to the Terrain Explorer.

   The elevations and differences for points 1186–1189 are:

   1186 - 735.26 Dif -2.5

   1187 - 735.21 Dif -3.25

   1188 - 735.24 Dif -3.5

   1189 - 735.18 Dif -2.35

8. Rebuild the surface. Select Existing in the Explorer, then click the right mouse button, select Build (surface), select OK to build the surface, and again select OK to exit the Done dialog box to return to the Explorer.

9. Next select the surface's name in the Terrain Explorer, Existing, click the right mouse button, cascade out the Surface Display menu, and select the Quick View routine to view the new triangulation. Minimize the Terrain Explorer to view the new triangulation. The new triangulation should triangulate the headwall correctly.

EXERCISE

Edit Fault

The Edit Breakline routine is similar to the PEDIT command of AutoCAD. The routine starts with the first vertex and marks the current vertex with an X. The Next and Previous options move the X to different vertices on the breakline.

Currently, the Swale breakline out of the headwall does not include point 135. The Edit Breakline routine allows you to add new vertices to the breakline definition.

1. Redraw the screen to remove the ghost triangles.

2. Restore the Terrain Explorer.

3. In the Terrain Explorer, select Breaklines, click the right mouse button, and select Edit Breakline (see Figure 5–23). Select the swale breakline. The routine displays an "X" at the beginning of the breakline. Point 135 represents a new vertex on the breakline after point 190. The new vertex is inserted between the beginning

point and the next point on the breakline. Type the letter **I** to set the insert option. Change the prompt to point number by using the point number dot toggle (.p). Identify point 135 as the new point. Accept the elevation by pressing ENTER. The elevation displayed is from the entry in the point database. Type **X** and press ENTER twice to exit the routine. When you exit the routine, it updates the external fault data file.

4. Rebuild the surface. Select Existing in the Explorer, then click the right mouse button, select Build (surface), select OK to build the surface, and again select OK to exit the Done dialog box to return to the Explorer.
5. Next select the surface's name in the Terrain Explorer, Existing, click the right mouse button, cascade out the Surface Display menu, and select the Quick View routine to view the new triangulation. Minimize the Terrain Explorer to view the new triangulation.
6. Redraw the screen to remove the ghost triangles.
7. Restore Terrain Explorer and dismiss it by clicking the "X" in the upper right corner or by selecting the Manager menu and then Exit.
8. Save the drawing.

EXERCISE

UNIT 3: EDITING A SURFACE

Editing the TIN occurs after the general approval of the surface. After processing all the surface data, it is time to review critical areas on the surface or inspect for spurious triangles. These false triangles occur when the TIN Modeler is unable to create the correct triangulation from all of the original data. As mentioned in the last unit, TINs from points need breaklines to preserve lineations. Not all problems are obvious, and the resulting surface will have triangles with diagonals that are wrong (see Figure 5–19). These triangles need to have their diagonals flipped to the second solution. The TIN Modeler may create spurious triangle legs around the periphery of the site. These legs need to be deleted because they may adversely affect contour generation. They may cause the contours to wrap or become chevrons at the edge of the surface. If there are places where point data is missing, the points can be edited into the surface. You can add or remove point data with the surface editing tools. You must edit the lines and points representing the surface with tools from the Edit Surface cascade of the Terrain menu. Editing the lines representing the surface with generic AutoCAD commands does not affect the surface definition.

When you edit a surface, you are editing a copy of the surface in memory. The edits you make are immediately a part of the surface copy. You must save the surface to make the changes permanent. This is done in the TME or with the Save Current Surface routine of the Terrain menu. If you decide not to keep or want to undo your changes, the process of undoing you editing is difficult. It would be better discard the changes and start the editing process over again. If you do not want to keep the changes you made to a surface, close the surface by selecting the surface name in the TME, pressing the right mouse button, and selecting Close. The routine asks you about the pending edits and when you answer no to the save changes prompt the edits are discarded.

The Edit Surface cascade contains three diagonal editing routines (see Figure 5–41): Add Line, Delete Line, and Flip Face. You must exercise caution with the Delete Line routine because it voids the elevation information between the four points that make up the two triangles sharing the common diagonal. If you delete a diagonal TIN line in the interior of a TIN, a hole occurs in the surface. A TIN hole manifests itself as a blank area in a 3D grid or as contours that disappear and reappear along a TIN line. Once you delete a TIN line, the only way to restore its elevations is to rebuild the surface using the Build (Surface) routine. Rerunning the Build (Surface) routine voids all editing of the surface. The Add Line routine of Edit TIN will not restore the elevation information to the four points that the Delete Line voided.

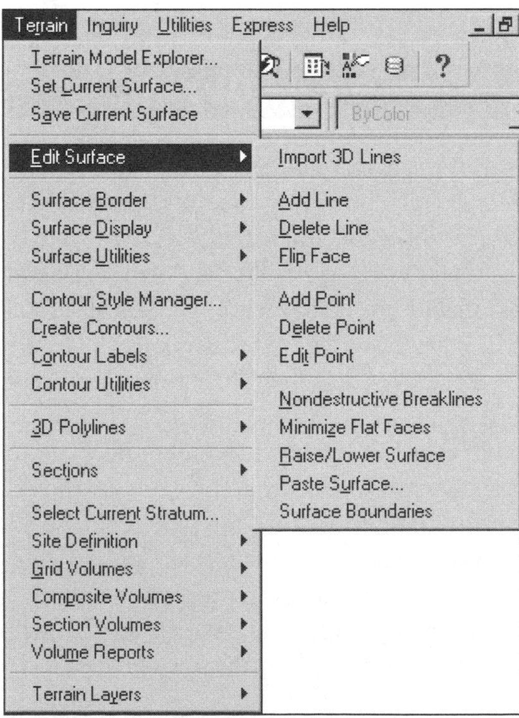

Figure 5–41

The cascade also contains three point editing tools: Add Point, Delete Point, and Edit Point. These routines manipulate the points in a surface. The Add Point routine allows you to add points to a surface that were not a part of the original data set.

As mentioned earlier, if you rebuild a surface at any time, you destroy all of the editing done to that surface. LDD tracks the changes you edit into a surface and allows you to apply them when rebuilding the surface, so all is not lost. However, if rebuilding the surface includes new data or data not included originally, the results of applying the edit history will be unpredictable. LDD will warn you when you attempt to rebuild a surface after it has been edited.

FLIP FACE

After evaluating the data, you may find that some diagonals in the surface are wrong. You may want to change an incorrect diagonal; that is, flip the diagonal connection. The easiest routine to do this task is Flip Face (see Figure 5–41). The routine flips the diagonal to the other two points in the group of four. As a result, the new diagonal changes the interpretation between the four points (see Figure 5–19).

ADD LINE

You can use a second routine to change diagonals, Add Line (see Figure 5–41). With the Add Line routine, you place the new line across the existing diagonal and when you exit the routine, it replaces the old diagonal with the new one.

If you have a surface with a very small number of points, you may have fewer triangles than you anticipated around the periphery of the surface. This occurs because of an algorithm within the Terrain Modeler attempting to limit the number of spurious periphery triangles. You may have to manually add the triangles in to make the surface valid.

DELETE LINE

If you want to void surface triangles and their elevations, use the Delete Line routine (see Figure 5–41). The need to delete lines occurs mainly on the periphery of the surface. Since the TIN Modeler tends to create spurious triangles around the edge of the surface, you need to eliminate these triangles because they do not result from correct interpretations. The Delete Line routine deletes the elevations from the side of the triangles facing the deleted line. The sides of the lines that face away from the deleted line retain their elevation data.

The Delete Line routine can create a problem within a surface: holes. A TIN hole is a group of TIN triangles within the body of the TIN that have no valid elevation. The deleting of interior TIN lines results in the voiding of the elevations within the body of the TIN. The Add Line routine does not heal this elevation void. A symptom of a hole is contours that abruptly stop at a point defined by a TIN line. The only way to restore the elevations to the hole is to rebuild the surface.

SURFACE BOUNDARIES

You can also control the spurious triangles and triangulation by drawing a polyline boundary around areas defining the surface. This boundary line accomplishes two goals: the control of spurious triangles and the exclusion of surface data. Also, a boundary can define a pond, lake, or building footprint you want to exclude from triangulation.

ADD A POINT

The Add a Point routine allows you to add new point data directly into a surface (see Figure 5–41). The point can be from the point database or one that you manually locate and assign an elevation.

DELETE POINT

The Delete Point routine deletes a point from the surface (see Figure 5–41). If the point is from the point database, you specify the point number and the routine removes the point from the surface. Otherwise, you identify the point with a selection and the routine deletes the point from the surface.

EDIT POINT

The Edit Point routine lets you modify the elevation of the point within the surface (see Figure 5–41). The routine modifies individual points or a group of selected points.

EXERCISE

Exercise

After you complete this exercise you will be able to:

- Use the Flip Face routine
- Use the Delete Line routine

FLIP FACE ROUTINE

The easiest method for changing the diagonals of a surface is the Flip Face routine. This routine automatically repositions the diagonal to the next pair of points.

1. Execute a Zoom Extents to view the current drawing, TERRAIN of the surfaces project.
2. Click the Layer icon and open the Layer dialog box. Select the layers Existing-SRF-FLT, CL, LOT, and RDS and freeze them. Select OK to exit the Layer dialog box.

3. If there are any points in the drawing, remove all of the points. Insert back into the drawing point 214, 11, and 179.
4. From the Edit Surface cascade of the Terrain menu, select Import 3D Lines. This brings the surface triangles into the drawing as objects you can select and edit.
5. Use the Zoom to Point routine, and zoom to point 214 using the height of 175.
6. Use the Flip Face routine to change a few surface diagonals.
7. Restore or open the Terrain Explorer, select Edit History, and view the log of surface edits.
8. Minimize the Terrain Explorer.

DELETE LINE ROUTINE

The Delete Line routine is for the periphery lines of the TIN. The TIN Modeler occasionally generates spurious TIN lines or connects points with questionable associations at the edge of the data. The Delete Line routine voids the surface information that is a part of the deleted TIN Lines. The routine voids the elevations from adjacent lines facing the deleted line.

1. Use the Zoom to Point routine, and zoom to point 11 using the height of 250.
2. Use the Delete Line routine of the Edit Surface cascade of the Terrain menu to delete the two triangles to the northeast of point 11. Use Figure 5–42 as a guide.

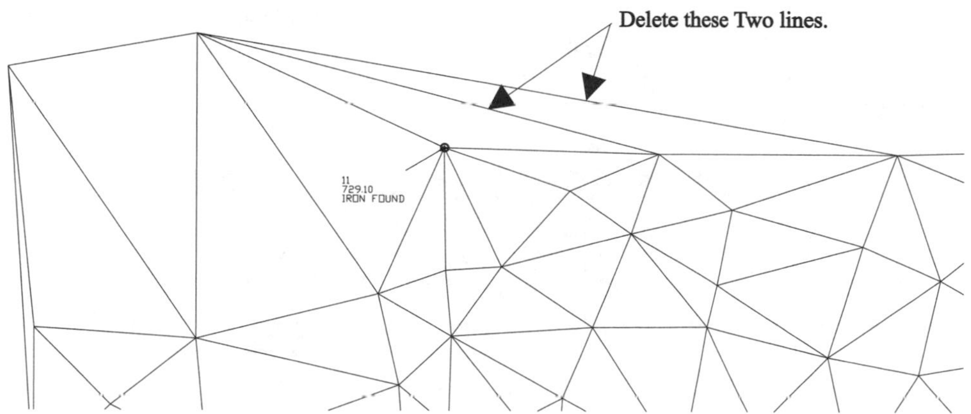

Figure 5–42

3. Use the Zoom to Point routine, and zoom to point 179 using the height of 250.
4. From the Edit Surface cascade of the Terrain menu, select the Delete Point routine. To be able to delete point 179, a point in the point database, toggle the routine into point number mode with the *.p* dot toggle. When prompted for the point number, you enter **179**. The routine deletes the point and rebuilds the surface. Press ENTER to exit the routine. The surface has a new solution based upon the deletion of point 179.

EXERCISE

5. Use the Add Point routine from the Edit Surface cascade of the Terrain menu. This routine allows you to add new or deleted points into the surface. Add point 179 back into the surface. Because you left the last routine in point number mode, this routine will start in point number mode. When the routine asks you to verify the elevation of point 179, press ENTER to accept the value.
6. Toggle out of point number mode by typing **.p** and pressing ENTER.
7. While you are still in the Add Point routine, select a point north and west of point 179. The next prompt is for an elevation of the new point. Use the elevation of **735** for the new point. The routine adds the new point and its elevation into the surface.
8. Exit the Add Point routine by pressing ENTER. The point is added to the surface, but not to the point database.
9. Restore Terrain Explorer to view the edit history. If you cannot see the point deletions and additions press the F5 key on the keyboard to refresh the history panel. After viewing the Edit History, close the Terrain Explorer.
10. Perform a Zoom Extents to see the entire surface. Then use the dynamic zoom tool to show more space around the surface triangles.
11. Reopen the Terrain Explorer and rebuild the surface without using any of the edit history. The Build (Surface) routine issues a dialog box warning you about the potential loss of edits done to the surface (see Figure 5–43). Select Yes to continue, make sure the toggle for Edit History is off in the Build Existing dialog box, and then select OK to build the surface. Select the OK button to dismiss the Done dialog box.

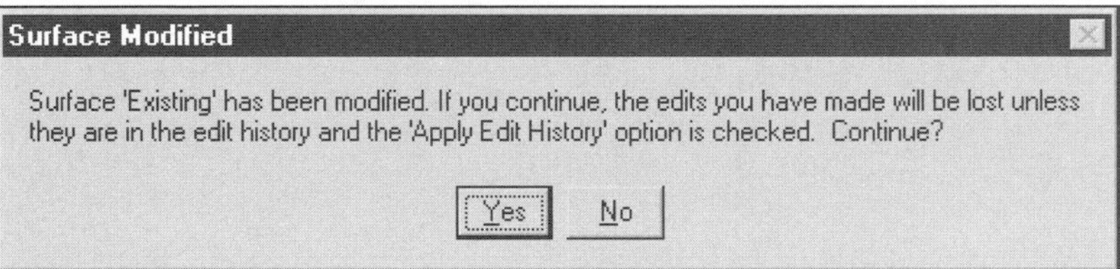

Figure 5–43

12. Minimize the Terrain Explorer.
13. Use the AutoCAD ERASE command to erase everything on the screen.
14. Use the Import 3D Lines routine of the Edit Surface cascade to import the current TIN as objects. The surface should be back to its original state.
15. Save the drawing.

EXERCISE

SURFACE BOUNDARIES

Another way of deleting peripheral triangles is to create a surface boundary. A surface boundary will also exclude points from a surface. So the surface boundary routine can act like delete points.

1. Draw a boundary line around the surface similar to the one found in Figure 5–44. Use the POLYLINE command of AutoCAD to draw the line. You will need to toggle off object snaps, object tracking, and polar to draw the polyline.

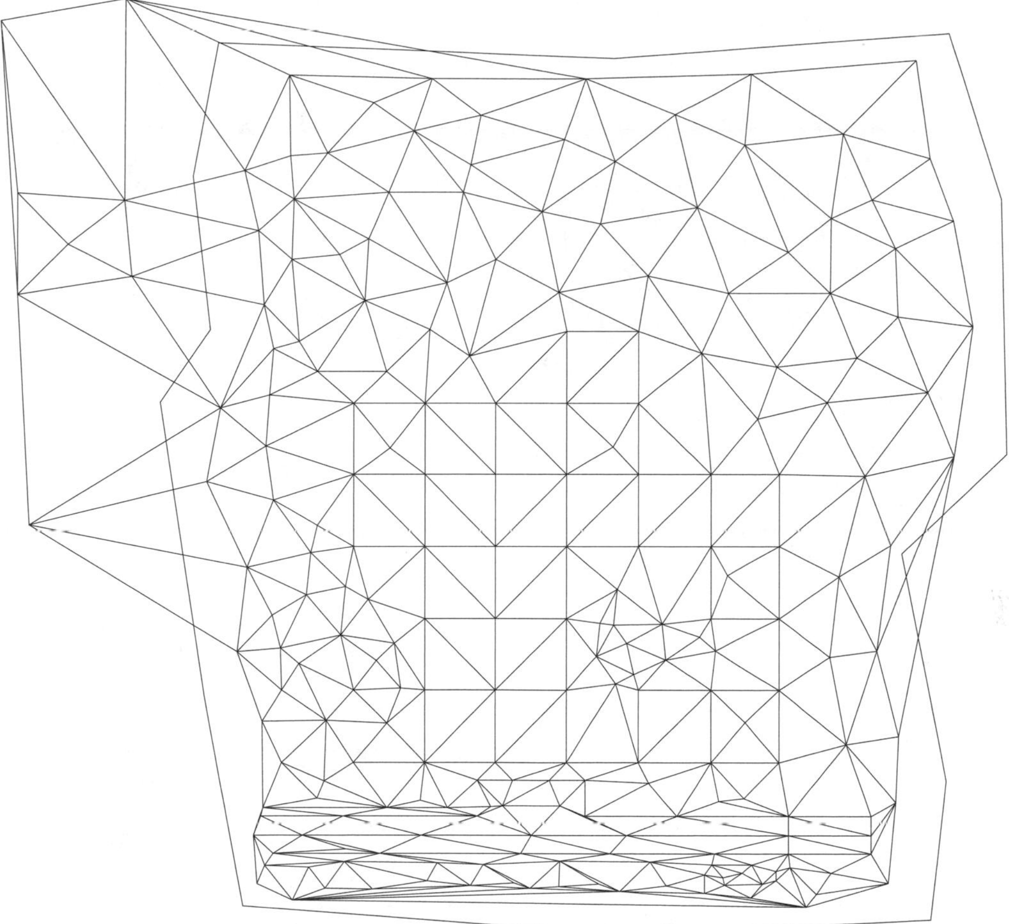

Figure 5–44

2. Before running the routine, use the Surface Layers routine to erase the triangles on the screen.

3. Use the Surface Boundaries routine of the Edit Surface cascade of the Terrain menu, select the polyline, and answer the prompts for the boundary. The routine asks if you want to add a boundary, then if you want to delete any existing ones, if the data within the boundary is to show or be hidden, and then if any triangles are to be calculated to the boundary line. The last prompt is for viewing the new surface. We want to exclude the points to the northwest of the main body of points. The resulting surface should be like Figure 5–45. The following is the sequence of questions and answers in the Boundary routine.

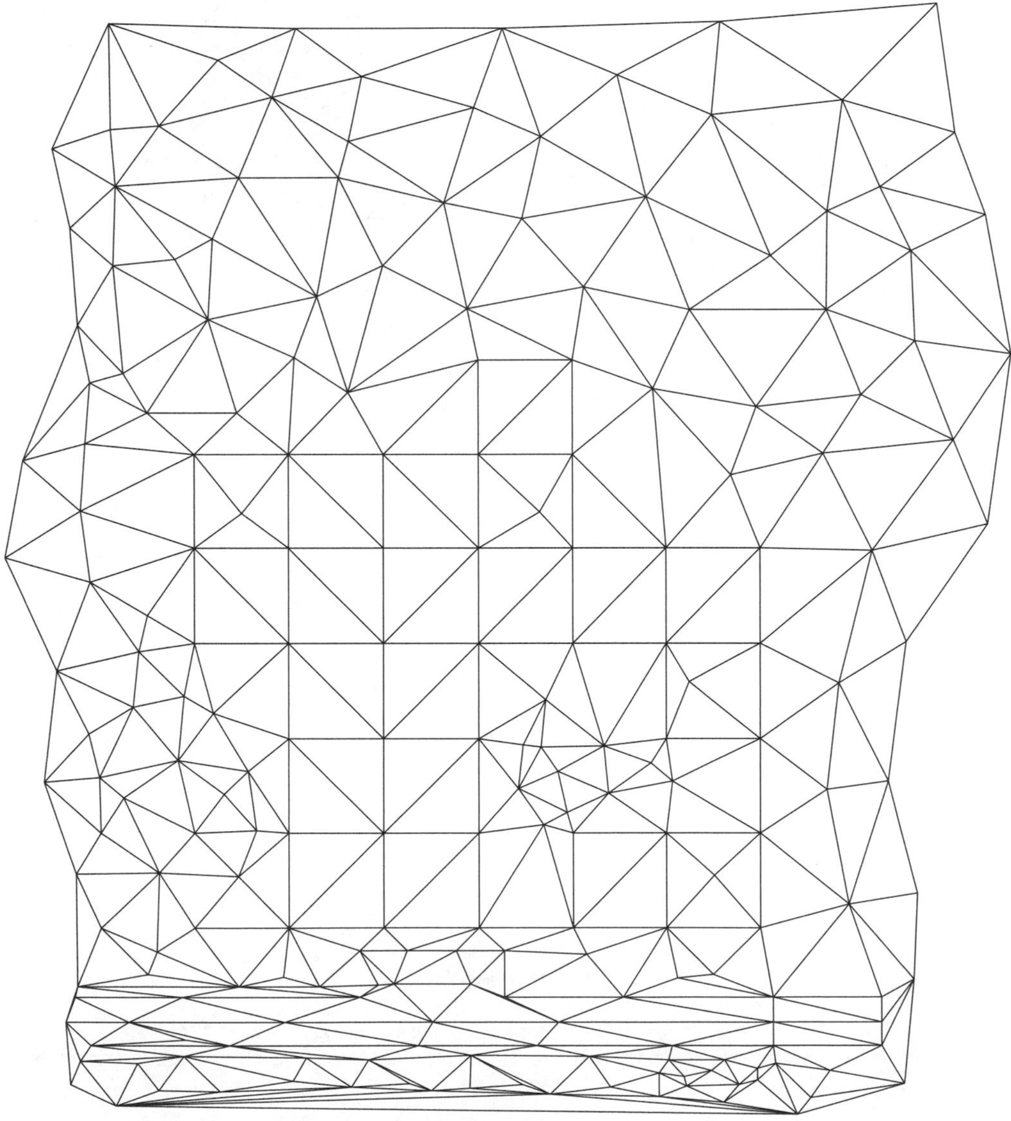

Figure 5–45

```
Command:
RemoveAll/Add <Add>: (press ENTER)
Remove all existing boundary definitions (Yes/No) <No>: y
Select polyline for boundary: (select the polyline)
Boundary definition [Show/Hide] <Show>: (press ENTER)
Make breaklines along edges (Yes/No) <No>: (press ENTER)
Select polyline for boundary: (press ENTER)
View/Review surface (Yes/No) <Yes>: (press ENTER)
Creating View ...400
Command:
```

4. Use the Zoom to Point routine and zoom to point 38 with a height of 200.
5. From the Edit Surface cascade of the Terrain menu use the Import 3D Lines routine to place lines representing the surface into the drawing.
6. Use the Delete Lines routine from the Edit Surface cascade of the Terrain menu to remove the lines identified in Figure 5–46.

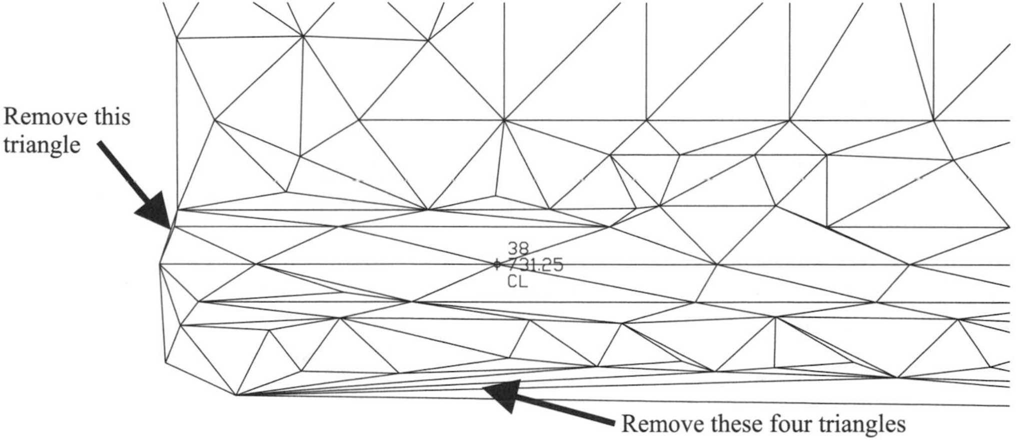

Figure 5–46

7. Use the Zoom to Point routine and zoom to point 35 with a height of 200.
8. Use the Delete Lines routine to remove the lines identified in Figure 5–47.
9. Use Zoom Extents to view the entire surface.
10. Use Zoom Dynamic to show more display area.

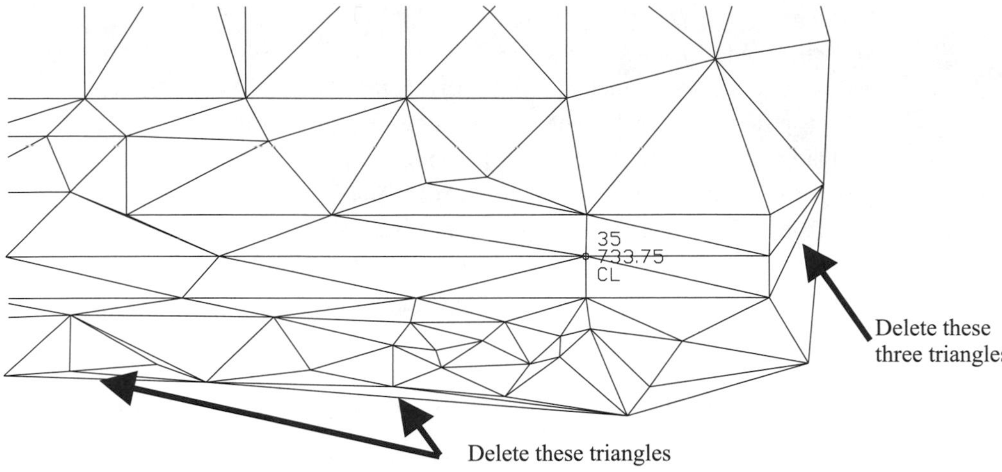

Figure 5–47

11. Create a Named View, surface, from the current display. Use the Named View routine from the View menu.
12. Close the Terrain Explorer.
13. From the Terrain menu, select the Save Current Surface routine to save the Current surface and its edits.

LDD automatically saves the surface when you build it. LDD does *not* automatically save the surface when you edit it. You must manually save the surface when you edit to keep the changes.

14. Use the AutoCAD ERASE command to remove everything from the display.
15. Save the drawing.

UNIT 4: SURFACE ANALYSIS

A surface contains two types of information: elevations at the corners of and slopes across the faces of the triangles. The surface display routines portray this information in several different forms.

Each analysis method portrays the surface as 2D or 3D AutoCAD objects. Also, the routines can display specific ranges of elevations or slopes when you are viewing the data. By assigning each range a color value, you can differentiate the ranges. The surface display routines place the objects on specific layers in the drawing. You can specify up to 16 ranges for slopes and elevations. The surface display routines have their own settings dialog box that was a part of the Unit 1 discussion and exercise of this section. Each routine will display its current settings while it runs.

If you have several surfaces, you may want to have more than one surface's information present in the same drawing. Each time you import a range group, the Display routine asks about erasing the previously imported entities. If the answer is Yes, the routine erases the previous group of objects before placing the new ones. The use of a layer name prefix allows multiple surfaces to import their data into the drawing without disturbing previously imported data. The prefix usually is the name of the surface appended to the default layer name. For example, when you want to see the grids for two surfaces on different layers, place a prefix in the Layer Prefix box of the Surface Display Settings dialog box. By placing an asterisk (*) followed by a dash (-) in the Layer Settings area of the Surface Display Settings dialog box, you set the prefix for the layer names. This setting indicates that the prefix for the layer names is the surface name followed by a dash followed by the default layer name; for example, EXIST-SRF-RNG15 or PROP-SRF-GRID, and so on.

SURFACE STATISTICS OF TERRAIN EXPLORER

The Extended Surface Statistics area of Terrain Explorer gives you general information about the current surface (see Figure 5–48). The report indicates the minimum and maximum size of triangles, number of triangles, and the calculated total 2D and 3D surface area. At the bottom of the statistics area is a report on the slopes on the surface. The report includes minimum, maximum, and mean values of slopes on the surface. The report represents the slopes as percentage grade values, having a range from 0 to infinity. The minimum and maximum values in the report give the range of the slopes in the surface. The mean characterizes the average slopes of the surface. The mean when compared to the minimum and maximum slopes indicates whether there is any skewing in the distribution of slopes on the surface.

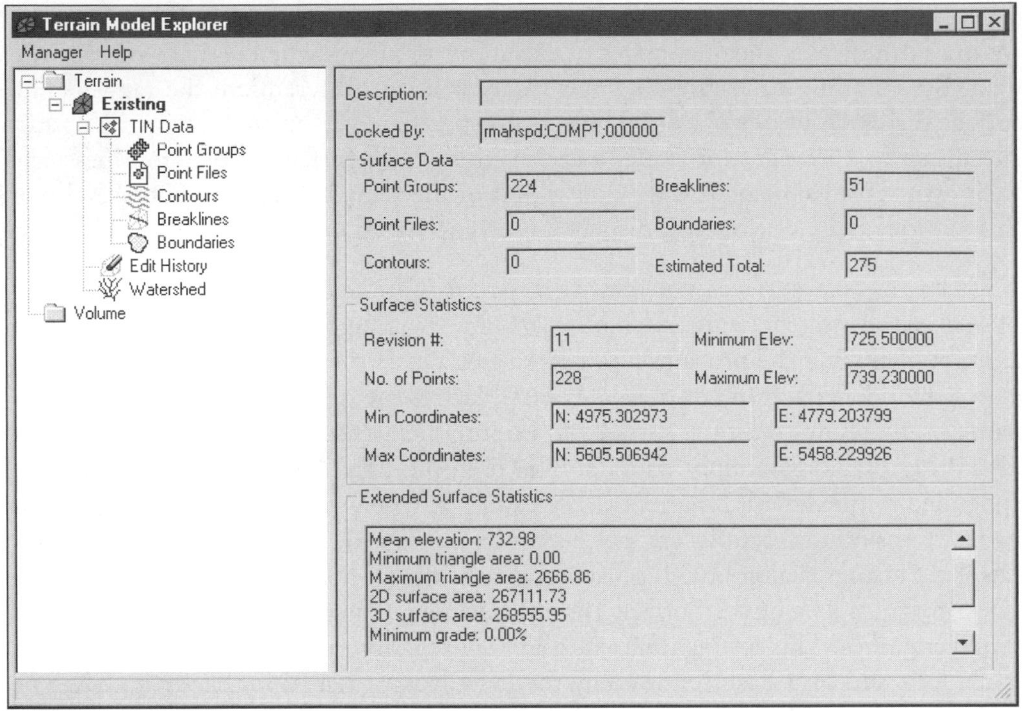

Figure 5–48

SLOPES

There are three types of slope display routines in the Surface Display cascade, 2D Solids, 3D Faces, or Polyface (mesh) (see Figure 5–49). The type of object you create depends upon the intended use of the information. Each routine displays a dialog box for controlling the settings for layer prefixes and a number of values. The routine includes the option of creating a legend.

ARROWS

The above-mentioned slope display routines display their information as triangles; however, there is another component to their slope, direction. When you view an area of large change, high slopes, is it a hill or a valley that is creating these slopes? The Arrow routine creates an arrow in the drawing that points down slope across the face of the triangle. The combination of arrows and triangular outline shows dramatically the distribution of slopes on the surface and allows you to differentiate between hills and valleys on the surface. This routine is found only in the Surface Display cascade of the Terrain menu.

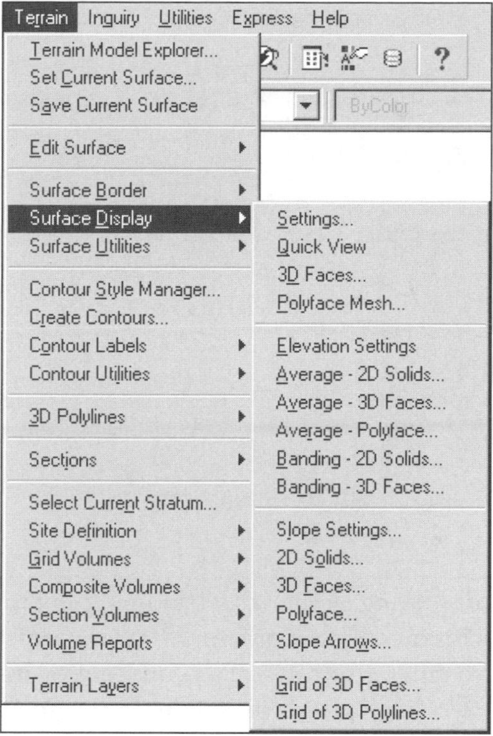

Figure 5–49

ELEVATION

Each corner of a triangle represents a known elevation. The elevation tools represent the elevations on the surface in one of two ways: first, by averaging the elevations of the triangle corners, and second, by slicing the triangle by the range elevations (see Figure 5–49).

AVERAGE

The routines in this group average the elevations found at the corners of a triangle and assign the entire triangle to a range group based upon its average elevation. The routines of this group display the results in one of three ways, 2D Solids, 3D Faces, or Polyface (mesh) objects. The type of object you create depends upon the intended use of the information. Before creating the objects, each routine displays a settings dialog box controlling the layer prefixes and a number of values. The routine includes the option of creating a legend.

An example of how an Average routine works is the following. The elevations at the corners of a triangle are:

```
721.25
723.34
724.58
```

The mean elevation for the entire triangle is 723.06.

If the ranges are:

```
Range 1 718.0 – 720.0
Range 2 720.0 – 722.0
Range 3 722.0 – 724.0
Range 4 724.0 – 726.0
```

The routine assigns the entire triangle to range #3.

BANDING

The second method of displaying elevations is banding. Banding calculates where the range elevation(s) cut across a triangle's surface. The routine assigns all or a portion of the triangle's face to each range group. Basically, this routine uses the same process as the contour generator. The routines of this group display the results in one of three ways: 2D Solids, 3D Faces, or Polyface (mesh) objects. The type of object you create depends upon the intended use of the information. Before creating the objects, each routine displays a settings dialog box controlling the layer prefixes and a number of values. The routine includes the option of creating a legend.

GRID OF 3D FACES

The Grid of 3D Faces routine creates a grid representing the surface elevations (see Figure 5–49). The grid presents a better view of the surface than triangles because of its regular spacing. The grid can be viewed with AutoCAD rendering tools, LDD's Object Viewer, or exported for use in 3D Studio VIZ. The routine has the ability to exaggerate the relief of the surface.

A grid can cover all or a portion of the surface. The advantage of defining a subarea of a surface is your ability to specify even smaller cell sizes. Each corner of the mesh represents a surface elevation. The smaller mesh size allows for greater detail to be visible in the surface. The introduction of an exaggeration factor accentuates the high and low elevations of the surface, just as when viewing the elevations of the surface triangles.

Each cell in the grid can have one, two, or four faces. A cell with more faces is able to show large changes in relief better than a cell with one face. For smooth or rolling surfaces, use cells with one face. If the surface has large changes in elevation over short distances, it is best to use two or four face cells.

SURFACE ELEVATION EXAGGERATION

When using any of the elevation evaluation tools, you can exaggerate the elevations on the surface. LDD uses the vertical factor as an exaggeration value. The vertical factor in the routine's settings is a multiplier that exaggerates elevations that vary from the base elevation. The farther away from the base elevation, the more the routine exaggerates the elevation.

For example:

```
Base elevation = 720.0
Point #1 elevation = 721.0
Point #2 elevation = 738.0
Vertical factor = 4.0
```

The following method of exaggeration is:

```
(Point elevation - Base elevation) * Vertical factor + Base elevation.
(Point #1 = 721.00)
721.0 - 720.0 = 1.0
1.0 * 4.0 = 4.0
720.0 + 4.0 = 724.0
(Point #2 = 738.00)
738.0 - 720.0 = 18.0
18.0 * 4.0 = 72.0
72.0 + 720.0 = 792.0
```

The farther away an elevation is from the base elevation, the greater the distance the elevation is from the base elevation.

SECTIONS

There are two section methods for analyzing a surface (see Figure 5–50). The first method looks only along a line, at one surface, and is semidynamic. This is the Create Section View routine. The second method creates section data that is imported into the drawing and is static. This is the Define Sections routine group. A section may be a cross section or profile view of a surface.

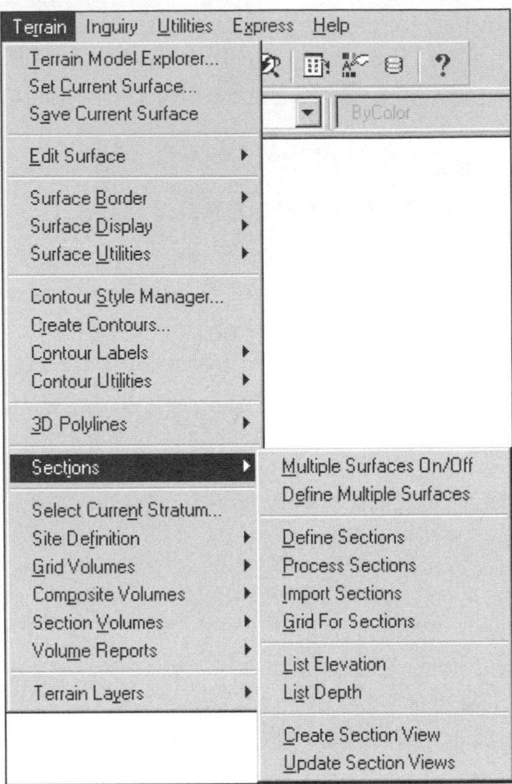

Figure 5–50

CREATE SECTION VIEW

The Create Section View routine displays a dialog box showing the elevations along a linked line. If you move the line, the object updates the section displayed in the dialog box. Deleting the line deletes the dialog box. The drawback to the routine is that it can show only one surface. You can save the currently displayed section to a WMF file or copy and paste it to another document.

DEFINE SECTIONS GROUP

The Define Sections Group routine creates data that you can import later to create sections. These sections are not dynamically linked to any lines in the drawing. The routine creates sections between any two points you select on the surface. The sections can display more than one surface and have an offset and elevation grid over the section.

SURFACE UTILITIES

The Surface Utilities menu contains three routines that aid in evaluating a surface. The routines are Water Drop, Line of Sight, and Fly By (see Figure 5–51).

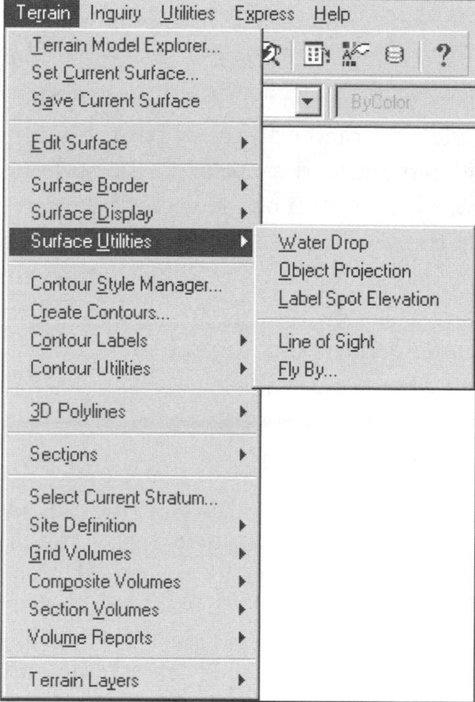

Figure 5–51

WATER DROP

The Water Drop routine traces the path of a water drop as it travels over a surface. The routine will stop at depressions or low spots in the surface. The path of the water drop is similar to the analysis from slope arrows.

LINE OF SIGHT

The Line of Sight routine creates a viewing point on the surface. The routine prompts you to select camera and target points and an elevation for both. After you set the paramcters for viewing, the routine displays the site with the Vpoint (orthographic) or Dview (perspective) routine of AutoCAD. The Line of Sight routine allows for the redefinition of the camera and target point.

FLY BY

The Fly By routine views the site from a moving camera or a camera panning a polyline path. First you create a polyline whose elevation is similar to the surface's elevation and is totally within the site area. After drawing the polyline, you start the routine. There are three methods of viewing the surface. The first is following the polyline as a camera path. The second is a fixed target. This method uses the polyline as the path of the camera as it views a stationary target. The third method fixes the camera and uses the vertices in the polyline as moving target points.

After drawing the polyline, you run the routine and select the polyline. In the Fly By dialog box, you set the viewing method (follow path, fix target, or fix camera), the height of the target and camera, and a prefix to the slide names. If you opt for a camera following a path, the direction of the vector between the vertices points the camera. This means that the polyline needs to point towards the site; otherwise, the routine views the empty areas around the site. The routines display the scenes either as a Vpoint (orthographic) or Dview (perspective) views. There are options to shade, hide, or show a wire frame surface and control the type of output. The output is either a slide or a Named view.

MULTIPLE SURFACES

Multiple surfaces permit you to simultaneously view the data from a group of surfaces. The only opportunity to view multiple surface data is in the Define Sections routine.

Generally, the surface display routines work with only one surface at a time. The reason for this is that during the process of generating data, creating or editing a TIN, or creating contours, only one surface is active. The only time it is appropriate to view several surfaces together is when you look at the surfaces in profile (long section) or cross section.

For you to be able to use multiple surfaces, more than one surface must be defined in the project. Then you complete two steps. The first step is toggling on Multiple Surfaces, and the second is creating a list of surfaces with the Define Multiple Surfaces routine (see Figure 5–50). The surface list contains the surfaces you want to work with. When you use the Define, Process, and Import Sections routines, they will draw a profile or section for each surface that intersects the section. Each surface section will be on its own layer whose name is the same as the surface it represents. You can assign a different color to each surface in the section by setting the color in the Layer dialog box.

INQUIRY MENU ROUTINES

TRACK ELEVATION

The Track Elevation routine displays the elevations under the cursor at the coordinate display of the AutoCAD interface (the lower left corner of the screen) (see Figure 5–52).

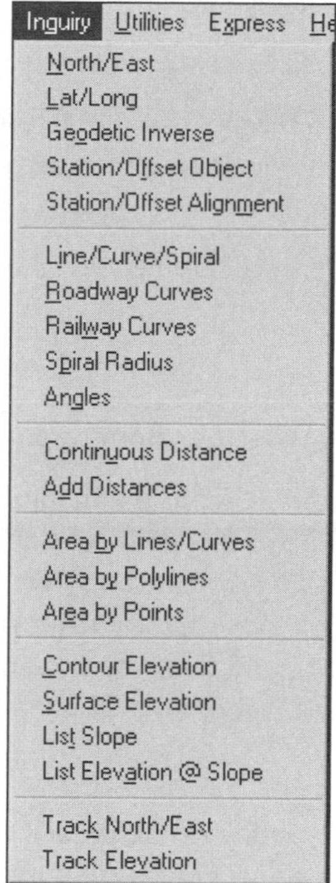

Figure 5–52

SURFACE ELEVATION

The Surface Elevation routine returns the northing/easting and elevation of a point you select from the screen.

Exercise

After you complete this exercise, you will be able to:

- Evaluate the site statistics
- Use surface analysis tools
- View slope and elevation characteristics on a surface
- Create a legend to create presentation drawing
- View the surface as a 3D grid
- Use the Fly By tools to view the surface

ELEVATION AND SLOPE REPORTING

1. If you are not in the TERRAIN drawing, open the drawing.
2. If there are triangles on the screen, use the AutoCAD ERASE command to remove them, or use the Surface Layer routine of the Terrain Layers cascade of the Terrain menu.
3. Import the surface triangles as a 3D line with the Import as 3D Lines routine of the Edit Surface menu.

Elevation Tracking

1. Use the Zoom to Point routine and zoom to point 107 with a height of 250.
2. Use the Track Elevation routine from the Inquiry menu to view the surface elevations. The routine displays the elevations at the bottom left of the screen where the coordinates would normally be. Press the SPACEBAR or ENTER to exit the routine. This routine will not cancel even if you start another command. You must first exit this command and then start the next routine. If the routine prompts you for a surface to use, select Existing.

Surface Elevation

3. Use the Surface Elevation routine of the Inquiry menu to select points to view their coordinates and elevations. Press ENTER to exit the routine.

SLOPE ANALYSIS

The current surface statistics for slopes, found in the Extended Surface Statistics area of Terrain Explorer, are adversely influenced by the headwall. The values for the Existing surface should be more reasonable in mean and maximum slope values. The slopes for a shear face push the maximum slope towards infinity and pull the mean towards a higher value.

The mean slope value divides the range of slopes in half. If the range is 0 to 80 and the mean is 40, this indicates that half of the triangle slopes are less than 40 and the other half are above 40. If the slope range is 0 to 100 and the mean is 80, this indicates that half of the slopes are less than 80 and the remainder are above 80, which signifies a rough (high slope) surface. If the slope range is 0 to 60 and the mean is 5.5, this indicates that half of the slopes are less than 5.5 and the remainder are between 5.5 and 60. This would indicate a smooth or rolling surface. The report gives you a quick look at the distribution of surface slopes.

The Extended Statistics Report about slopes indicates whether to use automatic or user-defined ranges. If the distribution of slopes is fairly even over the surface, use Auto-Range. If the distribution is skewed, the ranges need biasing towards the more frequent slopes. The headwall slopes distort the range and average of the Existing surface. If we were to remove the headwall, the slopes' values would change dramatically.

1. Restore the view, surface, by selecting the Named Views routine of the View menu.
2. Use the Surface Layer routine of the Terrain Layer cascade to delete all of the surface triangles.

3. Open the Terrain Explorer and view the Surface Statistics for the Existing surface. If they have not been computed, select the surface name, click the right mouse button, and select Calculate Surface Statistics. On a piece of paper write down the minimum, maximum, and average slope values and compare them to the values noted earlier. There should be a significant change in the maximum and mean slope values.
4. Close the Terrain Model Explorer.
5. From the Surface Display flyout of the Terrain menu, run the 2D Solids routine below Slope Settings to view the distribution of slopes. Add the layer prefix of *- (asterisk dash) to the settings if not set already. Toggle off Create Skirts. Set the number of range to 8, select Auto-Range to define the slope ranges, and select OK to set the minimum and maximum range values. If the colors of the ranges are not set, set them to the values shown in Figure 5–53. Select OK to exit the Ranges portion of the routine. Again select OK to exit the routine's dialog box and answer Yes to erase any preexisting lines. The routine displays a dialog box reporting the area of each range. There are only two ranges used, the first and last. The last range contains a small triangle disturbing the slope report. Select OK to exit the dialog box, and the routine draws the slopes.

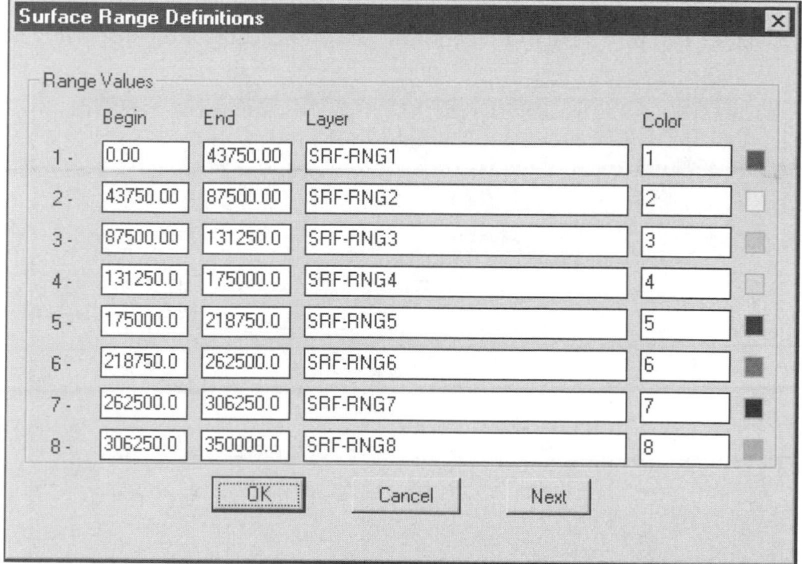

Figure 5–53

If you look at the first set of surface statistics before the development of the headwall, the current surface uses a range of 0 to 300 with an average slope around 6 percent. This distribution of slopes is a skewed distribution towards the low end of the slope values. As a result you need to define a user range of slopes that emphasizes the lower end of the distribution.

6. Rerun the Slopes 2D Solids routine and assign user-defined range values (see Figure 5–54). You replace the values by double clicking in the begin and end range value boxes and typing in the new values. Use the following range values. Select OK to exit the Surface Range Definitions settings and the Surface Elevation Shading Settings dialog box. Then answer Yes to the Erase previous objects question, view the new values in the report dialog box (their values have changed significantly), and then select OK to exit the report dialog box.

   Range 1   0.00 - 2.00
   Range 2   2.00 - 4.00
   Range 3   4.00 - 6.00
   Range 4   6.00 - 8.00
   Range 5   8.00 - 10.00
   Range 6   10.00 - 12.00
   Range 7   12.00 - 50.00
   Range 8   50.00 - 300.00

The routine creates a report about the values of each range. The report lists the range, beginning and ending range values, the area, and the percentage of the surface within the range. These values can be a part of a legend the routine creates after displaying the ranges.

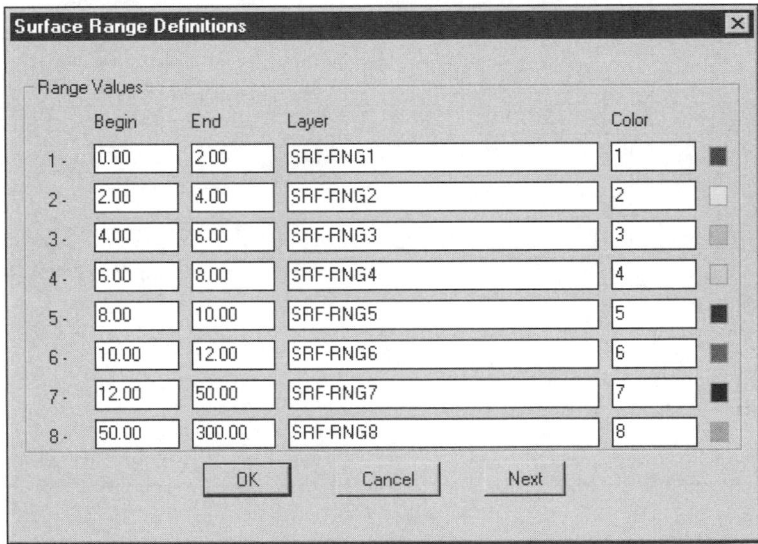

Figure 5–54

7. Rerun the 2D Solids routine from the Surface Display cascade and in the Range Statistics dialog box, create a legend. Select the Legend button and from the report values, create a legend (see Figure 5–55). The values may already be set.

Title — Surface Slopes

Set the following values:

Column 1: Color

Column 2: Beginning Range

Column 3: Ending Range

Column 4: Percentage

Column 5: Area

Column 6: None

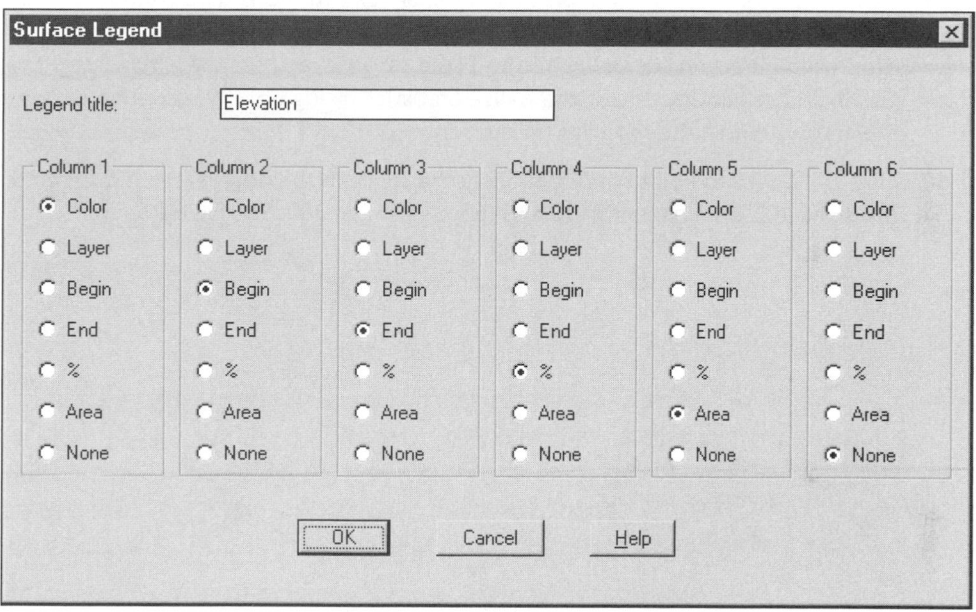

Figure 5–55

8. Choose OK in the Legend dialog box and place the legend at the right bottom of the screen. LDD centers the legend at the selected point.
9. Inspect the results. List a triangle entity. Notice that the entity is on a layer prefixed to the name of the surface.

The high slope values along the side of the roadway reflect the presence of a ditch and swale, features that were discussed earlier. Also note the high slope values in the northeast. Viewing only slope values does not indicate whether the slopes are on the side of a hill or valley. When viewing slope data, you cannot discern what surface features create the slopes.

To better visualize the surface slopes, it would be better to see the direction of slope as well as the amount of slope. LDD uses a slope arrow on the surface of the triangle to point down slope (direction of overall elevation change). The Arrow routine then "shows" both direction of change and amount of slope on the triangle face. In valleys or swales the arrows will point towards each other. If the arrows point away or radiate from a point, there is a high spot or point.

10. Run the Slope Polyface routine of Surface Display cascade to create a polyface mesh instead of 2D solids. Use the same ranges. Choose OK to exit the routine's main dialog box and answer Yes to erase the range views. Select OK to exit the Range Statistics dialog box and the routine creates hollow triangles. Notice the Legend is not erased by rerunning the routine.

11. Run the Slope Arrows routine and add the arrows to mesh. The routine is in the Surface Display menu. Use the same settings as before by selecting OK in the main dialog box and answering No to Erase range layers, set the arrow scale to 10.00, and select OK to exit the Range Statistics dialog box. The routine draws the arrows within the triangles on the screen.

12. Use the Zoom to Point routine to view the swale in the northeastern quarter of the surface. Zoom to point 126 and use a height of 300. See Figure 5–56.

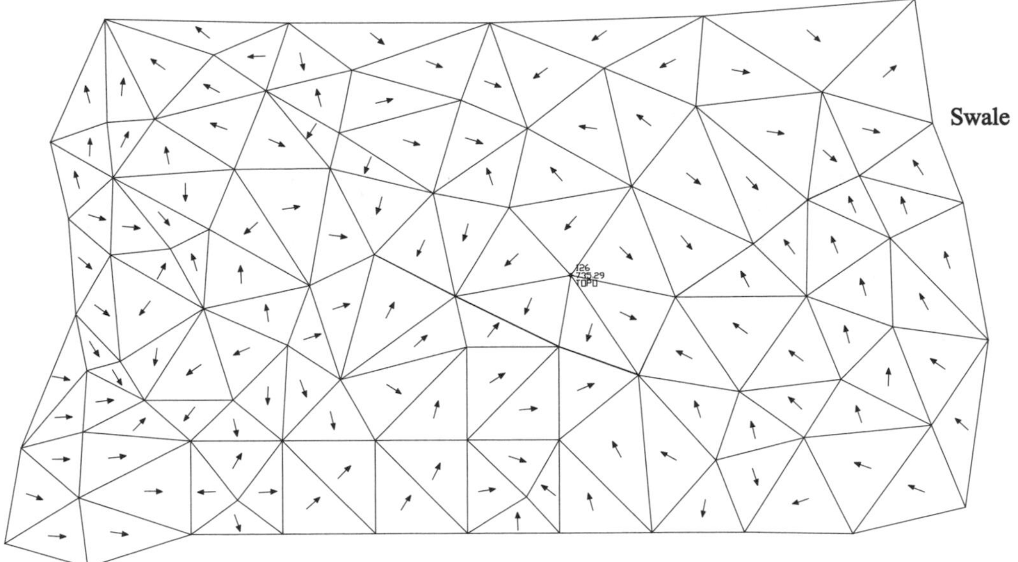

Figure 5–56

13. Use the Zoom to Point routine to view the hill in the southwestern quarter of the surface. Zoom to point 61 and use a height of 300. See Figure 5–57.

With the down slope arrows, you can see that the feature in the northeast section of the drawing is a swale. Arrows converge in areas of lower elevations, that is, a ditch or swale.

Arrows diverge from a point that is a high spot. All surrounding elevations are lower and the arrows point away from the high point.

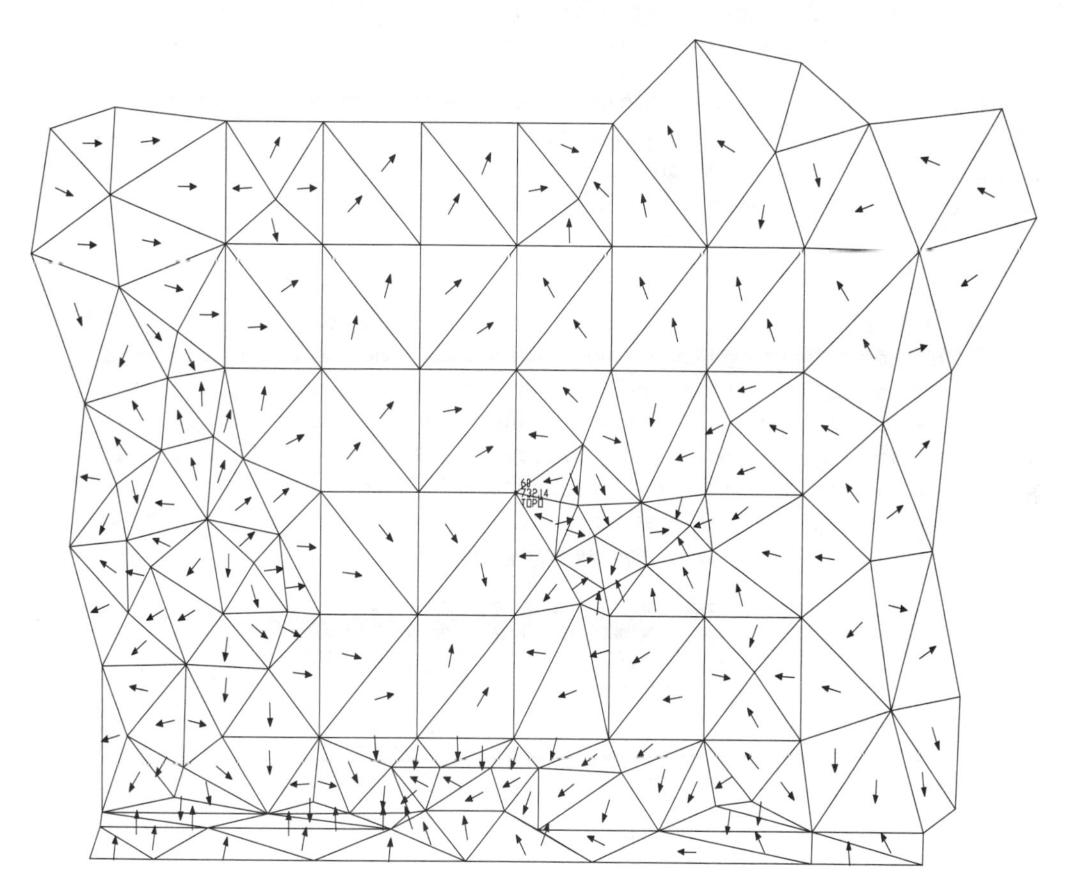

Figure 5–57

ELEVATION ANALYSIS

Average Elevation

The average elevation routines place the entire triangle into a range. This method can produce a very jagged display of elevations.

1. Restore the Named View, surface, to view the entire surface.
2. Erase the Slope Legend from the drawing.
3. Use the Range Layers routine of the Terrain Layers cascade to erase the objects from the screen.

4. Select Terrain Model Explorer from the Terrain menu. Select the Existing surface name and review the minimum and maximum elevations and the mean elevation for the surface in the Extended Surface Statistics area. The elevation distribution for the surface is about 14 feet from the lowest to the highest.

The mean elevation is a little larger than half of the elevation range, which means that there are a few more elevations in the higher ranges than in the lower ranges. The difference between the mean and the midpoint is not too extreme. From these values, assume a normal distribution of elevations and let the Elevation routines automatically create the range groupings.

5. Close the Terrain Model Explorer.
6. Use the Average–2D Solids routine of Surface Display to create a map of surface elevations. Toggle off Create Skirts. In the main dialog box, Surface Elevation Shading Settings, set the numbers of ranges to 5, set the ranges by selecting Auto-Range, and select OK in the minimum and maximum elevations dialog box. The Surface Range Definitions dialog box contains the colors from the slope ranging routines for the new elevation ranges. The dialog box should look like Figure 5–58. Exit both dialog boxes by selecting the OK button, erase all range data, and exit the Report dialog box by again selecting the OK button.

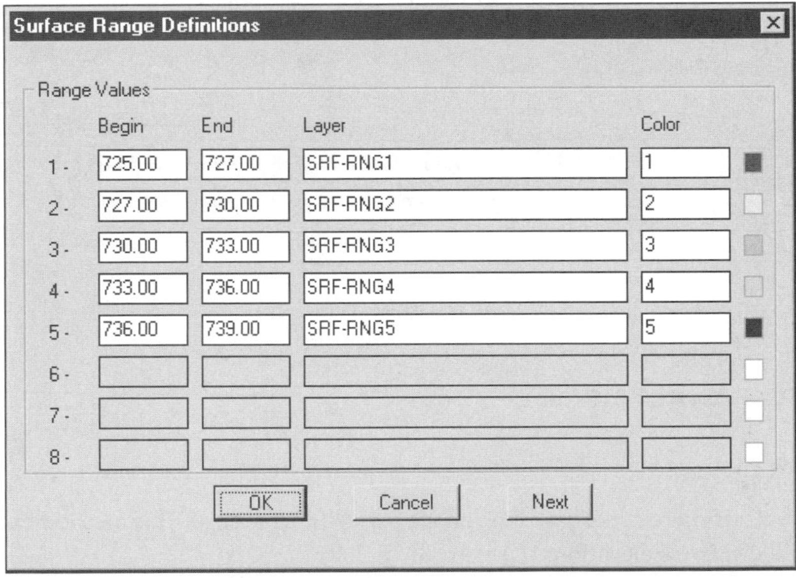

Figure 5–58

The area listed in the Report dialog box incorrectly estimates the areas because entire triangles are assigned to a range. This is not always the case. Many times only a portion of a triangle is in one or another elevation range. To be able to show and compute correctly the distribution of elevations over the surface, you need to use the banding routines of Surface Display.

Banding Elevation

The second method of displaying elevations is banding. Banding calculates where the range elevation(s) intersects the surface of the TIN triangle. The routine assigns all or a portion of the triangle face to each range group. Basically, this routine shows the contour interval on the surface. In the current range grouping, we will be looking at a contour interval of 3 feet.

1. Use the Banding–2D Solids routine to create a view of the site elevations. Use the same number of ranges and the same colors by choosing OK in the Surface Elevation Shading Settings dialog box. Answer Yes to the Erase existing layers question and review the new areas in the Report dialog box. Select OK to exit the Report dialog box and the routine creates a new display.

This result is a 3-foot contour, solid filled map of the surface.

DTM SECTIONS

The creation of surface sections is a three-step process: defining the section lines, processing the sections to sample for elevations along the line, and importing the sections into the drawing. The routines that define the process are found in the Sections cascade of the Terrain menu (see Figure 5–50).

1. Use the AutoCAD ERASE command to erase everything from the screen.
2. Use the 2D Polyline routine of the Surface Border cascade of the Terrain menu to draw a border for the surface.
3. Use the Define Sections routine of the Sections cascade to define one group of sections having three (3) sections. The routine first prompts you for the group name and then asks you to name and define the section's location. The two points used to define a section must be within the boundary of the surface. An example of the Define Sections routine is:

```
Command:
Group Name: G1
Section Name: S1
First point: (select start and endpoints of section)
Second point:
Section Name: S2
First point: (select start and endpoints of section)
Second point:
Section Label: S3
First point: (select start and endpoints of section)
Second point:
Section Name: (press ENTER to stop this group)
Group Name: (press ENTER to stop this group)
Command:
```

EXERCISE

4. The next step is to process the sections with the Process Sections routine of the Sections cascade. Since there is only one group and one surface, the processing is fast. The routine sets the group, G1, as the current group. If there were more than one group defined, the routine would prompt you to identify the current group.
5. Next, pan a point on the right side of the screen to the left side of the screen. The result should be a clear display area.
6. Import the sections with the Import Sections routine of the Sections cascade. Press ENTER to accept the datum layer and the vertical factor of 10 defaults. Use the same datum elevation (720) for all three sections. Place the first section at the bottom of the screen and the remaining sections above the first.
7. After placing the sections in the drawing, list the surface line in the section. The routine places the line on a layer that is the same name as the surface.
8. Use the Grid for Sections routine of the Sections cascade to add a grid to the sections using the default layer, an elevation offset of 5, and a horizontal offset value of 50. The datum block is the text that is to the left of the section datum line.
9. Use the List Elevation and List Depth query tools to query elevations within the section.
10. After exiting the listing routines, erase the sections and their grid from the screen.
11. Restore the view, surface, from the Named Views dialog box.
12. Save the drawing.

EXERCISE

3D VIEWING OF A SITE

The Elevation and Banding 3D Face routines allow you to create 3D models from the surface data. The routines create a skirt around the edge of the surface. The elevation of the base of the skirt is the base elevation setting, and the exaggeration of a surface's relief is the result of the vertical factor (see Figure 5–59).

1. Rerun the Banding–3D Faces routine of the Surface Display cascade, set the vertical factor to 3.5, set the base elevation to 720 (the bottom of the skirt), and toggle on Create Skirts. Use the same ranges. Exit the dialog box by selecting OK, answer Yes to delete any previous border and range layers, and then select OK to exit the Report dialog box. The routine builds what looks like the same surface representation.
2. Select all of the objects on the screen to activate their grips. Click the right mouse button to display the shortcut menu. Select Object Viewer from the shortcut menu. This displays the LDD Object Viewer, which is similar to the Orbit Viewer of AutoCAD (see Figure 5–60). The viewer allows you to spin the model, pan and zoom your view, and shade the surface. Try viewing the erosional feature from the northeast, shade the surface, and view the surface from different perspectives.

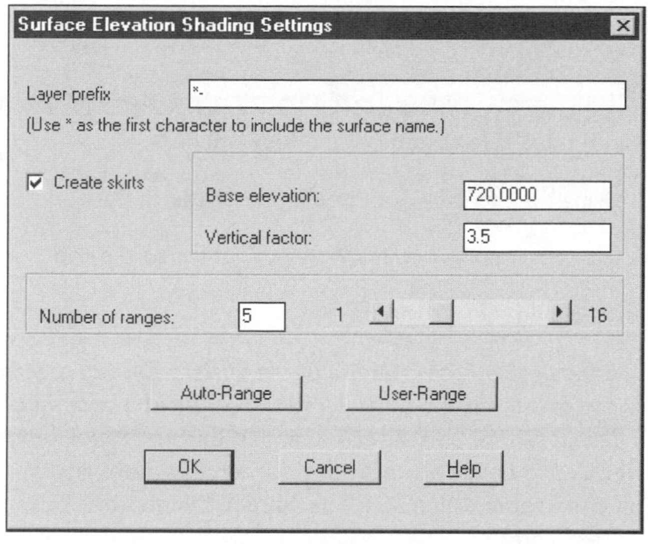

Figure 5–59

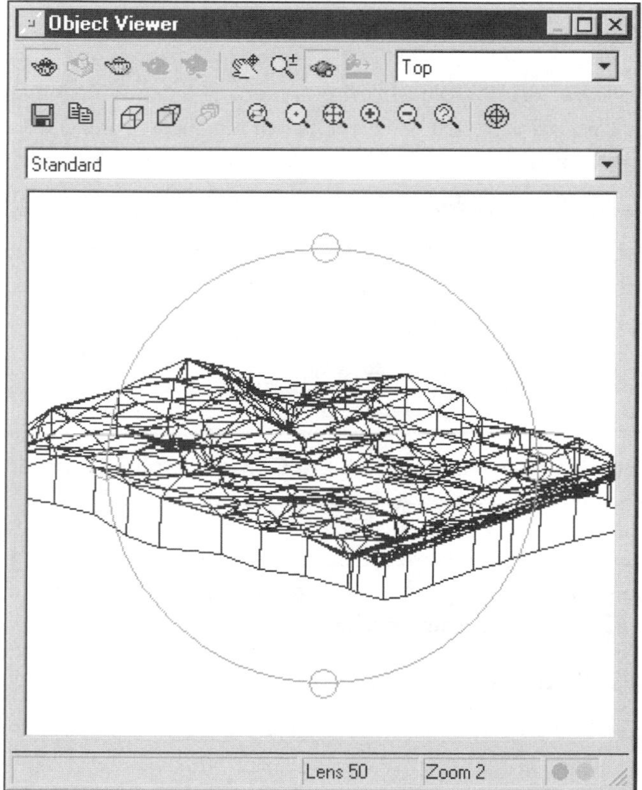

Figure 5–60

EXERCISE

3. Exit the Object view by clicking the "X" in the upper right corner of the dialog box.

4. Try viewing the surface with the Orbit Viewer of AutoCAD. Try different shadings from the Shade cascade of the View menu.

5. Set the drawing back to 2D wireframe in the Shade cascade of the View menu.

6. If needed, use the Top view of 3D Views to return to the top view of the surface.

7. Restore the Named View, surface.

8. Rerun the Banding–3D Faces routine of the Surface Display cascade, set the vertical factor to 1.0, set the base elevation to 720 (the bottom of the skirt), and toggle on Create Skirts. Use the same ranges. Exit the dialog box by selecting OK, answer Yes to delete any previous border and range layers, and then select OK to exit the Report dialog box. The routine builds what looks like the same surface representation.

9. Save the drawing.

Fly By

The Fly By routine views the site from a moving camera or a camera panning a path. First you create a pline whose elevation is near the elevation of the surface and is totally within the site area. After drawing the pline, start the Fly By routine. Depending upon the method of viewing (path, fixed target, or fixed camera), the routine views the site at each vertex on the polyline.

After you select the polyline, the routine displays an options dialog box. These options allow you to set the method of viewing, the height of the target and camera, the type of view, the type of shading, and a prefix to the slide names.

1. Start in the north central portion of the surface and draw a polyline with eight vertices similar to the one in Figure 5–61.

2. Use the Edit Elevation routine from the Contour Utilities cascade of the Terrain menu, and edit the elevation to 745.

3. Select the Fly By routine from the Surface Utilities cascade of the Terrain menu. Select the polyline when prompted by the routine. Set the options as shown in Figure 5–62 and create a series of slides.

4. Use the Vslide routine to view the slides.

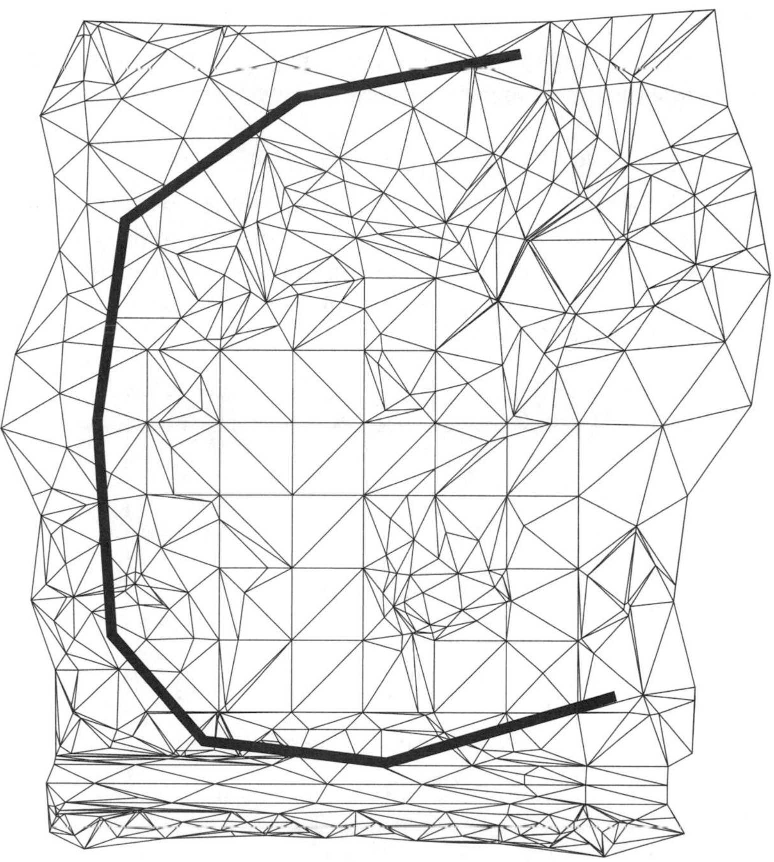

Figure 5–61

Surface Fly By

Fly By Type
- ( ) Follow path
- (•) Fixed target
- ( ) Fixed camera

Viewpoint Type
- ( ) Vpoint
- (•) Dview

View Type
- (•) Shade
- ( ) Hide
- ( ) None

Camera height: 750.000

Target height: 730.000

Output prefix:

Output Type
- (•) Slide
- ( ) View

OK | Cancel | Help

Figure 5–62

Line of Sight

The Line of Sight routine creates a single view of the surface from a camera and target point. The view can be either an orthographic or Dview (perspective) view.

1. Erase the polyline from the drawing.
2. Run the Line of Sight routine from the Surface Utilities cascade of the Terrain menu. Select a point for the camera and assign it an elevation of 740. Next select a point for the target and assign it an elevation of 730. The routine then displays the Surface Line of Sight dialog box (see Figure 5–63). Select the Vpoint button to view the ortho view of the site.

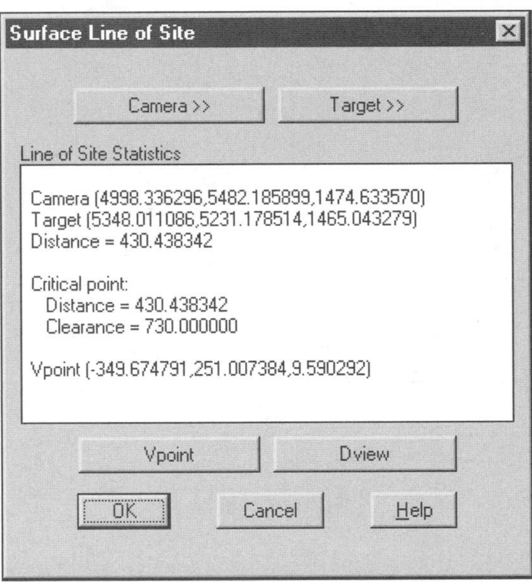

Figure 5–63

3. Return the drawing back to plan view and try Line of Sight from other positions and values.

3D Grid

The 3D Grid routine displays the surface as a mesh rather than as triangles. The advantage of the routine is the ability to specify a small mesh size. As a result, the mesh portrays clearly the details not well seen as triangles.

1. Return the drawing to plan view.
2. If the shading is not set, set it to 2D wireframe in the Shade cascade of the View menu.
3. Set the current view to surface in the Named View routine of the View menu.

4. Erase the objects from the screen. You can use either the AutoCAD ERASE command or the Border and Range Layer routines from the Terrain Layers cascade of the Terrain menu.
5. Save the drawing.
6. Use the 2D Polyline routine of the Surface Border cascade to import the surface boundary. The boundary will give you an area context when selecting points.
7. Run the Grid of 3D Faces routine from the Surface Display cascade of the Terrain menu.

The first prompt is for a rotation angle. The reason for this is that the surface shows best when the grid is rotated along the axis of the surface. In this example the rotation is 0.

8. Press ENTER to accept the zero rotation.

The next prompt is for a starting point. The default is AutoCAD's 0,0. The site exists near the 5000,5000 coordinate area. Select a point that is just outside of the west and south boundaries of the surface.

Next the routine prompts for an *M* distance. The terms *M* and *N* relate to a rotated axis system. If the site we are defining were rotated 45 degrees, the site would no longer refer to *X* or *Y* coordinates; thus LDD refers to this potentially rotated system as *M* and *N*.

9. Set the *M* and *N* spacing to 10.

The final prompt is for an upper right point. Select a point above and to the right of the TIN boundary.

The routine draws a box representing the grid boundary. The surface should be inside the box. In the lower left corner of the box should be a small square. This square represents a 10 by 10 grid cell.

If the box does not enclose the surface border, type **Y** for Yes to redefine the grid limits. If the surface is within the box, you accept the grid limits by pressing ENTER to accept No.

10. Press ENTER to accept No.

The routine responds by presenting a dialog box containing the current settings.

11. Set the base elevation to 720.0, the vertical scale to 3.5, and toggle on Create Skirts. Use Figure 5–64 as a guide for the settings. Select OK to exit the dialog box, answer Yes to erase any previous grid and border objects, and then the routine builds the grid.
12. View the grid from different angles with the Object Viewer. Select the grid on the screen to activate the grips, click the right mouse button to view the shortcut menu, and select Object Viewer.

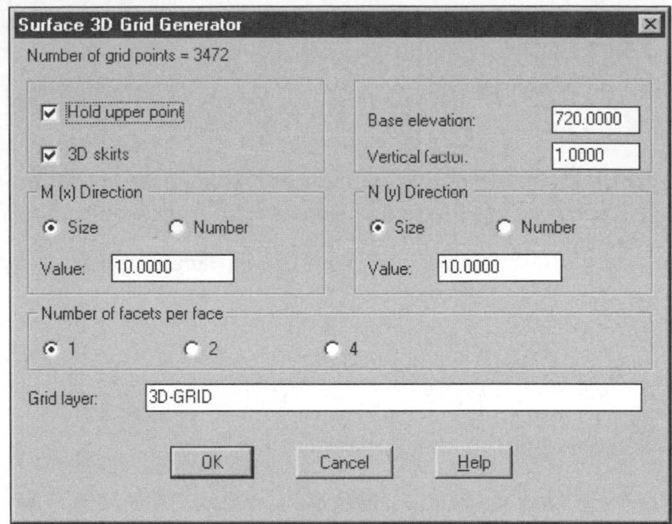

Figure 5–64

13. Rerun the 3D Grid routine. All you need to do is press ENTER at the prompts until you arrive at the Surface 3D Grid Generator dialog box. Reset the vertical factor to 1.0. Select OK to exit the dialog box and answer Yes to both erasing the border and grid layers. You now have a grid representing the true elevations of the surface.

3D Surface Intersections

With a 3D Grid surface, you can view interactions of other surfaces with the current surface, for example, a FEMA flood plain definition. If the flood elevation is 727.00 feet, how much of the current surface would be affected by the flood waters?

1. Create a new layer, WATER. Assign the layer the color blue (5) and make it the current layer.
2. Set the elevation to 727.00 with the AutoCAD ELEVATION command.
3. Use the 3D Face routine to draw a face around the surface. Select four points around the perimeter of the grid.
4. Set the elevation back to 0.00 with the AutoCAD ELEVATION command.
5. Set the shade mode to Flat Shaded in the Shade cascade of the View menu.
6. Select the 3D Face to activate its grips. With the 3D Face grips active, click the Properties icon to display the Properties dialog box.
7. In the Properties dialog box, change the elevation of the 3D Face to 730.00, 731.00, and 732.00. Each time you change the elevation, more of the water hides the surface.

EXERCISE

8. Close down the Properties dialog box by clicking the "X" in the upper right corner of the dialog box.
9. Set the shade mode to 2D wireframe in the Shade cascade of the View menu.
10. Erase all of the objects on the screen.
11. Save the drawing.

UNIT 5: SURFACE ANNOTATION

Land Development Desktop annotates a surface with contours. The Create Contours routine controls the interval and layers of the contours. A contour style controls the general display characteristics of the contour. A label representing the elevation of the contour is a part of the contour style definition.

A contour object represents a single continuous elevation on a surface. The contour object is similar to a 3D polyline. The Banding Elevation routines of Surface Display display elevations as an area of the surface based upon ranges. The range breaklines are lines representing the range elevation traveling across the triangle faces and in essence are contours. The Create Contour routine creates the contour object and calculates where it crosses a leg or face of a triangle.

CREATE CONTOUR

In the Create Contour routine, you specify several values controlling the creation of contours including the contour interval, exaggeration, and the layers for the contours (see Figure 5–9). You can produce polylines with a constant elevation instead of contours with this routine. The Contour Style area identifies the current contour style.

CONTOUR STYLES

The contours produced by the Create Contour routine reflect the current contour style settings. A contour style affects the appearance, text style, and placement. The current style will be used when creating the next set of contours.

CONTOUR APPEARANCE

The Contour Appearance panel of contour styles contains settings controlling the contour and label properties and the smoothness of the contour (see Figure 5–10). The contour display properties have to do with the display of grips and contour width. The label display properties control the appearance of grips with label, labels only, or no labels. Labels are a property of a contour line. By toggling on No Labels, you can prevent a contour from displaying a label.

The bottom portion of the panel controls the smoothness of the contour. The Add Vertices option adds more vertices to the contour to simulate curves. If you are toggling on this method, you need to set a smoothing factor. The Spline method uses the spline method found in PEDIT. When the contours are close together and contain sharp turns, the results may be crossing contour lines. Because of potentially crossing contours, you need to be careful when using the Spline option.

TEXT STYLE

The Text Style panel controls the text style used in the label of a contour (see Figure 5–11). The panel sets the text style, size, and color. If you select a fixed height text style, the height box is grayed out, because the style sets its own height. The middle portion of the panel controls the color of the text and any prefix and/or suffix added to the elevation. The panel also sets the precision of the label.

LABEL POSITION

The Label Position tab contains values affecting the orientation, readability, and bordering of labels (see Figure 5–12). The orientation portion sets the label above, below, or on a contour. If you select the label to be above or below the contour, you need to set the offset factor or if the label is on the contour, whether the label breaks the contour or not.

The Readability portion of the panel controls whether the label responds to a rotation in a layout (paper space) or the labels read left to right when facing a set of contours going up slope.

The last area of the panel controls the placement of a border around the label. The options are None, Rectangular, and Round Corners.

MANAGE STYLES

The Manage Styles panel shows the defined contour and the current style. The panel displays a list of currently defined styles from the *Contours* subfolder of the *Data* folder (see Figure 5–13).

CONTOUR LABELS

LDD annotates contour lines in two ways: at the end or in the interior of the contour. The routines are in the Contour Labels cascade of the Terrain menu (see Figure 5–65). Within each method of annotating contours are options to annotation by individual or groups of contours. All labeling routines use the contour style assigned to the contours to determine what to display and how to behave when labeled.

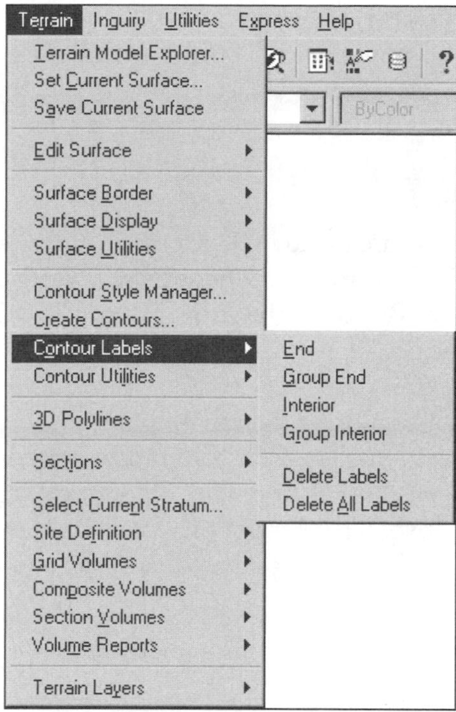

Figure 5–65

END AND GROUP END

The End and Group End routines label the ends of contours. The End routine does individual endpoints and the Group End routine labels multiple endpoints. The routines prompt for an offset distance and direction from the contour end and a rotation angle of the contour label.

INTERIOR AND GROUP INTERIOR

The Interior and Group Interior routines label the interior of contours. The Interior routine labels individual points within the contour. The Group Interior routine labels the contour at a specific point and at a user-defined spacing interval along the interior of the contour. The Group Interior routine uses an optional repeat value to label the same contour multiple times. For example, if the label repeat is set to 250, this tells the labeling routine to add a new label to the contour every 250 feet along its length.

DELETE LABEL AND DELETE ALL LABELS

With the elevation label as a property of the contour, the only way to delete the labels is to change the contour style definition or to use the label erasing routines. The Delete Label routine deletes labels from individual contours and the Delete All Labels deletes all contour labels.

LABEL SPOT ELEVATIONS

The Label Spot Elevations routine of the Surface Utilities cascade creates a leadered text object annotating the elevation of a selected point. The text is static and does not respond to possible changes of elevation at the point it labels.

EDITING CONTOURS

The contour editing routines are found in the Contour Utilities cascade of the Terrain menu (see Figure 5–66). The routines apply to site design situations. The Edit Elevations, Edit Elevations By Layer, and Edit Datum Elevations routines allow you to change a contour's elevation. The Edit By Layer routine selects all the contours on a single layer for editing. After you select a contour, the routine selects all the contours on the layer for editing. The routine then highlights each contour and prompts for the new elevation of the contour. The Edit Datum Elevations routine adjusts a group of contours by specifying an absolute or relative amount of change. The change value is either positive or negative. The Assign Elevation routine is an automated Edit Elevation routine. The Zero Elevation routine scans a selection set for 0 (zero) elevation contours or polylines.

Section 6 covers the contour editing routines in detail.

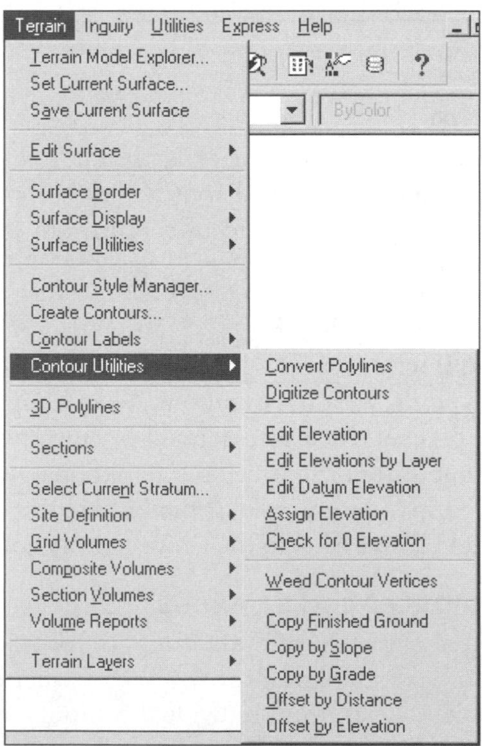

Figure 5–66

ANALYZING CONTOURS

The Contour Elevation routine of the Inquiry menu allows you to view the elevation of a contour or 3D polyline without having to read an AutoCAD listing.

Exercise

EXERCISE

After you complete this exercise, you will be able to:

- Create contours
- List and evaluate contour elevations
- Label the contours

CREATING AND LABELING CONTOURS

1. Create a new contour style. Select the Contour Style Manager routine from the Terrain menu. After setting the following values and referring to Figures 5–10, 5–11, 5–12, and 5–13, select OK to exit the dialog box.

Contour Appearance

Contour Display:

Contour and Grips: ON

Label Display:

Labels and Grips: ON

Smoothing Options:

Add Vertices: ON

Value: 5

Text Style

Text Style: L120

Color: 30

Precision: 0

Label Position

On Contour

Break Contour for Label: ON

Manage Styles

Style Name: Mystyle

2. Run the Create Contour routine found on the Terrain menu, and after setting the following values in the dialog box (see Figure 5–9), select OK to exit the Create Contours routine.

```
Interval
Normal: 1
Major: 5
Vertical Factor: 1
Contour Style: MyStyle
```

3. Answer Yes to delete any contours. The routine creates the surface contours. Since you have been using a layer prefix of asterisk dash (*-) in previous commands, the Create Contour routine continues using the prefix. The contours are now on the layers EXISTING-CONT-MNR and EXISTING-CONT-MJR.
4. Assign colors to the default layers, EXISTING-CONT-MNR and EXISTING-CONT-MJR in the layer dialog box.
5. List out some contour elevations using the Contour elevation routine from the Inquiry menu.

Contour End Annotation

1. Zoom into the northeastern quarter of the surface.
2. Use the End routine to annotate a few contours. After annotating a few contours, Undo the annotation.
3. Use the Group End routine to annotate a few contours. The routine asks you to drag a line across the contour ends, then prompts for a text rotation angle (use 15 degrees). After you annotate the contours, press ENTER to exit the routine. After annotating the contour ends, Undo the annotation.
4. Use the Interior routine to annotate the interior of a contour. After exiting the routine, use the AutoCAD UNDO command to undo the annotation.
5. Use the Group Interior routine to annotate a group of contours. After dragging a line over the contours, set the repeat interval to 250. The repeat interval places additional labeling along the contour every 250 feet. After placing the labels, exit the labeling routine.
6. Select a contour to display its grips. The contour and the label display grips so you can edit the location of either the contour or the label. Remember, if you displace the contour, it no longer represents the surface data. If you recreate the contours, the contour will return to its original location. Select the label grip and move the location of the label. Notice that there is a label every 250 feet along the length of the contour.
7. Use the Delete Labels routine from the Contour Labels cascade of the Terrain menu. The routine asks you to identify a contour and then to select a point nearest to the label you want to erase. After deleting a few labels, exit the routine by pressing ENTER.
8. Restore the Name View, surface.
9. Save the drawing.

UNIT 6: SURFACE POINT TOOLS

The Points menu contains surface point creation routines in the Create Points–Surface cascade (see Figure 5–67). There are three routines: Random Points, On Grid, and Along Polyline/Contour. The Random Points routine places a point into the drawing whose elevation is from the current surface. The On Grid routine places points in a user-defined spaced grid whose elevations are from the current surface. The Along Polyline/Contour routine places points along a polyline/contour whose elevations are from the current surface.

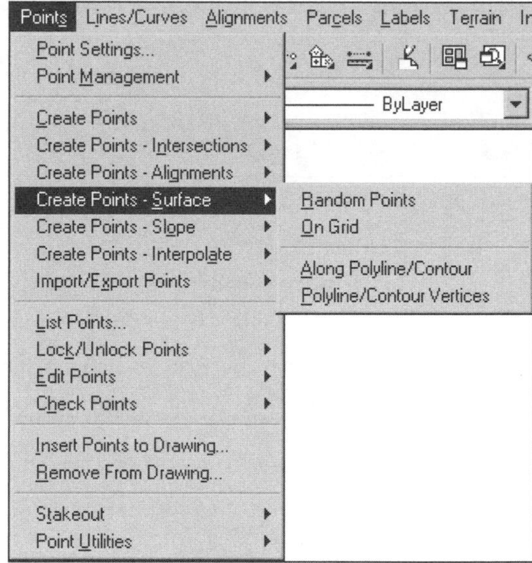

Figure 5–67

Exercise

After you complete this exercise, you will be able to:

- Set random points whose elevations are from a surface
- Set points on a grid whose elevations are from a surface
- Set points on a 3D polyline or contour whose elevations are from the object

RANDOM POINT

1. Run the Point Settings routine from the Points menu. In the Create tab, set the current point number to 500 and set descriptions as automatic with a value of Unit 6. See Figure 5–68. Select OK to exit the dialog box.

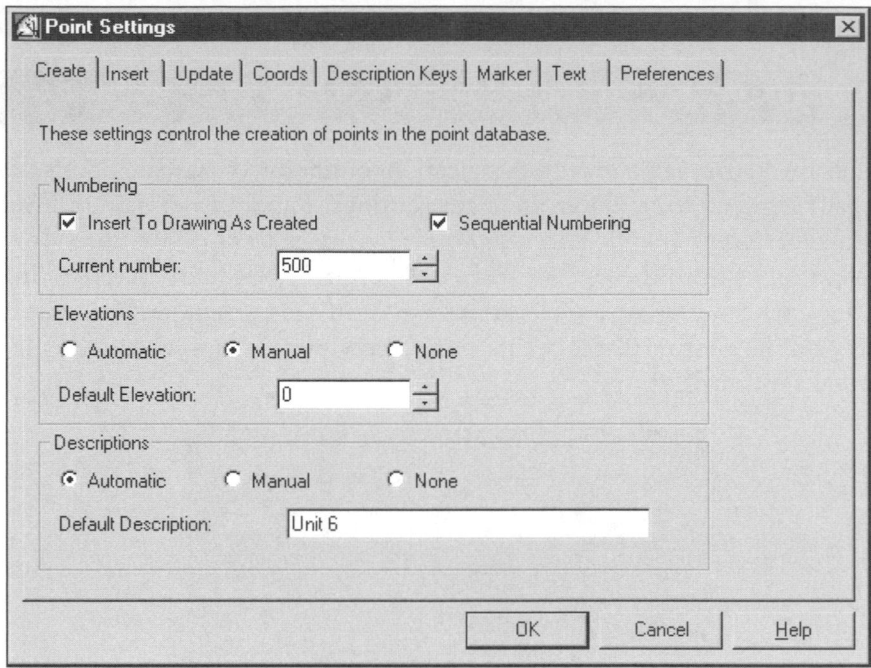

Figure 5–68

2. Select the Random Point routine from the Create Points–Surface cascade of the Points menu. Place about five points into the drawing and exit the routine.
3. Zoom in to view the points.
4. Erase the points from the drawing.
5. Do a Zoom Previous to see the entire surface.

ON GRID

The On Grid routine places points at a regular grid spacing. The routine prompts you for a rotation angle and base point for the grid. Next, the routine prompts a spacing interval for *X* and *Y* axes and then asks for the upper right corner of the grid. After you select the upper right corner, the routine responds by drawing a box to verify the grid orientation and location. If the box is correct, you press ENTER or type Yes to restart the defining process again. After you establish the corners of the grid, the routine uses the first point number at the end of the point database as the seed point number and places the points into the drawing. The elevations for the points are the elevations found on the current surface at the grid locations.

1. Define a grid that has no rotation and is within the boundary of the surface. Set a spacing of 50 feet in *X* and *Y* and select an upper right corner within the boundary of the TIN.

2. Zoom in to view the points.
3. Use the Remove Points from Drawing routine and select the points for removal.

ALONG POLYLINE/CONTOUR

The Pline routine places points at each vertex of a polyline whose elevation is from the current surface.

1. While zoomed in, draw a polyline within the bounds of the surface.
2. Use the Convert from 2D Polyline routine to convert the 2D polyline into a 3D polyline.
3. Use the Edit 3D Polyline routine and assign different elevations to the polyline vertices.
4. Run the Along Polyline/Contour routine of the Create Points–Surface cascade to create the point at each vertex.
5. Use the Remove Points from Drawing routine and select the points for removal.
6. Erase the polyline from the drawing.
7. Use the Erase routine from the Edit Points cascade and erase the number range from 500-1000.
8. Save the drawing.

This completes the discussion of the basic data groups in Land Development Desktop. The data groups—points, lines and curves, and surfaces—are the foundations on which projects are designed. These elements must be accurate and organized in the drawing and the project. If these elements contain errors, these errors will affect volumes, hydrological calculations, roadway design decisions, and grading strategies.

The next section explains how to create a new surface definition for the existing surface from contours and reviews some of the basic surface design tools found in LDD.

Site and Parcel Volumes

After you complete this section, you will be able to:

- Create a surface from contours, points, and breaklines
- Develop, modify, and create design contours
- Develop, modify, and create design 3D polylines
- Calculate Grid, Section, and Composite volumes
- Evaluate Grid, Section, and Composite volumes
- Calculate Grid and Composite parcel volumes

SECTION OVERVIEW

The calculation of volumes is a critical part of site design. You may need to calculate the volume of earthworks for a subdivision, stockpile, landfill, or Superfund cleanup site. Whatever the situation, a volume calculation is necessary. In Land Development Desktop, the routines for the volume calculation process are found on the lower portion of the Terrain menu. The process is a simple three-step procedure. First, name the two surfaces to compare; then, define the calculation area; and finally, calculate the volumes. LDD has three methods of calculating a volume: Grid, Section, and Composite.

UNIT 1

The volume you generate is between two surfaces. The first unit of this section is an overview of the surface design process and the calculation of volumes.

UNIT 2

The volume settings affect the volume calculation methods, the layers used in the process, and the annotation. These settings allow the user to create standards for the office and to bring a measure of consistency to drawing files. These settings are the topic of the second unit of this section.

UNIT 3

The calculation of volumes is between two surfaces. Generally, one surface represents existing conditions and the other is a design (modified) surface. This is by no means the only scenario. However, in this section, the assumption is an existing and design surface. Land Development Desktop offers the user several types of surface design tools. These tools include manipulating and creating points, contours, and 3D polylines. The most powerful tool for surface design in LDD is the 3D polyline. The 3D polyline provides surface elevation data and controls the triangulation process as a breakline. The design tools and creating a surface are the focus of the third unit of the section.

UNITS 4, 5, AND 6

The fourth, fifth, and sixth units cover the Grid, Section, and Composite volume calculation process, the evaluation of the results, and the annotation of the resulting values. The volume tools of LDD allow for an extensive review of the data and volume numbers.

UNIT 7

The last unit of the section covers the parcel volume methods in Land Development Desktop. The Grid and Composite methods are the only methods supported for subsite volumes. The evaluation of the results and the annotation of the volume values are the same as those covered in Units 4 and 5 of this section.

UNIT 1: THE SITE VOLUME PROCESS

The methods the designer uses to create a site depend upon the demands made of the site design. The eventual use of a site may be a subdivision, landfill, or park. The designer has to transform the site from its current condition to the final design. This transformation may involve raising and lowering areas within the site. To measure the amount of change is the task of the Volumes portion of the Terrain menu.

In a site design scenario, an initial topography creates the data for the first surface. The second surface is a designed surface based upon the original surface. This is not always the case. The scenario could be an assumed initial elevation for the first surface and the second is a stockpile of material. In essence a volume is just a comparison between two sets of surface data. In the case of a site design, the amount of material to move to create the new surface is of extreme interest. The Site Volume process represents the creation of surfaces and the estimation of volumes between them. This process is a set of tasks at the bottom of the Terrain menu (see Figure 6–1).

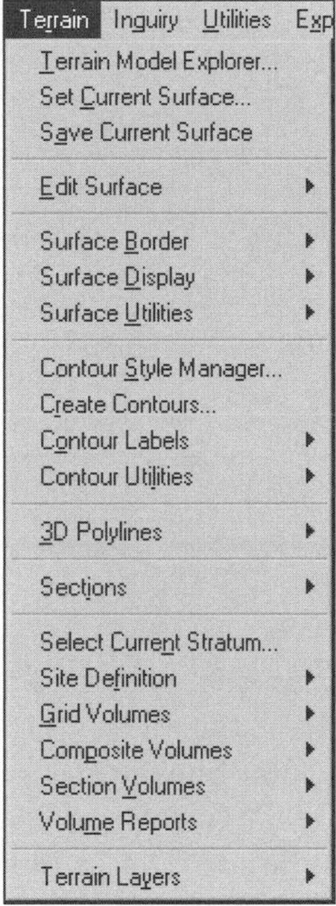

Figure 6–1

The Site Volume process begins with data for a surface (see Figure 6–2). The surface data comes from any combination of three sources: point, breakline, and contour. In order to calculate a volume, there must be two surfaces. After you create the two surfaces, the next step in the process is to define a stratum (see Figure 6–3). The stratum is an alias for the two volume surfaces. The volume routines calculate a volume between the two surfaces listed in a stratum.

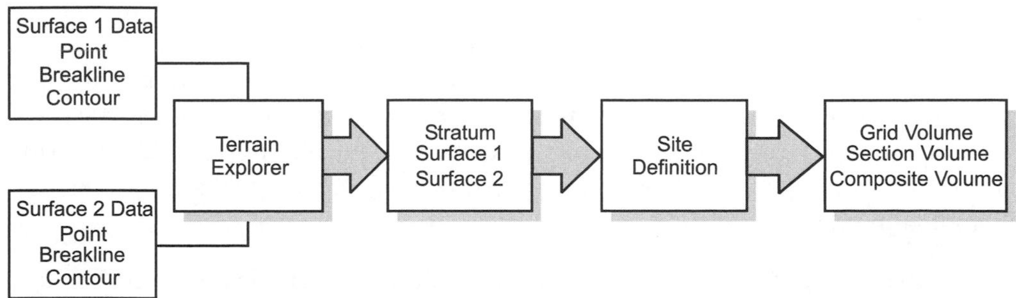

Figure 6–2

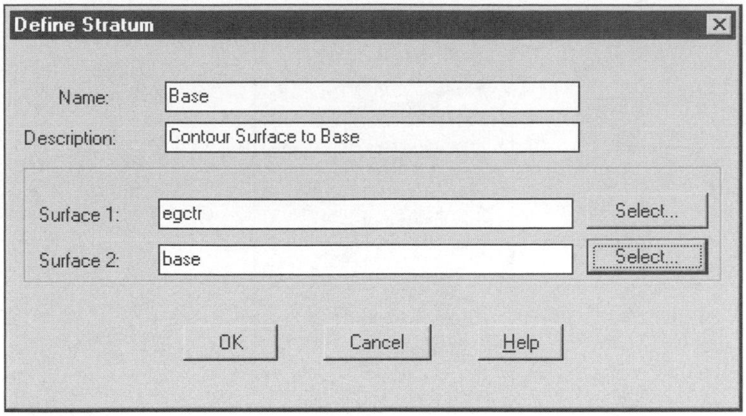

Figure 6–3

After you set the stratum, the next step is setting the sampling rules for the site (Volume Site Settings). The Volume Site Settings routine sets the grid and section sample spacing. You access the Volume Site settings from either the Drawing Settings dialog box or the Site Definition flyout of the Terrain menu. After you define the site settings, the next task is locating the site (Define Site routine of the Site Definition flyout). A site defines a volume calculation area. A site can be within or outside of the border of one or both surfaces.

If the surfaces are rotated, it may be necessary to rotate the site definition to match the rotation of the surfaces. When rotating a site, be aware that the upper right corner *must* be to the right and above the lower left corner of the site. During the site defining process, the routine will show a box representing the site and will draw a square representing the cell size or spacing in the lower left corner of the site box. If this square is outside of the box, you *must* redefine the site. A square outside of the box indicates that the upper right point is to the left and above the lower left corner. If you proceed with the calculation of volumes without correcting the site definition, the results will be unpredictable.

Finally, you use one or a combination of the three volume estimation methods to calculate a volume. Each volume method is an estimation process and each has its own strengths and weaknesses. The three methods for calculating volumes are Grid, Composite, and Section. The Section method uses either the average end or prismoidal model.

The Grid and Composite methods calculate a volume in a single step. The Section method uses two steps. The first step creates the cross sections and the second step calculates the volume.

The Grid Volume method will not evaluate any prism (a triangle portion of a grid cell) that has none or only one surface elevation at any one of its corners (see Figure 6–4). In the Section volume method, the routine will not sample for a volume until the current section has two surface elevations to compare. The Composite method develops a new surface whose triangles are from the combination of triangles of the two surfaces and samples only where there are two surface elevations at each triangle corner.

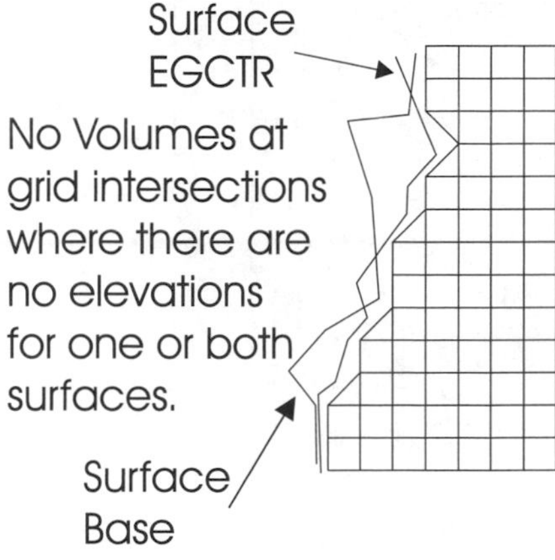

Figure 6–4

Because of the above conditions, a site (the area of interest) can be larger than both surfaces and still produce valid volume estimations. The Volume Generator scans the site area for the conditions used by the volume method, and it is only when the conditions are met that the Generator will begin to calculate volumes.

There is an intimate relationship between surfaces and volumes. Several critical elements influence the results of surface volume calculations. Each volume method calculates a volume differently. Each method is totally dependent on the quality of surface data. The better the quality of surface data, the better the volume results.

The best methods for moderating errors in volume calculations are consistent surface design methodologies, consistent data densities, and an awareness of the strengths and weaknesses of each volume calculation method. What is a good methodology for creating a design and what is the correct density of data are questions open to debate. What you need to develop is a consistent approach to the development of surface data. You must evaluate the numbers that a volume routine generates with the "ball park" value manually or mentally figured.

After calculating volumes, you can import into the drawing representations of the grid, composite, and section data that produces the volume calculations. You can represent volume data as surfaces, cut and fill contours, ticks and labels, or sample sections.

The Volume Reports cascade of the Terrain menu contains the routines that create reports to the screen or printer or tables in the drawing representing the resulting volume calculations (see Figure 6–5).

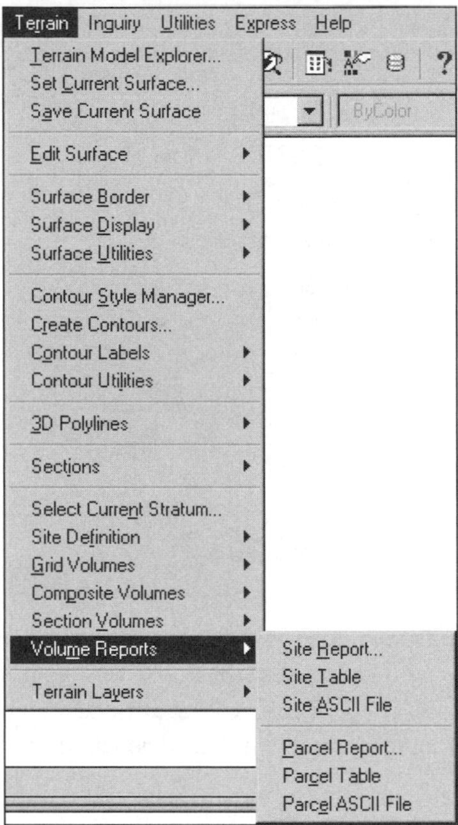

Figure 6–5

The Grid method can import grid ticks that represent the cut or fill values at the intersection points on the grid. The routine labels each tick (cell corner) with the elevation difference of the stratum surfaces.

The Grid Volumes and Composite Volumes routines create surfaces whose elevations are the differences in elevation between the two stratum surfaces. These surfaces become data for cut and fill contours representing depths of cut and fill.

The Section method creates cross sections that can be imported into the drawing. The Section Volumes cascade of the Terrain menu contains routines that insert the cross sections individually, all, or a page at a time.

Volumes calculated from contour data surfaces limit the effect of the design changes on the volume. The second surface design usually starts by copying the contours of the first surface to another layer. These copied contours become the starting point for the design of the second surface. When you use the original contours to design the second surface, the undisturbed contours limit the volume calculations to the first undisturbed contour beyond the design. This is because the volume between all unchanged contours is 0 (zero). The effect of the contours controlling the triangles allows you to radically change a surface and have only a volume extending out to the first undisturbed contour.

The blending of the second surface design into the first surface is best accomplished with these copied contours. Contours are one of the most effective means of creating a design surface, besides 3D polylines. The ease of blending a design into existing contours is straightforward (connecting the endpoints of new to existing contours), and areas undisturbed by the design process will produce no volume.

Contour Surface Rules:

- You should generate contours for each surface with the same contour style.
- When generating contour data for a surface, you should process the contour data with the same weeding and supplementing factors.

When you generate the contours for surfaces, they should have the same style. By varying the style, you can change the smoothing factor and possibly change the location of the contours. Copying the original contours to a new layer and then manipulating them for the new design fulfills this condition.

Different weeding and supplementing factors will produce a difference in surface data. If you create a difference in data between surfaces, there will be a change in the results of volume calculations.

UNIT 2: SITE VOLUME SETTINGS

All of the volume settings are found in the lower portion of the Edit Settings dialog box of the Drawing Settings routine on the Projects menu (See Figure 6-6). Remember that these settings come from the selected prototype when you first defined the project. You can edit the settings for the prototype so they are consistent from project to project. You can edit the settings in a project and copy them to a prototype to change or update its values. LDD also allows you to access these settings from the various flyout menus of the Terrain menu.

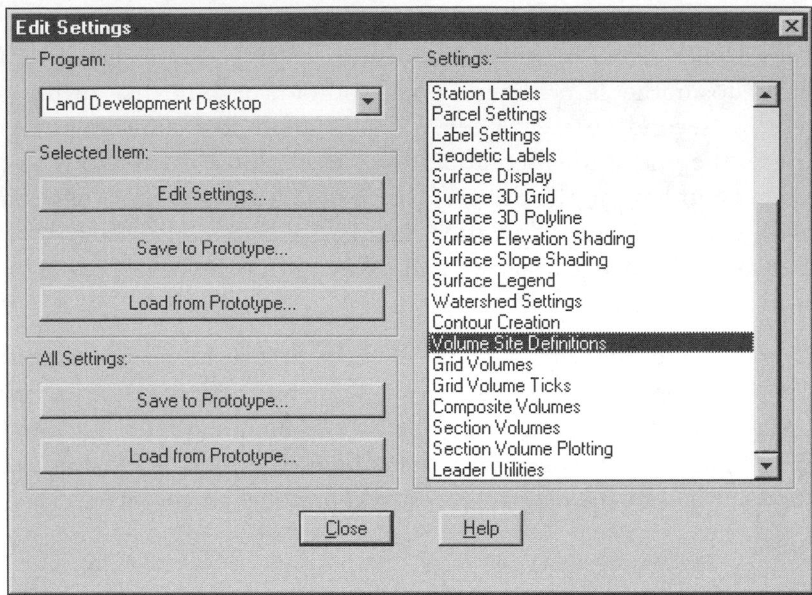

Figure 6–6

VOLUME SITE SETTINGS

The Volume Site Settings dialog box sets the values for the sampling spacing, site, volume labeling text styles, and the layers used in labeling volumes in the drawing (see Figure 6–7). The *M* and *N* direction portion of the dialog box sets whether the value box contains the size of or the number of cells. The grid volume method uses both *M* and *N* values to create the cell size or number when calculating a volume. The Section method uses either the *M* or *N* direction value to sample the surfaces in the direction that is perpendicular to the longest axis of the site. When you use the Section method, the value is the distance between sections. The reason for using *M* and *N* is that the sampling grid or section slices can be rotated to better sample the site. If rotated, the axes of the grid no longer represent the *X* or *Y* direction. To avoid confusion with *X* and *Y* directions, the axes of the site are referred to as *M* and *N*.

The Site Labeling section sets the text style when labeling the site. If you select a text style that has a variable height, for example standard, the routine uses the last used height. If you select a text style that is a fixed height style, the style will define the height of the text.

The Volume Labeling section sets the text style, the suffix, and the precision for the volume report routines.

The Site Layer setting specifies the layer used by the annotation routines.

The last setting toggles on the automatic placement of the site label into the drawing.

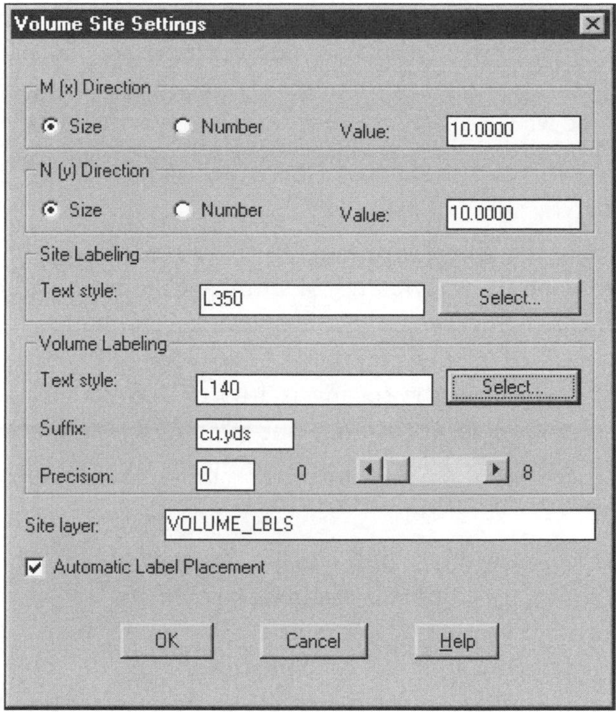

Figure 6–7

GRID VOLUME

The Grid Volume defaults set the initial values for the Grid volume calculations and adjustments (see Figure 6–8).

Grid Volume Settings
Elevation Tolerance
Minimum difference: 0.000
Grid Volumes Corrections
Cut factor: 1.000
Fill factor: 1.000
Grid Volumes Output
None
SDF
CDF
OK
Cancel
Help

Figure 6–8

The first value in the dialog box is the Elevation Tolerance. This is the minimum difference in elevation between the two surfaces that are included in a volume calculation. If the difference in elevation is set at 0.05, the volume generator does not include any grid node whose difference in elevation is less than 0.05 into the volume calculation. This setting tells the volume generator that the volume from the cell is not significant to the site volume.

The middle portion of the dialog box contains the Grid Volume adjustment values. The adjustments are for fluffing (swelling) or shrinkage (compaction) of the estimated volume. The cut factor is the swelling or increase of volume due to the removal of overburden. The cut factor is a number that is always one (1.0) or greater. A swelling factor of 5 percent means that the setting is 1.05 or 105 percent of the calculated volume. The fill factor is the loss of volume due to compaction and is a setting that is always one (1.0) or less. A fill factor of 5 percent means that the setting is 0.95 or 95 percent of the calculated volume.

The last setting is a toggle for a type of report for the grid volume. The report is either a space-delimited (SDF) or a comma-delimited (CDF) ASCII file. The report groups the corners of the grid cell into two volumetric triangles or prisms. The structure of the report is the cell's column and row, each prism's cut and fill volume, and the total net volume for the cell. Positive values mean fill and negative values mean cut. This is a meticulous method of reviewing how the grid routine calculates its volume.

GRID VOLUME TICKS

The Grid Volume Ticks defaults are for a tick and its label representing the difference in elevation between surfaces at a corner of a cell (see Figure 6–9). If the first surface is above the second surface at the cell corner, the label will be a negative number. If the first surface is below the second surface at the cell corner, the label will be a positive number. The first situation represents a cut and the second represents a fill. The defaults in the dialog box apply to both cut and fill ticks and their labels.

The Tick and Label Interval settings indicate how often to label the grid corner. A value of 1 labels each corner, a value of 2 labels every other corner, and so forth.

The Tick and Label Size settings control the plotted size of the tick mark and text.

The Grid Volume Ticks routine creates four layers. Two of the layers are for cut and two layers are for fill. Whether cut or fill, one of the two layers contains the tick and the other contains the labeling text.

The final default group, Label Position, controls the location of the text at the tick mark. The default is to place the text on the tick. Other choices for placing ticks are right, left, below (bottom), or above (top) the tick.

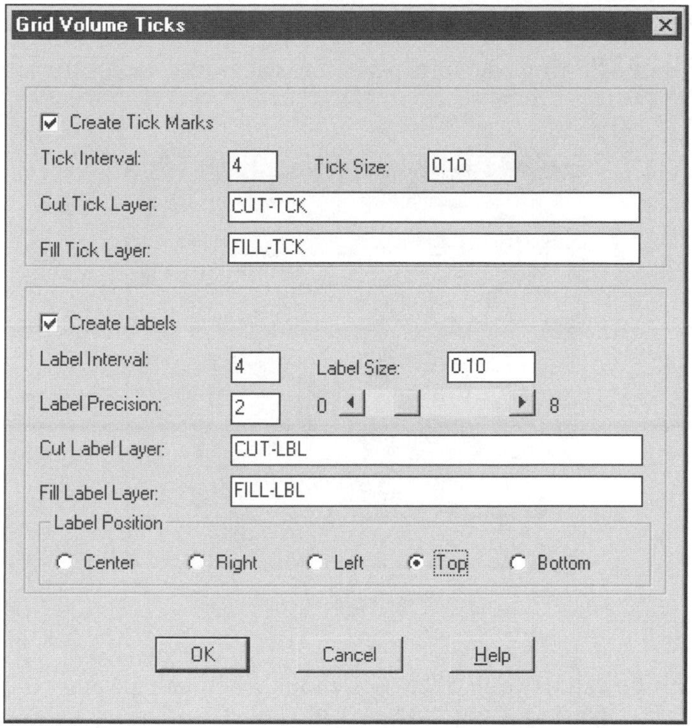

Figure 6–9

COMPOSITE VOLUME

The settings in the Composite Volume Settings dialog box affect the volume calculation and correction factors for this method (see Figure 6–10). The settings are the same as the Grid Volume settings, except for the report. The Composite method does not create a report. See the discussion on Grid Volume above.

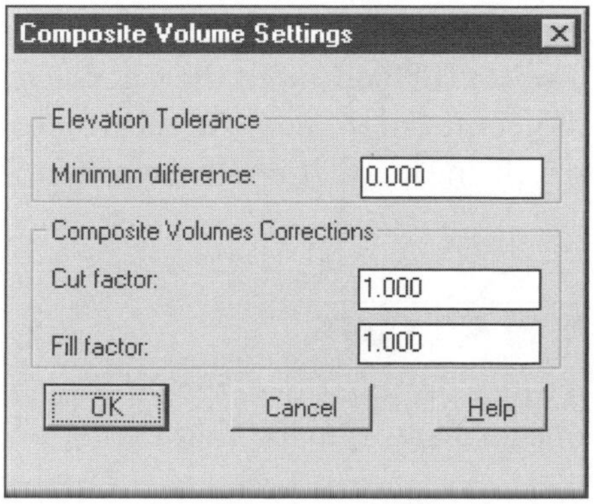

Figure 6–10

SECTION VOLUME

The Section Volume Settings defaults pertain to the initial values for section volumes (see Figure 6–11).

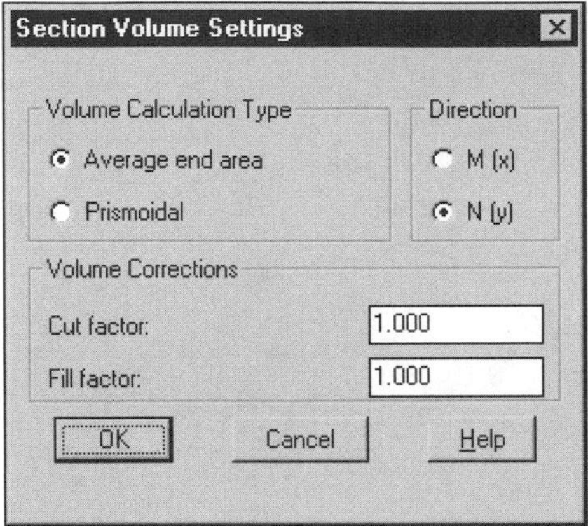

Figure 6–11

The Section routine can use one of two methods when calculating volumes: average end area and prismoidal. There is a distinct difference between the two methodologies.

The upper right portion of the Section Volume Settings dialog box sets the direction of sampling the two surfaces. When sampling in the *M* direction, the Section Volume routine slices the surface parallel to the *X*-axis (possibly rotated) of the site. If sampling the site in the *N* direction, the Section Volume routine slices the surface parallel to the *Y*-axis (possibly rotated) of the site.

The last group, Volume Corrections, affects the final volume report. The adjustments are for fluffing (swelling) or shrinkage (compaction) of the estimated volume. The cut factor is the swelling or increase of volume due to the removal of overburden. The cut factor is a number that is always one (1.0) or greater. A swelling factor of 5 percent means that the setting is 1.05 or 105 percent of the calculated volume. The fill factor is the loss of volume due to compaction, and is a setting that is always one (1.0) or less. A fill factor of 5 percent means that the setting is 0.95 or 95 percent of the calculated volume.

SECTION VOLUME PLOTTING

The Cross Section Plot Settings defaults affect the plotting of volume cross sections. You can import the sections into any drawing that is a part of a project.

The initial dialog box, Cross Section Plot Settings, toggles on or off the basic components and their corresponding layers in a basic cross section (see Figure 6–12).

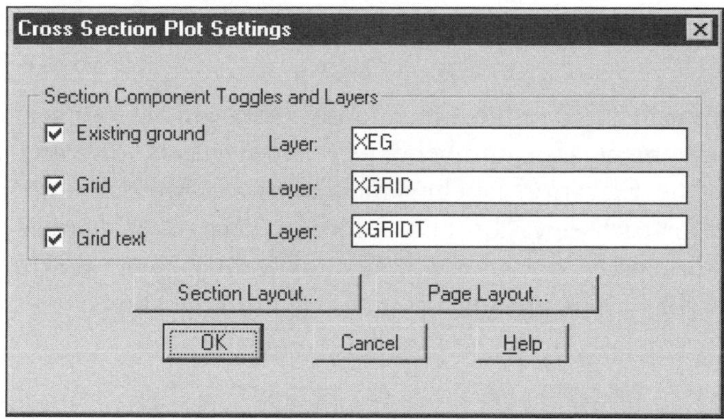

Figure 6–12

At the bottom of the dialog box are two buttons that display additional settings dialog boxes.

SECTION LAYOUT

The Section Layout dialog box sets the values for the annotation of the grid on a volume section (see Figure 6–13).

Section Layout
Offset
Increment: 25.000
Label interval: 2
Label precision: 0
Elevation
Increment: 2.000
Label interval: 2
Label precision: 0
Section Height
Rows below datum: 2
Rows above maximum: 0
OK
Cancel
Help

Figure 6–13

The upper left portion defines the values for the interval and annotation of cross-section offsets. The Offset increment is the number of feet between each vertical offset line. The Offset labeling increment sets the labeling frequency for offsets; 1 means each offset line, 2 means every other offset line, and so forth. The Offset precision is the number of decimal places in the offset label. The initial values are for offsets of 10 feet, labeling every other offset (2) and the offset distance is a whole number (0 precision).

The upper right portion defines the interval and annotation of cross-section elevations. The Elevation increment is the number of feet between each horizontal elevation line in the section. The Elevation labeling increment sets the labeling frequency for elevations; 1 means each elevation line, 2 means every other elevation line, and so forth. The Elevation precision sets the number of decimal places for the vertical label. The initial values for elevation are elevations every 2 feet, labeling every other elevation (2), and the elevation is a whole number (0 precision).

The bottom portion of the dialog box sets the number of elevation lines above and below the "normal" cross section.

PAGE LAYOUT

The Page Layout defaults set the sheet size and margins, the horizontal and vertical spacing between cross sections, and the number of pages stacked before starting a new column of cross sections (see Figure 6–14). The default page size is a 24 by 36-inch sheet of paper, with a 2-inch margin on the left side and a 1-inch margin on the remaining edges.

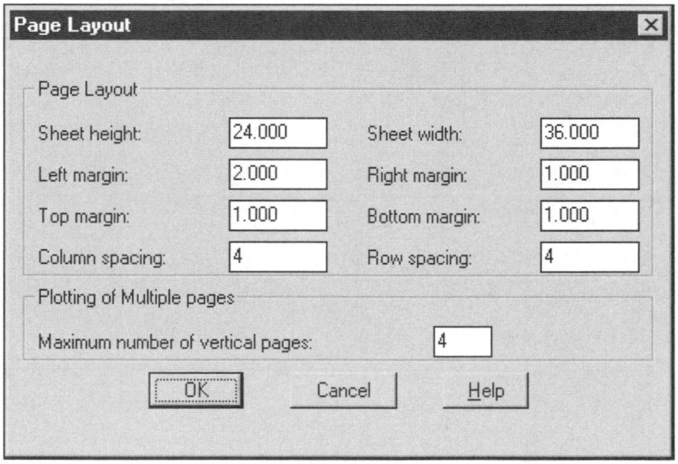

Figure 6–14

At the bottom of the Page Layout area are the settings for the column and row spacing between cross sections. The width and height of each space is the horizontal and vertical distance of a cross-section cell.

At the bottom of the dialog box are the settings for the maximum number of sheets in a stack of cross-section pages.

Exercise

After you complete this exercise, you will be able to:

- Review the Earthworks settings

1. If you are not in the TERRAIN drawing of the Surfaces project, open the file.
2. Select the Drawing Settings routine from the Projects menu to display the Edit Settings dialog box (See Figure 6-6).

VOLUME SITE DEFINITIONS

1. Select the Volume Site Definitions item from the list to display the Volume Site Settings dialog box. Set the following values. Use Figure 6–7 as a guide. After setting the values, select OK to exit the dialog box.

   M (x) Direction

   Size: Toggle ON Value: 10

   N (y) Direction

   Size: Toggle ON Value: 10

   Site Labeling

   Text Style: L350

   Volume Labeling

   Text Style: L140

2. Select Grid Volumes from the list to display the Grid Volume Settings dialog box. Set the Elevation Tolerance to 0.00. Refer to Figure 6–8. Select OK to exit the dialog box.
3. Select Grid Volume Ticks to display the Grid Volume Ticks dialog box. Set the following values. Refer to Figure 6–9. Select OK to exit the dialog box.

   Tick Interval: 4

   Label Interval: 4

   Label Precision: 2

   Label Position: Top

4. Select the Composite Volumes item to view the values in the Composite Volume Settings dialog box. Make sure the Elevation Tolerance is set to 0.00. Refer to Figure 6–10. Select OK to dismiss the dialog box.
5. Select Section Volumes to view the settings in the Section Volume Settings dialog box. Refer to Figure 6–11. Select OK to dismiss the dialog box.

6. Select Section Volume Plotting to view the Cross Section Plot Settings dialog box. Toggle on all of the components at the top of the dialog box. Refer to Figure 6–12.
7. Select the Section Layout button in the Cross Section Plot Settings dialog box to view the current values. Set the offset increment to 25 feet. Refer to Figure 6–13. Select OK to return to the Cross Section Plot Settings dialog box.
8. Select the Page Layout button to view the page layout settings. Refer to Figure 6–14. The current sheet is a 24x36-inch sheet of paper with a 2-inch binder margin on the left and 1-inch margins on the remaining edges. The column and row spacing is 4 cells. Cell measurements are from the offset and elevation values in the Section Layout settings. Currently a cell is 25x2 feet in size. The size of the cell on the paper is a function of the plotting scale. Select OK to return to the Cross Section Plot Settings dialog box.
9. Select OK to exit the Cross Section Plot Settings dialog box.
10. Select Close to dismiss the Edit Settings dialog box.
11. Save the drawing.

UNIT 3: BASIC SURFACE DESIGN TOOLS

In Land Development Desktop, there are three basic tool groups for designing new surfaces. The first group is point tools. The routines in this group create new points from existing points, surfaces, or 3D objects. These tools are found in the Create Points, Create Points–Surface, Slope, and Interpolate cascades of the Points menu.

The second group of tools creates new contours from existing contours. The tools create new contours by offsets at a grade or slope to a known elevation or distance. These tools are found in the Contour Utilities cascade of the Terrain menu.

The last set of tools creates 3D polylines. The routines create 3D polylines from existing points, known elevations, or surfaces. These tools are found in the 3D Polylines cascade of the Terrain menu.

If you have purchased the Civil Design add-on, you have one more designing tool, the grading object. The grading object is the topic of a later section of this book.

POINT DATA

CREATE POINTS

The Create Points cascade contains two routines to create points from Polyline/Contour vertices (see Figure 6–15). The first routine creates points whose elevations you enter, and the second takes the elevations from the polylines.

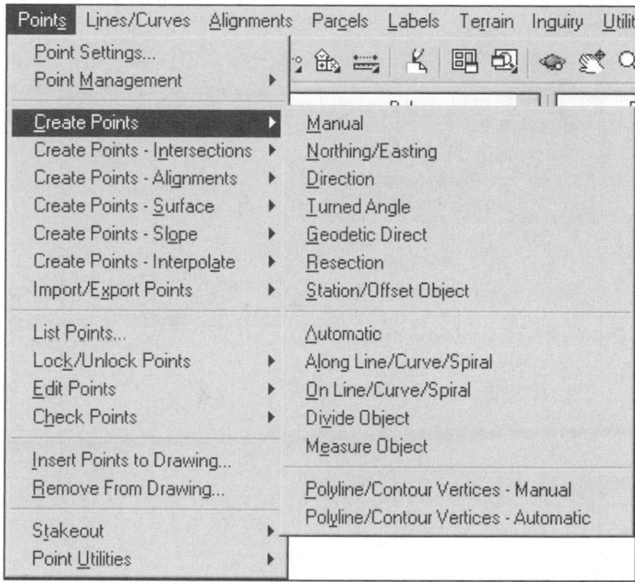

Figure 6–15

Polyline/Contour Vertices–Manual

The Polyline/Contour Vertices–Manual routine places points at each vertex of a polyline/contour, but prompts you for or wants verification of the elevation at the vertex. This routine works with 2D and 3D polylines. This routine allows you to locate points you need to use while designing a surface or to identify critical points along the polyline/contour. The elevation of the point is assigned by you or pulled from the 3D object.

Polyline/Contour Vertices–Automatic

The Polyline/Contour Vertices–Automatic routine places points at each vertex of a polyline/contour. The elevation for the point is from the elevation of the vertex of the polyline/contour. The routine assumes that the object is a 3D object with elevations at each vertex. By placing the points on the 3D Polyline, you are able to send the coordinates and elevations to a data collector for field staking.

CREATE POINTS–SURFACE

The routines of the Create Points–Surface cascade place points in the drawing whose elevation represents surface elevations (see Figure 6–16). The Random Points routine creates points when making a selection on the screen. The elevations of the points are elevations from the current surface. The On Grid routine creates points in a grid pattern whose elevations are from the current surface. By creating points whose elevation is from a surface, you are able to send the elevations to a data collector for field staking.

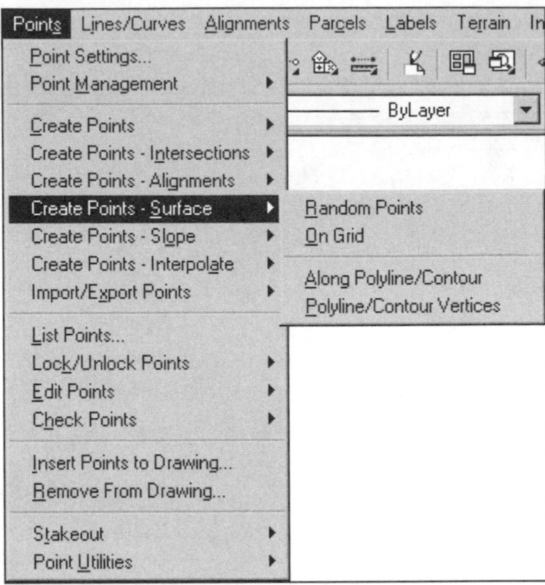

Figure 6–16

Along Polyline/Contour

The Along Polyline/Contour routine places points on the objects at a user-specified distance. The elevations of the points are determined by the elevations found on the current surface along the course of the polyline/contour. The distance between points is set by the user either as a specific distance or a selection of two points along the polyline/contour.

Polyline/Contour Vertices

The Polyline/Contour Vertices routine places points at the vertices of the objects. The elevations of the points are determined by the elevations found on the current surface along the course of the polyline/contour.

CREATE POINTS–SLOPE

The Create Points–Slope cascade contains three point routines (see Figure 6–17). All of these routines create new points that represent a design based upon a slope relative to existing points. Examples of this strategy are designing drainage around house pads, slopes around edges-of-pavement, and slope design around a parking lot.

High/Low Point

The High/Low Point routine places a point at the intersection of a grade or slope from two existing points. The slope or grade is along the line projecting from each point toward their intersection. This intersection maybe important to a surface design and as such needs to be a part of the point database. Eventually, this point will be staked out in the field and its coordinates come from the design drawing.

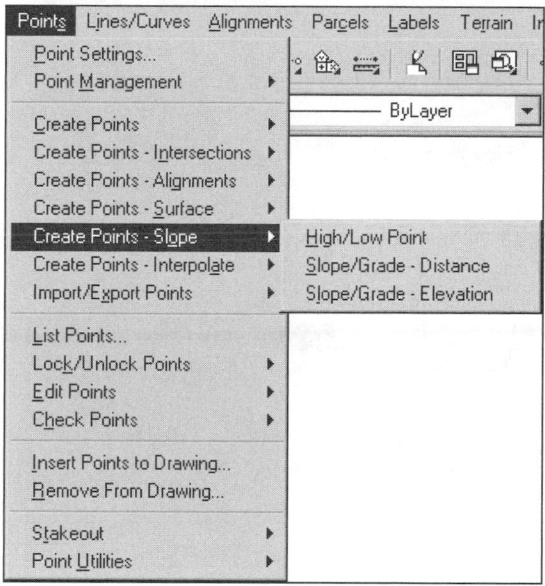

Figure 6–17

Slope/Grade Distance and Slope/Grade Elevation

The Slope/Grade Distance or Slope/Grade Elevation routines place point objects at an interval for a preset distance, at a slope or grade from that point. For example, starting at point 100, whose elevation is 723.34 feet: From point 100, traveling a distance of 75 feet at an azimuth of 285 degrees with a slope of 4 to 1, the routine places four intermediate points and an ending point 75 feet away. The five new points reflect the rise of elevation (4:1) and direction of travel.

The Slope/Grade Elevation routine places points from an existing point at a grade/slope, direction, and distance. You can specify an elevation override that places the new points before or after the existing point depending upon whether the elevation override is higher or lower than the initial point's elevation.

CREATE POINTS–INTERPOLATE

The routines in the Interpolate cascade place new points into the drawing by slopes and grades (see Figure 6–18). The routines work with existing points or AutoCAD objects. After selecting the beginning and ending points, you define the method for calculating the intermediate points. You can define a distance along the line between the points (a fixed incremental distance) or specify a number of points over a specific distance. You can identify points that are perpendicular to or intersecting with a line or figure. These routines allow you to create point data dense enough so you may not need to create breaklines to produce triangulation correctly representing the slopes on a surface.

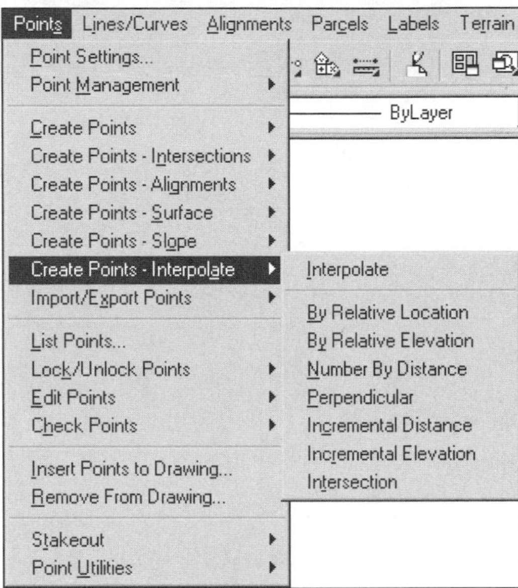

Figure 6–18

Interpolate

The Interpolate routine places a specific number of new points between two existing point objects. The elevations assigned to the new points reflect the slope and distance between the two existing points.

By Relative Location

The By Relative Location routine creates a new point as a user-specified distance from the first of two known points. The new point can be offset from the line connecting the two controlling points. The routine allows you to redefine the elevation of the second point as a function of a grade/slope, difference, or a user-specified elevation.

By Relative Elevation

The By Relative Elevation routine is the same as By Relative Location except the location of the new point is set by a user-specified elevation that occurs between the two initial points. The new point can be offset from the line connecting the two controlling points. The routine allows you to redefine the elevation of the second point as a function of a grade/slope, difference, or a user-specified elevation.

Incremental Distance

The Incremental Distance routine places points into the drawing based upon a user-specified interval distance measured from the first control point. The new points can be offset from the line connecting the two controlling points. The routine allows you to redefine the elevation of the second point as a function of a grade/slope, difference, or a user-specified elevation.

Incremental Elevation

The Incremental Elevation routine is the same as the Incremental Distance routine except the user specifies an increment of elevation to set the points. The new points can be offset from the line connecting the two controlling points. The routine allows you to redefine the elevation of the second point as a function of a grade/slope, difference, or a user-specified elevation.

Number By Distance

The Number By Distance routine creates a user-specified number of points between two controlling points. The new points can be offset from the line connecting the two controlling points. The routine allows you to redefine the elevation of the second point as a function of a grade/slope, difference, or a user-specified elevation.

Intersection

The Intersection routine allows you to place points where selected objects intersect or where they would intersect if extended. The new points can be offset from the line connecting the two controlling points and the intersecting object. The routine allows you to redefine the elevation of the second point as a function of a grade/slope, difference, or a user-specified elevation.

Perpendicular

The Perpendicular routine places a point on an existing object where a projected point is perpendicular to that object. The new points can be offset from the line connecting the two controlling points. The routine allows you to redefine the elevation of the second point as a function of a grade/slope, difference, or a user-specified elevation.

CONTOUR DATA

You can create contours for a design by manipulating existing contours or sketching in new contours. After you draft in the new contour locations, the Terrain Surface Modeler converts the contours into data for a surface.

CONTOUR UTILITIES

The Contour Utilities cascade contains tools for the creation of new contours from existing ones (see Figure 6–19). The new contours are a result of sketching or the offsetting of existing contours at a grade/slope or distance. These routines are best used in the creation of contours representing a slope from a house pad, roadway edge, or drainage design.

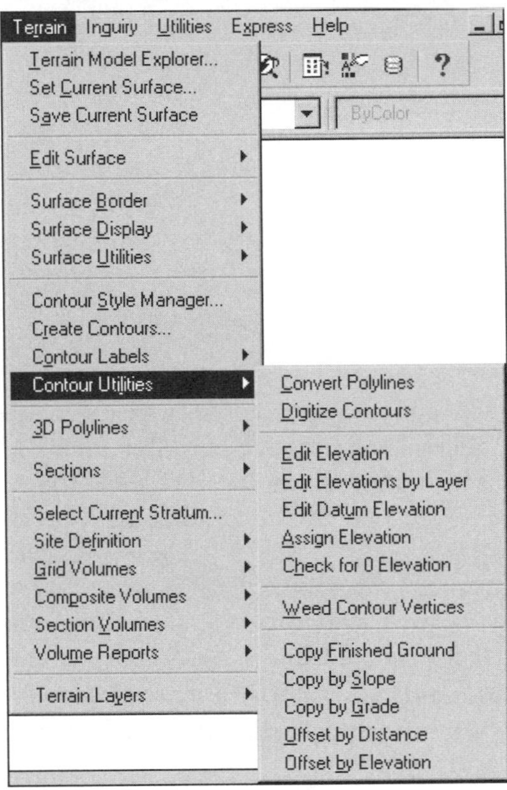

Figure 6–19

The top portion of the menu contains routines for drawing new contours. The Digitize Contours routine prompts for an elevation and assigns the elevation to the polylines you draw with the routine. Other routines enable you to draw polylines and then use the Edit Elevation, Edit Elevations By Layer, or the Assign Elevation routines to give the polylines the appropriate elevation. The Check for 0 (zero) Elevation routine scans a selection set of polylines for polyline with a 0 (zero) elevation. If there are any zero elevation polylines, the routine highlights the polyline and prompts you for the correct elevation.

The bottom portion of the menu contains routines that create new polyline/contours from existing ones.

Copy Finished Ground

The Copy Finished Ground routine copies a selection set of contours onto a new layer. The designer can start the design process by using the original contours. This routine prevents the design process from destroying the original set of contours.

Copy By Slope

The Copy By Slope routine creates new contours that are offset from the original contour representing a specific slope. For example, offsetting a contour at a 4:1 slope creates a new contour one-foot higher and four feet from the original contour.

Copy By Grade

The Copy By Grade routine creates new contours that are offset from the original contour representing a specific grade. For example, offsetting a contour at a 25 percent grade creates a new contour one-foot higher and four feet from the original contour.

Offset By Distance

The Offset By Distance routine creates new contours over a specific distance from an existing contour. The routine prompts for a grade to apply to the calculation of new contours, the distance to cover, the contour interval, and layer names for major and minor contour intervals. After you enter the information, the routine creates the contours to reach the specified distance.

Offset By Elevation

The Offset By Elevation routine creates new contours to a user-specified elevation from an existing contour. The routine prompts for a grade to apply to the calculation of new contours, the elevation to attain, the contour interval, and layer names for major and minor contour intervals. After you enter the information, the routine creates the contours to reach the specified elevation.

3D POLYLINES

The 3D Polylines cascade of the Terrain menu contains tools to crcatc and edit 3D polylines (see Figure 6–20). These tools create 3D polylines representing elevations or slopes.

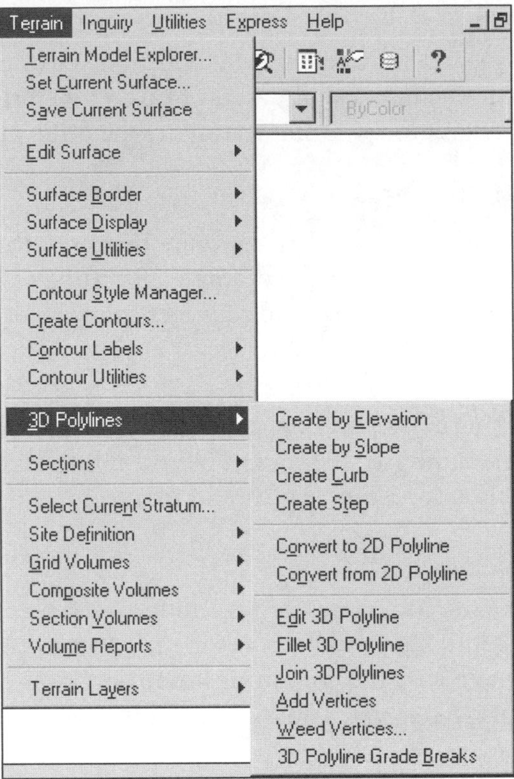

Figure 6–20

Create By Elevation

The Create By Elevation routine produce 3D polylines by specific elevations, surface elevations, or by transitioning between elevations along the 3D polyline. The routine allows for curves, transitions of elevation over several vertices, and assigning elevations to vertices from surface elevations. The routine does not create a curve in the 3D polyline, but a series of chord segments.

Create By Slope

The Create By Slope routine uses grades or slopes to calculate the elevation of the next vertex on the polyline. The user sets the elevation of the first point. After selecting the next and each succeeding point, the routine assigns an elevation based upon a grade or slope that occurs between the two selected points.

Create Curb

The Create Curb routine creates a new 3D polyline by offsetting from an existing 3D polyline. The routine prompts for the direction of the offset, the distance of the offset, and the elevation or difference in elevation from the original elevation for each vertex of the new polyline. If you enter an elevation, each vertex along the offset polyline is set to that elevation. If you enter a difference in elevation, the routine adds the value (the value can be a negative value) to all vertices of the new offset polyline.

Create Step

The Create Step routine creates a new polyline by offsetting from an existing polyline and applies a user-specified elevation or difference to each vertex of the offset polyline. The difference between this routine and the Create Curb routine is that the Create Step routine prompts for an elevation or difference in elevation at each vertex along the offset polyline.

There are several 3D Polyline editing tools in the 3D Polylines cascade.

Convert to 2D Polyline

The Convert to 2D Polyline routine creates a 2D polyline from a 3D polyline.

Convert from 2D Polyline

The Convert from 2D Polyline routine creates a 3D polyline from a 2D polyline.

Edit 3D Polyline

The Edit 3D Polyline routine edits existing 3D polylines, much like the PEDIT command of AutoCAD. The routine has an option to change the elevations found at each vertex of a 3D polyline.

Fillet 3D Polyline

The Fillet 3D Polyline routine adds a fillet to a polyline. The routine does not place a curve into the polyline, but places user-specified chord lengths instead.

Join

The Join routine allows you to join two 3D polylines into a single 3D polyline. The elevation must be the same at the joining point of the two polylines.

Add Vertices

The Add Vertices routine places new vertices to a 3D polyline at a user-specified distance.

Weed Vertices

The Weed Vertices routine removes redundant vertices from a 3D polyline.

Exercise

Two surfaces must be defined in a project before you can calculate a volume. There is now one surface in the Surfaces project and a second surface needs to be developed. The scenario for this exercise is the clearing of the site, filling in and leveling of elevations, and calculating a site (overall volume). The question is how much material needs to be moved to accomplish this goal.

This exercise has two parts. The first part is creating a surface whose data is from the contours of the existing surface. This is done for two reasons. The first reason is to compare the calculated volumes between the existing surface as a surface of points and a surface from contours and the design surface. The second reason is to review the potential problems of calculating a volume between surfaces create with points and contours.

The second part of the exercise creates the design surface. The design of the second surface sets the stage for the volume calculations.

After you complete this exercise, you will be able to:

- Create a surface from contours
- Design a site with points, contours, and 3D polylines

SURFACE FROM CONTOURS

1. Open the drawing file Terrain from the Surfaces project. The drawing is from the exercises of Section 5.
2. Restore the Named View, surface.
3. Use the Create Contours routine from the Terrain menu to create contours for the existing surface. Use Figure 6–21 as a guide in setting the contour values. The Mystyle of contours was developed in Section 5.

If you did not do the exercise in Section 5, use the *terrainv.dwg* drawing found on the CD that accompanies this book. Start a new drawing, terrain, and assign it to a new project. Accept the default point database definition and select i40 setup. After defining the new drawing and the project, use the AutoCAD insertion routine to insert the *terrainv.dwg*. Insert the drawing at 0,0,0, with a scale of 1, no rotation angle, and explode the file. The drawing will have the existing contours showing on the screen.

4. Open the Terrain Model Explorer.
5. Select Terrain at the top of the surface tree, click the right mouse button, and select Create New Surface from the shortcut menu. A new surface, surface1, appears on the list.
6. Select the name Surface1, click the right mouse button, and select Rename. Rename the surface EGCTR.

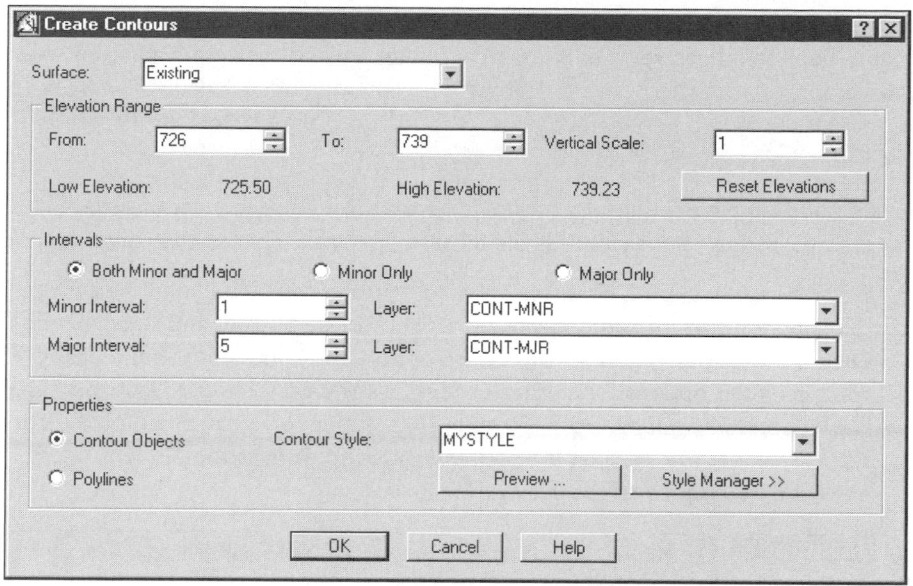

Figure 6–21

7. Create contour data for the surface using the contours in the drawing. Select the "+" (plus sign) under EGCTR to view the data types for the surface. Select the contours, click the right mouse button, and select Add Contour Data. Make sure Create as Contour Data is on. Use a weeding factor 15.0 for length and 4.0 for angles. The supplementing factors should be 30.0 and 0.25. Use Figure 6–22 as a guide. Select OK to exit the dialog box. Use the Layer option to select the contours from the screen to generate the surface data. Select one contour from each layer to identify that layer as surface data (*Existing-cont-mnr* and *Existing-cont-mjr*). Press ENTER to exit the selection process. The routine prompts for a layer name. This prompt allows the routine to scan layers that are Off or Frozen. Press ENTER to exit the prompt and reenter the Terrain Explorer.

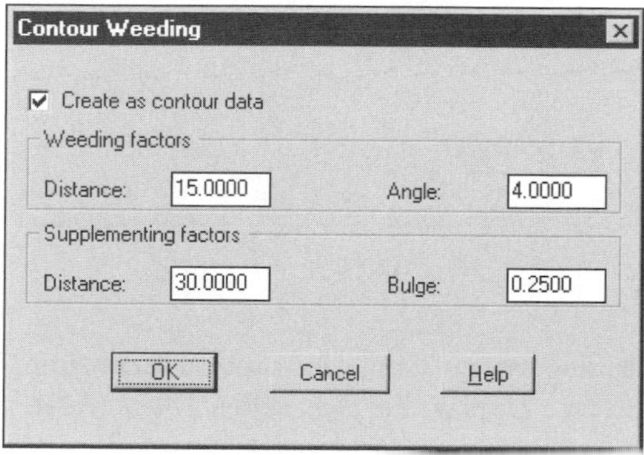

Figure 6–22

As discussed in Unit 2 of Section 5, the weeding factor removes redundant vertex data when the distance between three vertices is less than 15 feet and the deflection angle between the first and second vector is less than 4 degrees. The routine adds a calculated vertex to the contour data if the length between contour vertices more that 30 feet or if there is an arc segment with a bulge greater than 0.25.

The weeding and supplementing factors should be the same for every set of surface data the designer creates because consistency is the most important aspect of the volumes process.

8. Build the surface, EGCTR, using only the contour data. You build the surface by selecting the surface name, clicking the right mouse button, and selecting the Build option. After you select the Build option, the routine displays a dialog box containing the options for surface building. Toggle on Contour Data and Minimize Flat Triangles Resulting From Contour Data. Refer to Figure 6–23 for the correct settings. Select OK to start the processing and to dismiss the Done process dialog box to return to the Terrain Explorer.

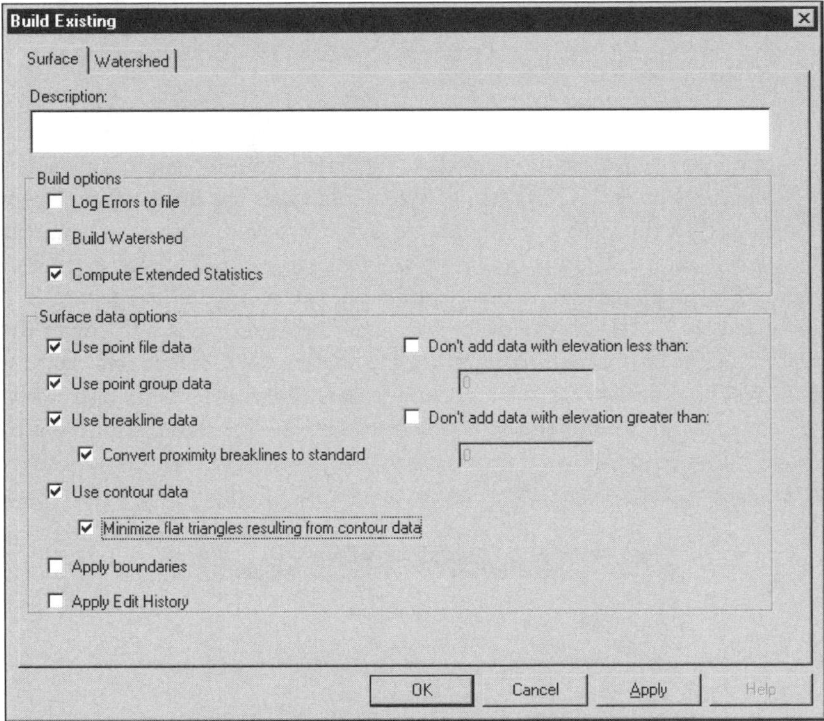

Figure 6–23

9. To view the resulting triangulation, select the surface name, click the right mouse button, select Surface Display, and then select Quick View.
10. Minimize the Terrain Explorer to view the surface triangles.
11. Redraw the screen to remove the Quick View triangles.

12. Restore the Terrain Explorer and review the point count for the new surface. This value is found in the middle right portion of the main surface panel. Compare this number to the number of points in the Existing surface. The EGCTR surface has many more triangles than the Existing surface.
13. Close the Terrain Explorer.
14. Save the drawing.
15. Copy the Existing contours to a new layer, EXCONTOUR. Use the Copy Finished Ground routine of the Contour Utilities cascade of the Terrain menu. Select and copy all of the contours.
16. Freeze the Existing contour layers, EXISTING-CONT-MJR and EXISTING-CONT-MNR.
17. Use the Layer dialog box to make EXCONTOUR the current layer.

DESIGNING THE SECOND SURFACE

The immediate area of concern is the northeast corner of the site, where the first priority is filling in the swale. Since the goal is to fill in the swale, the designer needs to indicate its removal. What indicates the filling in of a swale? If fill is brought into the site, the elevations found in the swale will rise. As the material fills the swale, the lower elevations recede towards the northeastern part of the site. In effect the fill pushes the contours from the interior of the site towards the exterior.

To create these changes in the drawing, remove the contours that are in the interior of the site and redraw new contours reflecting the filling in of the swale (see Figure 6–24).

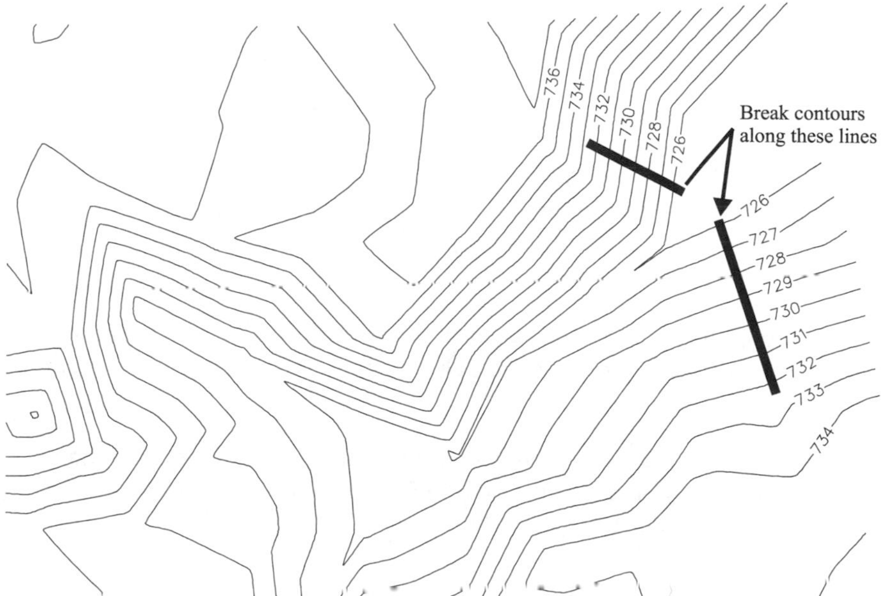

Figure 6–24

Northeast Section:

1. Use the Zoom to Point routine and zoom to point 111 with a height of 275. If there are no points in the drawing, use the ZOOM CENTER command with the coordinates of 5295,5440 and a height of 275.

2. Break the contours using Figure 6–24 as a guide.

3. Draw the new contour for the 726 elevation with the AutoCAD PLINE command. Remember to use the object snap of endpoint to connect the pline to the contour. Use Figure 6-25 as a guide for drawing the new contours.

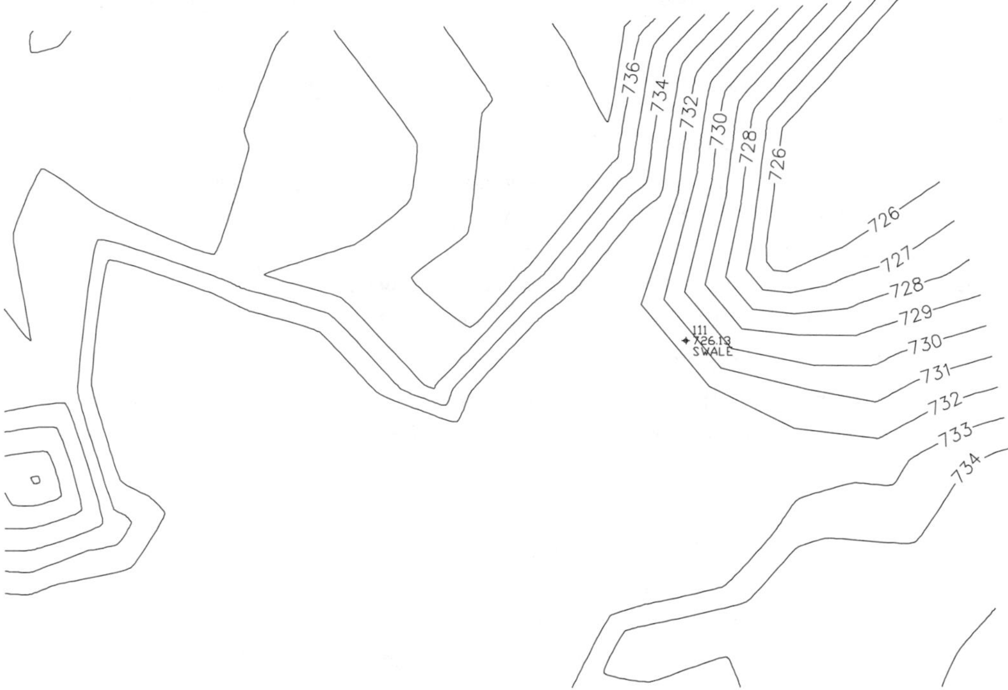

Figure 6–25

4. Check the elevation of the new contour with the Contour Elevation routine found on the Inquiry menu. The polyline should have the same elevation as the contour.

5. If the elevation of the new polyline does not match the elevation of the contour, edit the new polyline segment with the Edit Elevation routine of the Contour Utilities cascade and change its elevation to 726.

If you want the spacing of the new contours to represent a specific slope or grade, use the Copy at Slope or Grade routines of the Contour Utilities menu. The first contour is in place and now offset the next six contours at a set slope.

6. Use the Copy by Slope routine to create the remaining six contours. Set the interval to 1, the run to 8, and the rise to 1. Offset the new polylines to the southwest. To connect the polylines to the existing contours, activate the grips of the new polylines and the contours. Stretch the first and last vertices of the new polylines to connect to the end vertices of the broken contours. You may have to zoom in to better see the polylines and their connections to the contours.

Northwest Section:

7. Use the Zoom to Point routine and zoom to point 122 with a height of 300. If there are no points in the drawing, use the ZOOM CENTER command with the coordinates of 5070,5380 and a height of 300.

The next modification is on the northwest side of the new flat area. Create a cut or flattening out of a high spot by erasing some of the existing contours and breaking others (see Figures 6–26 and 6–27).

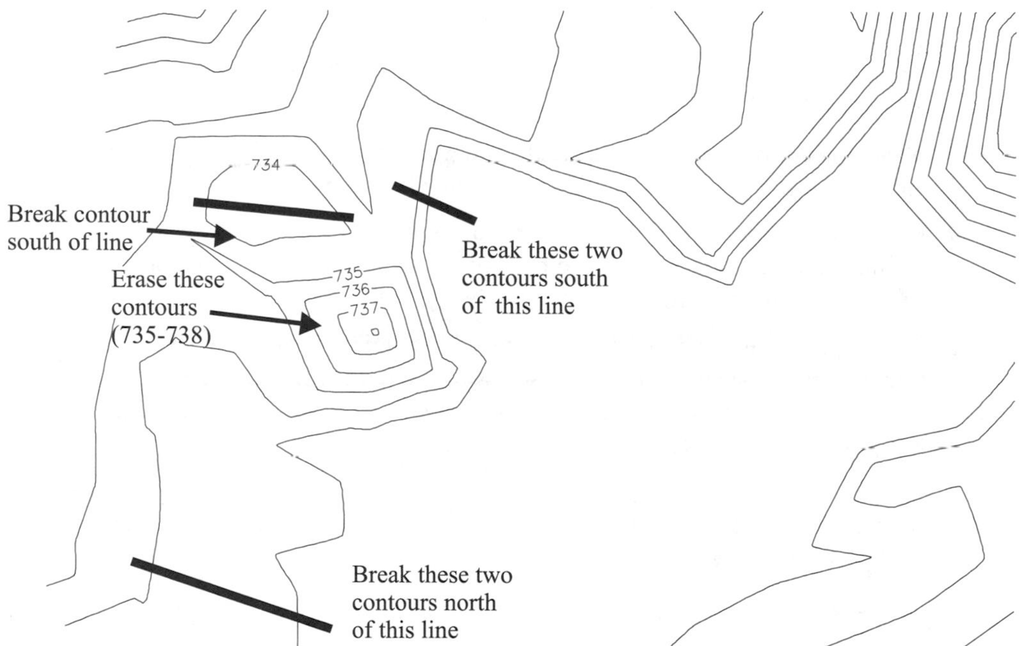

Figure 6–26

EXERCISE

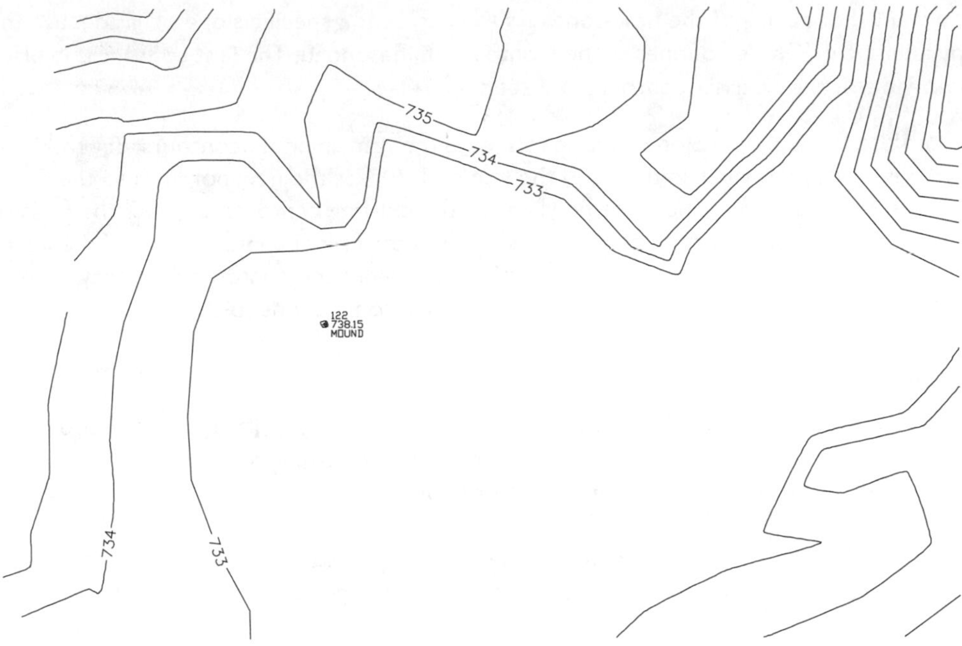

Figure 6–27

8. Erase the hill contours on the northwest of the new flat area—the contours around point 122.
9. Break the two contours on the south and north side of the flat area at the indicated locations (elevations 733 and 734). Use Figure 6–26 as a guide.
10. Break off the southern portion of the closed polyline. The 734 contour will connect on the southern end and connect from the eastern end to the existing 734 contour to the right.
11. Before you draw the contours make sure EXCONTOUR is the current layer. Use the AutoCAD POLYLINE command with the endpoint object snap to create the new segments connected to the existing contours. Use Figure 6–27 as a guide in creating the new polylines.
12. List the elevations of the new polylines—they should be the correct elevation. If the new polylines are not the right elevation, use the Edit Elevation routine to assign the correct elevations.
13. Save the drawing.

Southeast Section:

The next area of concern is the southeast side of the site. The flat interior area needs to expand to the southeast (see Figure 6–28 and 6–29). This means lowering the elevations in this area or moving some of the contours towards the southeast to the edge of the surface. You also need to erase the contour representing the pond.

EXERCISE

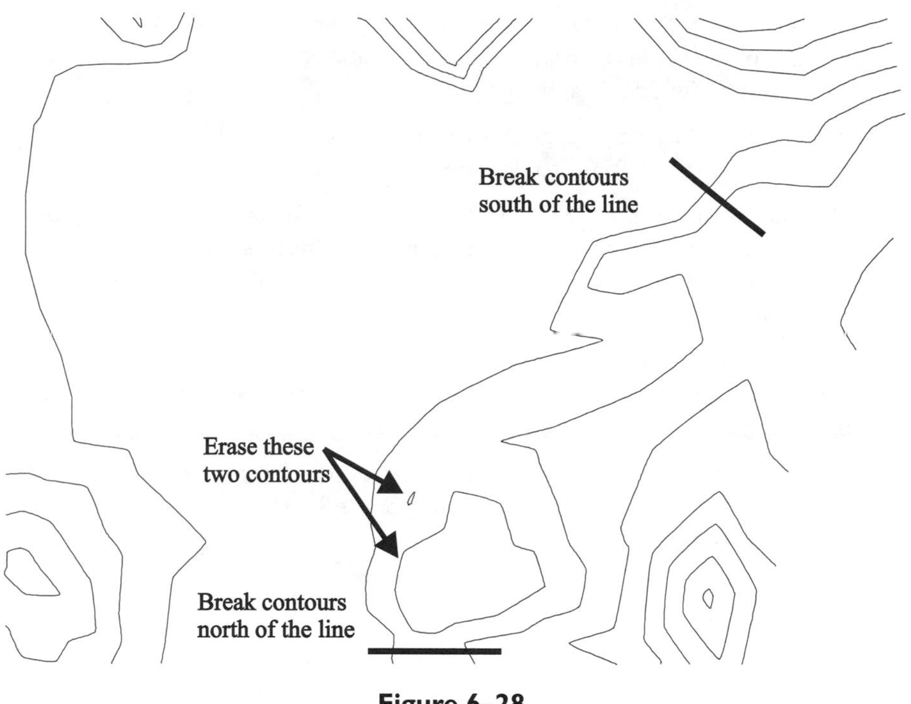

Figure 6–28

Figure 6–29

14. Use the Zoom to Point routine and zoom to point 67 with a height of 325. If there are no points in the drawing, use the ZOOM CENTER command with the coordinates of 5300,5285 and a height of 325.
15. Erase the two closed contours that represent the pond and another low area adjacent to the pond. Refer to Figure 6–28.
16. Break the two contours (733 and 734) and redraw their location. Refer to Figure 6–28. Remember to use the object snap of endpoint to set the connection of the pline to the existing contours. The new polylines should have the same elevation as the contours.
17. If the new polylines have an elevation of 0 (zero), use the Edit Elevation routine of the Contour Utilities menu and set them to the correct elevations (733 and 734). Refer to Figure 6–29.
18. Restore the view surface to view the impact of the changes on the surface. Your changes will show a flat area in the center of the surface (see Figure 6–30).

Figure 6–30

The problem with the current design is that the center of the site is flat and water will not drain off the site. One way to drain water from the site is to design a swale in the flat interior.

By placing points in the flat area, you create a shallow swale that moves any water from the interior of the site out to the northeast. The use of points is only one option. You can also use 3D polylines to create linear elevation designs.

Designing a Swale

19. Use the Zoom to Point routine and zoom to 113 with a height of 350. If there are no points in the drawing, use the ZOOM CENTER command with the coordinates of 5200,5300 and a height of 350.

20. Use the Point Settings routine of the Points menu to set the following values. Select OK to exit the dialog box after setting the following values in the Create tab:

```
Numbering
     Current Point Number: 500
Elevations
     Manual: ON
Descriptions
     Automatic: ON
The default description: SWCNTRL.
```

21. Manually set the following three points into the drawing using the Manual routine of the Create Points cascade (see Figure 6–31). Before you place the points make sure to toggle off object snaps.

Point Number	Elevation
500	732.90
501	732.40
502	732.10

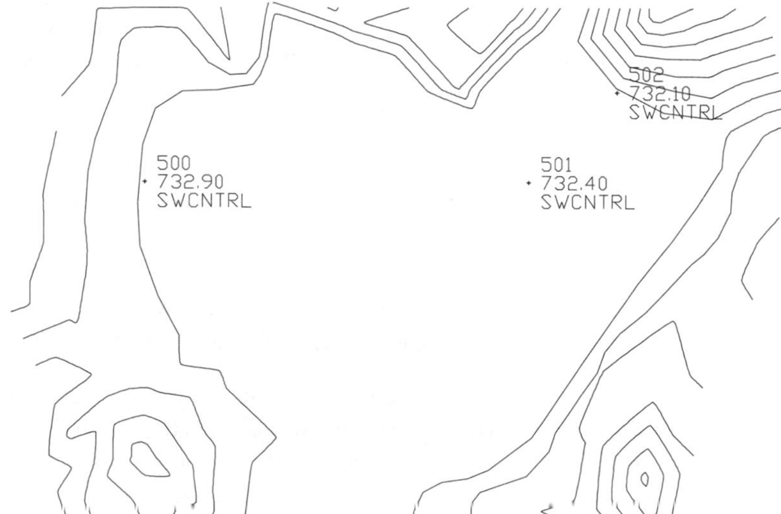

Figure 6–31

EXERCISE

The three new points indicate the upper, mid, and endpoints of a swale. To complete the swale design, place points between these controlling points by using the Interpolate routine. The routine places a specified number of points between two controlling points. Each intervening point has a different elevation that reflects a straight slope calculation between points.

The question is why place intervening points? The wide gap between each point along the swale may let the Surface Modeler create triangles that cross the path of the swale. By doing so, the Surface Modeler indicates a higher elevation crossing the swale as opposed to the linear relationship between the three points defining the swale. If the TIN Modeler connects a triangle between the contours on either side of the flat area, the implication would be an elevation of 733 along the triangle leg. This interpretation ignores the elevations along the swale points (732.9–732.4). To control this potential problem, you have two options: to place points between the controlling points of the swale, or to place a breakline between the controlling points of the swale. In either case, the purpose is to make the TIN Modeler create triangles that follow the swale.

The Interpolate routine of Create Points–Interpolate places points between two existing control points. The elevations of the new points are a straight slope calculation of a change in elevation over a distance. The new points prevent the TIN Modeler from connecting points across the swale, which means that the surface triangles will respect the swale definition.

A 3D polyline will design the southern branch of the swale. The 3D polyline will become a standard breakline for the design surface.

EXERCISE

22. Use the Interpolate routine of the Create Point–Interpolate cascade of the Points menu. The routine prompts for the elevation precision of the new points. Set the precision to 2. Then the routine prompts for the first and last points (see Figure 6–31). The routine responds with the distance and difference in elevation between the two selected points. Next the routine asks for the number of points to place between the two points. Finally, the routine prompts for the description of the new points. Use the description to SWCNTRL. Press ENTER to exit the routine. The point pairs are:

500–501 four intermediate points

501–502 three intermediate points

23. Save the drawing.

The last segment of the swale will be a 3D polyline. The polyline will have an elevation of 732.9 at its southern end and the same elevation as point 501 at its northern end.

24. Run the Create by Elevation routine found in the 3D Polylines cascade of the Terrain menu. The routine prompts for a from point starting point. Use the dot toggle to put the routine into point graphics mode (.g). Select point 501 as the starting point of the 3D polyline. The routine will echo back the elevation of point 501 (see Figure 6–31). Press ENTER to accept the elevation. The last point on the 3D polyline does not reference any existing point. You need to toggle the routine out of graphics mode by again typing the dot toggle (.g). Select the last point from the screen between the two 733 contours. The routine prompts with the elevation of 0.00 (zero). Enter the elevation of 732.90, press ENTER to use the value, type **X**, and again press ENTER to exit the routine. Use Figure 6–32 as a guide for creating the 3D polyline. The following is the command sequence.

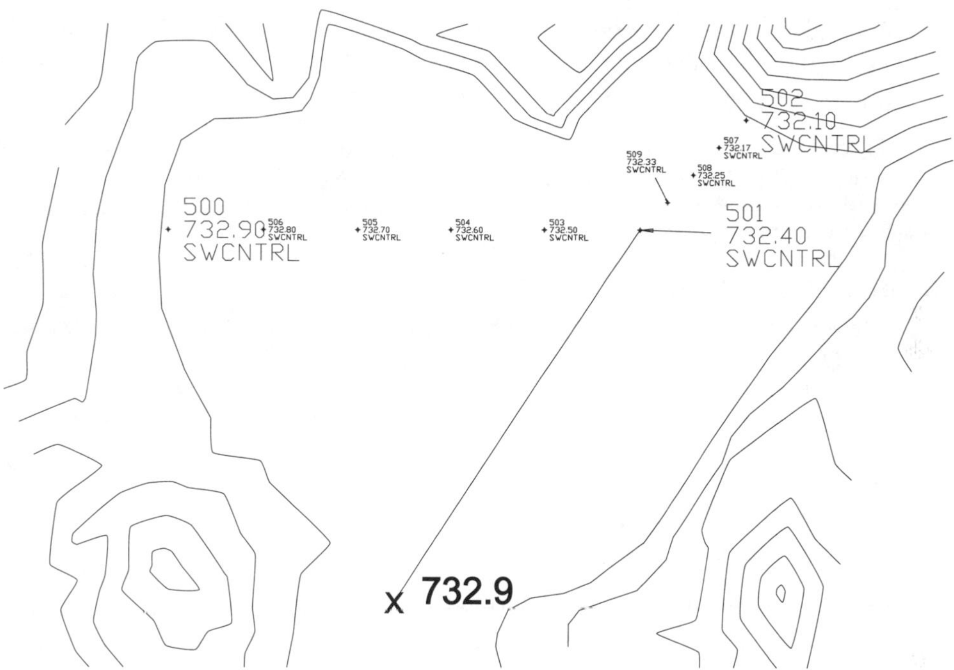

Figure 6–32

```
Command:
From point (or Entity): .g
From point (Entity):
 >>Select point object: (select point 501)
Starting elevation (or Dtm) <732.4>: (press ENTER to accept)
To point (eXit/Dtm/Curve/Undo/Entity/Transition):
 >>Select point object: .g (toggling out of Graphic mode)
To point (eXit/Dtm/Curve/Undo/Entity/Transition): (select the
   location)
Elevation <0.00>: 732.9 (enter the elevation)
Slope: 494.99:1, Grade 0.20 used for segments.
To point (eXit/Dtm/Curve/Undo/Entity/Transition): x
Command:From point: .g
```

The 3D polyline currently has only two vertices: the northern and southern ends. These two points are the only two data points for the swale. The 3D Polylines cascade of the Terrain menu contains an Add Vertices routine to add vertices to 3D polylines. The routine calculates an elevation for each new vertex by a straight slope calculation.

25. List out the polyline to view the elevations at each vertex. There are only two vertices listed.
26. Run the Add Vertices routine of the 3D Polylines cascade of the Terrain menu. When the routine prompts for a distance, enter **10** feet and answer Yes to erase the old 3D polyline. Finally, press enter to exit the routine.
27. Again list the new polyline to view the new vertices and their elevations. The new elevations are between the elevations of the southern and northern ends.
28. Save the drawing.

BUILDING THE SECOND SURFACE

A data set now exists for a second surface, which consists of existing contours, new contour segments, a 3D polyline, and points.

1. The points for the new surface have to be in a group before they are passed to the Terrain Explorer. Run the Point Group Manager routine of the Point Management cascade of the Points menu. Create a point group, base, in the Advanced area and group the points by the point number range of 500–600. The source for the points will be all points.
2. Run the Terrain Explorer routine from the Terrain menu.
3. Select the word *terrain* at the top of the Explorer, click the right mouse button, and select Create New Surface. The routine places a new surface in the list, surface1.
4. Select Surface1, click the right mouse button, select rename, and rename the surface Base.
5. Select the "+" (plus sign) under the base surface name to expose the data list.
6. Select Point Groups, click the right mouse button, select Add Group, and select Base from the list. You may have to click the list arrow to view the point groups to make Base the current group.
7. The next step is to define the 3D polyline as a standard breakline. Use the By Polyline routine from the Explorer. Select Breaklines, click the right mouse button, and select By Polyline. The routine hides the Explorer, Name the breakline SWCNTRL, select the polyline, and answer Yes to deleting the old polyline.
8. Select Contours in the Explorer, click the right mouse button, and select Add Contour Data. The routine displays the Contour Weeding dialog box. Make sure that contour data is on and use the same weeding (15.0 for length and 4.0 for

angles) and supplementing (distance 30.0 and bulge 0.25) factors. Select OK to dismiss the dialog box.

9. There is only one layer containing design contours, EXCONTOUR. Use the Layer option to select a contour from the screen. Press ENTER twice after selecting the contour to exit the routine and reenter the Explorer.

10. Run the Build (Surface) routine to create the Base surface. Select the surface name, click the right mouse button, and select Build (surface). The routine displays the Build base dialog box. Set the values as shown in Figure 6–33. Select the OK button to build the surface. Select OK again to dismiss the Done dialog box and reenter the Terrain Explorer.

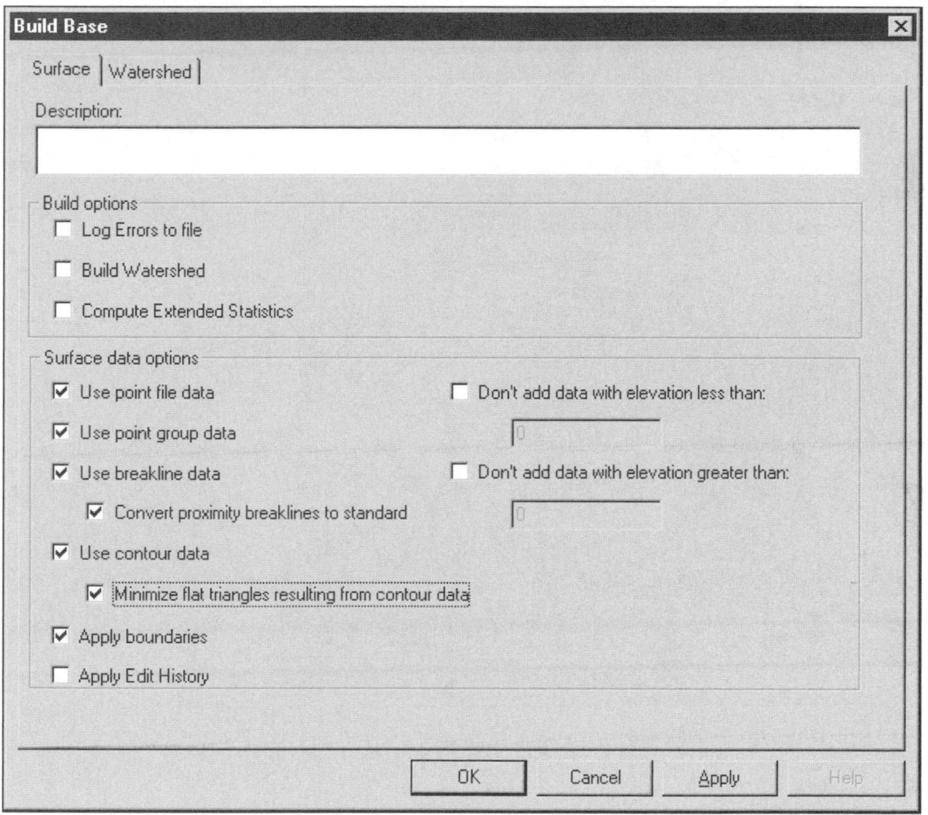

Figure 6–33

11. Select the Quick View routine from the Surface Display cascade to see the interpretation of the surface. Select the surface name, click the right mouse button, select Surface Display, and select Quick View. There should be no need for a breakline where the points define the swale. Notice how the vertices along the 3D polyline act as points defining the southern swale. Use Figure 6–34 as a reference.

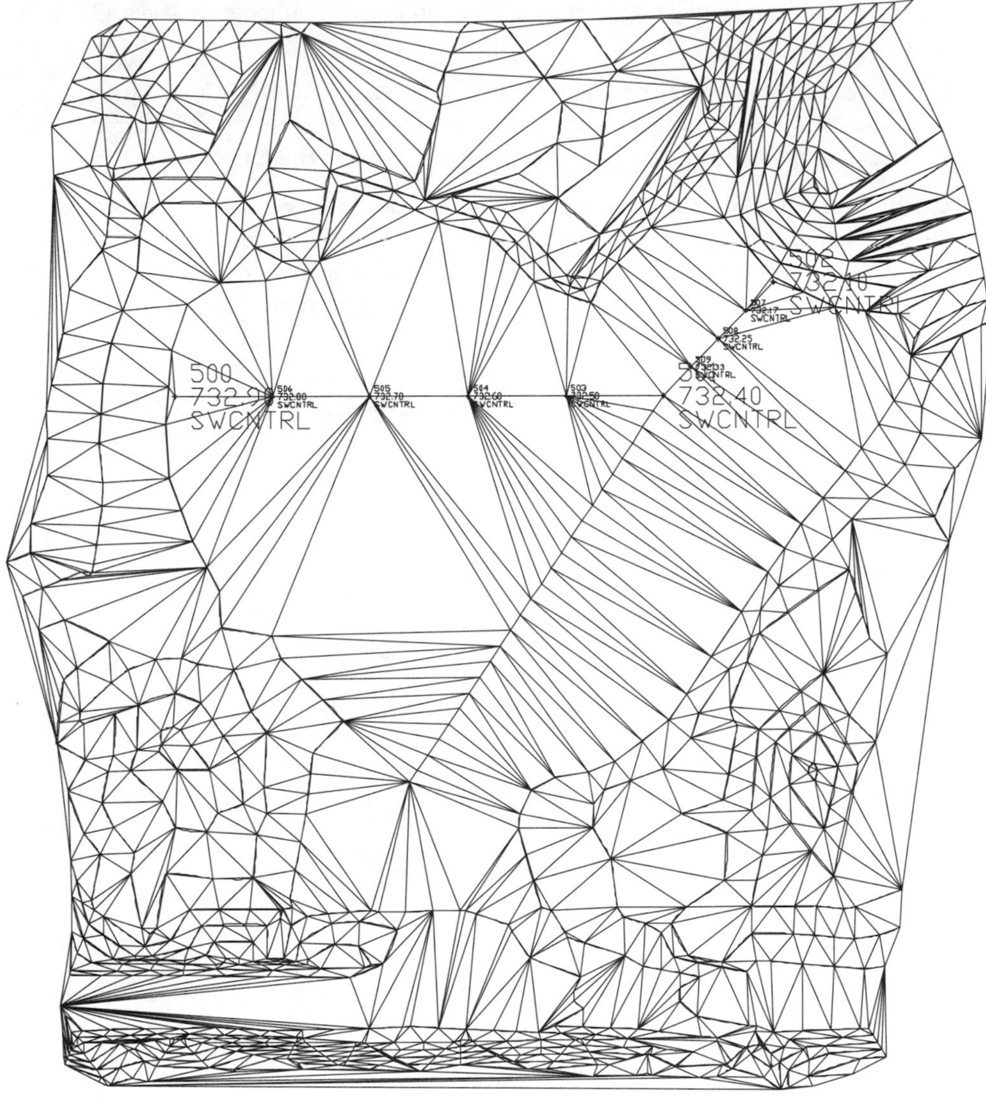

Figure 6–34

12. Redraw the drawing to remove the quick view TIN lines.

13. Use the Remove Points from Drawing routine and remove points 500–550 from the drawing

14. Erase the 3D polyline breakline from the drawing.

15. In the Layer dialog box, make layer 0 (zero) the current layer. Freeze layer EXCONTOUR. Select OK to exit to the drawing. There should be nothing on the screen.

16. Close the Terrain Explorer.
17. Restore the Named View, surface, to set the display.
18. Save the drawing.

EXERCISE

UNIT 4: CALCULATING A GRID VOLUME

Each volume method in Land Development Desktop calculates a volume differently. The Grid method slices the stratum in both the *M* and *N* direction. Then the volume calculator checks for an elevation from both surfaces at each corner of the cell. If there are elevations at all four corners of the cell, the Volume Generator uses the entire cell. If there are not valid elevations at one of the four corners in the cell, the volume generator does not include that prism (one-half of the cell) in the calculation (see Figure 6–4). If you vary the sample spacing, smaller or larger cells, in the Site definition, the routine may calculate different volume totals. The varying of the sampling interval will affect the area the volume method samples to determine the total volume.

The single greatest problem you face, outside of bad data, is how often to sample a site. You should sample a site with an interval similar to the survey interval. In some cases the spacing can be as large as 50 feet or as small as one-half foot. To sample the site with a 20-foot grid is to sample the interpretations between the 50-foot spacing and to miss the one-half foot points. If there are bays and peninsulas at the edges of the site and the sample spacing is too large, the Volume Generator may exclude those bays or peninsulas because the spacing excludes those areas. The Site Definition spacing should sample the greatest amount of area with the fewest number of points. This optimum spacing varies based upon the size, shape, and relief found on the site. Different sampling intervals in some cases will produce different volumetric results. The differences can be severe or minor. If the differences in the results are severe, you should question the spacing of the sampling or the surfaces.

One method of evaluating the volume calculation is an exhaustive report created by the Grid Volume routine. The report reviews the elevations and quantities on a prism basis for each cell. The report is a toggle in the Grid Volume Settings dialog box (see Figure 6-8). A large site would produce a huge file. Another method of evaluation would be to import the border of the most restrictive surface and compare how closely the grid matches the border. If there is a wide gap between the grid and the border, this would indicate the need to reduce the cell size (see Figure 6–4). If the grid is close to the border and extremely dense, you may be able to try a larger-sized cell.

The grid volume calculation process creates a volume surface. The Terrain Explorer displays this surface not as a terrain surface, but as a volume surface (see Figure 6–35).

Terrain Explorer displays volume surfaces at the bottom of the Explorer. Explorer places a grid icon in front of the name of the grid surface.

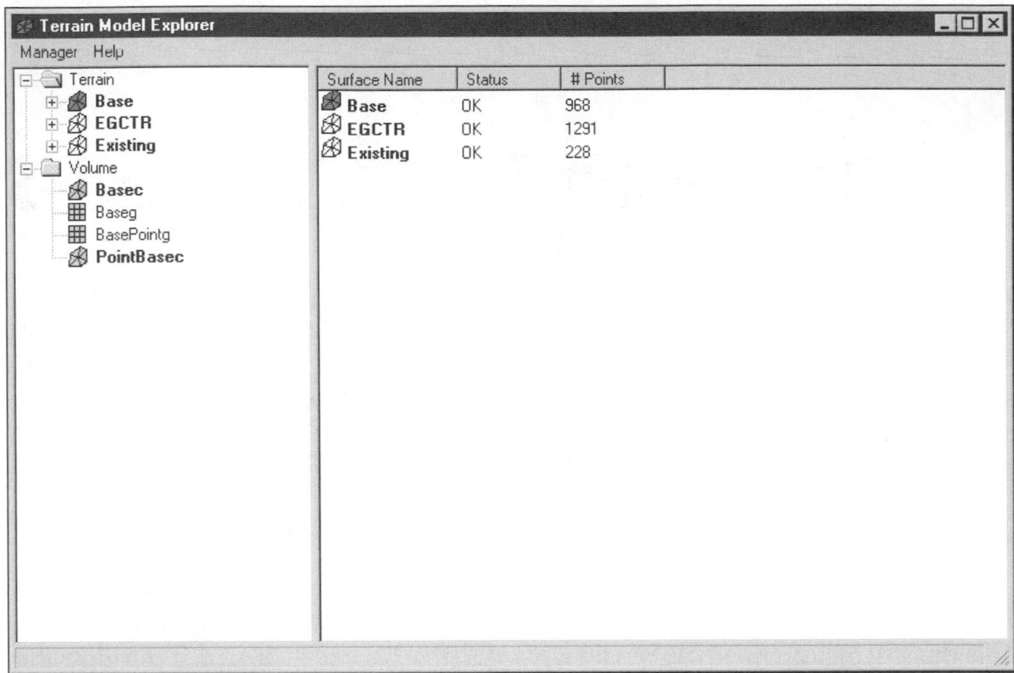

Figure 6–35

The site volume reports in the Volume Reports cascade apply to each volume method (see Figure 6–5). The routines are Site Report, Site Table, and Site ASCII File.

The Site Report routine creates a report representing the final volume values. The report identifies site, stratum name, and surfaces, the cut, fill, and net amount, and the methods used to calculate the volume.

The following is an example of the report:

Site Volume Table Unadjusted				**Cut**	**Fill**	**Net**	
Site	Stratum	Surf1	Surf2	Yards	Yards	Yards	Method
Base10	Base	EGCTR	base	1128	4038	2910 (F)	Grid
				1150	4058	2908 (F)	End area
				1419	4033	2615 (F)	Composite

The Site Table routine places the above report into the drawing.

The Site ASCII File routine creates an ASCII comma-delimited file (CDF). The file contains the site name, stratum name, the names of surface 1 and 2, the method, the total cut, the total fill, and the net amount. This file allows you to use the LDD volume numbers in other programs.

The following is an example of the report:

```
base10,base,egctr,base,Grid,1063.864548,3654.106785,2590.242237
Base10,base,egctr,base,End
area,1278.116453,3891.819258,2613.702805
base10,base,egctr,base,Composite,1100.567626,2703.206510,1602.638884
```

Exercise

After you complete this exercise, you will be able to:

- Calculate a Grid Earthworks Volume
- Review the volume numbers
- Annotate the volume numbers

EXERCISE

CALCULATING A GRID VOLUME

Step 1—Define a Stratum

The Site Volume Process is found in the lower portion of the Terrain menu. The first task is defining a stratum. A stratum is an alias for the two volume surfaces. In this exercise you want to compare the Existing surface (surface 1) to the Base surface (surface 2) (see Figure 6–36). The first volume calculation involves the surfaces, *egctr* and *base*. As you select each surface for a stratum, the routine displays a dialog box reviewing the statistics of each surface (see Figure 6–37). Remember that the surface *existing* is from points and breaklines and *egctr* is a surface of contours with some points and breaklines. The surface data for the *base* surface comes from the *egctr* surface contours. So, the volume comparison should be between *egctr* and *base*.

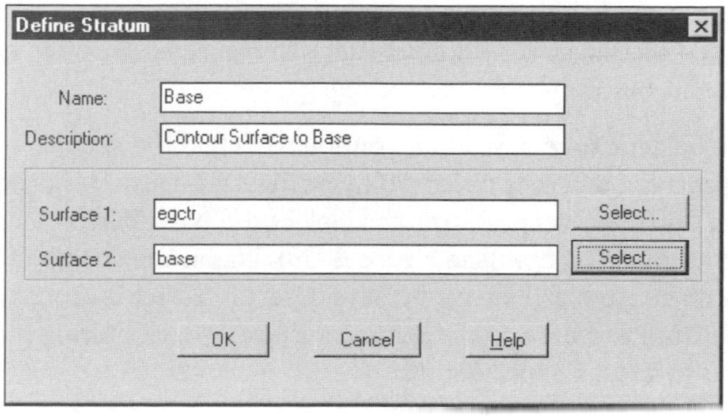

Figure 6–36

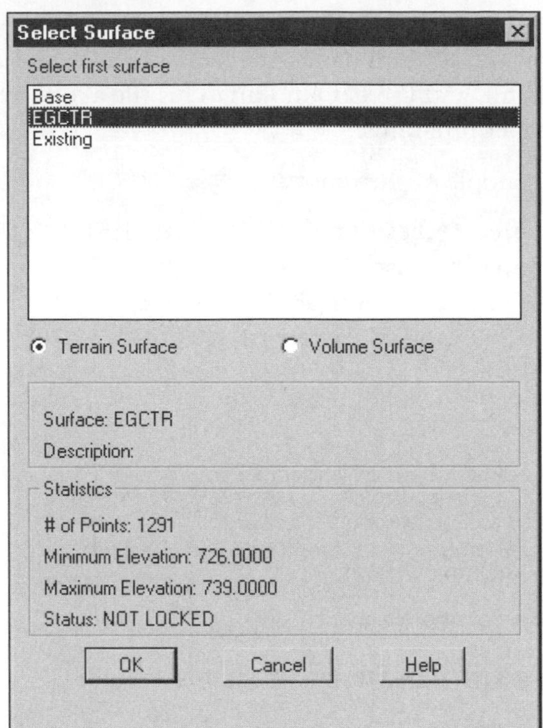

Figure 6–37

The second stratum defined compares the Existing surface to the Base surface. This is to evaluate the difference, if any, between different types of surface data and their effect on the resulting volume.

1. This exercise continues in the Terrain drawing of the Surfaces project.

2. Use the Select Current Stratum routine from the Terrain menu to define the first stratum. Name the stratum Base. Enter the description of Contour Surface to Base. Choose the Select button to identify the surface names. The first surface is *egctr* and the second surface is *base*. Refer to Figure 6–36. After entering the data, select the OK button to exit the routine.

2. Rerun the Select Current Stratum routine to define the second stratum. The routine displays a different dialog box (see Figure 6–38). This dialog box shows only when there are defined stratums. Select the New button to display the Define Stratum dialog box (see Figure 6–36). Name the stratum PointBase and describe the stratum as Existing to Base. Use the Select buttons to set Existing as the first surface and Base as the second surface. After entering the data, select the OK button to exit the routine.

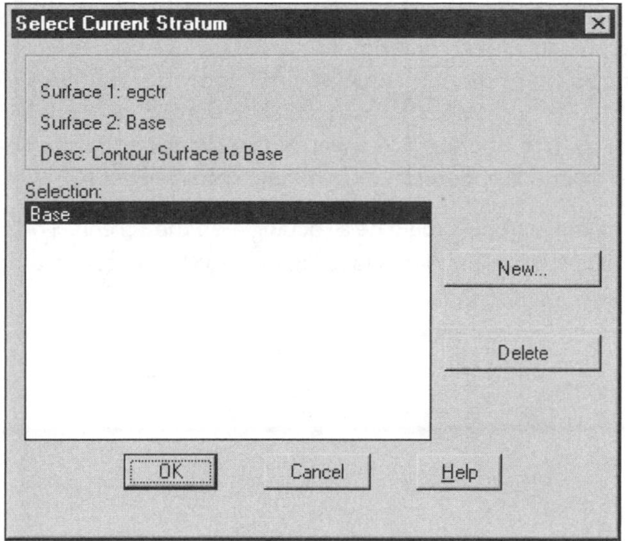

Figure 6–38

Step 2—Setting the Site Defaults

The next step is setting the Site Settings (see Figure 6–7). The first default is the size or number of the grid points.

1. Run the Site Settings routine of the Site Definition cascade of the Terrain menu and set the values as shown in Figure 6–7. Make sure the cell size is set to 10.
2. Exit the Volume Site Settings dialog box by selecting the OK button.

Step 3—Define a Site

The next step is to define a site. The base point of the site is the lowest left point outside both of the surfaces' boundaries. The site does not have to be within the boundaries of both surfaces because the volume calculations occur only where the volume routine finds an elevation for both surfaces.

1. Use the Named Views routine of the view menu to set the surface view as current. Select OK to exit the dialog box.
2. Use the Set Current Surface routine of the Terrain menu and set the *egctr* surface as the current surface. Then use the 2D Polyline routine of the Surface Border cascade of the Terrain menu to import the *egctr* boundary.

3. Define the site by using the Define Site routine of the Site Definition cascade of the Terrain menu. Use a rotation of 0 (zero) for the site. Select a point at the lower left bottom of the screen (AutoCAD coordinates 4850,4950). Confirm the spacing of 10 for both *M* and *N* axes. Select an upper right corner of the site outside the surface boundary. Answer No to changing the site line and answer Yes to the delete old boundary lines. Name the site Base10.

The result of running the routine should be a rectangle on the screen. The rectangle is larger than the TIN boundary for both surfaces (see Figure 6–39). The cell at the lower left shows a 10x10 cell. Your site will not show this cell.

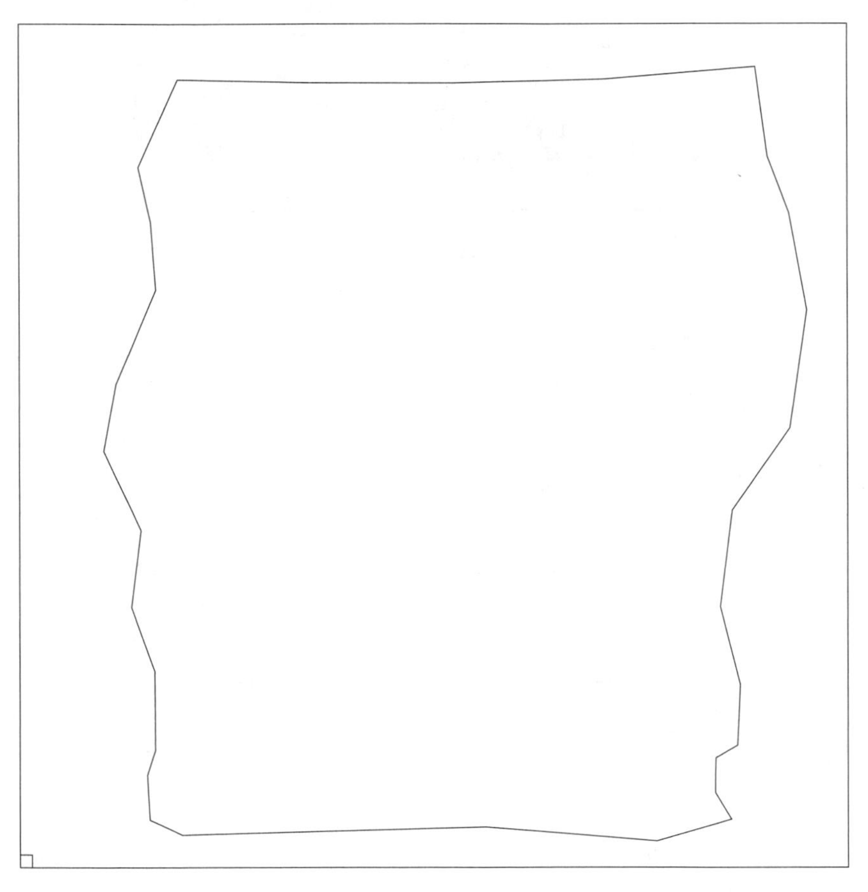

Figure 6–39

Step 4—Calculate a Grid Volume

The Grid method lays a mesh, the site definition, over the surfaces and at each intersection of the mesh the routine locates a difference in elevation between the surfaces. It is from these differences in elevations at the intersection points that the routine calculates a volume.

EXERCISE

1. The current stratum is the PointsBase stratum, the last stratum defined. Rerun the Select Current Stratum routine to set the current stratum to Base.
2. Run the Calculate Total Site Volume routine from the Grid Volumes cascade of the Terrain menu. The routine displays the Site Volume Librarian dialog box (see Figure 6–40).

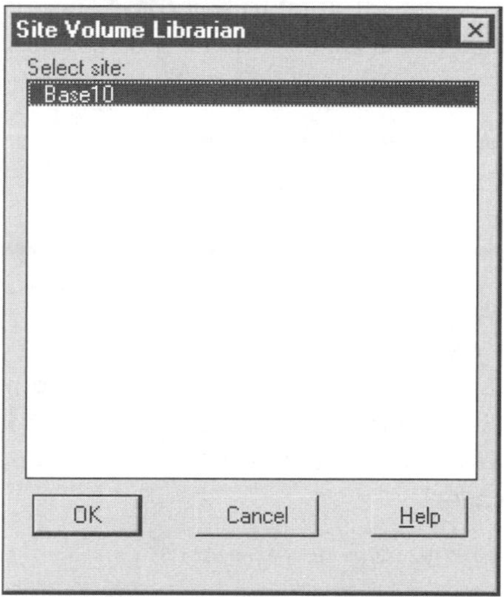

Figure 6–40

3. Select Base10 in the Site Volume Librarian and select OK to exit the dialog box.
4. The routine displays the Grid Volume Settings dialog box. Make sure the Grid Tolerance is set to 0.00 and the report toggle is set to None. Select OK to exit the dialog box.
5. The next dialog box prompts for a surface name used by the Grid Volume generator (see Figure 6–41). Name the surface, *baseg*; that is, *base* grid surface. After entering the surface, select OK to exit the dialog box. The routine calculates a volume. You may have to toggle to the AutoCAD text screen to view the results. Your results should be similar to the following:

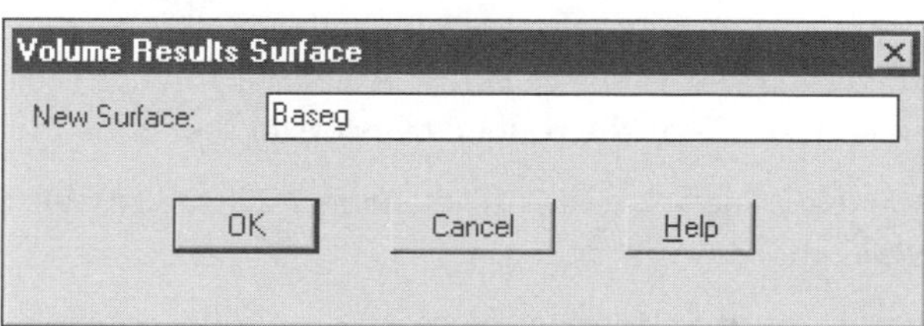

Figure 6–41

EXERCISE

```
Current stratum: Base
Site name = Base10
Cut = 1392 cu.yds   Fill = 4012 cu.yds
Net = 2620 cu.yds FILL
```

6. Redo steps 1 through 5 to calculate a volume for the PointBase Stratum. You need to change the current stratum to PointBase and then select Calculate Total Site Volume to calculate the volume. Name the volume surface BasePointg. Your results should be similar to the following:

```
Current stratum: Point Base
Site name = Base10
Cut = 1606 cu.yds   Fill = 4387 cu.yds
Net = 2781 cu.yds FILL
```


EVALUATING GRID VOLUMES

1. Open the Terrain Explorer and notice that at the bottom of the surface list are two new surfaces, Baseg and BasePointg. Each surface has a grid icon in front of its name. This icon indicates that the surface is a volume surface resulting from the grid volume process.
2. Select each surface name to review the statistics.
3. Select the Baseg surface, click the right mouse button, select surface display, and select Quick View. This shows the grid that calculates the volume.
4. Minimize the Terrain Explorer.
5. Select the Set Current Surface routine from the Terrain menu and select the EGCTR surface.
6. Run the 2D polyline routine of the Surface Border cascade to view the grid to border gap. Some of the surface did not get sampled, because of the spacing and location of the site.
7. Set Baseg as the current surface using the Set Current Surface routine in the Terrain menu. You need to select the volume surface toggle to see the list of volume surfaces (see Figure 6–42).
8. View the surface elevations using the Elevation Tracking routine from the Inquiry menu. You may have to move the Terrain Explorer to view the elevations. The elevations show in the area where AutoCAD displays its coordinates. After viewing the elevations, exit the routine by pressing ENTER or the SPACEBAR.
9. Erase the boundary line and site outline and redraw the screen. The screen should now be empty.
10. Run the Create Contours routine to create cut and fill contours. Use Figure 6–43 as a guide for setting the values. Set the Minor interval to 0.5 and the Major interval to 2.5. Select OK to create the contours.

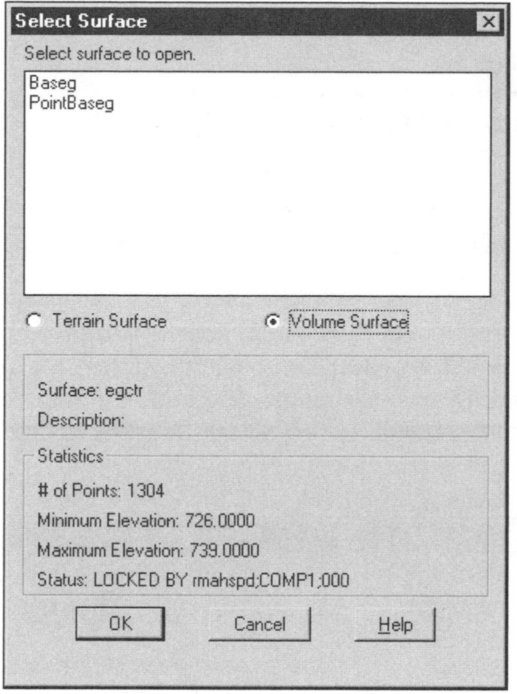

Figure 6–42

Create Contours
Surface: Baseg
Elevation Range
From: -5 To: 5.5 Vertical Scale: 1
Low Elevation: -5.07 High Elevation: 5.61 Reset Elevations
Intervals
Both Minor and Major Minor Only Major Only
Minor Interval: 0.5 Layer: CONT-MNR
Major Interval: 2.5 Layer: CONT-MJR
Properties
Contour Objects Contour Style: MYSTYLE
Polylines Preview ... Style Manager >>
OK Cancel Help

Figure 6–43

11. Erase the contours from the screen.
12. Save the drawing.

ANNOTATING GRID VOLUMES

Grid Volume Ticks

The Grid Volume Ticks routine creates ticks and places a label at each tick representing the difference in elevation between the two surfaces. Each tick marks a cell corner in the grid.

1. Run the Grid Volume Ticks routine from the Grid Volumes cascade of the Terrain menu. The routine displays the Volume Site Librarian dialog box. Select the Base10 site and then select OK to exit the dialog box. The next dialog box is the Grid Volume Ticks dialog box (see Figure 6–44). After setting the following values, select OK to exit the dialog box, answer Yes to erase any existing grid ticks, and the routine creates the ticks and labels.

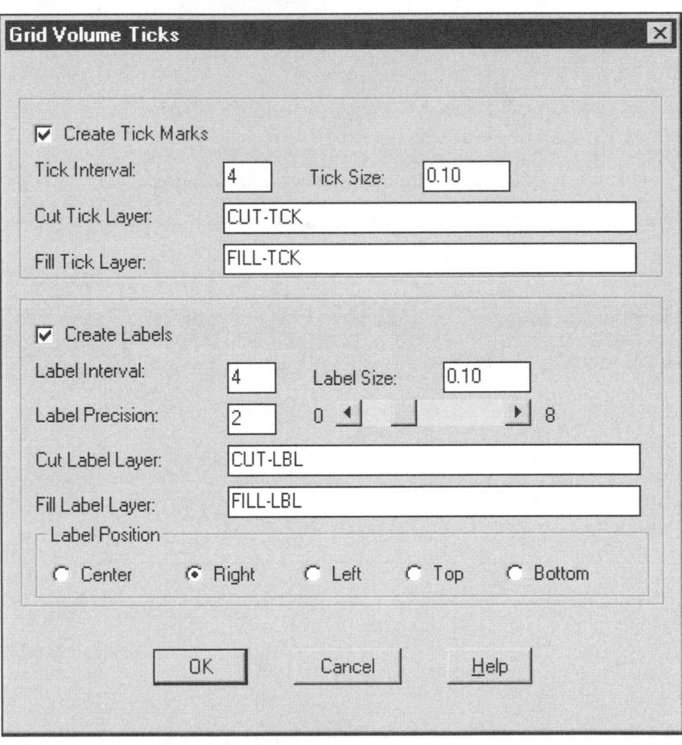

Figure 6–44

```
Tick Spacing: 4 (every fourth grid intersection)
Label Spacing: 4 (every fourth grid intersection)
Label Precision: 2
Label Position: Top
```

EXERCISE

The other defaults remain the same.

2. After the routine creates the ticks and labels, assign colors to the following layers: *CUT-TCK*, *CUT-LBL*, *FILL-TCK* and *FILL-LBL*.

The ticks portray the cutting and filling done on the site. They also show the amount of cut and fill in the site at every other grid intersection point (see Figure 6–45).

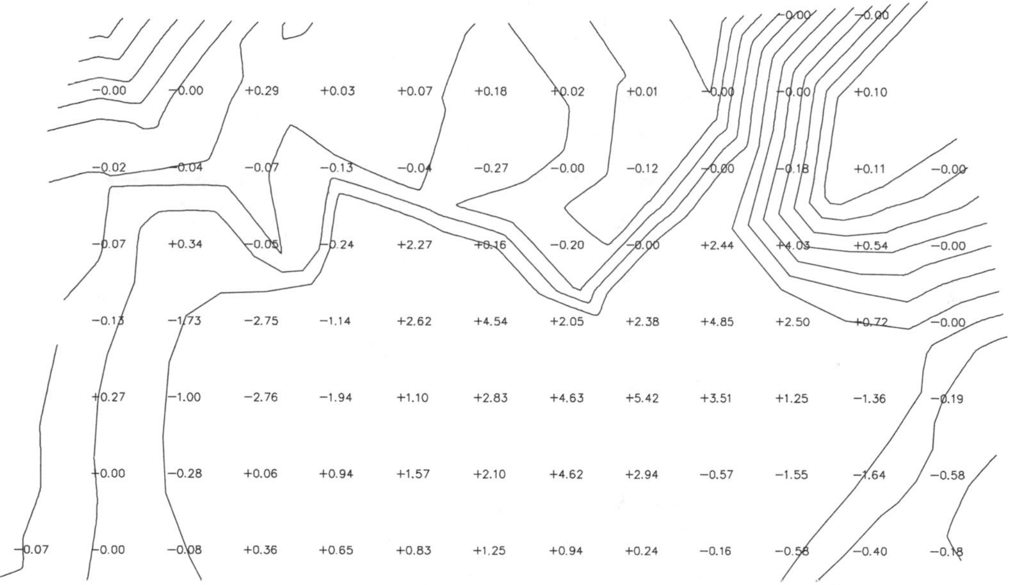

Figure 6–45

3. Erase the ticks and labels from the screen.
4. Save the drawing.
5. If you have zoomed in to view the ticks and labels, reset the surface view as the current view of the drawing.

UNIT 5: CALCULATING SECTION VOLUMES

The second method of calculating a volume is by the Section method (see Figure 6–46). The Section method samples by slicing across the surfaces in a single direction. The spacing value for the site sets the slicing interval for sampling the surfaces. As discussed with grid volumes, sampling at different intervals may produce different volume estimates.

The generating of Section volumes is a two-step process. The first step is sampling and constructing the sections from the surface data (Sample Sections), and second step is calculating a volume total (Calculate Volume Total). The Section volume routine reports the calculated volume to the text screen.

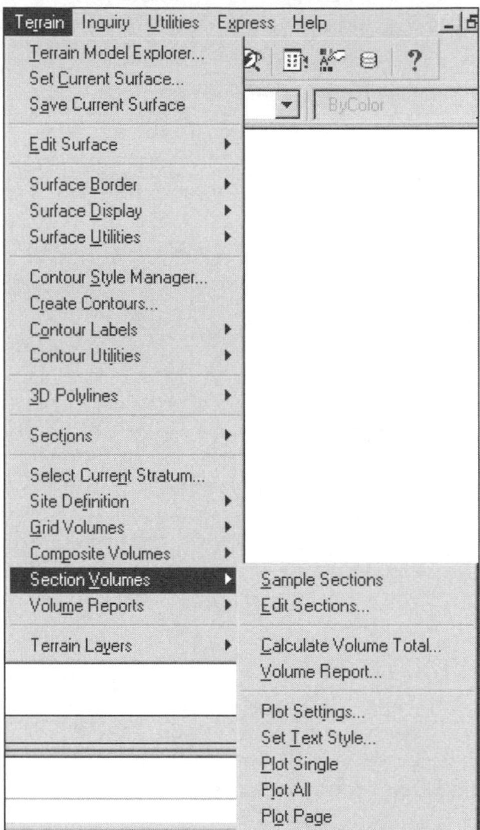

Figure 6–46

While running both steps of the volume routine, you encounter the Section Volume Settings dialog box (see Figure 6–11). The top portion of the dialog box toggles which method to use when calculating the Section volume. The two methods for calculating a section volume are average end area and prismoidal.

The middle portion of the dialog box sets the direction of section sampling. The two directions are *M* and *N*. If the site is not rotated, *M* is the same as the *X* axis and *N* is the same as the *Y* axis. If the site is a square, sampling in either the *M* or *N* direction suffices. If the site is longer in one axis, the best strategy is sampling perpendicular to that axis. For example, if the site is longer in *X* than *Y*, sample the site by slicing in the *Y* or *N* direction.

The last set of values in the dialog box is adjustments to the volume for cut (fluff) and fill (compaction). These values are multipliers of the final volume value. Since fluffing will increase the amount of cut volumes, its value is always 1.0 or greater. If the fluff factor is 5 percent, then the cut value is 1.05, 105 percent of the volume results. The compaction value means that there will be a loss of volume, so its value is always 1.0 or less. If there is a compaction rate of 5 percent, then the fill value is 0.95, 95 percent of the volume results.

Land Development Desktop has two methods to review section volume calculations. The first method is to view the cross sections with the Cross Section editor. The second is a set of routines that import volume sections into the drawing.

You can import the sections into a drawing by one of three methods. The first method is by individual sections in which the routine prompts for each station to import. After you set the station, the routine imports the section into the drawing.

The second method is to import all the sections. This method brings in all the sections in a semiorganized manner. The horizontal and vertical spacing of the page layout control how the sections appear on the screen.

The final way to import the sections is in a page format. The Page routine follows the Layout defaults of the Section Utilities area of the Section defaults.

A site volume may involve several sections. When importing the sections, the process may generate several pages of sections. Before importing all the sections at one time, you should import one page of sections to review the effects of the spacing settings. If the spacing defaults need to change, you have only one page of sections to undo before adjusting the settings in the Layout dialog box.

The site volume reports in the Volume Reports cascade apply to each volume method. The routines are Site Report, Site Table, and Site ASCII File.

The Site Report routine creates a report representing the final volume values. The report identifies site, stratum name, and surfaces, the cut, fill, and net amount, and the methods used to calculate the volume.

The following is an example of the report:

Site Volume Table Unadjusted				**Cut**	**Fill**	**Net**	
Site	Stratum	Surf1	Surf2	Yards	Yards	Yards	Method
Base10	Base	EGCTR	base	1128	4038	2910 (F)	Grid
				1150	4058	2908 (F)	End area
				1419	4033	2615 (F)	Composite

The Site Table routine places the above report into the drawing.

The ASCII routine creates an ASCII comma-delimited file (CDF). The file contains the site name, stratum name, the names of surface 1 and 2, the method, the total cut, the total fill, and the net amount. This file allows you to use the LDD volume numbers in other programs.

The following is an example of the report:

```
base10,base,egctr,base,Grid,1063.864548,3654.106785,2590.242237
Base10,base,egctr,base,End
area,1278.116453,3891.819258,2613.702805
base10,base,egctr,base,Composite,1100.567626,2703.206510,1602.638884
```

Exercise

After you complete this exercise, you will be able to:

- Calculate a Section Earthworks Volume
- Review the volume numbers
- Import volume cross sections
- Annotate the volume numbers

CALCULATING SECTION VOLUMES

1. Continue in the terrain drawing of the Surfaces project and use the Select Current Stratum routine from the Terrain menu to set the current stratum to Base.

2. Use the Sample Sections routine of the Section Volumes cascade of the Terrain menu to create the volume cross sections. When the routine displays the Site Librarian, select the Base10 site and select the OK button to exit the dialog box. Use the Average End method and sample in the direction as indicated in the dialog box. Do not apply any volume adjustment factors. The routine exits the dialog box and samples the surfaces at the 10-foot slicing interval.

3. Calculate a section volume using the Calculate Volume Total routine of the Section Volumes cascade of the Terrain menu. When the routine displays the Site Librarian, select the Base10 site and select the OK button to exit the dialog box. Use the Average End method and sample in the direction as indicated in the dialog box. Do not apply any volume adjustment factors. The routine exits the dialog box and samples the surfaces at the 10-foot slicing interval.

The two steps, creating the sections and calculating the volume, are almost exactly alike. The two routines prompt for the same values and display the same dialog boxes.

4. Note the volume values. Compare the values to grid totals for the Base Stratum. Your results should be similar to the following values.

```
Current stratum: Base
Site name = Base10
Passing through sections determining the strata conditions...
Station: 6+30
Displaying strata report for stratum: Base
Cut: 1416 cu.yds   Fill: 4033 cu.yds
Net: 2616 cu.yds (FILL)
Volume calculations done. Press any key to continue...
```

5. Use the Set Current Stratum routine to set PointBase as the current stratum.

6. Repeat steps 2 and 3 to calculate a volume for the PointBase stratum.

7. Note the volume values. Your values should be similar to the following.

```
Current stratum: PointBase
Site name = Base10
Passing through sections determining the strata conditions...
Station: 6+30
Displaying strata report for stratum: PointBase
Cut: 1687 cu.yds   Fill: 4465 cu.yds
Net: 2778 cu.yds (FILL)
Volume calculations done. Press any key to continue...
```

8. Compare the values to the grid totals for the PointBase stratum. The point base stratum is the comparison of a point surface to a design surface from contours.

EXERCISE

ANALYZING CROSS SECTIONS

Land Development Desktop has two methods to review section volume calculations. The first method is to view the cross sections with the Cross Section editor. The second is a set of routines that import volume sections into the drawing.

Edit Sections

1. Use the Select Current Stratum routine of the Terrain menu to set Base as the current stratum.

2. Select the Edit Sections routine from the Section Volumes cascade. Select the Base10 site from the Site Librarian and select OK to exit the dialog box. The routine displays another dialog box that displays the elevations and offsets for cross sections of each surface (see Figure 6–47). You may have to move up several stations to view the beginning of the volume data.

You can modify the data used for the calculation of the volume in this dialog box. However, the dialog box gives you a useful tool in evaluating the section data.

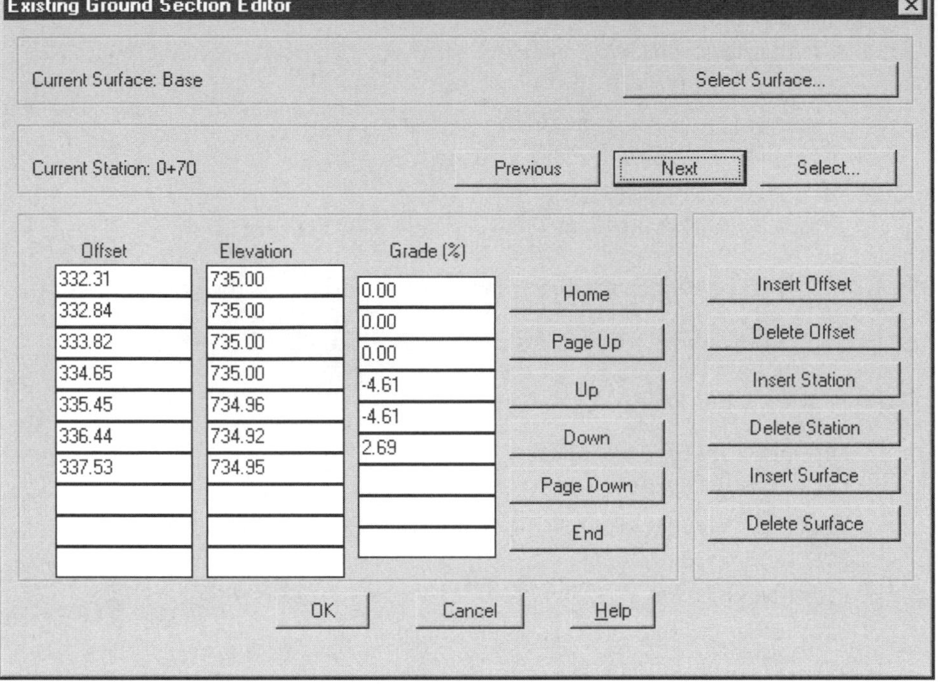

Figure 6–47

Viewing Volume Sections

1. Start a new drawing and save the changes to the current drawing. Name the new drawing *sitesec* (see Figure 6–48). Assign the new drawing to the surfaces project and use the *aec_i.dwt* template file.

2. In the Load Settings dialog box, select the LDD1-40 setup. If you do not have the setup, load the *i40* setup file. After selecting the setup, select the Load button and then the Finish button to dismiss the dialog box. Select OK to dismiss the Finish dialog box.

3. Click the Save Drawing icon in the upper left to save the drawing.

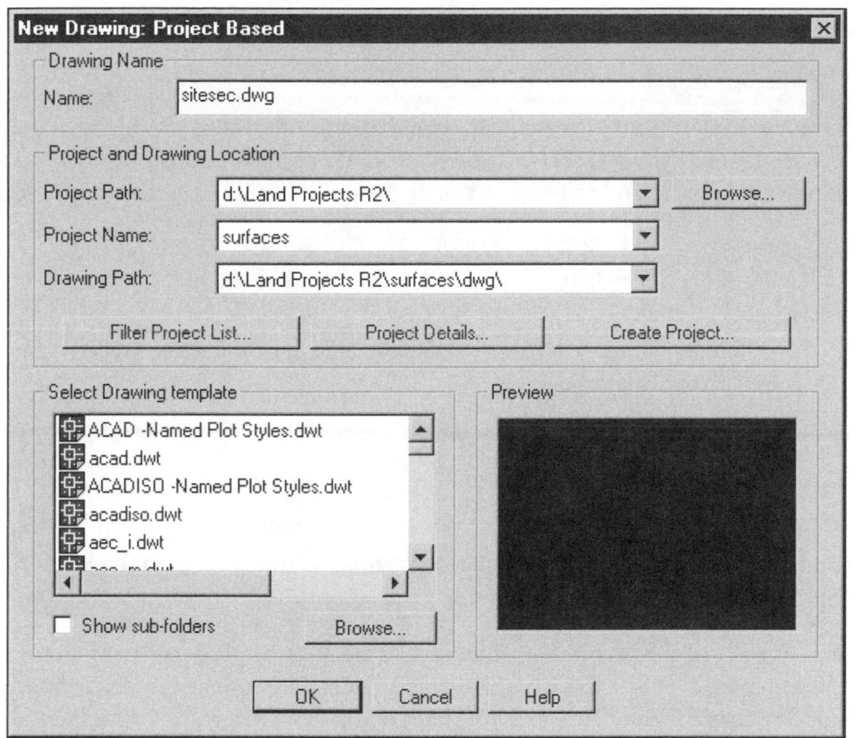

Figure 6–48

Importing Single Cross Sections

The first method of importing the sections is by individual section stations. The Plot Single routine of the Section Volumes cascade of the Terrain menu imports one section at a time. The routine first prompts for the site, then the station, and finally insertion point of the section. After placing the section, the routine starts the prompting for another station until you press ENTER twice to exit the routine.

1. Use the Set Current Stratum routine to set the Base stratum as the current stratum.
2. Run the Plot Single routine, which is in the Section Volumes cascade of the Terrain menu. Select the Base10 site and select OK to exit the dialog box. The routine prompts for a station. Remember, the Edit Sections routine did not display any cross sections until around station 70.
3. Place a few sections into the drawing starting around station 120 and press ENTER twice to exit the routine.
4. Run the Layer command and assign different colors to the XEG-Base and XEG-Egctr layers.

5. Zoom in to view the sections.

The routine places the lines representing the surfaces on layers prefixed with XEG-.

6. Use the Back option of the UNDO command to undo everything just done. You can do this by typing **UNDO** at the AutoCAD command prompt, Typing the letter **B** for back, and answering Yes to the question about undoing everything.

```
Command: undo Enter the number of operations to undo or
[Auto/Control/BEgin/End/Mark/Back] <1>: b
This will undo everything. OK? <Y> y
ZOOM GROUP GROUP GROUP QSAVE GROUP AECCSETUPNEWDWG GROUP .LAYER
  GROUP .LAYER
GROUP
Everything has been undone
```

Importing All of the Cross Sections

The second method is the All method. The routine follows the settings in the Section Layout defaults for the spacing between each section.

1. Run the Plot All routine. The routine is in the Section Volumes cascade of the Terrain menu. Select the Base10 site and select OK to exit the dialog box. The routine asks for the cross-section sheet origin. Select a point in the lower left of the display. The routine places all of the cross sections into the drawing.
2. Execute a Zoom Extents to view all of the sections.
3. Use the ZOOM command to zoom in to view the sections.

The routine places the lines representing the surfaces on layers prefixed with XEG-.

4. Use the Back option of the UNDO command to undo everything just done. You can do this by typing **UNDO** at the AutoCAD command prompt, then typing the letter **B** for back, and answering Yes to the question about undoing everything.

```
Command: undo Enter the number of operations to undo or
[Auto/Control/BEgin/End/Mark/Back] <1>: b
This will undo everything. OK? <Y> y
ZOOM GROUP GROUP GROUP QSAVE GROUP AECCSETUPNEWDWG GROUP .LAYER
  GROUP .LAYER
GROUP
Everything has been undone
```

Importing a Page of Cross Sections

The last method imports sections in organized sheets. The Plot Page method can import one or more pages of sections at a time. The spacing between sections within the page is from the setting in the Page Layout Settings dialog box.

1. Run the Plot Page routine. The routine is on the Section Volumes cascade of the Terrain menu. Select the Base10 site and select OK to exit the dialog box. Import multiple pages into the current drawing and answer Yes to import them into this drawing. Start with section 0+00 (0.00) and use the AutoCAD coordinates of 0,0 as the insertion point. The routine places the page into the drawing. The sections could use a different spacing.
2. Run the Zoom Extents routine to view all the pages.
3. Zoom in to view some of the sections.
4. Save the drawing.
5. Reopen the *terrain* drawing and run the Named Views routine of the View. Set the Surface view as the Current view and select the OK button to exit the routine.

EXERCISE

UNIT 6: COMPOSITE SURFACE VOLUME

The Composite volume calculation process creates a volume surface. The elevations of the resulting surface are the differences in elevations between the two surfaces. The surfaces' triangles are a composite of the triangles from both surfaces. This routine is the most comprehensive volume calculation method.

The Terrain Explorer displays this surface not as a terrain surface, but as a volume surface. Terrain Explorer displays volume surfaces at the bottom of the Explorer. Explorer places a composite surface icon in front of the name of the grid surface.

The site volume reports in the Volume Reports cascade apply to each volume method. The routines are Site Report, Site Table, and Site ASCII File.

The Site Report routine creates a report representing the final volume values. The report identifies site, stratum name, and surfaces, the cut, fill, and net amount, and the methods used to calculate the volume.

The following is an example of the report:

Site Volume Table Unadjusted				**Cut**	**Fill**	**Net**	
Site	Stratum	Surf1	Surf2	Yards	Yards	Yards	Method
Base10	Base	EGCTR	base	1128	4038	2910 (F)	Grid
				1150	4058	2908 (F)	End area
				1419	4033	2615 (F)	Composite

The Site Table routine places the previous report into the drawing.

The Site ASCII File routine creates an ASCII comma-delimited file (CDF). The file contains the site name, stratum name, the names of surface 1 and 2, the method, the total cut, the total fill, and the net amount. This file allows you to use the LDD volume numbers in other programs.

The following is an example of the report:

```
base10,base,egctr,base,Grid,1063.864548,3654.106785,2590.242237
Base10,base,egctr,base,End
area,1278.116453,3891.819258,2613.702805
base10,base,egctr,base,Composite,1100.567626,2703.206510,1602.638884
```

Exercise

After you complete this exercise, you will be able to:

- Calculate a Composite Earthworks Volume
- Review the volume numbers

COMPOSITE SURFACE VOLUME

1. Use the Set Current Stratum routine from the Terrain menu to set Base as the current Stratum.
2. Calculate a Composite volume with the Calculate Total Site Volume routine. The routine is in the Composite cascade of the Terrain menu. Identify the site as Base10, select OK to exit the Site Librarian, and name the resulting surface, Basec. The settings for the Composite volume routine should be like those found in Figure 6-10. Your results should be similar to the following:

```
Current stratum: Base
Site name = Base10
Cut = 1419 cu.yds   Fill = 4033 cu.yds
Net = 2615 cu.yds FILL
```

3. Restore the Terrain Explorer and view the surface Basec with the Quick View routine of the Surface Display. After viewing the surface, redraw the screen to remove the lines.
4. Run the Create Contours routine to generate cut and fill contours for the Basec surface. These contours should look smoother than the contours for the Baseg surface. After inspecting the contours, erase the contours from the drawing.

5. Use the Set Current Stratum routine to set PointBase as the current Stratum.

6. Calculate a Composite volume with the Calculate Total Site Volume routine. The routine is in the Composite cascade of the Terrain menu. Identify the site as Base10, select OK to exit the Site Librarian, and name the resulting surface, PointBasec. The settings for the Composite volume routine should be like those found in Figure 6-10. Your results should be similar to the following:

```
Current stratum: PointBase
Site name = Base10
Cut = 1684 cu.yds   Fill = 4445 cu.yds
Net = 2761 cu.yds FILL
```

7. Restore the Terrain Explorer and view the surface PointBasec with the Quick View routine of the Surface Display. After viewing the surface, redraw the screen to remove the lines.

Volume Output—All Volume Methods

1. Use the Site Report routine to create a volume report. The routine displays an adjustment dialog box (see Figure 6–49). Select the OK button to exit the dialog box. The routine then displays the report dialog box (see Figure 6–50).

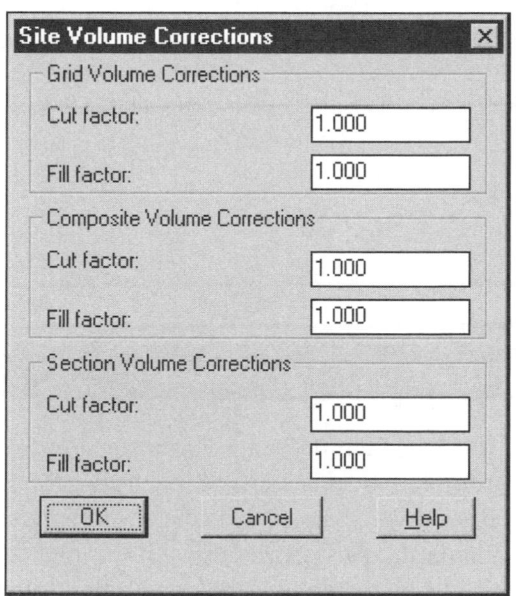

Figure 6–49

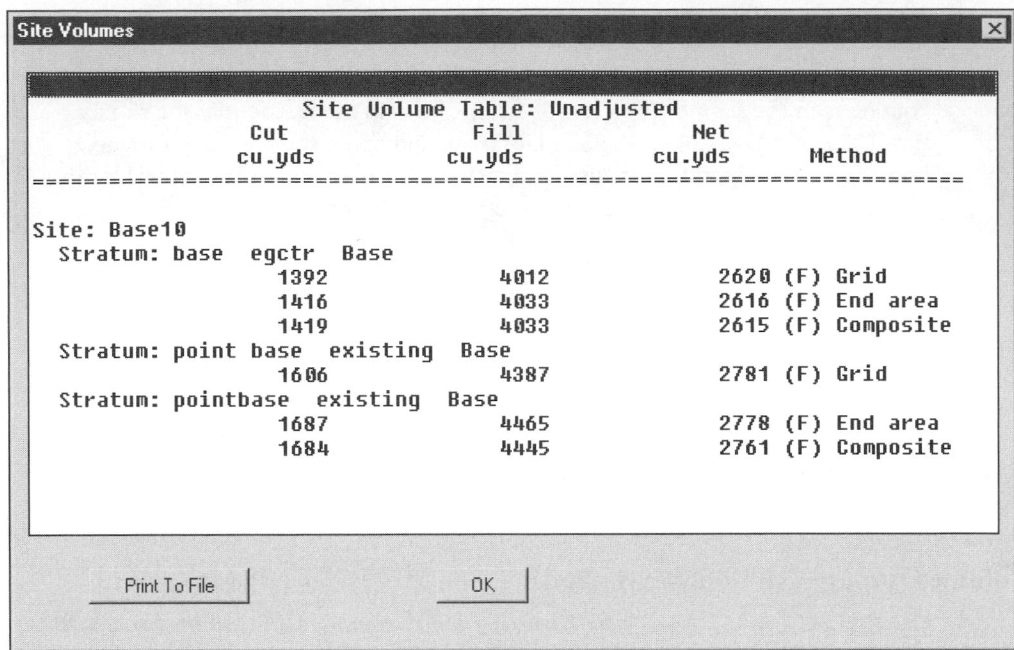

Figure 6–50

2. Place a volume table into the drawing. Use the Site Table routine in the Volume Reports cascade of the Terrain menu. The table cascades to the right and down from the insertion point. Zoom in to view the table. Undo the zoom and the table insertion.
3. Save the drawing.

UNIT 7: PARCEL VOLUMES

As the name implies, a parcel volume is a volume for a subarea smaller than the overall site. In a subdivision, a parcel is a lot, or in a golf course, the parcel could be the first hole. The Parcel volume routines calculate the impact of a design on a smaller scale. The process for calculating parcel volumes is the similar to that of site volumes (see Figure 6–51). The additional steps are the drawing and defining of parcels. All parcels defined in a project from the routines in the Parcels menu are automatically a parcel volume location.

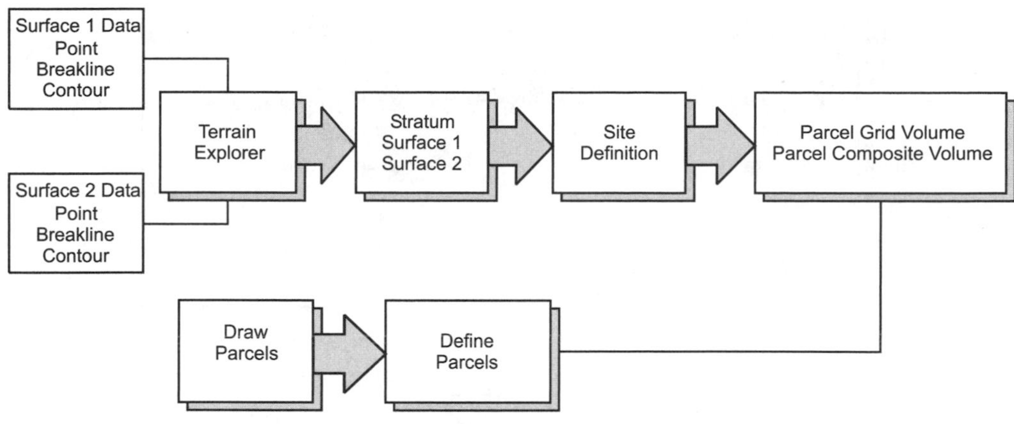

Figure 6–51

LDD requires you to define the parcels before you can calculate parcel volumes. The only methods of calculating a volume are Grid and Composite. The parcel volume routines use the same settings as the site volume routines. Both parcel volume routines display a dialog box that allows you to choose which parcels to calculate a volume for.

Exercise

After you complete this exercise, you will be able to:

- Calculate a Grid parcel volume
- Calculate a Composite parcel volume

PARCEL VOLUME

1. Thaw the Lot layer to view the parcels in the project. Isolate the lines on the Lot layer.
2. Run the Layer command and create a new layer area. Assign the layer a color and make it the current layer. Select the OK button to exit the Layer dialog box.
3. Run the Bpoly routine of AutoCAD to create a parcel polylines. Select points in the interior of eastern lots. The routine creates a polyline on the current layer. Press ENTER to exit the routine.
4. In the Layer dialog box freeze the Lot layer and exit the dialog box.
5. Run the Parcel Settings routine and set the values in the dialog box to match those found in Figure 6–52. After setting the values select the OK button to exit.
6. Define the Parcels with the Define By Polylines routine.

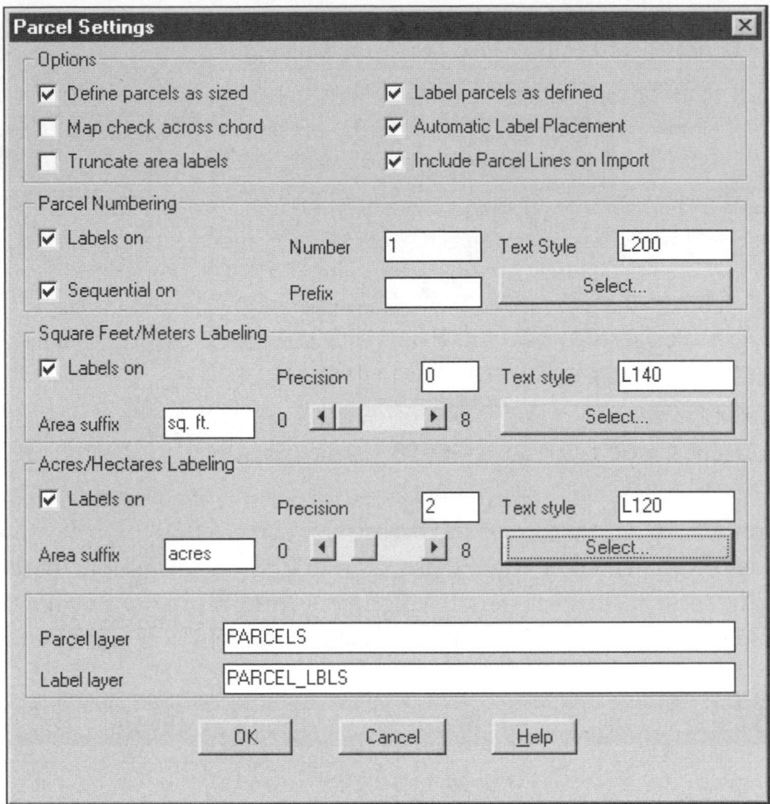

Figure 6–52

Calculating Parcel Volumes–Grid Method

1. Set the current stratum to Base with the Select Current Stratum routine from the Terrain menu.

2. Calculate parcel volumes for all of the lots by running the Calculate Parcel Volumes routine of the Grid cascade. The routine first displays the Site Librarian. Select Base10 and select OK to exit the dialog box. Next the routine displays the parcel list. Select all parcels and then select OK to exit and calculate the parcel volumes. Use the default grid volume settings. The routine places the volumes in the text screen of AutoCAD. Your results should be similar to the following:

```
Current stratum: Base
Site name = Base10
Cut = 1 cu.yds    Fill = 1 cu.yds
Net = 0 cu.yds CUT
```

```
Cut = 0 cu.yds   Fill = 16 cu.yds
Net = 16 cu.yds FILL
Cut = 149 cu.yds   Fill = 100 cu.yds
Net = 50 cu.yds CUT
Cut = 485 cu.yds   Fill = 439 cu.yds
Net = 47 cu.yds CUT
Cut = 0 cu.yds   Fill = 1756 cu.yds
Net = 1756 cu.yds FILL
Cut = 136 cu.yds   Fill = 348 cu.yds
Net = 212 cu.yds FILL
Cut = 118 cu.yds   Fill = 39 cu.yds
Net = 79 cu.yds CUT
Cut = 1 cu.yds   Fill = 11 cu.yds
Net = 11 cu.yds FILL
```

3. Change the current stratum to PointBase and calculate the parcel volume for all lots.

Calculating Parcel Volumes–Composite Method

1. Change the current stratum to Base and calculate a Composite Parcel Volume. Use the default composite volume settings. Your results should be similar to the following:

```
Current stratum: Base
Site name = Base10
Cut = 1 cu.yds   Fill = 1 cu.yds
Net = 0 cu.yds FILL
Cut = 0 cu.yds   Fill = 16 cu.yds
Net = 16 cu.yds FILL
Cut = 153 cu.yds   Fill = 101 cu.yds
Net = 52 cu.yds CUT
Cut = 492 cu.yds   Fill = 445 cu.yds
Net = 47 cu.yds CUT
Cut = 0 cu.yds   Fill = 1761 cu.yds
Net = 1761 cu.yds FILL
Cut = 142 cu.yds   Fill = 350 cu.yds
Net = 207 cu.yds FILL
Cut = 121 cu.yds   Fill = 42 cu.yds
Net = 79 cu.yds CUT
Cut = 1 cu.yds   Fill = 13 cu.yds
Net = 12 cu.yds FILL
```

2. Change the current stratum to PointBase and calculate a Composite Parcel Volume for this stratum.
3. View the Parcel Volume report by running the Parcel Report of the Volume Reports cascade of the Terrain menu.

This ends the section on site volume calculations. In a later section, we will return to the problem of site design and explore new tools in the Civil Design add-in product.

No matter which method you use, all of the calculation methods should produce a similar volume. If they do not, this is an indication that something is wrong. You need to check the site spacing and how it impacts the coverage of the surfaces by each volume method. For example, when you evaluate the results of the grid method, if there are wide gaps between the grid and the edge of the design, the site needs to be redefined to a smaller spacing.

Consistency is the keyword—consistent methods of developing surfaces and creating surface data for the Terrain Explorer.

The Roadway Design Process—Horizontal Alignments

After you complete this section, you will be able to:

- Edit the setting for a roadway centerline
- Draw a roadway centerline
- Edit a roadway centerline
- Analyze a roadway centerline
- Annotate a roadway centerline
- Create new points and objects relative to the centerline

SECTION OVERVIEW

This section and the next two sections cover the roadway design process of Land Development Desktop. The reason for dividing this topic into three sections is that the horizontal alignment phase of design is in LDD and the profile and cross-section phases are in an add-in to LDD, Civil Design. Each phase—horizontal alignment, profile, and cross section—has its own separate menus. Each menu contains tools to create, analyze, edit, and annotate the design focus.

The roadway design process ultimately produces roadway volume estimation, plan, profile, and cross sections of a design. The designer has flexible templates, control over ditches, superelevation calculations, benching, match slope control, and a host of spiral design options.

The roadway design process follows a design sequence from concept to the viewing of the design calculations. The roadway design process reduces the three-dimensional aspects of the roadway design problem into its 2D components (see Figure 7–1). This means projecting the 3D roadway design onto planes showing the road design as 2D. This is exactly what occurs when you design a roadway on paper. The paper is the 2D plane on which the designer represents the respective portions of the 3D roadway design problem.

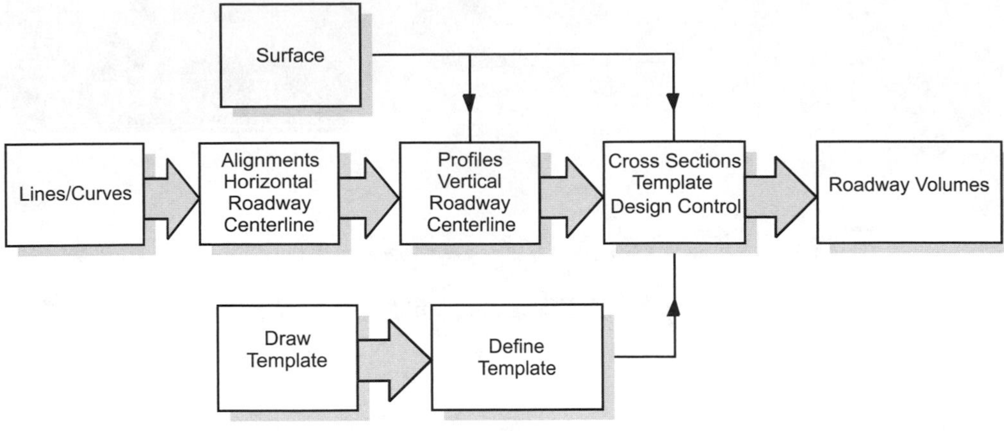

Figure 7–1

The roadway horizontal design tools are in the Alignments menu (see Figure 7–2). The vertical aspects of the roadway are on the Profiles menu. The cross-sectional and template view of the design is in the Cross Sections menu. Both the Profiles and Cross Sections menus are in the Civil Design add-in. When building a vertical frame of reference for profiles and cross sections, the routines read data from the surfaces stored within the project. When the design calls for a roadway template, the designer can use one from a library of templates or develop one just for the project.

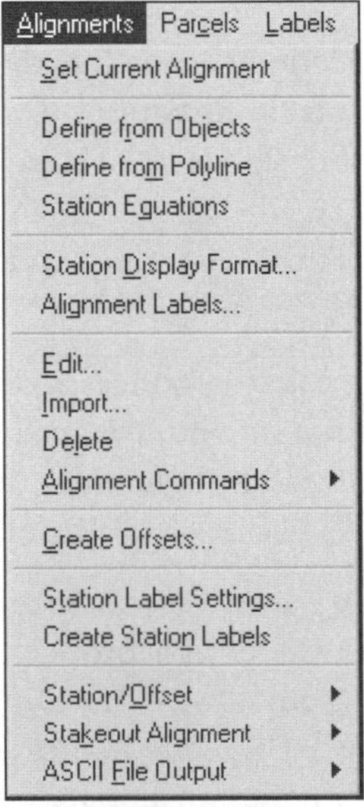

Figure 7–2

The design process is not overly flexible. You cannot design the horizontal roadway centerline and then jump to setting the template into the design. The interaction of the template depends upon the vertical design developed in profile and cross sections. What occurs at any point in the process depends upon what occurred before the current point in the process.

Rarely is a design a single pass through the design process. All of the initial values of the design elements give the designer a starting point from which to optimize their final values. To modify a design or element in the design means returning to that point in the process, changing the values or elements, and working forward from that point towards developing a new volume.

LDD creates external data files containing the road design data. These external data files allow anyone to review and/or edit the roadway data when working on the project.

UNIT 1

The settings used by the horizontal alignment are the first unit of this section. There are a number of layers, initial values, and styles to define and modify. As with other settings in LDD, these settings can be saved to a prototype, and when you select the prototype for a new project, the settings become a part of the project.

UNIT 2

A centerline in Land Development Desktop is a set of lines, curves, and maybe spirals. The tools of the Lines/Curves menu provide tools in addition to those in generic AutoCAD. After you draw the centerline, it must be defined. Defining in LDD is a process of creating an external data entry. The entry in the external file is a snapshot of the current conditions on the screen. If you edit the objects on the screen that are a part of a roadway centerline, you must redefine the centerline to update the external file. The use of the line and curve tools and the definition of a roadway centerline are the focus of Unit 2.

UNIT 3

Unit 3 covers the analysis of a centerline's design numbers. Most engineers design by calculations or design criteria. Lines and arcs on the LDD screen do not tell the designer the whole story of their values. These numbers must be queried by listing routines or reviewed in the Horizontal Editor.

UNIT 4

There are many times when the design changes or there are station equations on the centerline. The Horizontal Editor is the most powerful analysis and design-editing tool. The reason for this is that the Editor displays all of the essential data the designer needs to make a correct decision about a design change. The Alignments menu of LDD has a routine that adds station equations to a centerline of a roadway. The Horizontal Editor and the Station Equations tools are the subjects of the fourth unit.

UNIT 5

The annotation of the centerline is the topic of Unit 5. There are two basic styles of annotation. The first style is labeling parallel to and offset from the roadway. The second style has labels perpendicular to and on the centerline.

UNIT 6

The last unit reviews routines that create new objects that reference the centerline. These routines create lines, curves, points, and stakeout reports.

UNIT 1: ROADWAY SETTINGS

Land Development Desktop controls the settings for horizontal alignments. The settings are found in the Edit Settings dialog box of the Drawing Settings routine from the Projects menu (see Figure 7–3). The initial settings are from the selected prototype when defining the project. The settings are extensive in that they control design values, layers, annotation styles, text styles, and new object creation settings.

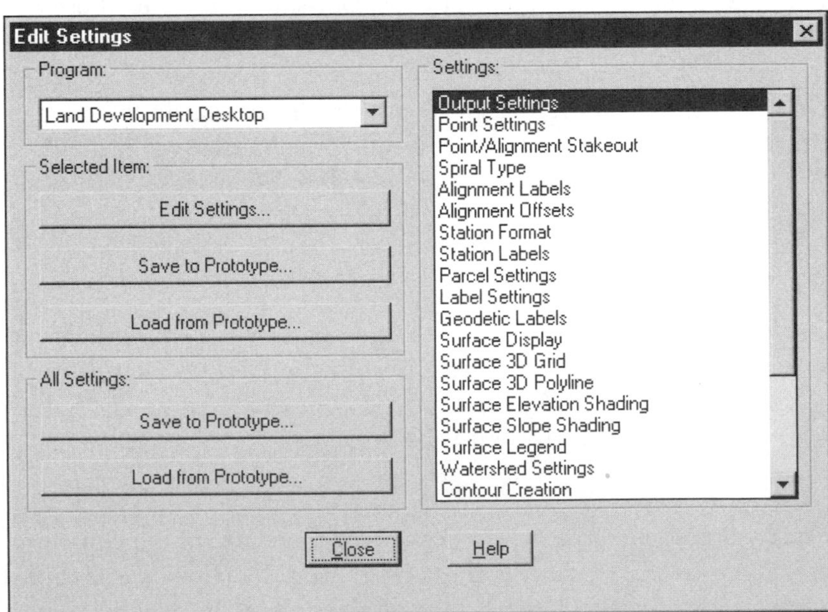

Figure 7–3

ALIGNMENT LABELS

The Alignment Labels Settings dialog box controls the labeling layer prefix and prefixes to information that appears on an annotated centerline (see Figure 7–4). The layer prefix at the top of the dialog box is the same as the layer prefix of surfaces. If you place an asterisk (*) in the box, each layer used by the alignment routines will be prefixed by the name of the current alignment. A dash after the asterisk is to separate the alignment name from the root layer name.

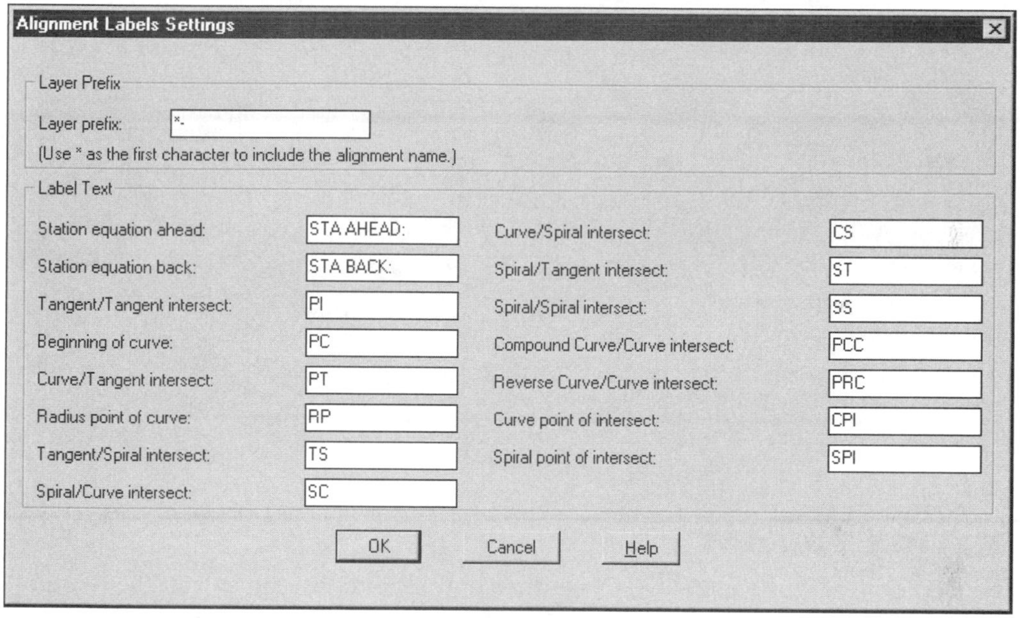

Alignment Labels Settings

Layer Prefix

Layer prefix: *-

(Use * as the first character to include the alignment name.)

Label Text

Station equation ahead:	STA AHEAD:	Curve/Spiral intersect:	CS
Station equation back:	STA BACK:	Spiral/Tangent intersect:	ST
Tangent/Tangent intersect:	PI	Spiral/Spiral intersect:	SS
Beginning of curve:	PC	Compound Curve/Curve intersect:	PCC
Curve/Tangent intersect:	PT	Reverse Curve/Curve intersect:	PRC
Radius point of curve:	RP	Curve point of intersect:	CPI
Tangent/Spiral intersect:	TS	Spiral point of intersect:	SPI
Spiral/Curve intersect:	SC		

OK Cancel Help

Figure 7–4

ALIGNMENT OFFSETS

The Alignment Offset Settings dialog box sets the layer, distance, and side for line/curves offset from a defined centerline (see Figure 7–5). You can run this routine as many times as you like. You need only change the distances, layers, and sides. If the layers are not present in the drawing, the routine creates the layers using AutoCAD default values.

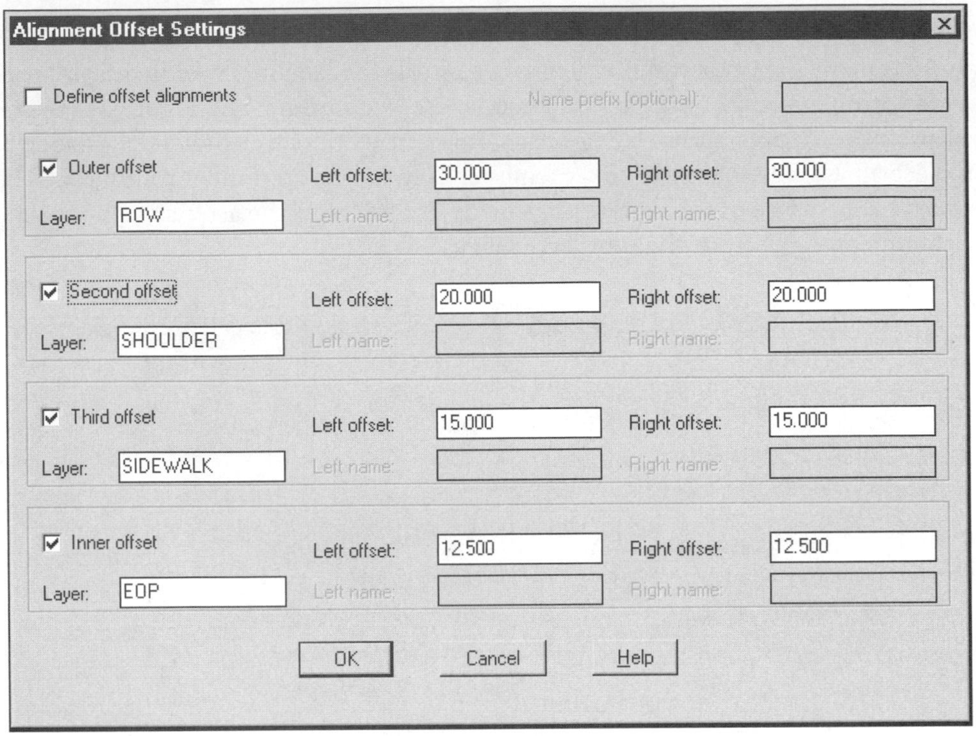

Figure 7–5

STATION FORMAT

The Edit Station Format dialog box controls how a centerline appears in the routines of Land Development Desktop (see Figure 7–6). The top portion of the dialog box displays the effects of the settings in the middle and bottom areas of the dialog box.

The middle portion, Numeric Format Options, controls showing leading 0 (zeros) in the station (01+00 instead of 1+00 for a base 1000 stationing system), whether negative stations have brackets, and whether a station has no precision when it is an even station. The right side of the dialog box sets the number of decimal places, the minimum width of a station, and the character used for a decimal.

The bottom portion of the dialog box toggles on the station system, the character used to separate the digits, and the base number for the stationing system. LDD can display stations as straight decimal values instead of the traditional "+" (plus) system, or you can choose to use a different character instead of the plus sign (+). The base stationing affects how the stations will appear on the screen. If the base number is 1000, a 300 station appears as 03+00 if the toggle to show leading 0 (zeros) is on. These settings make LDD able to handle imperial and metric jobs and their different methods of displaying stationing values.

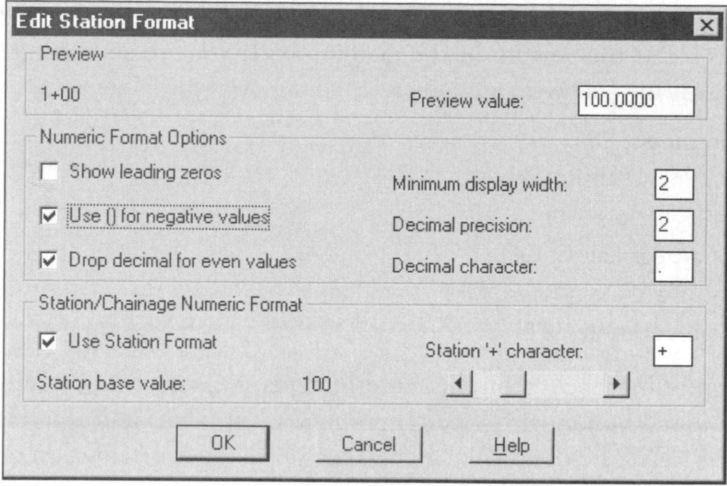

Figure 7–6

STATION LABELS

The Alignment Station Labels Settings dialog box controls the layers, location, and orientation of centerline labeling (see Figure 7–7). The top portion of the dialog box toggles on or off elements to label, and if this setting is toggled on, what layers to use in the drawing.

Alignment Station Label Settings

Station labels — Layer STALBL
Station point labels — Layer STAPTS
Station equations labels — Layer STAEQU
Perpendicular labels
Stations read along road
Plus sign location

Station label increment 100.00
Station tick increment 50.00
Station label offset 15.00

OK Cancel Help

Figure 7–7

By default the labeling is parallel and offset from the roadway centerline. The distance of the offset is set at the bottom of the dialog box. If you are stationing with offset parallel labels, set the following defaults:

Station Labels: ON
Station Point Labels: ON
Station Equation: ON
Perpendicular Labels: OFF
Stations Read along Road: OFF
Plus Sign Location: OFF

If you want centerline labeling on the centerline, perpendicular to the centerline, and at the station mark, you must toggle on perpendicular labels and plus sign location (see Figure 7–8). The Perpendicular Labels toggle changes orientation of the labels. If the Plus sign location is not on, the routine offsets the labels from the station point along the centerline of the alignment. When stationing with perpendicular labels, use the following defaults:

Station Labels: ON
Station Point Labels: ON
Station Equation: ON
Perpendicular Labels: ON
Stations Read along Road: OFF
Plus Sign Location: ON

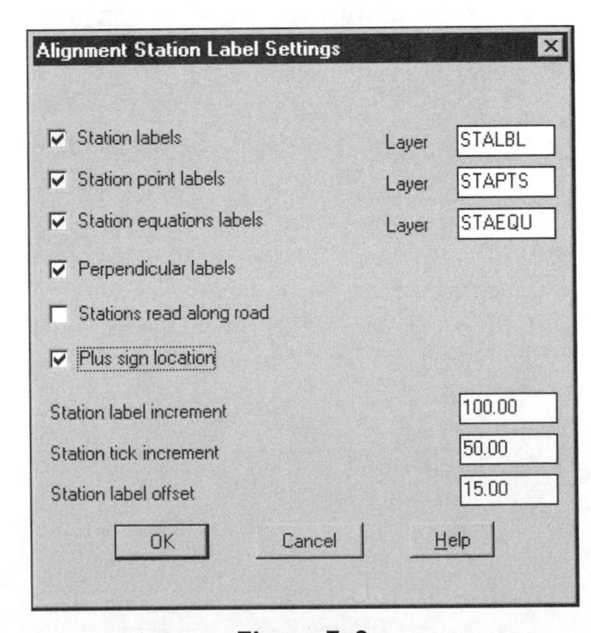

Alignment Station Label Settings
Station labels Layer STALBL
Station point labels Layer STAPTS
Station equations labels Layer STAEQU
Perpendicular labels
Stations read along road
Plus sign location
Station label increment 100.00
Station tick increment 50.00
Station label offset 15.00
OK Cancel Help

Figure 7–8

The Stations Read Along Road toggle controls the orientation of the station labels. On some alignments, stationing may end up being upside down. To prevent the upside down stationing, toggle off Station Read Along Road. This toggle ensures that the text is always right side up along the length of the centerline no matter how convoluted the centerline.

The Plus Sign Location toggle controls the location of the station label. With the Plus Sign Location toggle on, it forces all station labels, whether parallel or perpendicular, onto the centerline.

The Station Label Increment has to be a multiple of the Station Tick Increment. If the Station Label Increment is 100, the tick increment can be 5, 10, 20, 25, or 50.

CURVE TABLES

Land Development Desktop creates curve/spiral segments from a superelevation table (see Figure 7–9). You can use the American Association of State Highway and Transportation Officials (AASHTO) tables as the basis for a roadway design. There are additional tables for Canadian and Malaysian road designs (see Figure 7–10). You can edit the table entries if you use different standards. When you create a curve with a table routine, the routine displays a default speed table. If the table is not the correct one, you select a different table while the routine is running. After setting the correct table, you select the speed and radius. The routine creates a curve with a radius from the table and if appropriate, attaches a spiral into and out of the arc.

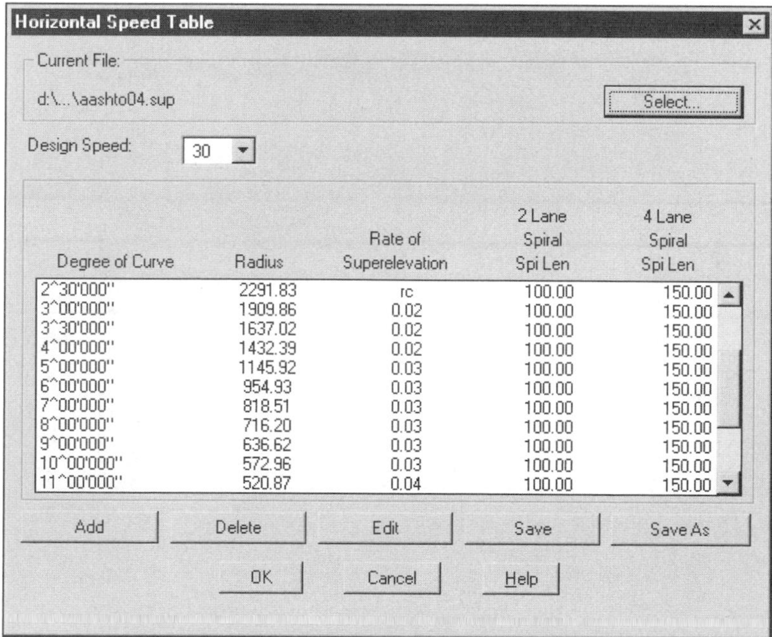

Figure 7–9

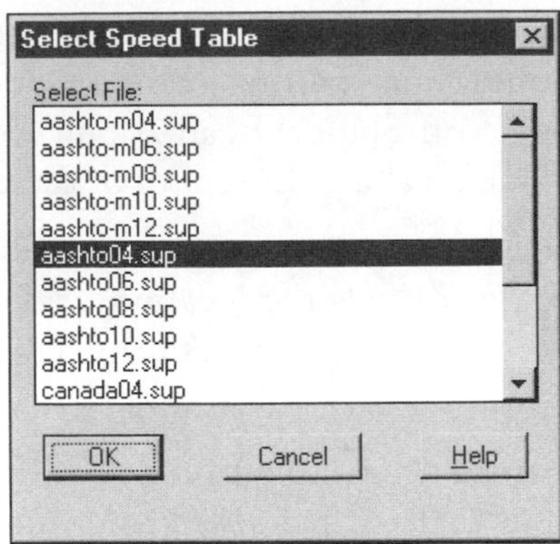

Figure 7–10

Exercise

After you complete this exercise, you will be able to:

- Know and set the different settings for alignments

STARTING A NEW DRAWING AND PROJECT

1. Start a new drawing, Roadway, and assign it to a new project, Roadway (see Figures 7–11 and 7–12). Use the LDDTR2 prototype. Describe the project as the Exercise for Section 7 and add the keyword Roadway in the Keyword box. Select the OK button to exit the Project Details and New Drawing dialog boxes. The routine creates a new project and copies the settings from the prototype to the project.

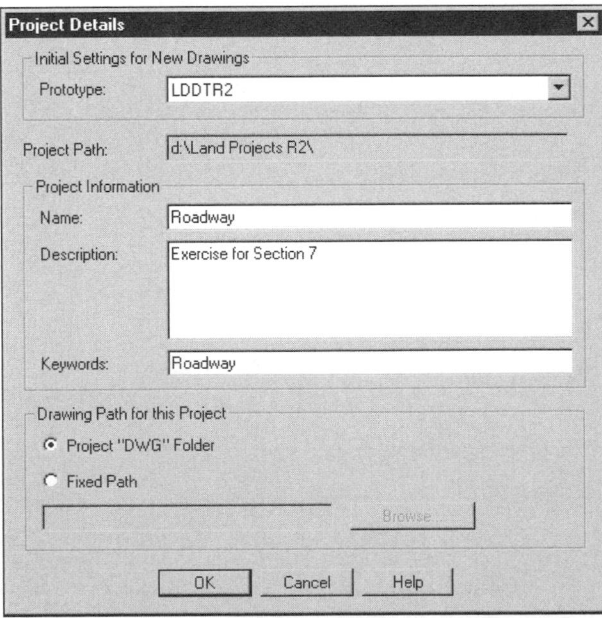

Figure 7–11

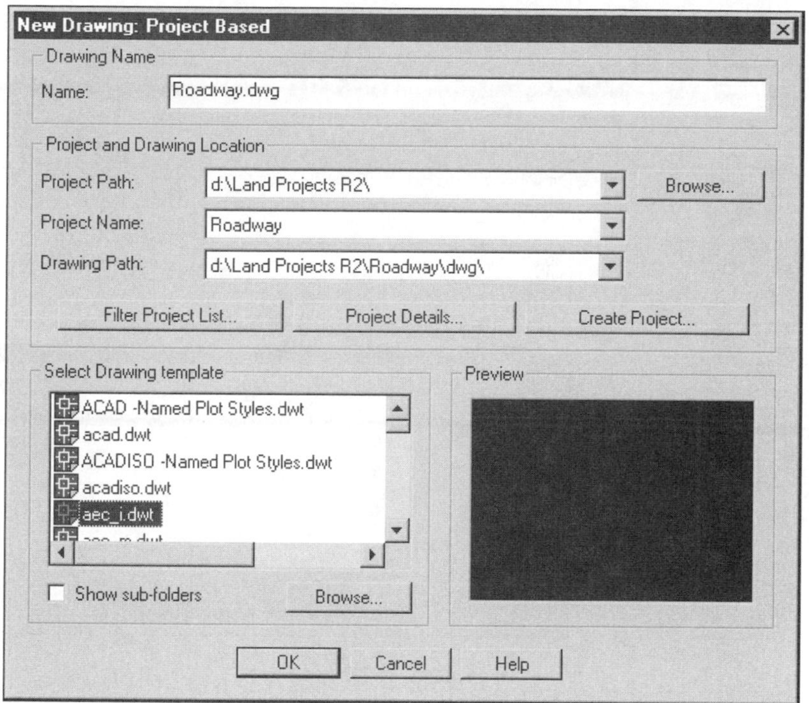

Figure 7–12

2. The next dialog box asks for the default values for the point database (see Figure 7–13). Select OK to exit the dialog box.

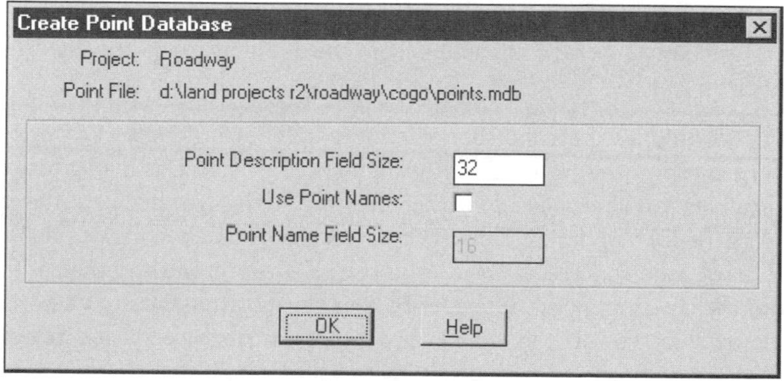

Figure 7–13

3. The next dialog box asks you to set up the drawing scale and associated values. Select the LD1-40 setup file, then select the Load button (the upper right of the dialog box), and finally select the Finish button at the lower right of the dialog box (see Figure 7–14).

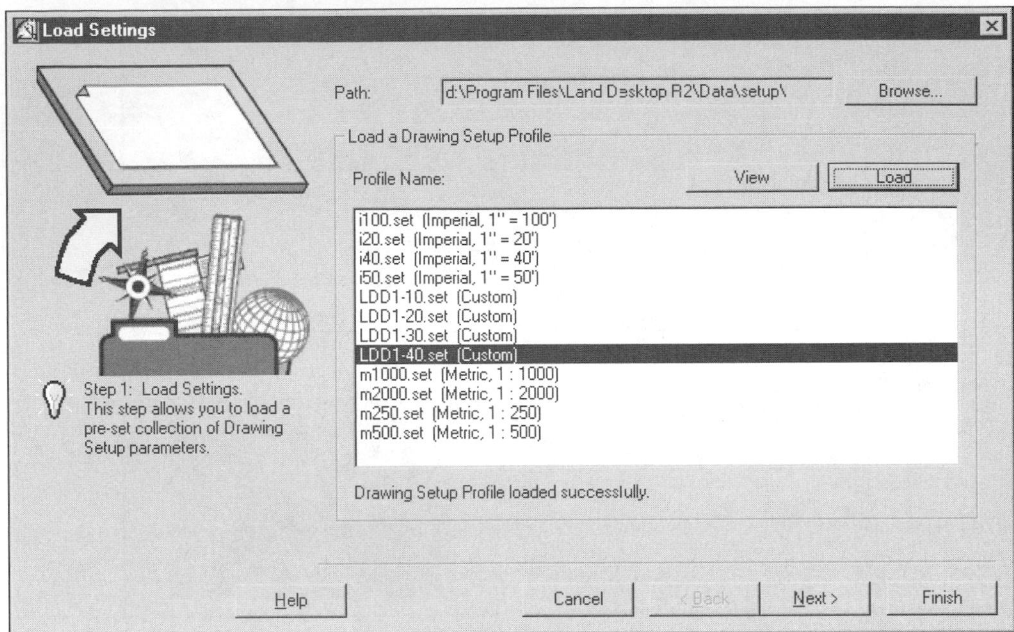

Figure 7–14

4. After you exit the Drawing Setup dialog box, the review settings dialog box displays. Review the values and then select the OK button to dismiss the dialog box.

5. Save the drawing.

ALIGNMENT SETTINGS

6. Select the Drawing Settings routine from the Projects menu to view the alignment settings.

7. Select Alignment Labels and then select the Edit Settings button to view the current settings (see Figure 7–4). As mentioned earlier, this dialog box contains the prefixes for labels placed into the drawing. The Layer prefix at the top is the same as the Terrain layer prefix. This prefix places labels on layers defined by their associated centerline. When you relabel a centerline and the routine asks to delete the labels, only the appropriate labels are erased. Enter an asterisk followed by a dash to set the layer prefix (*-), if it is not already set. Select OK to exit the dialog box.

8. Select Alignment Offsets and then select the Edit Settings button to view the settings (see Figure 7–5). Toggle on all of the offsets. Select OK to exit the dialog box.

9. Select Station Format and then select the Edit Settings button to view the settings (see Figure 7–6). This dialog box formats stationing when LDD prompts you

about the centerline. Toggle on Use () (parentheses) for negative stationing. After setting the toggle, select the OK button to exit the dialog box.

10. Select Station Labels and then select the Edit Settings button to view the settings (see Figure 7–7). The default method of labeling is parallel and offset from the centerline. This will be changed later in another exercise. Select OK to exit the dialog box.
11. Select the Close button to exit the Edit Settings dialog box.

EXERCISE

UNIT 2: CREATING A ROADWAY CENTERLINE

The roadway design process starts with lines and arcs in the drawing file. A polyline can also be a centerline. The only drawback to using a polyline is that the stationing of the centerline corresponds to the direction of the polyline segments. If a centerline contains a spiral segment, LDD represents spirals as a polyline with extended entity data. Do not edit spirals with the PEDIT command, because you may lose the extended entity data LDD places on the object.

To create centerlines, simply use one of the Define routines, Objects or Polyline. Each Define routine prompts you for a starting reference point, name, description, and starting station. The reference point is the starting point of the line segment. If for some reason another point along the line is the starting point (of the stationing), you select that point with an AutoCAD object snap. You can select a point away from the centerline, and the routine will find a point on the centerline perpendicular to the point you select.

The last step in defining the centerline is filling in the required information in the Define Centerline dialog box.

LINES/CURVES MENU

Section 3 of this book covers the line and curve tools of Land Development Desktop. The Lines/Curves menu contains several tools for creating and transcribing written descriptions of lines and curves that may appear in a project (see Figure 7–15). Many times the lines that define centerlines of a roadway are parallel and offset from boundary lines defining a project. The curve segments may be fillets you place between the tangent lines.

It makes no difference how the lines and curves are developed as long as they are correctly represent the centerline or the roadway.

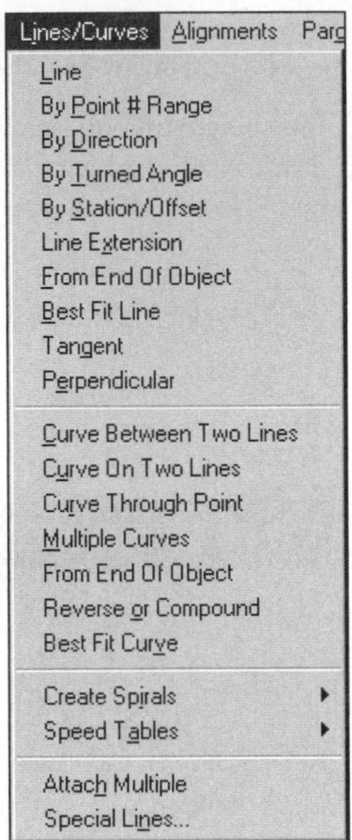

Figure 7–15

ALIGNMENTS MENU

The Alignments menu contains all of the routines you need to create, analyze, edit, and annotate roadway centerlines (see Figure 7–2). Seven routines from the menu complete these four basic tasks (see Figure 7–16).

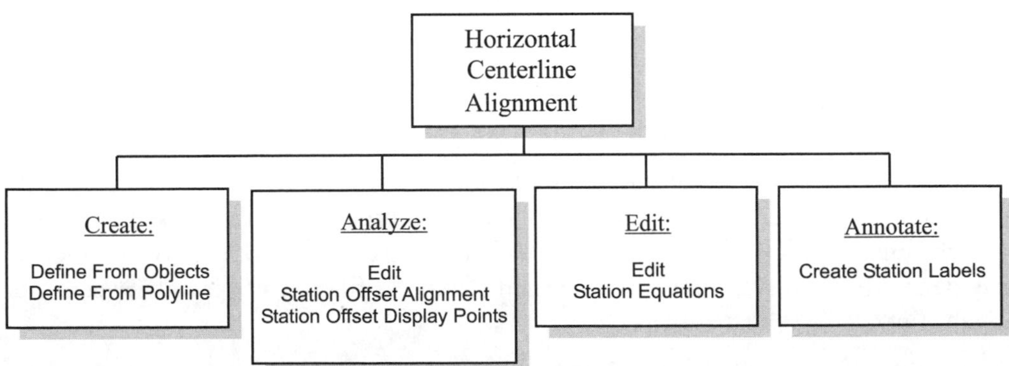

Figure 7–16

The defining routines take the line/curve objects you select and create alignment definitions. Generally, alignments are roadway centerlines, but the Civil Design templates are flexible. Using supplemental horizontal alignments, transition alignments, creates the horizontal stretch of a template. LDD, however, lists horizontal and transitional alignment together. You must be aware of which alignments are centerlines and which are transitional.

CENTERLINE ALIGNMENTS

The centerline consists of lines and curves or polylines in the drawing. After you draw the centerline lines and curves, the Define from Objects or Polyline routine creates the centerline. The routines defining alignment create an entry in an external database file representing the centerline. The definition gives you access to the design numbers that the LDD objects represent in the drawing.

The external database is an access database and is a record locked file (see Figure 7–17). This means that several people can define and edit alignments as long as they are not trying to work on the same centerline. LDD places the main alignment database in the *Align* folder of the project. Each alignment has a folder below the main folder containing profile and cross-section data. The subfolder is the same name as the alignment.

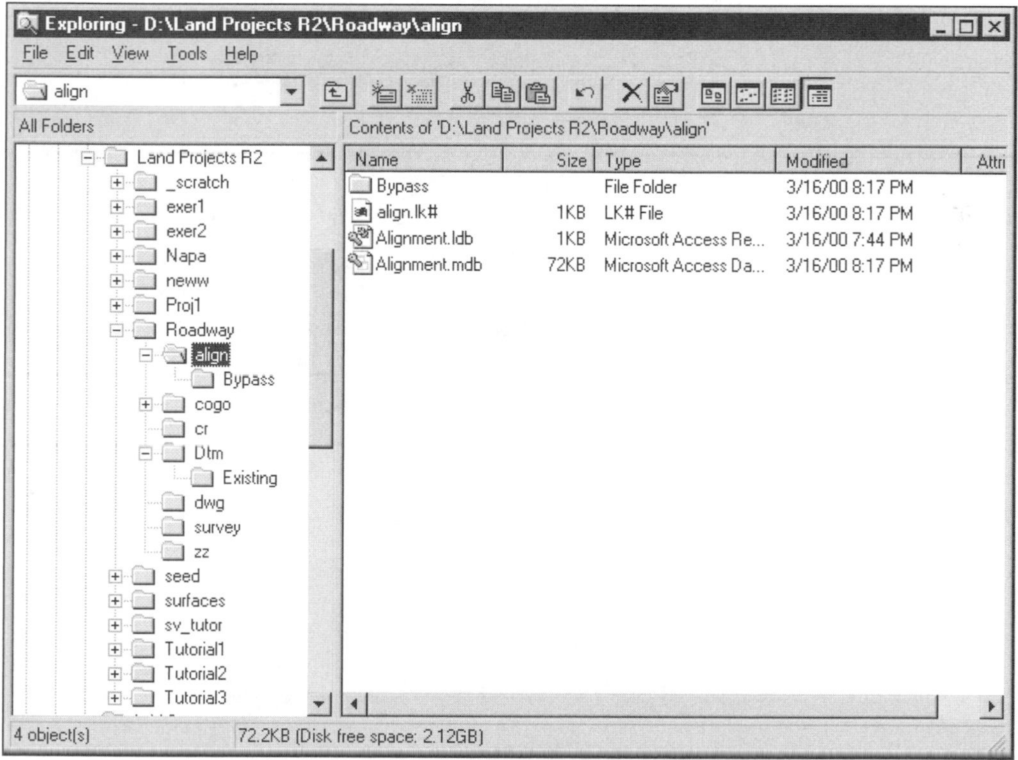

Figure 7–17

SELECTING A CURRENT ALIGNMENT

A drawing may have several lines and curves representing different horizontal alignments. Objects representing an alignment can be in one or several drawings at the same time even though there is only one definition entry for all occurrences of an alignment. When you are working in a drawing, only one alignment can be current at any one time. If the project contains several alignments, the designer must be sure that the correct alignment is the current alignment. This is because the current alignment is the focus of all of the routines you use. To change the focus of the routines, you must change the current alignment. To change the current alignment, you will use the Select Current Alignment routine in the Alignments menu most of the time.

If you are in a drawing session and you select an alignment routine, the routine will force you to select a current alignment if one is not set. The default method of setting the current alignment is to select an object in the drawing representing the alignment. If you are unable to select the correct object, the routine has two additional methods of identifying the current alignment. In the first optional method, you start by clicking the right mouse button and selecting an alignment from a list inside a dialog box (see Figure 7–18). In the second option, you cancel out of the dialog box and enter the number of the alignment at the command prompt.

Also, whenever you define a new centerline, it becomes the current alignment.

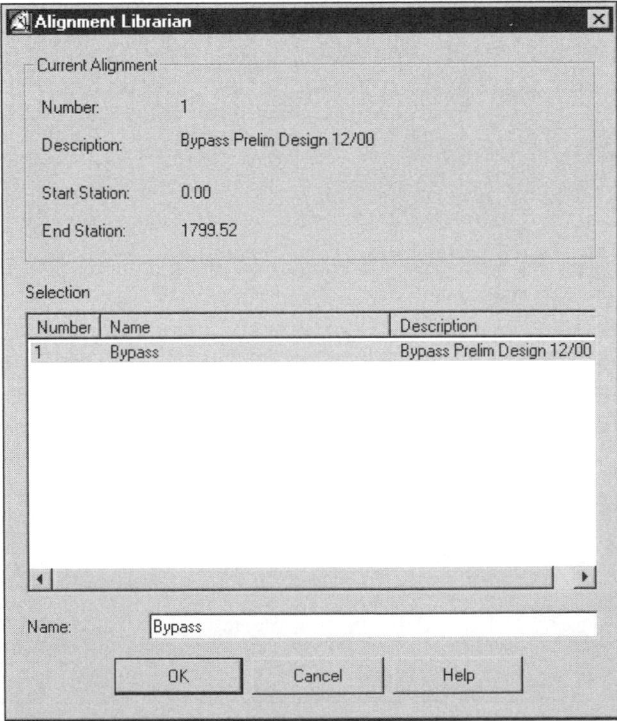

Figure 7–18

EXERCISE

Exercise

After you complete this exercise, you will be able to:

- Draw the roadway centerline
- Define a roadway centerline

DRAW THE HORIZONTAL CENTERLINE

1. If you are not in the Roadway drawing, open it now.
2. Currently the drawing is empty. From the CD, insert the Road drawing. Insert the drawing at 0,0,0, with a scale factor of 1, a 0 (zero) rotation, and explode the drawing as you insert it into the current drawing. Use the ZOOM EXTENDS command to view the entire site.
3. The drawing contains contours and a point. The point does not automatically become a part of the project. You must add the point into the project point database. Run the Modify Project routine of the Check Points cascade. Toggle on Add Unregistered Points to Project Database and select OK to exit the dialog box. Refer to Figure 7–19 for the Check Points settings.

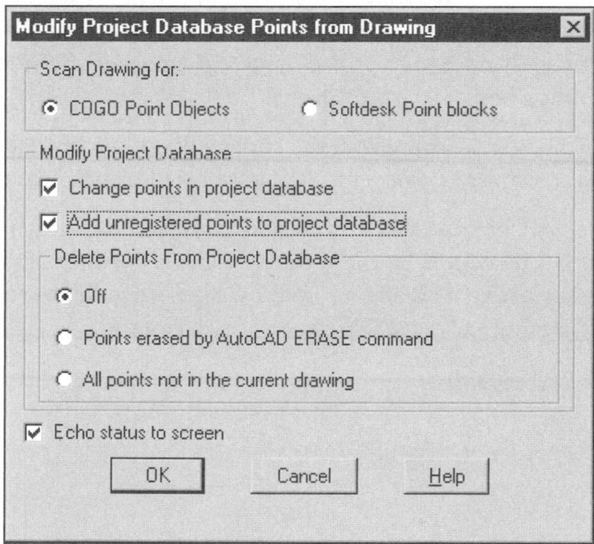

Figure 7–19

4. Open the Terrain Model Explorer and create a new surface. Select Terrain at the top of the upper left panel, click the right mouse button, and select Create New Surface. A new surface, surface1, appears in the Terrain tree. If the surface1 surface does not appear, select the "+" (plus sign) to expand the Terrain tree to view the surface name list.

5. Select the terrain surface, surface1, click the right mouse button, then select Rename, and rename the surface to *Existing*. After renaming the surface. select the "+" (plus sign) under the Existing surface name to expose the surface data list.

6. Select Point Groups, click the right mouse button, and select Add Point Group. The routine displays a dialog box indicating that there are no point groups and asks if you want to run the Group Manager routine to create a group (see Figure 7–20). Select Yes to call up Group Manager.

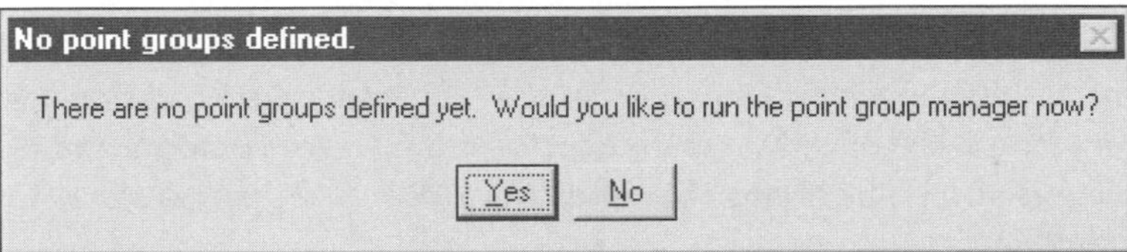

Figure 7–20

7. In the Group Manager dialog box, click the Create Point Group icon to display the Create Group dialog box. Enter **Spot** as the name of the group and select the Build List button to view the Point List dialog box. Toggle All Points to identify all points in the drawing as members in the point group. Select OK to exit the Point List and the Create Point Group dialog boxes and reenter the Group Manager. Click the "X" in the upper right corner of the Manager to dismiss the dialog box.

8. Again select the Point Group from the surface data list, click the right mouse button, then select Add Point Groups, and select the Spot group from the list.

9. Select Contours from the data list, click the right mouse button, and select Add Contour Data. The routine displays the weeding/supplementing dialog box. Set the weeding factors to 15.0 and 4.0 and the supplementing factors to 20.0 and 0.25 (see Figure 7–21). Select OK to exit the dialog box.

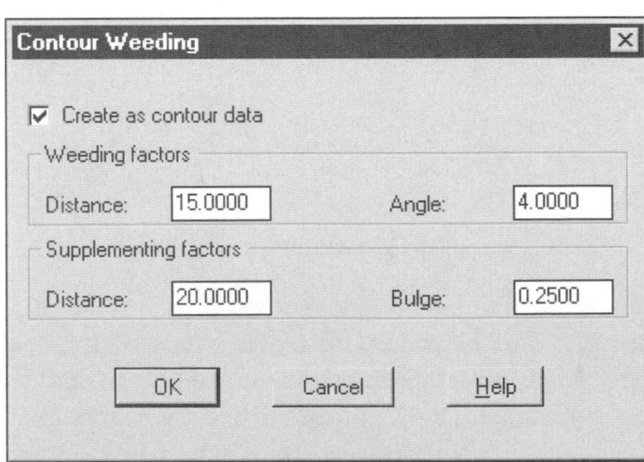

Figure 7–21

EXERCISE

There are two layers with contours, Existing-Cont-Mjr and Existing-Cont-Mnr. These layers also contain 3D polylines. The Terrain Modeler does not differentiate between contours and 3D polylines. The only difference between contours and 3D polylines is that the contour has the same elevation at each vertex, and the 3D polyline has potentially different elevations at each vertex. They both act as breakline surface data.

10. At the Select objects prompt, use the Layer method. The default method is shown in greater angle brackets. If the option is set to Entity, type the letter **L** and press ENTER to change the mode or just press ENTER to go to the selecting process. Select one object from each layer. The routine should echo back the layer name of each selected layer. After you select the two objects, press ENTER twice to exit the selection process. As you reenter the Terrain Modeler, the contours are evaluated and a report about weeding and supplementing is put on the command prompt.

11. Run the Build surface routine to create the surface. Select the surface name, click the right mouse button, and select Build. The routine displays the Build Options dialog box. Refer to Figure 7–22 for the settings and when you are finished, select the OK button to create the surface. Select the OK button of the Done dialog box to reenter the Terrain Explorer.

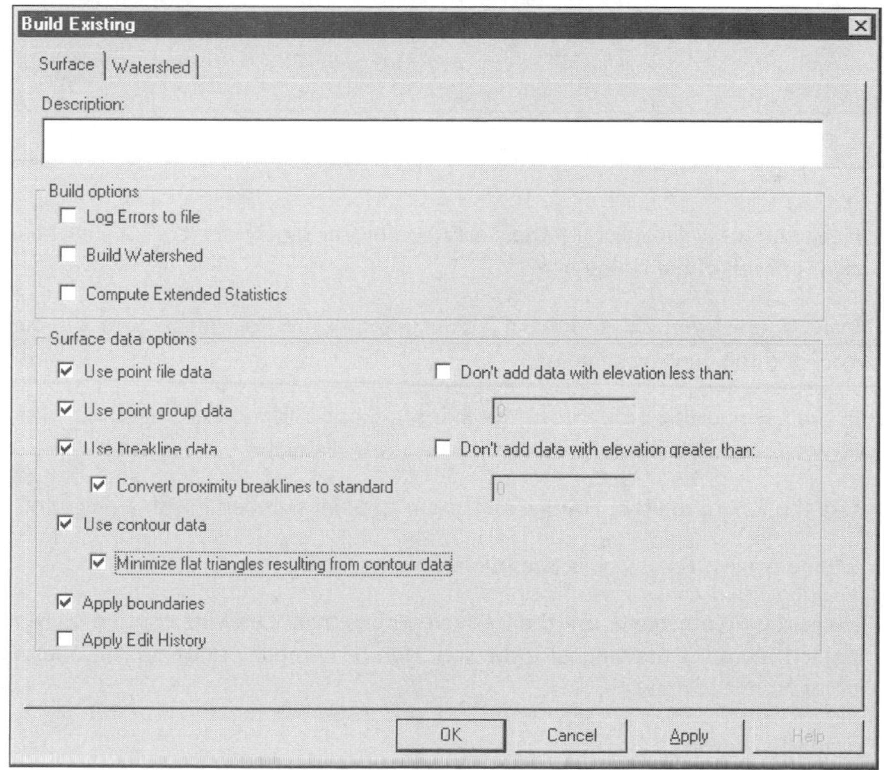

Figure 7–22

EXERCISE

12. Check the elevations listed in the Explorer. They should be between 704 and 741 (see Figure 7–23).

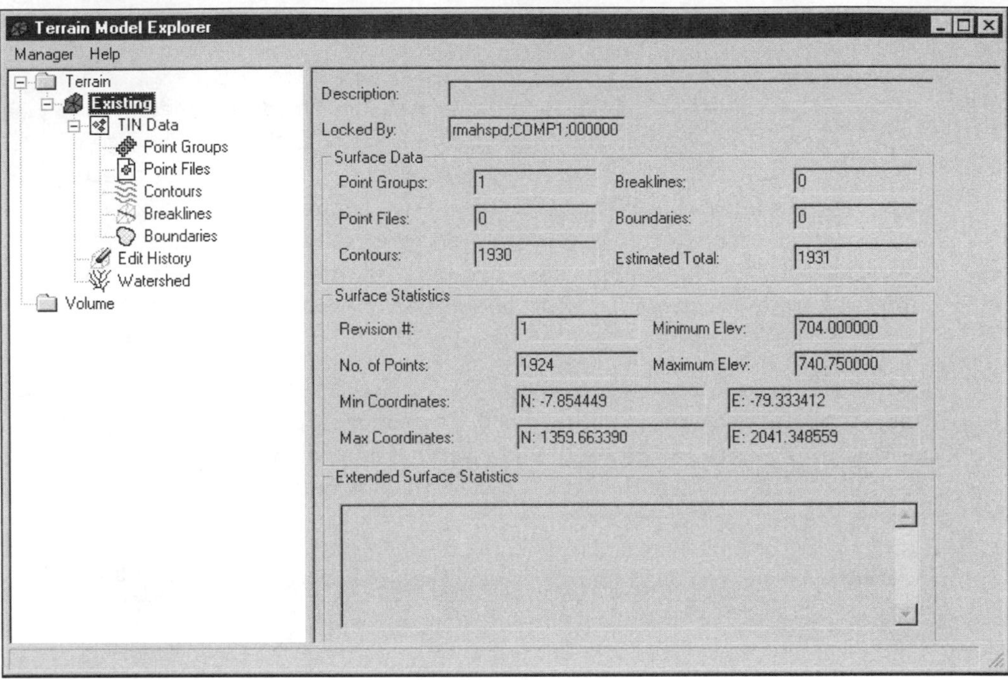

Figure 7–23

13. If the surface is OK, dismiss the Terrain Explorer by clicking the "X" in the upper right corner of the dialog box.

14. Make a new layer, CL, and make it the current layer. Assign the layer the color cyan and the linetype of center.

15. In the Layer dialog box, freeze the Existing-Cont-Mjr and Existing-Cont-Mnr layers. Select the OK button to exit the Layer dialog box.

16. Use the Zoom to Point routine and zoom to point number 1 with a height of 1650.

17. Define a view, Roadway, in the Named Views dialog box.

18. From the Points menu, use the Remove Points from Drawing routine to remove point 1 from the drawing. Use the selection or number option to remove this point from the drawing.

19. Save the drawing.

Drawing the Roadway Tangent Lines

The new roadway centerline will start as a northing/easting coordinate on the eastern side of the drawing.

1. Draw the centerline of the roadway with the By Direction routine from the Lines/Curves menu. The first two segments are designed by bearings and the last segment is an azimuth. Toggle the By Direction routine into northing/easting mode by typing the dot toggle *.n*. This changes the prompt from Starting Point: to Starting Point: Northing:. Refer to Figure 7–24 and the table below for directions and distances. After drawing the last tangent, press ENTER twice to exit the routine.

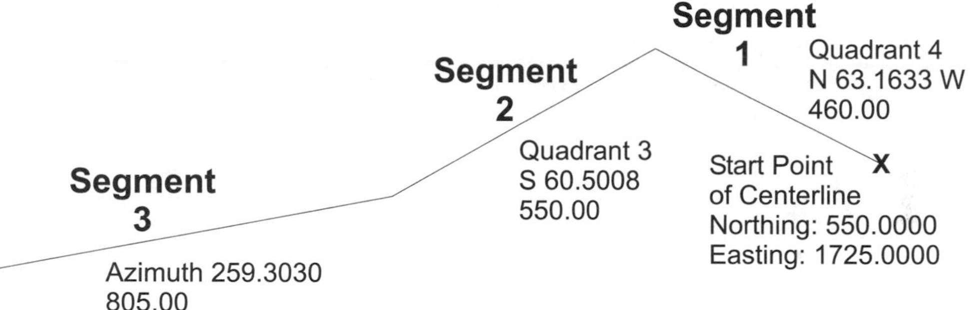

Figure 7–24

	Northing	Easting	
Start of Segment 1 –	550.00	1725.000	
	Distance	Quadrant	Direction
Segment 1 –	460.00	4	63.1633
Segment 2 -	550.00	3	60.5008
	Distance	Azimuth	
Segment 3 -	805.00	259.3030	

The following is the command sequence.

```
Command:
COGO Line by direction
Starting point: .n
Starting point:
 >>Northing: 550
 >>Easting: 1725
```

```
Angular units: Degrees/Minutes/Seconds (DD.MMSS)
Quadrant (1-4) (Azimuth/POints): 4
Bearing (DD.MMSS): 63.1633
Distance: 460
Angular units: Degrees/Minutes/Seconds (DD.MMSS)
Quadrant (1-4) (Azimuth/POints): 3
Bearing (DD.MMSS): 60.5008
Distance: 550
Angular units: Degrees/Minutes/Seconds (DD.MMSS)
Quadrant (1-4) (Azimuth/POints): a
Azimuth (DDD.MMSS): 259.3030
Distance: 805
Angular units: Degrees/Minutes/Seconds (DD.MMSS)
Azimuth [Bearing/POints]:
Perimeter = 1815
Starting point:
 >>Northing: .n
Starting point:
```

2. Use the Curve Between Two Lines routine from the Lines/Curves menu to place two arcs at the Point of Intersections (PIs) of the centerline by. Use Figure 7–25 as a reference for your work. The first arc is 125 feet in radius, and a tangent length of 150 defines the second arc.

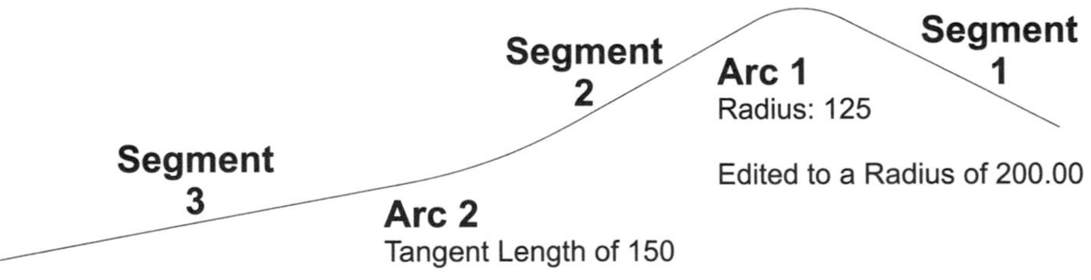

Figure 7–25

The routine prompts for the selection of the each tangent line (segments 1 and 2). After you select the two segments, the routine responds with a list of options. To enter a radius, press ENTER and enter the radius, 125. The routine creates the arc and then prompts for the next two segments. Select segments 2 and 3. When the routine displays the option list, enter **t** for tangents, press ENTER to choose the option, and enter the length of tangent, 150. The length is a distance along segment 2 you travel before entering the second arc. The following is the command sequence for the routine.

```
Command:
COGO Curve by two known tangents
Select first tangent: (select segment 1)
Second tangent: (select segment 2)
FACTOR [Length/Tangent/External/Degree/Chord/Mid/MDist/<Radius>]:
  (press ENTER to select the radius option)
Radius: 125
Select first tangent: (select segment 2)
Second tangent: (select segment 3)
FACTOR [Length/Tangent/External/Degree/Chord/Mid/MDist/<Radius>]: t
Tangent length: 150
Select first tangent: (press ENTER to exit the routine)
Command:
```

The resulting curves are not the final values. Rather than erasing the curves and refilleting the tangents, we will adjust the radius in the Horizontal Alignment Editor after defining the lines and curves as a roadway centerline.

3. Save the drawing.

EXERCISE

DEFINE THE CENTERLINE

1. Define the horizontal alignment by using the Define from Objects routine in the Alignments menu (see Figure 7–2). The routine starts by prompting for the starting point of the centerline. Select near the southern end of segment 1 to define the starting point. A red blip X appears at the endpoint of the segment nearest to your selection point. After you select the first segment, select the remaining segments. It does not matter how or in what order you select the remaining objects.

LDD places a red X at the end of the first segment to indicate the assumed starting point of the alignment. This is the default reference point of the alignment. If the start (reference) point of the alignment is another point away from the endpoint of the line, select a new starting (reference) point for the alignment. You should use an AutoCAD object snap to define the new start (reference) point. If the endpoint of the segment is also the starting point of the alignment, just press ENTER at the reference point prompt.

If you identify a point-of-beginning away from the end of the segment, the Define routine assigns lower stations from the new starting point to the endpoint of the segment. If you set the reference point 50 feet down the line and define this point as the beginning station 0 (zero) LDD assigns negative stations to the portion of the line from the reference point to the end of the line; that is, -0+10, -0+20, and so forth.

```
Command:
Select entity: (select segment 1 near its eastern end)
Select objects: 1 found (select the remaining centerline objects)
Select objects: 1 found, 2 total
Select objects: 1 found, 3 total
Select objects: 1 found, 4 total
Select objects:
Connecting entities: ....
Done!
Select reference point (Enter for start): (press ENTER)
```

The routine presents a dialog box for the entry of alignment values (see Figure 7–26). Enter a name, Bypass, and a description, Bypass Prelim Design 12/00, about the centerline. After entering the values, select the OK button to exit the dialog box and routine. When you exit the dialog box, the routine displays a report about the alignment.

```
————————— ALIGNMENT DATA —————————
Description: Bypass Prelim Design of 12/00
Name: Bypass                  Number: 1            Length: 1799.52
Starting station: 0+00        Ending station: 17+99.52
Superelevation data created.
```

2. Save the drawing. This save preserves the drawing and the horizontal alignment definition in the alignment database.

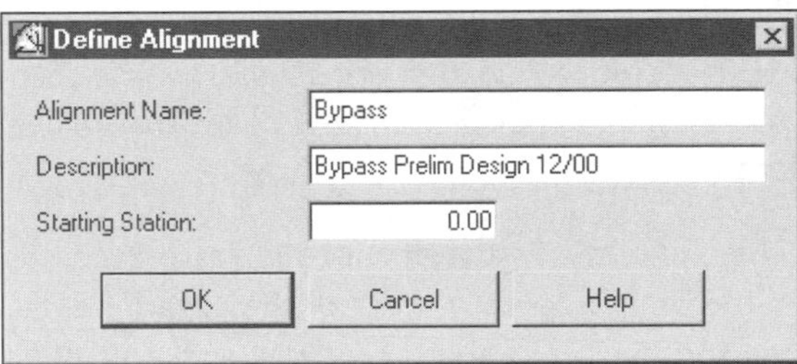

Figure 7–26

UNIT 3: CENTERLINE ANALYSIS

ANALYZING A HORIZONTAL ALIGNMENT

There are two types of analysis when you are working with alignments. The first is analyzing the alignment itself and the second is analyzing objects related to the alignment.

When analyzing the alignment itself, the Edit routine of the Alignments menu displays the horizontal design numbers represented by the objects in the LDD drawing. Many times the designer is not interested in what is on the screen, but is interested in the design criteria controlling the design itself. The only way to "see" these numbers is to open the Horizontal Editor and view the calculated values (see Figure 7–27) of the alignment. The Horizontal Editor is the topic of the next unit of this section.

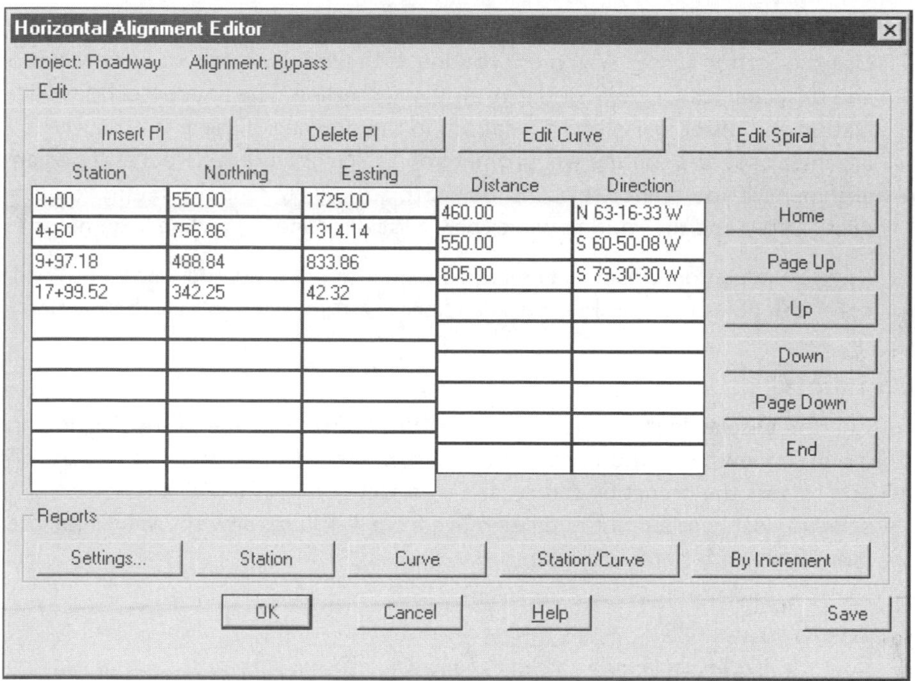

Figure 7–27

The second method is analyzing objects relative to an alignment. The first routine, Station Offset Alignment from the Inquiry menu, lists the stations and offsets of selected points relative to the centerline. The routine allows the dot toggles (*.n*, *.g*, and *.p*) to reference points from the database or northing/easting coordinates.

The listing routine Line/Curve/Spiral displays information similar to the information found in Horizontal Editor. This routine places a report about the selected objects on the text screen of LDD.

The last routine for analysis relative to an alignment is the Display Points routine of the Station Offset cascade of the Alignments menu. This routine displays the station and offset of points relative to the current centerline. This report can be sent to a file named in the Output Settings dialog box of the Drawing Settings routine of the Projects menu.

EXERCISE

Exercise

After you complete this exercise, you will be able to:

- Analyze the roadway centerline
- Analyze objects relative to a roadway centerline

STATION OFFSET ALIGNMENT

1. Use the Station Offset Alignment routine of the Inquiry menu to review station and offset values for points relative to the centerline. The routine echoes back station and offset for selected locations in the drawing. If there is no current alignment set, the routine will prompt you to select an object representing an alignment. If you cannot select an object to set the current alignment, you can press ENTER and the routine will display a list of define centerline alignments.
2. Toggle the Station Offset Alignment routine into point number mode (.p) and enter the point number 1. Type **.p** and press ENTER twice to exit the routine.

Display Points

3. Run the Display Points routine of the Station/Offset cascade of the Alignments menu to view the station and offset of a selected set of points. The routine allows you to sort the report by station and to select points as either a range or selection set. Use a point number range from 1-100 to view the values for the points currently in the drawing.

Line/Curve/Spiral

4. Use the Line/Curve/Spiral routine to list out the different segments of the alignment. Even though the routine displays the reports to the text screen, the reports contain the same design numbers as the Horizontal Editor. The numbers are rounded to the settings of the drawing.

Horizontal Editor

5. Run the Edit routine from the Alignments menu. The routine displays a dialog box that contains the tangent line definition of the alignment (see Figure 7–27).
6. Select into the station 4+60 box and select the Edit Curve button. This dialog box displays the design numbers of the roadway (see Figure 7–28). Select the Next button at the bottom of the dialog box to display the next curve's values.
7. Select the OK button once to dismiss the curve data and again to exit the Horizontal Editor. If there is a dialog box asking to save the changes, select No to dismiss the dialog box and discard the changes.
8. Save the drawing.

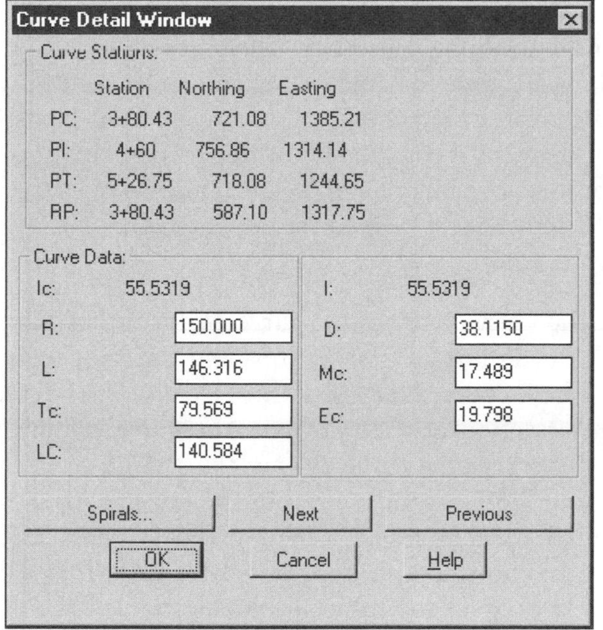

Figure 7–28

UNIT 4: CENTERLINE EDITING

EDITING A HORIZONTAL ALIGNMENT

After you have defined and evaluated the centerline, you can modify the alignment in the Horizontal Editor. The alignment definition is a snapshot of the objects selected during the defining process. If you use AutoCAD commands to change any of the alignment's graphical entities, you must redefine the alignment to update the external definition. You can modify the alignment in the Editor. When you save the changes and exit the dialog box, the routine will modify the objects in the drawing to match the changes made in the editor.

The Edit routine of the Alignments menu displays the Horizontal Editor (see Figure 7–27). The middle portion of the dialog box contains the stationing, coordinates, bearings, and distances of the tangent lines of the centerline. The stationing is the location of the PI (Point-of-Intersection) of the centerline tangents. In LDD a curve cannot be the first or last segment of a centerline. You can change the northing/easting values for the PI in the dialog box. When you change the values and press ENTER, the dialog box recalculates the design values for the current alignment.

At the top of the dialog box are buttons that add or delete centerline PIs or buttons that call a curve or spiral editing dialog box. The buttons on the side of the Horizontal Editor dialog box allow you to maneuver up and down the PI list if it is extensive. The bottom buttons create different reports about the current centerline. The Station report is a tangent segment report, the Curve report contains curve/spiral data, the Station/Curve report is a combination of the previous two reports, and the last report is a user-defined Increment report about the current alignment. The Settings button sets the format of the reports.

When you select the Curve button, the curve editor displays (see Figure 7–28). At the top of this dialog box are the stations and northings/eastings of the PC, PI, PT, and RP. These codes stand for Point of Curve (the beginning point of a curve), PI (the intersection of the two lines the curve is tangent to), Point of Tangency (the point where the curve ends at the second line segment), and Radius Point (the center point of the curve). In the lower portion of the dialog box are values of other characteristics of the curve. These characteristics include radius, tangent length, chord length, mid ordinate, and other values. The editor allows you to edit any one of the values in this portion of the dialog box. When you edit the value and press ENTER on the new number, the dialog box will update to display the new design numbers.

The Spirals button displays the spiral editor (see Figure 7–29). This editor displays the same station and northing/easting values as found in the curve editor. The lower portion displays all of the pertinent information for a spiral. The editor allows you to edit any one of the values in the lower portion of the dialog box. When you edit the value and press ENTER on the new number, the dialog box will update to display the new design numbers.

Some alignments have changes in stationing at points along their path. LDD calls this point a station equation. At this point, the stationing of the alignment changes and from this point the stationing may increase or decrease. LDD allows you to place several station equations on any one alignment.

The Import and Delete routines of the Alignments menu allow you to place alignment definitions into the drawing or to remove them from the drawing or the project or both. Both of these routines work with one alignment at a time. The Multiple Selections routine of the Alignment Commands cascade of the Alignments menu allows you to select more than one alignment and to either delete them from the drawing and/or external file, or to import them into the drawing (see Figure 7–30). If you are deleting the alignment, the dialog box also allows you to delete the associated profile and cross section objects and files.

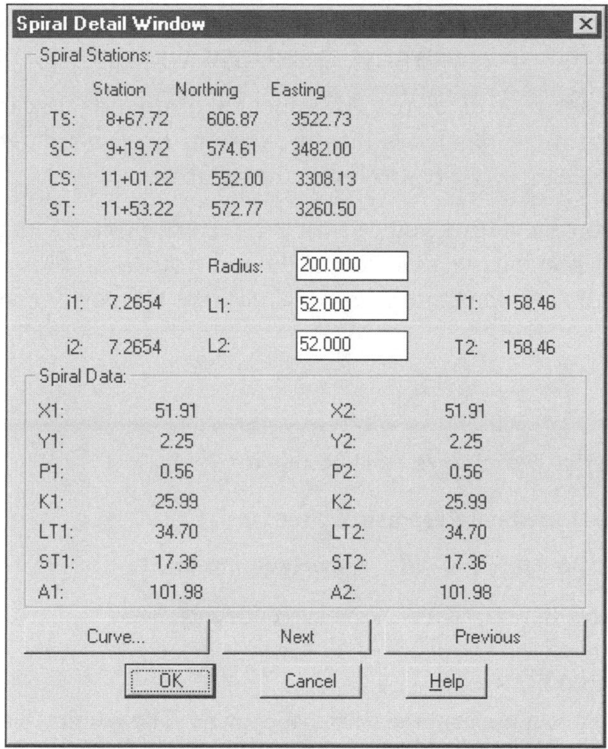

Figure 7–29

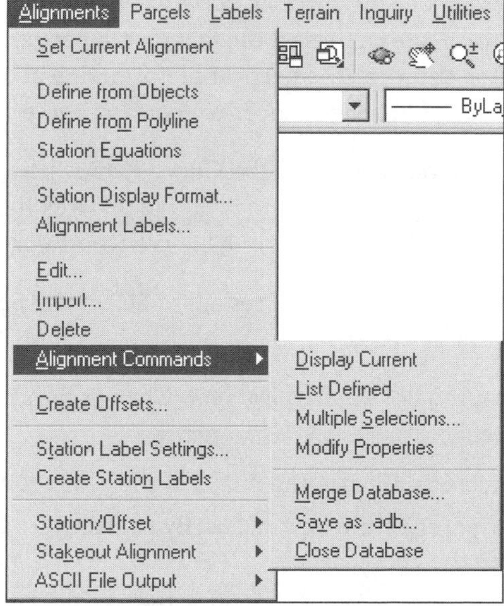

Figure 7–30

The Modify Properties routine of the Alignment Commands cascade allows you to change the color, layer, and linetype of the selected alignment.

The Merge Database routine of the Alignment Commands cascade allows you to combine other alignment databases into the current project's database. The Close Database routine releases your control over the database.

The Save as adb routine allows you to save the current project's alignment database into the format of previous versions of LDD and Softdesk. This is only necessary when passing the data to someone who does not have the current software.

Exercise

After you complete this exercise, you will be able to:

- Use the Horizontal Alignment Editor
- Review and edit curves in a roadway centerline
- Write and view a report on the current centerline

HORIZONTAL EDITOR

1. Run the Edit routine from the Alignments menu. The routine displays the Horizontal Editor (see Figure 7–27).

To view the curve data, select either a station, northing or easting cell of a segment that has a curve (station 4+60). Remember there are no curves at the first or last entry of this dialog box. After placing the cursor in the cell, select the Edit Curve button in the upper right of the dialog box. The curve editor shows all the pertinent information about the curve. The Next button at the bottom of the curve editor moves to the next curve on the roadway.

To change the radius of a curve, select into the Curve Radius box (R:) and edit the value. After you change the value and press ENTER, the routine updates the associated curve data. You can edit any of the curve values to change the curve (see Figure 7–31).

If the alignment has spirals, the same procedures apply. Select the Spiral button, view, and if necessary, edit any of the spiral values.

2. Edit the curves on the horizontal alignment. Change the radius of each arc to the following values:

   Curve 1 (Station 4+60): Change the radius of the first arc to **200.00**.

   Press ENTER after entering the new radius. By pressing ENTER, you force the dialog box to recalculate the curve values.

   Select the Next button to view the data for the second curve.

   Curve 2 (Station 9+97): Change the radius of the second arc to **350.00**.

Press ENTER after entering the new radius. By pressing ENTER, you force the dialog box to recalculate the curve values.

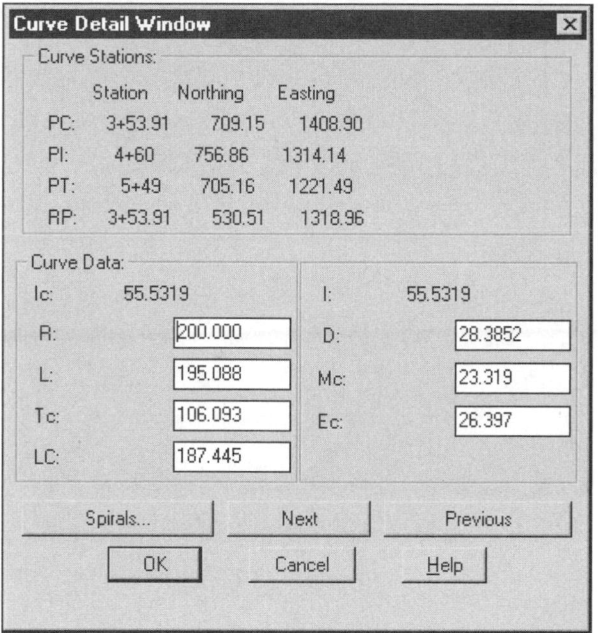

Figure 7–31

3. Select the OK button to exit the curve editor.
4. Select OK to exit the Horizontal Editor. When the Editor prompts you to save the changes, select Yes.

When you exit the editor and save the changes, the routine automatically removes the old centerline and imports the new definition into the drawing.

5. Use the Line/Curve/Spiral routine found on the Inquiry menu to view the new line and arc segment values. You may have to flip to the text screen to view the reported values. Press ENTER to exit the routine.
6. Rerun the Horizontal Editor routine (Edit in the Alignments menu) and select the settings button at the bottom left of the Editor. The Settings button displays the Output Settings dialog box. Set the settings using Figure 7–32 as a guide. Select the OK button to exit the Output Settings dialog box.
7. Create a Station/Curve report. The routine displays a report dialog box (see Figure 7–33). Press ENTER to accept the beginning and ending stations for the report. Review the values in the report. Select the OK button to exit the dialog box.

EXERCISE

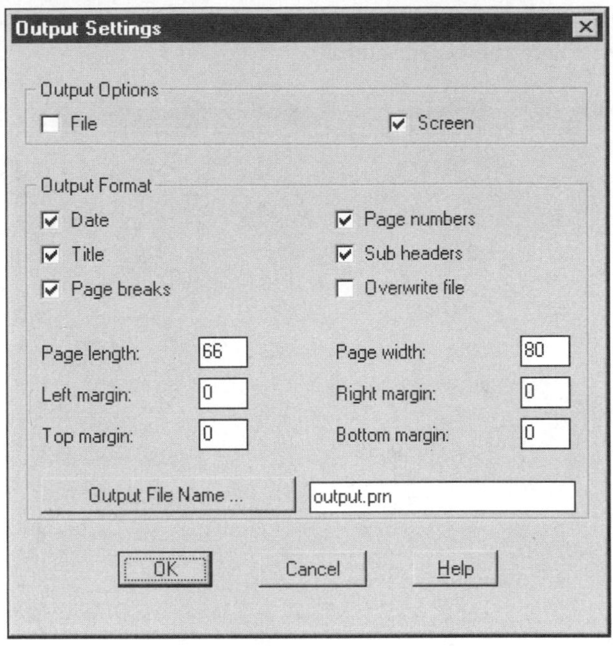

Figure 7–32

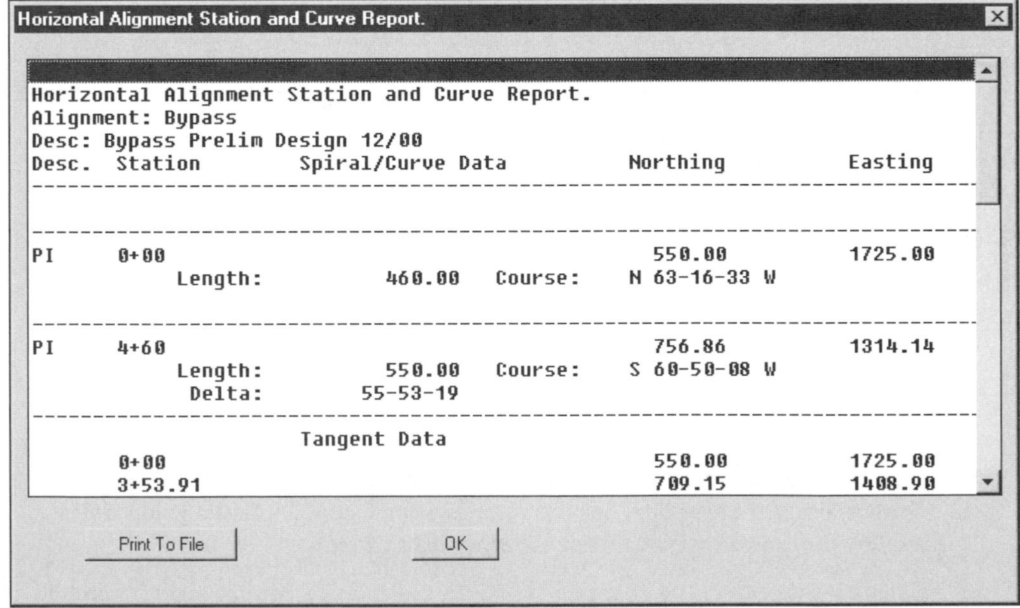

Figure 7–33

8. Select the By Increment button and generate an increment report, using an increment of 50 feet. Press ENTER to accept the beginning and ending stations for the report, and enter **50** at the increment prompt. The routine displays a report dialog box (see Figure 7–34). Review the values in the report. Select OK to exit the report.

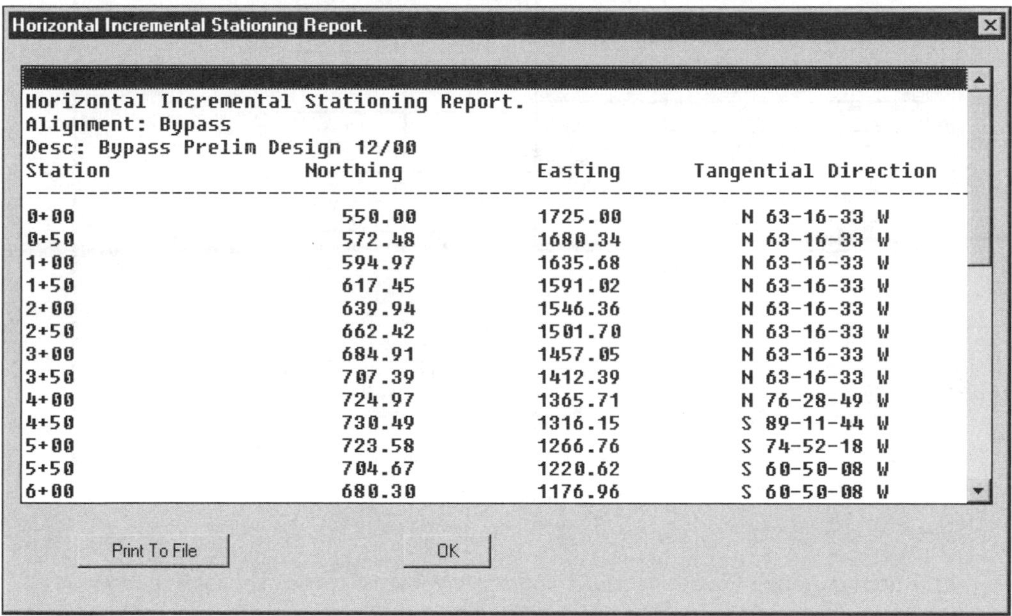

Horizontal Incremental Stationing Report.

Horizontal Incremental Stationing Report.
Alignment: Bypass
Desc: Bypass Prelim Design 12/00

Station	Northing	Easting	Tangential Direction
0+00	550.00	1725.00	N 63-16-33 W
0+50	572.48	1680.34	N 63-16-33 W
1+00	594.97	1635.68	N 63-16-33 W
1+50	617.45	1591.02	N 63-16-33 W
2+00	639.94	1546.36	N 63-16-33 W
2+50	662.42	1501.70	N 63-16-33 W
3+00	684.91	1457.05	N 63-16-33 W
3+50	707.39	1412.39	N 63-16-33 W
4+00	724.97	1365.71	N 76-28-49 W
4+50	730.49	1316.15	S 89-11-44 W
5+00	723.58	1266.76	S 74-52-18 W
5+50	704.67	1220.62	S 60-50-08 W
6+00	680.30	1176.96	S 60-50-08 W

Print To File   OK

Figure 7–34

9. Select the OK button to exit the Horizontal Editor.

10. Save the drawing.

STATION EQUATIONS

1. Draw a line that represents the location of the station equation. Use the By Station/Offset routine in the Lines/Curves menu. Press ENTER twice to exit the routine after using the following values to create the line.

Station	Right Offset	Left Offset
1350.00	12.00	-12.00

2. Use the Zoom Window routine to view the intersection.

3. Before adding the equation to the roadway, run the Edit routine of the Alignments menu to view the current stationing (see Figure 7–35). The roadway stationing ends around 18+00. Select the OK button to exit the editor

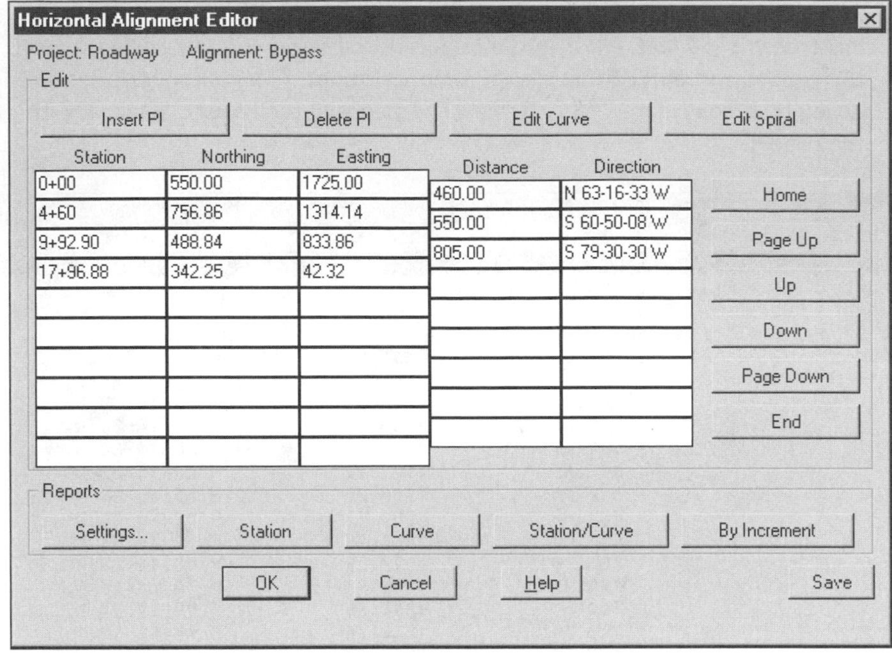

Figure 7–35

4. Run the Station Equations routine of the Alignments menu. The intersection between the alignment and the crossing line is the location of the equation. The stationing will change at this point and it will increase from that point towards the end of the roadway. The first step is to tell the routine what action to take, in this case Add. Then you identify the intersection with an AutoCAD object snap. The routine responds with the current station of the intersection point. Press ENTER to accept the station 1350. The next two prompts set the station, 2000, and whether the stations increase or decrease from this point. Set the stations to increasing.

The routine prompts as follows:

Command:
Alignment Name: Bypass   Number: 1 Descr: Bypass Prelim Design of 12/00
Starting Station: 0.00 Ending Station: 1796.88
Starting Station: 0.00
Select operation [Clear/Add/eXit/Modify/Delete] <eXit>: **a**
Select point on alignment for station equation: _int of *(use an AutoCAD object snap)*
Enter station back <1350.00>: *(press* ENTER*)*
Enter station ahead: **2000**
Select stationing order [Increase/Decrease] <Increase>: *(press* ENTER*)*
Starting Station: 0.00

```
EQUATION     STATION-BACK     STATION-AHEAD      ORDER
____        ______          ______        ____
       1          1350.00          2000.00    INCREASING
Select operation [Clear/Add/eXit/Modify/Delete] <eXit>: x
```

5. Save the drawing.

The Create Station Labels routine places annotation on the current alignment. The annotation reflects the settings of the Station Label Settings dialog box (see Figure 7–7).

6. Run the Edit routine of the Alignments menu to view the new stationing of the roadway design. Select the OK button to exit the dialog box.
7. Use the Named View routine to restore the view Roadway.
8. Save the drawing.

EXERCISE

UNIT 5: CENTERLINE ANNOTATION

By adding a station equation, you change the stations along the roadway from the equation point. The overall length of the centerline does not change. However, the alignment reports will indicate the new ending station.

You must relabel the alignment if you edit the alignment design numbers. There is no association between the text and the objects they label. The routine uses regular text and layers defined in the Station Label Settings dialog box.

The Station Label routine uses the same layers for all alignments in the drawing. When executing, the routine asks about deleting all existing labeling. If you answer Yes, the routine erases all labels on the station label layers. This may not be correct if you have one alignment labeled and you are now doing the second. To avoid this problem, the labeling routines look to see if a layer prefix is set in the Alignment Label Settings dialog box (see Figure 7–4). If there is a prefix, the labeling routine uses the prefix to create the layer names. If the current alignment is Bypass and layer prefix is set to *- (an asterisk and a dash), the layer name for point labels would be BYPASS-STAPTS.

The Stationing Defaults create two types of stationing parallel to and perpendicular to the roadway centerline. The Default mode is parallel to the roadway. See the discussion in Unit 2 of this section about the stationing settings.

The Labeling routine of the Station/Offset cascade of the Alignments menu labels selected points in the drawing with a station/offset label relative to the current alignment.

Exercise

After you complete this exercise, you will be able to:

- Station the roadway centerline
- Modify the station labeling values
- Label Stations/Offset of points relative to the current alignment

STATIONING A CENTERLINE

1. Set the Station Label settings to the following values to produce parallel labeling. Use Figure 7–36 as a guide for the settings. Run the routine Station Label Settings from the Alignments menu to set the values. The following is the command sequence.

   Station Labels: ON

   Station Point Labels: ON

   Station Equation: ON

   Perpendicular Labels: OFF

   Stations Read along Road: OFF

   Plus Sign Location: OFF

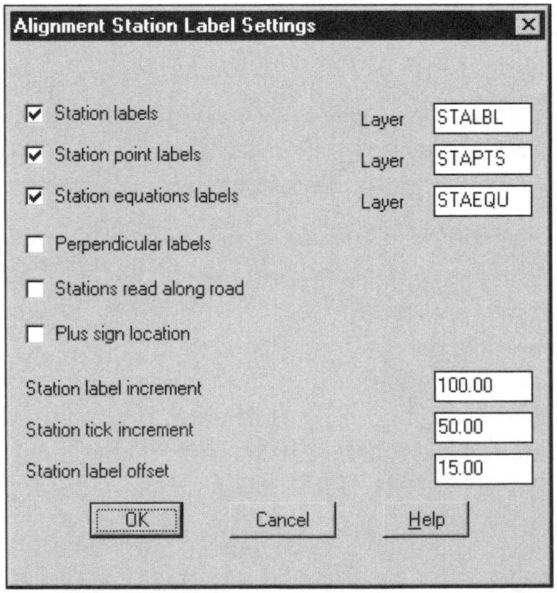

Alignment Station Label Settings

- [x] Station labels — Layer: STALBL
- [x] Station point labels — Layer: STAPTS
- [x] Station equations labels — Layer: STAEQU
- [ ] Perpendicular labels
- [ ] Stations read along road
- [ ] Plus sign location

Station label increment: 100.00

Station tick increment: 50.00

Station label offset: 15.00

OK | Cancel | Help

Figure 7–36

2. Use the Create Station Labels routine of the Alignments menu to label the roadway with parallel labeling.

```
Command:
Alignment Name: Bypass  Number: 1  Descr: Bypass Prelim Design of 12/00
Starting Station: 0.00  Ending Station: 2446.88
Beginning station <0>: (press ENTER)
Ending station <2446.88>: (press ENTER)
Delete existing stationing layers (Yes/No) <Yes>: (press ENTER)
Erasing entities on layer <Bypass-STALBL> ...
Erasing entities on layer <Bypass-STAEQU> ...
Erasing entities on layer <Bypass-STAPTS> ...
No entities found on selected layers!
Command:
```

3. Zoom in and pan around the drawing to view the labeling.

4. Restore the Roadway Named View to see the entire project area.

5. Set the Station Label settings to the following values to produce perpendicular labeling. Use Figure 7–37 as a guide for the settings. Run the routine Station Label Settings from the Alignments menu to set the values. As before, press ENTER to accept the beginning and ending station and to erase any previous labels.

Station Labels: ON

Station Point Labels: ON

Station Equation: ON

Perpendicular Labels: ON

Stations Read along Road: OFF

Plus Sign Location: ON

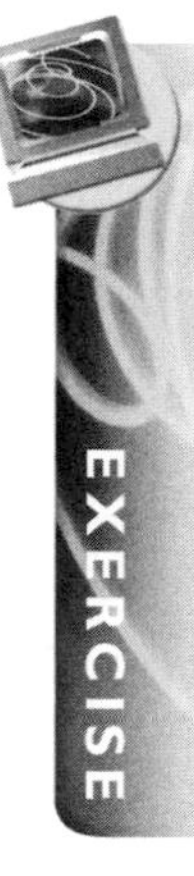

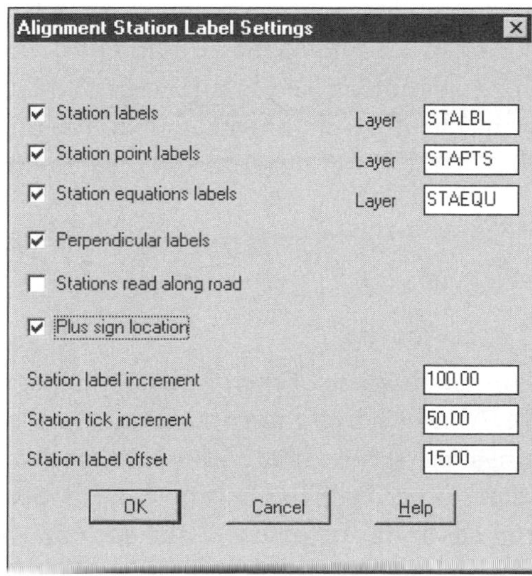

Figure 7–37

6. Use the Create Station Labels routine to label the roadway with perpendicular labeling.
7. Zoom in and pan around the drawing to view the labeling.
8. Use the LAYER command to freeze the labeling layers. The layer will have the BYPASS prefix.
9. Restore the Roadway Named View to see the entire project area.
10. Save the drawing.

LABEL–STATION/OFFSET

1. Label some station/offsets by running the Label routine of the Station/Offset cascade of the Alignments menu.
2. Undo station labeling.
3. Save the drawing.

UNIT 6: CREATING NEW DATA RELATIVE TO AN ALIGNMENT

LDD provides several tools that set points and create new objects in reference to the current alignment, and create reports from the current alignment.

CREATE OFFSETS

A drawing may contain several lines that are parallel and offset from the centerline alignment. The Create Offsets routine of the Alignment menu creates parallel and offset lines from a defined alignment (see Figure 7–5). You can offset up to eight parallel lines (four on each side of the alignment) and do the procedure several times. All you need to do is change the offset distances and layer names. The routine is in the Alignments menu.

POINT TOOLS

CREATE POINTS–ALIGNMENT

The Create Points–Alignment cascade of the Points menu has several point creation tools (see Figure 7–38). Almost all of the routines use the current point settings found in the Point Settings dialog box. If a routine does not use the current settings, it will prompt for the values it needs or the routine may supply its own descriptions. In some cases even if the elevation toggle is on, the routine will not prompt for an elevation. The Create tab of the Point Settings dialog box sets the current point number and prompting for elevations and descriptions.

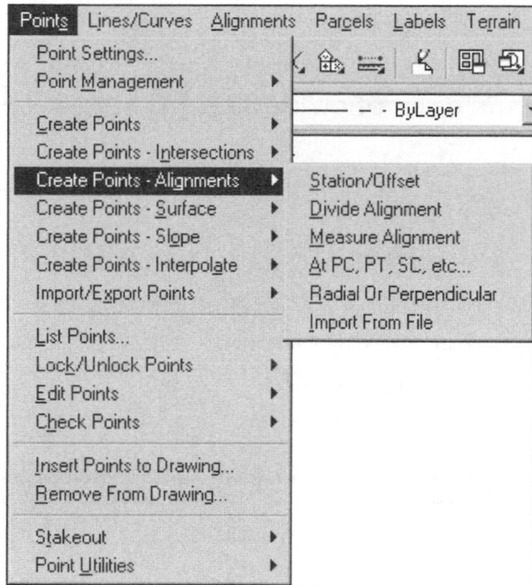

Figure 7–38

Measure Alignment and Divide Alignment

The Measure Alignment and Divide Alignment routines place points on the alignment either at a user-specified distance or divide the alignment into a specified number of segments.

Station/Offset

The Station/Offset routine places points at specific stations and offsets along the centerline. When the routine starts, it asks for the starting point number, then prompts for the station and offset. If elevations and description are on in Point Settings, the routine prompts for those values. The routine continues prompting for different offsets for the same station until you press ENTER at the offset prompt. This tells the routine to prompt for a new station and then to repeat the offset prompting.

Import From File

The Import From File routine reads an external file and places points in the drawing representing the coordinates in the external file. The file formats the routine reads are:

```
Station, Offset
Station, Offset, Elevation
Station, Offset, Rod, Hi
Station, Offset, Description
Station, Offset, Elevation, Description
Station, Offset, Rod, Hi, Description
```

The Rod and Hi are for level surveys. The Rod is the prism elevation and the Hi is the instrument height.

Radial/Perpendicular

This routine places points on the centerlines that are radial or perpendicular to the centerlines. If elevations and description are on in Point Settings, the routine prompts for those values.

At PC, PT, SC, etc

The AT PC, PT, SC, etc routine places points at critical design points along the centerline. The critical points are PCs (point of curve beginning), PTs (point of curve ending), SC (point of spiral curve intersection), RP (radial point), CC (compound curve points), and spiral curve intersections.

CREATE POINTS–INTERSECTION

The Create Points–Intersection cascade of the Points menu contains routines that place points at intersections of direction, distances, other alignments, and other objects in the drawing (see Figure 7–39).

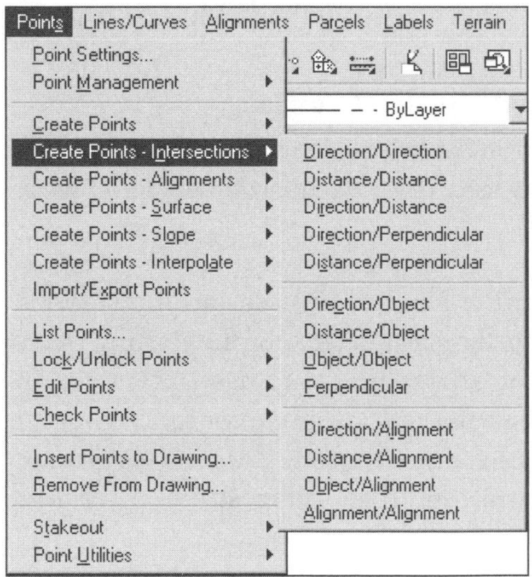

Figure 7–39

Direction/Alignment

This routine places points that are at the intersection between an alignment and a direction from a known point. The intersection can be an offset from both or either the direction or alignment.

Distance/Alignment

The Distance/Alignment routine creates points at the intersection of a distance from a known point and an alignment. There can be an offset from the alignment.

Object/Alignment

This routine creates an intersection point at the intersection of an object (line, circle, or spiral) and an alignment. Both the object and the alignment can have an associated offset.

Alignment/Alignment

The Alignment/Alignment routine creates a point at the intersection of two alignments. There can be an offset from both or either alignments.

STAKEOUT ALIGNMENT

The Stakeout Alignment cascade of the Alignments menu creates a report or file that a field crew can use to stake out an alignment (see Figure 7–40).

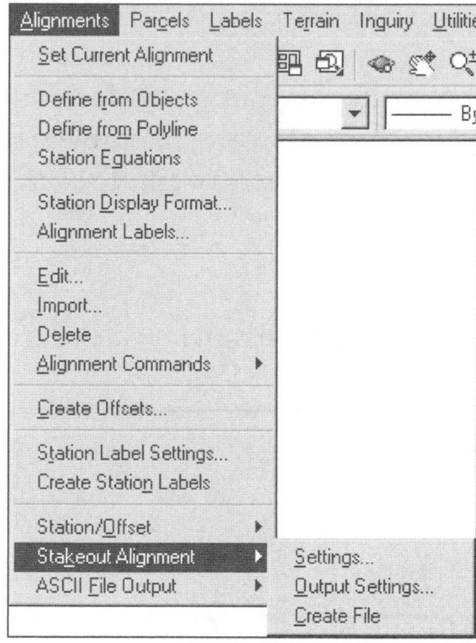

Figure 7–40

Output Settings

The Output Settings routine sets the defaults for printed reports (see Figure 7–32). The settings include the name of the file, header values, and page size.

Settings

The Settings routine displays the Stakeout Settings dialog box, which sets the turned angle method used by the crew when staking out the centerline (see Figure 7–41). The default method is right turned angles. You can use deflection angles or bearings to create the report.

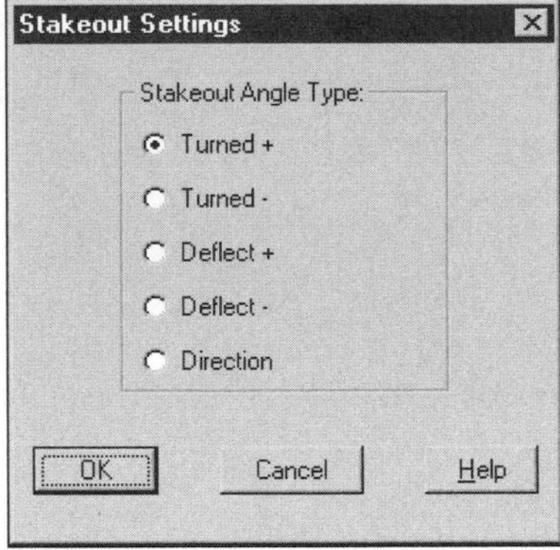

Figure 7–41

Create File

The Create File routine creates the stakeout report for the staking of the centerline. The routine prompts you for two points, the instrument and backsight point, the beginning and ending stations, and if there is an offset needed for the points. The offset can be a positive or negative value.

ASCII FILE OUTPUT

The ASCII File Output cascade of the Alignments menu creates a text file containing the definition of the centerline (see Figure 7–42).

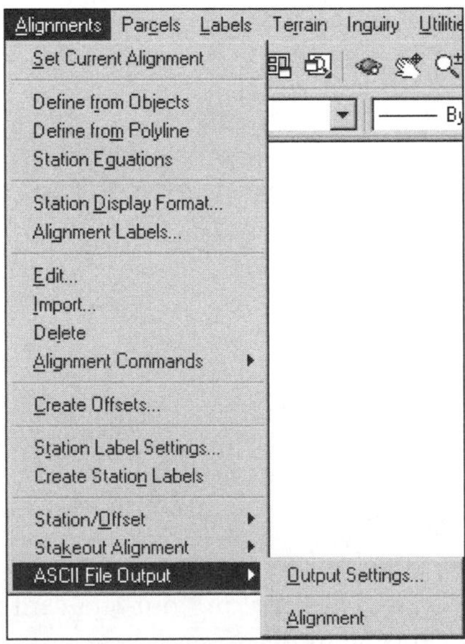

Figure 7–42

Output Settings

The Output Settings routine sets the default values for the current report (see Figure 7–32).

Alignment

The Alignment routine creates the alignment report. The report allows you to use the alignment data of LDD in another program that can read the file. The structure of the file has numbers appearing at the beginning of each line. The number indicates the type of data that is in the remainder of the line. The number 0 indicates a line, 1 indicates a curve, and 3 indicates a station equation. The internal station is the actual station and the external station is a new station value if the station occurs after a station equation. The format of the file is the following:

Alignment name, number, starting station (internal),length description

0,internal sta, external sta,N 1,E 1,N 2,E 2,Dist.,Dir.

1,internal sta, external sta,BC N,BC E,CC N,CC E,EC N,EC E,Length,Radius,Delta

2,internal sta, external sta,BS N,BS E,SPI N,SPI E,ES N,ES

E,Length,Theta,Radius,A,Offset,External,spiral type, direction defined

3,length along alignment (including starting station), external sta,type

The report for the Bypass alignment is the following:

```
# AutoCAD Land Development Desktop Alignment Output 2.0
Bypass,1,0.000000,1796.882340
Bypass Prelim Design 12/00
3,1350.000000,2000.000000,0
0,0.000000,0.000000,550.00000000,1725.00000000,709.15064248,1408.89639203,353.907358,2.675168
1,353.907358,353.907358,709.15064248,1408.89639203,530.51428418,1318.95723765,705.15921539,
   1221.49366497,195.088056,-200.000000,0.975440
0,548.995414,548.995414,705.15921539,1221.49366497,516.87694727,884.11075501,386.364388,3.650608
1,935.359802,935.359802,516.87694727,884.11075501,822.50557691,713.54950282,478.35708612,777.28188305,
   114.065508,350.000000,0.325901
0,1049.425310,1049.425310,478.35708612,777.28188305,342.25075568,42.32128623,747.457030,3.324707
```

Exercise

After completing this exercise, you will be able to:

- Create offsets from the roadway centerline
- Set points along and on alignment
- Create a stakeout report about the alignment
- Create an ASCII report about the alignment

CREATING OFFSETS FROM AN ALIGNMENT

1. If you are not in the Roadway drawing, open the Roadway drawing of the Roadway project.

2. Create the following offsets by running the Create Offsets routine in the Alignments menu (see Figure 7–43). Set the following values to create the offsets:

	Distance	**Layer**
Outer Offset	33.00	ROW
Second Offset	32.00	SIDEWALK
Third Offset	27.00	SIDEWALK
Inner Offset	13.50	EOP

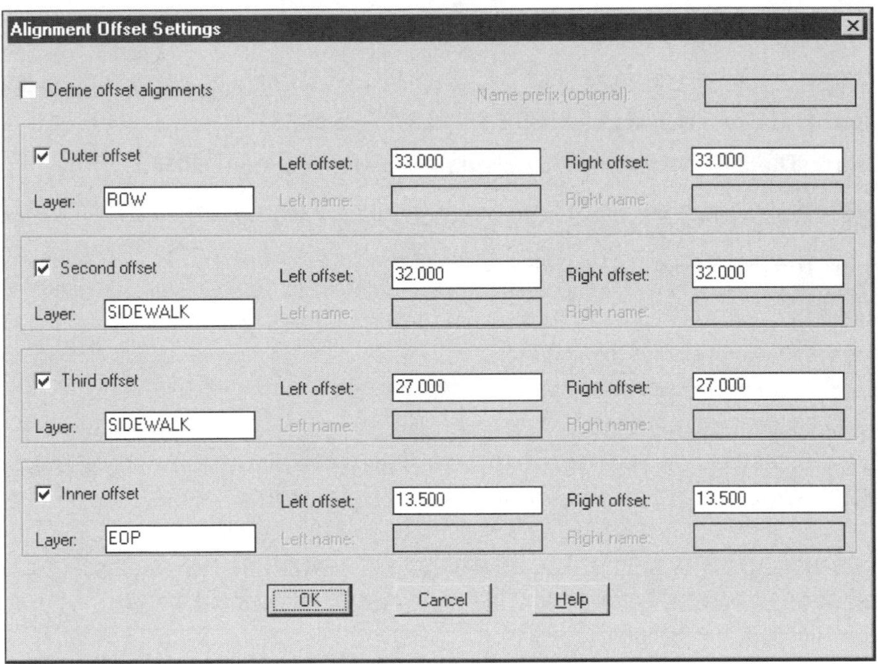

Figure 7–43

3. When the offset finishes running, use the Layer command to freeze the ROW, SIDEWALK, and EOP layers.

Placing Points Along a Defined Alignment

4. In the Create Panel of the Point Settings routine of the Points menu set the current point number to 50. In the same panel set the Elevations radio button to none. After setting the values select OK to exit the dialog box.

5. Run the At PC, PT, SC, etc. routine of the Create Points – Alignment cascade of the Points menu to create points on the alignment. The routine prompts for the beginning and ending station and the current point number to use. Press ENTER at each prompt to run the routine.
6. View the resulting points in the List Points routine of the Points menu. Use the All Points option. Notice that each point has a description that indicates its function in the alignment.

Importing Points from File

7. Read in the file *road1.nez*. Use the Import From File routine in the Create Points–Alignment cascade of the Points menu. The routine first asks for the location of the point file, then the format of the file (in this case format 5), the separator (comma or space—use 1 for a comma), the number indicating an invalid elevation, and finally whether there are any invalid stations and offsets.

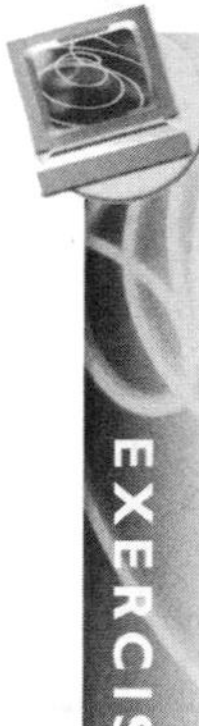

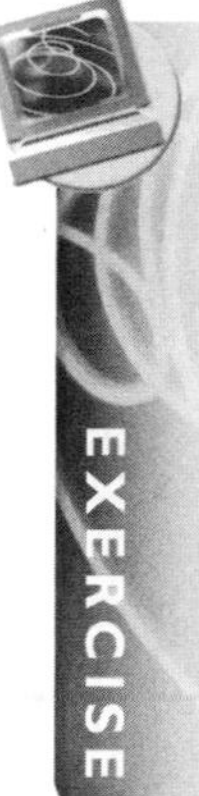

The format of the file is: station, offset, elevation, and description. The following is an excerpt from the file.

```
500,12.50,723.34,bc
515,13.50,724.73,tc
```

The following is the prompting of the routine.

```
Command:
Available File Formats.
___________
  1.  Station, Offset
  2.  Station, Offset,  Elevation
  3.  Station, Offset,  Rod, hi
  4.  Station, Offset,  Description
  5.  Station, Offset,  Elevation, Description
  6.  Station, Offset,  Rod, hi, Description
Enter file format [1/2/3/4/5/6]: 5
Available File Delimiters.
_____________
  1. Comma <,>.
  2. Space < >.
Enter file delimiter [1/2] <2>: 1
Invalid elevation <-99999>:
Invalid station/offset <none>:
Import complete!
```

Display Points

8. Run the Display Points routine of the Station/Offset cascade of the Points menu. Sort the points by station and select the point number range of 1–200.
9. Remove all of the points from the drawing. Use the Remove Points from Drawing routine of the Points menu. Select the All Points option.

Staking Out a Centerline

10. Run the Settings routine from the Stakeout Alignment cascade of the Alignments menu. Set the angle type to Turned Angle +.
11. Use the Create File routine of the Stakeout Alignment cascade of the Alignments menu to create a stakeout report for the Bypass alignment. Use point 61 as the occupied point and point 1 as the backsight point. Accept the default beginning and ending stations, set the offset distance to 20.00 feet, and the spacing to 25 feet.

Creating an Alignment Report

12. From the Alignments menu, run the Alignment routine. The routine first prompts for the directory of the file. Press ENTER to accept the *Align* directory of the project. The next prompt is for the file name. Use the file name of *road.txt* and press ENTER to create the file.
13. Use the Windows Notepad and open the file. The file is in the Align directory of the project.
14. Save the drawing.

EXERCISE

The horizontal alignment is the first step in designing a roadway in Land Development Desktop. Desktop contains tools that create, analyze, edit, and annotate the centerline in plan view. There are additional tools that create points and other objects in relation to the alignment in the drawing.

The next two steps, profiles and cross sections, occur in an add-in product, Civil Design. The next section discusses profiles and the tools this add-in provides to you.

The Roadway Design Process—Profiles and the Vertical Alignment

After you complete this section, you will be able to:

- Edit the setting for a roadway profile
- Draw a vertical roadway centerline
- Define a vertical roadway centerline
- Edit a vertical roadway centerline
- Analyze a vertical roadway centerline
- Annotate a vertical roadway centerline
- Create new points relative to the vertical centerline

SECTION OVERVIEW

This section covers the second phase of the roadway design process, profiles. The profile phase of roadway design is the domain of the Civil Design add-in to Land Development Desktop. LDD has tools only for the plan (horizontal) design.

The previous section, Section 7, covered the horizontal alignment or plan view of the roadway. This section reviews the side view or profile view of the roadway design. The profile view is the first to use vertical (surface) data (see Figure 8–1). The vertical data can come from either a surface or a file. It is in profile view that you start to develop a feel for the impact of the design on a volume result. The height of the roadway above or below the existing ground elevations gives you visual feedback as to what amount of earthworks needs to be done to build the road design.

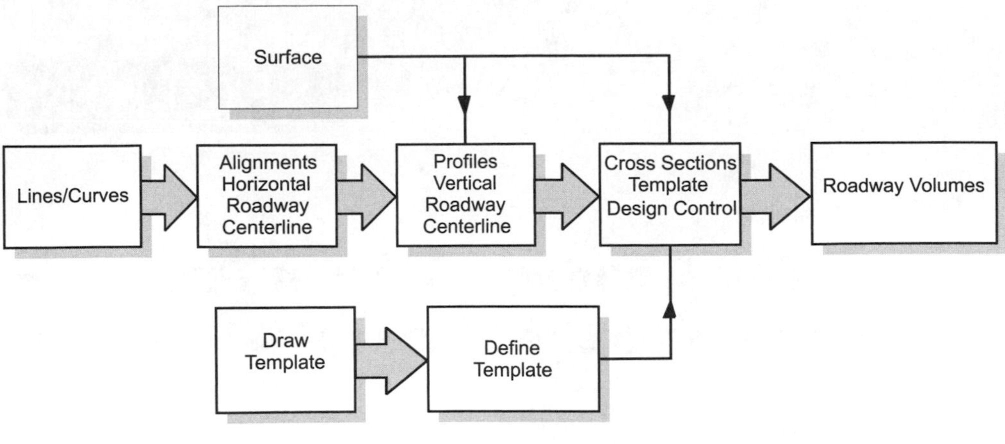

Figure 8–1

UNIT 1

The Civil Design add-in represents the vertical roadway centerline as a vertical alignment. In the drawing, the centerline is a series of lines and arcs. The arcs represent vertical curves, so there is more data associated to the arc than just a radius. Vertical curve designs depend upon whether the curve is on the crest of a slope or at the bottom of a slope. There are different design criteria for each. The Civil Design settings are the initial values for the curve computations and are the subject of the first unit of this section.

UNIT 2

The second unit of this section covers the steps needed to create a profile. A profile is the result of sampling in order to determine what elevations exist along the horizontal centerline of the road design. After you create a profile in the drawing, Civil Design offers you several tools to evaluate stations and elevations within the profile.

UNIT 3

Within the context of the profile, the designer creates the elevations the template will use as it moves from the beginning to the end of the roadway. For Civil Design, this is the vertical alignment. After the alignment has been designed, Civil Design has several tools to evaluate and edit the vertical alignment. Creating the vertical alignment is the topic of the third unit of this section.

UNIT 4

The analysis and editing of a vertical alignment is the focus of the fourth unit of the section. This unit covers the creation of a vertical alignment ASCII report similar to the one for horizontal alignments.

UNIT 5

The fifth unit of this section reviews the annotation of profiles and vertical alignments.

UNIT 6

The last unit covers the creation of points from the vertical design.

UNIT 1: PROFILE SETTINGS

The profile defaults found in the Edit Settings dialog box control the names and values used by various profile routines (see Figure 8–2). If a layer prefix is set in the Profile Labels Settings dialog box, the layers that appear in all profile dialog boxes will be prefixed by the current alignment's name in the drawing. A layer prefix is an asterisk (*), the name of the current alignment, followed by a separator, which is either a dash (-) or underscore (_). If the current alignment is Maple, the prefix of *- (asterisk dash) creates a prefix of Maple- for each layer applicable to profile.

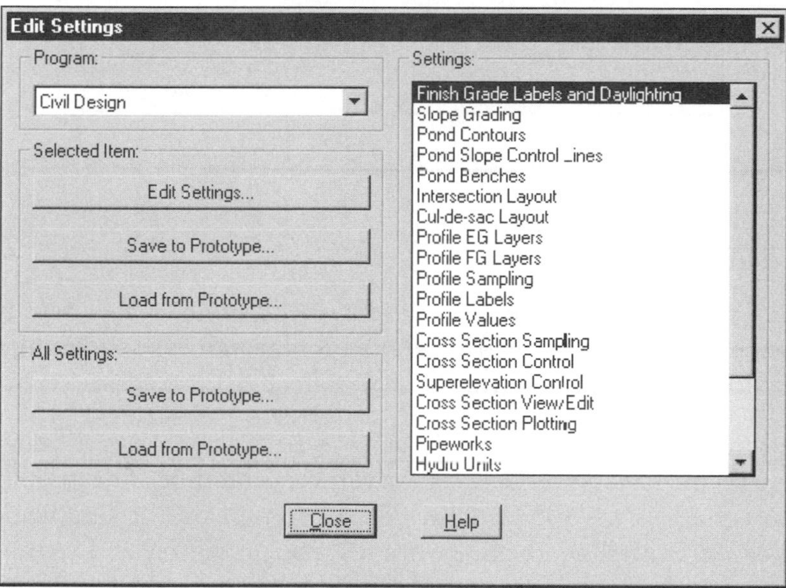

Figure 8–2

PROFILE EG LAYERS

The Profile EG Layers choice displays the Existing Ground Layer Settings dialog box, which sets the names for layers used by various profile routines representing existing ground conditions (see Figure 8–3). The letter P indicates that these are profile layers and the EG indicates that these layers represent existing ground conditions. If these layers are not present in the drawing, the routine will create layers with standard desktop defaults (white and continuous).

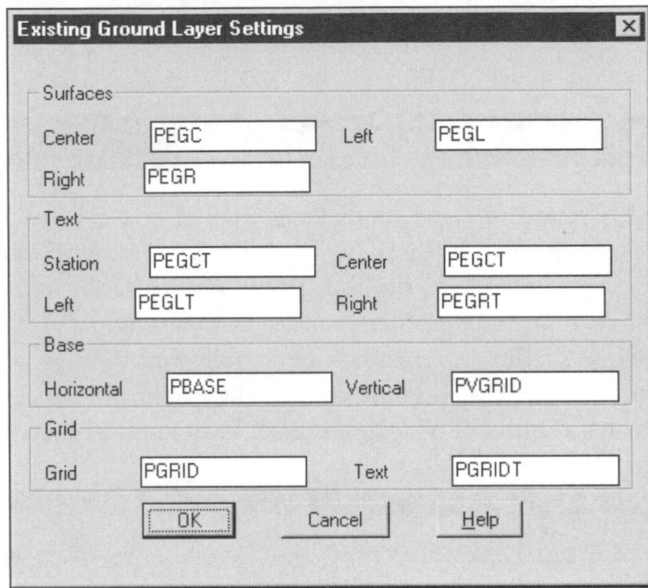

Figure 8–3

PROFILE FG LAYERS

The Profile FG Layers choice displays the Finished Ground Layer Settings dialog which, sets the names for layers used by various profile routines representing proposed ground conditions (see Figure 8–4).

The top portion of the dialog box sets the layer names for the vertical alignment object layer and the annotation layer for the vertical centerline. The most important layer in this group is the PFGC layer. When you create the line work for the proposed vertical centerline, the line work must be on this layer. The routine that defines the vertical alignment turns off all of the layers except for PFGC during the defining process.

The lower portion of the dialog box sets the right and left side layer names for transitional vertical alignments. A transitional alignment allows you to control the vertical behavior of a roadway template.

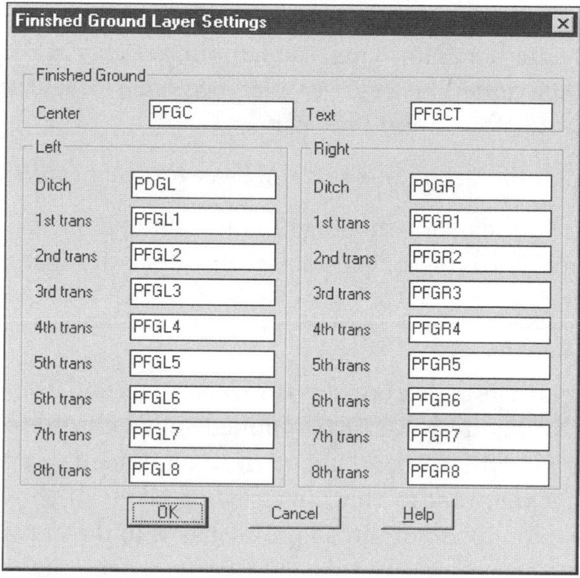

Figure 8–4

PROFILE SAMPLING

The Profile Sampling Settings dialog box controls the sampling of surface elevations for the profile, the sampling for a right and left profile, and the layer name for profile sample lines (see Figure 8–5). As you design the vertical for the centerline, you can use the results of the right and left sampling to develop a sense of the impact of the design on earthwork results.

Profile Sampling Settings

Sample offset tolerance 0.500

Sample Lines

Import

Layer PROSAM

Left and Right Sampling

Sample left/right

Sample left offset 12.000

Sample right offset 12.000

OK Cancel Help

Figure 8–5

The Sample Offset Tolerance setting forces the sampling to occur more frequently around curves in a centerline. Normally, the sampling occurs at the intersection of a surface triangle and the centerline. A lower tolerance value forces the profile sampling routine to sample more often around the curve.

The middle portion of the dialog box sets the layer for the profile sampling line(s).

The bottom portion of the dialog box toggles on right and left side sampling. After you have toggled on this sampling, you can set the offset distance of the sampling.

PROFILE LABELS

The Profile Labels Settings dialog box sets the layer prefix and user-defined labels for profile annotation (see Figure 8–6). As mentioned above, the prefix of *- prefixes the EG and FG layers with the name of the current alignment. The Profile Labeling routine will ask you if you want to erase any existing labels. If you answer Yes to the question and the layer prefix is not set, all profile labels in the drawing will be erased. If you answer Yes to the question and the layer prefix is set, only the profile labels of the current alignment are erased.

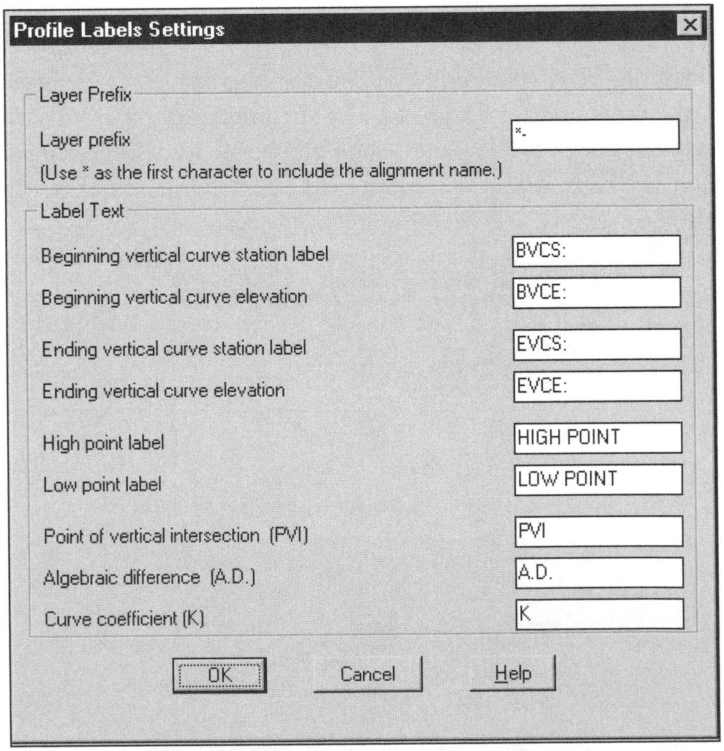

Figure 8–6

PROFILE VALUES

The Profile Values Settings dialog box sets values for four critical functions found in profiles (see Figure 8–7). At the top of the dialog box, in the Stationing Increments area, are settings for the frequency of sampling for tangent lines and vertical curves. Included in this area is the spacing for elevation grid lines.

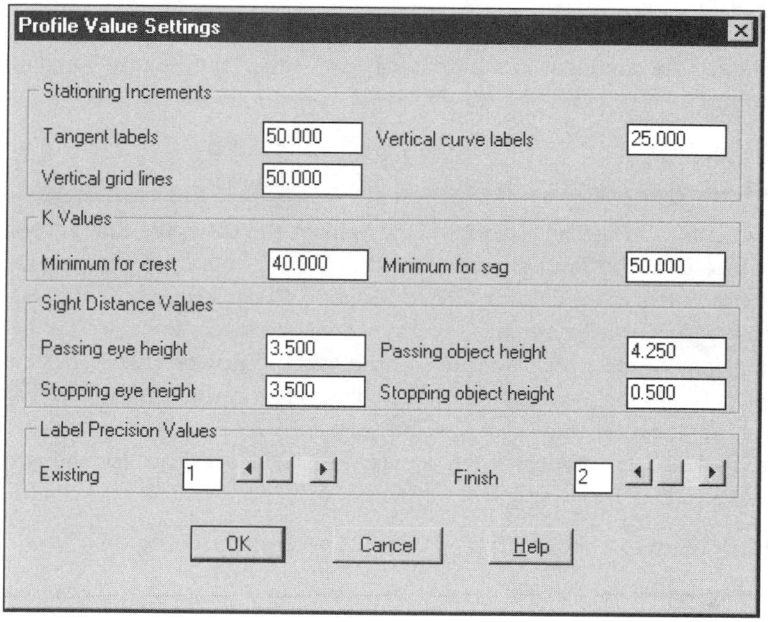

Profile Value Settings

Stationing Increments

Tangent labels 50.000 Vertical curve labels 25.000

Vertical grid lines 50.000

K Values

Minimum for crest 40.000 Minimum for sag 50.000

Sight Distance Values

Passing eye height 3.500 Passing object height 4.250

Stopping eye height 3.500 Stopping object height 0.500

Label Precision Values

Existing 1 Finish 2

OK Cancel Help

Figure 8–7

The next area in the dialog box, K values, controls the minimum K values for crest and sag vertical curves. A K value is the horizontal distance along a vertical curve to create a change of 1 percent in grade on the vertical curve.

The Sight Distance Values area sets default heights for safety calculations in the design of vertical curves. These values allow you to evaluate the quality of the design to the code requirements of the job.

The Label Precision Values area sets the default precision for existing and proposed elevations. Generally, existing elevations are to one decimal place and proposed elevations are to two decimal places.

EXERCISE

Exercise

After you complete this exercise, you will be able to:

- Review the profile settings

PROFILE SETTINGS

Profile EG Layers

1. This exercise continues in the Roadway drawing of the Roadway project from Section 7.
2. If the menu is not the Civil Design menu, change to the Civil Design menu by using the Menu Palettes routine of the Projects menu.
3. Select Profile E(xisting) G(round) Layers from the list in the Edit Settings dialog box (the Drawing Settings routine of the Projects menu), and then select the Edit Settings button. The Existing Ground Layer Settings dialog box displays (see Figure 8–3). Civil Design uses several predefined layers for a profile. You can change any of the layer names presented in the dialog box. The Surfaces settings are the layers for the existing surface's line work in the profile. The Text layer is the annotation for the elevations and stations. The Base layers are for the bottom line work and annotation of the profile. The Grid layers are for the grid you may put over a profile.
4. Select OK to exit the Existing Ground Layer Settings dialog box.

Profile FG Layers

5. Select Profile F(inished) G(round) Layers from list in the Edit Settings dialog box, and then select the Edit Settings button. The Finished Ground Layer Settings dialog box includes two layers at the top (see Figure 8–4). The PFGC layer is most important because it is the layer on which Civil Design draws the vertical centerline alignment. The layer PFGCT is for annotation. The remaining layers are for drawing and importing transitional vertical alignments. Civil Design identifies the transitional alignment by the layer it is on in the drawing.
6. Select OK to exit the Finished Ground Layer Settings dialog box.

Profile Sampling

7. Select Profile Sampling from the list in the Edit Settings dialog box, and then select the Edit Settings button. The Profile Sampling choice opens the Profile Sampling Settings dialog box containing the default values for sampling (see Figure 8–5). The Sample Offset Tolerance value forces the sampling routine to sample more frequently around curves. The middle portion of the dialog box toggles the importing of sample lines and the layer they use in the drawing. The bottom portion of the dialog box toggles the sampling of the right and left side of the roadway. When you toggle on Left and Right Sampling, the offset distances need to be set.

8. Select the OK button to exit the Profile Sampling Settings dialog box and to return to the Edit Settings dialog box.

Profile Label Settings

9. Select Profile Labels from the Edit Settings list and then select the Edit Settings button. The Profile Label Settings dialog box opens (see Figure 8–6).

In a complex design there may be times when a drawing will have more than one profile. If there is more than one profile, they will be sharing the same layers in the drawing. The Create Profile routine prompts for erasing any objects on the profile layers. When there are several profiles using the same layers, the result of saying "yes" would be disastrous.

Civil Design allows for the assignment of a prefix to the layer names of a profile. The prefix is the name of the associated horizontal alignment. To set the prefix, place the prefix characters in the Layer prefix box. An asterisk (*) indicates the name of the current alignment that prefixes the layer names. A dash or underscore (- or _) separates the prefix from the layer name. The entry in the dialog box is either *- or *_. These entries create layer names such as NICHOLS-EGC or NICHOLS_EGCT.

10. Place an *- (an asterisk followed by a dash) in the Layer Prefix box (see Figure 8–6). The remainder of the dialog box identifies labeling notation for critical points on a vertical alignment.
11. Select the OK button to exit the Profile Label Settings dialog box and to return to the Edit Settings dialog box.

Profile Value Settings

12. Select Profile Value from the list in the Edit Settings dialog box, and then select the Edit Settings button to display the Profile Value Settings dialog box (see Figure 8–7). The Stationing Increments area sets the values for the stationing of the roadway alignment. Normally, you label stations at 50-foot intervals. However, set the tangent labels and vertical grid line to 50 and the vertical curve labels to 25. This will label more elevations along the vertical curve at the base of the profile.
13. The K Values section provides the minimum K value for crest or sag vertical curves. Set the minimum crest to 40 and the sag to 50.
14. The Sight Distance Values area sets the initial values used in vertical curve calculations. Some vertical curve calculations have sight distance, eye heights, and object height values in their formulas, and this is the area in the dialog box where you set those values.
15. The last section of the dialog box sets the precision for labeling existing and proposed profile elevations. The current settings are OK for the project.
16. Select the OK button to exit the Profile Value Settings dialog box and to return to the Edit Settings dialog box.
17. Select the Close button to exit the Edit Settings dialog box.

UNIT 2: PROFILES

The development and use of profiles requires a sequence of two steps. The first step is sampling the elevations along the centerline, and the second is creating the profile. The two steps are found in the Existing Ground and Create Profile cascade of the Profiles menu.

ELEVATION SAMPLING

Before you can create a profile, the horizontal centerline must be defined. After defining the centerline, you need to know what elevations are along its path. To determine what elevations are along the centerline, Civil Design uses a sampling process. Simply, as the sampling routine travels down, and at specific intervals on the alignment, it determines what elevations exist on the centerline. Sampling takes place more frequently on curves. The sampling process always samples at the PCs and PTs of the curves.

Civil Design has four methods of sampling (see Figure 8–8), which are available on the Existing Ground cascade. The first method is Sample From Surface. This method is by far the simplest. The Sample From Surface routine gleans elevations from a surface under the centerline. You can sample to the right and left of the centerline to view ditch or other elevations offset and parallel to the roadway.

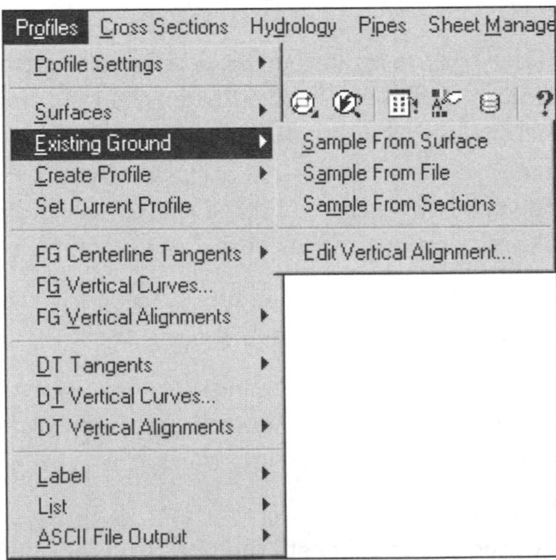

Figure 8–8

The second method, Sample From File, samples elevations by reading an external file that has a station number and its elevation. The Sample From File method samples only one surface and only along the centerline. This method is ideal for creating profiles from an alignment sketched on a contour map.

The Sample From File routine requires a specific file format. The file has to be a space-delimited ASCII file with the first field representing the station and the second representing the elevation at that station. The following is an example of a data file for the Sample From File method.

```
0.00      723.33
32.83     724.78
50.00     732.31
```

The third method, Sample From Sections, reads a cross-section data file for the elevations along the centerline. The data file contains elevations for stations and offsets along the centerline. The routine reads the file and creates the station and elevation data needed for the profile.

The fourth method, Edit Vertical Alignment, samples the existing elevations from the station and elevation data you type into the Vertical Alignment Editor (see Figure 8–9). After you enter the station and elevation values in the editor and confirm the values upon exiting, the routine samples the roadway centerline elevations.

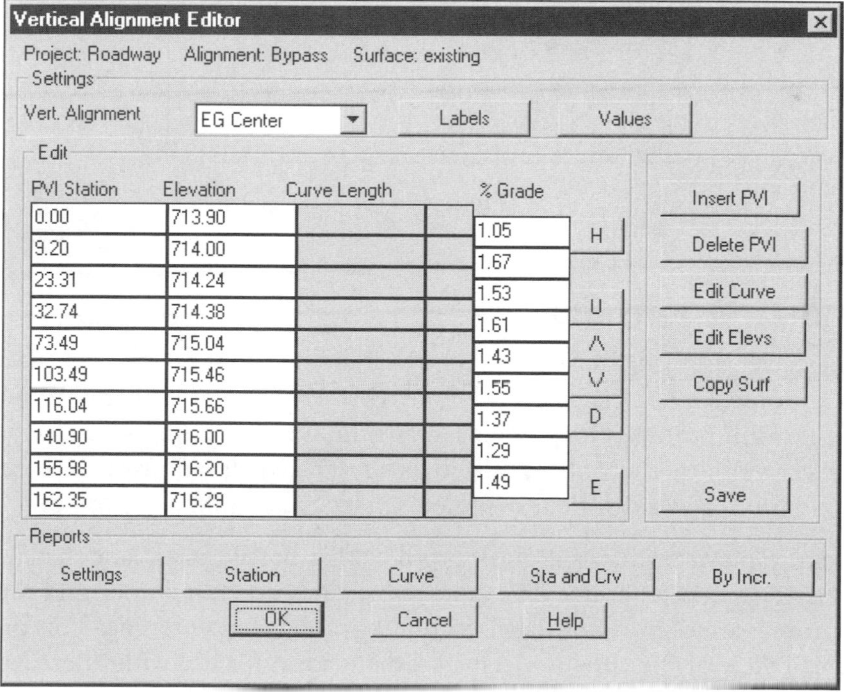

Figure 8–9

The Sample From File, Sample From Sections, and Edit Vertical Alignment methods cannot sample to the right or left of the centerline. They do require an elevation for the beginning and ending stations of the centerline alignment. If there is no elevation for the end of the alignment, the Create Profile routine will truncate the profile at the last elevation along the centerline.

If you have multiple surfaces and want them to be a part of the profile, you must complete two steps: first, toggle on Multiple Surfaces and then identify the surfaces with the Select Multiple Surfaces routine. These steps are found in the Surfaces cascade of the Profiles menu (see Figure 8–10). The only method that can sample multiple surfaces is the Sample From Surfaces routine.

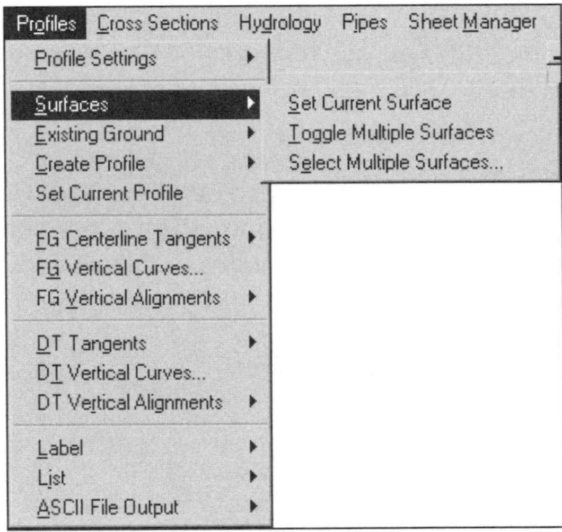

Figure 8–10

CREATING A PROFILE

After you have sampled the station and elevation information, the next step is to create the profile. Civil Design creates three types of profiles (see Figure 8–11). The first type is a full profile. This profile contains lines representing surface elevations along the centerline, a stationing grid and labels, labeled elevations at the base of the station grid lines, and a datum or base elevation of the profile.

The second type of profile is a quick profile. This profile is a scant sketch of the elevations along the centerline. The quick profile has no elevation or grid markings and no datum elevation. It only labels stationing along the centerline. This profile is a thumbnail view of the situation. This profile is ideal for use with Sheet Manager when creating plan and profile sheets.

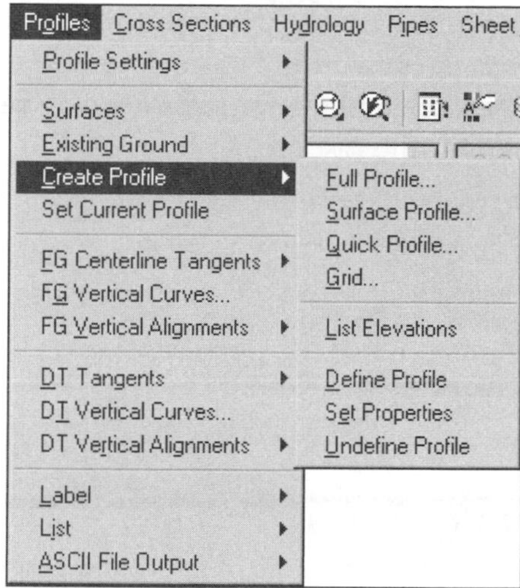

Figure 8–11

The last type of profile is the surface profile. This routine creates surface lines within an existing profile where you are using multiple surfaces. When working with multiple surfaces, create one full profile and then within the context of the full profile, create each additional surface's profile. For example, if a project has three surfaces—eg, base, and pile1—you would make one (1) full profile for eg and two (2) additional surface profiles for base and pile1.

ANALYZING A PROFILE

Civil Design provides a single tool to analyze the elevations and station within a profile. This tool is found in the Create Profile cascade of the Profiles menu (see Figure 8–11). The routine, List Elevations, echoes back the station, the elevation of the selected point, and the elevation of the existing ground at the point.

EDITING A PROFILE

The editing of a profile is limited to two routines, Define Profile and Set Properties. When the Create Profile routine executes, it places a block into the drawing identifying the location of the profile. If you should move the profile with the AutoCAD MOVE command, the block is not updated and any routine referencing the profile will return false values. The Define Profile routine will update the block to the new location of the profile so that the listing routines will report the correct elevations.

The Set Properties routine displays a dialog box allowing you to change the construction of the current profile (see Figure 8–11).

When you want to erase a profile, erasing the objects from the screen is not enough. The Create Profile routine creates an information block in the drawing that the ERASE command will not delete. To delete the information block, run the Undefine Profile routine found in the Create Profile cascade.

Exercise

After you complete this exercise, you will be able to:

- Sample the elevations along the centerline
- Create a profile

SAMPLE THE EXISTING GROUND

1. If you are not in the Roadway drawing of the roadway project, open the drawing now. Make sure you have the Civil Design menu loaded. Use the Menu Palettes routine of the Projects menu to load the Civil Design menu.
2. Select the Set Current Surface routine of the Surfaces cascade of the Profiles menu to set the existing surface as the current surface.
3. Sample the elevations along the Bypass centerline using the Sample From Surface routine of the Existing Ground cascade from the Profiles menu. If there is no current alignment, the routine will prompt you to select an object representing the Bypass alignment. If you cannot select an object representing the alignment, press ENTER to display a list of defined alignments. Select Bypass from that list. When the routine displays the Sample Offset Tolerance dialog box, make sure the tolerance is set to 0.5. Select OK to exit the dialog box. Press ENTER twice to accept the beginning and ending station values. After you accept the stations, the routine samples the surface.
4. Run the Edit Vertical Alignment routine of the Existing Ground cascade from the Profiles menu to view the elevations along the centerline. The routine displays the Vertical Alignment Editor showing the stations and elevations along the centerline (see Figure 8–9). The D button pages down the listing and the U button pages up. The H button displays the top row and the E button displays the last row of profile data. Select OK to exit the dialog box.

CREATE A PROFILE

1. Use the Zoom Center routine of the Desktop and use the coordinates of 2000,0 and the height of 2000.
2. Create a full profile using the Full Profile routine of the Create Profile cascade. The routine displays the Profile Generator dialog box. The dialog box identifies the beginning and ending stations of the profile and whether the profile is right-to-left or left-to-right (lower stations to higher stations). The Datum entry sets

the base elevation of the profile. The routine finds the nearest 5 elevation and then subtracts 10 to create the datum elevation. There may be reasons to lower this value if subsurface utilities are to be a part of the profile. Enter a datum of **680**. Refer to Figure 8–12 to set the values. After setting the values, select the OK button to exit the dialog box.

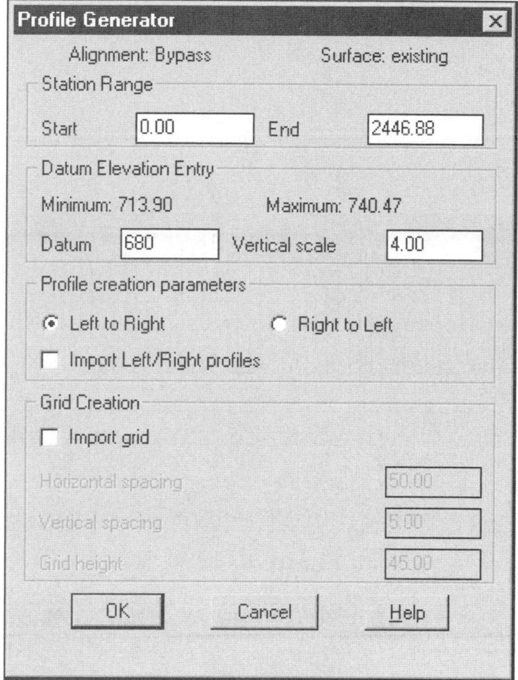

Figure 8–12

3. Select a point at the lower left side of the screen. Use Figure 8–13 as a guide to place the profile. When the routine prompts about deleting the profile layers, answer Yes because this is the first time there is nothing to erase. Notice that all the layer names have a prefix of Bypass.

4. All of the layers are white for the profile because the Create Full Profile routine created several new layers. Assign the layers the following colors using the Layers Properties Manager (see Figure 8–14). After setting the values, select OK to exit the layers dialog box.

   Green - BYPASS-PEGC
   Yellow - BYPASS-PBASE
   Cyan - BYPASS-PEGCT
   Gray - BYPASS-PVGRID

5. Save the drawing.

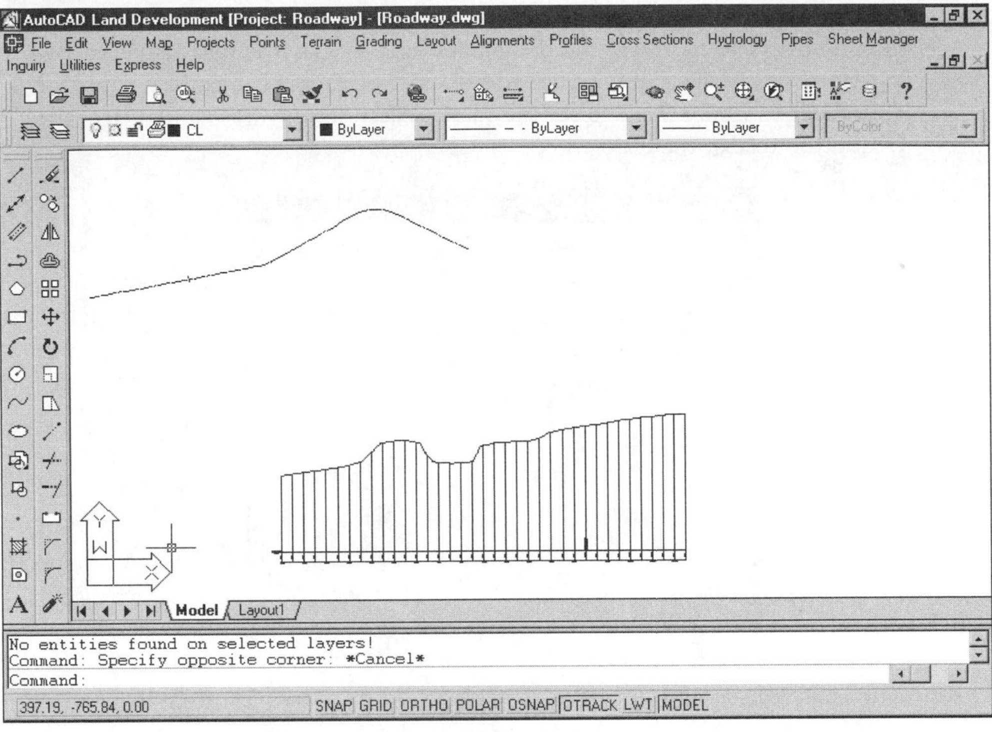

Figure 8–13

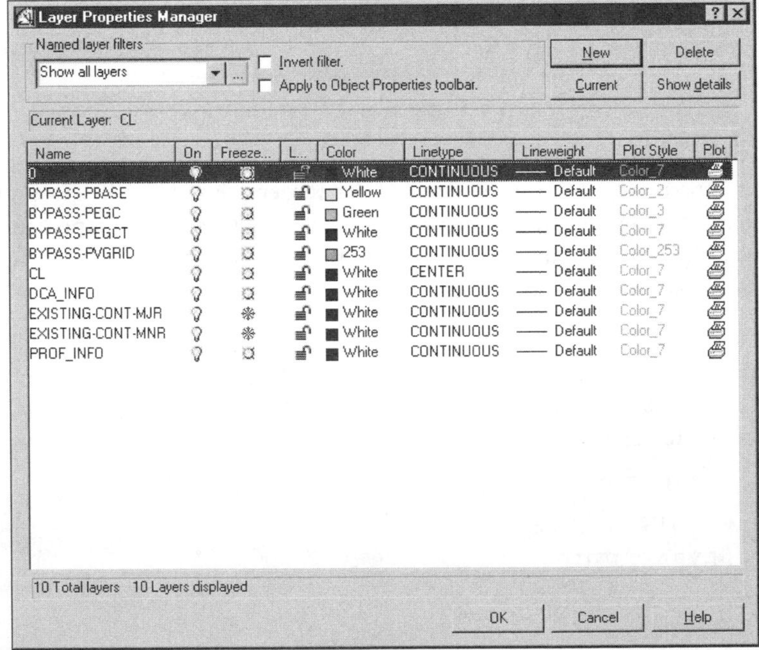

Figure 8–14

EXERCISE

LISTING ELEVATIONS IN A PROFILE

1. Use the List Elevations routine of the Create Profile cascade to report station, surface, selection point elevations, and the difference in elevations between the selected point and the existing ground. Press ENTER twice to exit the routine.

UNIT 3: CREATING A VERTICAL ALIGNMENT

CREATING VERTICAL TANGENTS

After you have created the profile, the next step is drawing the vertical tangent lines and designing vertical curves where necessary. All of the routines to create, analyze, edit, and annotate the vertical alignment are in the Profiles menu. The routines that accomplish these tasks are listed in Figure 8–15. The lines and curves representing the alignment are the vertical path the road template travels over the length of the centerline. After you have drafted these objects, the next step is to define them as the vertical alignment for the roadway. The defining process creates an external file entry that contains the information found on the screen. This definition is a snapshot of the objects on the screen. If you edit the objects with Desktop tools, you must redefine the alignment definition to match what is on the screen.

All of the routines you use to create a vertical alignment force you to use specific numbers. An alternative to using specific grades, slopes or elevations at stations along the profile is to simply draw in the vertical tangents, define the vertical alignment from the tangent lines, and edit the vertical alignment in the Vertical Editor. While in the Vertical Editor you can adjust PVI stations, elevations, and add in the needed vertical curves.

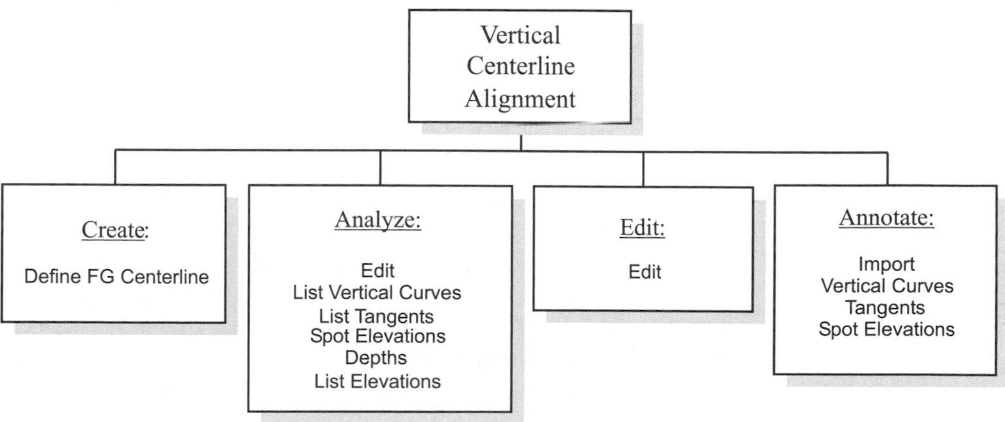

Figure 8–15

Before creating the vertical line work, you need to set the correct current layer. The Define FG Centerline routine isolates the PFGC layer, assuming the vertical line work exists on the layer. If the Profile layer prefix is toggled on, the layer name would be BYPASS-PFGC. To establish the correct current layer for the vertical work, run the Set Current Layer routine in the FG Centerline Tangents cascade of the Profiles menu. The routine sets the PFGC layer as the current layer.

Civil Design has two tools with which to create tangents for the proposed roadway centerline: Crosshairs @Grade and Create Tangents (see Figure 8–16). The Create Tangents routine of the FG Centerline Tangents cascade provides the greatest flexibility. The routine starts by locating a starting point at the beginning of the profile. The starting point is either a specific station or a user-selected point within the profile. If you specify a station, the routine offers the elevation of the station as a default value with an opportunity to change the elevation if so desired. If you select a point within the profile, the routine prompts with the station and elevation of the selected point and allows for the changing of either value.

After you set the starting point of the tangent, the Create Tangents routine presents four options to create the opposite end of the tangent. The first option is to specify the station of the endpoint of the tangent. After you specify the station, the routine prompts for the elevation it has to attain. The routine then draws a tangent from the starting point and elevation to the ending point and elevation.

The second method of creating tangents is by specifying a grade to follow from the starting point. The routine starts by prompting for a grade, and then the routine prompts for a length to travel to the next station or uses a selected point within the profile to determine the end station. After you enter all of the data, the routine draws the tangent line.

The third method is to select a point within the profile, and the routine will use the station and elevation of the point selected as the end of the tangent.

In the last method, if you select the length option, the routine asks for a grade and draws the tangent line.

The Crosshairs @Grade routine rotates the crosshairs to match a specific grade or slope. With the crosshairs rotated, you can draw tangents within the profile representing the grade or slope. The Crosshairs @Grade routine draws lines on the current layer.

If you use the Create Tangents routine of the FG Centerline Tangents cascade, the routine will automatically place the line work on the layer PFGC. After drawing the line work, the routine will reset the current layer to the previous settings.

EDITING VERTICAL TANGENTS

You can adjust the tangents by one of four methods. The first is to erase the lines and rerun the Create Tangents or Crosshairs @Grade routine. The second method is to use the Change Grade 1 or Change Grade 2 routines. The next method is using the Move PVI routine. The Change Grade and Move PVI routines are found in the FG Centerline Tangents cascade of the Profiles menu (see Figure 8–16). The last method is by changing the numbers in the Vertical Alignment Editor after you have defined the vertical centerline. After defining the vertical alignment, you have access to all the numerical values of the tangent lines.

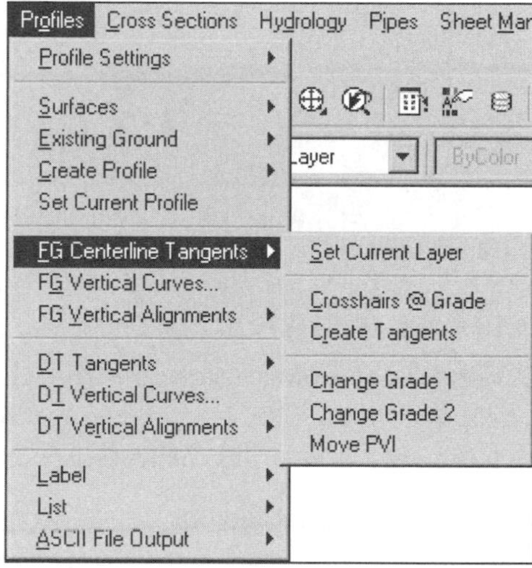

Figure 8–16

CREATING VERTICAL CURVES

A roadway design may need vertical curves where the vertical tangents intersect. A vertical curve allows the traffic to transition from one grade to another. Civil Design provides several vertical curve tools (see Figure 8–17). The vertical curve creation routines address three basic situations. The first situation is for crests. The methods of constructing these vertical curves include passing sight and stopping sight distance. The second situation is for sags. This group of methods include headlight and rider comfort conditions. The last group of methods includes length of vertical curve, minimum K value, passing through a point or selecting a high point, and a grade break.

The grade break method is used where there is no need for a vertical curve between two different tangent slope lines.

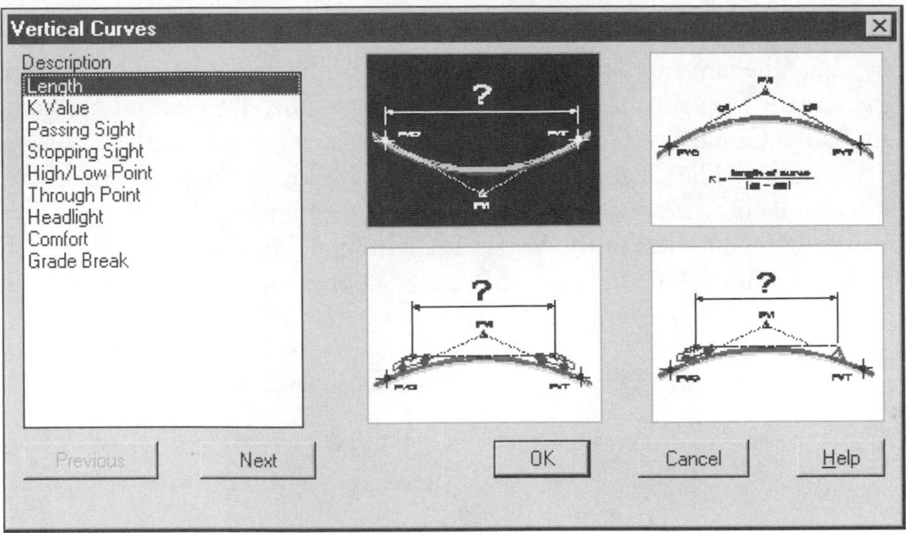

Figure 8–17

DEFINING A VERTICAL ALIGNMENT

After you have drawn the tangents and vertical curves within the profile, the next step is the defining of the vertical alignment. The Define FG Centerline routine is in the FG Vertical Alignment cascade of the Profiles menu (see Figure 8–18).

The defining process takes the desktop data and adds the vertical design values to the alignment resource files. From this definition, the designer can review the numerical values of the design. You can review the numerical values in the Vertical Alignment Editor.

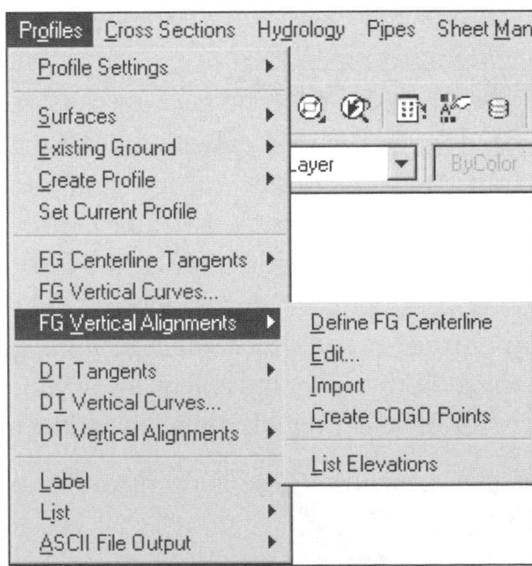

Figure 8–18

EXERCISE

Exercise

After you complete this exercise, you will be able to:

- Create the vertical tangent lines
- Create a vertical curve
- Define a vertical alignment

DRAW PROPOSED VERTICAL CENTERLINE

1. Continuing in the Roadway drawing of the Roadway project, use the Create Tangents routine of the FG Centerline Tangents cascade of the Profiles menu to draw the vertical centerline tangent lines. The routine will draw the lines on the correct layer. Start the tangent at station 0.00 and use its existing elevation. The last tangent endpoint is an object snap at the right side of the profile. Use the following values for the vertical alignment.

Station	**Elevation**	
0+00 – 6+00	728.00	(6+00)
6+00 – 24+46.88	Endpoint of Existing Surface in Profile	

```
Command:
Alignment Name: Bypass  Number: 1  Descr: Bypass Prelim Design 12/
  00
Starting Station: 0.00  Ending Station: 2446.88
Start Station: 0+00     Existing Elevation: 713.9032
Select point (or Station): s
Station <0>: (press ENTER to accept)
Elevation <713.9>: (press ENTER to accept)
Station: 0+00   Elevation: 713.9032
Select point [Station/eXit/Undo/Length]: s  (the s is for Station)
Enter station: 600       (the station is 6+00)
Select point [Grade/Elevation/Undo/eXit]: e  (the e is for eleva-
  tion)
Elevation <713.9>: 728
Station: 6+00   Elevation: 728.0000  Last Grade: 2.3495
Select point [Station/eXit/Undo/Length]: _endp of   (object snap)
Station: 24+46.88   Elevation: 740.4672  Last Grade: 1.0416
Select point [Station/eXit/Undo/Length]:
Command:
```

EXERCISE

VERTICAL CURVES

1. Create a vertical curve by K value. Select the K Value routine from the FG Vertical Curves dialog box of the Profiles menu. The routine prompts you to choose the in and out tangent lines. Select the two tangents in the profile. Use the K value of 150 for the initial vertical curve at station 6+00. After the routine returns the curve length, change the curve length to 200, and then press ENTER twice to accept the vertical curve length and to exit the routine.

```
Command:
Alignment Name: Bypass  Number: 1  Descr: Bypass Prelim Design 12/
  00
Starting Station: 0.00  Ending Station: 2446.88
Select incoming tangent: (select the first tangent line)
Select outgoing tangent: (select the second tangent line)
Minimum K: 150
Calculating.  Please wait.
Length of curve <196.17>: 200
The final K value = 152.93
Select incoming tangent: (press ENTER to Exit)
```

DEFINE THE VERTICAL ALIGNMENT

1. Define the vertical alignment by using the Define FG Centerline routine of the FG Vertical Alignments cascade from the Profiles menu. The routine turns off all of the layers in the drawing except for BYPASS-PFGC. Select the first tangent line (left tangent segment) near its left endpoint. After selecting the initial segment, select the remaining tangents and curves (crossing window). Press ENTER to stop selecting objects and to exit the definition routine. The routine indicates each PVI on the alignment with a yellow "X."

UNIT 4: ANALYZING AND EDITING A VERTICAL ALIGNMENT

After you have defined the vertical alignment, Civil Design offers several tools to evaluate and edit them. The routines of the FG Vertical Alignments cascade of the Profiles menu define, import, and edit vertical alignments in the project (see Figure 8–18).

VERTICAL ALIGNMENT EDITOR

The best analysis and editing tool for vertical alignments is the Vertical Alignment Editor. The editor displays the numerical data of the current vertical alignment (see

Figure 8–19). Like the Horizontal Alignment Editor, the Vertical Alignment Editor allows for the manipulation of the elevations and/or stations of PVIs and vertical curves of the alignment.

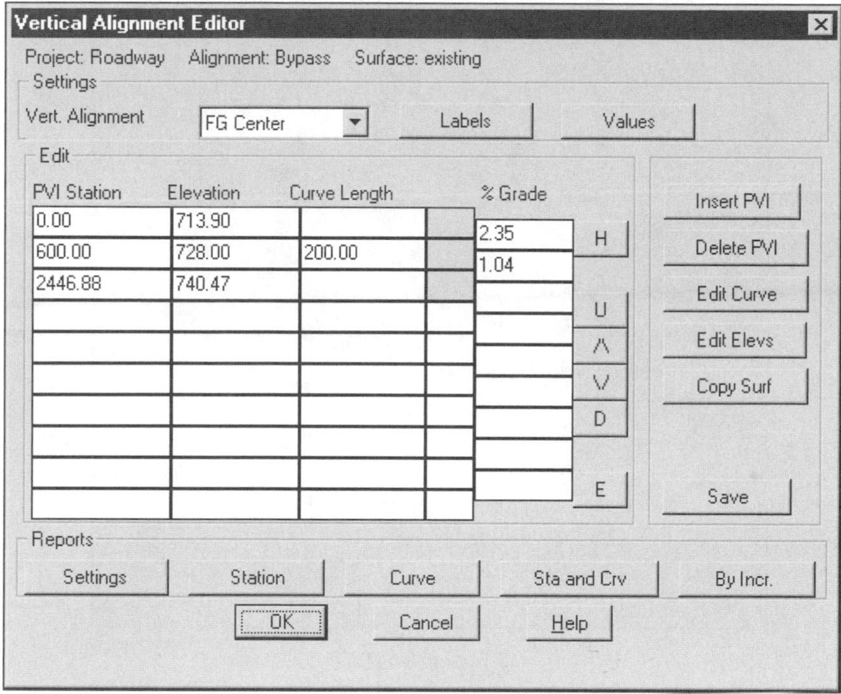

Figure 8–19

The Vertical Editor reports on the statistics of a vertical alignment. There are three types of reports generated by the Vertical Editor. The first report type is only a tangent or curve data report. The second report is a combined tangent and curve data report. The third report is a user-specified incremental report along the centerline.

Before viewing or editing the vertical curve data, you must identify the PVI station by selecting the cell for the PVI station. Then select the Edit Curve button to view the curve data dialog box (see Figure 8–20). The dialog box displays the current values for the curve. You can edit or create a vertical curve by editing any value in the lower portion of the dialog box. After changing a curve value, press ENTER to see the editor update all the pertinent vertical curve data.

When you exit the Vertical Editor and save the changes, the editor does not import the new vertical alignment into the drawing. This is the exact opposite of what happens when you exit the Horizontal Editor. To replace the old vertical alignment with the new definition, run the Import routine of the FG Vertical Alignments cascade. The

Import routine asks about replacing the existing line work in the profile and annotating the vertical alignment definition. If you want the annotation, answer Yes, and the routine annotates the tangents, vertical curves, and station elevations.

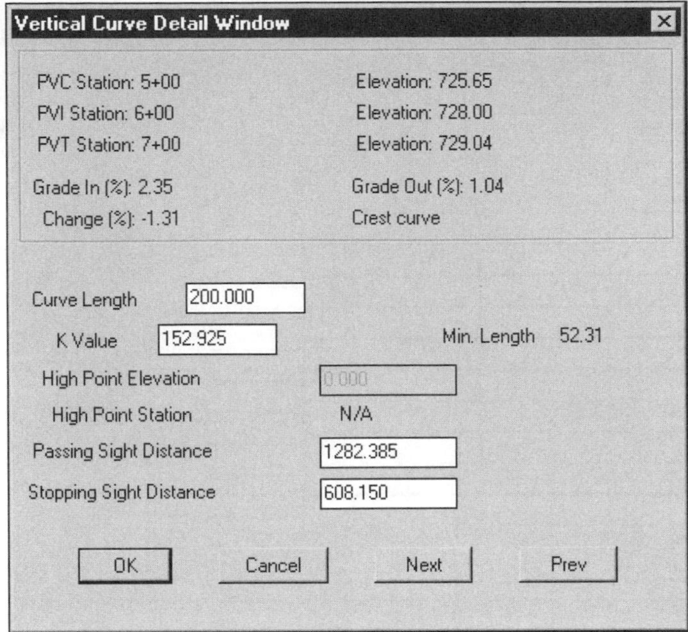

Figure 8–20

ANALYSIS TOOLS

Other listing tools for elevations within the profile are in the List cascade of the Profiles menu.

VERTICAL CURVES

The Vertical Curves routine reports the design numbers of a selected vertical curve to the Desktop text screen.

TANGENTS

The Tangents routine reports the design numbers of a selected tangent to the Desktop text screen.

SPOT ELEVATIONS

The Spot Elevations routine reports the station and elevation of a point selected in the profile.

DEPTHS

The Depths routine reports the vertical distance between two points selected in the profile.

LIST ELEVATIONS

The List Elevations routine reports the station, elevation of the selected point, the elevation of the vertical alignment normal to the selected point, and difference in elevation between the selected point and the normal to the vertical centerline.

Exercise

EXERCISE

After you complete this exercise, you will be able to:

- Review the design numbers of a vertical alignment
- Use different listing routines on a profile
- Edit the vertical centerline

PROFILE LISTING ROUTINES

1. Use the AutoCAD ZOOM command and create a full screen view of the profile.
2. Create a named view, Profile, using the Named View routine of the View menu.

Vertical Curves

3. Use the Vertical Curves routine of the List cascade of the Profiles menu to report the design values of the vertical curve. When the routine prompts you to select the vertical curve, select the vertical curve in the profile. The routine displays a report in the text window of AutoCAD. You may have to flip to the text screen to view the report. After reviewing the report, press ENTER to exit the routine.

Tangents

4. Run the Tangents routine of the List cascade of the Profiles menu to report the design values of the vertical alignment tangents. Press ENTER to exit the routine. You will have to flip to the text screen of the Desktop to view the report.

Spot Elevations

5. Use the Spot Elevations routine of the List cascade of the Profiles menu to report the station and elevation of a selected point in a profile.

Depths

6. Use the Depths routine of the List cascade of the Profiles menu to report the vertical difference between two selected points in the profile. You may have to toggle off object snaps to pick points that are not endpoints. Press ENTER to exit the routine.

List Elevations

7. Use the List Elevations routine of the List cascade of the Profiles menu to report the elevations and stations of selected points and their related vertical centerline elevation.

VERTICAL ALIGNMENT EDITOR

The Vertical Alignment Editor displays the numerical data of the vertical alignment (see Figure 8–19). Before editing or viewing the vertical curve data, identify the PVI station by selecting the cell for the PVI station. Then select the Edit Curve button to view the curve data dialog box (see Figure 8–20). Adjust the vertical curve by editing any value of the vertical curve. After changing a curve value, press ENTER to see the editor update all the pertinent vertical curve data.

1. Edit the vertical alignment by using the Edit routine of the FG Vertical Alignments cascade of the Profiles menu.
2. Select the station 6+00 cell that has the vertical curve. After selecting the cell, select the Edit Curve button. This will display the vertical curve data dialog box.

   Change the vertical curve length to 250.00 feet. Press ENTER to view the data changes.

   Change the K value to 175. Press ENTER to view the data changes.

   Change the K value to 100. Press ENTER to view the data changes.

   Change the vertical curve length to 250.00 feet. Press ENTER to view the data changes.
3. Keep the vertical curve length to 250 and select the OK button to exit the curve data dialog box.
4. Exit and confirm the editor changes by clicking on the OK and Save buttons.

When you exit the Vertical Editor, the routine does not update the vertical alignment in the drawing. You must use the Import command of the FG Vertical Alignments cascade to see the new entities on the screen. Before importing the entities onto the screen, the routine prompts about annotating the new and removing the old alignment entities.

5. Import the alignment, but do not annotate the line work. Use the Import routine of the FG Vertical Alignments cascade of the Profiles menu. Answer No to the annotation of the alignment and answer Yes to the erasure of previous alignment objects.
6. Save the drawing.

The current alignment is only preliminary.

7. Reenter the Edit routine of the FG Vertical Alignments cascade and create a Station and Curve report. The routine prompts for the beginning and ending stations. Press ENTER twice to accept the stations, and then the routine displays a Tangent and Curve report for the vertical centerline alignment.
8. Select the OK button twice to exit the report and Editor dialog boxes.
9. Save the drawing.

UNIT 5: ANNOTATING A VERTICAL ALIGNMENT

Civil Design provides tools to label information from the profile and vertical alignment. There are two sets of tools. The first tool set is the Import routine of FG Vertical Alignments. The Import routine will label all tangent segments and vertical curves when importing a vertical alignment into a drawing (see Figure 8–18). The second set of tools is in the Label cascade of the Profiles menu (see Figure 8–21). This cascade has tools that label individual elements of a profile or vertical alignment.

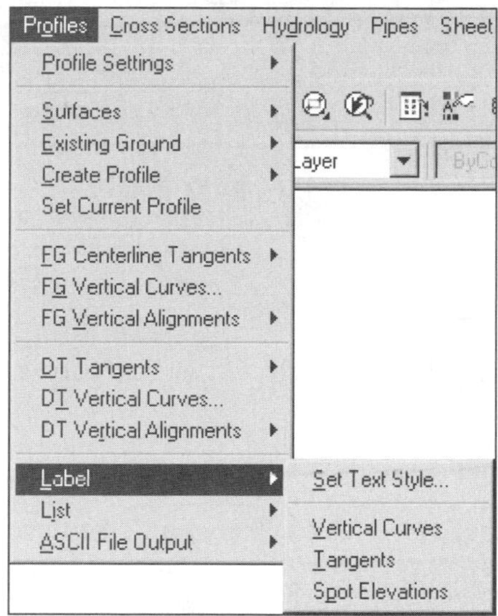

Figure 8–21

SET TEXT STYLE

This routine sets the text style for the annotation routines. The routine calls the Text Style dialog box of the Desktop to define or set current a text style (see Figure 8–22). The Leroy styles (L prefixed) are a fixed height text style group. Their name identifies the plotting height of the text on a piece of paper. All other styles need to have an appropriate text height set before using them for annotation.

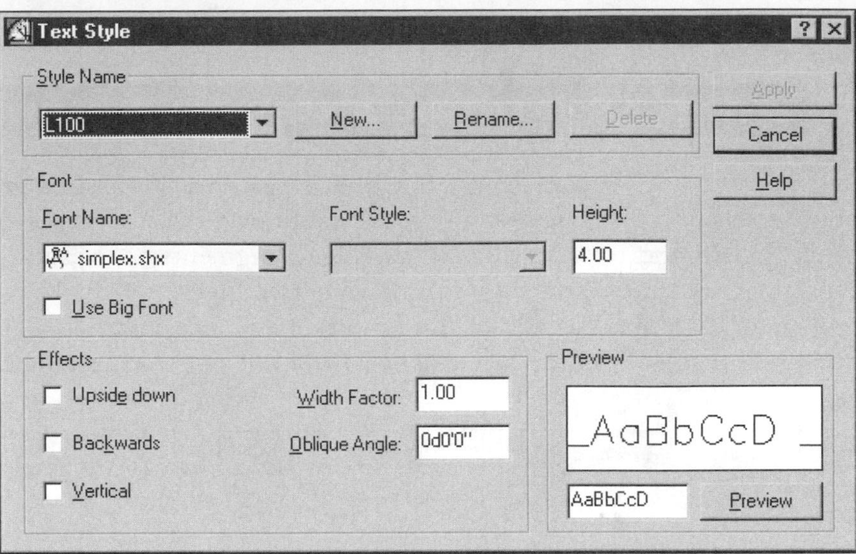

Figure 8–22

VERTICAL CURVES

The Vertical Curves routine labels user-selected vertical curves of the vertical alignment.

TANGENTS

The Tangents routine labels user-selected tangents of the vertical alignment.

SPOT ELEVATIONS

The Spot Elevations routine labels user-selected points with their profile elevations. The routine prompts you to select a point within the profile and if necessary to create a leader from that point to the location of the label. The routine places the label and leader on the current layer.

Exercise

After you complete this exercise, you will be able to:

- Manually annotate a vertical alignment
- Annotate spot elevations within the profile

ANNOTATING VERTICAL ALIGNMENTS

1. Run the Set Text Style routine of the Label cascade from the Profiles menu. Set the current text style to L120. After setting the style, select the CANCEL button to exit the dialog box.
2. Select the Vertical Curves routine from the Label cascade of the Profiles menu. The routine prompts for the two tangents in and out of the vertical curve. Select the two tangent lines. After you select the tangents, the routine prompts for the vertical curve. Select the arc segment in the profile. You may have to zoom in to be able to select the arc segment. After you select the arc segment, the routine labels the vertical curve above the profile. The critical point codes come from the Profile Label Settings.
3. After labeling the vertical curve, press ENTER to exit the routine.
4. Select the Tangents routine from the same cascade menu. The routine prompts you to select a tangent. After you select a tangent, the routine labels the grade on the tangent line.
5. After labeling the tangent's curve, press ENTER to exit the routine.
6. Select the Spot Elevation routine from the Label cascade of Profiles. The routine prompts for a label point. This is the station and elevation that will be labeled by the routine. After you select the point, the prompt changes to selecting leader points. Select one point to set the end of the leader, select a second point to turn and end the leader, and press ENTER to label the point.
7. Undo back and remove all of the labeling.
8. Save the drawing.

UNIT 6: CREATING POINTS FROM A VERTICAL ALIGNMENT

This unit covers the routine that creates points from the vertical alignment definition. The routine is the Create COGO Points routine of the FG Vertical Alignment cascade of the Profiles menu (see Figure 8–18). The routine displays the Centerline Point Output dialog box, identifying the current alignment, beginning and ending stations, and the interval for placing the points (see Figure 8–23). The dialog box also sets the description for the points placed along the centerline. The elevations of the points come from the elevations of the vertical alignment. The routine places the points on the horizontal alignment.

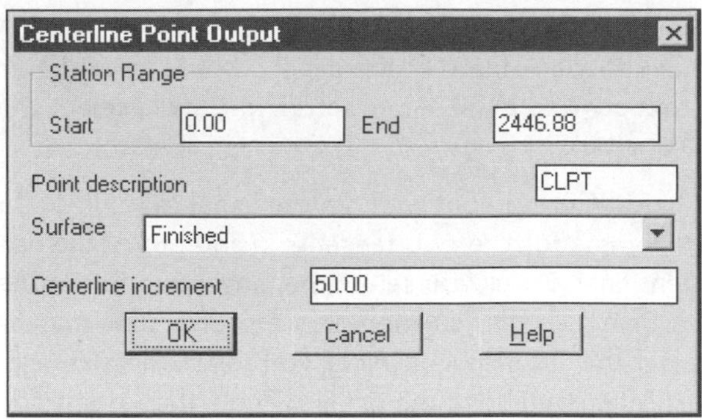

Figure 8–23

Exercise

After you complete this exercise, you will be able to:

- Place points along a vertical alignment

CREATING POINTS FROM VERTICAL ALIGNMENTS

1. Select the Create COGO Points routine from the FG Vertical Alignment cascade of the Profiles menu. The routine displays the Centerline Point Output dialog box. Select OK to exit the dialog box and set the current point number to 75. The routine places points along the horizontal alignment (plan view) representing the elevations of the vertical alignment.
2. Run the List Points routine from the Points menu and toggle on all points to list. Scroll down to view the point list after point number 75. All of the points have a CLPT description. Select OK to exit the dialog box.
3. Erase the point number range of 75–300 from the drawing. Use the Erase routine of the Edit Points cascade of the Points menu. Use the Number option to specify the point range.
4. Use the Pack Point Database routine of the Point Utilities cascade of the Points menu to resize the point database. The routine first reports the current state of the point database. Press ENTER to continue the process. Next the routine displays a dialog box (see Figure 8–24) informing you that it will now remove all empty records from the database. Select the Yes button to continue.
5. Save the drawing.

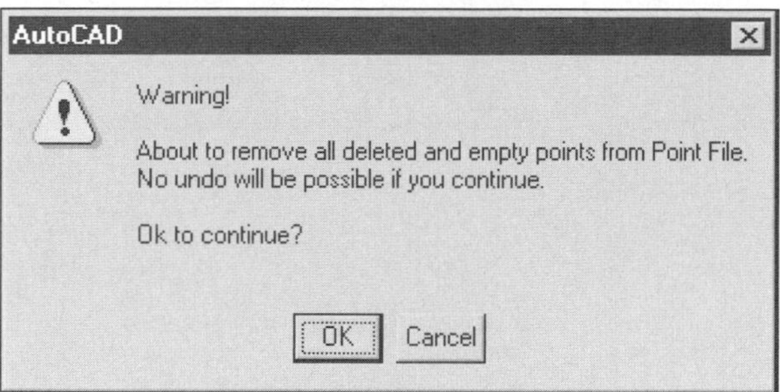

Figure 8–24

This completes the discussion of profiles in Civil Design. A profile is the second major phase in the process of road design. The first was the horizontal alignment. The next phase of the process is to look to the right and left of the centerline and view the interaction of a template and the alignments. This interaction produces the critical result, a volume. Section 9 covers the creation, evaluation, and editing of roadway templates and designs.

SECTION 9

The Roadway Design Process—Cross Sections and Volumes

After you complete this section, you will be able to:

- Edit the settings for a roadway cross section
- Sample for cross-section elevations
- Create a template
- Set values in and the use of the Edit Control dialog box
- Review the roadway volumes
- Edit a template
- Edit a design
- Import design cross sections
- Create plan and profile sheets with Sheet Manager
- Create section sheets with Sheet Manager
- Annotate the design results
- Create new points from the resulting roadway design

SECTION OVERVIEW

This section is the last section covering the basics of roadway design. The first two roadway design sections, 7 and 8, cover the creation of the horizontal centerline alignment, profiles, and the vertical centerline alignment. This section covers the cross-section view of a roadway design (see Figure 9–1). The cross-section view of the design includes a template representing a cross section of the surfaces of a roadway. A template can be complex, simple, or dynamic with the use of transitional alignments.

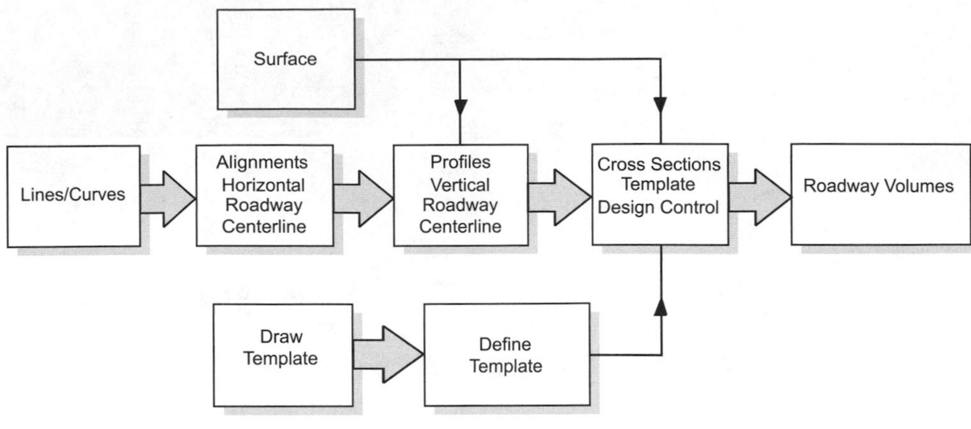

Figure 9–1

The goal of the design will vary from project to project, but generally the goal is not to have to move more spoil material than necessary to create the road. The interaction of the template with the necessary ditches, slopes, and grades makes the resulting design a process of evaluating designs and editing them to understand the impact the changes make to the overall volumes. The editing of the design may take you to the very beginning of the design process, the horizontal alignment. Or, the modifications can occur in profiles or in the cross-sections area. Wherever the editing takes place, you must move forward from that point to the final processing of the design. If you change the vertical alignment, you do not need to resample the cross sections, but you have to reprocess the design. If you change the horizontal alignment, you will need to resample and rebuild the profile and cross sections to view the impact of the changes.

After processing the cross sections, the Civil Design add-in calculates a volume for general design grading and template material volumes. On the Cross Sections menu, you will find tools for creating, evaluating, and annotating the volume totals.

UNIT 1

The settings used by this phase of the design are the focus of the first unit of this section. The Civil Design settings control layer names, sampling rates, and slopes used in the development of a road design.

UNIT 2

The first step in working with cross sections is the sampling of a surface to create the cross sections. There are three methods of sampling for cross sections. These methods are Sample From Surface, Sample From File, and Edit Sections. The process of sampling the surfaces is the focus of the second unit of the section.

UNIT 3

The third unit of the section covers the drawing, defining, and editing of a template. As mentioned above, the template can be simple or complex. The template can have subassemblies added to it or be a collection of surfaces. You can use subassemblies to develop a library of templates that vary by the attached type of curbs and cut and fill slopes. Also, transitional alignments can manipulate the template in the horizontal and vertical directions. When editing a template, you can locate and define the behavior of these transition points and indicate where the alignments attach to the template.

UNIT 4

The evaluation of all of the design elements, vertical centerline, template, and slopes occurs in the Edit Design Control dialog box of the Design Control flyout of the Cross Sections menu. The Edit Design Control dialog box is in essence a design processor. When you exit the dialog box, the routine evaluates all of the design values and attachments and creates the cross sections and volume results. The Edit Design Control dialog box is the focus of the fourth unit of this section.

UNIT 5

The review of the resulting volumes and cross sections is the topic of Unit 5. This unit covers viewing, editing, and evaluating the cross sections and the design volume results.

UNIT 6

The sixth unit of this section reviews the editing of the design needed to fine-tune it for volumes. This unit will also cover the importing of critical points on a template into the profile to aid in the design review process.

UNIT 7

After you have refined the design, the Civil Design add-in offers several tools to create section and plan and profile sheets. Unit 7 covers the importing of cross sections into the drawing. This process is similar to the importing process for site volume sections.

UNIT 8

The use of Sheet Manager for plan and profile and sections sheets is the topic of the eighth unit of this section.

UNIT 9

The last unit covers the creation of new points and objects from a final design. This includes creating points, 3D models, and pasting the design into the existing surface to create a final design surface.

UNIT 1: CROSS SECTION SETTINGS

The Cross Section Settings for roadway design control layers, initial design criteria, and the plotting of cross sections. The Sheet Manager menu contains routines to create a different set of cross sections. You access all of the Cross Section settings in the Drawing Settings Dialog box of the Projects menu.

CROSS SECTION SAMPLING

You can access the Section Sampling Settings dialog box from the Drawing Settings routine of the Projects menu. The Section Sampling Settings dialog box has four sections (see Figure 9–2). Each section controls a different aspect of the sampling process. The top portion of the dialog box sets the right and left offset distance. This distance is the limit of the design impact. Most of the time the impact of the design ends before reaching this offset distance. However, the design can go no further than this distance from the centerline.

Section Sampling Settings

Swath Widths

Left 75.000 Right 75.000

Sample Increments

Tangents 50.000 Curves 25.000 Spirals 25.000

Additional Sample Control

PC's/PT's TS-SC's/CS-ST's

Alignment start Alignment end

Save sample list Read sample list

Add specific stations

Sample Lines

Import Layer XSSAM

OK Cancel Help

Figure 9–2

The Sample Increments area sets the frequency of sampling on tangents, curves, and spirals.

The Additional Sample Control area sets user-defined critical design points. These points are PCs and PTs of curves, Spiral-Curve points, the beginning and end of the alignment, and user-selected points from the plan view of the road or entered stations.

The Sample Lines area at the bottom of the dialog box toggles on the importing of lines representing the cross sections and creates the layer for the lines if the layer does not exist.

CROSS SECTION CONTROL

In the Control Editor dialog box, you can access a number of other dialog boxes for controlling templates, ditches, slopes, and benches (see Figure 9–3). The Template Control button displays the Template Control dialog box. This dialog box sets the template and datum to use when processing the design (see Figure 9–4).

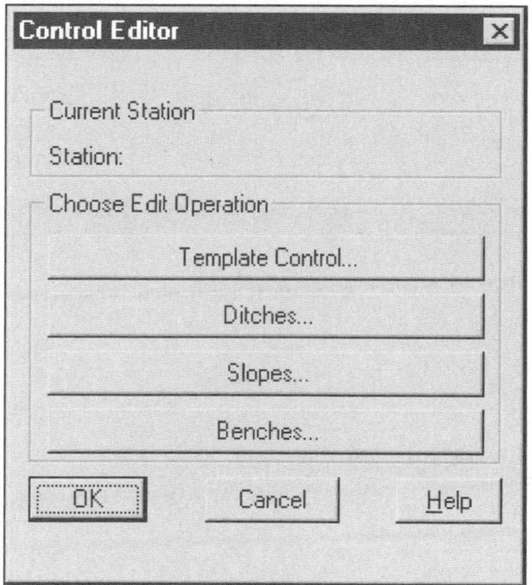

Figure 9–3

The Ditches button displays the Ditch Control dialog box (see Figure 9–5). This dialog box sets the parameters for a ditch along the edge of the road. A ditch can be on the right and/or left side, in cut or fill or both, and can be a fixed distance or slope determined distance from the centerline. You must toggle on a minimum of two parameters in the dialog box to make the ditches occur in the cross sections.

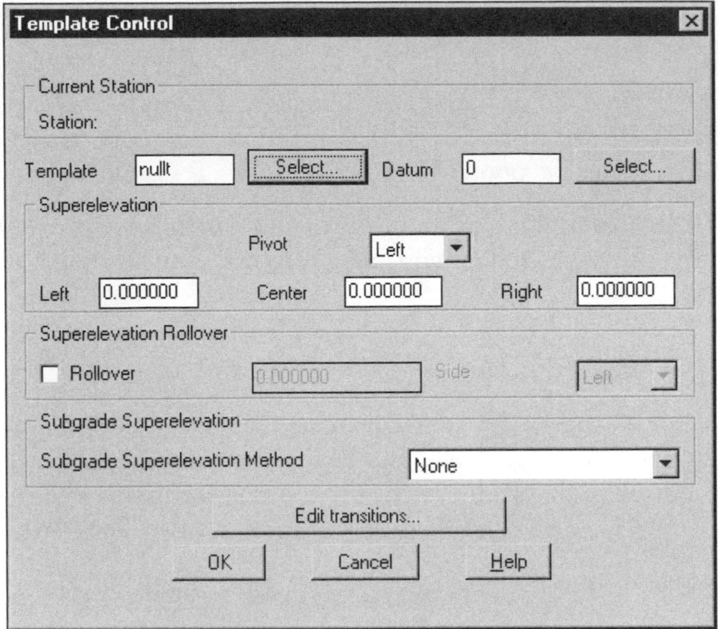

Figure 9–4

Figure 9–5

The Slopes button displays the Slope Control dialog box, which sets the type and amount of slopes to use when attempting to daylight the right and left end of the template to the existing ground (see Figure 9–6). The upper area of the dialog box controls daylighting (on or off), the type of slope to use, and amount of slope. The bottom portion of the dialog box defines the offset distance for the right-of-way (ROW). The Hold ROW toggle allows the design to daylight at the ROW whether or not there is a solution before encountering the ROW. Civil Design allows the ROW to be an alignment. This means that the ROW offset distance can vary based upon the definition of the alignment.

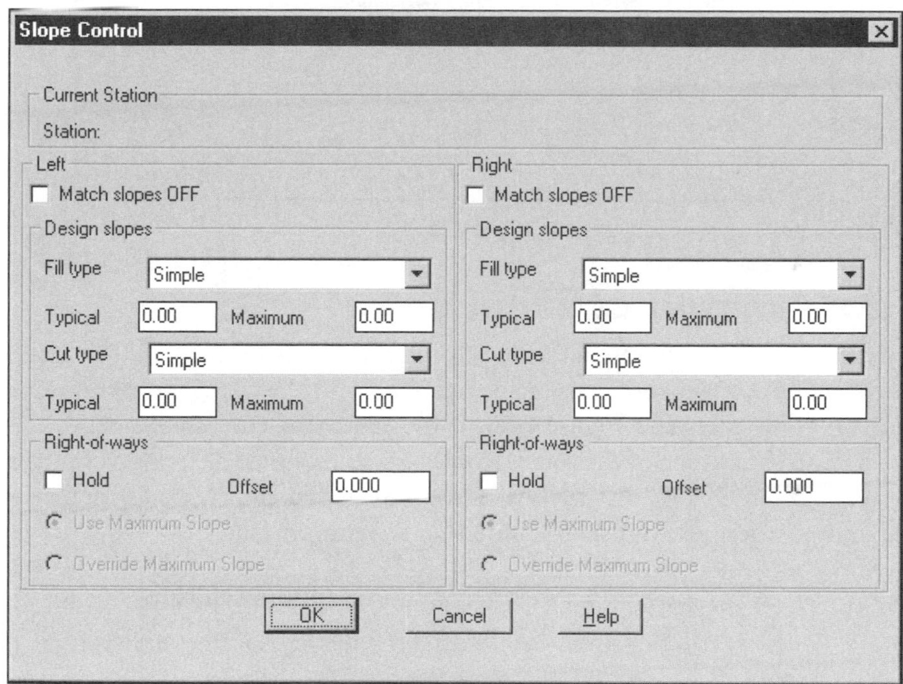

Figure 9–6

The Benches button displays the Bench Control dialog box, which toggles on and sets the values for right and left side roadway benches (see Figure 9–7). The benches can be different for cut and fill situations.

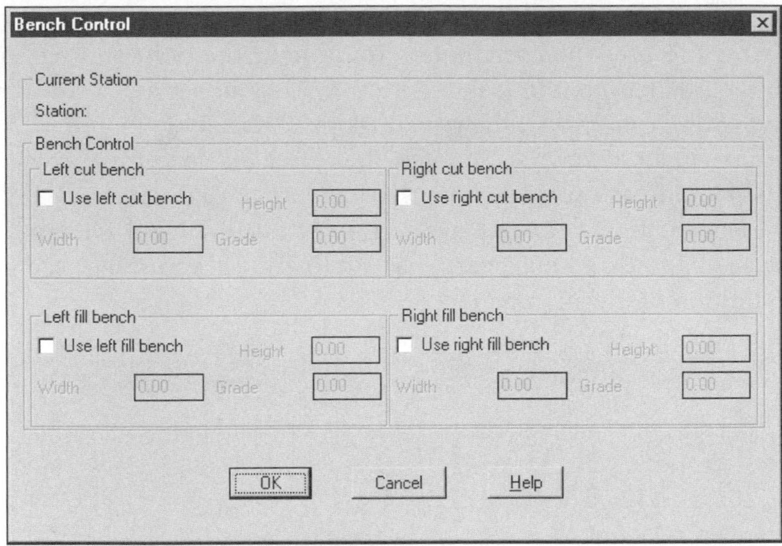

Figure 9–7

SUPERELEVATION CONTROL

You can access the Superelevation Curve Settings dialog box from the Drawing Settings routine of the Projects menu. This dialog box sets control values that are not standard superelevation numbers (see Figure 9–8). If the centerline curves were designed with the Create Curves routine of the Speed Tables cascade of the Lines/Curves menu, the superelevation numbers are on the curve. If the curves were not created with the Create Curves routine, these settings set the initial superelevation values.

Superelevation Curve Settings
Curve Edit Information
Start sta 0.00 End sta 0.00 Method C
E value 0.060000 Direction Right Rollover 0.000000
Transition In
Runout 50.000
Runoff 100.000
% runoff 100.000
Transition Out
Runout 50.000
Runoff 100.000
% runoff 100.000
Subgrade Superelevation Method None
Transition In
Distance 0.00
Transition Out
Distance 0.00
OK Cancel Help

Figure 9–8

CROSS SECTION VIEW/EDIT

The Cross Section View/Edit choice in the Edit Settings dialog box opens the Template View Settings Editor. These settings affect the display of cross sections during an interactive editing session (see Figure 9–9). The Toggles area at the top of the dialog box toggles on and sets the colors for different elements of the cross section.

The Grid Values area in the middle of the dialog box sets the offset, elevation, and precision for values that make up the cross section. At the bottom of the dialog box, in the Miscellaneous Values area, are settings that control the vertical scale factor, text size, and zoom in and out factor.

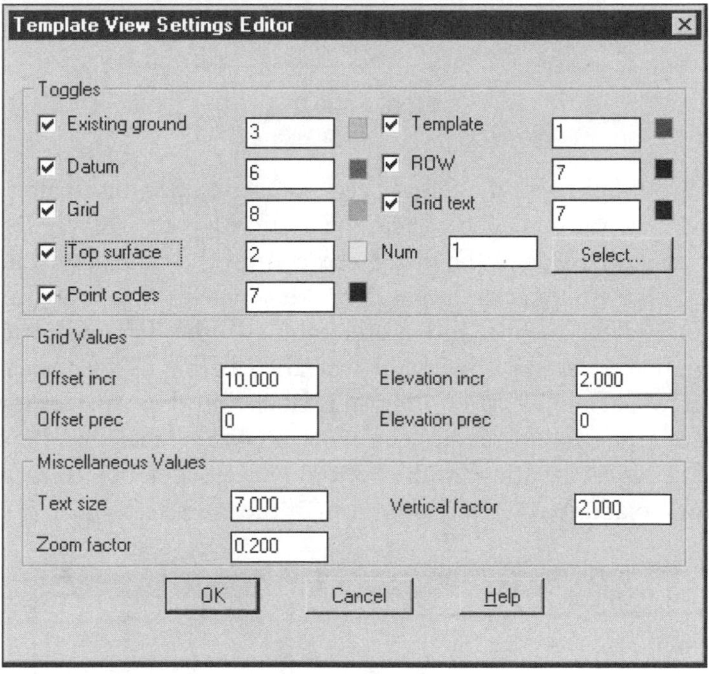

Figure 9–9

CROSS SECTION PLOTTING

The Cross Section Plotting settings occupy three dialog boxes. The first dialog box, Cross Section Plotting Settings, controls the appearance and layer for different elements in the cross section (see Figure 9–10). The layers for all of the elements are prefixed with the letter “X”.

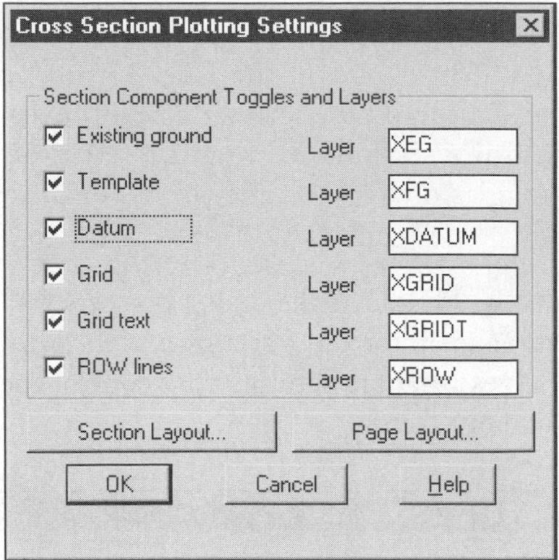

Figure 9–10

The Section Layout button in the Cross Section Plotting Settings dialog box calls the Section Layout dialog box, which affects the section layout (see Figure 9–11). The settings at the upper left side of the dialog box control the offset interval, its labeling, and annotation. The settings at the upper right side of the dialog box are values affecting the elevation interval, its labeling, and its annotation. Both offset and elevation settings include an increment value. This increment sets the labeling frequency. A value of one means that each offset and elevation interval is labeled in the cross section. A value of two means that every other offset and elevation interval is labeled in the cross section and so forth. The last two settings at the bottom of the dialog box control the number of elevation rows to add to the top and bottom of the cross section.

Section Layout

Offset incr	10.000	Elevation incr	2.000
Offset lbl incr	2	Elevation lbl incr	2
Offset prec	0	Elevation prec	0
FG lbl prec	2	EG lbl prec	1
Rows below datum	2	Rows above max	1

OK Cancel Help

Figure 9–11

The third dialog box, Page Layout, contains the settings for the definition of a plotting sheet of cross sections (see Figure 9–12). The two settings at the top of the dialog box set the size of the sheet. The next four settings control the no-plot margins for the sheet. The next two settings control the horizontal and vertical spacing between the cross sections. The spacing is in cross-section cell size. The last setting limits the number of stacked pages to four. After importing four sheets of sections, the routine starts a new column of sheets.

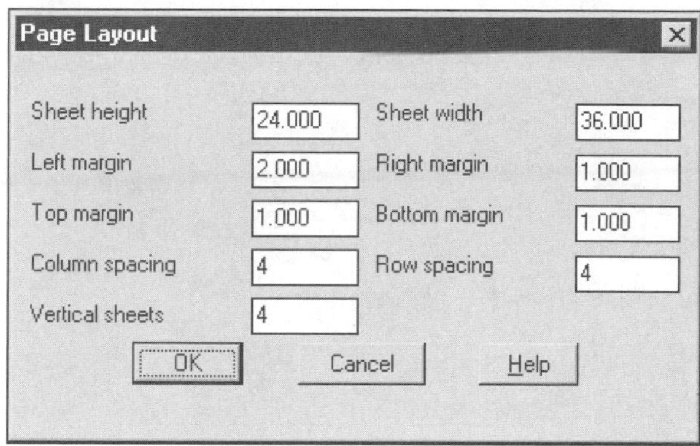

Figure 9–12

SHEET MANAGER SETTINGS

In the Edit Settings dialog box, the Sheet Manager Settings choice displays the Settings dialog box, which controls several aspects of a section, plan, profile, or plan and profile sheet (see Figure 9–13). The Style Database is a path to a folder containing sheet styles. Civil Design creates four different libraries of sheets. You can add any number of libraries and sheet styles to the folder structure.

The Layers group names the three basic layers for a sheet style. The sheet style contains several elements; however, they all reside on these three layers.

The next set of values, Sheet Options, sets the Layout name for the sheets, whether the match lines are model space objects, whether the plan view of the sheet is set by the length of profile, and whether there is an offset for the profile. A negative profile offset starts the profile view before the alignment starts. A positive profile offset starts the profile view after the alignment starts.

The Generate Sheets and Return to Model Space toggle returns control to model space after creating a sheet series.

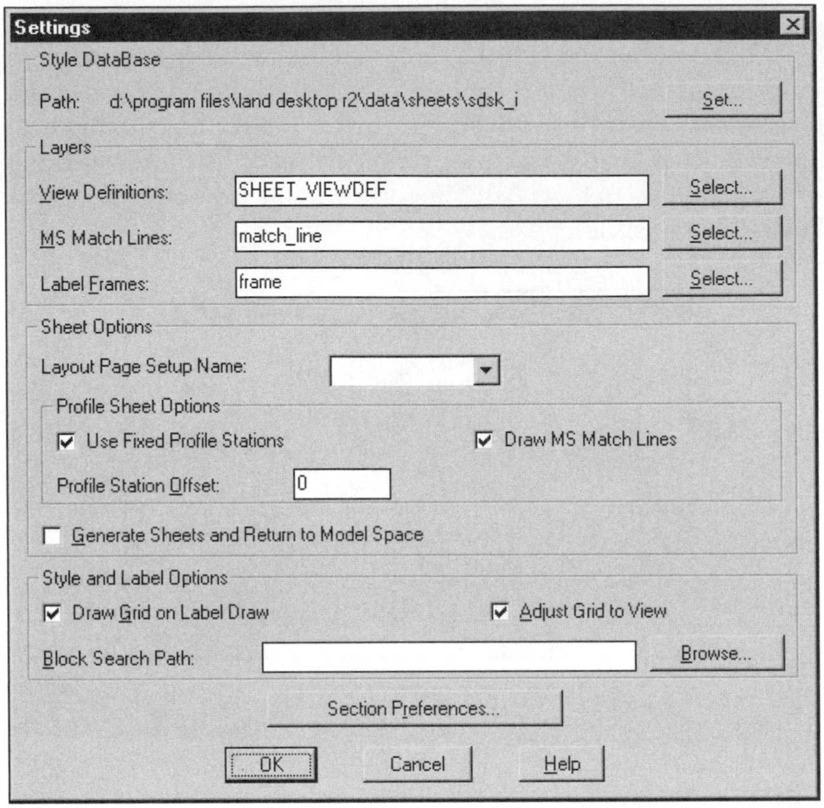

Figure 9–13

The Style and Label Options area controls how grids are drawn in a sheet, their alignment to their associated frame, and a new path to block objects used in annotating a sheet. The Draw Grid on Label Draw toggle forces the sheet generator to recreate the grid when the labeling is placed into the sheet. The Adjust Grid to View toggle forces the grid to align to the labeling if the labeling interval should change.

SECTION PREFERENCES

The Section Preferences button calls the Cross Section Preferences dialog box of Sheet Manager (see Figure 9–14). The Cross Sections Sheet Options area contains several basic settings for the creation of sections. Sections have an independent horizontal and vertical scale from the profile and the drawing. An invisible grid can overlay the sheet, and you can snap sections to the grid. The Section Sheet Border Spacing area controls the margins around the sheet edges. The Internal Section Spacing area sets the amount of space between each section. The last settings of this group define the horizontal and vertical progression of stations.

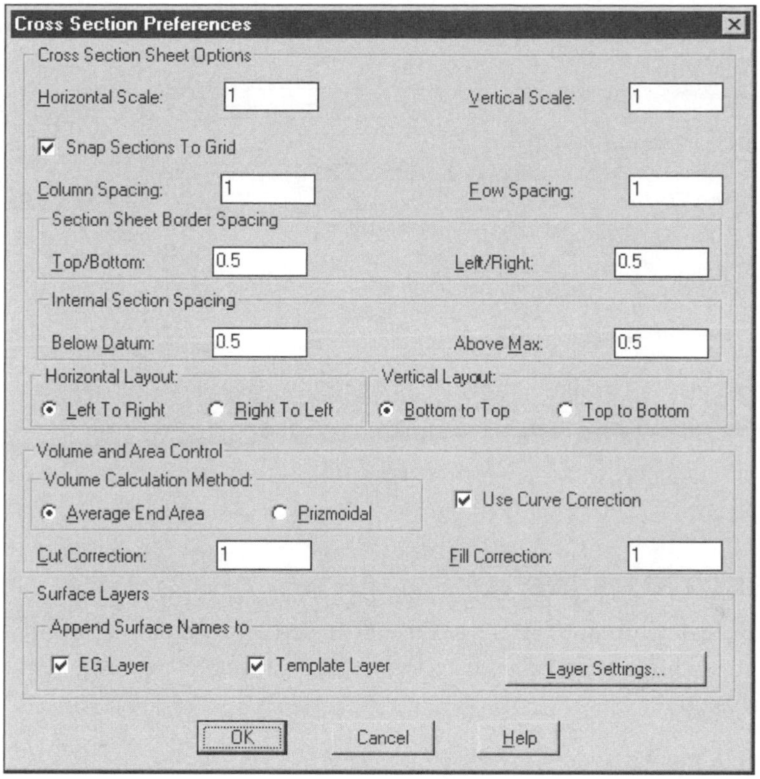

Figure 9–14

The Volume and Area Control settings set the method for calculating the roadway volume and the type of corrections. The default method of volume calculation is average end. The optional method is the prismoidal method. The first correction is for going around curves. As the design travels around a curve, the inside distance between cross sections is less than the outside distance between sections. This toggle compensates for the difference in distances. The help topic for this toggle explains how the volume calculation compensates for the template rounding a curve. The last values are the cut and fill corrections. The cut correction accounts for an increase in volume because of the removal of overburden. The cut correction value is always 1 or greater. The fill correction accounts for a decrease in volume due to the compaction of material. The fill correction value is always 1 or less.

The Surface Layers area of the dialog box allows you to append the name of the project to the layers of the section drawings. Sheet Manager uses all paperspace layers when creating cross sections. The Layer Settings button displays a list of the current layer name for the sections (see Figure 9–15). You can edit and change the names of the layers used by the Sheet Manager cross sections.

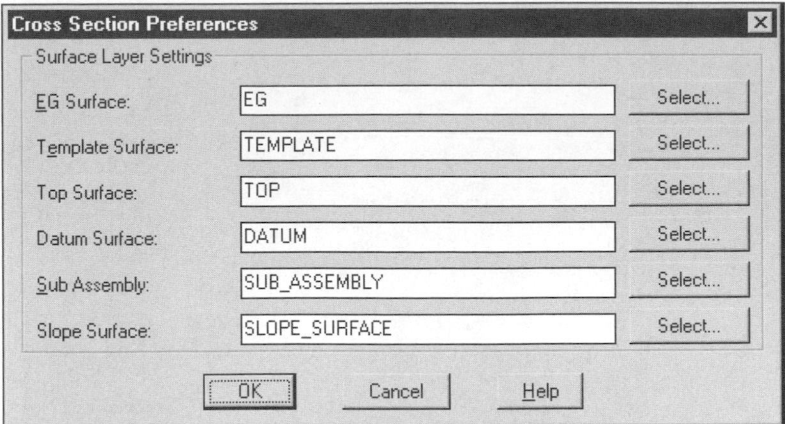

Figure 9–15

CROSS SECTIONS MENU

The Cross Sections menu contains a number of settings that affect templates. These settings control the location of the template path and the two tables, material and pcodes (points codes).

SET TEMPLATE PATH

The Set Template Path routine sets the location of the template library. By default the path is to the templates area of the *Data* folder of Land Development Desktop. The routine displays the Template Path dialog box, which defines the path to the library and the folder of the specific template library (see Figure 9–16). Civil Design installs two libraries, Imperial (adtpl_i) and Metric (adtpl_m).

Template Path
User Preferences Cross Section Templates root path
Root path:
d:\program files\land desktop r2\data\tplates\
Path: adtpl_i Browse...
Template path:
d:\program files\land desktop r2\data\tplates\adtpl_i\
OK Cancel Help

Figure 9–16

MATERIAL CODES TABLE

When defining a template, Civil Design requires you to assign a material to each surface in the template. The prompt for the material is from a list of materials stored with Civil Design. The Edit Material Table routine of the Template cascade displays the Material Table Editor with the materials list (see Figure 9–17). You can edit, add, and delete from the list.

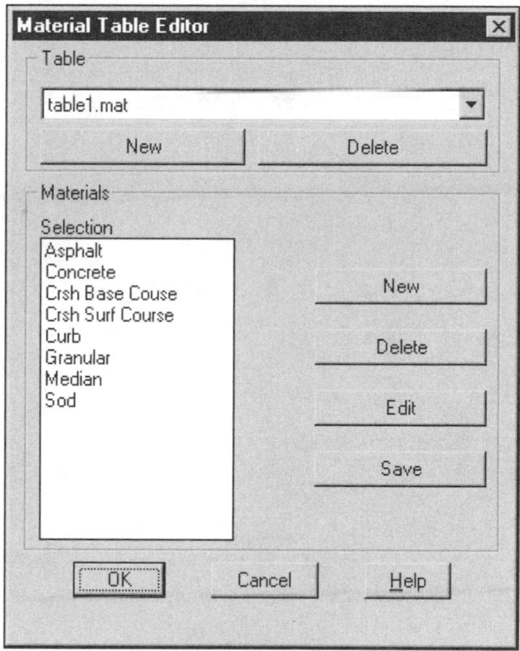

Figure 9–17

PCODES TABLE

The Edit Point Code Table routine of the Templates menu displays the Point Code Table Editor (see Figure 9–18). P(oint) Codes are critical points on a template. Civil Design reserves the first 24 pcodes for its own use. You can add codes after 25. The New button allows you to add new Pcodes to the Pcode list. The delete button deletes existing Pcodes and the Edit buttons allows you to edit the definition of existing Pcodes. The New and Delete buttons at the top of the dialog box allow you to create and delete existing Pcode files.

When you define a template, it has only two pcodes assigned to three of its critical points. One point is the finished-ground-reference point and the two remaining points represent the right and left template connect-out point. You can place additional points on the template by editing the template. When editing the template, you select the critical point on the template and then select the appropriate code from the Pcode dialog box.

When importing points in to a drawing representing the roadway design, you can use the pcode description for the descriptions of the points.

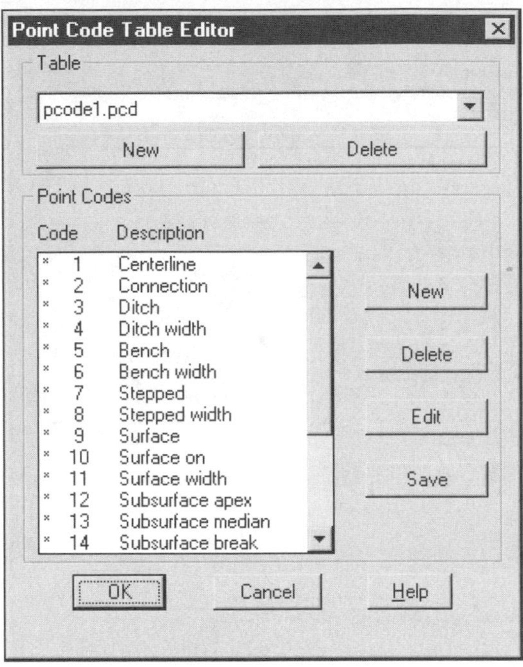

Figure 9–18

EXERCISE

Exercise

After you complete this exercise, you will be able to:

- Know and review the basic cross-section settings

CROSS SECTION SETTINGS

1. If you are not in the Roadway drawing of the Roadway project, open the file with Land Development Desktop.
2. If needed, change to the Civil Design menu. You change the menu by selecting Menu Palettes from the Projects menu, then selecting the Civil Design R2 menu, and then selecting the Load button at the bottom of the Menu Palette Manager (see Figure 9–19).

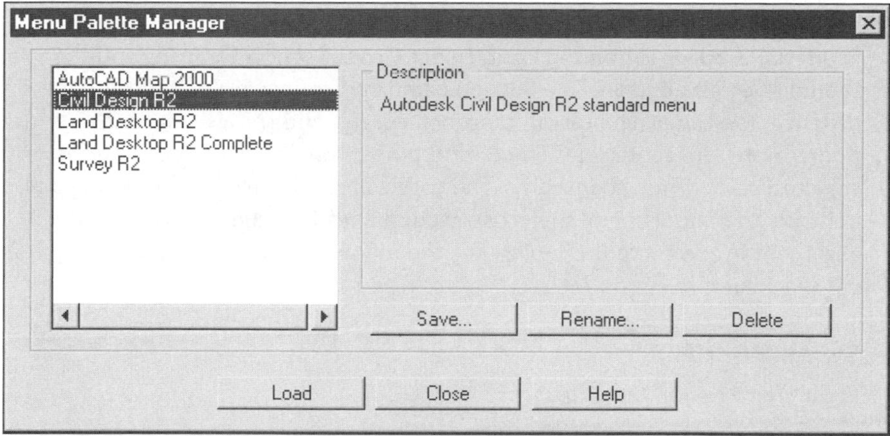

Figure 9–19

3. Select Drawing Settings from the Projects menu to display the Edit Settings dialog box (see Figure 9–20).

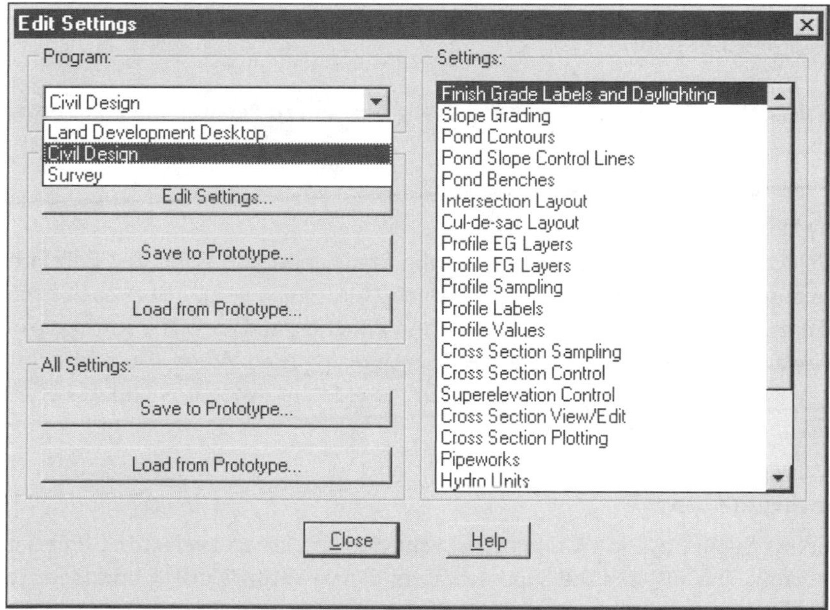

Figure 9–20

4. Select the drop list of products at the top left of the dialog box and change the product to Civil Design.

Cross Section Sampling

5. From the Settings list on the right, select Cross Section Sampling and then select the Edit Settings button. The Section Sampling Settings dialog box sets the offset distance, the sampling interval (tangents, curves, and spirals), and other critical points along the centerline. The critical points can be entered as stations or selected points from the display. The toggle at the bottom of the dialog box controls the importing of the cross-section lines into the plan view of the alignment in the current drawing. Set the following values and use Figure 9–2 to set the values.

   Swath Width:

   Right: 75 Left: 75

   Sample Increments:

   Tangent: 50 Curve: 25 Spiral: 25

   Addition Sample Control:

   Toggle ON all except for Read Sample List

   Sample Lines:

   Import Lines: ON

6. Select the OK button to exit the dialog box and to reenter the Edit Settings dialog box.

Cross Section Control

7. Select Cross Section Control from the Settings list and select the Edit Settings button. This displays the Control Editor (see Figure 9–3). Select each button, Template Control, Ditches, Slopes, and Benches, and view the settings. After reviewing the settings, cancel out of each dialog box. When you have completed a reviewing all of the settings, select OK to exit the dialog box and to return to the Edit Settings dialog box.

Superelevation Control

8. Select Superelevation Control from the Settings list and select the Edit Settings button. This displays the Superelevation Curve Settings dialog box (see Figure 9–8). The values in the dialog box define the superelevation process for curves not designed with speed tables. Each curve along a centerline will require its own set of values. When you are finished reviewing the settings, select OK to exit the dialog box and to return to the Edit Settings dialog box.

Cross Section View/Edit

9. Select Cross Section View/Edit from the Settings list and select the Edit Settings button. This displays the Template View Settings Editor (see Figure 9–9). The editor sets the initial values for the View/Edit (cross sections) routine. The View/ Edit routine is an interactive cross section editor. Use Figure 9–9 as a guide in setting the values. When you have completed reviewing and setting the values, select OK to exit the dialog box and to return to the Edit Settings dialog box.

Cross Section Plotting

10. Select Cross Section Plotting from the Settings list and select the Edit Settings button. This displays the Cross Section Plotting Settings dialog box (see Figure 9–10). This dialog box toggles on and sets the layers for elements in a cross section. Toggle on all of the elements.

Section Layout

11. Select the Section Layout button to display the Section Layout dialog box. This box sets the elevation and offsets values for the cross section. Select the OK button to accept the values and to dismiss the Section Layout dialog box.

Page Layout

12. Select the Page Layout button to display the Page Layout dialog box. This box sets the sheet size, margins, spacing, and placement in the drawing. Select the OK button to accept the current values and again select the OK button to exit the Cross Section Plotting Settings dialog box.

Sheet Manager Settings

13. Scroll to the bottom of the list, select Sheet Manager Settings from the list, and select the Edit Settings button. This displays the Settings dialog box (see Figure 9–13). This dialog box controls the creation of plan, profile, and plan and profile sheets.
14. Select the Section Preferences button at the bottom of the dialog box to view the settings for Sheet Manager cross sections (see Figure 9–14). Select the Layer Settings button to view the layer names used in generating cross sections. Select the OK button to exit the layer settings dialog box and to reenter the Cross Section Preferences dialog box.
15. After reviewing the section settings, select the OK button once to exit the Cross Section Preferences dialog box and to return to the Settings dialog box.
16. Select the OK button to exit the Settings dialog box and reenter the Edit Settings dialog box.

17. Select the Close button to exit the Edit Settings dialog box.
18. Select the Set Template Path routine of the Cross Sections menu. The routine displays a dialog box for creating the template path (see Figure 9–16). The top portion of the dialog box defines the path to the library as installed by Civil Design. The middle portion defines the actual template library folder. The bottom portion displays the entire template path.
19. Select the Edit Material Table routine from the Templates cascade of the Cross Sections menu. The routine displays the current material list (see Figure 9–17). The Material Table Editor allows you to add, edit, create a new table, and delete existing materials. Select the OK button to exit the dialog box.
20. Select the Edit Pcode Table routine from the Templates cascade of the Cross Sections menu. The routine displays the current pcode list (see Figure 9–18). The Point Code Table editor allows you to add, edit, create a new table, and delete existing pcodes. Select the OK button to exit the dialog box.

UNIT 2: CROSS SECTION SAMPLING

The next stage of the design process is viewing the existing ground to the right and left of the centerline. This activity moves to the Cross Sections menu of Civil Design see Figure 9–21). In this stage, the designer creates the cross sections of the design. On each of these cross sections, the design processor places a roadway template. From a connecting point out of the template, a daylight slope searches for an intersection with the existing ground surface. The slope intersects the existing ground surface as either a cut or a fill slope. Civil Design calls this point (the intersection between a slope and the existing ground surface) a daylight or catch point.

Civil Design has three methods to sample for cross-section elevation. These methods are in the Existing Ground cascade of the Cross Sections menu: Sample From Surface, Sample From File, or Edit Sections (see Figure 9–22). By far the easiest method is Sample From Surface. This routine can sample more than one surface when developing the elevations for the cross sections.

If there are multiple surfaces in the project and the surface to sample is not the current surface, run the Set Current Surface routine from the Surfaces cascade of the Profiles menu. If you have toggled on Multiple Surfaces in the Profiles menu, the Sample From Surfaces routine of the Existing Ground cascade of the Cross Sections menu samples the same surfaces found on the surface list from Profiles.

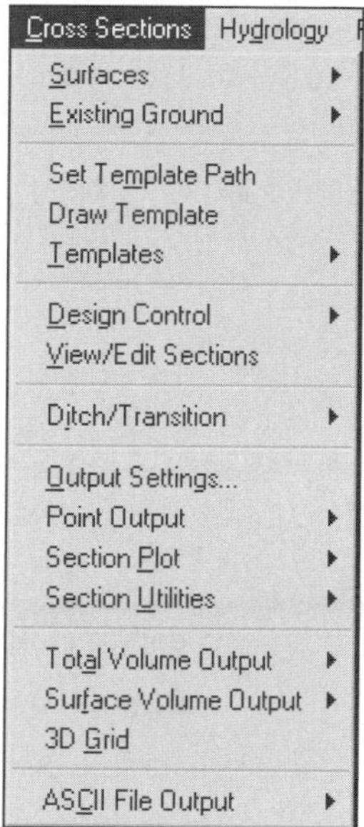

Figure 9–21

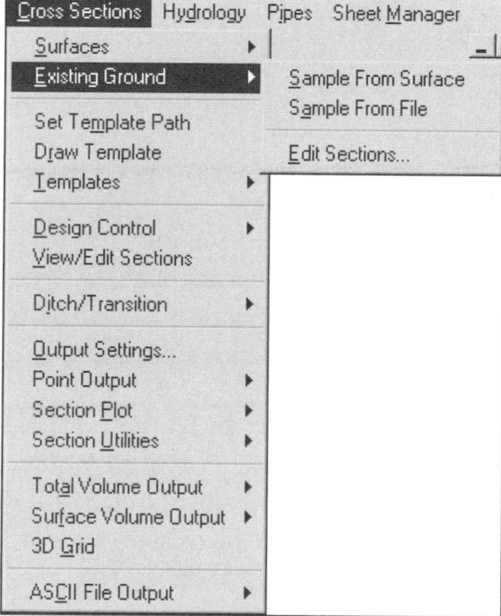

Figure 9–22

The Sample From File method reads a file that lists station, offsets, and elevations for one or more surfaces. A negative offset is an offset to the left of the centerline and a positive offset is to the right of the centerline. After reading the file with the Sample

From File routine, Civil Design considers the surface sampled. The file format is rigid and documented in the Help file. The following is the file format and an example of the resulting file.

```
0.000000                          station number
S Existing                        surface name
-75.0000 720.0000 offset and elevation
-45.0000 718.54                   "      "
4.8594 721.0000                   "      "
25.2905 714.6810 "      "
75.0050 715.2940 "      "
E                                 end of station data
50.000000                         station number
S Existing                        surface name
-75.0000 726.2699
-33.4758 729.8359 offset and elevation
39.9956 728.4365 "      "
40.0000 726.2699 "      "
50.2744 720.9167 "      "
75.9956 713.4365 "      "
E                                 end of station data
```

The last method of sampling is by the Edit Sections method. The process is simply entering offset and elevation values for each station along the roadway. You can enter data for more than one surface within the Section Editor. After you save and exit the editor, Civil Design considers the cross sections sampled.

The next question to settle on is: how far to the right or left of the roadway centerline can the roadway impact? Most times the design daylights to the design surface before the ROW or end of the cross section. If the design does not daylight before the end of the section, how then does the design processor handle this condition? When sampling the surface(s) for sections, the sampling routine uses a swath width (a user-defined offset distance from the centerline) to determine width of a cross section. The swath value sets the farthest left and right offset distance a roadway design can have. If the design processor calculates a daylight point outside the swath width, the routine forces the daylight point to the intersection of the top surface and the end of the cross section. Civil Design calls this slope pinning. In this situation, you have no control over the resulting daylight slope.

The second method is holding a right-of-way (ROW) offset instead of the sample swath width. If the right-of-way is 33 feet, the design cannot go any farther than 33 feet from the centerline of the road. The design processor ignores any solutions before the ROW and daylights only to the ROW intersection with the design surface.

Many times, however, the ROW is not a constant offset from the centerline. To handle ROWs whose distance varies from the centerline, Civil Design has a right and left ROW horizontal alignment. If you assign the alignments to the design processor (in the Edit design Control dialog box), the design processor will account for the horizontal location of the ROW alignments when calculating the cross sections.

Exercise

EXERCISE

After you complete this exercise, you will be able to:

- Sample the elevations for cross sections

SAMPLING ELEVATIONS

1. Pan the drawing so you can see the horizontal roadway centerline. Create a named view, Roadway, with the Named View routine of the View menu.
2. Run the Sample From Surface routine in the Existing Ground cascade of the Cross Sections menu. The routine displays the Section Sampling Settings dialog box.
3. If the routine prompts for an alignment, either select a segment of the centerline in the display or click the right mouse button and select the alignment (Bypass) from the list to set the current alignment.
4. If the routine prompts you to select a surface to sample, select the existing surface from the surface list.
5. Set the following values and toggles in the Section Sampling Settings dialog box, if they are not set already. Use Figure 9–2 as a guide for setting the values.

   Swath width: 75 Feet Left and Right

   Sample Tangents: 50

   Curves and Spirals: 25

   PCs/PTs: ON

   Alignment Start: ON

   Alignment End: ON

   Save Sample List: ON

   Add Specific stations: ON

Exit the dialog box by selecting the OK button. Press ENTER to accept the beginning and ending stations.

```
Command:
Select alignment:
Alignment Name: Bypass  Number: 1    Descr: Bypass Prelim Design of 12/00
Starting Station: 0.00  Ending Station: 2446.88
Beginning station <0>:
Ending station <2446.88>:
Enter the Specific stations (do not use 1+00; use the decimal equivalent).
Enter critical station (or Point): 275.50
Enter critical station (or Point): 540.75
Enter critical station (or Point): 2015.25
Enter critical station (or Point): 2305.10
```

Select three points along the roadway centerline.

Toggle to Point mode by entering the letter p.

```
Enter critical station (or Point): p
Select critical station point (or Station): (select a position along
   centerline)
Select critical station point (or Station): (select a position along
   centerline)
Select critical station point (or Station): (select a position along
   centerline)
Select critical station point (or Station): (press ENTER to exit)
```

The routine processes the stations and samples the elevations along the roadway.

```
Sampling terrain data from the surface.
Scanning Cross section Input.
Current surface: existing
Group: BYPASS Section: 1796.882340
Starting station: 0+00         Ending station: 24+46.88
You have sampled sections for 1796.88 feet of alignment.
```

UNIT 3: THE TEMPLATE

After sampling the cross-section data, the final preliminary to roadway design processing is drawing and defining a road template. The Templates cascade of the Cross

Sections menu contains all of the tools for designing, defining, and editing templates (see Figure 9–23).

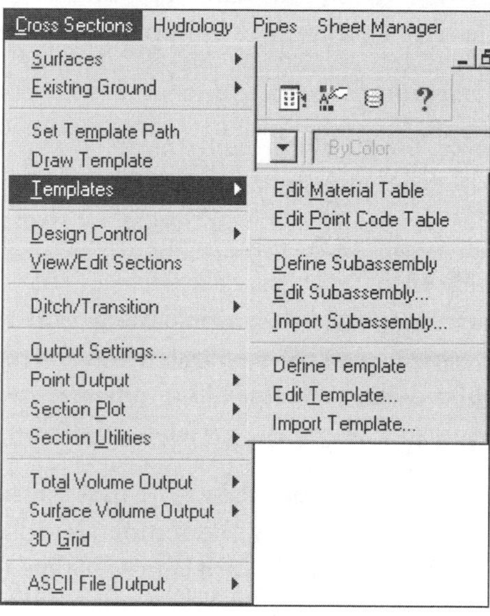

Figure 9–23

The Civil Design template is simply a set of surfaces (A) or a template with central surfaces and subassemblies (B) (see Figure 9–24). Subassemblies represent peripheral template elements, shoulders and curbs. You may have templates with the same central surfaces, but with different shoulder and curb specifications due to job conditions. The subassembly library quickly allows you to define new templates with minimum drafting. Roadway shoulders can be subassemblies that change depending upon the template being in either a cut or fill situation.

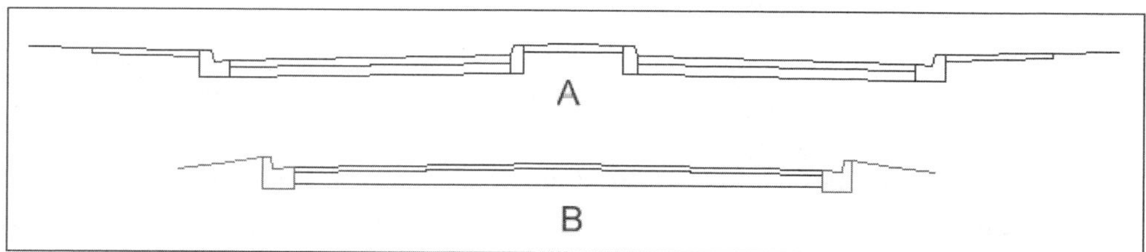

Figure 9–24

The template is a set of surfaces. You create template surfaces by one of two methods. The first method involves drawing the template in place. The second method involves drawing, moving, or selecting additional surfaces while editing the template.

A template can be symmetrical or asymmetrical. In a symmetrical template, the right side of the template is a mirror image of the left. If the template is symmetrical, you need to draw only the left side of the template. The defining process will mirror the template to the right side. An asymmetrical template requires the drawing of both the left and right sides of the template since they are different from each other.

If the template is symmetrical, the inner surfaces, surfaces closing across the centerline, are open and the peripheral surfaces, curb, sidewalks, and so on, are closed polylines. The Define Template routine of the Templates cascade mirrors the left side (the drawn side) of the template to the right side closing the central surfaces. For example, when you draw the bituminous or granular base surfaces that cross the centerline of a symmetrical template, they are open in the initial left portion of the template drawing. The Define Template routine closes the surfaces when mirroring the template. The curb, sidewalk, and shoulder surfaces are closed. Each of the peripheral surfaces close on their respective sides and not across the centerline to complete closure.

There are three methods of drawing a template. The first method is using the Desktop POLYLINE command. The command draws the template surface polylines with more flexibility than Civil Design's Draw Template routine. The greatest disadvantage of the POLYLINE command of the Desktop is that it is unable to draw a polyline with a grade or slope. You have to calculate values for the endpoints before drafting the polyline segments or you have to create controlling geometry before drafting the template. So even though the POLYLINE command is more flexible, it does not use the terms or calculations of a roadway cross section.

The second option is drawing lines representing the different surfaces of the template and then joining them into polylines with the Pedit routine of Desktop. This allows for a sculpted template, but you must remember that the polylines need to travel in a counterclockwise direction for template definition. Also, the polylines of one surface can lie over the top of the polylines of another surface in the template. So you must have lines over lines to have enough for the final template polylines.

The last option is using the Draw Template routine of the Cross Sections menu. The Draw Template routine uses grades and slopes to calculate the location of vertices along a polyline. A negative slope or grade (-2:1 or -50%) goes down and a positive slope or grade (2:1 or 50%) goes up. The slope or grade goes up or down relative to a starting point of the polyline segment. Civil Design refers to the distance a slope or grade travels as an offset or horizontal distance. There are other options for drawing surface segments. Those options are relative distance and selecting points.

The convention of left as negative and right as positive is a part of the Draw Template routine. When describing segments that go left, the offset is negative. When describing segments that go to the right, the offset is positive. When changing elevations, a negative elevation change or negative slope goes down in elevation. When changing elevations, a positive elevation change or positive slope goes up in elevation. These are standard roadway conventions.

Template and Subassembly Rules

- Each element is a 2D polyline.
- Draw all elements counterclockwise.
- If the template is symmetrical, central elements (surfaces) are open and peripheral elements are closed.
- If the template is symmetrical, draw only the left side of the template. The definition process will produce the right side.
- If the template is asymmetrical, all elements are closed 2D polylines.
- If the template is asymmetrical, draw the entire template as counterclockwise surfaces and elements.

During the Define Template routine, the routine prompts for a material for each surface in the template. The Define Template routine also prompts you to identify a number of critical points on a template. The first critical point is the connect-out point. A connect-out point can have two meanings. The first type of connect-out point is where the slope to find daylight connects to the end of template. The second type of connect-out point is where subassemblies connect to the central surfaces of the template or where a shoulder subassembly attaches to a curb subassembly.

The second critical point is a set of datum points used in calculating volumes. The next critical point is a set of points defining the top of the template. The template defining routine does not create the Top-of-Template points. The definition of the template top occurs in the editing of a template. The final critical point is the finish-ground-reference-point. This point is where the template attaches to the vertical alignment.

When you define a template, the finish-ground-reference point and the connect-out points are automatically assigned point codes. You can add other point codes to the template that indicate critical points you want to import into the drawing.

DEFINING A TEMPLATE

The Define Template routine of the Templates cascade creates a snapshot representation file of the template on the screen. This external template representation is a file in a folder pointed to by the template path. The template becomes a part of the template library. All users and projects can see and use templates from the template library.

There are two types of templates. The first template uses subassemblies. This template starts as central surfaces bookended by subassemblies. The second type of template uses no subassemblies and can be complex in construction and use. In the second type of template, all template areas are 2D polylines.

The Define Template routine prompts the user for four critical points on the template: the Final Grade Reference Point (the point on the template that connects to the Vertical Centerline Alignment), the connect-in and connect-out points, and the datum points. The Connect In and Out points setting tells the Define Template routine how to attach the subassembly to the central surfaces of the template and/or to attach a subassembly to other subassemblies. The outermost connect-out point is used to establish the start of the ditch or the starting point for the daylight slope.

The datum line is the limit of excavation or the limit of fill to create the foundation of the template.

DEFINING A SUBASSEMBLY

The Define Subassembly routine of the Templates cascade (see Figure 9–23) creates a snapshot representation file of the subassembly on the screen. This external representation is in the folder pointed to by the template path. The subassembly becomes a part of the template library. All users can see and use the subassembly.

The Define Subassembly routine prompts the user for three critical points on the subassembly: the connect-in and connect-out points and the datum points. The Connect In and Out points tell the Define Template routine how to attach the subassembly to the central surfaces of the template and/or to attach a subassembly to other subassemblies. The outermost connect point is used to establish the start of the ditch or the starting point for the daylight slope.

The datum line is the limit of excavation or the limit of fill to create the foundation of the template.

EDITING A TEMPLATE

The process of defining a template does not completely finish the template definition. The template needs a top-of-template, but the defining process does not create one. To be able to create a top-of-template, you must edit the template. The Edit Template routine of the Templates cascade adds the top-of-template definition as well as other critical items to a template. These include points (pcodes), transition control points, and superelevation zones.

When you start the Edit Template routine, the routine displays a dialog box listing the currently defined templates. You select the template you want to edit, click on the

OK button, and then routine prompts you for an insertion point. The insertion point places a copy of the template into the drawing for you to edit.

EDIT SURFACE

The Edit surface (Edsrf) option of the Edit Template routine allows you to add a surface, delete a surface, modify a surface, change the material assigned to a surface, and to edit the pcodes on a template. The following is the command sequence for the Edit surface option (Edsrf) and Addsurf option. The Modify option can delete or add vertices to an existing surface within the template.

```
Command:
(After selecting the template from a dialog box)
Pick insertion point:
Edsrf/SAve/eXit/ASsembly/Display/SRfcon/Redraw <eXit>: e
Addsurf/Delsurf/Modify/MName/Points/Redraw/eXit <eXit>: a
Draw/Move/Select <Select>:
```

Add Surface (Addsurf)

The Add Surface option adds a new surface to an existing template by drawing the new surface within the context of the edited template, by moving an existing polyline into the template, or by selecting a polyline already present in the edited template.

Modify

The Modify option allows you to edit the 2D polyline defining a portion of the template. The routine is similar to the PEDIT command of the Desktop.

MName

The MName option allows you to change the material assigned to a surface within the template or subassembly.

Points (Pcodes)

The Points option of the Edit Template routine list allows you to add point codes to a template. By default a template has two critical points identified, the final grade centerline and the outer connect-out point. The point codes are stored in an external file that displays when using this option (see Figure 9–18). Civil Design reserves the first 24 slots. You can add your own point description to the list after the reserved codes. When importing points, the routine has an option to import points based upon the point code values.

SRFCON

The SRfcon option of Edit Template allows you to redefine critical points on the template element. The routine edits the Datum, Connect points, and Top-of-surface.

The SRfcon option also has a DIsplay option that allows you to view the current element definitions. This is very handy when you want to view the results of your defining and/or editing process. The following sequence shows the entry into and options inside of the SRfcon routine. To select an option, type a letter(s) code and press ENTER, and the routine will change the prompts. This can be a confusing process, so be careful while editing the template.

```
Command:
(After selecting the template from a dialog box)
Pick insertion point:
Edsrf/SAve/eXit/ASsembly/Display/SRfcon/Redraw <eXit>: sr
Connect/Datum/Redraw/Super/Topsurf/TRansition/eXit <eXit>:
```

ASSEMBLY

The Assembly option displays the subassembly dialog box. This allows you to change the associated right and left curbs and cut and fill shoulders on a template.

```
Command:
(After selecting the template from a dialog box)
Pick insertion point:
Edsrf/SAve/eXit/ASsembly/Display/SRfcon/Redraw <eXit>: as
```

DISPLAY

The Display option allows you to view the critical definitions in a template. The option displays the datum, connect points, pcodes, top surface definition, transition control points, and transition types. Each item is highlighted in a different color and line type.

```
Command:
(After selecting the template from a dialog box)
Pick insertion point:
Edsrf/SAve/eXit/ASsembly/Display/SRfcon/Redraw <eXit>: d
Display [Datum/Connect/Points/Super/SHoulder/Topsurf/TRansition/eXit/Redraw/
   TType]
<eXit>:
```

EDITING A SUBASSEMBLY

The process of defining a subassembly does not completely finish the subassembly definition. The subassembly needs a top-of-subassembly, but the defining process does not create one. To be able to create a top-of-subassembly, you must edit the subassembly. The Edit Subassembly routine of the Templates cascade of the Cross Sections menu also allows you to redefine the datum and connect points of the subassembly.

SRFCON

The SRfcon option of the Edit Subassembly routine allows you to redefine critical points on the element. The routine edits the Datum, Connect points, and Topsurface. The SRfcon option also has a DIsplay option that allows you to view the current element definitions. This is handy when you want to view the results of your defining and/or editing process. The following sequence shows the entry into and options within the SRfcon routine. To select an option, type a letter(s) code and press ENTER, and the routine will change the prompts. This can be a confusing process, so be careful while editing the template.

```
Command:
(After selecting the template from a dialog box)
Pick insertion point:
Delete/Insert/Next/SAve/Previous/SRfcon/Move/Redraw/eXit <Next>: sr
Edit [Datum/Connect/Topsurf/eXit/DIsplay/Redraw] <eXit>: t
Pick top surface points (left to right): (select a point)
Pick top surface points (left to right): (select a point)
Pick top surface points (left to right): (select a point)
Pick top surface points (left to right): (select a point)
Pick top surface points (left to right): (press ENTER to exit)
Display [Datum/Connect/Topsurf/eXit/Redraw] <eXit>: (press ENTER)
Edit [Datum/Connect/Topsurf/eXit/DIsplay/Redraw] <eXit>: (press ENTER)
Delete/Insert/Next/SAve/Previous/SRfcon/Move/Redraw/eXit <Next>: x
Save subassembly [Yes/No] <Yes>: (press ENTER)
Subassembly name <curb612>: (press ENTER)
Subassembly exists. Overwrite [Yes/No]: y
```

HORIZONTAL AND VERTICAL MANIPULATION OF THE TEMPLATE

The Civil Design template is flexible. You can manipulate the template by using transitional alignments (additional horizontal and/or vertical alignments). Transitional alignments attached to template control points and affect the motion of the template. To help control the transition alignments, Civil Design assigns the control points characteristics. These characteristics determine the behavior of the template as it responds to the motion of the control points. The design processor of the Edit Design Control knows the location of these control points at each cross section whether there is an attached alignment or not.

Initially, you may have to run the template along the centerline without any influence from horizontal or vertical alignments. If you define these critical transition points on the template, but do not assign them any alignments, the design processor will calculate their location for the entire length of the roadway. Civil Design provides tools to import these points as lines representing their path into plan and/or profile view (see Figure 9–25).

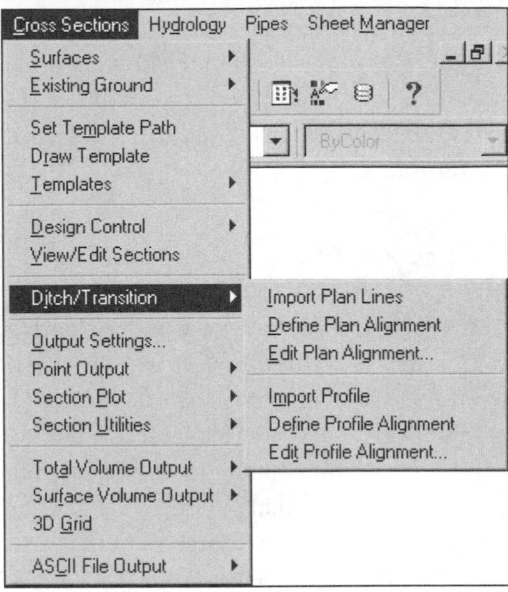

Figure 9–25

Some of the critical points of the template design are automatically available as Desktop objects; for example, the ditch base. First you process the design and them import the path of the point into the drawing as horizontal (plan) or vertical (profile) lines representing the ditch base. After editing the line work to better represent the final design, you define the edited line work as transitional alignments. Finally, you attach the alignments to the template in the Edit Design Control dialog box to produce the desired design behavior. There can be eight (8) horizontal and vertical control points on the template and two (2) additional control points for ditch control.

If a control point controls a zone on a template, you have to define a transition zone. For example, you want a median curb to disappear as the design transitions from a median divided highway into a two-lane road. The template has to have a transition zone containing the curb. Thus, a zone is a pair of related points on the template. One point is the control point and the other point defines the farthest point of the zone.

You can assign alignments to the control points in the Edit Design Control dialog box. Section 10 of this book covers transitioning and transitional alignments.

Exercise

After you complete this exercise, you will be able to:

- Draft and define roadway subassemblies
- Draft and define a roadway template
- Edit roadway subassemblies and a template

TEMPLATE DRAWING SETUP

1. Start a new drawing. Save the changes to Roadway before exiting to the New Drawing Project Based dialog box (see Figure 9–26). Name the drawing CDTPL.

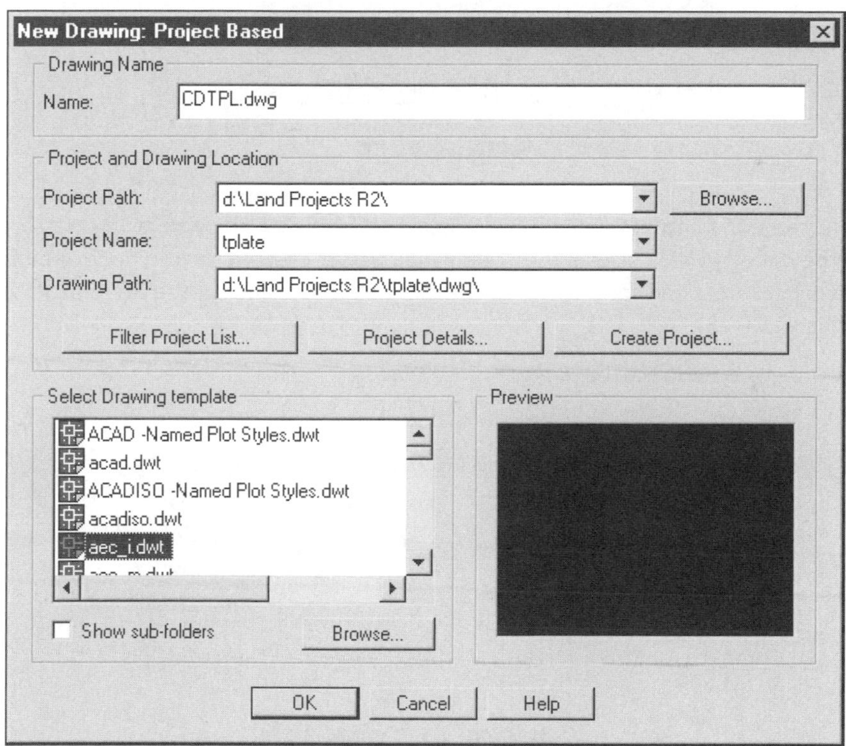

Figure 9–26

2. Assign the drawing to a dummy project named tplate (see Figure 9–27).
3. Accept the default point database settings by selecting the OK button.

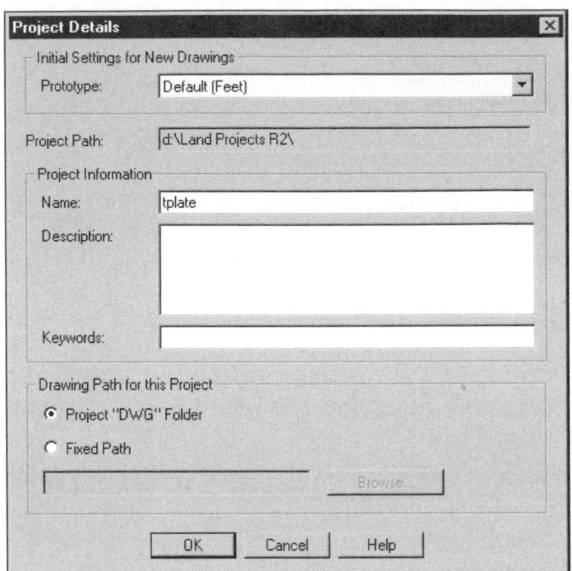

Figure 9–27

4. In the Load Settings dialog box, select the Next button twice at the bottom of the dialog box to display the scale panel. Select custom for the horizontal and vertical scale and set both to 1. Set the following values in the Scale panel (see Figure 9–28).

```
Horizontal scale: Custom   Value: 1
Vertical scale: Custom   Value: 1
```

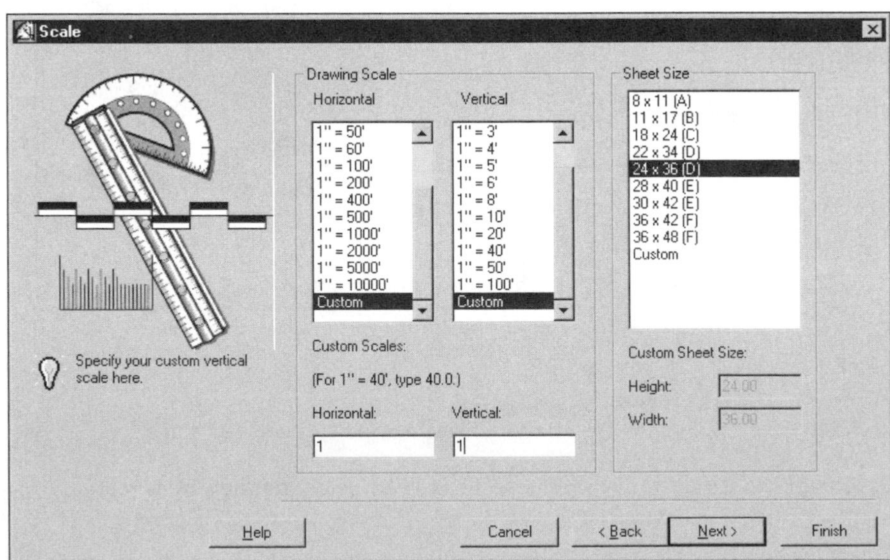

Figure 9–28

5. Select the Next button until the dialog box displays the Text Style panel. Select the text style Leroy, then the Load button, and finally set L100 as the current style (see Figure 9–29).

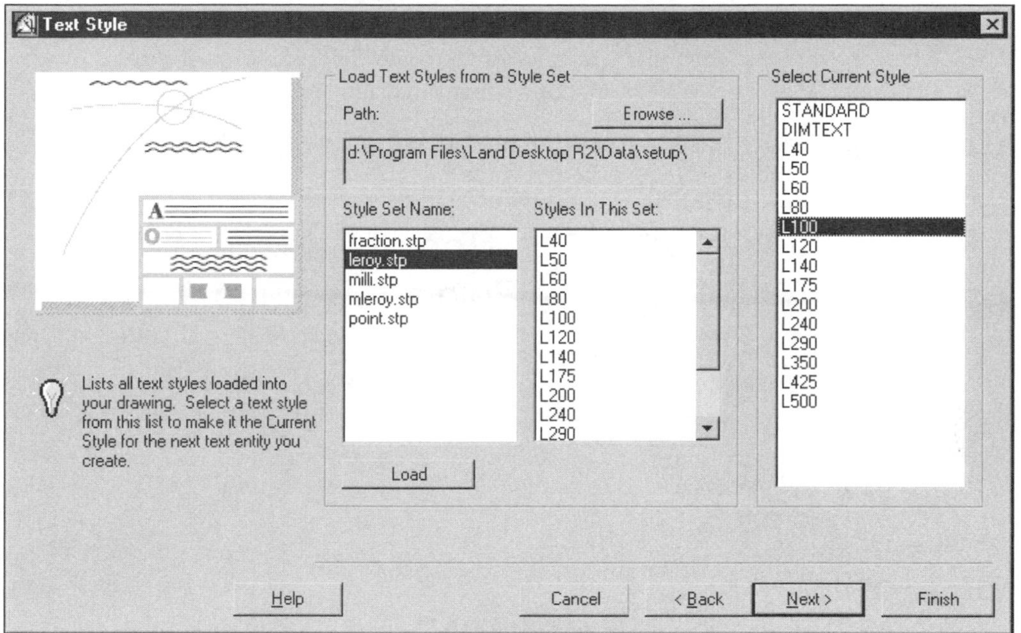

Figure 9–29

6. Select the Finish button to exit the Load Settings dialog box.

7. Select the OK button to dismiss the Settings Review dialog box.

Drawing Subassemblies

The peripheral portions of the template are subassemblies, curbs and shoulders. The definitions of different curbs and shoulders create a library of subassemblies. During the template defining process, you can add subassemblies to the central surfaces. All templates do not have to have subassemblies.

Remember, when drawing the 2D polylines for the template, draw all of them counter clockwise. When drawing the subassembly polyline segments, the Draw Template routine refers to segments that go to the right or left as offsets. A positive offset draws a polyline segment to the right and a negative offset draws a polyline segment to the left.

When describing vertical changes in the subassembly without a slope or grade, the Draw Template routine describes the change as a change of elevation. A change of 0.5 (positive) indicates that the next endpoint is one-half foot higher than the current position. A change of -0.5 (negative) indicates that the next endpoint is one-half foot lower than the current position.

Drawing the Curb Subassembly

The curb has a flow area 12 inches wide with a barrier width and height of 6 inches. The curb is a closed object (see Figure 9–30).

1. Set a running object snap of endpoint and intersection by clicking the right mouse button while the pointer is over the osnap toggle. Select Settings and set the two object snaps. Select the OK button to exit the dialog box.
2. Start the Draw Template routine of the Cross Sections menu. Select a point on the screen and use Figure 9–30 as a guide.

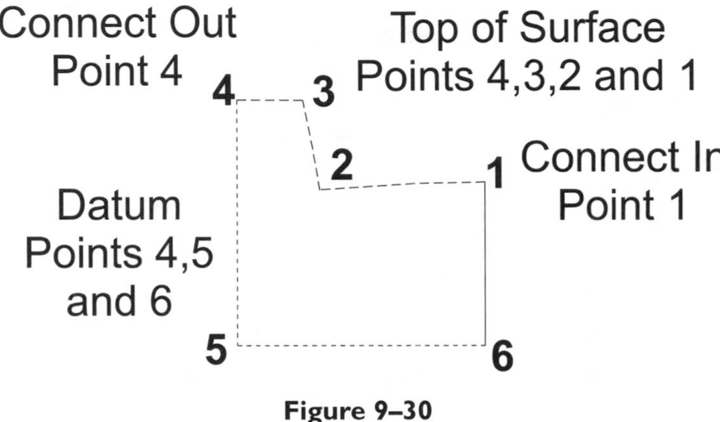

Figure 9–30


```
Starting point: (select a point on the screen)
```

All the curb endpoints are a relative distance from this starting point. The last segment of the curb closes to the first point.

If the routine prompts for something other than relative mode, toggle into Relative mode by typing the letter **r**.

```
Select point (Relative/Grade/Slope/Close/Undo/eXit): r
Change in Offset (Grade/Slope/Close/Points/Undo/eXit): -1
Change in elev: -0.05
Change in Offset (Grade/Slope/Close/Points/Undo/eXit): -0.1
Change in elev: .55
Change in Offset (Grade/Slope/Close/Points/Undo/eXit): -.4
Change in elev: 0
Change in Offset (Grade/Slope/Close/Points/Undo/eXit): 0
Change in elev: -1.5
Change in Offset (Grade/Slope/Close/Points/Undo/eXit): 1.5
Change in elevation: 0
```

The last endpoint is the closing of the curb polyline. The letter C, close, toggles on the Close option. The Draw Template routine automatically exits when you use the Close option.

```
Change in Offset (Grade/Slope/Close/Points/Undo/eXit): c
Starting Point:
```

Drawing the Shoulder Subassembly

The shoulder is 4 feet wide with an initial slope of -6:1 for 2 feet and -10:1 for 2 feet. The slopes are to draw water away from the back of curb (see Figure 9–31).

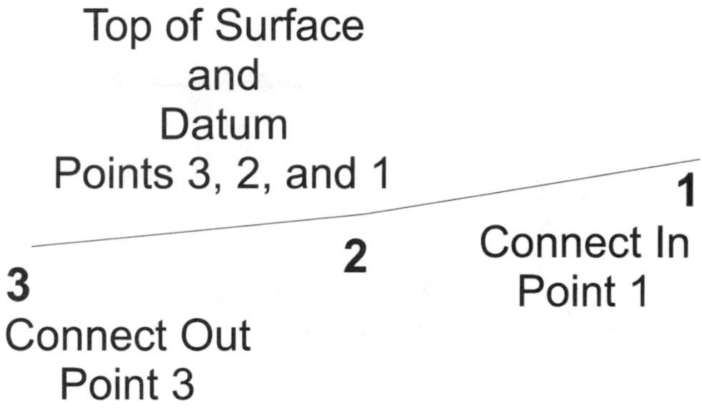

Figure 9–31

1. While still in the Draw Template routine, draw the shoulder subassembly.

```
Starting Point: (select a point on the away from the curb screen)
```

If the routine prompts for something other than slope mode, type the letter **s** to toggle the routine into Slope mode.

```
Change in Offset (Grade/Slope/Close/Points/Undo/eXit): s
Slope (3 for 3:1) (Relative/Grade/Points/Close/Undo/eXit): -6
Change in offset: -2
Slope (3 for 3:1) (Relative/Grade/Points/Close/Undo/eXit): -10
Change in offset: -2
Slope (3 for 3:1) (Relative/Grade/Points/Close/Undo/eXit): x
Starting point: (press ENTER to exit)
```

The subassembly polylines do not have to be closed objects.

Critical Points on a Subassembly

A subassembly has three critical points (see Figure 9–32). The first critical point is the connect-in point. The connect-in point attaches to the connect-out point of the central surfaces of

EXERCISE

the template or the connect-out point of an adjacent subassembly. Depending upon the situation, the connect-out point of a subassembly functions in one of two ways. If the subassembly is the innermost subassembly, the connect-out point is the connect-in point of the next subassembly. If the subassembly is the outermost subassembly, the connect-out point connects the template to the daylight slope or ditch design for the roadway. For example, the connect-in point of the curb connects to the central surfaces and the connect-out point of the curb connects to the connect-in point of the shoulder.

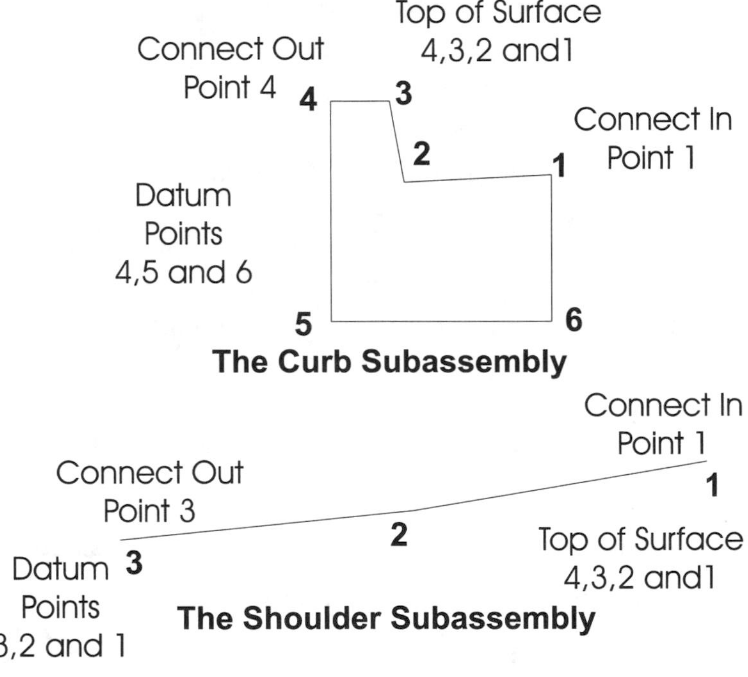

Figure 9–32

After the subassembly connection points, the next important set of points is the datum points. These points are a part of the volume calculations of the subassembly. Identify the points from left to right along the edge of the subassembly.

The final set of critical points for a subassembly is the top. The top of a subassembly, whether open or closed, is a straight line between the connect-in and connect-out points. If you want something other than a straight line definition, redefine the top definition by editing the subassembly with the SRfcon (Surface control) tool. This step will be done later in this exercise.

Defining the Subassemblies

Refer to Figures 9–30 and 9–31.

1. Run the Define Subassembly routine in the Templates cascade of the Cross Sections menu. Define the curb and shoulder subassemblies. The material assignments are from the Materials Table Editor dialog box.

Defining the Curb

The routine prompts:

```
Pick current surface connect point in: (select upper right point of curb -
  vertex 1)
(The material dialog box displays) (select Curb from the dialog box)
Pick entity (must be a 2D polyline): (select curb)
Pick current surface connect point out: (select upper left point of curb
  vertex 4)
Pick current surface datum points: endp (select vertex 4)
Pick current surface datum points: endp (select vertex 5)
Pick current surface datum points: endp (select vertex 6)
Pick current surface datum points: (press ENTER)
Save subassembly (No/<Yes>): (press ENTER)
Enter subassembly name: curb612
Define another subassembly (No/<Yes>): (press ENTER)
```

Defining the Shoulder

```
Pick current surface connect point in: (select right endpoint of shoulder -
  vertex 1)
(The material dialog box displays) (select Sod from the dialog box)
Pick entity (must be a 2D polyline): (select shoulder)
Pick current surface connect point out: (select left endpoint of shoulder -
  vertex 3)
Pick current surface datum points: endp (select vertex 3)
Pick current surface datum points: endp (select vertex 2)
Pick current surface datum points: endp (select vertex 1)
Pick current surface datum points: (press ENTER)
Save subassembly (No/<Yes>):
Enter subassembly name: shdr
Define another subassembly (No/<Yes>):n
Command:
```

Editing Subassemblies: Adding Top of Surface

If you want to have a top of surface that is not a straight line between the connect-out and connect-in points, you must edit each subassembly. The Surface Control option (Srfcon) of Edit Subassembly changes the top-of-surface definition. Even though the template has a default top-of-surface (straight line), the Display option of Edit Template may not show any top-of-surface definition.

EXERCISE

1. Make sure that the running osnap is endpoint and intersection.
2. Run the Edit Subassembly routine of the Templates cascade of the Cross Sections menu. Select the curb subassembly. Use Figure 9–30 as a guide.

```
Pick insertion point: (select a point in the editor)
```

The routine shows the connect-in point (X) and the datum of the subassembly (purple dashed line).

```
Delete/Insert/Next/Save/Previous/SRfcon/Move/Redraw/eXit <Next>: sr
Edit (Datum/Connect/Topsurf/eXit/DIsplay/Redraw) <eXit>: T
Pick top surface points (left to right): endpoint (vertex 4)
Pick top surface points (left to right): endpoint (vertex 3)
Pick top surface points (left to right): endpoint (vertex 2)
Pick top surface points (left to right): endpoint (vertex 1)
Pick top surface points (left to right): (press ENTER)
Edit (Datum/Connect/Topsurf/eXit/DIsplay/Redraw) <eXit>: x
Delete/Insert/Next/Save/Previous/SRfcon/Move/Redraw/eXit <Next>: x
Save subassembly (Yes/NO) <Yes>: (press ENTER)
Subassembly name <curb612>: (press ENTER)
Subassembly exists. Overwrite? (No/Yes): Y (press ENTER)
Command:
```


This redefines the top of the curb subassembly.

Redefine the top of the shoulder subassembly. Refer to Figure 9–31.

2. Again, run the Edit Subassembly routine. Select the shoulder subassembly.

```
Pick insertion point: (select a point in the editor)
Delete/Insert/Next/Save/Previous/SRfcon/Move/Redraw/eXit <Next>: sr
Edit (Datum/Connect/Topsurf/eXit/DIsplay/Redraw) <eXit>: T
Pick top surface points (left to right): endpoint (vertex 3)
Pick top surface points (left to right): endpoint (vertex 2)
Pick top surface points (left to right): endpoint (vertex 1)
Pick top surface points (left to right):(press ENTER)
Edit (Datum/Connect/Topsurf/eXit/DIsplay/Redraw) <eXit>: x
Delete/Insert/Next/Save/Previous/SRfcon/Move/Redraw/eXit <Next>: x
Save subassembly (Yes/NO) <Yes>: (press ENTER)
Subassembly name <shoulder>: (press ENTER)
Subassembly exists. Overwrite? (No/Yes): Y (press ENTER)
Command:
```

This redefines the top of the shoulder subassembly.

This process of defining and editing to define a top-of-surface applies to every subassembly and template you define. The defining process does not create a top surface for a template. You must edit the template and the subassembly to define a top surface.

Drawing a Template

When drawing a symmetrical template, draw only the left side of the template (see Figure 9–33). The defining process mirrors the left side to the right to create the remaining half of the template. As a result, the core surfaces are open surfaces that the mirroring process closes. The curb and shoulders surrounding the central surfaces of the template are added to the template definition when defining the template, if using subassemblies.

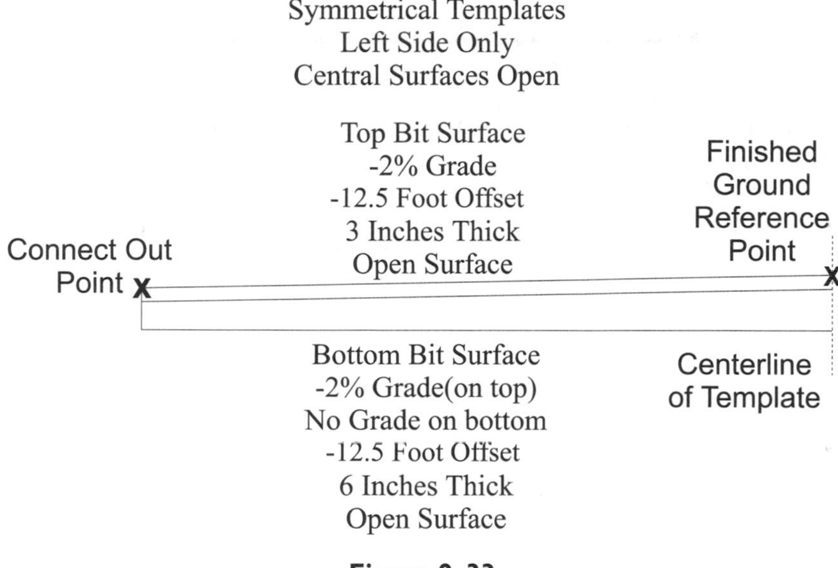

Figure 9–33

When describing vertical motion in the template without a slope or grade, the Draw Template routine describes the motion as a change of elevation. A change of 0.5 (positive) indicates that the next endpoint is one-half foot higher than the current position. A change of -0.5 (negative) indicates that the next endpoint is one-half foot lower than the current position.

Drawing Template Surfaces

The template for the current exercise has two (2) central surfaces: Top Bit and Base Bit (see Figure 9–33). The Top Bit surface is 3 inches thick and bottom bit is 6 inches thick.

Top Bit Surface

1. While still in the template drawing, select the Draw Template routine from the Cross Sections menu.

The routine prompts for a starting point. Select a point in the center of the screen.

```
Starting point: (select a point)
Change in Offset (Grade/Slope/Close/Points/Undo/eXit): g
```

The routine responds with a list of options about how to get to the next point. If the prompt is not for grade, type the letter **g** to toggle into Grade mode.

The next point is 12.5 feet to the left at a grade of -2%.

```
Grade (%) (Relative/Slope/Points/Close/Undo/eXit): -2
Change in offset: -12.5
```

The next endpoint is 3 inches straight down. A grade or a slope can describe this movement, but with large numbers. The best option for moving straight up, down, right, or left is relative.

Toggle into Relative mode by typing the letter **r**. The change of offset is 0 and the change of elevation is -.25 of a foot.

```
Grade (%) (Relative/Slope/Points/Close/Undo/eXit): r
Change in Offset (Grade/Slope/Close/Points/Undo/eXit): 0
Change in offset: -.25
```

The return to the center of the roadway completes the surface segment. Remember, this is a central surface so it is an open surface. The Define routine will close the surface when the routine mirrors the surface to the right side. The return to the centerline is by a 2% grade for 12.5 feet to the right (a positive offset). Toggle on Grades and enter the following:

```
Change in Offset (Grade/Slope/Close/Points/Undo/eXit): g
Grade (%) (Relative/Slope/Points/Close/Undo/eXit): -2
Change in offset: -12.5
Grade (%) (Relative/Slope/Points/Close/Undo/eXit): (press ENTER)
Starting point:
```

After you press ENTER, the routine prompts for a new starting point.

If you make mistakes, there is an Undo option within the command. The Undo removes the last segment, but keeps the Draw Template routine running. If you inadvertently exit the routine, you can start over or start where the routine left off. Remember to join the pline segments together with the PEDIT command to make them a single 2D polyline.

Bottom Bit Surface

Even though the bottom of the Top Bit surface is the top of the Bottom Bit surface, the template must have the Bottom Bit surface as its own polyline. The bottom segment of the Top Bit surface provides a starting and a second point for the top of the second surface. You

EXERCISE

must draw the Bottom Bit polyline over the Top Bit polyline segment. So the first two selections for the second surface are by Desktop object snaps snapping to vertices on an existing 2D polyline (see Figure 9–34).

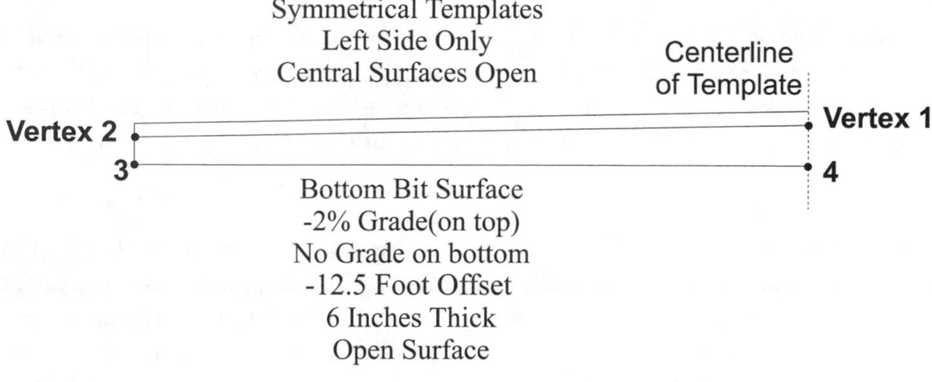

Figure 9–34

2. While still in the Draw Template routine draw the bottom bit surface.

Verify if the routine is prompting for a starting point, if not rerun the Draw Template routine. Select the bottom centerline endpoint of the top surface.

```
Starting point: endpoint (select at vertex 1)
```

If, after you choose the starting point, the routine prompts with something other than point, toggle into Pointing mode by typing the letter **p** and pressing ENTER. An example of this is the following:

```
Grade (%) (Relative/Slope/Points/Close/Undo/eXit): p
Select point (Relative/Grade/Slope/Close/Undo/eXit): endpoint (select at
  vertex 2)
```

The second point is the lower outside endpoint of the Top Bit surface.

The next endpoint is a straight change of elevation, 6 inches. Toggle into Relative mode with the letter r.

```
Grade (%) (Relative/Slope/Points/Close/Undo/eXit): r
Change in Offset (Grade/Slope/Close/Points/Undo/eXit): 0
Change in offset: -.5
```

Most likely there is no crown to the bottom of the Bottom Bit surface. The return to the centerline is a straight change of offset and no change in elevation. The offset change is 12.5 feet and the elevation change is 0.

EXERCISE

```
Change in Offset (Grade/Slope/Close/Points/Undo/eXit): 12.5
Change in offset: 0
Change in Offset (Grade/Slope/Close/Points/Undo/eXit): (press ENTER)
Starting point: (press ENTER to exit the routine)
```

If you make mistakes, there is an Undo option within the command. The Undo removes the last segment, but keeps the Draw Template routine running. If you inadvertently exit the routine, you can start over or start where the routine left off. Remember to join the two pline segments together with the PEDIT command to make them a single 2D polyline.

Defining a Template

The Define Template routine identifies several critical points on the template (see Figure 9–35). The first critical point is the Finished Ground Reference point. This is the template point that the design processor pins onto the horizontal and vertical alignments as the template travels along the road. The second critical point is the connect-out point, which is where the search for daylight begins or the ditch design parameters take over. This point is the end of the template's influence and the beginning of either a slope search for the top surface (the catch points or daylight points) or where the design processor starts creating the ditch and then searches for daylight. The last critical point to define is the datum line points. The datum line represents the bottom limits of excavation for the template.

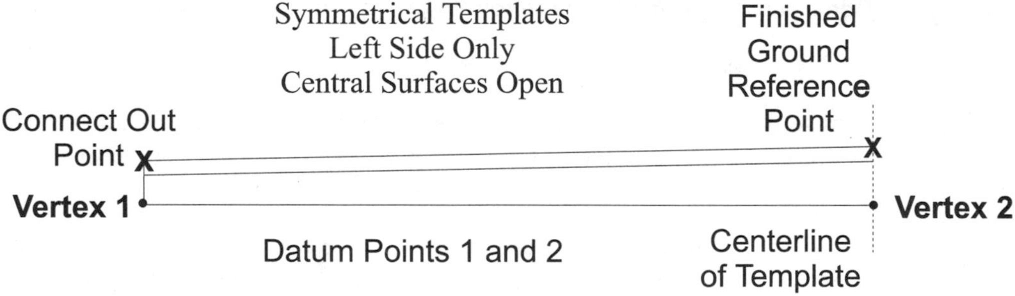

Figure 9–35

The finished ground reference point is the point where the template attaches to the vertical and horizontal alignments (seen as one by the design processor). In the current exercise this is the right upper endpoint of the Top Bit surface. The finished ground reference point can be anywhere on a template. The only condition is that the point represents the location of the horizontal and vertical alignment locations of the template (a single point).

The connect-out point is the outer top vertex of the Top Bit surface. It is at this point that the curb subassembly attaches to the central surfaces.

EXERCISE

The datum points start at the bottom outside vertex to the bottom right hand endpoint of the Bottom Bit surface. This line indicates the amount of excavation needed to build this portion of the template. Refer to Figure 9–35 for the location of these points.

The final step in creating a template is to define the template. A Define routine in Civil Design produces external data. The Define Template routine asks for several pieces of data. The routine first prompts for the location of the finished ground reference point and whether the template is symmetrical or asymmetrical. The Define Template routine then asks you to select the 2D polyline representing the template. After you identify the surfaces that make up the template, the routine asks for the surface material type, the connect-out point (where either the ditch attaches, or the subassemblies attach, or where the template influence ends and slopes to daylight begin), and the datum points (limits of excavation).

1. Start the Define Template routine located in the Templates cascade of the Cross Sections menu.

The first question is the location of the finished ground reference point. Select the upper right endpoint of the Top Bit surface.

```
Select finished ground reference point: endpoint (select upper right vertex)
```

The next series of prompts are for the surfaces in the template. The first question asks if the surfaces are symmetrical. The answer is yes for all the surfaces.

The next question is to identify the 2D polylines that represent the surfaces. Make sure, when selecting the two (2) entities, to select the objects where the objects do not overlap. Places to select the entities are the top segment of the Top Bit surface, the bottom segment of the Bottom Bit, and the backside of the curb.

```
Command:
Pick finish ground reference point:
Is template symmetrical [Yes/No] <Yes>: (press ENTER)
Select template surfaces...
Select objects: 2 found, 2 total
Select objects:
```

The Define Template routine highlights the base bituminous surface and asks whether the surface is normal or subgrade. Press ENTER on normal. After you press ENTER to accept normal, the routine displays the Surface Material Names dialog box.

The material list does not include Base Bituminous, so select the New button and enter the Base Bituminous material name (see Figure 9–36). After defining and identifying the material, select the OK button to exit the dialog box.

EXERCISE

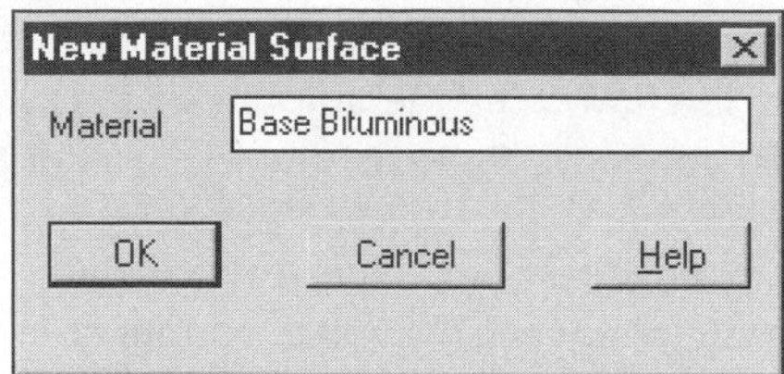

Figure 9–36

```
Surface type [Normal/Subgrade] <Normal>:
```

The Define routine then highlights the asphalt surface and asks whether the surface is normal or subgrade. Press ENTER on normal. After you press ENTER to accept normal, the routine displays the Surface Materials Names dialog box.

The material names list includes Asphalt. Select the material and select the OK button to exit the Surface Material Names dialog box (see Figure 9–37).

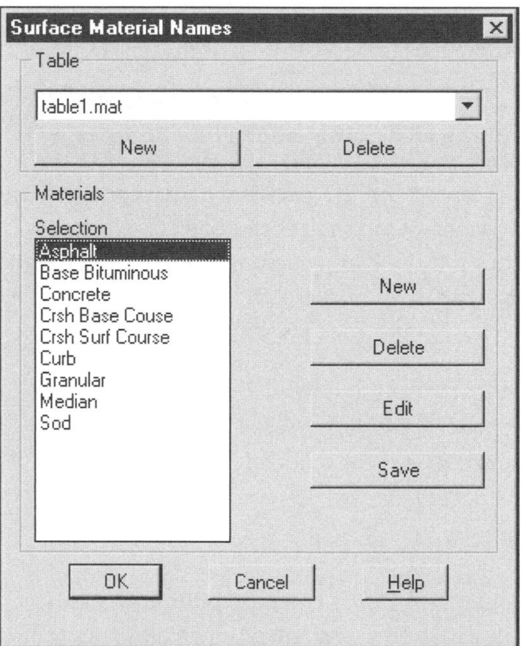

Figure 9–37

```
Surface type [Normal/Subgrade] <Normal>:
```

The next critical point to define is the connect-out point. This is where a subassembly connects to the central surfaces. If the template does not have any subassemblies, the connect-out point becomes one of two things: The connect-out point is either where slopes start searching for daylight or where the ditch design connects to the template (see Figure 9–35).

Identify the outer upper vertex of the Asphalt surface as the connect-out point.

```
Pick connection point out:
```

The last set of points is the datum or excavation limits line. The routine expects these points in order from left to right (see Figure 9–35).

```
Datum number <1>:
Pick datum points (left to right):
Pick datum points (left to right):
Pick datum points (left to right):
```

After defining the surfaces, the Define Template routine displays the Subassembly Attachments dialog box prompting you for the template subassemblies. The dialog box indicates that there are currently no subassemblies assigned by showing nulls in the boxes. In the current example, the template has a curb and a cut and fill shoulder subassemblies. The shoulder subassembly is both a cut and fill subassembly (see Figure 9–38).

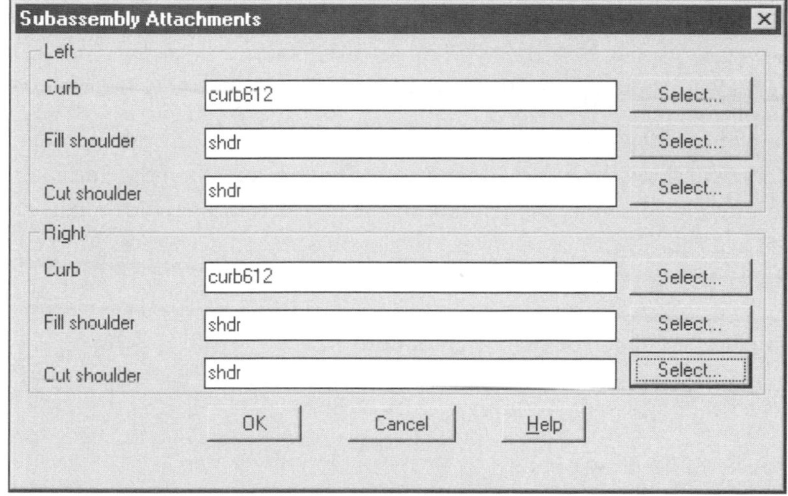

Figure 9–38

Assign Curb612 to the right and left curb subassembly.

Assign Shdr to the right and left cut and fill shoulder subassembly.

EXERCISE

A template can have different shoulder subassemblies for cut and fill situations. A template can also have different curbs and shoulders for the left and right sides.

```
Save template [Yes/No] <Yes>:
Template name: STD2L
Define another template [Yes/No] <Yes>: n
```

The template is now a part of your template library.

Editing and Viewing Critical Points on a Template

To view the entire template and its datum and connect points, use the Edit Template routine of the Template cascade.

1. Erase all of the objects from the screen and regenerate the screen. After defining the template, the original line work is not needed.
2. Select the Edit Template routine of the Templates cascade. The routine displays a dialog box listing the defined templates. Select the STD2L template from the list of templates. The dialog box contains a viewer that shows a distorted view of the template. Exit the dialog box by clicking on OK.

Select an insertion point in the middle of the Graphics Editor.

```
Command:
(After selecting the template from a dialog box)
Insertion point: (select a point in middle of the screen)
```

The Edit Template routine changes and/or adds to the template. For the moment, however, the Edit routine will show the critical points on the template. When the Edit routine inserts the template into the drawing, the routine marks the datum line with a dashed line. The subassemblies are red. The subassemblies are blip objects and do not respond to any object snap choices.

The Edit Template routine prompts for an option.

```
Edsrf/SAve/eXit/ASsembly/Display/SRfcon/Redraw <eXit>:
```

3. Choose the Redraw option.

```
Edsrf/SAve/eXit/ASsembly/Display/SRfcon/Redraw <eXit>: r
```

The Redraw option removes the datum from the screen. To be able to view the datum and critical points on the template, select the Display option by typing **d**. The Display option returns another list of choices including displaying the datum, the connect points, redrawing the image, or exiting Edit. The finished ground reference point is a connect point, which is a connect point to the alignments. Exercise the Display option. Display the Datum, the connect points, points, and top surface. Even though the routine says there is no defined top surface, it will draw a straight line from connect point to connect point. Use the Redraw option to clear the template before displaying another item from the list.

```
Edsrf/SAve/eXit/ASsembly/Display/SRfcon/Redraw <eXit>: d
Display [Datum/Connect/Points/Super/SHoulder/Topsurf/TRansition/eXit/Redraw/
  TType]
<eXit>: d
Datum number <1>:
Display
[Datum/Connect/Points/SHoulder/Super/Topsurf/TRansition/eXit/Redraw/TType]
<eXit>: r
Display
[Datum/Connect/Points/SHoulder/Super/Topsurf/TRansition/eXit/Redraw/TType]
<eXit>: c
Display
[Datum/Connect/Points/SHoulder/Super/Topsurf/TRansition/eXit/Redraw/TType]
<eXit>: r
Display
[Datum/Connect/Points/SHoulder/Super/Topsurf/TRansition/eXit/Redraw/TType]
<eXit>: p
Display
[Datum/Connect/Points/SHoulder/Super/Topsurf/TRansition/eXit/Redraw/TType]
<eXit>: r
Display
[Datum/Connect/Points/SHoulder/Super/Topsurf/TRansition/eXit/Redraw/TType]
<eXit>: t
Top surface number <1>:
Top surface <1>, not found.
Display
[Datum/Connect/Points/SHoulder/Super/Topsurf/TRansition/eXit/Redraw/TType]
<eXit>: r
Display
[Datum/Connect/Points/SHoulder/Super/Topsurf/TRansition/eXit/Redraw/TType]
<eXit>: (press ENTER to exit the display options)
```

EXERCISE

Defining the Top Surface

The Define Template routine defines the top of template as a straight line between the connect-out points of the left and right side of the template. If applying the STD2L template to the design data and then viewing the cross sections, the viewing routine will complain that there is no top of surface 1 in the template. You must edit the template to define a top of surface. The top of surface travels from left to right along the upper vertices of the central surfaces of the ADV2L template (see Figure 9–39).

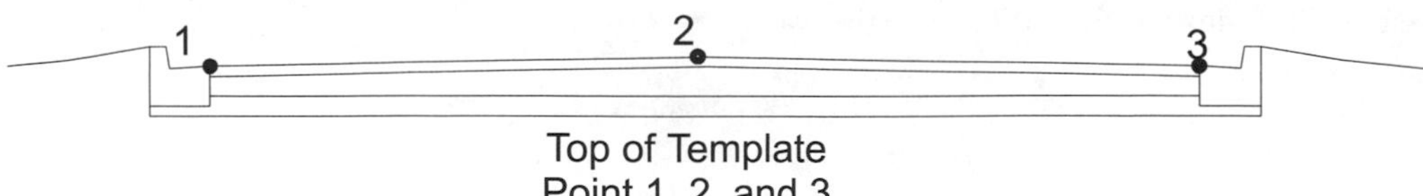

Figure 9–39

```
Edsrf/SAve/eXit/ASsembly/Display/SRfcon/Redraw <eXit>: sr
Connect/Datum/Redraw/Super/Topsurf/TRansition/eXit <eXit>: t
Top surface number <1>:
Pick top surface points (left to right): (select vertx 1)
Pick top surface points (left to right): (select vertx 2)
Pick top surface points (left to right): (select vertx 3)
Pick top surface points (left to right): (press ENTER to exit)
Connect/Datum/Redraw/Super/Topsurf/TRansition/eXit <eXit>:
Edsrf/SAve/eXit/ASsembly/Display/SRfcon/Redraw <eXit>:
Save template [Yes/No] <Yes>:
Template name <std2l>:
Template exists. Overwrite [Yes/No]: y
```

4. Reopen the Roadway drawing in the *DWG* directory of the Roadway project. Save the changes to the template drawing.

UNIT 4: PROCESSING A ROADWAY DESIGN

The final step of the roadway design process is identifying and processing all of the elements that affect the volume calculations. This step is done in the Edit Design Control routine of the Design Control cascade of the Cross Sections menu, which displays the Design Control dialog box. In this dialog box, you identify the template, the template's datum, daylight slopes, ROW conditions, ditch parameters, and benching conditions.

THE DESIGN CONTROL DIALOG BOX

The Edit Design Control routine of the Design Control cascade of the Cross Sections menu is the final step in the design process (see Figure 9–40). The Edit Design

Control routine calculates the daylight points and the cross-section volumes of the design. In the Design Control dialog box, the designer assigns the template, the beginning and ending station for evaluation, the daylight slopes, ditch control, benches, and the horizontal and vertical transition alignments (see Figure 9–41).

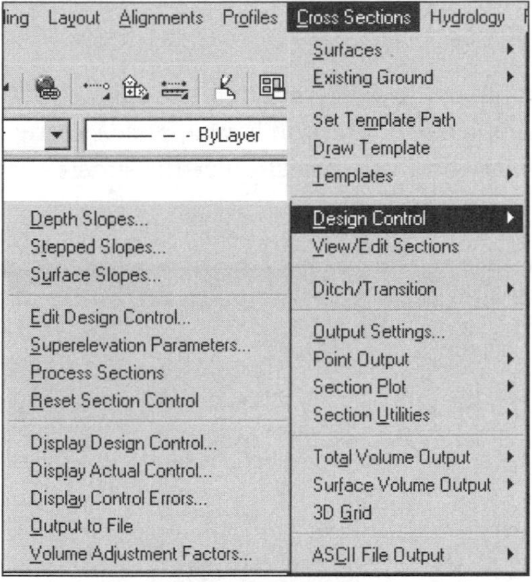

Figure 9–40

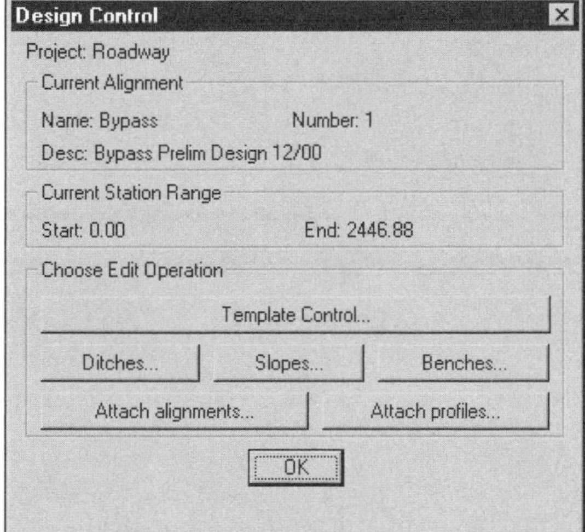

Figure 9–41

Civil Design provides additional tools to vary the daylight slopes depending upon predefined conditions: Depth Slopes, Stepped Slopes, and Surface Slopes. These tools can vary daylight slopes by the depth of either cut or fill or by the type of material, and introduce multiple benches (see Figure 9–40).

TEMPLATE CONTROL

The Template Control dialog box sets the template and datum for evaluating roadway design volumes (see Figure 9–42).

The settings at the bottom of the dialog box apply to superelevation criteria. The only time these values become active is after you have applied superelevation control to the alignment. These settings allow you to customize the superelevation parameters for a range of stations along the centerline.

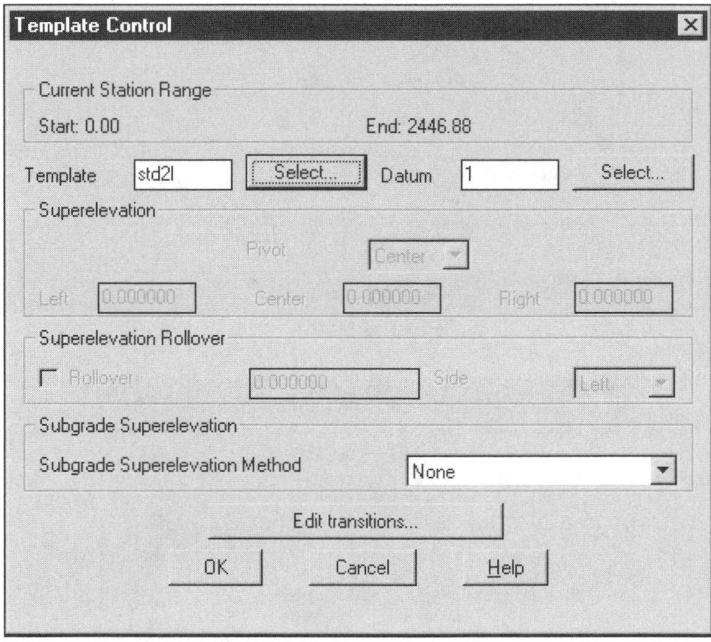

Figure 9–42

DITCHES

Ditch Design Parameters

The Ditches button of the Design Control dialog box displays a dialog box that sets several parameters that affect the resulting ditch design (see Figure 9–43).

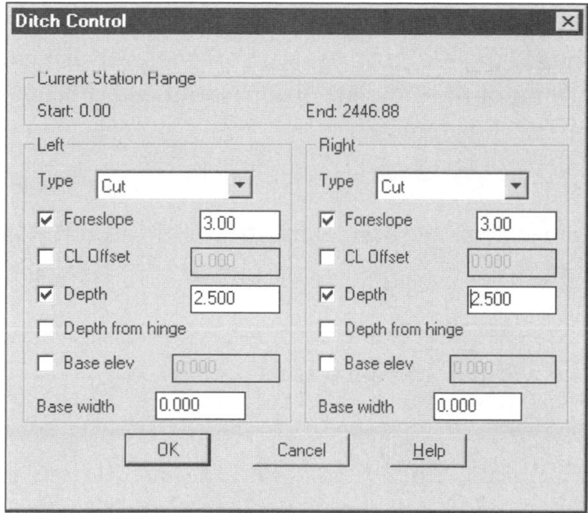

Figure 9–43

Ditch Rules:

- Ditches are active only after setting two design parameters.
- When assigning transition alignments to control ditch depths, toggle Base Elevation on.

Ditch Types

The top portion of the dialog box controls the type and side of a ditch. The type of ditches are Cut only, Fill only, Cut and Fill, and No ditches.

Foreslope

This is the value of a slope from the template to the base of the ditch. The foreslope value is always a down slope run and is a positive value. An upslope with a positive value will give you a berm.

CL Offset

This is the distance from the centerline to the base of the ditch. The CL offset is a constant value. This "fixes" the ditch at a constant distance from the centerline.

Depth

This is the distance from a point on the template to the bottom of the ditch. A positive value creates a ditch and a negative value creates a berm.

Civil Design measures the depth of the ditch from one of two points on the template. The first is the finished ground reference point. The second is the hinge point. A hinge point is the connect-out point of a subassembly or template surface.

Depth from Hinge

The Depth from Hinge (Toggle) toggles on or off the measuring of the ditch depth from a connect-out point of the template or subassemblies. This toggle activates only after toggling on the depth parameter.

Base Elevation

This value is set or reflects the base elevation of the ditch when using vertical transition alignments.

Base Width

The base width is the width of the ditch at its bottom.

SLOPES

The Slopes button calls the Slope Control dialog box. These settings affect the daylight slopes that blend the design back into the design surface (see Figure 9–44). The routine prompts for the cut and fill slope values in the settings dialog box. The user specifies an optimal slope value (typical) and highest slope value acceptable (maximum).

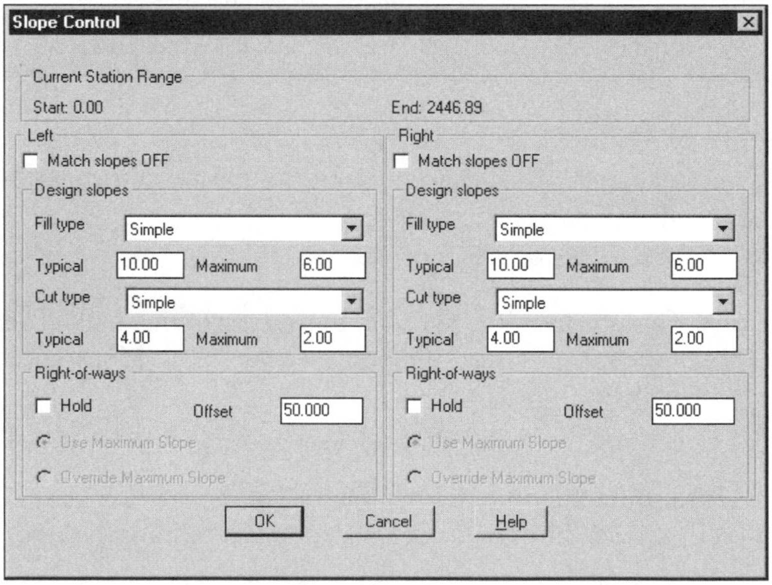

Figure 9–44

The uses of simple slopes limits the effects of the design to the swath width of the cross section in cut and fill cross sections. In Civil Design there are two additional slope types, depth and stepped, for fill situations. The same two types of slope control are found in the cut slope settings plus one new slope setting, surface. The parameter changes the slope used for daylighting based upon which surface the design is cutting through. These control settings are the subject of Section 11.

ROW

The bottom of the dialog box sets the display of a ROW line in the cross section. By default the ROW is a fixed offset distance from the centerline of the roadway. If the ROW offset value varies over the length of the roadway, you can define a ROW alignment to control the location of the ROW in a cross section.

Hold ROW

If the Hold ROW toggle is off, the design processor can use the width of the cross section (from the end of the ditch to the end of the section).

If the Hold Right-of-Way toggle is on, this fixes the daylight solution to the intersection of the daylight surface and the ROW offset distance. The Hold ROW toggle forces the design processor to solve the daylight point in a set order. The order creating the solution is by first using the typical slope, then the maximum slope, and finally pinning the slope to ROW.

BENCHES

The Benches button displays the Bench Control dialog box (see Figure 9–45). The settings in this dialog box affect the occurrences and characteristics of benches along the roadway. Benches can occur in cut or fill or both cut and fill. The toggles allow you to have the benching on the right side of the template behave differently than benching on the left side. Some of the parameters you set affect the width of the bench, its height, and its slope.

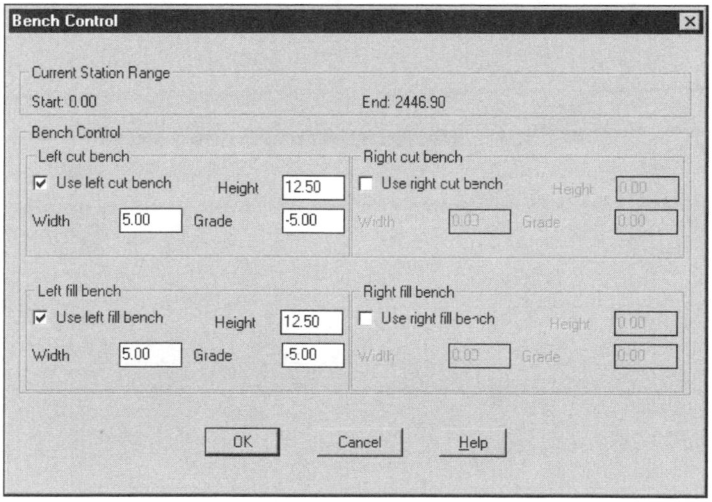

Figure 9–45

After you complete this exercise, you will be able to:

- Merge roadway design elements in the design processor
- Assign a template
- Review ditch settings
- Review slope settings

PROCESSING A ROADWAY DESIGN

The Edit Design Control routine of the Design Control cascade of the Cross Sections menu binds together the horizontal and vertical alignments, the cross sections, and template elements. The Design Control dialog box sets the template, datum, ditch parameters, transition control, and the daylight slopes for cut and fill. When you exit the Design Control dialog box, the design processor uses these elements along with the beginning and ending stations of the alignment to evaluate them into a volume.

1. Run the Edit Design Control routine of the Design Control cascade of the Cross Sections menu. If there is no current alignment, select an object that represents the Bypass alignment, or press ENTER to display the Alignment Librarian (see Figure 9–46). You can then select the Bypass alignment from the list of defined alignments.

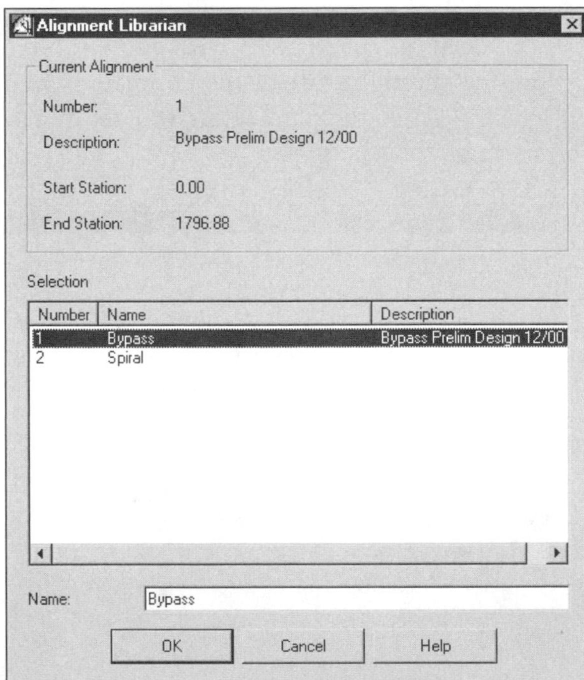

Figure 9–46

2. After you select the current alignment, the routine then displays a dialog box asking you to confirm or change the beginning and end stations for the current run of the routine. Dismiss this dialog box by selecting the OK button. The Edit Design Control routine then displays the Design Control dialog box, which supplies values to the design processor.

The Design Control dialog box displays the current alignment, its affected stations, and a group of buttons. It is from this dialog box that you choose and apply the initial data for the roadway design. The Template Control, Ditches, Slopes, Benches, and Transition alignments that affect the roadway design are set in the subdialog boxes displayed by selecting the appropriate buttons (see Figure 9–41).

3. Select the Template Control button. This displays the Template Control dialog box (see Figure 9–42). Currently there is no template selected for this segment of the roadway. Choose the Select button to display the template library. Select the STD2L template from the library list and then select the OK button to exit the librarian.

4. Select the Datum button to display the Datum Librarian (see Figure 9–47). A template can have more than one datum, and this is where you select which datum to use. Select the OK button to exit the Datum Librarian and Template Control dialog boxes and to return to the Design Control dialog box.

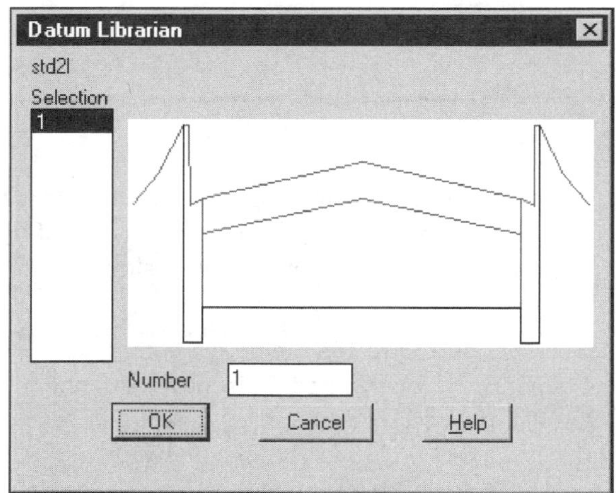

Figure 9–47

5. Select the Slopes button to display the Slope Control dialog box (Figure 9–44). Set the following values in the dialog box. After setting the slope values, select the OK button to exit the dialog box and to return to the Design Control dialog box.

EXERCISE

Design Slopes

	Left	Right
Fill Type	Simple	Simple
Typical	10:1	10:1
Maximum	6:1	6:1
Cut Type	Simple	Simple
Typical	4:1	4:1
Maximum	2:1	2:1
Right-of-way	Hold - OFF	
Offset	50	50

There is no benching for the current design.

Bench	All toggles OFF

Ditch Design Parameters

6. Select the Ditches button to display the ditch design parameters (see Figure 9–43). Set the following ditch parameters in the dialog box. After setting the values, select OK to exit the Ditch Control dialog box and to return to the Design Control dialog box.

	Right	Left
Type:	Cut	Cut
Foreslope:	3:1	3:1
Depth:	2.5	2.5

7. Exit the Design Control dialog box by selecting the OK button. When you select the OK button, the design processor processes the design. If there are any errors, the routine will display an error message dialog box before returning to the command prompt.
8. Select the View Errors button to see the design errors (see Figure 9–48). The errors are the same; right slope pinned. This means that the minimum and maximum slopes did not intersect the target surface before encountering the outer end of the cross section (75 foot offset or swath width). Select the OK button to close the Control Processing Errors, Section Processing Status, and Processing Status dialog boxes.
9. Civil Design records the design processor errors in a file. You can view the error file with the Display Control Errors routine of the Design Control cascade. You can print or send the errors to a named text file. Run the Display Control Errors routine to view the error report.
10. Select the OK button to exit the Errors dialog box.
11. Save the drawing.

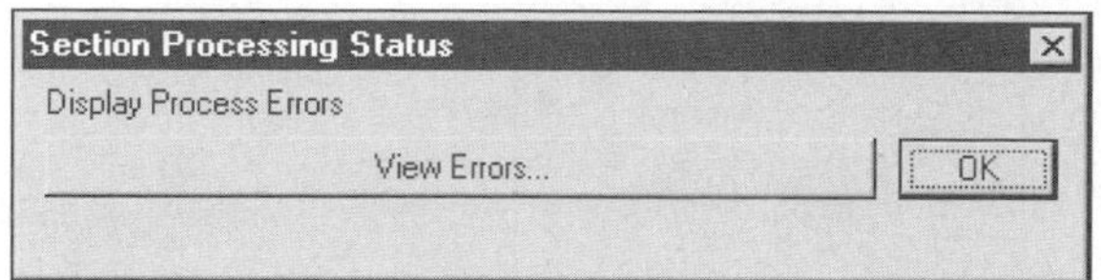

Figure 9–48

UNIT 5: REVIEWING DESIGN RESULTS

The Cross Sections menu of Civil Design contains several tools that allow you to view and evaluate the results from the volume calculations (see Figure 9–49). The first method is to view the resulting cross sections (View/Edit Sections). This allows you to decide whether the current settings visually accomplish your design goals. The second method is viewing the volumes from the processing of the design. This review has three different methodologies. The first methodology is simply viewing the numbers from the calculations. The Total Volume Output cascade contains three routines that create an overall volume (see Figure 9–50). The second method is in the same cascade, but allows you to view a graphical representation of the volume and how it is arrived at over the length of the road design, the Mass Haul diagram. The Mass Haul diagram reports the balancing of the design as well as the portions of the design that create or use material (cut and fill). The last methodology is to plot the cross sections in the drawing (see Figure 9–51).

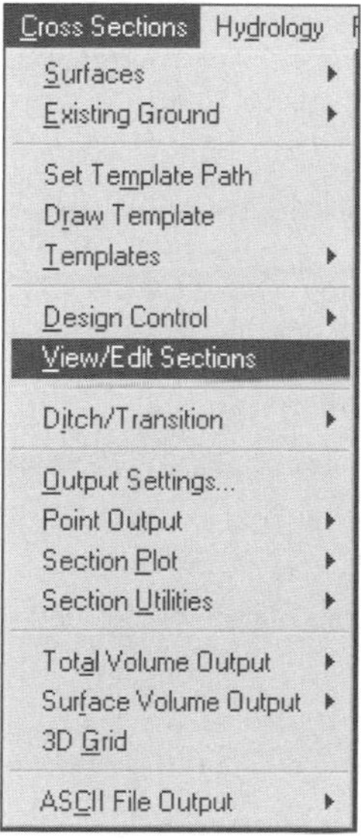

Figure 9–49

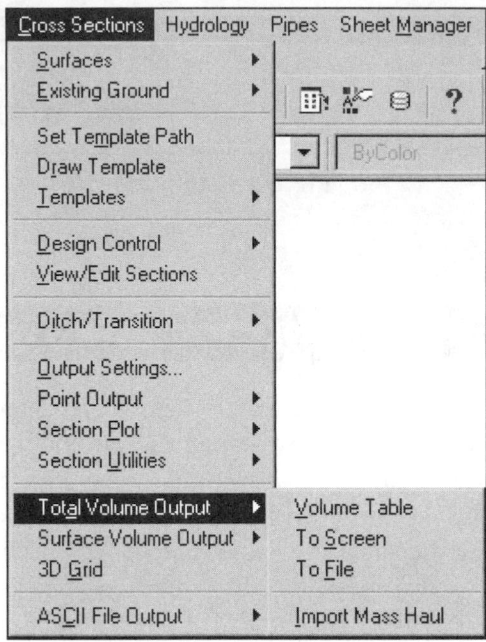

Figure 9–50

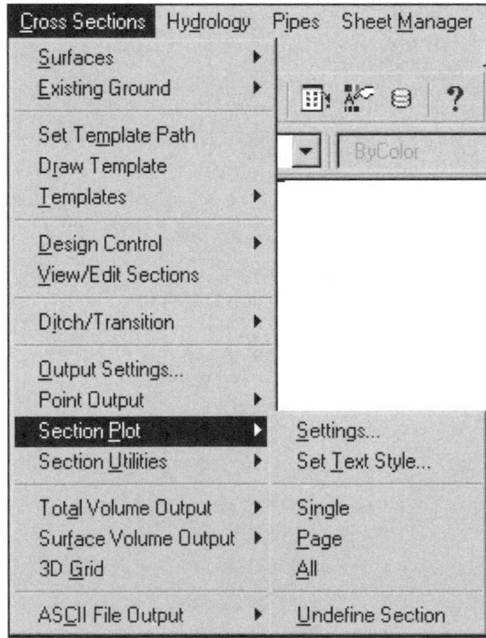

Figure 9–51

VIEW/EDIT SECTIONS

The View/Edit Sections routine of the Cross Sections menu displays the resulting cross sections as transient images on the screen. Options within the routine allow for the review of Design and Actual settings. Other options control the zoom amount, the vertical exaggeration, and colors of the cross-section elements. In the View/Edit Sections routine, there are tools that query and edit different aspects of a cross section as well as routines to identify slopes, elevations, and offsets within a cross section. Other tools can edit the slopes in the section. Any editing done at a station within the View/Edit Sections routine is only for the current section. Each operation has a submenu within the View/Edit Sections routine. You must pay attention to which submenu is active. It has a tendency to be confusing.

The View/Edit Sections routine displays the current station, the alignment name, the beginning and ending station for the review, and the current cross section station. The routine then displays an option list that indicates by a capital letter which character invokes the option. The default option is at the end of the list and is enclosed in right and left angle brackets.

If there are any error messages for the current cross section, the View/Edit Sections routine will display the error message. An example of displaying an error message is:

```
Warning:  Station: 7+00   Right slope pinned to end
Sta: 7+00   Section View & Display commands
Actual/Design/Edit/Id/Next/Previous/eXit/Sta/View/Zoom <Next>:
Warning:  Station: 7+50   Right slope pinned to end
Sta: 7+50   Section View & Display commands
Actual/Design/Edit/Id/Next/Previous/eXit/Sta/View/Zoom <Next>:
```

When you first enter the View/Edit Sections routine, the routine displays the alignment name, description, and the beginning and ending stations. After displaying the alignment information, the routine displays the main prompt line. You select an option by typing the capitalized letter and pressing ENTER. Below is an example of the routine's first prompt.

```
Command:
Alignment Name: Bypass Number:1  Descr: Bypass Prelim Design of 12/00
Starting Station: 0.00  Ending Station: 2446.88
Sta: 0+00   Section View & Display commands
Actual/Design/Edit/Id/Next/Previous/eXit/Sta/View/Zoom <Next>: a
```

ACTUAL

The Actual option displays the Actual Control Parameters dialog box, containing the actual slopes and cut and fill areas of the current cross section (see Figure 9–52).

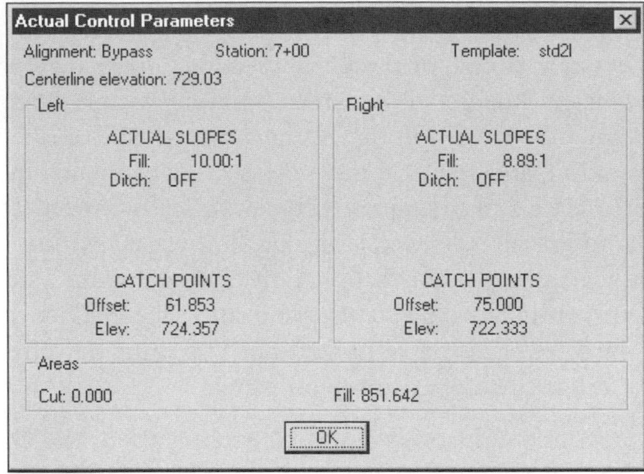

Figure 9–52

DESIGN

The Design option displays the Design Control Parameters dialog box, containing the slope, ROW, and bench settings (see Figure 9–53). If there is any benching or transitioning occurring, the buttons at the bottom display additional dialog boxes containing information on these influences.

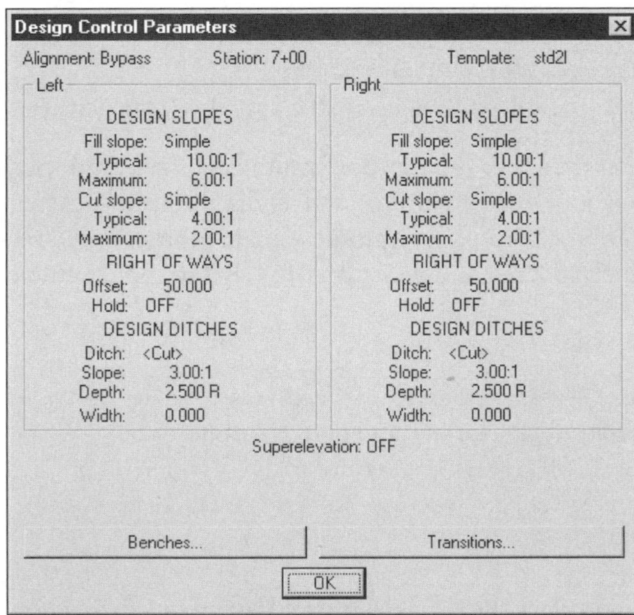

Figure 9–53

EDIT

The Edit option displays a new list of options that contains a new tool to edit the Design Control or Transition control used on the cross section. The ditch option displays a new list of options to edit ditch and slopes present in the current cross section. Below is an options list for Edit.

```
Actual/Design/Edit/Id/Next/Previous/eXit/Sta/View/Zoom <Next>: e
Sta: 1+00   Section Edit & Display commands
Actual/Control/Ditch/Id/eXit/Transition/Undo/Zoom <eXit>:
```

Control

The Control option displays Control Editor dialog box that changes the Design Control settings for the current cross section (see Figure 9–54).

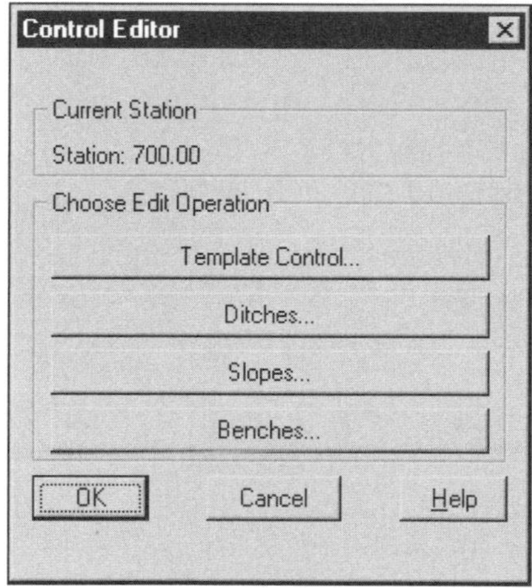

Figure 9–54

Ditch

The Ditch option displays a new list of options that modify the slope or location of critical design points in the cross section.

```
Actual/Control/Ditch/Id/eXit/Transition/Undo/Zoom <eXit>: d
Sta: 1+00   Section Edit & Display commands
Actual/Control/dSlope/dElev/dWidth/dPos/Id/Mslope/eXit/Undo/Zoom <eXit>: x
```

Actual, Control, Id, Undo, Zoom, and eXit are the same routines found in the main option list.

DSlope. The dSlope option graphically edits the foreslope of the ditch, the slope between the connect-out point of the template and the base of the ditch. After you select two points in the editor to define a new foreslope, the option returns the selected slope value. If you accept the slope or enter a different value, press ENTER and the option changes the foreslope for the section.

```
First ditch foreslope point: (select a point)
Second ditch foreslope point: (select a point)
Resultant slope (s:1) <2.5>: (press ENTER to accept)
```

dElev. The dElev option graphically edits the elevation of the ditch. Select a point in the section for the new ditch elevation and the option returns the selected elevation. If you accept the elevation or enter a different value, press ENTER and the routine changes the ditch's elevation for the section.

```
Select ditch elevation point: (select a point)
Resultant elevation <723.63>: (press ENTER to accept)
```

dPos. The dPos option graphically changes the offset and depth of the ditch. After you select a point in the editor, the routine returns the offset and elevation of the selected point. If you accept the values or enter a different value, press ENTER and the routine will modify the offset and depth of the ditch.

```
Select ditch position: (select a point)
Resultant offset  Right <30.09>: 33 (press ENTER to accept)
Resultant depth <3.74>: 4.00 (press ENTER to accept)
```

dWidth. The dWidth option graphically edits the base width of the ditch. Select two points and the option returns the distance ditch width. If you accept the distance or enter a new value, press ENTER and the routine changes the width of the ditch.

```
First ditch width point: (select a point)
Second ditch width point: (select a point)
Resultant width <2.96>: (press ENTER to accept)
```

Mslope. Mslope edits the match slope from the ditch to the daylight surface. This option edits only one side of the template at a time.

```
First match point: (select a point)
Second match slope point: (select a point)
Resultant slope (S:1) <2.53>: (press ENTER to accept)
```

ID

The Id option reports the station and offset of selected points within the cross section. The option does not allow you to use an object snap when selecting points in the cross section.

```
Select point:
Offset: -41.74, Elev: 711.78
Select point:
Offset: 29.72, Elev: 711.52
```

NEXT/PREVIOUS/EXIT/STATION

These options, Previous, Next, and Sta allow you to move forwards, backwards, and to a specific station in the cross section list.

VIEW

When you select the View option, the response is the Template View Settings Editor, containing settings that affect the display of the cross section (see Figure 9–55).

The top portion of the dialog box toggles on and off the visibility of cross-section elements and sets their color when displayed.

The middle portion of the dialog box sets the grid offset and elevation interval and the precision used in annotating the cross section.

The bottom portion controls the text size, the zoom increment, and the vertical exaggeration factor for the section.

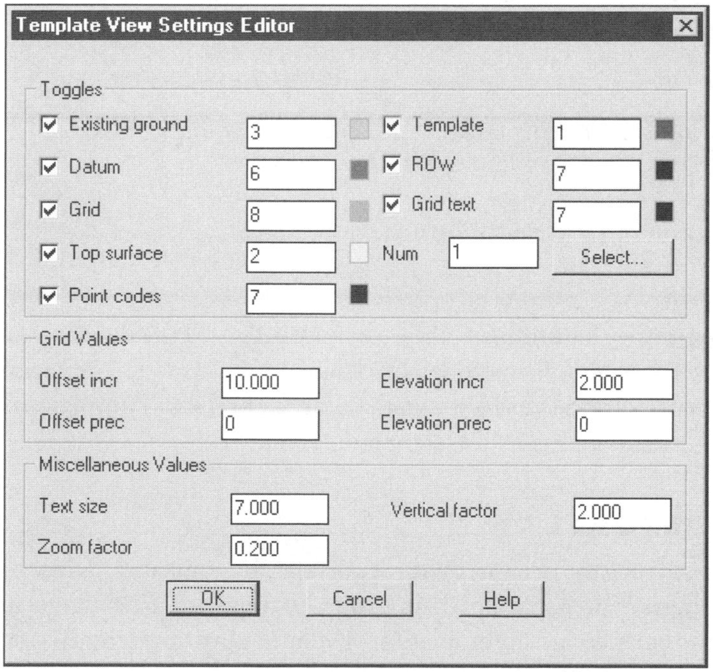

Figure 9–55

ZOOM

The Zoom option displays a list of options.

```
Sta: 0+00    Section Zoom commands
Zoom [All/In/Out/Point/eXit/View/Window] <In>:
```

All

The All option resets the screen to a centered full view of the cross section and drops out of the Zoom submenu.

In/Out

The In and Out zoom options use the zoom increment found in the View dialog box. This increment reduces or expands the view by a factor of 0.2, if 0.2 is the current setting.

Point

This option prompts the user to define a new center point of the displayed cross section.

Window

The Window option prompts you to select two points to isolate a section of the cross section.

View

The View option sets the display defaults in the View dialog box.

EXit

The eXit option returns to the main view option list.

POINT CODES

The View option toggles on the point code numbers of critical template points. The current template has two point codes by default, Final Grade Centerline and Connect Out (internal surfaces and shoulder) points. The Edit Template routine allows you to add additional points to the template.

VOLUME REPORTS

Civil Design provides the designer with reports of the total and surface volumes. The Mass Haul diagram is a part of this group of reports. The diagram is a graph representing the balance and accumulation of volumes over the length of the centerline. The diagram shows the volume's balance at any point along the design and as to whether the design is generating cut or fill volumes. The diagram helps in making decisions about free haulage.

The reports are in the Total Volume Output cascade of the Cross Sections menu (see Figure 9–50). There are routines to send the volume reports to the screen, a table within the drawing, or to a file or printer. The volumes represent the amount of material between the template datum line and the top surface.

The volume reports use two types of adjustments. The first adjustments are for Cut and Fill. With the removal of overburden, the material may swell in volume. This is a cut adjustment made to the volume and it increases the overall volume. If the material is transported to a new location and deposited, the addition of more material and the weight of the machinery will compact the material lying underneath. This is a Fill adjustment made to the volume and it decreases the overall volume. The reports apply the adjustments as multipliers.

The second type of adjustment is made to curves along the centerline. When turning a curve, the distance between cross section is less on the inside of the curve than on the outside of the same curve (see Figure 9–56). This adjustment makes allowances for the difference in spacing between sections. The line work intersecting the centerline indicates program and user critical points. These points include PCs, PTs, Spiral intersections, and other user defined points.

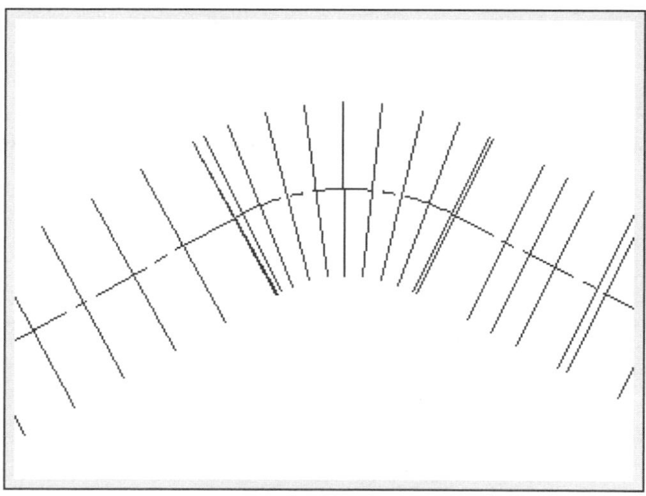

Figure 9–56

MASS HAUL DIAGRAM

The most useful report of Total Volume Output is the Mass Haul diagram. The diagram portrays the balance and location of cut and fill volume accumulation as a chart. This chart allows evaluation of the design and free haulage potential within the design (see Figure 9–57).

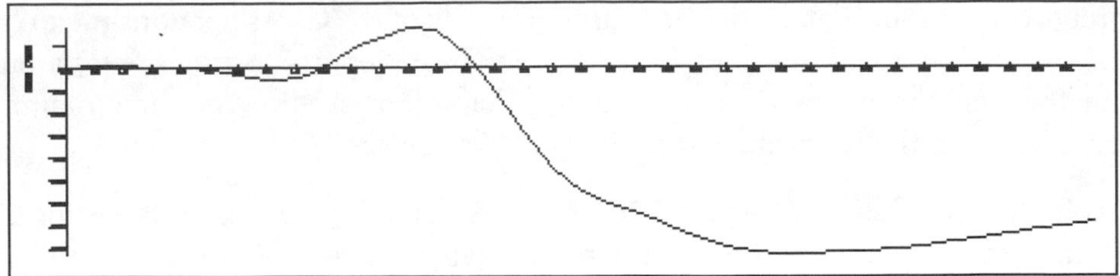

Figure 9–57

SURFACE VOLUMES

The Template Surface routine of the Surface Volume Output cascade generates a report containing the surface volumes of the template (see Figure 9–58). The report is a file report. If the template has Top Bit, Bottom Bit, Granular Base, and Curb surfaces, the report generates a volume for each surface. The report indicates the first surface, but omits the names of any other surfaces in the report. To see the individual surface names as a part of the report, you need to toggle on Subheaders in the Output Settings dialog box of the Drawing Settings routine. The totals for the surfaces are at the end of the report.

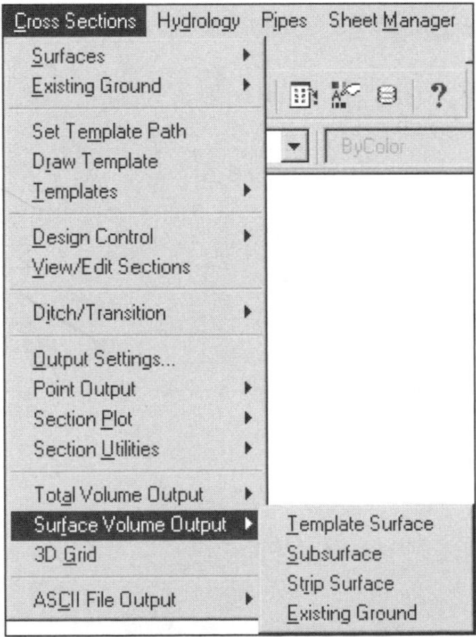

Figure 9–58

This report does not prompt for volume adjustment values because the routine looks for a file containing adjustment factors from a list of surface names. The Volumes

Adjustment Factors routine of the Design Control cascade modifies the values in the file. Since bituminous, concrete, and granular materials have different shrinkage factors, Civil Design allows for differences with the Factors file. After finalizing the design, you are able to use and create surface volumes.

SUBSURFACE

The Subsurface routine of the Surface Volume Output cascade of the Cross Sections menu creates a report on the subsurface volumes. The volume is for surfaces below the topmost surface. The project has to have Multiple Surfaces on before being able to generate this routine. This report is for surfaces that are not a part of the template.

STRIP SURFACE

The Strip Surface routine writes a volume report to a file representing the volume of material removed from the top surface. An example would be removing 6 inches of topsoil from the existing ground surface and then calculating a volume. The design processor assumes that no surface stripping has occurred when calculating a volume. The routine calculates the volume of removed material between either the ROW or the Daylight points and the named surface.

SECTIONS

The routines of the Section Plot and Section Utilities cascades import, list, and label cross sections. See Unit 7 for a discussion of these routines.

Exercise

EXERCISE

After you complete this exercise, you will be able to:

- Use the cross-section viewer
- View the cross sections
- Review the design results
- Review the options within the viewer
- Generate volume reports
- Generate a surface volume report
- Create a Mass Haul diagram
- Evaluate the design

The View/Edit Sections routine of the Cross Sections menu displays the cross-section and template along the centerline. The view of these elements is temporal, in that the lines defining the cross section and template are not actual Desktop objects, but only blip images. Since they are blip images, you cannot use object snaps to select points from the screen.

VIEW/EDIT SECTION

View Option

1. Start the View/Edit Sections routine of the Cross Sections menu. If the routine prompts you to select a centerline, either select an object representing the horizontal centerline or click the right mouse button to display the alignment list and select the Bypass alignment.

The initial prompt from the start of the view command is the following:

```
Actual/Design/Edit/Id/Next/Previous/eXit/Sta/View/Zoom <Next>:
```

2. Run the View option by typing the letter **v** and pressing ENTER. The option displays the Template View Settings Editor . Set the following values and toggles (see Figure 9–55). After setting the values, select the OK button to view the new cross section display.

Toggles:

Toggle ON all elements

Grid Values:

Offset:	Increment	Precision
	10.0	0
Elevation:	Increment	Precision
	2.0	0

Miscellaneous

Text Size: 7.0

Zoom Factor: 0.2

Vertical Exaggeration: 2.0

Actual and Design Options

3. View the Actual values for the current section (see Figure 9–52). Type the letter **a** and press ENTER, and the option displays a dialog box containing the current section's values. Select OK to exit the Actual Control Parameters dialog box.

4. View the design values for the current section (see Figure 9–53). Type the letter **d** and press ENTER, and the option displays a dialog box containing the current section's values. Select OK to exit the Design Control Parameters dialog box.

Previous, Next, and Sta Options

5. Set the Next option by typing the letter **n**. Go to the next few stations by pressing ENTER on the Next option.

6. Set the Previous option by typing the letter **p**. Go to the previous few stations by pressing ENTER on the Previous option.

7. Use the Station option by typing the letter **s** to go to a specific station. The routine responds with a station prompt. Remember to enter stations as decimal values.

```
Actual/Design/Edit/Id/Next/Previous/eXit/Sta/View/Zoom <Next>: s
Enter station: 600
Sta: 6+00   Section View & Display commands
```

8. Go to station 600 with the Sta option.

9. Use the Next option (the default value) to view sections 650 and 700.

When you are view the section for station 700, the View/Edit Sections routine issues the error message, right slope pinned at end. This error message means that the slope creates a daylight point beyond the 75 foot limit (in this case the swath width). To handle this condition, the design processor pins the slope at the intersection of the offset and the top surface.

```
Actual/Design/Edit/Id/Next/Previous/eXit/Sta/View/Zoom <Next>:
Warning:  Station: 7+00   Right slope pinned to end
Sta: 7+00   Section View & Display commands
Actual/Design/Edit/Id/Next/Previous/eXit/Sta/View/Zoom <Next>:
```

10. View the actual data for this cross section. The dialog box does not display the error message (see Figure 9–52).

Zoom Option

```
Actual/Design/Edit/Id/Next/Previous/eXit/Sta/View/Zoom <Next>: z
Zoom (All/In/Out/Point/eXit/View/Window) <In>:
```

11. Run the Zoom option. Zoom in by pressing ENTER.

12. Set the Zoom option to Out and press ENTER to expand the view of the cross section.

13. Use the Point option to center a different point of the cross section on the screen.

14. Use the Zoom All option to exit the Zoom submenu and to return to the main View/Edit Sections prompt.

EXERCISE

ID Option

15. Run the Id option. Select points around the template. The routine reports the offset and elevation of the selected points. Remember, you cannot use object snaps to select points. Press ENTER to exit the Id option.

16. If you are not at Station 7+00, use the Sta option to set the current section to 7+00.

Edit option

17. Run the Edit option by entering the letter **e** and pressing ENTER. The Edit option returns a new series of prompts. Any editing done is only for the current cross section. If you need to modify several stations, you must edit one station at a time.

```
Actual/Design/Edit/Id/Next/Previous/eXit/Sta/View/Zoom <Next>: e
Sta: 7+00   Section Edit & Display commands
Actual/Control/Ditch/Id/eXit/Transition/Undo/Zoom <eXit>: c
Warning:  Station: 7+00   Right slope pinned to end
```

18. Run the Control option and set the typical fill slope to 2:1 for the right side. Confirm the change and exit. The routine changes the slope for the section.

19. Use the Undo option to remove the change from Control.

This is not the only place in which to edit the daylight slope. In the current situation, fill, our design control turns ditches off. The Ditch option of the Edit mode manipulates the match slope of the nonexistent ditch. If manipulating the match slope (mSlope) with the Ditch option, the routine interactively defines a new daylight slope; however, it turns on ditching for the section. Thus, when you are defining a new match slope and ditches are off, your choice is to use the Control option to edit the slope value for the cross section. If you decide to use the tools in the Ditch option, you turn on ditching where there is to be no ditching.

20. Exit the Edit suboption list of the View/Edit Sections routine by typing the letter **x** and pressing ENTER.

21. Use the Sta(tion) option to set the current station to 4+50.

22. Run the Edit option by entering the letter **e** and pressing ENTER. The prompts change to the Edit prompts.

23. Change to the Ditch options list by entering the letter **d** and pressing ENTER.

```
Actual/Control/Ditch/Id/eXit/Transition/Undo/Zoom <eXit>: d
Sta: 4+50   Section Edit & Display commands
Actual/Control/dSlope/dElev/dWidth/dPos/Id/Mslope/eXit/Undo/Zoom
   <eXit>: x
```

The ability to control graphically the daylight slope (mslope) and other aspects of the ditch is in the Ditch option of Edit. All the changes made apply only to the current section. If the changes are to take place over a number of sections, you must edit one section at a time. It may be more practical to use transition control to place the ditch in a more desirable horizontal and vertical position. If that is the case, you need to edit only the critical stations and import them into plan and/or profile view. You then use the critical stations to draw the location of the ditch in plan and/or profile. After drafting the desired behavior, you define transition alignments from the line work and attach them to control the ditch.

24. Use the Mslope option to modify the right match slope. You start the process by typing the letter m and pressing ENTER. The routine prompts you to select two points to define a new match slope. After you select two points, the routine responds with the slope and asks if it is OK or if you want to enter a new slope value. If the slope is OK, or if you enter a new slope, press ENTER to change the slope.

```
Actual/Control/Ditch/Id/eXit/Transition/Undo/Zoom <eXit>: d
Sta: 4+50   Section Edit & Display commands
Actual/Control/dSlope/dElev/dWidth/dPos/Id/Mslope/eXit/Undo/Zoom <eXit>: m
First match slope point:
Second match slope point:
Resultant slope (S:1) <6.84>: 6.5
```

25. Use the Undo option to undo the change to the cross section.

26. Exit the cross section viewer by typing the letter **x** and pressing ENTER. You will have to do this about three times.

27. Save the drawing.

EXERCISE

VOLUME REPORTS

Upon processing the design data, you need to review the results. The review is necessary to evaluate whether the design achieves the goals the designer has in mind. If the design passes the review, the process moves on to the creation of reports, sections, and a representation of the calculations. If the design does not pass review, the design process must return to an earlier step in the process to make corrections and/or additions.

Total Volume to Screen

28. Run the To Screen routine of the Total Volume Output cascade of the Cross Sections menu to view the overall volume of the roadway.

```
Command:
Alignment Name: Bypass Number: 1 Descr: Bypass Prelim Design of 12/00
Starting Station: 0.00  Ending Station: 2446.88
Volume computation type [Prismoidal/Avgendarea] <Avgendarea>:
Use of curve correction [Yes/No] <Yes>:
Use of volume adjustment factors [Yes/No] <Yes>: n
Beginning station <0>:
Ending station <2446.88>:
```

The routine calculates volumes by prismoidal or average end area methods. The routine prompts for curve sampling corrections and for volume adjustment factors. The cut factor is a value of 1.0 or larger and the fill factor is a value of 1.0 or less. Finally, specify the beginning and ending station of the report. Press ENTER to exit the report.

Mass Haul Diagram

The most useful report of Total Volume Output is the Mass Haul diagram. The diagram portrays the balance and location of cut and fill volume accumulation as a chart. This chart allows evaluation of the design and free haulage potential within the design (see Figure 9–57).

29. Create a Mass Haul diagram for the design results using the Import Mass Haul routine of the Total Volume Output cascade of the Cross Sections menu. You may have to pan to place the diagram above the profile and below the roadway centerline.

```
Command:
Alignment Name: Bypass Number: 1 Descr: Bypass Prelim Design 12/00
Starting Station: 0.00  Ending Station: 2446.88
Volume computation type [Prismoidal/Avgendarea] <Avgendarea>:
Use of curve correction [Yes/No] <Yes>:
Use of volume adjustment factors [Yes/No] <Yes>: n
Pick insertion point:
Beginning station <0>:
Ending station <2446.88>:
Vertical scale (cu. yds.) <5000>:
```

The current Mass Haul diagram indicates that the design starts out balanced, but the design from its midpoint to the end produces a need for more fill material. It would seem desirable to have a design that balances. The diagram also indicates that the design produces a great need for fill at the midway point. If you lower the central portion of the vertical design, the design lessens the demand for fill and better balance the overall volume.

30. Undo enough times to remove the Mass Haul diagram.

31. Create a view, Mass Haul, using the Named View routine of the View menu.

32. Save the drawing.

EXERCISE

UNIT 6: EDITING A ROADWAY DESIGN

At this moment the question is: "Is the design sufficient to continue?" If the answer is yes, the process moves on to the generation of volume reports, cross sections, and points.

If the answer is no, where does the redesign begin? The point at which to start over depends on where you make the changes to the design. If the plan design changes, the process defined in sections 7, 8, and 9 needs to start over again. This means producing a new centerline, new profile from sampling along the centerline, new tangents and vertical curves within the profile, and new sections from sampling along the new centerline path. If the plan design is correct, does the vertical need to change? If only the vertical needs to change, the changes occur in the profile area. Or, is the solution a matter of a wider swath or a different template? Then the changes start in the Cross Sections menu. Whatever you decide, the decision to change a design necessitates a return to a previous point in the design process.

Whatever the decision is, the Civil Design tools allow for changes. However, the tools all require you to follow the rigid design process methodology from the point of the changes to the processing of the design in the Edit Design Control routine.

The first step for correcting the current design is to add a new surface to the template. The base granular surface was left off the template. After you have redesigned the template, the next step is to modify the vertical alignment to better balance the volumes produced by the design. The first modification to the design will be adding a new tangent segment and a second vertical curve to the middle of the vertical alignment. This will reduce the need for fill material to help balance the site. If this adjustment does not quite balance the design, the last adjustment will be to edit the elevations of the PVIs (Point-of-Vertical Intersection) to help the balancing process.

Exercise

After you complete this exercise, you will be able to:

- Edit a template
- Add a surface to a template
- Add a second datum to the template
- Insert a new PVI
- Modify the PVI elevations in profile
- Reprocess the design after modifying the elements

EDITING THE TEMPLATE

The office decides that the template needs to change. The change to the template is adding a granular base for the asphalt. The granular base is a central element and is symmetrical around the center of the template. As a result, the granular base is an open element when drawn. The mirroring of the surface by the defining process will close the surface.

After you add the new granular surface, the old datum is no longer valid. You will either have to redefine the template datum points or define the new datum as datum 2 of the template. These changes will affect the datum of the subassembly.

Adding the Granular Surface

The decision to add to the template means running the ADdsurf option of the Edit Template routine. The ADdsurf option prompts for the method of adding a surface. The options are Draw, Move, or Select. Move assumes an existing 2D polyline in the editor that you move into position on the template. The Select option assumes the new surface is in its proper location in relation to the template. You only have to select the 2D polyline to add it to the template. The last option, the one to use for this exercise, is Draw. We will draw the new surface in place and define the surface (see Figure 9–59).

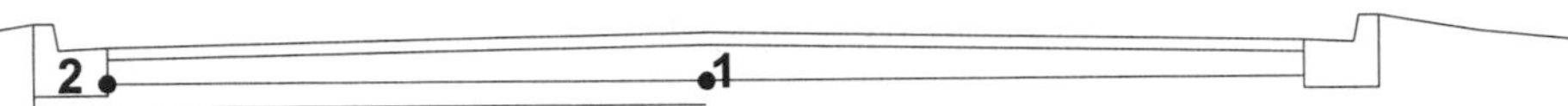

Figure 9–59

EXERCISE

1. Open the drawing CDTPL of the tplate project. Save any changes to the roadway drawing.
2. In the CDTPL drawing, erase any entities from the screen.
3. Run the Edit Template routine of the Templates cascade on the Cross Sections menu. The routine presents the template library dialog box. Select the STD2L template. Insert the template in the center of the screen.

```
Command:
(After selecting the template from the Libraian)
Pick insertion point:  (select a point on the screen)
```

The current prompt for the Edit Template routine is:

```
Edsrf/SAve/eXit/ASsembly/Display/SRfcon/Redraw <eXit>:
```

4. Run the ADdsurf option of the Edit Surface option list. First enter the letter **e** and press ENTER to change the prompts to edit mode. Then enter the letter **a** and press ENTER to start the Add Surface routine.

```
Edsrf/SAve/eXit/ASsembly/Display/SRfcon/Redraw <eXit>: e
Addsurf/Delsurf/Modify/MName/Points/Redraw/eXit <eXit>: a
```

The option returns a new prompt.

```
Draw/Move/Select <Select>: d
```

5. Run the Draw option. The option is just like the Draw Template command that drew the original template surfaces.

The starting point is the center endpoint on the bottom of the Bottom Bit surface (see Figure 9–59).

```
Starting point: endpoint  (select vertex 1)
```

The next endpoint along the granular surface is the bottom of the Bottom Bit surface. If the next prompt is not Select Point, type a **p** to return to Point mode.

```
Change in Offset (Grade/Slope/Close/Points/Undo/eXit): p
Select point (Relative/Grade/Slope/Close/Undo/eXit): endpoint (outer bottom
  of Bottom Bit surface - vertex 2)
```

The next endpoint is on the curb subassembly. Since the subassembly cannot be referenced with object snaps, you need to toggle into Relative mode from the current position. Toggle on Relative mode by typing an **r** at the option list prompt. There is no change in the offset from the current point, only a drop of .25 of a foot. The surface polyline then moves under the curb and then down to give depth to the granular surface. The surface ends back at the centerline, now 14.0 feet to the right, 12.5 feet for the roadway, and 1.5 feet for the curb.

```
Select point [Relative/Grade/Slope/Close/Undo/eXit]:
Select point [Relative/Grade/Slope/Close/Undo/eXit]: r
Change in offset [Grade/Slope/Close/Points/Undo/eXit]: 0
Change in elev: -.25
Change in offset [Grade/Slope/Close/Points/Undo/eXit]: -1.5
Change in elev: 0
Change in offset [Grade/Slope/Close/Points/Undo/eXit]: 0
Change in elev: -.25
Change in offset [Grade/Slope/Close/Points/Undo/eXit]: 14
Change in elev: 0
Change in offset [Grade/Slope/Close/Points/Undo/eXit]: x (press ENTER)
```

The Draw option now turns to the definition of the new surface. The first prompt is about the symmetry of the surface. Then after you press ENTER to select a material, the routine displays the Surface Materials Names dialog box and you select Crsh Base Course.

```
Surface type [Sym/Asym]: s
Press any key to select material name:
```

The Draw option mirrors the granular surface, and you return to the main prompt line of Edit Template.

```
Addsurf/Delsurf/Modify/MName/Points/Redraw/eXit <eXit>: (press ENTER)
Edsrf/SAve/eXit/ASsembly/Display/SRfcon/Redraw <eXit>: (press ENTER)
```

The current datum definition does not include the granular surface. You need to redefine the datum to include the granular surface.

Editing the Datum

The last task is to edit the datum of the template. Currently, the datum follows the bottom of the Base Bit surface, not the new granular surface. Either edit the datum to include the granular surface or define a second datum.

Rather than redefining the first datum, define a new datum. The second datum starts at the back of the curb just like the first datum. The datum, however, continues straight down to the bottom of the granular surface and over to the outer right side of the template.

6. Define a second datum for the surfaces.

```
Edsrf/SAve/eXit/ASsembly/Display/SRfcon/Redraw <eXit>: sr
Connect/Datum/Redraw/Super/Topsurf/TRansition/eXit <eXit>: d
Datum number <1>: 2
Pick datum points (left to right): endpoint (left top vertex of
  granular)
Pick datum points (left to right): endpoint (left bottom vertex of
  granular)
Pick datum points (left to right): endpoint (right bottom vertex of
  granular)
Pick datum points (left to right): endpoint (right bottom vertex of
  granular) (press ENTER)
Pick datum points (left to right): (press ENTER)
Connect/Datum/Redraw/Super/Topsurf/TRansition/eXit <eXit>: x
```

The next question is whether the template definition routine understands the fact that the datum of the curb is different in each datum. In the first datum, the datum for the curb goes down the back of the curb, across its bottom, and to the front of the curb. In the second datum, the datum just goes down the back side of the curb. To view the results of the definition, go into the Display option and view the different datums.

```
Edsrf/SAve/eXit/ASsembly/Display/SRfcon/Redraw <eXit>: r
Edsrf/SAve/eXit/ASsembly/Display/SRfcon/Redraw <eXit>: d
Display
[Datum/Connect/Points/Super/SHoulder/Topsurf/TRansition/eXit/
  Redraw/TType]
<eXit>: d
Datum number <1>: 1
Display[Datum/Connect/Points/SHoulder/Super/Topsurf/TRansition/
  eXit/Redraw/TType]
```

```
<eXit>: r
[Datum/Connect/Points/SHoulder/Super/Topsurf/TRansition/eXit/
  Redraw/TType]
<eXit>: d
Datum number <1>: 2
Display
[Datum/Connect/Points/SHoulder/Super/Topsurf/TRansition/eXit/
  Redraw/TType]
<eXit>:
Display
[Datum/Connect/Points/SHoulder/Super/Topsurf/TRansition/eXit/
  Redraw/TType]
<eXit>: x
Edsrf/SAve/eXit/ASsembly/Display/SRfcon/Redraw <eXit>:
```

The second datum does not look very good. The template does not understand that it must have two definitions to handle the current situation. If you are defining the Curb612 datum to be the back of the curb, the first datum definition becomes erroneous.

The second datum is all that is needed. The datum of the curb needs redefinition to match the second datum of the template. Since the current routine is Edit Template, no changes can be made to the subassemblies. You need to exit the Edit Template routine and run the Edit Subassembly routine.

7. Exit the Edit Template routine and save the changes to the template. Erase the entities from the screen.

```
[Datum/Connect/Points/SHoulder/Super/Topsurf/TRansition/eXit/
  Redraw/TType]
<eXit>: x
Edsrf/SAve/eXit/ASsembly/Display/SRfcon/Redraw <eXit>:
Save template [Yes/No] <Yes>: (press ENTER)
Template name <std21>: (press ENTER)
Template exists. Overwrite [Yes/No]: y (press ENTER)
Command:
```

Editing a Subassembly

To fix the errors in the datum, the curb subassembly needs to be edited. Editing the subassembly is much like editing a template. The Edit Subassembly routine is in the Templates cascade of the Cross Sections menu.

1. Run the Edit Subassembly routine in the Templates cascade of the Cross Sections menu.

The routine presents the Subassembly Library dialog box.

EXERCISE

2. Select the curb612 subassembly and place it in a open area in the drawing.

```
Command:
Pick insertion point: (select a point on the screen)
Delete/Insert/Next/SAve/Previous/SRfcon/Move/Redraw/eXit <Next>:
```

The changing of the datum is the domain of the Surface Control option.

```
Delete/Insert/Next/SAve/Previous/SRfcon/Move/Redraw/eXit <Next>: sr
Edit [Datum/Connect/Topsurf/eXit/DIsplay/Redraw] <eXit>: d
Pick datum points (left to right): endpoint (top back of the curb)
Pick datum points (left to right): endpoint (bottom back of the curb)
Pick datum points (left to right): (press ENTER to exit)
```

3. View the changes to the datum with the Display option of Edit Subassembly. Redraw the subassembly before viewing the new datum definition. This new definition will not be correct for datum 1, but it will be correct for datum 2.

```
Edit [Datum/Connect/Topsurf/eXit/DIsplay/Redraw] <eXit>: di
Display [Datum/Connect/Topsurf/eXit/Redraw] <eXit>: r
Display [Datum/Connect/Topsurf/eXit/Redraw] <eXit>: d
```

4. Exit the Display option and editing routine and save the edited subassembly.

```
Display [Datum/Connect/Topsurf/eXit/Redraw] <eXit>:
Edit [Datum/Connect/Topsurf/eXit/DIsplay/Redraw] <eXit>:
Delete/Insert/Next/SAve/Previous/SRfcon/Move/Redraw/eXit <Next>: x
Save subassembly [Yes/No] <Yes>:
Subassembly name <curb612>:
Subassembly exists. Overwrite [Yes/No]: y
```

Changing the Datum in Design Control

The new datum has to become a part of the design criteria. By setting datum 2 as the current datum in the Template Control dialog box of the Edit Design Control routine, you change the volume calculation for the template and design. After you make the change, the design processor will reprocess the design data.

1. Open the Roadway drawing of the Roadway project. Save the changes to the CDTPL drawing.
2. Run the Edit Design Control routine of the Design Control cascade of the Cross Sections menu.

The routine presents you with a prompt for an alignment. Press ENTER at the prompt, select the Bypass alignment from the Alignment Librarian dialog box, and select OK to exit the dialog box.

3. Select the OK button to accept the beginning and ending stations of the alignment for this session.
4. First select the Template Control button and then choose the Select button to the right of the Datum number.

The routine displays the Datum Librarian. Notice that Datum 1 cuts across the bottom of the curb to meet the inner surface datum.

5. View datum numbers 1 and 2. Select datum number 2 and exit the dialog box by selecting the OK button (see Figure 9–60). Select the OK button to exit the Template Control dialog box.

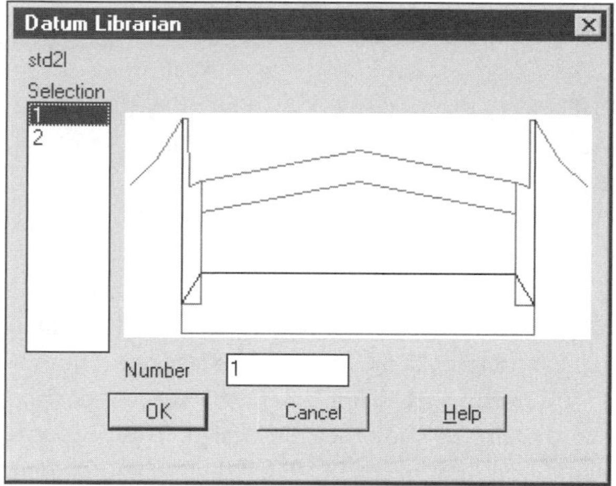

Figure 9–60

6. Select OK to exit the Design Control dialog box to process the design. The design still has errors. The design is still pinned to the right side. Select OK to exit the Section Processing Status and Process Status dialog boxes.
7. Use the View/Edit Sections routine of the Cross Sections menu to view the new template in the cross sections. The new surface and datum are now a part of the design's sections.

MODIFYING THE DESIGN

Upon further review, the vertical for the roadway centerline needs to change to balance the volumes. The changes involve adding a new tangent and vertical curve to the roadway. As a result of the changes, the design will demand less fill and create more cut. This will help balance the roadway better than the current design.

1. Rerun the To Screen volume report of the Total Volume Output cascade of the Cross Sections menu to view the overall volume of the roadway.

EXERCISE

END AREA VOLUME LISTING WITH CURVE CORRECTION

Station	Cut Area (sqft) Volume (yds)	Fill Area (sqft) Volume (yds)	Cut Cumulative Volume (yds)	Fill Cumulative Volume (yds)
22+02.69	114.80	0.00		
	199.59	0.00	3644.48	10498.20
22+50	113.02	0.00		
	205.77	0.00	3850.25	10498.20
23+00	109.21	0.00		
	20.61	0.00	3870.86	10498.20
23+05.10	109.01	0.00		
	189.48	0.00	4060.34	10498.20
23+50	118.87	0.00		
	204.99	0.00	4265.32	10498.20
24+00	102.51	0.00		
	154.23	0.86	4419.55	10499.06
24+46.88	75.13	0.99		

The volume report of the current design indicates the need for fill. The amount of cut material does not equal the amount of fill material (the last two values from the Cumulative Cut and Fill columns). This means that to implement the current design requires the builder to import material into the site to construct the design. The proposed changes may help balance the design (even the amount of cut and fill). If these changes do not even the balance, you may have to raise or lower the elevation of the two central PVIs to adjust the volume calculation. The changing of PVI elevations will produce more cut or fill and again help balance the volumes.

The decision to change the vertical aspects of the road means returning to the Profiles menu. In the Profiles menu, Civil Design provides tools to edit the vertical alignment. In this case, add a new tangent and vertical curve to the design.

2. Exit the report and restore the profile view.

There are two methods to edit the vertical alignment. The first is to redraw the tangents and create the new vertical curves within the profile. After redrawing the vertical design, you have to redefine the vertical alignment. The second method is to directly edit the vertical alignment numbers in the Vertical Alignment Editor. After saving the changes, you need to import the new definition into the profile.

The Edit Grade 1 and Grade 2 tools of the FG Centerline Tangents cascade only work on tangents that do not have a vertical curve. The current vertical alignment tangents have a vertical curve, so the Edit Grade 1 and Grade 2 tools do not apply.

You will make the changes in the Vertical Alignment Editor.

The editor allows for the insertion of PVIs (Points-of-Vertical Intersection) into the alignment (see Figure 9–61). The Insert PVI button places a blank PVI cell before the current PVI in the dialog box. You set the current PVI by selecting into the cell of a station. After you insert the PVI, the next step is to assign the PVI a station, an elevation, and then place a vertical curve between the two new tangents.

3. Run the Edit routine of the FG Vertical Alignments cascade in the Profiles menu.
4. In the Vertical Alignment Editor, insert a new PVI before station 24+46.88. The new PVI is at station 10+50 and its elevation is 728.00. Select the station 2446.88 box and then select the Insert PVI button. An empty station box appears before 2446.88. Enter **1050** and select the adjacent elevation cell and make the elevation of station 10+50 728.00 (see Figure 9–61).

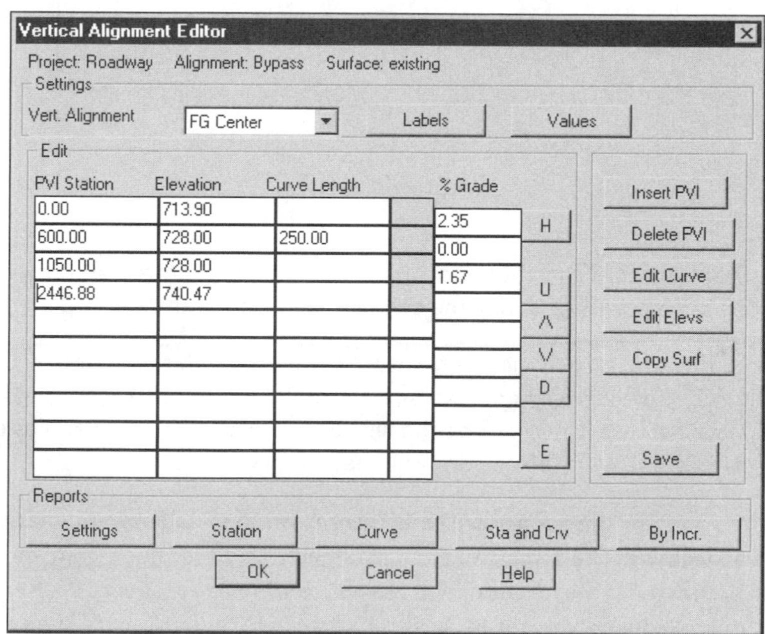

Figure 9–61

Notice that the dialog box changes the grades on the tangents adjacent to the new PVI.

5. Select the Edit Curve button and create a vertical curve with a length of 125. The Vertical Curve Detail Window appears. Select the Curve Length box and set the length to 125.00. Press ENTER on the new curve length. After you press ENTER in the Curve Length box, the new value forces the dialog box to calculate the vertical curve data (see Figure 9–62). Exit the Vertical Curve Detail Window by choosing the OK button.

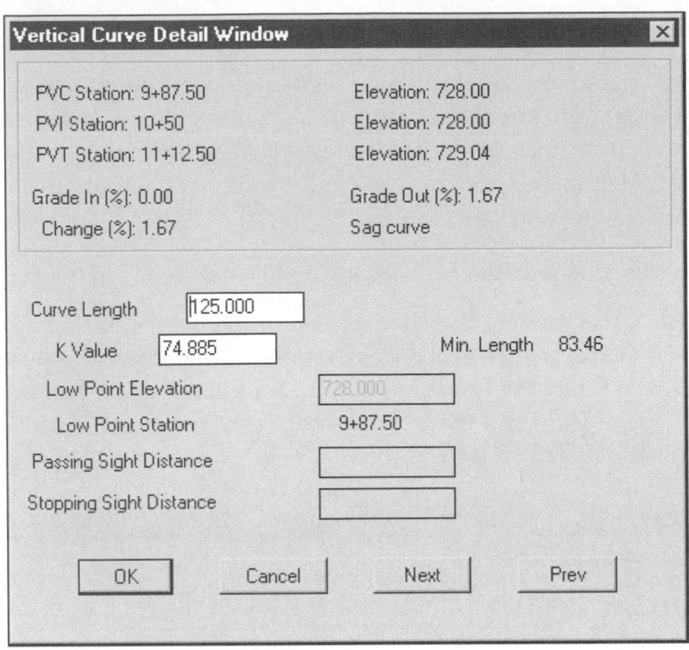

Figure 9–62

6. Exit the Vertical Alignment Editor by selecting the OK button and when the routine prompts about saving the editing session, answer Yes.

7. Import the new vertical alignment into the drawing with the Import routine found in the FG Vertical Alignments cascade of the Profiles menu. *Do not* annotate the alignment because it may not be the final design. However, answer Yes to erase the existing line work before drawing the new alignment.

In order for you to view the impact of these changes on the roadway volumes, the design data needs reprocessing. The template, ditch parameters, and its slope control remain the same. What has changed is the location of the template within a cross section. Rerunning the Edit Design Control routine will not process the changes. You must force the reprocessing of the design with the Process Sections routine of the Design Control cascade of the Cross Sections menu.

8. Run the Process Sections routine in the Design Control cascade of the Cross Sections menu. The routine prompts about reprocessing the design from the beginning to the end of the alignment. Choose the OK button to reprocess the design from beginning to end.

9. Choose the OK button to dismiss the Section Processing Status and Process Status dialog boxes to return to the command prompt. Some stations still have the daylight points pinned to the ROW.

10. View the new volume totals by running the To Screen routine in the Total Volume Output cascade of the Cross Sections menu.

END AREA VOLUME LISTING WITH CURVE CORRECTION

Station	Cut Area (sqft) Volume (yds)	Fill Area (sqft) Volume (yds)	Cut Cumulative Volume (yds)	Fill Cumulative Volume (yds)
22+02.69	233.00	0.00		
	384.73	0.00	7822.88	6176.60
22+50	206.15	0.00		
	354.77	0.00	8177.64	6176.60
23+00	177.00	0.00		
	33.16	0.00	8210.8	6176.60
23+05.10	174.16	0.00		
	280.88	0.00	8491.69	6176.60
23+50	163.65	0.00		
	265.84	0.00	8757.53	6176.60
24+00	123.45	0.00		
	172.41	0.86	8929.93	6177.46
24+46.88	75.13	0.99		

The results swing to an abundance of cut material. So now the design creates more extra material than it can use.

11. If you have not already done so, erase the old Mass Haul diagram and create a new diagram. You may have to pan down the screen to create the diagram.

```
Command:
Alignment Name: Bypass Number: 1 Descr: Bypass Prelim Design of 12/00
Starting Station: 0.00  Ending Station: 2446.88
Volume computation type [Prismoidal/Avgendarea] <Avgendarea>:
Use of curve correction [Yes/No] <Yes>:
Use of volume adjustment factors [Yes/No] <Yes>: n
Pick insertion point:
Beginning station <0>:
Ending station <2446.88>:
Vertical scale (cu. yds.) <1000>:
```

The new design parameters fail to balance the volumes. The design now has more cut than the roadway can use. The design needs to create less cut and more fill volumes to balance. To create less cut, you need to raise the elevation at station 10+50. The raising of the elevation will create a need for more fill.

EXERCISE

12. Erase the Mass Haul diagram.

13. Restore the profile view or pan to the profile.

14. Run the Edit routine of the FG Vertical Alignments cascade and raise the elevation of the PVI at station 10+50 to 729.50. Notice that the grades change entering and exiting station 10+50. Select into the elevation box for station 10+50 and change the elevation to 729.50.

15. Exit the editor by choosing the OK button and the Yes button to save the changes. Bring the new vertical alignment into the drawing with the Import routine found in the FG Vertical Alignments cascade. *Do not* annotate the alignment; it may not be correct. Answer Yes to erase the existing line work.

16. Run the Process Sections routine in the Design Control cascade of the Cross Sections menu. The routine prompts for the stations to reprocess. In this case we will reprocess from the beginning to the end of the alignment. Choose the OK button to start the reprocessing of the design.

17. Choose the OK button to dismiss the Section Processing Status and Process Status dialog boxes to return to the command prompt. Some sections still have their daylight slope pinned to the end of the section.

18. View the new volumes by running the To Screen routine in the Total Volume Output cascade of the Cross Sections menu.

The cut and fill volumes are close to being balanced.

19. Restore the view, Mass Haul, or pan to where you can draw the diagram and if you have not already done so, erase the previous Mass Haul diagram.

20. Create a new Mass Haul diagram (see Figure 9–63).

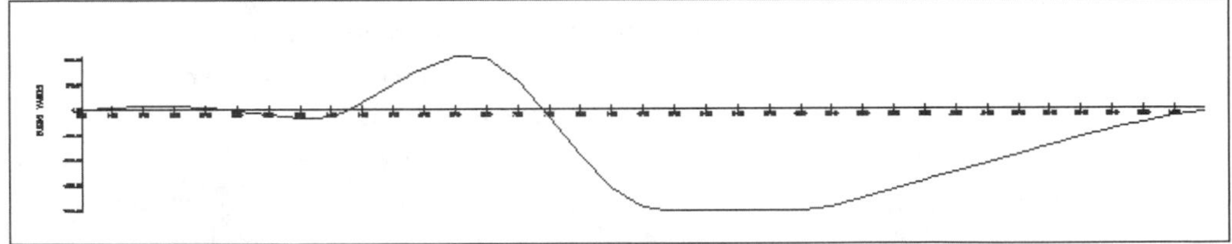

Figure 9–63

```
Command:
Alignment Name: Bypass Number: 1 Descr: Bypass Prelim Design of 12/00
Starting Station: 0.00  Ending Station: 2446.88
```

EXERCISE

```
Volume computation type [Prismoidal/Avgendarea] <Avgendarea>:
Use of curve correction [Yes/No] <Yes>:
Use of volume adjustment factors [Yes/No] <Yes>: n
Pick insertion point:
Beginning station <0>:
Ending station <2446.88>:
Vertical scale (cu. yds.) <1000>:
```

The design is close to balancing. Our work is finished.

The vertical alignment still is without annotation. You can annotate the alignment in one of two ways, manually or automatically. The first, manual, has labeling routines in the Label cascade of the Profiles menu (see Figure 9–64). You can label the tangents and the vertical curves by using the appropriate labeling routine and selecting the appropriate Desktop objects. The second way, automatic, labels all of the elements of the vertical alignment automatically. Automatic annotation occurs as an option when importing the vertical alignment into the drawing.

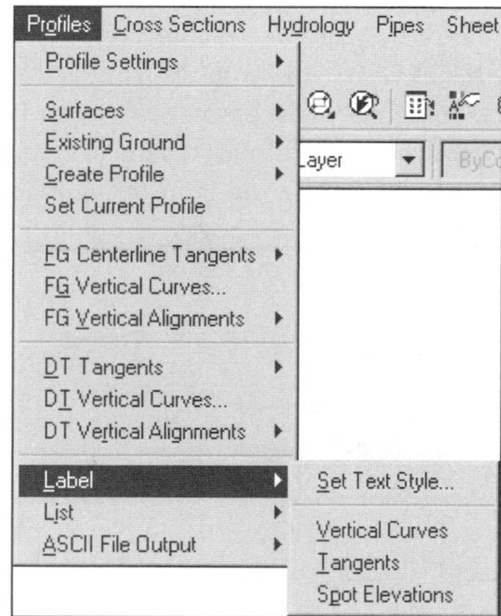

Figure 9–64

21. Restore the view, Profile, or pan to view the profile.

22. Run the Import routine in the FG Vertical Alignment cascade of the Profiles. Answer Yes to the annotation and Erase objects prompts. Compare your profile to Figure 9–65.

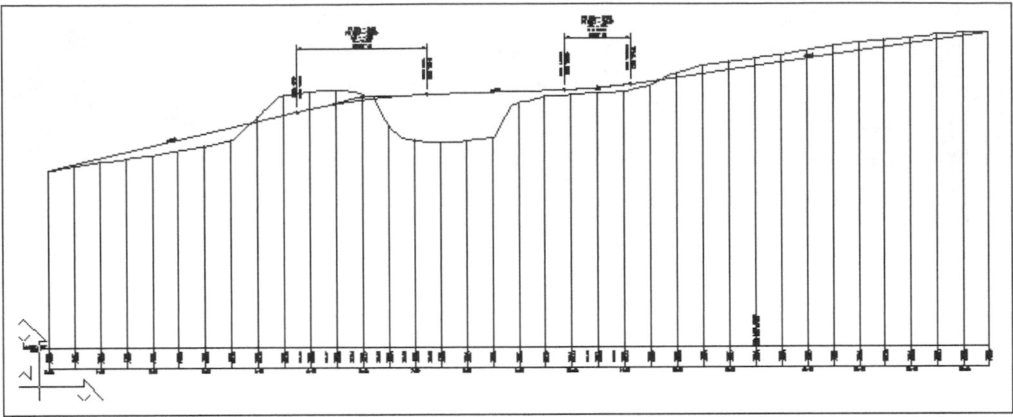

Figure 9–65

23. Save the drawing file.

Surface Volumes

The Template Surface routine of the Surface Volume Output cascade generates a report containing the volumes of surfaces within the template (see Figure 9–66). The report is a file report. If defining Top Bit, Bottom Bit, Granular base, and Curb surfaces, the routine generates a volume report for each surface. The report indicates the first surface, but omits the names of any other surfaces in the report. To see the individual surface names as a part of the report, you need to verify that the Subheaders toggle is on in the Output Settings routine. The totals for the surfaces are at the end of the report.

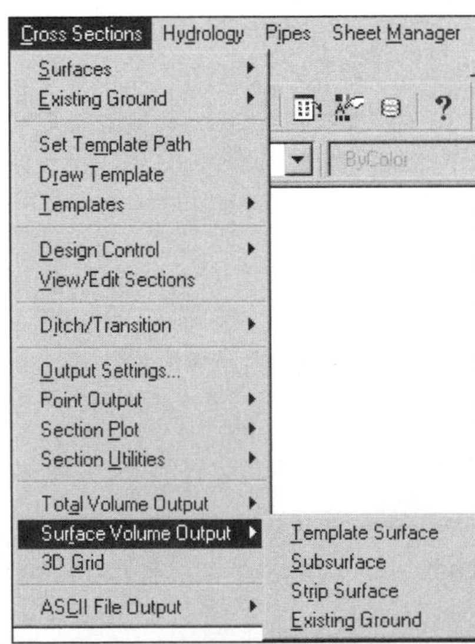

Figure 9–66

This report does not prompt for volume adjustment values because the routine looks for a file containing adjustment factors from a list of surface names. The Volume Adjustment Factors routine of the Design Control cascade modifies the adjustment factors in the file. Since the materials bituminous, concrete, and granular have different shrinkage factors, Civil Design allows you to specify different amounts of shrinkage for each surface in the template.

1. Run the Output Settings routine in the ASCII File Output cascade of the Cross Sections menu and toggle on Subheaders (see Figure 9–67).

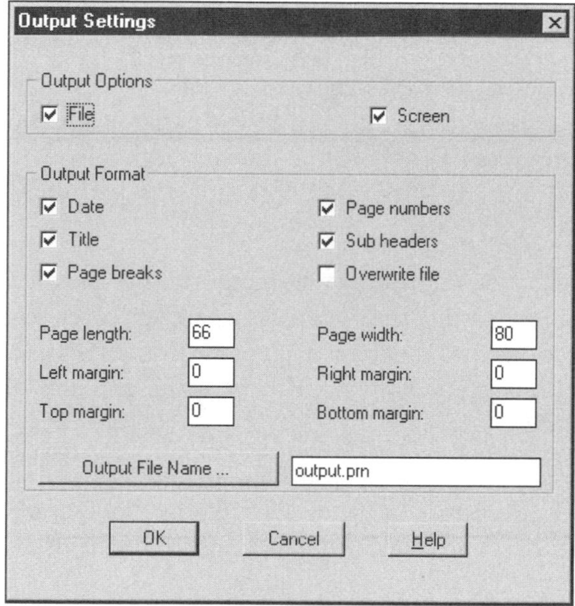

Figure 9–67

2. Run the Template Surface report. The routine places the file in the project folder. View the report with an editor.

```
Project:          Roadway                    Sat April 22 14:50:42 2000
Alignment:        Bypass
Start range:      0+00 to 24+46.88

    TEMPLATE END AREA VOLUME SUMMARY WITH CURVE CORRECTION
        Surface                                  Volume (yds)
   ------------------------------------------------------------
        Base Bituminous                          1039.86
        Asphalt                                  415.94
        Crsh Base Couse                          881.80
        Curb                                     225.94
        Sod                                      0.00
```

UNIT 7: CREATING CROSS SECTIONS

The routines that create and annotate the cross sections are in the Section Plot cascade and Section Utilities cascade of the Cross Sections menu. The sections created by the routines in the Section Plot cascade are more for the final design than for the evaluation of the design.

After you review the sections with the View/Edit Sections routine and decide to keep the design results unaltered, the next step is importing the cross sections into a drawing. You can import the cross sections into any drawing that is a part of the project. The Section Plot tools import sections into the drawing individually, as a page, or all (a semiorganized page but with no border). The sections can be plotted to their own drawing if you want.

The location of each section in the drawing is important to the routines of the Section Utilities cascade. The location of the sections is a part of a block placed in the drawing when placing the sections into the drawing. You must not move the sections around. If you move the sections in the drawing, the reporting and annotating routines will return incorrect values.

If you need to relocate the sections, you have to follow three steps: undefine the sections (the Undefine Sections routine of the Sections Plot cascade), erase the objects representing the sections, and reimport the sections into the drawing.

SECTION PLOT CASCADE

SETTINGS

This routine calls the Cross Section Plotting Settings dialog box. This is the same dialog box called in the Drawing Settings routine (see Unit 1 and Figures 9–10, 9–11, and 9–12).

TEXT STYLE

This routine calls the Desktop Text Styles dialog box and allows you to define and/or set current a text style used by the labeling routines of Section Utilities.

SINGLE

The Single routine plots individual cross sections at user-selected points in the drawing.

PAGE

The Page routine creates one or multiple pages of cross sections in the current drawing or they can be exported to their own drawing.

ALL

The All routine plots all of the cross sections to the current drawing without a page frame.

UNDEFINE SECTION

As a routine places a section into the drawing, a block accompanies it. This block contains several pieces of information about the section. If you move the section(s) and do not adjust the values in the related block, the listing/labeling routines of Section Utilities will report erroneous values. The only good way of moving the sections is to undefine them and reimport them at their new location in the drawing.

SECTION UTILITIES CASCADE

SELECT ROUTINES

The top portion of the Section Utilities cascade contains routines to select a current section (station) and then to zoom to it (see Figure 9–68).

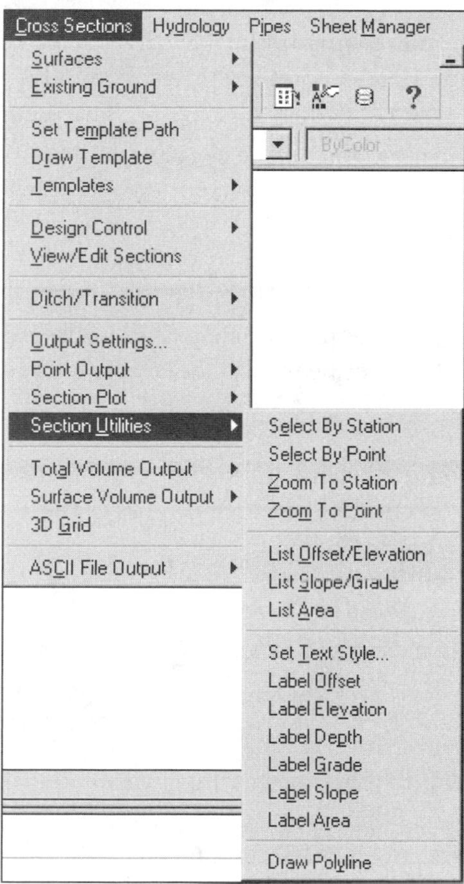

Figure 9–68

Select by Station

The Select by Station routine allows you to set the current station by entering the station number. The station number is a decimal value, for example, 100 not 1+00. The routine does not zoom to the cross section.

Select by Point

The Select by Point routine sets the current station by selecting in the area of the section representing the station. The routine does not zoom to the cross section.

Zoom to Station

The Zoom to Station routine zooms the display to a cross section (station). The routine prompts for the station number and the height of display to use when showing the section. The section is exaggerated, and you need to be aware of a reasonable display height. The sections of the next exercise range from 100 to 175 feet in height.

Zoom to Point

The Zoom to Point routine zooms the display to a selected point within a cross section (station). After you select a point within a section, the routine prompts for the height of display to use when showing the section. The section is exaggerated, and you need to be aware of a reasonable display height. The sections of the next exercise range from 100 to 175 feet in height.

SECTION LISTING ROUTINES

List Offset Elevation

The List Offset Elevation routine lists the station and offset of selected points within the current cross section. The routine reports the results in the command prompt area.

List Slope/Grade

The List Slope/Grade routine lists the slope and grade between two selected points within the current cross section. The routine reports the results in the command prompt area.

List Area

The List Area routine lists the area of selected points within the current cross section. Since there are objects in the drawing, you can use the Desktop object snaps to select the points representing the area. The routine reports the results in the command prompt area.

LABEL

The lower portion of the Section Utilities cascade of the Cross Sections menu contains routines that label specific points within the cross section. The labeling routines prompt for the location of a label in the current section. There are two methods for placing labels into a section, automatic and manual. The automatic method prompts for the orientation of the text and places it at the selected location. The manual method places text anywhere within the section and with a specific rotation angle for the text.

Set Text Style

This routine calls the Text Style dialog box of the Desktop.

Label Offset

This routine labels the offset of the selected point in the current cross section. The labeling portion of the routine prompts for a rotation angle and an auto(matic) or manual placement of the offset. The Auto option prompts you for another choice, random or linear. The Random option places the offset label at the point selected in the cross section. The Linear option prompts you to select a point that defines a line parallel with the datum and to which the label is parallel.

The Manual option creates a label and prompts you for the angle and location of the label.

Label Elevation

This routine labels the elevation of the selected point in the current cross section. See Label Offset for a description of the label placement options.

Label Depth

This routine labels the vertical depth of two selected points in the current cross section. See Label Offset for a description of the label placement options.

Label Grade

This routine labels the grade between two selected points in the current cross section. See Label Offset for a description of the label placement options.

Label Slope

This routine labels the slope between two selected points in the current cross section. See Label Offset for a description of the label placement options.

Label Area

This routine labels the area of a set of selected points defining a closed boundary. See Label Offset for a description of the label placement options.

Draw Polyline

The Draw Polyline routine is the same as the Draw Template routine. This routine allows you to draw new line segments within a cross section. The routine understands grades, slopes, and vertical exaggeration.

Exercise

After you complete this exercise, you will be able to:

- Import cross sections
- Use the routines of the Section Utilities menu

IMPORTING CROSS SECTIONS

1. Start a new drawing using the *aec_I* template drawing and name the drawing CDSEC (see Figure 9–69). Save the changes to the Roadway drawing. Save the changes to the Roadway drawing.
2. Assign the new drawing to the Roadway project.

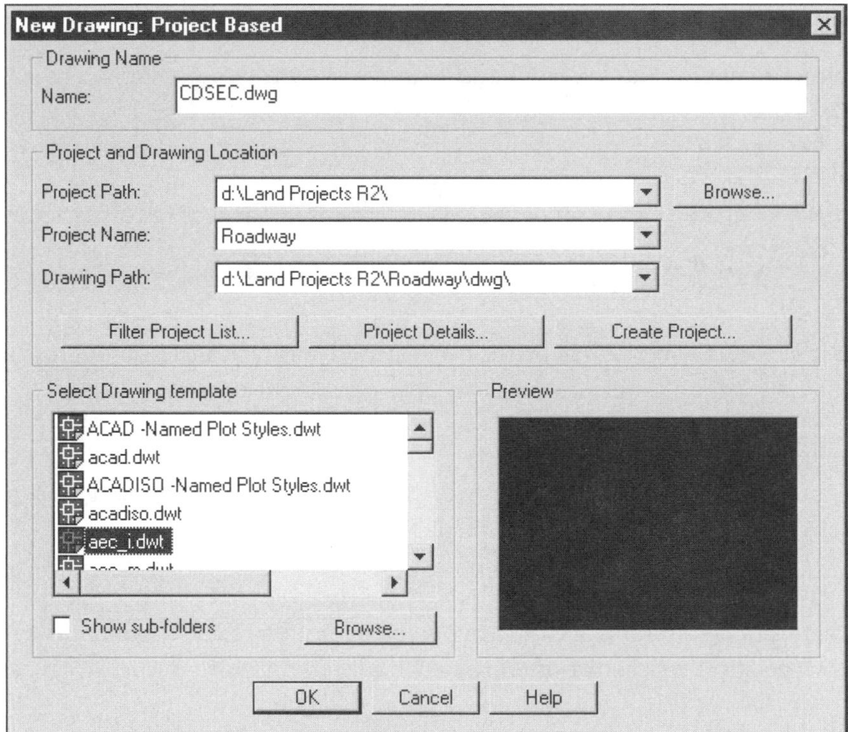

Figure 9–69

3. Load the LDD1-40 settings by selecting the LDD1-40 setup file (see Figure 9–70) in the Load Settings dialog box.

Before importing the sections, you need to review the section plot settings. The settings set values for the section layers, sheet sizes, sheet margins, and annotation.

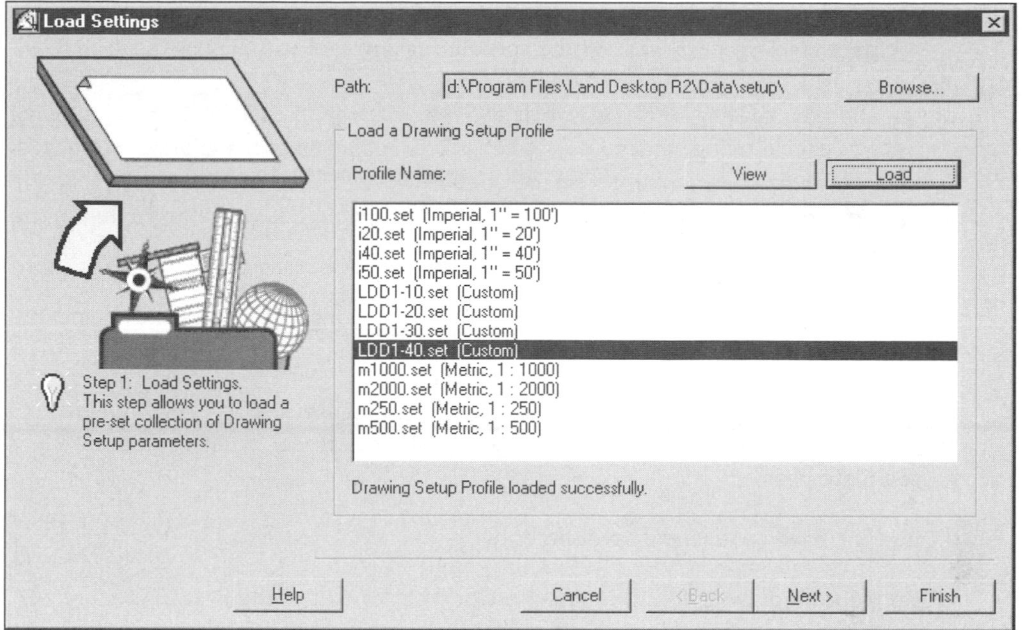

Figure 9–70

4. Rerun the Settings routine of the Section Plot cascade of the Cross Sections menu. The Cross Section Plotting Settings dialog box toggles on the elements of a cross section and sets the layers for these elements (see Figure 9–10). Make sure all of the toggles are on.
5. Select the Section Layout button to view the dialog box containing the settings for the offsets and elevations in the section. Use Figure 9–11 as a guide to setting the values.

The Offset and Elevation increments define the grid around each section. The default is every 10 feet horizontally and 2 feet vertically.

The Offset label and Elevation label increments set how often the labeling occurs on the grid. A value of 1 means each grid line has a label. A setting of 2 means every other grid line, and so forth.

The FG and EG label precisions set the precision for the existing and finished ground elevation label.

The Rows below and above the datum set the number of extra lines above or below the body of the grid. Generally, this value is zero (0).

6. Exit the Section Layout dialog box by selecting the OK button.

7. Select the Page Layout button to view the dialog box containing the settings for the page (see Figure 9–12). Set the column spacing to 4 and the row spacing to 4.

The column and row spacing setting controls the space between each cross-section. Civil Design refers to each spacing unit as a cell. A cell is one rectangle of the cross-section grid. Row spacing controls the number of spaces between each row of cross sections, the vertical spacing. Column spacing controls the number of spaces between each column, the horizontal spacing.

8. Exit the Page Layout and Cross Section Plotting Settings dialog boxes by selecting the appropriate OK buttons.
9. Review the template path for the drawing (see Figure 9–16). The template path should point to the *adtpl_l* folder. This is the folder where the std2l template resides.
10. Use the Save routine and save the drawing.
11. Reopen the drawing file. This sets the Undo back point of the drawing to a newly opened drawing file.
12. Use the Page routine of the Section Plot cascade of the Cross Sections menu to import a single page to view the results of the defaults. Import the sections into the current drawing from station 0.00 and place the page at 0,0. When the routine prompts for an alignment, press ENTER and select the Bypass alignment from the dialog box. The spacing of the sections is not very good in the sheet.
13. Undo the sections and the frame with the Back option of Undo. Undo back to the beginning of the drafting session.
14. Reenter the Settings routine of Section Plotting and set the column spacing to 8 and row spacing to 4 in the Page Layout dialog box.
15. Import a single page to view how the new values impact the page layout. Import the sections into the current drawing from station 0.00 and place the page at 0,0. The spacing looks good.
16. Undo the sections with the Back option of Undo. Use the Page routine to import the multiple pages of sections. The Undo does not affect the any changes you make to the Cross Section Plotting Settings.
17. After importing all of the sections, use Zoom Extents to view all of the cross sections.

EXERCISE

Section Utilities Menu

1. Use the AutoCAD Dist routine to measure the top to bottom of a few cross sections. This gives you an approximate height to give to the zoom to a section routine.
2. Use the Zoom to Station routine to view how Civil Design sets and zooms to the current cross section.

To be able to use the tools in the Section Utilities menu, you need to zoom to a section to see it better.

3. Use the Zoom to Point routine to set and view the current section. Try zooming to a few different stations.
4. Finally, use the Zoom to Station routine and zoom to station 7+00. Use the height of 200.

The second group of tools in the Section Utilities menu is listing tools (see Figure 9–68). These listing tools evaluate an offset and elevation, a slope or grade, or an area.

5. List some offsets and elevations, slopes and grades, and areas in the current section.

The last group of tools in the Section Utilities menu labels the values found in the section (see Figure 9–68). The labeling routines are nothing more than the individual routines grouped together in the Listing group of the Section Utilities menu.

6. Label some of the values in the cross section. If necessary, zoom to a new section and try other labeling routines. Label some of the values using the Manual method and others using the Auto methods. When using the Auto method, try using the Random and Linear option.
7. Open the Roadway drawing and save the changes to the CDSEC drawing.

UNIT 8: SHEET MANAGER

The generation and plotting of roadway design sheets is not an easy process. The process of defining match lines, rotating centerline data, border, and defining annotation settings is complex and lengthy. The Sheet Manager is a way of applying an office standard to a set of plots. The sheet styles of Sheet Manager become useable to any job requiring these types of sheets. When modifications are made to a style, they can be saved under the project so that the modifications do not affect the original sheet style or they can be placed in the library as a new style.

The Sheet Manager is a plan, profile, plan and profile, and section sheet generation utility. Sheet Manager is an alternative to the Plot Sections portion of the Cross Sections menu. The Sheet Manager has its own pull-down menu in Civil Design (see Figure 9–71). The routine generates the sheets from a sheet style. Civil Design supplies an initial sheet design, but it is up to you to define any other sheet styles.

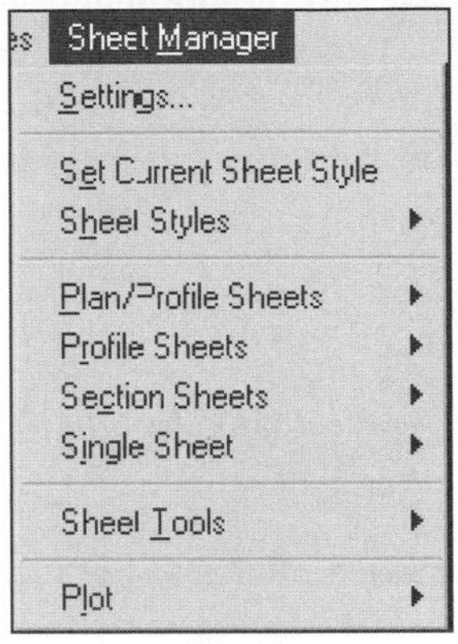

Figure 9–71

A sheet style contains one or two paper space viewports and several frames. The viewports present the Desktop objects representing the roadway design. The frames contain annotation definitions. Civil Design allows the stacking of viewports and frames to produce the final desired plot.

SHEET MANAGER SETTINGS

The Settings dialog box of the Sheet Manager controls several aspects of a plan, profile, or plan and profile sheet (see Figure 9–72). The Style Database is a path to a folder containing sheet styles. Civil Design creates four different libraries of sheets. You can add any number of libraries and sheet styles to the folder structure.

STYLE DATABASE

The top portion of the dialog box displays the path to the sheet library. When you select the Set button, Civil Design displays the Style Database Path dialog box (see Figure 9–73). The dialog box points to the installed location for the files.

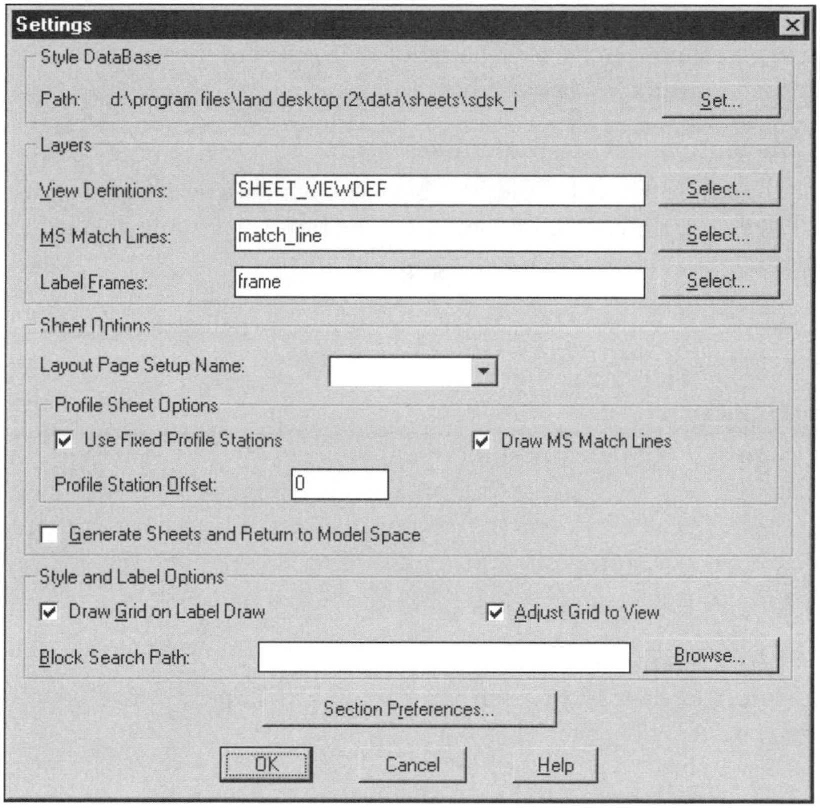

Figure 9–72

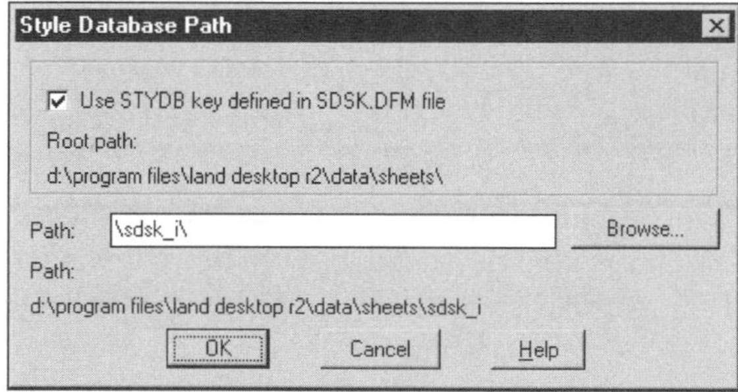

Figure 9–73

LAYERS

The Layers portion of the dialog box sets the initial layers needed for a sheet style. There are several other layers for a style; however, they are set in the creation of a sheet style.

SHEET OPTIONS

The next set of values to review is the sheet style options. There are two toggles in this group that control the Plan and Profile layout.

Fixed Profile Station and MS Matchlines

The first toggle, Use Fixed Profile Stations, sets which length controls the layout. With the toggle on, the controlling length is the profile stationing. With the toggle off, the alignment is the controlling length.

The second toggle sets the match lines in model space. The toggle does not have to be on to annotate the match lines.

Profile Station Offset

The next value in this group is the amount of offset to the first station of the profile. No value in the box means that the profile stationing starts at the edge of the viewport. A positive value means that the profile stationing starts after the start of the profile. A negative offset shows a space between the edge of the viewport and the start of the profile.

MODEL SPACE RETURN

The last toggle, Generate Sheets and Return to Model Space, returns you to model space upon generating the sheets.

STYLE AND LABEL OPTIONS

Draw Grid on Label Draw

The Draw Grid on Label Draw toggle redraws the grid any time the labels are drawn. When the toggle is off, the grid redraws only when generating the sheets or forcing the grid to be redrawn.

Adjust Grid to View

The Adjust Grid to View toggle controls the attachment of the grid to a frame. If the toggle is off, the grid places its line starting with the edge of the viewport. If the toggle is on, the grid aligns the first vertical grid line with the first vertical profile line or station.

Block Search Path

The Block Search Path box holds a path for blocks used in the profile that your custom blocks.

SHEET STYLE ANATOMY

The process of setting up and creating a sheet prototype is not intuitive. The process involves defining viewports, frames, and label attachments. Sometimes a frame can define more than one type of annotation. Some annotation elements do not blend well with others when present in the same frame. There is no documentation of the elements and their behaviors in combination with other elements. Much of the development of a prototype sheet is a trial-and-error process.

The best way to learn Sheet Manager is by dissecting the frame prototype supplied by Civil Design. The dissection allows you to view what elements combine and how to do the annotation stacking.

Since the task of Sheet Manager is complex, the settings and methods of developing a style are complex. You may find it necessary to traverse four levels of dialog boxes to make a single decision.

VIEWPORTS

When defining a sheet style with viewports, you must define the function of the object you design. When working with viewports, you must assign display responsibilities and scales to them. The Create Viewport routine draws the viewport and prompts for the responsibility and scale of the new viewport (see Figure 9–74). A viewport has one of two responsibilities viewing the plan or profile of the roadway segment. The Viewport category sets the responsibility and scale of an existing viewport.

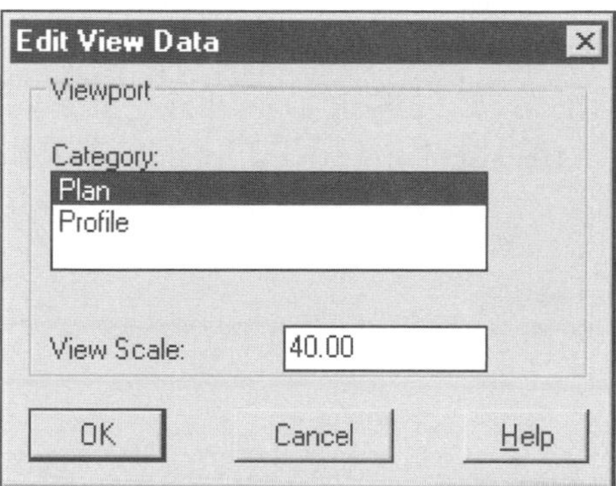

Figure 9–74

FRAMES

The Create/Edit Frame routine from the Sheet Styles cascade of the Sheet Manager menu creates and edits new and existing frames in a sheet style. The routine records the frame's coordinates and displays the Edit Frame Data dialog box. The Edit Frame Data dialog box contains the type of information that the frame displays as well as any other required information to make the annotation correct (see Figure 9–75).

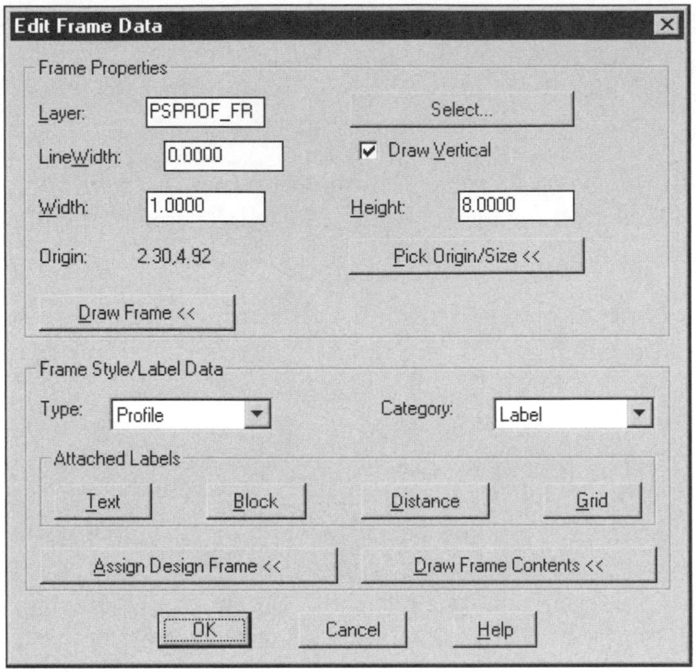

Figure 9–75

Frame Properties

The Frame Properties portion of the Edit Data dialog box controls the layer size and orientation of the frame in the sheet style (see Figure 9-75). In the case of the current frame, the following apply:

```
Layer: PS_PROF_FRAME
Line Width: 0.00
Width: 1.00
Height: 8.00
Origin: 2.30, 4.92
```

These pieces of information define the physical location and appearance of the frame. The frame is 1 by 8 inches in size and is drawn for vertical measurements.

Frame Styles/Label Data

Each frame contains data specifying what the frame annotates. The different options and values in this area of the Edit Frame Data dialog box set or modify the frame's characteristics.

Type. There are three types of frames: profile, plan, and section. This sets the basic ground rules for what can be a part of the frame (see Figure 9–75).

Category. A frame is one of four possible categories: label, view, table, and section (see Figure 9–75).

Label. The category label indicates that the frame will produce labels. The Type setting sets the kinds of annotation that appears within the frame. A vertical label frame annotates the profile elevations of the profile grid. A horizontal label frame annotates the station and station elevations along the length of the profile.

View. A View frame lies on top of a sheet viewport and allows annotation to be placed on top of the viewport. The View frame labels vertical tangent grades and design locations for profile viewports and stationing for plan viewports.

Table. A table frame is for cross sections only. This frame contains information that is not location-specific such as the area and volumes of the associated cross section.

Section. The section frame contains cross sections of the roadway design.

Edit Attached Text Labels

When you select the Text button in the Edit Frame Data dialog box, the selection displays the Edit Attached Text Labels dialog box (see Figure 9–76). This dialog box reports the assigned labeling responsibilities of the frame and the placement of the labels within the frame.

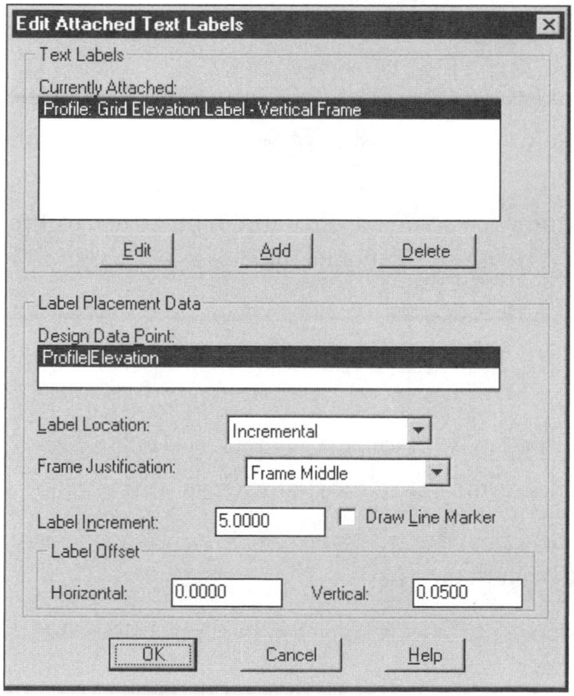

Figure 9–76

Currently Attached. The Currently Attached box in the Text Labels area displays what annotation set is attached to the frame. An annotation set contains the design point data that the set annotates. Some sets annotate only one data point. Other sets label more than one data point. Read the documentation of each annotation set very carefully.

The Text Labels area also contains three buttons, as follows.

Edit. When you select the Edit button. The Text Label Properties dialog box displays (see Figure 9–77). This dialog box sets the text style, size, justification, rotation, and layer. The Edit button at the bottom of the dialog box displays the definition of the label, the Edit Text Format dialog box (see Figure 9–78). This is where you can modify or create new labeling types.

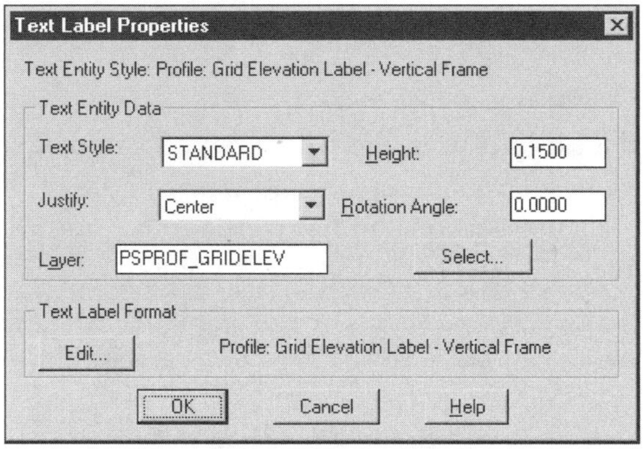

Figure 9–77

Add. The Add button allows you to add more annotation to the currently selected frame. A frame can have more than one annotation set assign to it.

Delete. The Delete button deletes a currently assigned annotation set to the frame.

Label Placement Data. This area of the dialog box describes the data annotated and its location.

Design Data Point. The Design Data Point box names the specific data point that is labeled out of the annotation set selected at the top of the dialog box.

Label Location. The Label Location box sets the methodology of the labeling to reflect the type of design data.

- Intersection—The label is a point in the design labeled in the frame, for example, PVI station.
- Incremental—The selected label is repetitive and within the boundary of the frame, for example, profile elevations.

- Design—The label is a point in the design labeled is in the viewport, for example, BVC station.
- Design Aligned—The label is aligned with the design element it labels, for example, the slope of a vertical tangent line in the profile view.
- Fixed—Creates a label at a fixed point on the frame, for example, a north arrow or project name.
- Design Incremental—Places labels on a design object at specified increment. This label works only with frame of view type, for example, alignment stationing at a 50-foot increment in the plan view of the roadway.

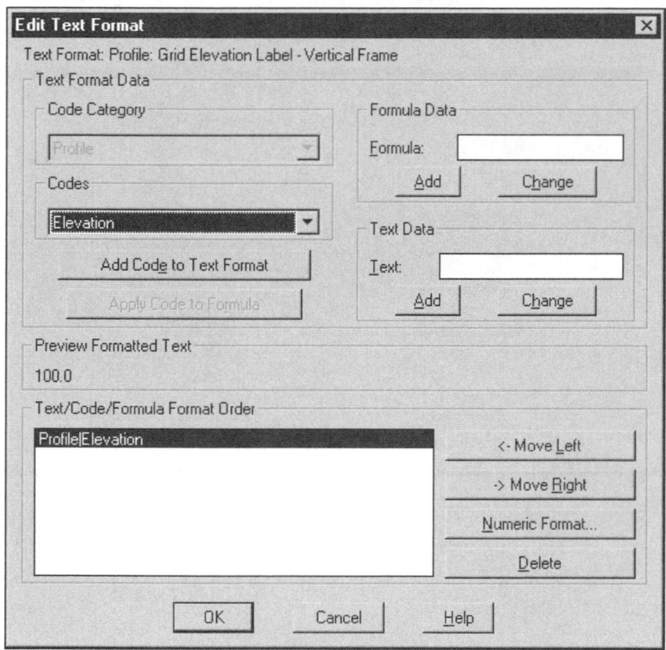

Figure 9–78

Frame Justification. Sets the justification of the label. There are three choices: Frame Left, Right, and Middle.

Label Increment. If the design point data is an incremental value, the value sets the increment that the label annotates.

Draw Line Marker. This toggle draws in the view frame a line from the frame label to the point in the view frame.

Label Offset. These values set the offsets for the text of the label in the frame. There are two types of offsets, horizontal and vertical. If there are more than one label type occurring in the frame, the offset allows you to change their positions so they do not overrun each other.

DEFINING A SERIES

After defining a sheet style, the next step is to create a sheet series. The Layout Sheet Series routine of the Plan/Profile Sheets cascade places a set of frames into the plan view of the alignment representing the location of the plan view of the sheet series. The first step is to name the series and the second step is to name the sheet style (see Figures 9–79 and 9–80). The Layout Sheet Series routine places frames sized from the scale in the plan viewport category. The horizontal scale has to be the same for both the plan and profile viewport.

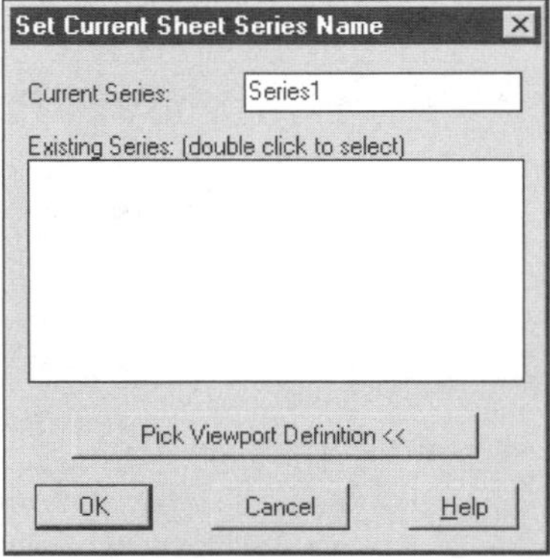

Figure 9–79

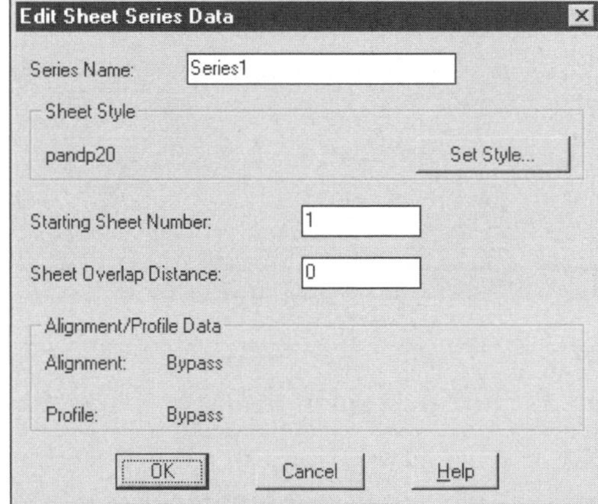

Figure 9–80

EDITING A SERIES

Once the series is laid out, the next step will be editing the location of specific frames and/or adjusting the datum elevation of the profile portion of the sheet. The Edit Sheet Layout routine displays a dialog box containing the stations of the plan view and allows for the moving and rotating of the plan view or the modifying of the datum elevation of the profile (see Figure 9–81). When editing the profile elevation, the routine displays a box enclosing the portion of the profile shown in the sheet (see Figure 9–82). The box will rise or fall in relation to the editing of the datum elevation.

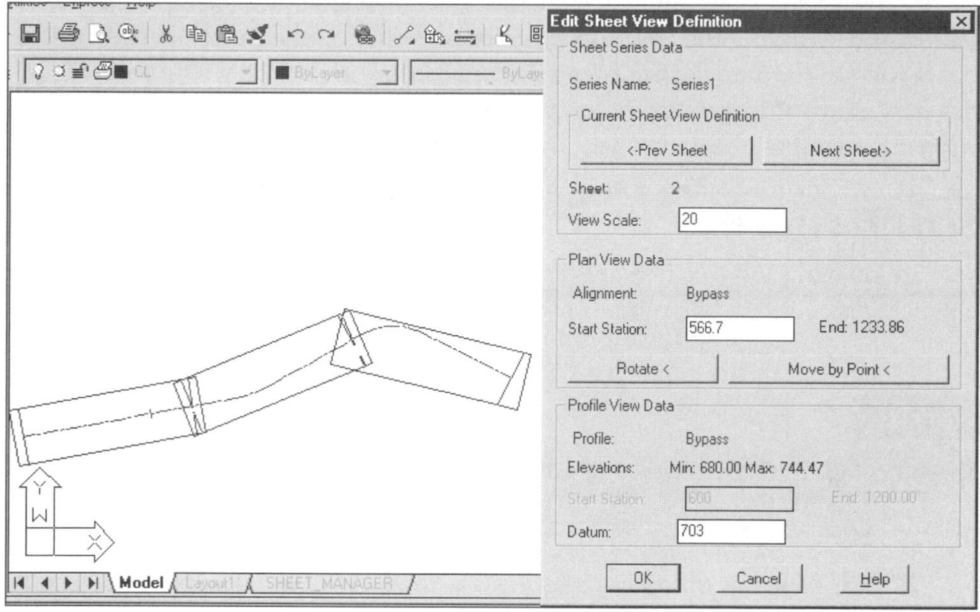

Figure 9–81

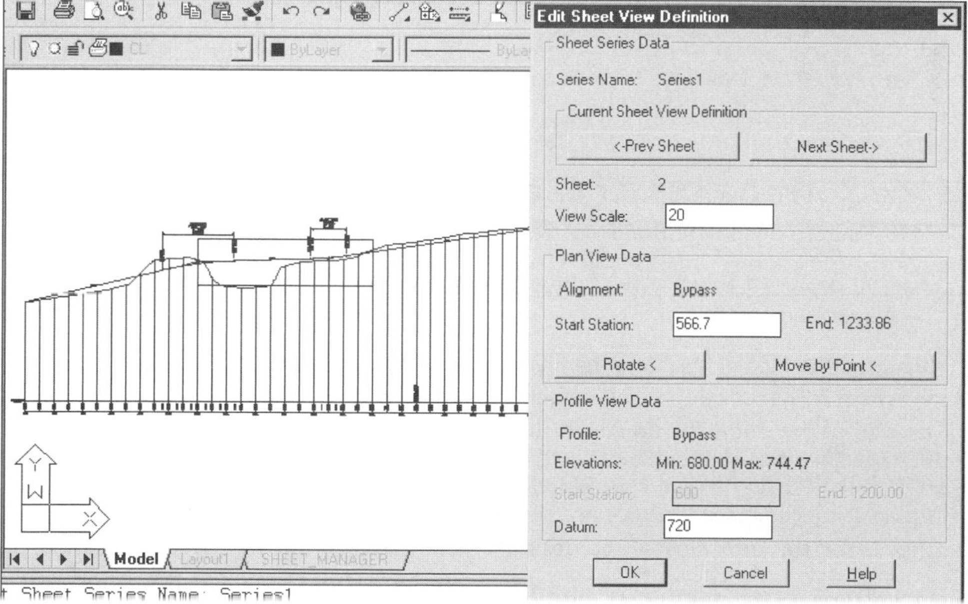

Figure 9–82

CREATING SHEETS

The Generate Sheet–Individual or Series routine creates the individual sheets or the entire sheet series. If running the Individual routine, you select the frame from plan view or specify the sheet number and the routine creates the sheet as a layout of Desktop. The name of the sheet is the series plus the number of the sheet (Maple + 001). If you are running the Series option, the routine creates all of the sheets for the series in the layout space of the Desktop. A sheet is a combination of a drawing file (dwg) and a database file (sdf–sheet database file).

EDITING SHEETS

There are no real tools to edit a generated sheet other than to modify the sheet style, the paper space, and/or the model space objects of the sheet.

EXERCISE

Exercise

After you complete this exercise, you will be able to:

- Load a sheet style
- Edit a sheet style
- Edit a viewport definition
- Define a plan and profile series
- Edit a plan and profile series
- Generate a plan and profile series
- Split a profile

THE ANATOMY OF A STYLE SHEET

1. If you are not in the Roadway drawing, open the drawing.
2. Run the Load Sheet Style routine found in the Sheet Styles cascade of the Sheet Manager menu.
3. Load the SDSKPLPR drawing file. Civil Design loads the sheet framework into a sheet manager layout. The default sheet size for the Layout may not match the size of the frame for the SDSKPLPR sheet.

The sheet is made up of two groups of objects. The first group is the paper space viewports. In this plan and profile sheet, there are two viewports. One viewport is for the profile and the other is for the plan view of the design. The function of the viewport is to display the Desktop objects of the design. The second group is the frames that define the annotation for the sheet. The function of the frame is to define the annotation and notes of the underlying design. This means that some frames are the same shape as a viewport and can lie over a viewport. There can be more than one frame over a viewport.

4. To view the properties of the viewports, select the Set Viewport Category routine from the Sheet Styles cascade of the Sheet Manager menu. Use a crossing window to view the values for the top and bottom viewports. Select the Cancel button to exit each dialog box. The routine returns a dialog box containing the properties of each viewport (see Figures 9–74). Within the viewport is the scale, 1" = 40'. Change the scale of both viewports to 1" = 20'.

Civil Design stores information defining the size, location, layers, type, category, and any label attachments within the frame.

5. Run the Create/Edit Frame routine from the Sheet Styles cascade of the Sheet Manager menu. Select the yellow box immediately to the left of the bottom viewport. The routine displays a dialog box containing the settings of the frame (see Figure 9–75).
6. To view the annotation definition within the frame, select the Text button in the current dialog box. The button displays another dialog box containing the type of annotation in the frame). The new dialog box contains the label attachment to the vertical box (see Figure 9–76). The bottom portion of the dialog box controls the type, location, and offsets of the label. The current attachment labels the profile grid elevations in a vertical frame at an increment of 5 feet. The text is frame middle justified with a vertical offset of 0.05 to center the text on the grid line.
8. Select Edit at the top of the dialog box to view the Text Label Properties dialog box (see Figure 9–77). This dialog box controls the text style, size, justification, and rotation angle.
9. Choose Cancel in all of the dialog boxes until you exit the routine.
10. Press ENTER to rerun the Create/Edit Frame routine. Select the horizontal box immediately below the vertical box on the left side of the sheet. This box is also a profile labeling box.
11. Select the Text button to view the details of the frame. This box is more complex, because it contains labeling for both the existing ground and proposed roadway elevations.
12. Select Profile:Elevation Label - EG Incremental (100.0).

The Profile:Elevation Label set has three design labeling points, but only two will work. The label is an elevation label, so the EG Center|Station is defined as the same as the EG Center|Elevation. This label type places an elevation label at a set increment in the frame. In this case the increment (50) is set in the Label Increment box in the lower portion of the box. The Grade Break label is not an incremental since the change of grade occurs at a point.

If you select the EG Center|Grade Break, the label would reference a point where the grade breaks on the existing ground profile.

13. Select Profile:Elevation Label - FG Tangent Incremental (100.00). This attached label produces annotation for the elevations along the tangent lines of the vertical alignment. The labeling occurs every 50 feet.
14. Choose Cancel button until you exit all of the dialog boxes.
15. Press ENTER again to rerun the Create/Edit Frame routine. Select the bottom frame of the sheet. Select the Text button to view the attachments. This frame contains the stationing of the profile. The labeling includes the profile and vertical curve beginning and ending stations. The labels for the vertical curve are intersection labels and the stationing of the profile is incremental.
16. Choose the Cancel button until you exit the dialog boxes.

Frames can overlay a viewport. In this case the frame is assigned to a different category: view. A view frame allows for the labeling of design elements within the viewport view of the profile. These elements include tangent grades, design locations, and piping information.

A frame can have more than one type of attachment. Besides the text attachment, a frame can have a block, distance, or grid attachment. You can use of the Control key while selecting the frame to help you cycle through the frames to select the correct frame.

17. Press ENTER to rerun the Edit routine and select the frame over the bottom viewport. In the main dialog box, select the Text button to view the labeling attachments. The frame labels the vertical curve information at the top of the frame.

The labeling style may contain more than one line of information. The style can create several lines of annotation. The ability to produce multiple lines of annotation is a result of creating a format that specifies what piece of data resides on which line of annotation. The format includes the label variables, text, and characters for the contents of each line ({}) and new line (\P). The braces enclose the text and variables of each line.

18. Select the label for vertical curve data and select the Edit button at the top of the dialog box. This displays the Text Label Properties dialog box.
19. Select the Edit button under text label format. This displays a dialog box that contains a multiple line format (see Figure 9–83).
20. Choose the Cancel button until you reach the Edit Frame Data dialog box and select the Block button. This displays the Edit Attached Block Labels dialog box. The current frame contains two block definitions (see Figure 9–84). Both are flag blocks marking the beginning and ending of the vertical curve.

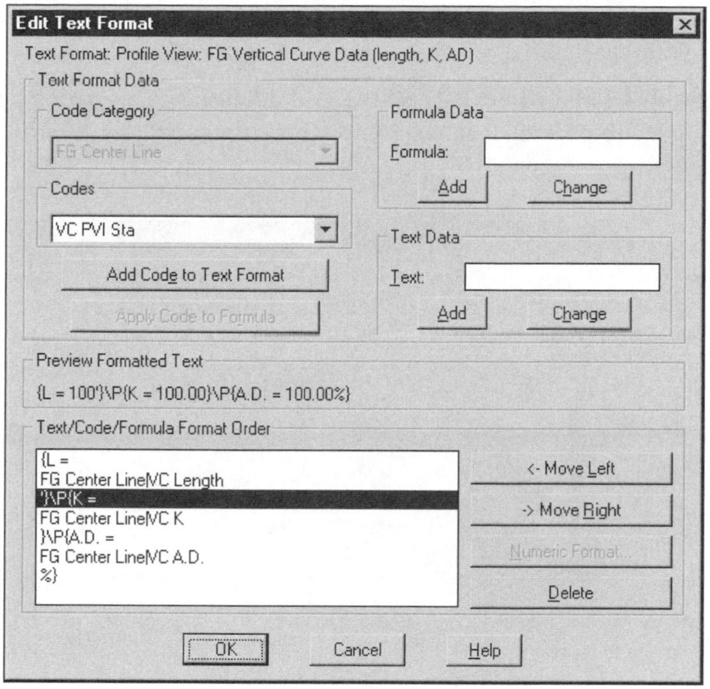

Figure 9–83

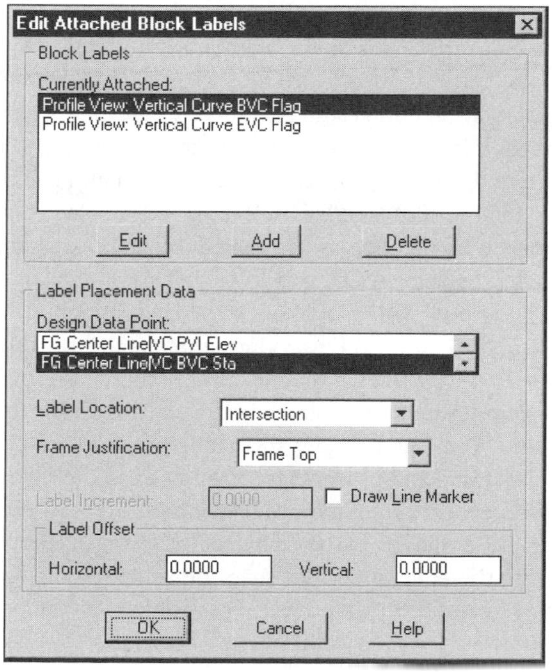

Figure 9–84

21. Choose the Cancel button to exit the block dialog box and to return to the Edit Frame Data dialog box.

22. Select the Grid button to display the grid definition. This dialog box indicates a grid one and quarter inch square for the frame profile view frame (see Figure 9–85).

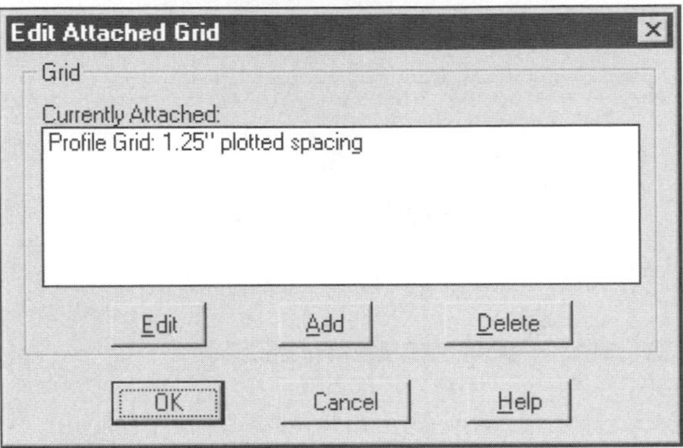

Figure 9–85

24. Choose the Cancel buttons in the boxes until you return to the Command prompt.

25. To add a station labeling frame to the plan view of the sheet, use the Create/Edit Frame routine from the Sheet Styles cascade of the Sheet Manager menu. The first prompt is to select or press ENTER to create a new frame. Press ENTER and the routine displays the Edit Frame Data dialog box. Set the frame Category to View. After setting the frame Category, select the Label button. The editor displays an empty dialog box. Select the Add button and select Plan:Station Labels–Incremental (0+00) (see Figure 9–86). Next, select the Edit button of the Select Styles dialog box to display the Text Label Properties dialog box. Set the text height to 0.2. Select the OK button to exit the Text Label Properties and Select Styles dialog box. You should be in the Edit Attached Text Labels dialog box. Set the Label Location to Design Incremental and set the increment to 50.0 (see Figure 9–87). Select the OK button to exit the current dialog box and to return to the Edit Frame Data dialog box. Select the Origin and Size button to define the Frame. The routine hides the dialog box and attaches a rubber band to the origin of the frame and prompts for the lower left (origin) point of the frame. Select the lower left corner of the plan viewport using the endpoint object snap. After selecting the lower left corner the prompt changes to selecting the upper right corner. Use the endpoint object snap and select the upper right corner of the plan viewport. After selecting the upper right point, the routine returns you to the Edit Frame Data dialog box. Select the Draw Frame button to draw the new Plan View frame.

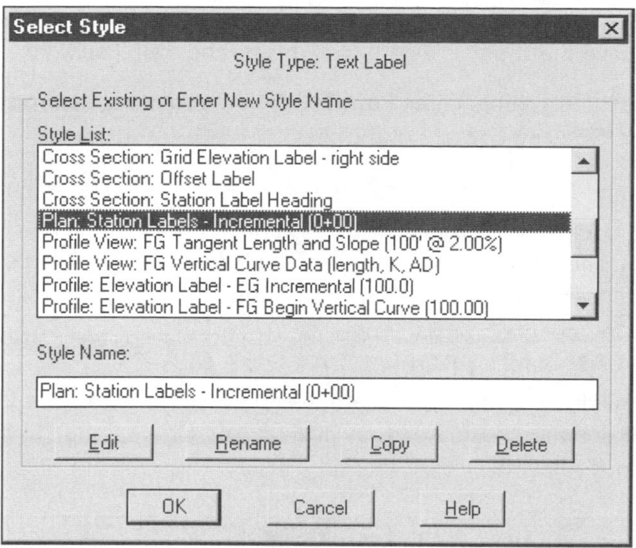

Figure 9–86

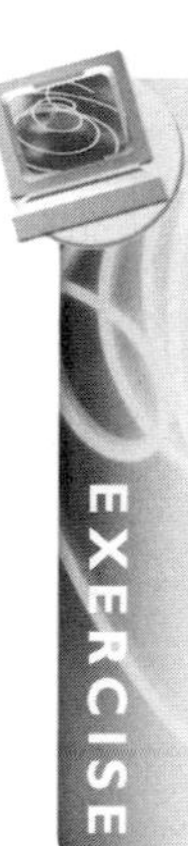

Edit Attached Text Labels
Text Labels
Currently Attached:
Plan: Station Labels - Incremental (0+00)
Edit
Add
Delete
Label Placement Data
Design Data Point:
Alignment|Station
Alignment|Match Line
Label Location: Design Incremental
Frame Justification: Frame Middle
Label Increment: 50
Draw Line Marker
Label Offset
Horizontal: 0.0000
Vertical: 0.0000
OK
Cancel
Help

Figure 9–87

26. Use the Save Sheet Style routine of the Sheet Styles cascade of the Sheet Manager menu to save the sheet as a new style. Name the style PandP20.

The routine prompts for a name. Enter **PandP20**. After you enter the name of the style, the routine queries the type and corners of the sheet.

```
Command:
Enter Type of sheet (Planprof/pRofile/Section) <Planprof>: (press ENTER)
Saving sheet: c:\program files\land desktop
  r2\data\sheets\sdsk_i\PandP20.dwg...
```

DEFINING A PLAN AND PROFILE LAYOUT SERIES

1. Start a new drawing and name the drawing CD120. Use the *aec_i* prototype and assign the drawing to the Roadway project (see Figure 9–88). Save the changes to the previous drawing.

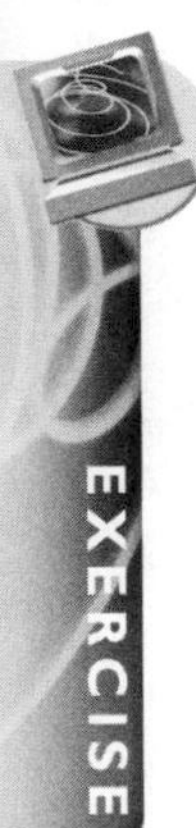

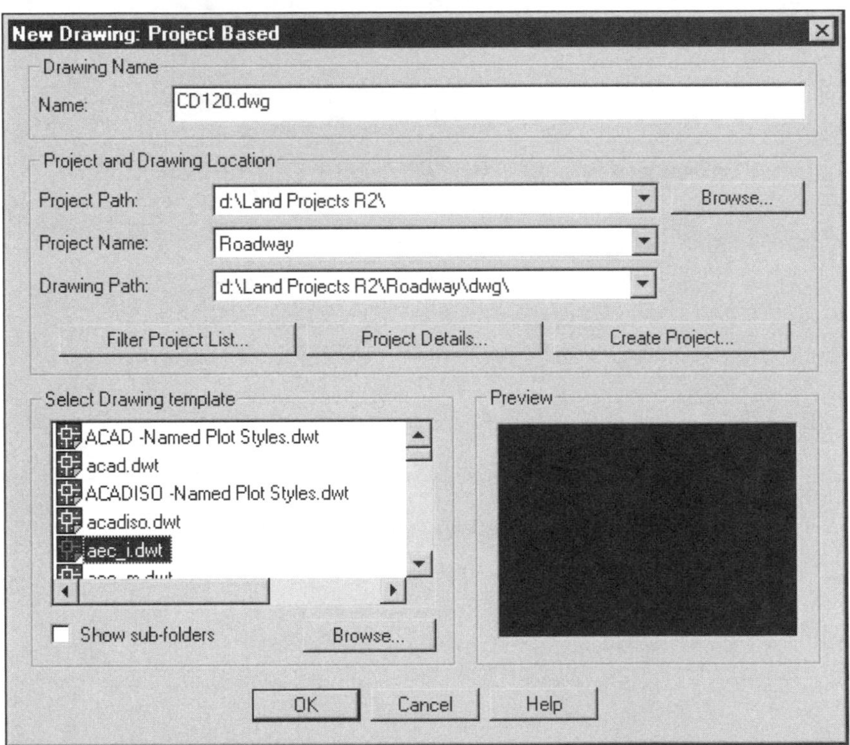

Figure 9–88

2. Select the LDD1-20 Drawing Setup file to set the drawing to a 1" - 20' horizontal and a 1" - 2' vertical scale (see Figure 9–89). Select the Finish button and the OK button to load the setup and dismiss the setup review dialog box.

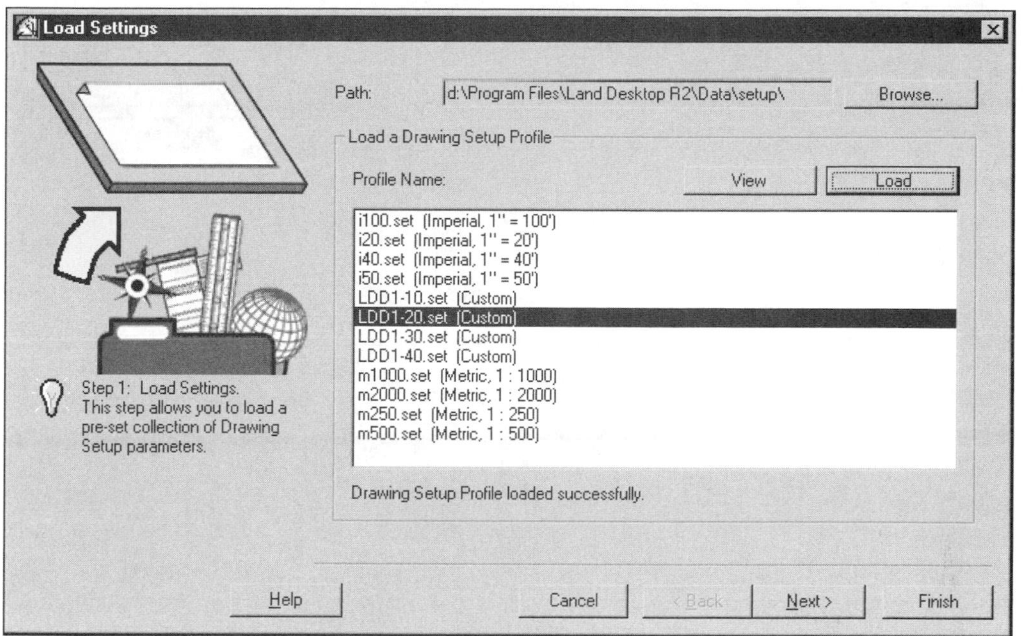

Figure 9–89

EXERCISE

3. Use the Save command and save the drawing.
4. Create a new layer, CL, and make the layer the current layer. Assign the layer the color of cyan and the linetype of center.
5. Use the Import routine of the Alignments menu to import the alignment into the drawing. When the routine displays the alignment list, select the Bypass alignment.
6. Use Zoom Extents to view the centerline. Use the Zoom command again with the .4x scale to view more of the area around the centerline (Zoom .4x).
7. Pan the centerline near the center top of the screen.
8. Make a new layer, MISC, and make it the current layer.
9. Set the profile layer prefix in the Labels and Prefix routine of Profile Settings (*-) of the Profiles menu.
10. Create a quick profile for the bypass alignment at the bottom of the screen underneath the alignment. You create the profile by using the Quick Profile routine of the Create Profile cascade of the Profiles menu. Use Figure 9–90 as a guide for placement of the profile.
11. Import the vertical alignment into the profile. Do not annotate the vertical alignment. Answer Yes to deleting any existing layers.

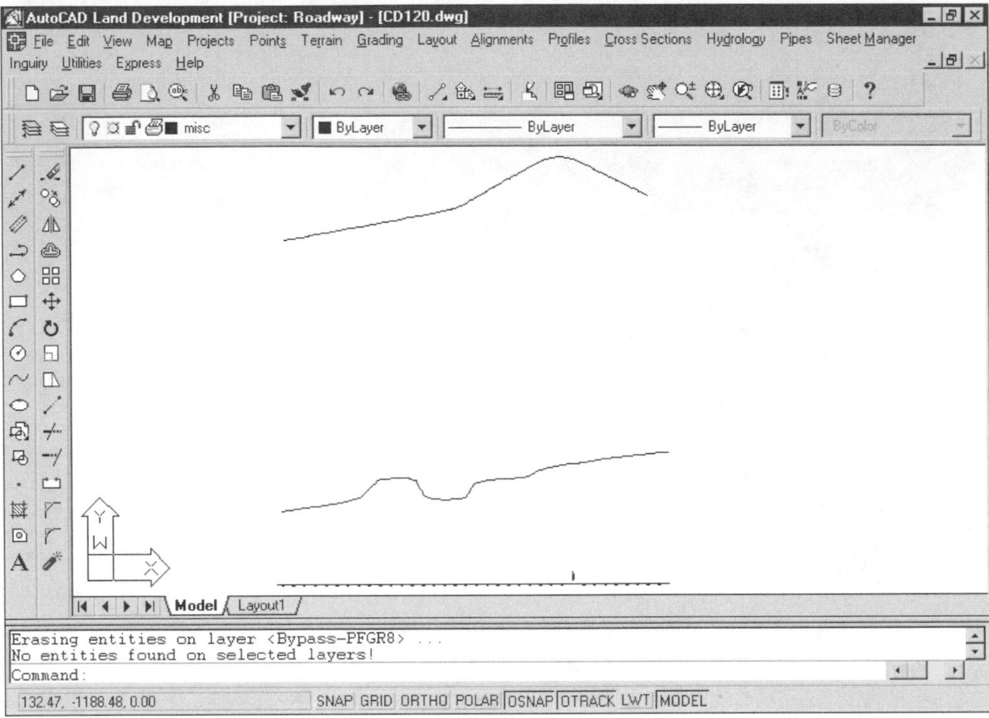

Figure 9–90

12. Pan the plan and profile to the left edge of the screen.
13. Save the drawing.

EXERCISE

CREATING THE SHEET SERIES

The process of creating the plan and profile sheets starts with a Sheet Series. The annotation of the alignment is 1" = 20' so it is correct for the plan view of our newly defined sheet style. The profile view will annotate the profile completely from paper space. The sheet style will use the viewport scale to size and locate the sheets. If the location or rotation is not correct, the Edit Sheet Series routine edits the rotation and location of a sheet.

1. Run the Layout Sheet Series routine from the Plan/Profile Sheets cascade of the Sheet Manager menu. The routine displays the Set Current Sheet Series Name dialog box. Enter the name of the series as PandP20 in the top text area of the dialog box (see Figure 9–91). After entering the name of the series, select the OK button to dismiss the dialog box.
2. The routine displays a new dialog box, Edit Sheet Series Data, containing the name of the series, the stations for the beginning and end of the sheets, section overlap, and at the middle of the dialog box a Set Style button to set the current sheet style. Select the button and change the sheet style to Pandp20 (see Figure 9–92).

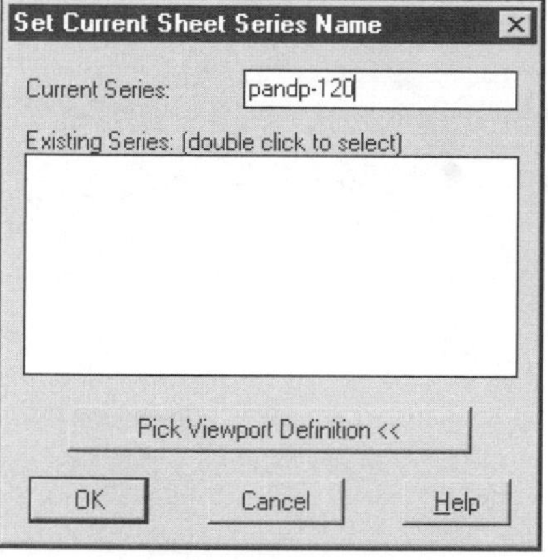

Figure 9–91

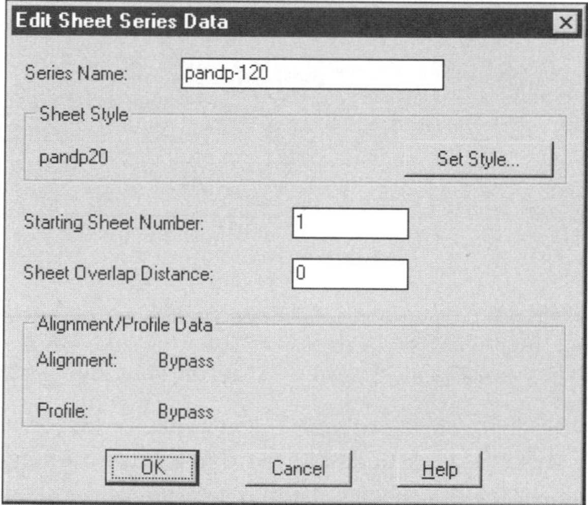

Figure 9–92

3. After setting all of the values, select the OK button to create the Sheet Series. The routine covers the alignment with yellow marker boxes for the plan views.

EDITING A SHEET'S LOCATION

The first sheet needs to be rotated to better display the centerline and to raise the datum of some of the profiles. Use the Edit Sheet Series routine of the Plan/Profile Sheets cascade of the Sheet Manager menu. Use the Rotate option of the dialog box to manipulate the rectangle representing the first section (eastern end of the centerline). The bottom portion of the dialog box shows the datum elevation and a box representing the profile view in the profile.

1. Run the Edit Sheet Series routine of the Plan/Profile Sheets cascade (see Figure 9–93).

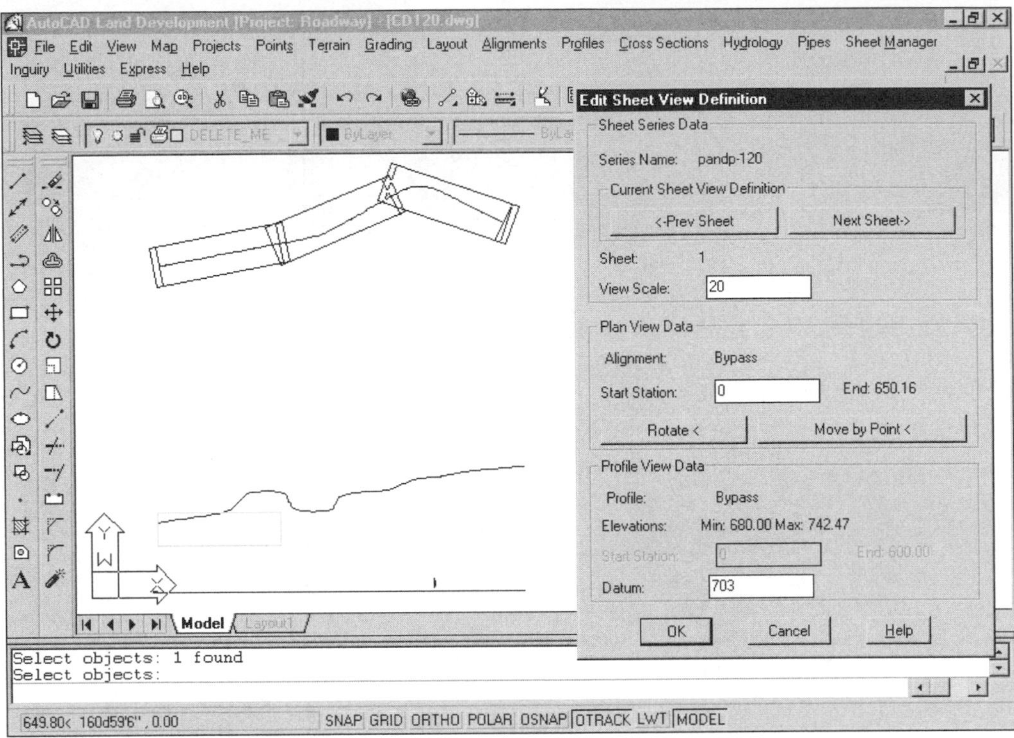

Figure 9–93

2. After rotating the first sheet, review the profile view location. Raising or lowering it will not help. We will split the sheet to better view the profile.
3. Select the Next Sheet button to view the next profile location. The datum elevation for sheet two is around 703 feet and locates the view well below the vertical alignment for this sheet. Change the datum of the second sheet to 717 and press ENTER to view the change of the profile view.
4. Select the Next Sheet button to view the next profile location. The datum elevation for sheet three is also 703 feet and locates the view well below the vertical alignment for this sheet. Change the datum of the second sheet to 727 and press ENTER to view the change of the profile view.
5. Select the OK button to save and exit the Edit Sheet Series routine.

GENERATING A PLAN AND PROFILE SHEET

After creating the Sheet Series and editing the location of the sheets, you can create one of the sheets to view the results. When you are satisfied with the results, the next step is to generate the series. Finally, the plotting of the series can be done individually or as a part of a batch plot process.

EXERCISE

1. Run the Generate Sheet–Individual routine of the Plan/Profile Sheets cascade of the Sheet Manager menu to create a PandP sheet. Select the plan view box of the second sheet of the series (see Figure 9–94).

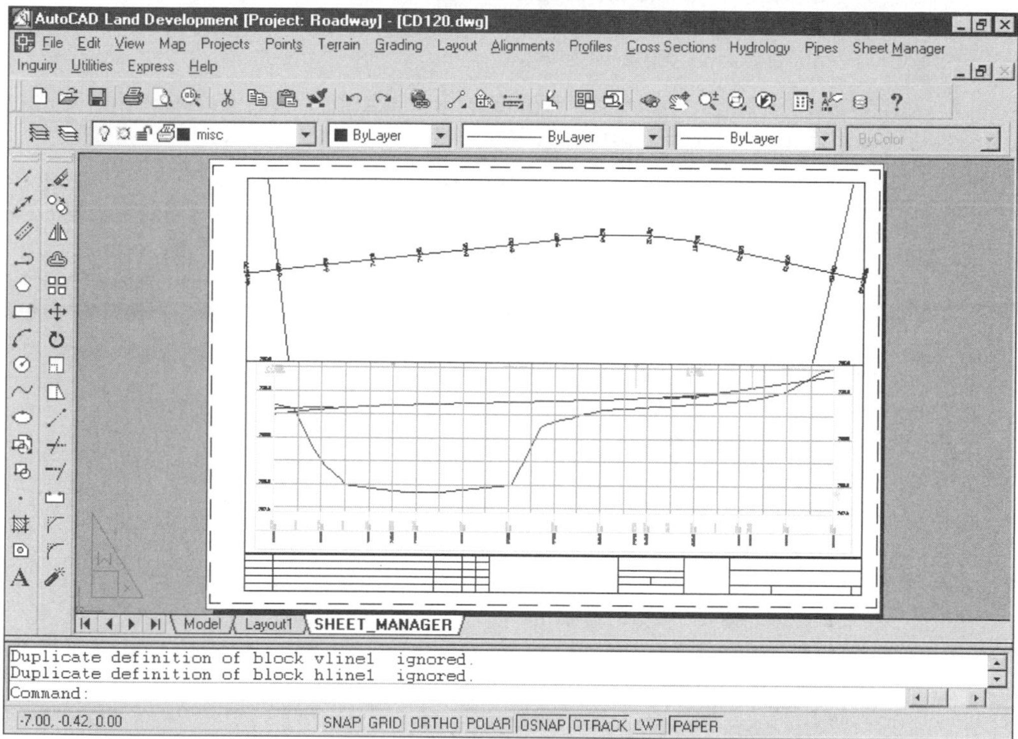

Figure 9–94

2. Create the entire series by running the Generate Sheet–Series routine of the Plan/Profile Sheets cascade of the Sheet Manager menu. When the routine displays the Set Current Sheet Series Name dialog box, select the OK button to create the series.

3. Use the Load Sheet–Individual routine of the Plan/Profile Sheets cascade of the Sheet Manager menu to load the first plan and profile sheet. Select *S001.dwg* to load the drawing file.

EDITING THE PROFILE VIEW

The current profile shows only a portion of the centerline vertical alignment. The amount of change is too much for the current vertical scale. To remedy this situation, the first step is to the raise the datum of the entire profile view and see how much of the profile is still cut off. If some of the profile still does not show, we can split the profile view into two portions with different datums showing the currently cut off portion. The Sheet Tools cascade of the Sheet Manager menu contains several tools for the manipulation of plan, profile, and section views.

1. The first step is to change to datum for the entire profile view of sheet 1. Use the Change Profile View Datum routine of the Sheet Tools cascade of the Sheet Manager menu. When the routine prompts you to select a profile viewport, use a crossing window to select the profile viewport. After selecting the viewport, press ENTER and enter **710** for the new datum. The routine adjusts the datum for the view and redraws the profile view.
2. Zoom in on the right half of the profile view. Make sure you can select the viewport and can see from stations 300 to the end of the profile.
3. Next run the Split Profile View routine of the Sheet Tools cascade of the Sheet Manager menu. Again use a crossing window to select the profile viewport. Make station 400 the splitting point. The routine creates two viewports from the single profile viewport. The new viewport is the viewport on the right-hand side. When the routine prompts you for a new datum elevation, enter the new datum of 715.

```
Command:
Select a Profile viewport to split:
Select a Viewport:
Select objects: Specify opposite corner: 4 found, 1 total
3 were filtered out.
Select objects:
Pick viewport split point (Station/MSpace/PSpace/eXit) <eXit>: s
Profile: Bypass  VP Starting sta: -0.00  VP End Sta: 600.00
Profile: Bypass  VP Starting sta: -0.00  VP End Sta: 600.00
Enter split station: 400
Profile: Bypass, Min Elev: 680.00, Max elev: 742.47, Datum Elev:
  710.00
Enter new Datum Elev <710.00> : 715
Labeling...
Alignment Name: Bypass    Number: 1      Descr: Bypass Prelim
  Design of 12/00
Starting Station: 0.00  Ending Station: 2446.88.........Done
The left viewport does not need adjusting.
```

4. Save the drawing.
5. Run the Save Sheet–Individual routine of the Plan/Profile Sheets cascade to save the sheet. If the routine prompts, answer Yes to overwriting the existing file.
6. If you have access to a plotter, plot the plan and profile sheets.
7. Use the Zoom Previous routine to view the entire plan and profile sheet.
8. Select the Sheet Manager Layout tab, click the right mouse button, and select Delete.
9. Save the drawing.

EXERCISE

DEFINING AND USING A CROSS SECTION SHEET STYLE

The process for creating cross sections is similar to the plan and profile process, except that the section generator looks for the horizontal and vertical scale in the Sheet Manager settings. First you set a sheet style, define a section sheet series, and then generate the sheets. The section sheet is different from the plan and profile style in that the section style contains no viewports, but is simply a frame. This frame expands and contracts to contain the roadway cross sections. Associated with the cross section is annotation that reports and calculates the cut and fill areas and the volume totals.

1. Review the current section settings of Sheet Manager. Open the Sheet Manager Settings dialog box, from the Settings routine of the Sheet Manager menu, and select the Section Preferences button at the bottom of the dialog box to display the section settings. Compare and, if necessary, change your settings to match the settings shown in Figure 9–95. After setting the values, select the OK button to exit and save the new settings.

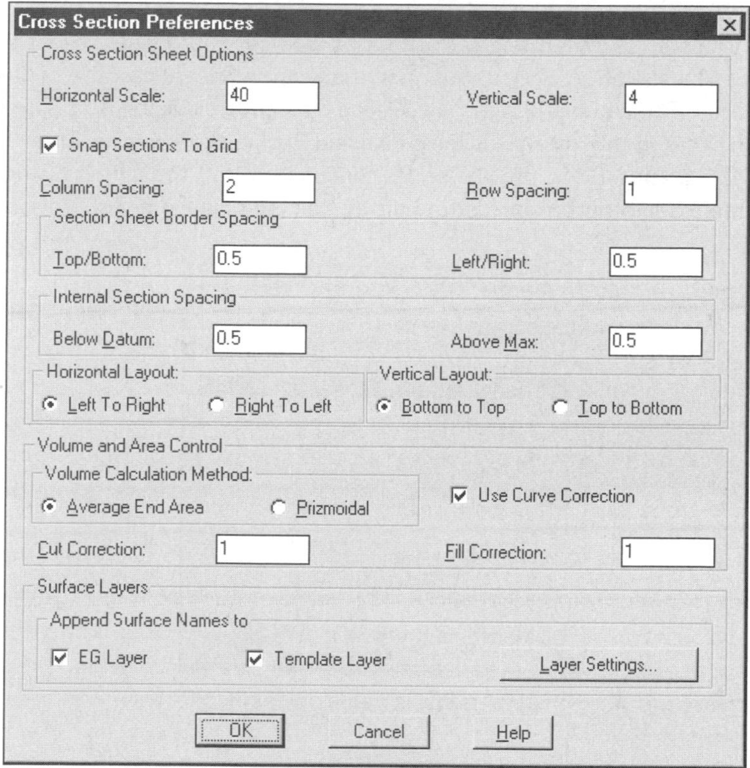

Figure 9–95

2. Set the current sheet style to SDSKSECT by running the Set Current Sheet Style routine of the Sheet Manager menu.

EXERCISE

3. Load the sheet style to view its format. Use the Load Sheet Style routine of the Sheet Styles cascade of the Sheet Manager menu.
4. Use the Edit/Create Frames routine to view the frame at the right of the section (the frame looks like a line).
5. Select the Text button in the Edit Frame Data dialog box.

The frame contains two attachments. The first attachment is the cut and fill area. Both of the attachments calculate and label their respective numbers.

6. To view the calculation formulas, first select which attachment to view. Select the Edit button and then, in the Text Label Properties dialog box, select the Edit button under the text label format to display the formula dialog box.
7. Choose the Cancel button in the dialog boxes to exit all of the dialog boxes.
8. Erase all of the sheet elements from layout and select the model tab to return to model space.
9. The first step to generate a section series is to set and define a sheet series. Run the Set/Define Sheet Series routine in the Section Sheets cascade of Sheet Manager. The routine displays the Define Sheet Series dialog box. The dialog box is the same as the one for defining Plan and Profile sheet series. Enter the series name as **BypassSect** into the series name box at the top of the dialog box, and choose the OK button to exit the dialog box (see Figure 9–96).

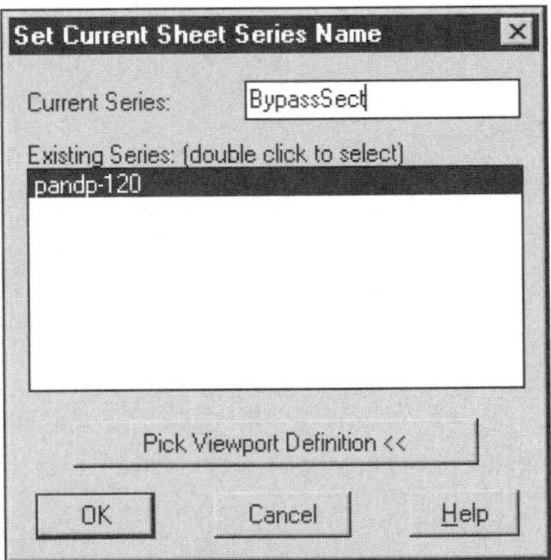

Figure 9–96

10. The last step is running the Generate Section Series routine to create the sections. The routine is in the same cascade as Set/Define. The routine displays the Edit Section Sheet Series Data dialog box with the sheet and section number starting values, starting and ending stations, and the current sheet style. Make sure the current style is SDSKSECT (see Figure 9–97). Choose the OK button in

the dialog box to exit and generate the sections. See Figure 9–98 for an example of the resulting cross-section sheet.

Edit Section Sheet Series Data

Series Name: BypassSect

Sheet Style

sdsksect Set Style...

Starting Numbers

Sheet Number: 1

Section Number: 1

Alignment Data

Start Station: 0

End Station: 2446.88

Alignment: Bypass

OK Cancel Help

Figure 9–97

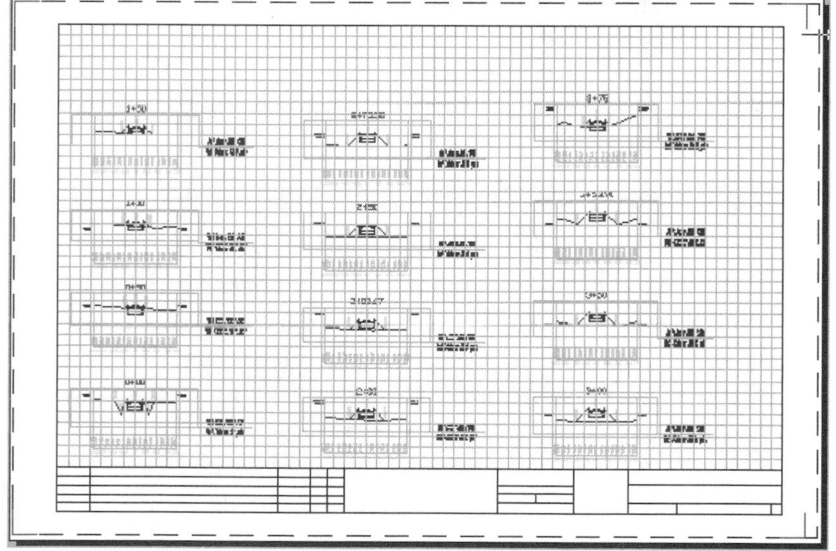

Figure 9–98

11. Call up the Layer dialog box and freeze the Series and matchline layers.
12. Save the drawing.

UNIT 9: CREATING POINTS AND A SURFACE FROM THE FINAL DESIGN

Civil Design allows you to create new points from the roadway design. These points represent the datum, the top of the design, daylight points, or individual point codes from the template definition. The Point Output cascade contains the routines for importing points into a drawing (see Figure 9–99). Once you have the points in the project, you can use the points as coordinates uploaded into a data collector to stake out the roadway.

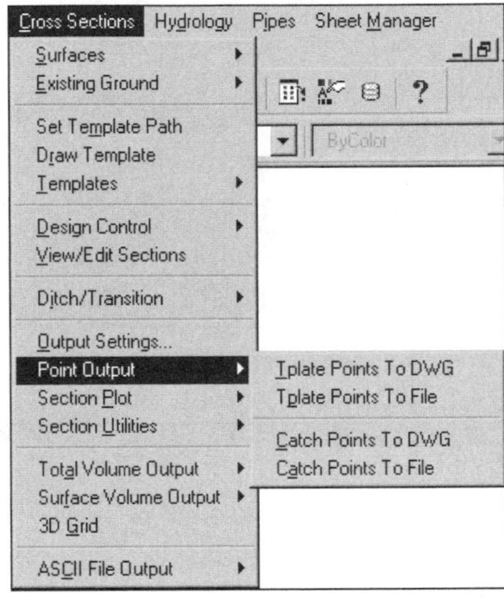

Figure 9–99

The merging of the design with the existing surface creates a new surface representing the results of the design. The 3D Grid routine of the Cross Sections menu creates a representation of the design out of 3D Faces. The Terrain Model Explorer converts the grid of 3D Faces into point data for a surface representation of the roadway design. After making a surface from the roadway 3D Faces, you paste the design surface onto the existing ground surface. After you paste the surfaces together, the next step is to use the Surface Saveas routine to save the surface under a new name. This prevents the destruction of the existing ground surface.

Exercise

After you complete this exercise, you will be able to:

- Import design points into the drawing
- Use the Section Utilities menu

- Create a roadway 3D grid
- Paste the roadway onto the existing ground surface

TEMPLATE POINTS

The Points Output cascade of the Cross Sections menu has tools for the importing of points that represent parts of the template. These parts include the top-of-surface, datum, and existing ground.

1. Reopen the Roadway drawing and save the changes to the previous drawing.

2. Restore the Roadway named view.

3. Make a new layer, DPTS, make it the current layer.

4. Run the Tplate (template) Points to Drawing routine of the Point Output cascade of the Cross Sections menu. Import the Top of surface 1 for the elevation of the points. Start with point number 1000. Answer Yes to importing the catchpoints.

The routine places into the drawing points whose elevations represent the top of the template. Each vertex on the top of the template is a point in the drawing. The points represent the catch point, back-of-curb, top-of-curb, flow line, EOP, and centerline.

5. Erase the points from the drawing with the Erase routine of the Edit Points cascade of the Points menu. Use the number range of 1000–2000.

Template Point—Catch Points

You can import the catch points and lines with the Catch Points to DWG routine in the Point Output cascade of Cross Sections menu. The Catch Point to DWG routine imports the points and lines representing the daylighting between the design and existing ground into the drawing.

6. Import the daylight lines and points into the drawing with the Catch Points to DWG routine of the Points Output cascade of the Cross Sections menu.

```
Import catch points (No/Yes) <Y>: y
Import daylight lines (No/Yes) <Y>: y
Beginning station <0.00>: (press ENTER to accept)
Ending stations <2446.88>: (press ENTER to accept)
Current point number <1864>: 1000
```

7. Erase the points from the drawing with the Erase routine of the Edit Points cascade of the Points menu. Use the number range of 1000–2000. Erase the Daylight lines from the drawing.

8. Rerun the Tplate Points to Drawing routine of the Points Output cascade of the Cross Sections menu and use the Pcode option. The routine displays the current list of pcodes. Select a few codes to import into the drawing and complete the routine by selecting the OK button. When the routine prompts for a current point number, enter **1000**.
9. Zoom in on the points to view their descriptions. The routine assigns point descriptions based upon the pcode list values.
10. Erase the points from the drawing with the Erase routine of the Edit Points cascade of the Points menu. Use the number range of 1000–2000. Erase the Daylight lines from the drawing.
11. Save the drawing.

MERGING DESIGN AND EXISTING SURFACES

1. Freeze the CL layer.
2. Run the 3D Grid routine of the Cross Sections menu to produce the roadway. Set the base elevation to 700 and the vertical scale to 1.0.

```
Command:
Alignment Name: Bypass  Number: 1 Descr: Bypass Prelim Design 12/00
Starting Station: 0.00  Ending Station: 2446.88
Beginning station <0>:
Ending station <2446.88>:
Surface points to import [Existing/Datum/Top]: t
Top surface number <1>:
Vertical scaling factor <1>:
Base elevation <0>: 700
```

3. Start up the Terrain Model Explorer. Select Terrain at the top left of the dialog box and click the right mouse button. Select Create New Surface.
4. A new surface, surface1, appears in the dialog box. Select Surface1, click the right mouse button, and rename the surface to Roadway.
5. Next collect the data for the surface. The Roadway surface needs to convert the 3D grid into point data. Select Point Files from the dialog box, then click the right mouse button and select Add Points from AutoCAD Objects. This displays a submenu; select 3D Faces (see Figure 9–100). The routine hides the Terrain Explorer and asks how to select the objects. Press ENTER to accept selecting by layer. Select an entity to identify the RDGRID layer.
6. Erase the 3D roadway grid.
7. Build the surface.
8. Select the surface Existing, click the right mouse button, and select Open (Set Current).
9. Minimize the Terrain Explorer.

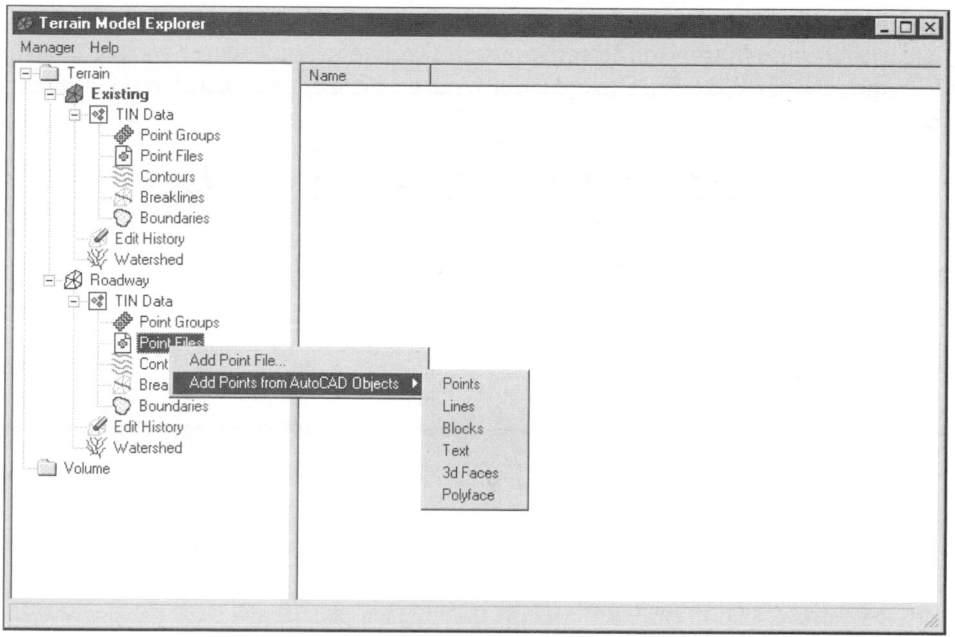

Figure 9–100

10. Use the Paste routine of the Edit Surface cascade of the Terrain menu to paste the roadway surface on to the Existing surface. The routine displays the Select Surface dialog box showing Existing as the current layer and prompting you to select which surface to paste on to Existing (see Figure 9–101).

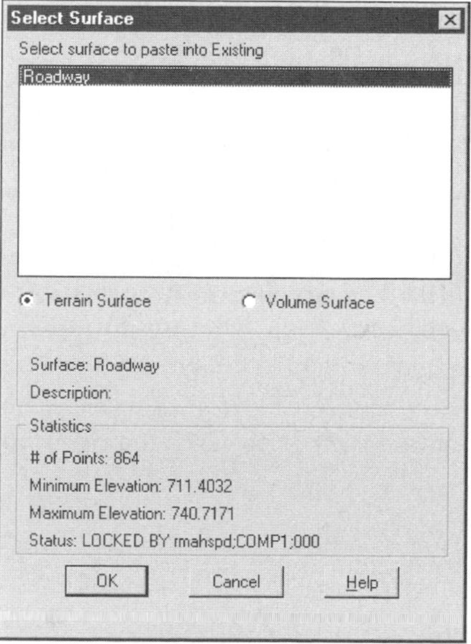

Figure 9–101

EXERCISE

11. Restore the Terrain Model Explorer and select the Existing surface. Click the right mouse button and select Saveas to save the new combined surface under the new name Design. Describe the surface as Existing and Roadway Together (see Figure 9–102).

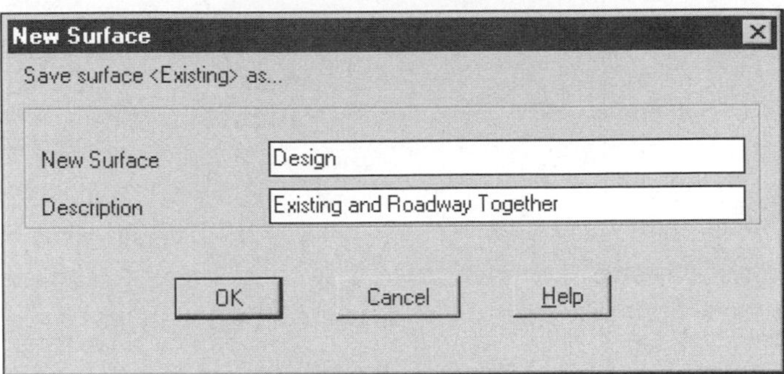

Figure 9–102

12. Close Terrain Model Explorer.
13. Save the drawing.

The visualization of the roadway is with the 3D Grid routine. Since the elements of the road are small, the grid needs to be rather small as well. Depending upon the type of computer and how large a 3D Grid can be, viewing and creating a 3D Grid can be quite frustrating. View only a small portion of the roadway instead of viewing the entire roadway.

Use the 3D Grid routine to define a portion of the surface for viewing by defining a smaller area of the surface for the grid routine. Define the grid area where the road crosses the depression in the middle of the site.

13. Import the Design surface boundary to determine the location of the grid site.
14. Run the Grid of 3D Faces routine in the Surface Display cascade of the Terrain menu. Use a rotation of 0 and select a lower left and upper right point within the boundary of the surface. Use an *M* and *N* spacing of 2.0, a base Elevation of 710.00, and a vertical factor of 3.5. Use Figure 9–103 as a guide for creating the grid.
15. Freeze the CL and SRF-Border layers.
16. Select the grid to activate their grips. Click the right mouse button and select view objects.
17. Erase the grid from the screen.
18. Save the drawing.

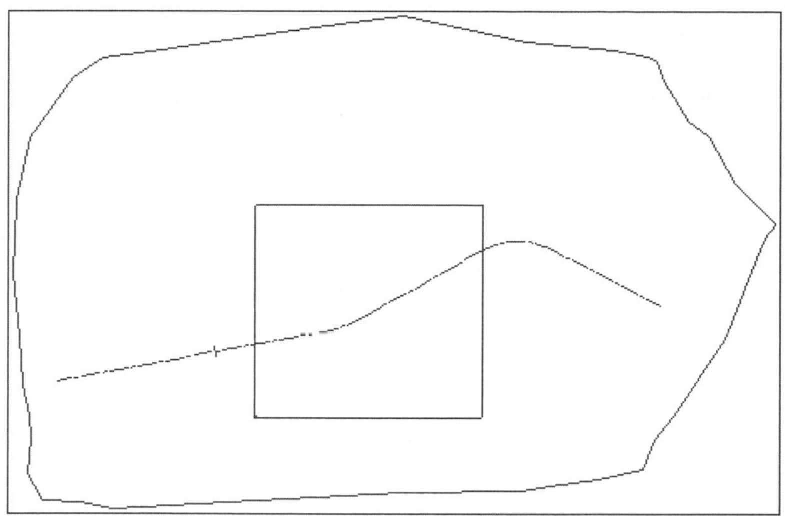

Figure 9–103

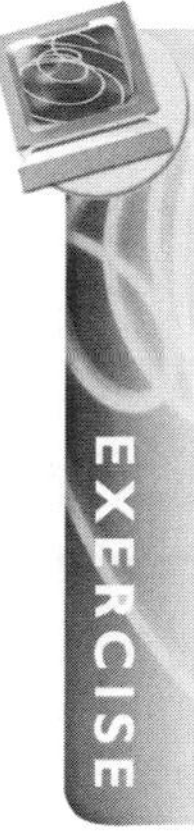

The roadway design process of Civil Design is lengthy and rigid; however, the process, once understood, is simple and straightforward. The steps in designing a roadway are the same as the manual methods used today. The design process uses two-dimensional views to solve a three-dimensional problem. The plan view is an orthographic view of the horizontal design. The profile view is a side view of the vertical along the centerline. The cross-section view is a view to the right and left of the design as you walk down the roadway.

As discussed in this section, not all designs are as simple and require you to manipulate points on the template or the ditch. Civil Design uses horizontal and vertical alignments to control the motion of a template. You can create these alignments in plan or profile view or in the horizontal and vertical editors. The next section covers the topic of the use of transitions in Civil Design.

Transition Alignments and Templates

After you complete this section, you will be able to:

- Define templates with different transition properties
- Define a template with multiple surfaces
- Define a template with a median
- Process template designs
- Import ditch and transition line work

SECTION OVERVIEW

The greatest advantage to road design in the Civil Design add-in is the ability to manipulate the horizontal and/or the vertical motion of a template. With this ability, the designer can accommodate the needs of almost any design.

Civil Design manipulates a template with alignments. Horizontal transition alignments control the left and right (offset) motion of a template. Vertical transition alignments control the up and down (elevation) motion of a template. A template can have up to eight horizontal and eight vertical transition points on each side of the template centerline. You can apply one horizontal and vertical alignment to a transition control point on a template.

A transitional alignment attaches to the template at the transition control point. The transition control point represents the location of the alignment on the template. In plan view, a horizontal alignment pulls or pushes the template transition control point away from or toward the centerline. The result of the pushing and pulling is the expansion or contraction of the template. An example of a roadway design horizontal transition is the road width transitioning from two to four lanes, and possibly back to two lanes. By using horizontal transition alignments, the Civil Design program manipulates the width of the road.

While manipulating the horizontal location of a transition control point, the design processor needs to calculate the elevation of the control point. If we assume that the edge-of-pavement point is a transition control point and if the template is to maintain a constant grade from the crown of the road to the edge-of-pavement, as the road widens, the elevation of the edge-of-pavement must lower to maintain the cross grade. As the road narrows, the elevation of the edge-of-pavement rises to maintain the cross grade. Thus, the design processor is calculating the elevations found at the transition control point (the edge-of-pavement) as the alignment manipulates only the horizontal location of the template (see Figure 10–1).

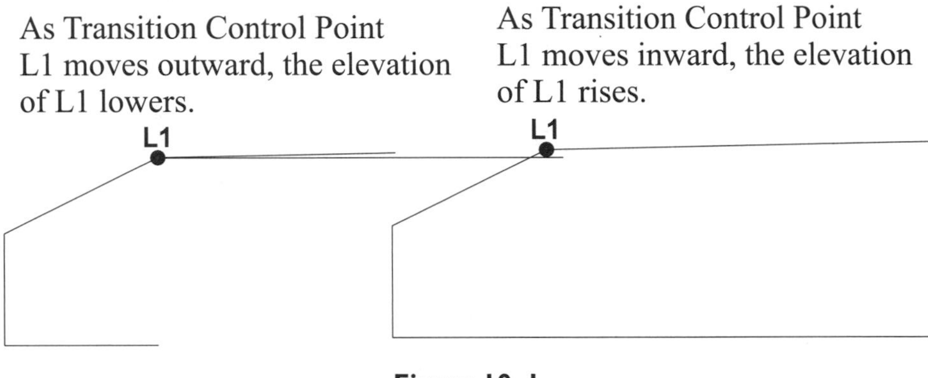

Figure 10–1

Initially, the designer may not have all the pertinent information to create the transition alignments. By processing the template under one set of design parameters, the output of the design processor provides Land Desktop objects from which to build the final transitional alignments.

The method of processing a preliminary design, modifying its results, and applying the new alignments to control a design applies equally to template transitions. You may initially define transition control as a way of getting data. Civil Design automatically calculates the location of each transition control point even if they are not attached to an alignment. Importing the control point data as lines in plan or profile view provides initial alignment data. After modifying the positions of the lines and defining them as transition alignments, you can then attach the new alignment definitions to the template. When you attach the transition alignments to the template control points, the template responds to the motion caused by the transition alignments.

An example of this is designing ditches. Ditch design can be fairly static as seen in the previous section. The ditch design made the assumptions that the roadway slopes defined the way the ditches drained and that the ditches occur only in cut situations. As the slopes along the road change, so do the ditch slopes. What if there is a specific way the ditches are to drain? You may need to manipulate the horizontal and vertical aspects of the ditch to accomplish this design goal. This manipulation would be accomplished through transitional alignments.

While running the template down the roadway, the design processor of Civil Design (the Edit Control routine) is calculating the location of the bottom of the ditch on both sides of the roadway. The design processor automatically calculates the vertical and horizontal locations of the ditch base. Routines in the Ditch/Transition cascade of the Cross Sections menu can import the ditch base calculations into plan and/or profile views of the roadway. The routines import the location of the ditch base as lines in the drawing. These lines become the basis for new alignments controlling the ditch base. After attaching the new alignments to the ditch, they manipulate the horizontal and vertical locations of the ditch base.

The connection between the alignment definitions and the Edit Design Control routine is static. When you are redefining an alignment, Edit Design Control does not receive the updated definition. You must manually reassign the plan or profile alignment to the template in the Design Control dialog box. When you do so, the Edit Design Control routine reads the new definition from the alignment database and correctly calculates the cross sections.

UNIT 1

The first unit of this section covers the transition control and region points and their properties. The location and assignment of properties are essential for the successful processing of a design. The region and control points can be the same or different points on the template. When they are separated, the region between them does not distort, but the surface towards the roadway centerline from the region point will distort.

UNIT 2

The processing and reviewing of the combinations of properties and control points is the focus of the second unit of this section. The incorrect combination of properties can produce unwanted results.

UNIT 3

Unit 3 reviews the location of region and control points as they pertain to templates with multiple surfaces. It is critical that the region point touches those surfaces you want to manipulate.

UNIT 4

The transitioning of a region is the focus of Unit 4. This unit works with a template that has a median. The design starts with the median, then collapses the median from the cross section, and returns the median to the cross section.

UNIT 5

Unit 5 reviews the importing of transition control into the profile and plan views of the design.

UNIT 6

The importing and manipulation of the ditch alignment is the focus of the last unit of this section.

UNIT 1: TRANSITION CONTROL AND REGION POINTS

There are two critical transition points on a template: the transition control and transition region points.

TRANSITION CONTROL POINT

A transition control point is where an alignment attaches to a template. Think of the control point as the eye of a needle and the alignment as the thread passing through the needle's eye. If not attaching an alignment to the control point, the Edit Design Control routine (the design processor) still calculates the control point's location along the roadway centerline. After completing the design processing, Civil Design provides tools that import the location of the control point into plan and/or profile views.

TRANSITION CONTROL POINTS

- A control point does not have to be on the template.
- The motion of the control point affects all the surfaces that touch or intersect a vertical line passing through the control point.

As the control point travels along the roadway, the design processor always knows the location of the transition control point on the template. This is true even if you are not using the control point. In fact, the control point may be there only to generate data for future horizontal and/or vertical alignments. After processing the design data, you can import the lines representing the continuous location of the control point into the drawing. After you have edited the lines to better represent design specifications, the next step is defining a new alignment from the lines to manipulate the transition control point.

When you select the Attach Alignments button in the main panel of the Edit Design Control routine, the Attach Alignments dialog box displays. In this dialog box, you attach the transition alignments to the template (see Figure 10–2).

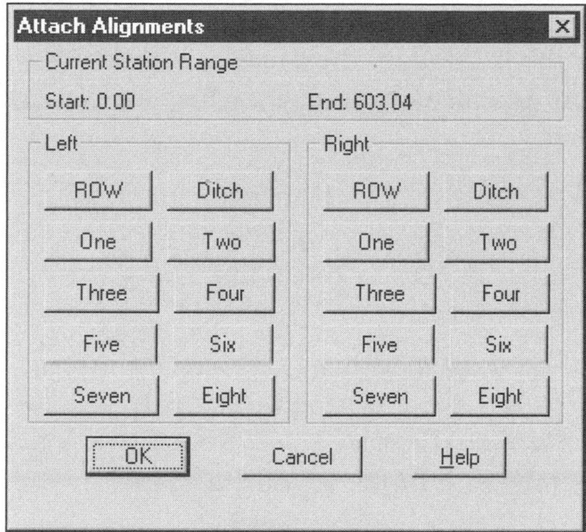

Figure 10–2

TRANSITION REGION POINT

A transition region point is the second defining point of the transition region. The transition region point can be to the right, left, or the same point as the transition control point. The only real condition is that the control point cannot be any closer to the centerline than the region point. All surfaces that intersect or cross a vertical line passing through the region point follow the motion of the control point.

TRANSITION CONTROL AND REGION POINTS

The transition control point works in tandem with a transition region point. In some situations, the transition control and transition region points are the same point on the template. They are the same point when a single point on the template manipulates all of the desired surfaces. When you are defining two separate positions, the intent is to have control over a specific region within the template.

The widening of a roadway from two to four lanes is an example of transition control and transition region points being the same point. The expansion and contraction of the template occur at the edge-of-pavement (a point on the template).

When you want to manipulate a zone within the template, the transition control and region points are separate points. An example of when you would need to separate the control and region points is a template with a median when you want to transition it to a roadway with no median. The control point is the edge-of-pavement and the region point is the intersection of back-of-curb and the end of the median (see Figure 10–3). The curb is in the region because it is to disappear when the road closes out the median. As the control point moves away from the centerline, the curb does not distort, but the median stretches to fill the new median width. When the control point moves towards the centerline, the median shrinks to accommodate its new smaller width. The curb and median disappear when the left and right control points are on top of each other.

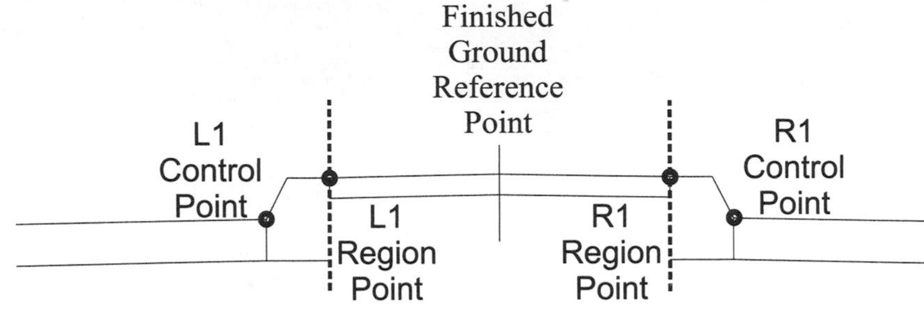

Figure 10–3

The Edit Template routine has an option that defines transition points on a template. The order for defining the transition points is first, the region point, and then, the control point. You need to know the location of these points before starting the editing routine, because if the location is defined incorrectly, the results are unpredictable.

TRANSITION PROPERTIES

A template transition region has three sets of behavioral properties: pinned or dynamic, constrained or free, and hold elevation or grade. The pinned or dynamic properties set applies only to the central surfaces of a template that cross over the centerline alignment. A pinned control point always stays on its side of the centerline alignment, never to cross. A dynamic control point moves freely across the centerline alignment. The constrained or free property set applies to any zone in a template. A constrained zone can only stretch away from the centerline. As a result, a constrained region holds a minimum width. A free zone reacts to the transition regardless of its original definition. The hold elevation or grade properties set applies to situations when using only horizontal transitions. A control point with the hold elevation property keeps its elevation relative to the Finish Ground Elevation no matter how the control point moves. These means that any grade drawn in the original template will change as this point moves. The hold grade property means that the control point will raise or lower its elevation to preserve the linked grade.

PINNED VS. DYNAMIC

Pinned

A region that has the property of pinned holds the inner vertex of the transition region. A pinned region should never cross the centerline of the roadway. A pinned region point always connects to the centerline of the alignment. If the region point crosses the centerline alignment, the region point draws a line from the region point to the centerline. If right and left region points cross the centerline, both region points attach a vertex to the centerline.

The line connecting the pinned region to the centerline is a result of the hold elevation or grade property. If you are setting the hold grade property, the line is at the grade between the point and the centerline elevation. If you are setting hold elevation, the design processor holds the elevation of the transition control point. The design processor holds the elevation to the difference between the control point and the finished ground reference point. If the control point is 3 inches below the reference point, the design processor ensures that the control point is always 3 inches below the reference point no matter how much the control point moves laterally.

Dynamic

The property of dynamic allows a transition control point to manipulate the template across the centerline alignment. A dynamic region holds all the grades of the templates' surfaces that cross the transition region. Holding a surface's grade allows regions to cross the centerline and maintain surface order. The crossing of the centerline is through the motion of right and left transition control points. The control points can move the template to different horizontal and/or vertical locations irrespective of the location of the centerline alignment. Once the template crosses the centerline, the horizontal and vertical aspects of the template depend almost entirely on the transition control and region points and not the centerline reference point.

The dynamic property affects only the central portions of the template (surfaces that cross the centerline of the template). If the transition control points intersect, the surfaces in the dynamic transition region will disappear. When a region disappears, the adjacent transition region becomes dynamic.

CONSTRAINED VS. FREE

Constrained

A constrained transition region point only moves away from the center of the template. This template property insures a minimum width to the centerline or next template control point.

Free

A free region can grow and shrink without restriction.

HOLD ELEVATION VS. HOLD GRADE

The last property set, hold elevation or grade, applies to templates that only have horizontal transitioning. The property informs the design processor how to handle the vertical aspects of the template.

Hold Elevation

The property of hold elevation sets the elevation of a transition control point. The design processor sets the elevation of a control point by the vertical distance between the original transition control point and the finished ground reference point in the template definition. For example, if the transition control point is 4 inches lower than the finished ground reference point, the control point will always be 4 inches lower than the current centerline elevation. The transition control point maintains the difference in elevation no matter how far the point moves to the right, left, or past the centerline alignment. If you apply the property of elevation to a transition control point, the grades within the region change to accommodate the new location of the transition control point.

Hold Grade

The grade property of a transition point changes the elevation of the transition control point to maintain the grade.

- The grade is always from the region point to a point on the template.

The properties of hold elevation or grade are moot if the design control includes a vertical alignment.

COMBINING PROPERTIES

The combination of transition region properties controls the behavior of a template as it crosses the centerline or merges into another transition point.

You need to evaluate the design and the behavior of the template before defining the template transition points. Depending upon the behavior of the zones, the template may have to contain different types of transition zones. Some zones may have the transition control and region points as separate locations on the template. Other transition zones may have both points as a single point on the template.

The successful transitioning of a template across a centerline occurs only when assigning the correct properties for each transition region. The following properties are for a crossing region point:

- Dynamic
- Free
- Hold Elevation or Grade

The combination of the three properties, dynamic, free, and hold elevation or grade, gives complete control over a template that crosses a centerline.

Exercise

EXERCISE

The first part of this exercise is the establishment of the design environment. The first task is starting a new drawing and project. After you establish a new drawing, the next step is to insert a drawing with the contours and lines representing the centerline and transition alignments. After you have completed building the surface, the next step is to define the centerline and the two transitional alignments MEDCL, R1-TRANS, and L1-TRANS.

The next step is to define four templates: CDTPFG, CDTPFE, CDTDFE, and CDTDCG.

After you complete this exercise, you will be able to:

- Create an existing ground surface
- Define a centerline
- Create a profile
- Draft and define the vertical design
- Sample for cross sections
- Draft and define a basic template
- Define four templates with different transition properties

STARTING A NEW DRAWING AND PROJECT

1. Start a new drawing, CDTRANS and use the aec_I template (see Figure 10–4).

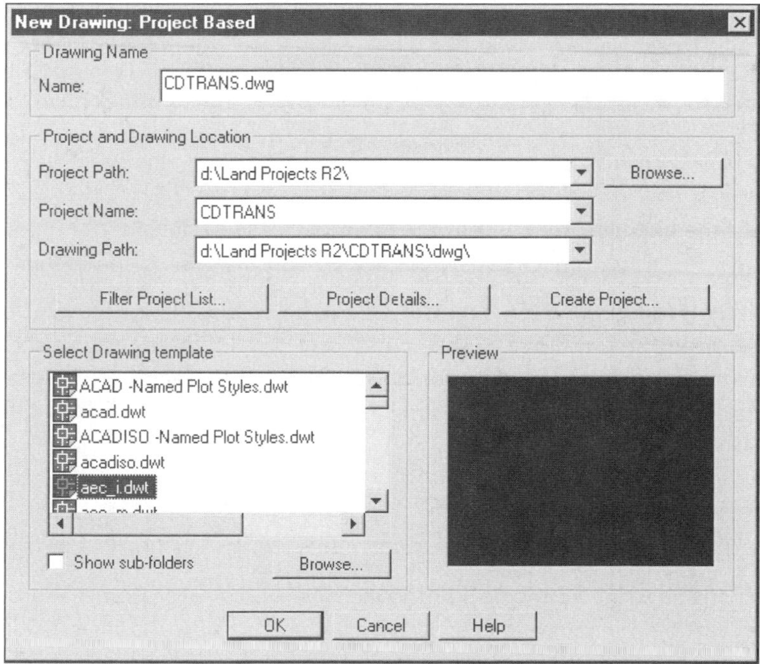

Figure 10–4

2. Create a new project, CDTRANS, and use the default Feet prototype to create the project (see Figure 10–5). Add the description of Civil Design Transitions to the project. After defining the project, select the OK button to dismiss the Project Details and New Drawing dialog boxes.

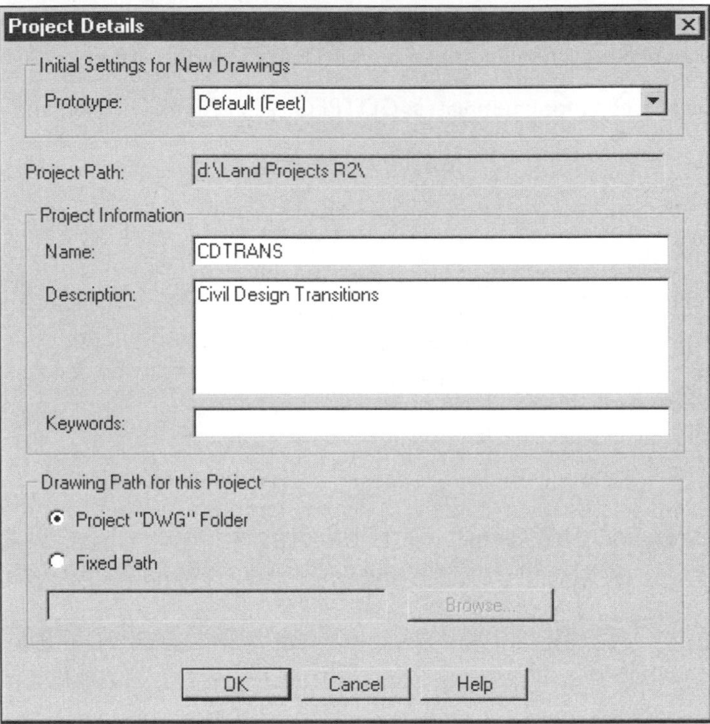

Figure 10–5

3. Select the OK button to accept the point database defaults.
4. When the Load Settings dialog box appears, set the following values for the drawing (see Figure 10–6). You will have to select the Next button at the bottom of the dialog box to move to the appropriate panel. Or, you can select the i50 setup and click on the load button to assign the setup values to the drawing. After setting the values, select the Finish button and then select the OK button to dismiss the Finish dialog box.

   Scale:

   Horizontal Scale: 50.0

   Vertical Scale: 5.0

   Zone:

   No Datum, No Projection

EXERCISE

Text Style:

Text Type: Leroy

Text Style: L100

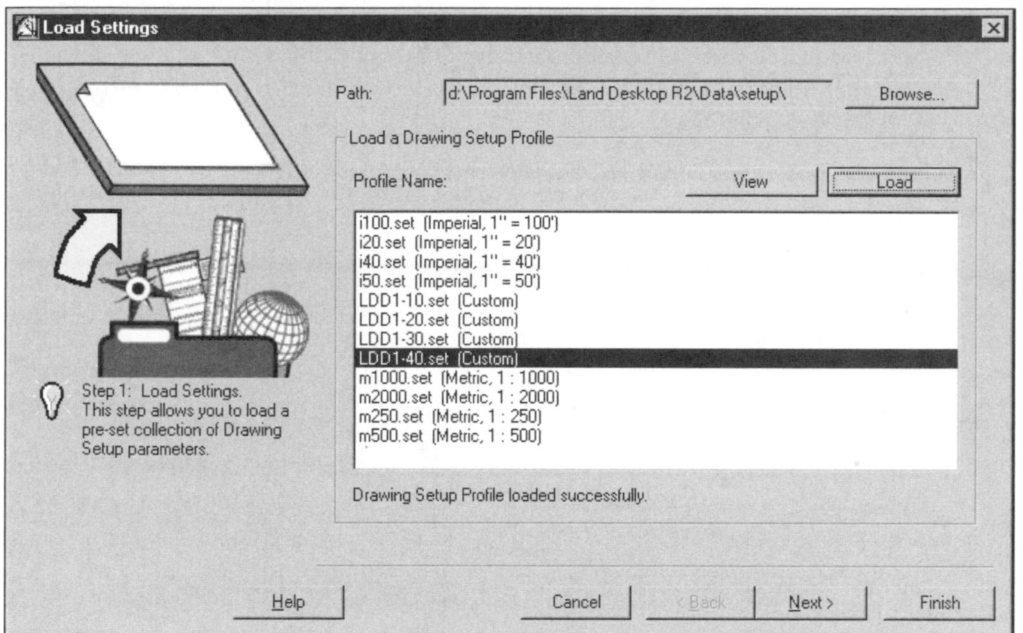

Figure 10–6

5. Insert the drawing CDTRANS from the CD that comes with the book. Insert the drawing at 0,0,0 and explode it upon insertion.

6. Do a Zoom Extents to view the entire drawing.

7. Use the Save routine of the File menu to save the file.

DEFINING A SURFACE

The next task is to create an existing ground surface. The surface is from contours in the drawing.

1. Create a surface from contours. Call up the Terrain Explorer by selecting Terrain Explorer from the Terrain menu. Select Terrain at the top left of the dialog box, click the right mouse button, and select Create New Surface. Click the plus sign (+) below Terrain to show the new surface (Surface1). Rename the surface to Existing by selecting Surface1, clicking the right mouse button, and selecting the Rename option.

2. Click the plus sign (+) to the left of the Existing surface to show the data list. To make a surface from contours, select Contours, click the right mouse button, and select Add Contour Data. The routine hides the Terrain Explorer and displays the Contour Weeding dialog box. Set the following values found below to create a denser triangulation of the contour data. Next the routine prompts for the method of selecting the contours. Press ENTER to accept the Layer method of selection. Select one contour on each layer (CONT-NML and CONT-HGH). Press ENTER twice to exit the contour data selection process and to return to the Terrain Model Explorer.

   Use the following weeding and supplementing factors.

Contour Data:	**ON**	
	Distance	Angle
Weeding:	**5.0**	**2.0**
	Distance	Bulge
Supplementing:	**10.0**	**1.0**

3. Build the surface using only contour data, and toggle on Minimize Flat Triangles Resulting from Contour Data. Select OK in the Done dialog box and close down the Terrain Explorer (see Figure 10–7).

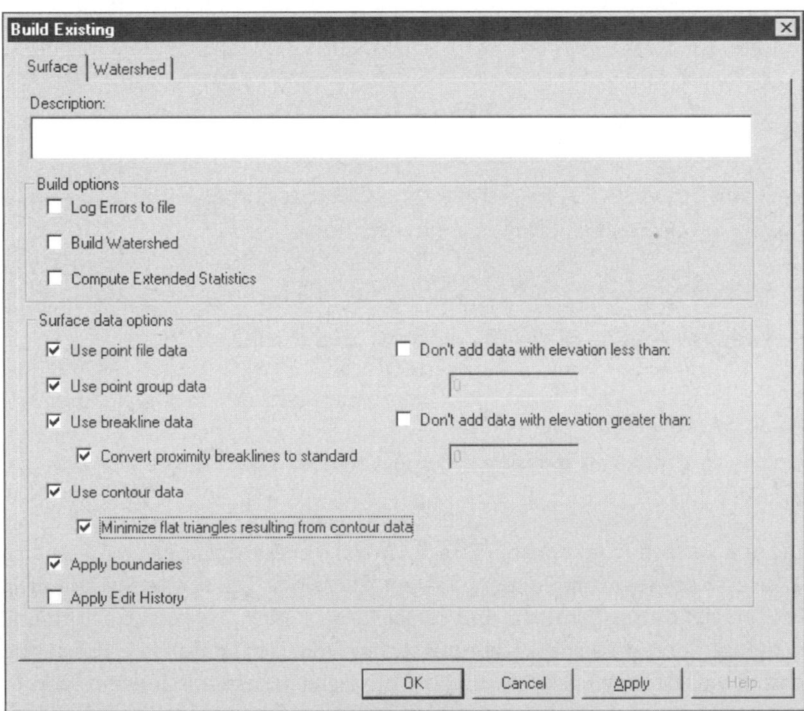

Figure 10–7

EXERCISE

4. Freeze the layers CONT-NML and CONT-HGH.

DEFINING THE ALIGNMENTS

The next task is to define the horizontal centerline and transitions. The line work for the centerline and the transition alignments is already in the drawing.

1. Use Zoom Center to view the objects for the exercise. Use the coordinates of 4500,4800 and a height of 650.
2. If the layers are not visible or frozen, thaw and turn on the TRANSCL, BSDK1L, and BSDK1R layers.
3. Define the alignment, Transc1 using the line work on the layer TRANSCL. The alignment consists of a single centerline segment found in the lower center of the drawing. The alignment starts at its southern end and the starting station is 0.00. When the routine asks about the centerline reference point, press ENTER to accept the default starting point. Describe the centerline as Centerline for transitional exercise. After setting the values, select the OK button to exit the dialog box and define the centerline.

The next task is to sample the elevations along the centerline. After you have sampled the centerline elevations, the next step is creating a full profile.

4. Set a layer prefix of *- in the Label and Prefix Settings routine in the Profile Settings cascade of the Profiles menu. After setting the prefix, select the OK button to exit the dialog box.
5. Sample along the centerline using the Sample From Surface routine of the Existing Ground cascade of the Profiles menu.
6. Pan the display so the centerline is at the left side of the screen.
7. Create a full profile with the Full Profile routine in the Create Profile cascade of the Profiles menu. When the routine displays the Create Profile dialog box, select the OK button to accept the defaults and to dismiss it. Select a point to the right of the centerline.
8. Draw the vertical alignment with the Create Tangents routine of the FG Centerline Tangents cascade of the Profiles menu. First draw the two tangent lines. The end of the vertical alignment (603.04) should be a point selected by the endpoint object snap. Use the following values to draw and create the tangent lines and vertical curve (see Figure 10–8).

Station	Elevation	Vertical Curve
0+00	690.84	
3+00	691.00	150.00
6+03.04	682.60	

EXERCISE

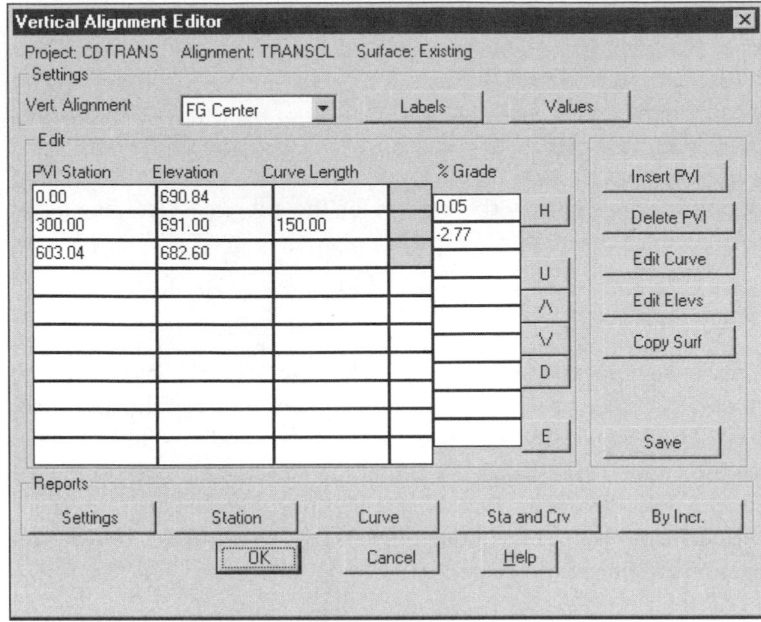

Figure 10–8

9. Use the Define FG Centerline routine of the FG Vertical Alignments cascade of the Profiles menu to define the vertical centerline.

10. Use the Import routine of the FG Vertical Alignments cascade of the Profiles menu to redraw the new vertical alignment with annotation (see Figure 10–9).

11. Sample the cross-section elevations along the roadway centerline. Use the Sample From Surface routine of the Existing Ground cascade of the Cross Sections menu. The sampling rate for tangent and curve sections is 10 feet at a swath of 50 feet. Sample the beginning and end of the alignment and Curve PCs and PT.

12. Define the following transitional alignments (see Figure 10–10). The line work for the alignment is on a layer with the same name as the alignments, so you may have to turn on or isolate the layers to be able to select the appropriate objects.

    - BSDK1L—Isolate the layer, BSDK1L (use the Layer dialog box) to define the alignment. The starting point of the alignment is the southernmost end of the line work. Press ENTER to accept the reference point and name the alignment BSDK1L. Describe the alignment as the Left Transitional Alignment and its starting station is 0.0.

    - BSDK1R—Isolate the layer, BSDK1R (use the Layer dialog box) to define the alignment. The starting point of the alignment is the southernmost end of the line work. Press ENTER to accept the reference point and name the alignment BSDK1R. Describe the alignment as the Right Transitional Alignment and its starting station is 0.0.

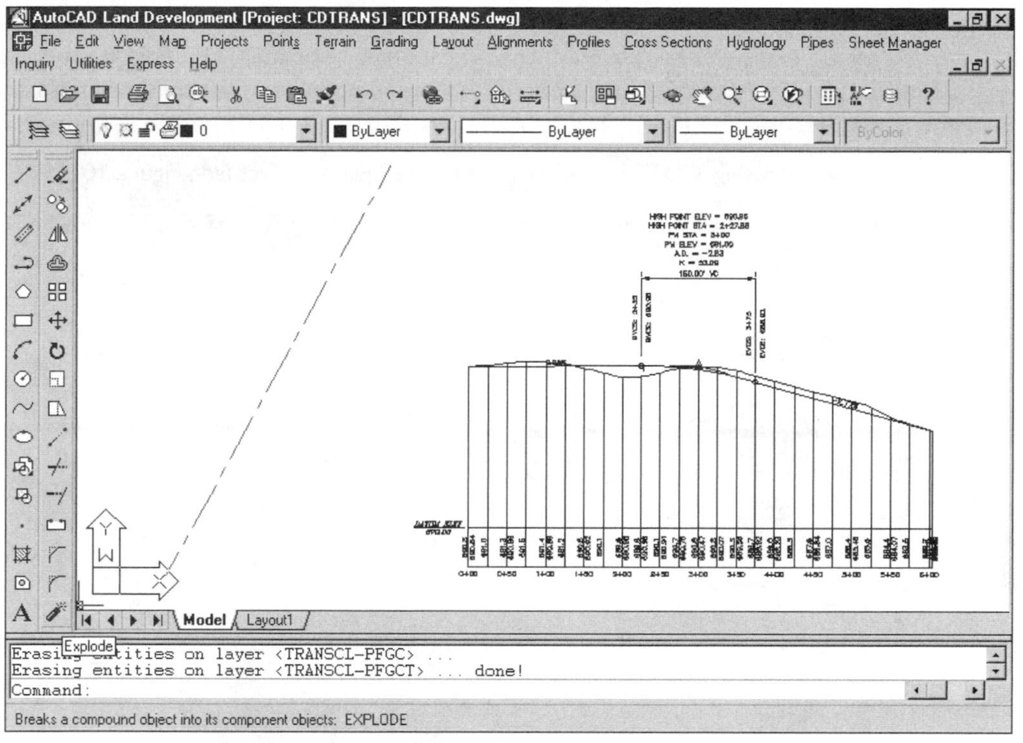

Figure 10–9

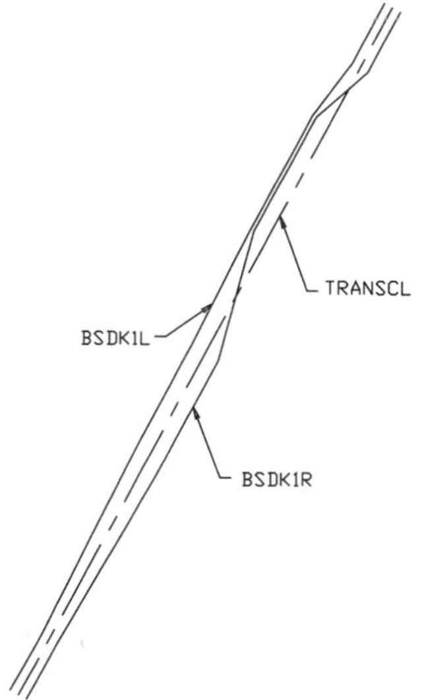

Figure 10–10

13. Click the Save icon and save the drawing.

DEFINING TRANSITIONING TEMPLATES

This portion of the exercise draws and defines transitioning templates. The Edit Template option contains routines that define the control and region points on the template and the transition behaviors.

1. Start a new drawing, CDTTPL, and assign it to the Tplate project (see Figure 10–11).

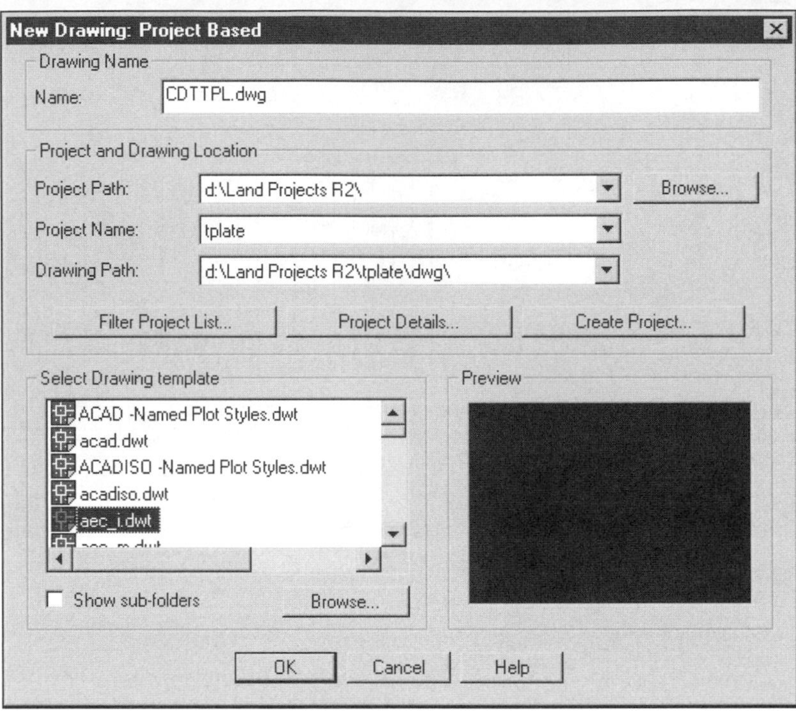

Figure 10–11

2. If this is a new project, select the OK button to accept and dismiss the Point Database dialog box.

3. If Tplate is not a new project, the Load Settings dialog box displays. Select the Next button to view and set values in the Scale, Zone, and Text panels of the Load Settings dialog box. You must select custom in the horizontal and vertical scale area to set the 1.0 scale. Use the following values (see Figures 10–12, 10–13, and 10–14).

   Scale:

   Horizontal Scale: 1.0

   Vertical Scale: 1.0

   Zone:

   No Datum, No Projection

Text Style:

Text Type: Leroy

Text Style: L100

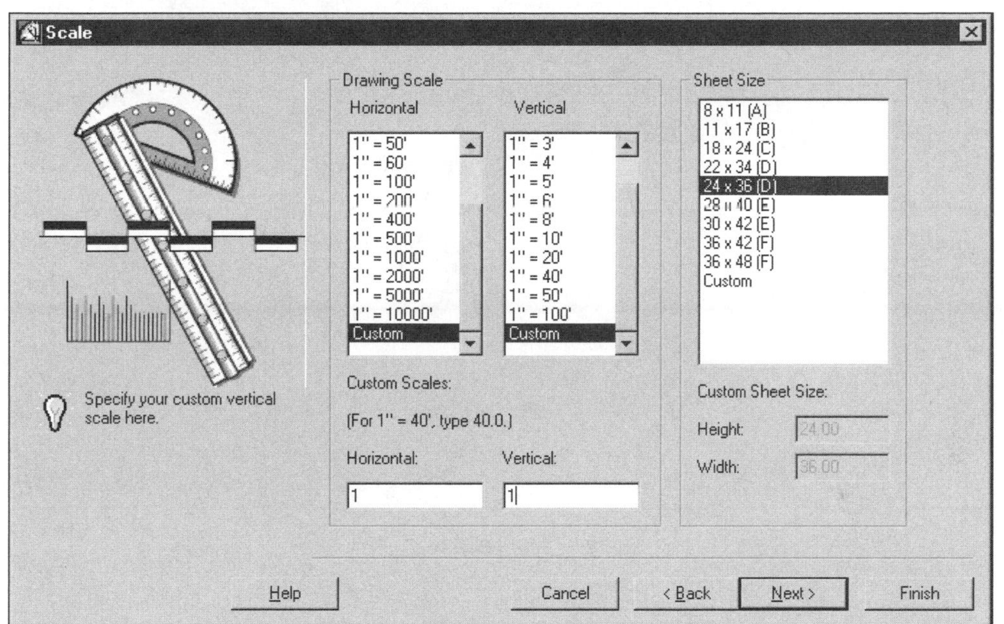

Figure 10–12

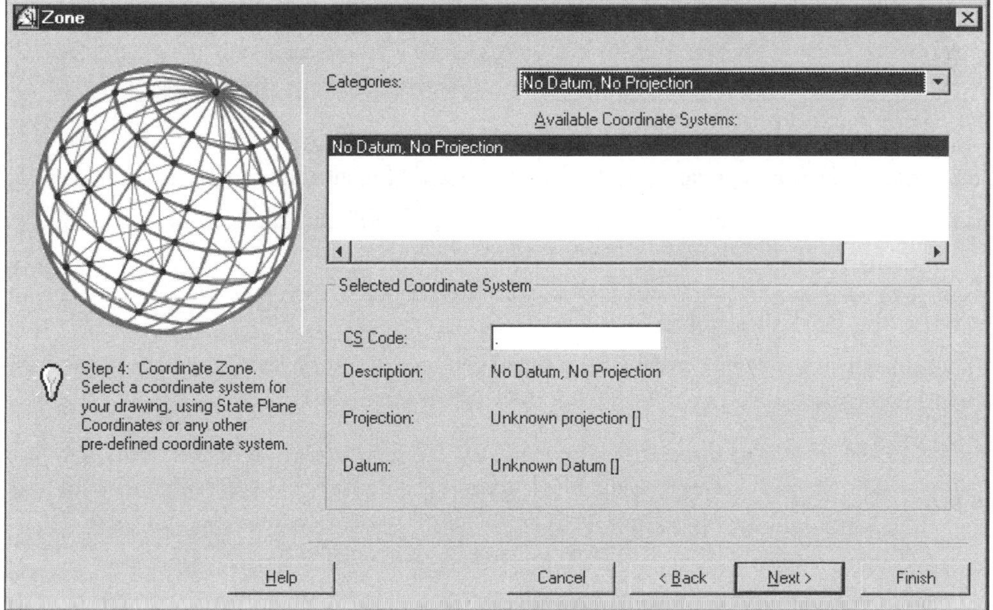

Figure 10–13

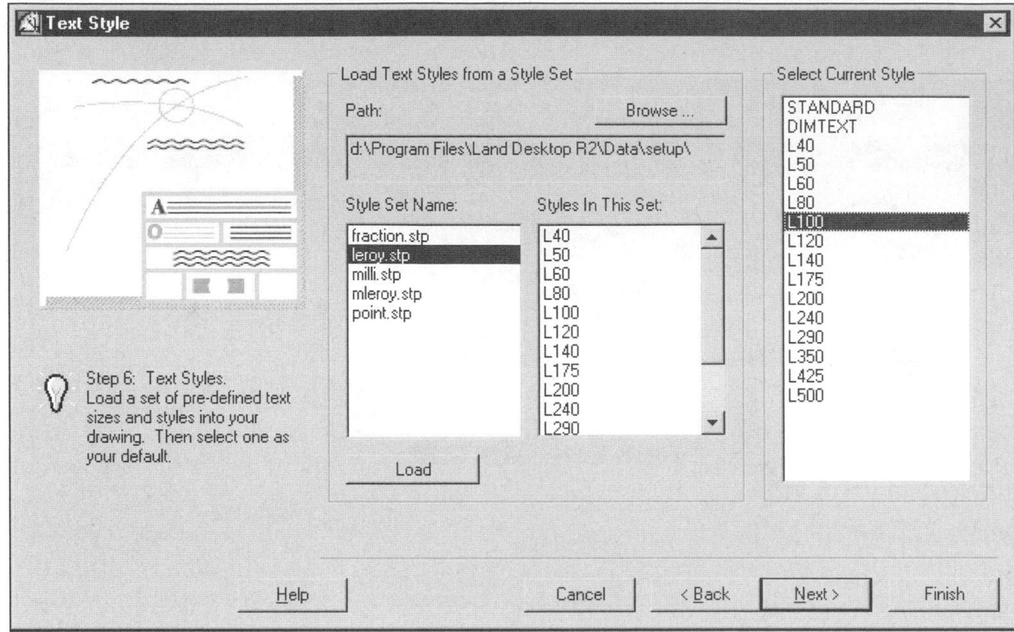

Figure 10–14

4. Click the Save icon to save the drawing.
5. Select the Object snap button and set the object snap to intersection and endpoint.

The transitional templates for this exercise are edited versions of a very basic template. By viewing the same template with different properties, you can view the varying effects of the differing combinations of properties.

The basic template needs a datum and top-of-surface edited into the template. After defining and editing the basic template, we will again edit the template, but save it under different names representing different combinations of properties.

Drawing the Basic Template

The basic template is symmetrical with a single surface. The template has one point for the transition control and region points, L1 and R1 (see Figure 10–15).

1. Create the basic template using the Draw Template routine of the Cross Sections menu. Start the template in a blank area at the center of the screen and draw the first segment as a -2% grade with an offset of 5 feet. The next segment is a -2:1 slope for 2 feet. The last segment is a relative change with an offset of 7 feet and a change of elevation of 0 feet.

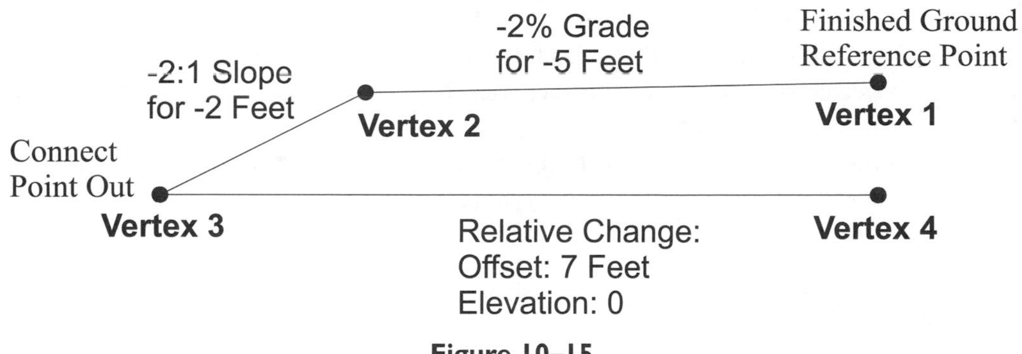

Figure 10–15

```
Starting point:
Slope (3 for 3:1) [Relative/Grade/Points/Close/Undo/eXit]: g
Grade (%) [Relative/Slope/Points/Close/Undo/eXit]: -2
Change in offset: -5
Grade (%) [Relative/Slope/Points/Close/Undo/eXit]: s
Slope (3 for 3:1) [Relative/Grade/Points/Close/Undo/eXit]: -2
Change in offset: -2
Slope (3 for 3:1) [Relative/Grade/Points/Close/Undo/eXit]: r
Change in offset [Grade/Slope/Close/Points/Undo/eXit]: 7
Change in elev: 0
Change in offset [Grade/Slope/Close/Points/Undo/eXit]: x
Starting point:
```

EXERCISE

2. Use the Define Template routine in the Templates cascade of the Cross Sections menu. The finished ground reference point is the upper right endpoint (vertex 1) of the template (see Figure 10–15). The surface is symmetrical and its surface is a normal surface. Select the polyline on the screen and when prompted by the dialog box, assign the material type as Asphalt. The connect point out is the lower left outermost endpoint of the template (vertex 3). The datum points are the lower left outermost endpoint (vertex 3) and the rightmost endpoint (vertex 4) of the bottom polyline. When the Subassembly dialog box appears, make sure the curb and shoulders are set to NULLS. Select the OK button to dismiss the dialog box and name the template, CDT.

The Define Template routine prompts as follows:

```
Command:
Pick finish ground reference point: (vertex 1)
Is template symmetrical [Yes/No] <Yes>: (press ENTER )
Select template surfaces...
Select objects: 1 found
```

```
Select objects:
Surface type [Normal/Subgrade] <Normal>:(press ENTER )
Pick connection point out:(vertex 3)
Datum number <1>:(press ENTER)
Pick datum points (left to right):(vertex 3)
Pick datum points (left to right):(vertex 4)
Pick datum points (left to right):
```

The Define Template routine presents the subassembly dialog box. Select OK to dismiss the box.

```
Save template (Yes/No) <Yes>:(press ENTER)
Template name: CDT
```

All the curbs and shoulders in the dialog box are nulls.

EXERCISE

Edit CDT Template

The CDT template needs a Top-of-Surface. This is only done in the editing of a template (see Figure 10–16).

1. Use the Edit Template routine of the Templates cascade of the Cross Sections menu to edit the CDT template. Select a point in the middle of the screen to insert the template. Use the sr option to change to the Surface Control option list, and choose the Topsurface option by typing the letter **t**. When the routine prompts for the surface number, press ENTER to accept 1. Select vertices 2, 1, and 5 to define the top of template (see Figure 10–16). After editing in the Top-of-surface, press ENTER until you have saved the template. When the routine prompts to overwrite the definition, type the letter **y** and again press ENTER .

```
Command:
Pick insertion point:
Edsrf/SAve/eXit/ASsembly/Display/SRfcon/Redraw <eXit>: sr
Connect/Datum/Redraw/Super/Topsurf/TRansition/eXit <eXit>: t
Top surface number <1>:
Pick top surface points (left to right): (select vertex 2)
Pick top surface points (left to right): (select vertex 1)
Pick top surface points (left to right): (select vertex 5)
Pick top surface points (left to right): (press ENTER)
Connect/Datum/Redraw/Super/Topsurf/TRansition/eXit <eXit>:
Edsrf/SAve/eXit/ASsembly/Display/SRfcon/Redraw <eXit>:
Save template (Yes/No) <Yes>:
Template name <cdt>:
Template exists. Overwrite [Yes/No]: y
```

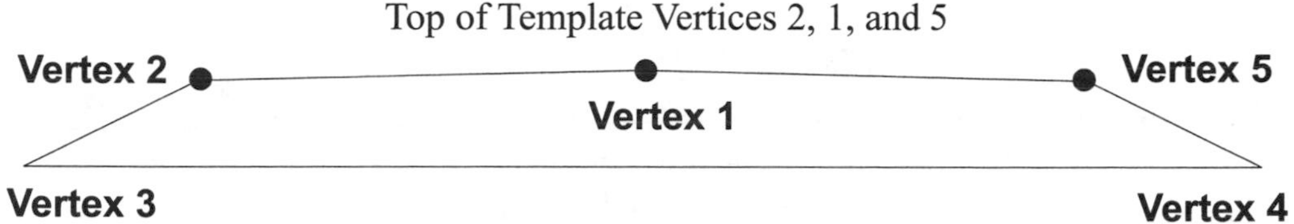

Figure 10–16

Defining the CDTPFG Template

The first template has the combination of the pinned, free, and hold grade property (CDTPFG) (see Figure 10–17). The pinned property says the transition region and control points do not cross their side of the centerline. If they do cross the centerline, they will draw lines to an elevation on the centerline of the cross section.

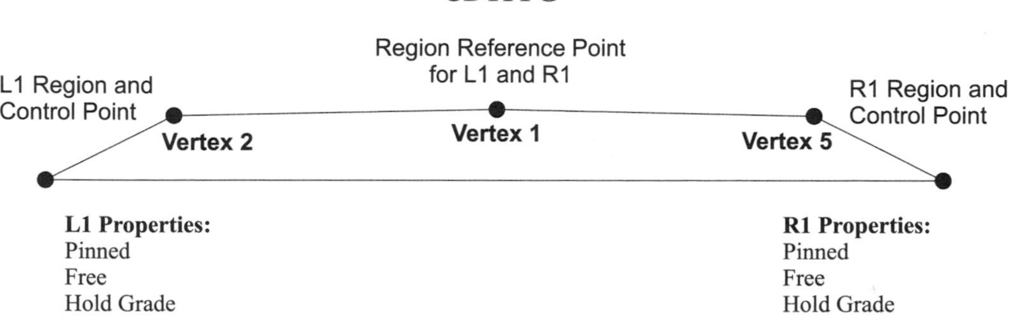

Figure 10–17

Editing CDT to Create CDTPFG

1. Erase any line work from the screen.

2. Run the Edit Template routine from the Templates cascade of the Cross Sections menu. Select the CDT template and place it in the middle of the screen. After placing the template into the drawing, type the letters **sr** and press ENTER to change to the Surface Control option list. Start the transition defining process by typing the letters **tr** and pressing ENTER to change to the transition definition process. Define vertex 2 as the first left transition point (L1). Vertex 2 is also the template control point—the point where the left transition alignment attaches to the template. Use vertex 1 and the grade reference point for the hold grade definition. Exit out of the left side transition routine and edit the right side of the template. Use vertex 5 for the region and control point (R1) and vertex 1 as the region reference point for the hold grade property.

```
Command:
Pick insertion point:
Edsrf/SAve/eXit/ASsembly/Display/SRfcon/Redraw <eXit>: sr
Connect/Datum/Redraw/Super/Topsurf/TRansition/eXit <eXit>: tr
Edit transition region [Left/Right/All/eXit] <eXit>: l
Edit left transition region [1/2/3/4/5/6/7/8/eXit] <eXit>: 1
Pick first left transition region point:    (select vertex 2)
Template surface transition [Dynamic/Pinned] <Pinned>:
Transition region type [Constrained/Free] <Free>:
Pick transition control point (RETURN for same):
Horizontal transition to hold [Grade/Elevation] <Grade>:
Pick transition reference point: (select vertex 1)
Edit left transition region [1/2/3/4/5/6/7/8/eXit] <eXit>:
Edit transition region [Left/Right/All/eXit] <eXit>: r
Edit right transition region [1/2/3/4/5/6/7/8/eXit] <eXit>: 1
Pick first right transition region point: (select vertex 5)
Template surface transition [Dynamic/Pinned] <Pinned>:
Transition region type [Constrained/Free] <Free>:
Pick transition control point (RETURN for same):
Horizontal transition to hold [Grade/Elevation] <Grade>:
Pick transition reference point: (select vertex 1)
Edit right transition region [1/2/3/4/5/6/7/8/eXit] <eXit>:
Edit transition region [Left/Right/All/eXit] <eXit>:
Connect/Datum/Redraw/Super/Topsurf/TRansition/eXit <eXit>:
Edsrf/SAve/eXit/ASsembly/Display/SRfcon/Redraw <eXit>:
Save template (Yes/No) <Yes>:
Template name <cdt>: CDTRPFG
```

3. Erase the line work from the screen.

By repeating the above sequence of steps starting with the editing of the CDT template and by assigning transition properties, you develop the needed template definitions. The only thing you need to do is change the transition properties.

Template Name	Zone Properties
CDTPFE	Pinned, Free, and Hold Elevation (see Figure 10–18)
CDTDFE	Dynamic, Free, and Hold Elevation (see Figure 10–19)
CDTDCG	Dynamic, Constrained, and Hold Grade (see Figure 10–20)

EXERCISE

CDTPFE

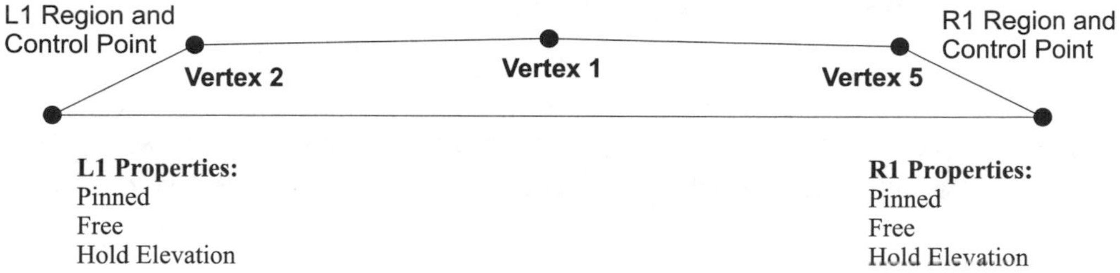

Figure 10–18

CDTDFE

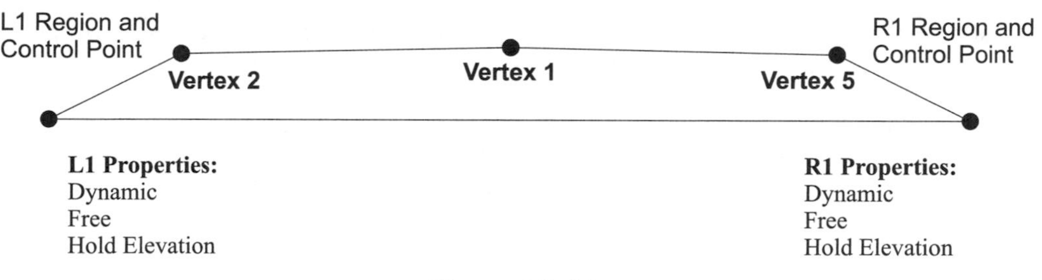

Figure 10–19

CDTDCG

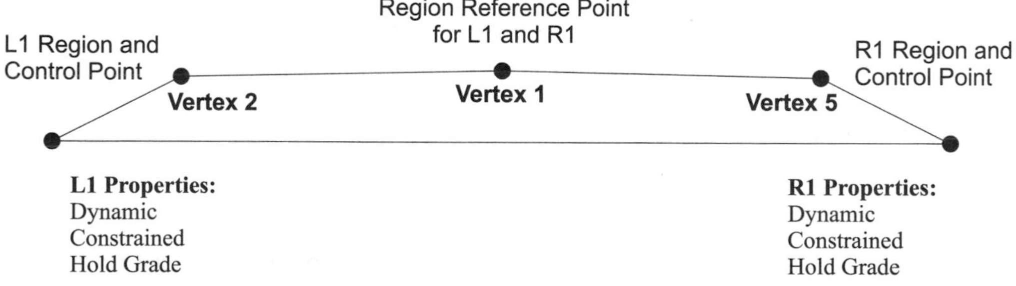

Figure 10–20

3. Before you exit the current drawing, you may want to Edit each template and display its transition and transition type definition. The Display option of Edit Templates allows you to view the current transition location and definition.

4. After editing, reviewing, and saving the new template definitions, open the Cdtrans drawing of the Cdtrans project. Save the changes to the template drawing.

UNIT 2: PROCESSING AND VIEWING TRANSITIONS

After defining the templates, it is a matter of specifying the template, the daylight slopes values, and attaching the first right and left alignments to the design. You assign these elements to the design in the Edit Design Control routine. When you exit the dialog box, the Edit Design Control routine processes the design and creates the template cross-section data. The next step is to view the cross sections that result from the transition properties and the location of the transition alignments.

After viewing the behavior of the template, all you need to do is rerun the Edit Design Control routine and change the template. This change causes the design processor to reevaluate the design and then creates new cross sections for you to view.

Each template will produce a different behavior as it travels down the centerline. It will become apparent what characteristics are for transitional alignments that cross the centerline as you view the distortion of the template.

Exercise

After you complete this exercise, you will be able to:

- Process the design with each template
- View the template's behavior within the cross sections

PROCESSING TRANSITIONS

Template CDTPFG

1. Use the Set Current Alignment routine from the Cross Sections menu to set theRanscl alignment as the current alignment.
2. Run the Edit Design Control routine in the Edit Control cascade of the Cross Sections menu, and accept the beginning and ending stations by selecting the OK button. The routine displays the Design Control dialog box.
3. Select the Template Control button to display the Template dialog box. Choose the Select button to view the Template Librarian dialog box and select CDTPFG from the list (see Figures 10–21 and 10–22). Select the OK button until you exit to the Template Control dialog box.

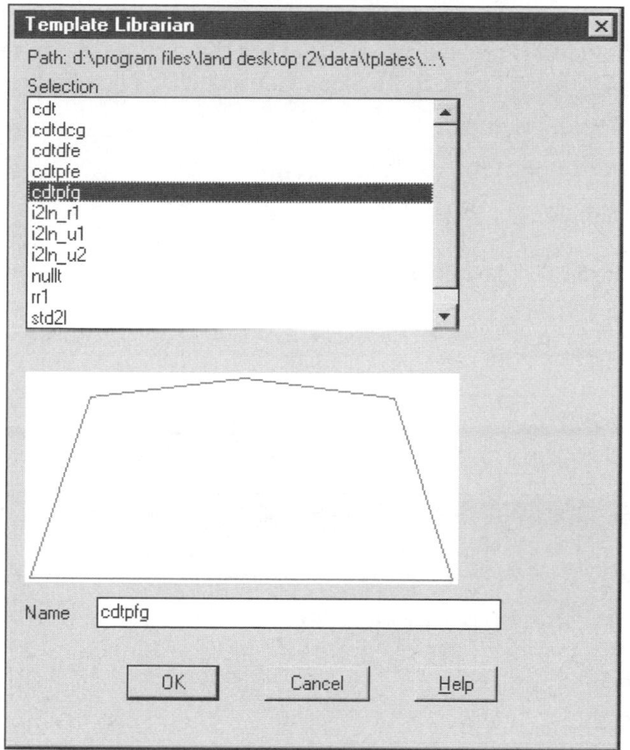

Figure 10–21

Template Control

Current Station Range

Start: 0.00 End: 603.04

Template cdtpfg Select... Datum 1 Select...

Superelevation

Pivot Left

Left 0.000000 Center 0.000000 Right 0.000000

Superelevation Rollover

Rollover 0.000000 Side Left

Subgrade Superelevation

Subgrade Superelevation Method None

Edit transitions...

OK Cancel Help

Figure 10–22

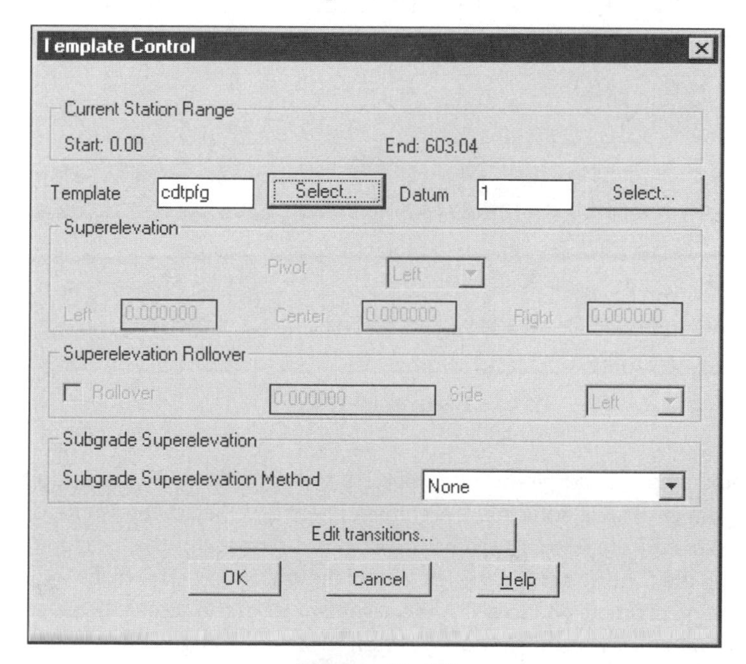

4. Select the Slope Control button to display the Slopes dialog box. Use the following values and review Figure 10–23 for the settings. After setting the values, select the OK button to exit the Slope Control dialog box and to return to the Design Control dialog box.

Left

Slope:	Simple	
	Typical	Maximum
Fill:	4:1	2:1
Cut:	5:1	3:1

Right-of-ways

Offset 33.00

Right

Slope:	Simple	
	Typical	Maximum
Fill:	4:1	2:1
Cut:	5:1	3:1

Right-of-ways

Offset 33.00

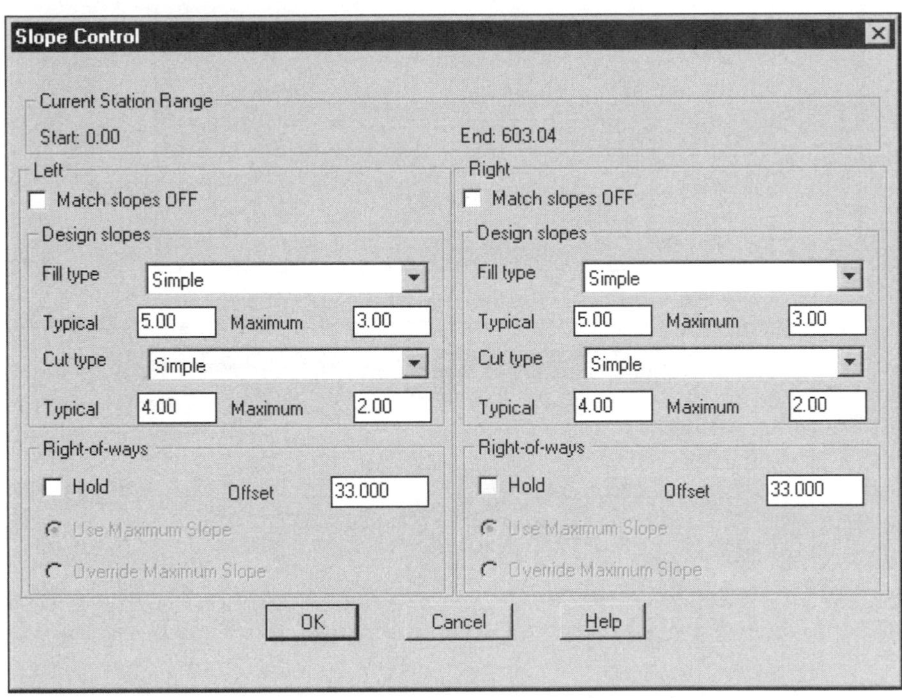

Figure 10–23

5. The next step is to assign the left (BSDK1L) and right (BSDK1R) transition alignments to the design. This is done in the alignment dialog box. Select the Attach Alignment button to view the Attach Alignment dialog box. First assign the Left One Alignment by selecting the One button on the left side of the dialog box. This hides the dialog box and the routine prompts you to select the alignment from the screen. If you cannot select the alignment, click the right mouse button to view the Alignment Librarian to select the BSDK1L alignment. The routine returns you to the Attach Alignment dialog box. Select

the One button on the right side of the dialog box to assign the BSDK1R alignment. This hides the dialog box, and the routine prompts you to select the alignment from the screen. If you cannot select the alignment, click the right mouse button to view the Alignment Librarian to select the BSDK1R alignment. Select the OK button to exit the Attach Alignment dialog box and to return to the Design Control dialog box.

6. Select the OK button to process the design and again select OK to exit the Process status dialog box. There should be no errors when processing this design.
7. To view the results, use the View/Edit Sections routine from the Cross Sections menu.

Reviewing CDTPFG Cross Sections

Notice that everything looks normal with the template while transitioning and moving down the roadway centerline. As long as each transition region point is on its respective side of the centerline, things seem normal. When the right transition control point crosses the centerline alignment, the region point draws a line from the template to the centerline. Since the template has the hold grade property, the line attaching the control point to the centerline alignment represents the grade referenced during the defining of the transition reference grade. Once the control point crosses back to its respective side of the centerline, the point behaves correctly (see Figure 10–24).

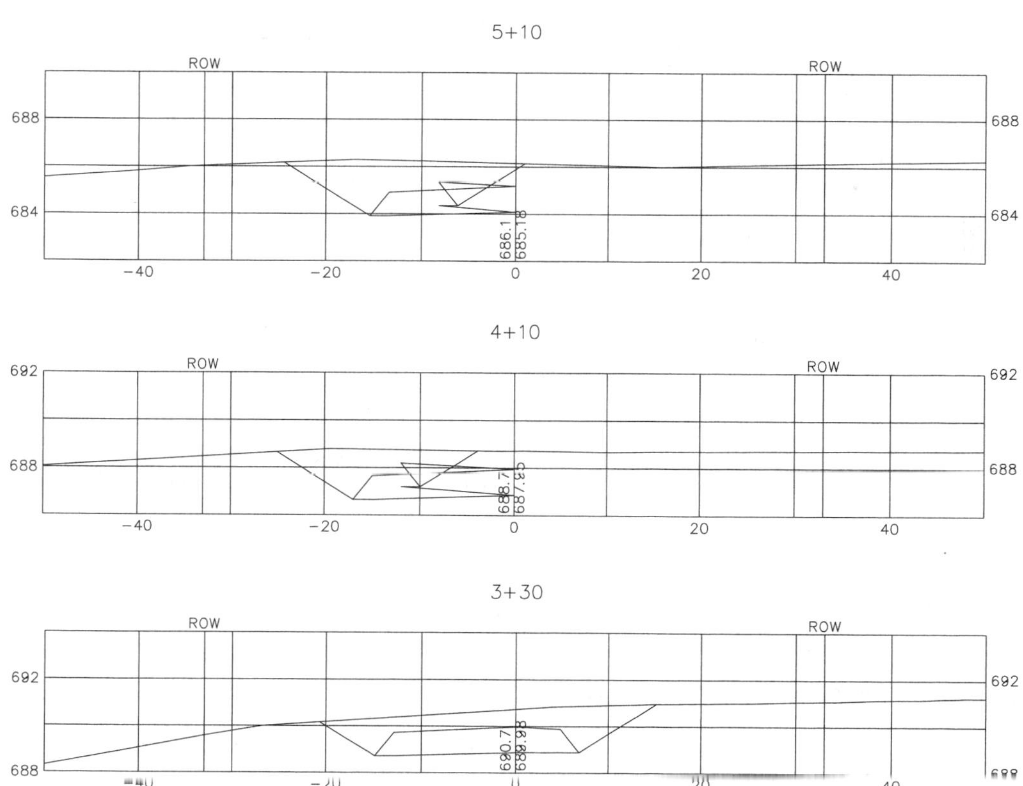

Figure 10–24

The conclusion from this run is simple. When manipulating the template across the centerline with transition control, the transition zone has to have the dynamic property, not pinned.

Any of the templates that have the property, pinned, will not successfully cross the centerline. The difference between CDTPFG and CDTPFE is that the first template connects to the centerline with a grade and the second template connects by an elevation.

Template CDTPFE

Since the new template uses the same slopes and transition alignments, you need only change the name of the template in the Template Control dialog box.

1. Change the template to CDTPFE using the Template Control dialog box of the Edit Design Control routine in the Edit Control Cascade of the Cross Sections menu (see Figure 10–25). Exit the both dialog boxes by selecting the OK button. The design processor processes the design with the new template.

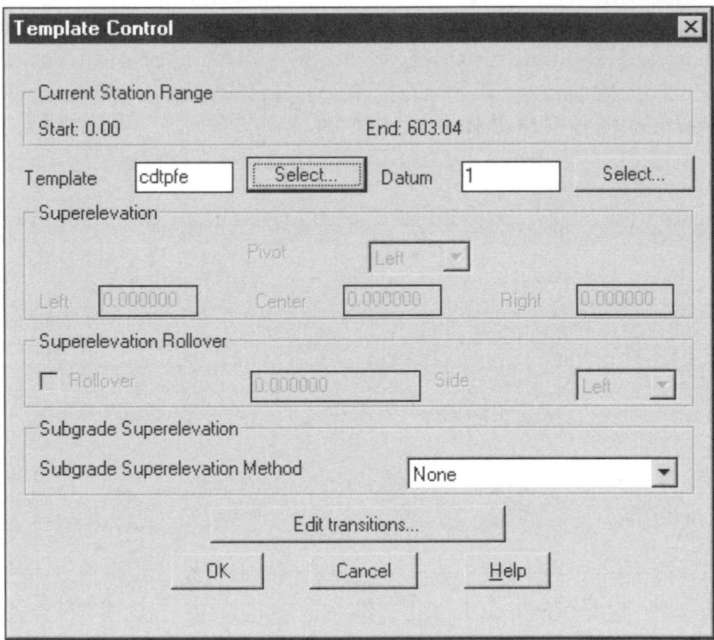

Figure 10–25

2. View the new template cross sections using the View/Edit Sections routine from the Cross Sections menu.

Reviewing CDTPFE Cross Sections

Notice that all is well as long as the control points remain on their side of the centerline. When the right control point moves across the centerline, it connects itself to the centerline alignment with a line segment. Since this template has the property of hold elevation, the line connecting the region point to the centerline alignment is a line that represents the difference in elevation you defined in the template (see Figure 10–26).

Again, the pinned region property is not for templates crossing the centerline alignment.

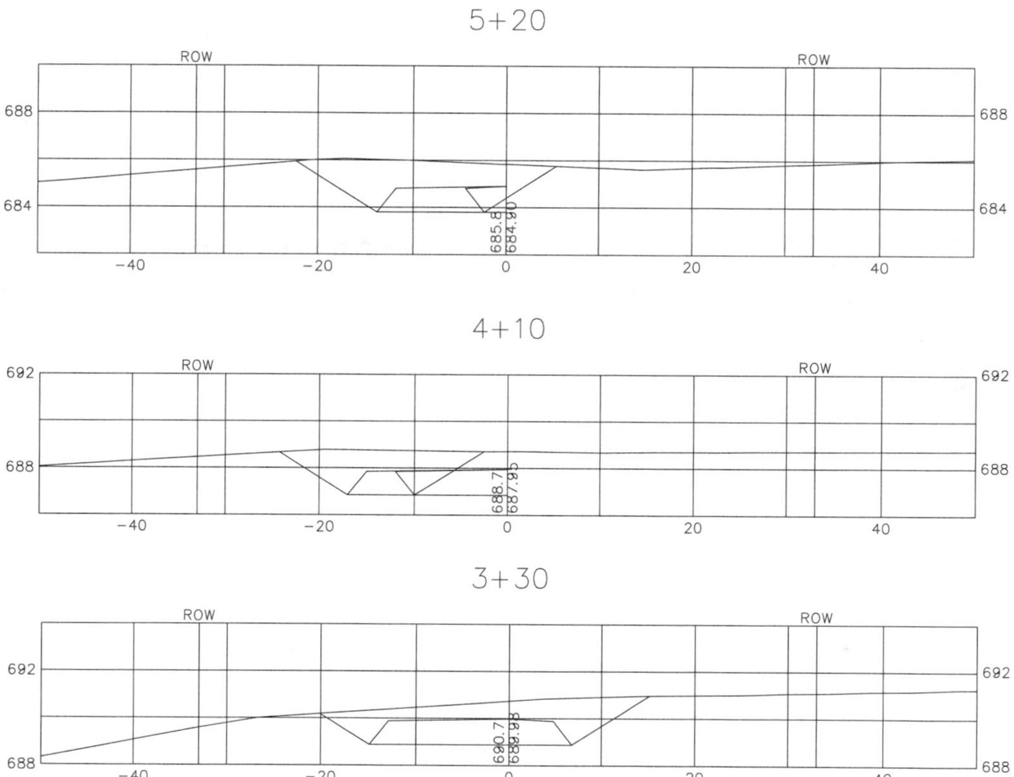

Figure 10–26

Template CDTDFE

Since the new template uses the same slopes and transition alignments, you need only change the name of the template in the Template Control dialog box.

1. Change the template to CDTDFE using the Template Control dialog box of the Edit Design Control routine (see Figure 10–27). Exit both dialog boxes by selecting the OK button. The design processor processes the design with the new template.

2. View the new template cross sections using the View/Edit Sections routine in the Cross Sections menu.

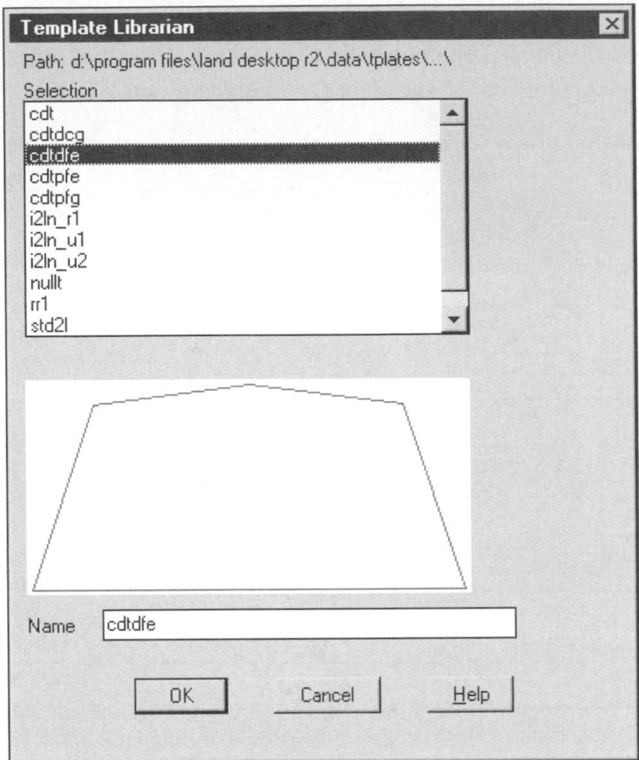

Figure 10–27

Reviewing CDTDFE Cross Sections

The transition of the control point across the centerline is smooth. The dynamic region property allows the template to be self-contained and not dependent on the centerline. The collapse of the transition regions occurs between the L1 and R1 regions. The property of hold elevation allows the region points to align with the elevation on the vertical alignment as the region point crosses the centerline alignment (see Figure 10–28).

Template CDTDCG

Since the new template uses the same slopes and transition alignments, you need only change the name of the template in the Template Control dialog box.

1. Change the template to CDTDCG using the Template Control dialog box of the Edit Design Control routine (see Figure 10–29). Exit both dialog boxes by selecting the OK button. The design processor processes the design with the new template.

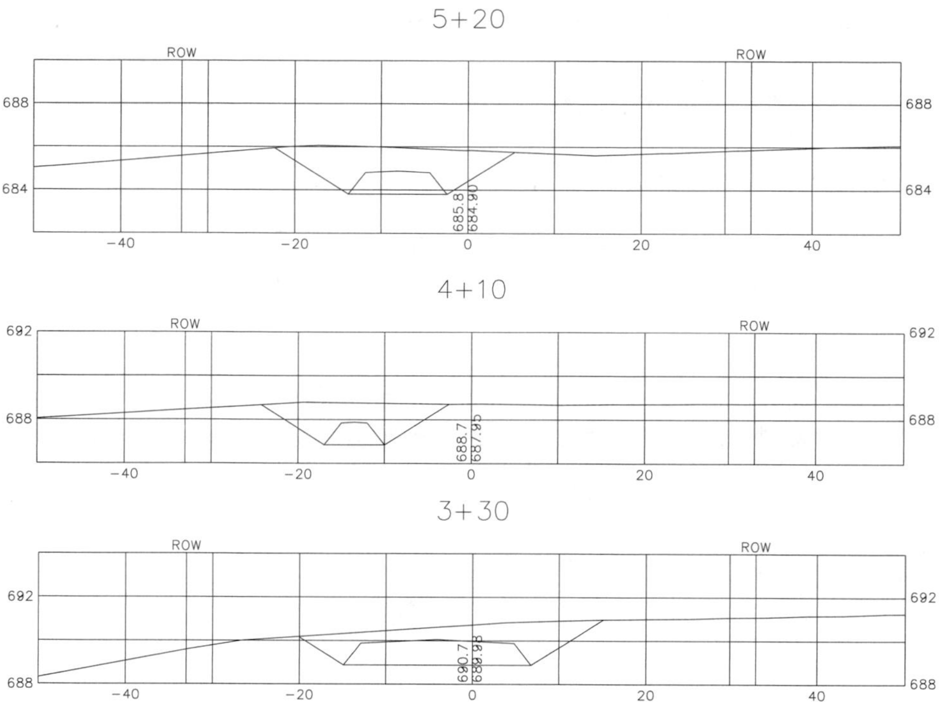

Figure 10–28

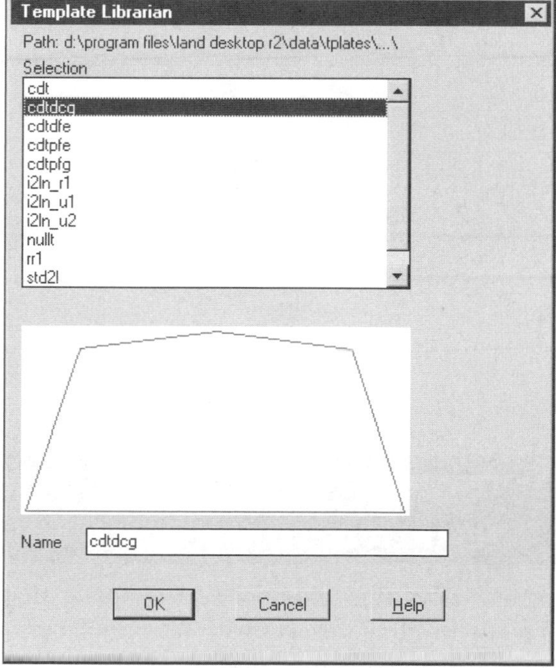

Figure 10–29

EXERCISE

2. View the new template cross sections using the View/Edit Sections routine from the Cross Sections menu.

3. Click the Save icon to save the drawing

Reviewing CDTDCG Cross Sections

Crossing the template over the centerline occurs without any problems. The property of constrained, however, prevents the shrinking of the template. When the transition alignments squeeze the template control points, the template maintains its minimum width (see Figure 10–30).

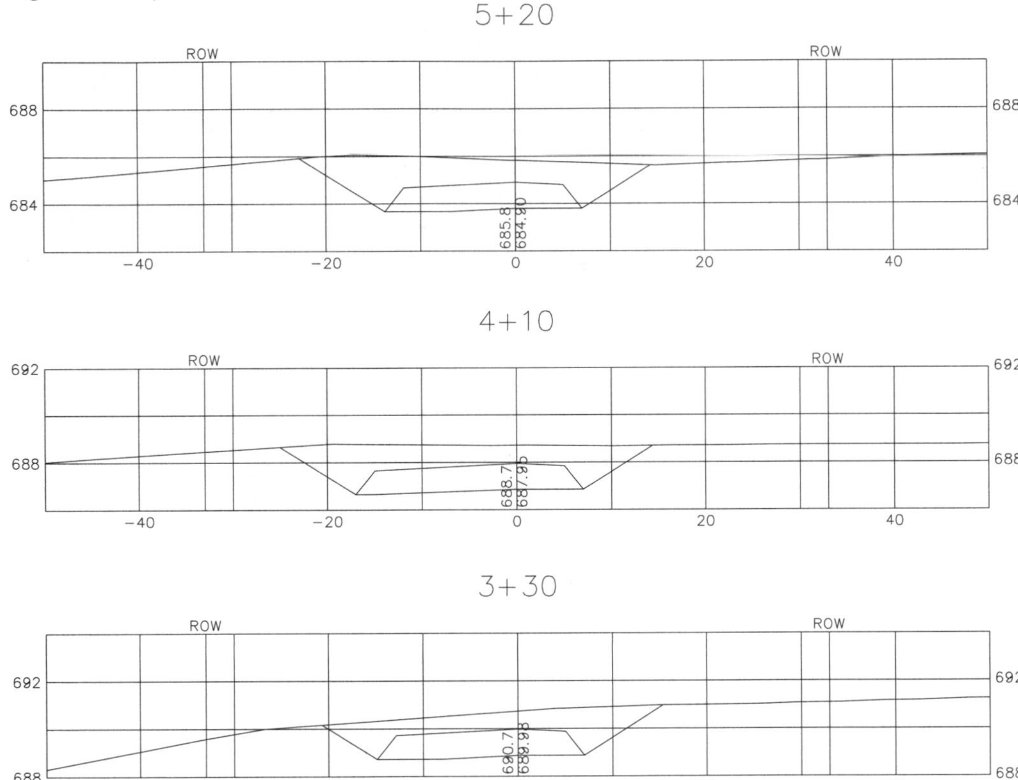

Figure 10–30

UNIT 3: TRANSITION CONTROL OF MULTIPLE TEMPLATE SURFACES

The next question is: how does this all apply to a template that has two or more surfaces? The answer for controlling surfaces within a template is that both the transitional control and region points control the surfaces that intersect a vertical line passing through the points.

To explore how multiple surfaces react in a template, create a new template similar to the CDT template. The new template will have three central surfaces. Two of the three surfaces cross the transition region point, which allows you to view the effects of the region point on adjacent and nonadjacent surfaces.

Exercise

The new template, CDT2, has the same shape as the CDT template. The new template, however, will have three surfaces, two of which are affected by the transition region point region (see Figure 10–31). As the template transitions, you can observe the effects the region point has on surfaces intersecting a vertical line passing through the transition.

After you complete this exercise, you will be able to:

- Define a template with multiple surfaces
- Process the design of the template
- View the behavior of the template surfaces within the cross sections

DRAWING THE CDT2 TEMPLATE

1. Open the drawing CDTTPL of the Tplate project.
2. Draw a new template with the Draw Template routine of the Cross Sections menu. Name the template CDT2. The template is symmetrical, so draw only the left portion of the template. Use the following values for the template and refer to Figure 10–31.

   Surface 1:

   Material: Asphalt

   The surface has a -2% grade for 4 feet and is 3 inches thick (.25).

```
Command:
Starting point:
Select point [Relative/Grade/Slope/Close/Undo/eXit]: g
Grade (%) [Relative/Slope/Points/Close/Undo/eXit]: -2
Change in offset: -4
Grade (%) [Relative/Slope/Points/Close/Undo/eXit]: r
Change in offset [Grade/Slope/Close/Points/Undo/eXit]: 0
Change in elev: -.25
Change in offset [Grade/Slope/Close/Points/Undo/eXit]: g
Grade (%) [Relative/Slope/Points/Close/Undo/eXit]: 2
Change in offset: 4
Grade (%) [Relative/Slope/Points/Close/Undo/eXit]:
```

Surface 2:

Material: Granular

The top of the surface 2 is the bottom of the surface 1. You can select vertices (the point option using the endpoint object snap) from surface 1 to define the top of surface 2. After selecting the last endpoint from surface 1, you draw the remainder of surface 2 from this point, starting with a grade of -2% for 1 foot. Then a -2:1 slope for 1 foot for the next portion. Next is a relative change of offset of 0 and a change of elevation -0.35. Finally, the last segment is a relative change of offset of 6 feet and a change of elevation 0.

```
Starting point:
Grade (%) [Relative/Slope/Points/Close/Undo/eXit]: p
Select point [Relative/Grade/Slope/Close/Undo/eXit]: (select the
   lower right point on surface 1)
Select point [Relative/Grade/Slope/Close/Undo/eXit]: (select the
   lower left point on surface 1)
Select point [Relative/Grade/Slope/Close/Undo/eXit]: (select the
   upper left point on surface 1)
Select point [Relative/Grade/Slope/Close/Undo/eXit]: g
Grade (%) [Relative/Slope/Points/Close/Undo/eXit]: -2
Change in offset: -1
Grade (%) [Relative/Slope/Points/Close/Undo/eXit]: s
Slope (3 for 3:1) [Relative/Grade/Points/Close/Undo/eXit]: -2
Change in offset: -1
Slope (3 for 3:1) [Relative/Grade/Points/Close/Undo/eXit]: r
Change in offset [Grade/Slope/Close/Points/Undo/eXit]: 0
Change in elev: -.35
Change in offset [Grade/Slope/Close/Points/Undo/eXit]: 6
Change in elev: 0
Change in offset [Grade/Slope/Close/Points/Undo/eXit]:
```

Surface 3:

Material: Crushed (Crsh)Base Course

The top of surface 3 is the bottom of the second surface. Use the point option to arrive at the 2:1 slope point on the second surface. At the outer left endpoint of the second surface, the Crushed Base Course continues with a slope of -2:1 for 1 foot. The side of the surface is a relative change of offset 0 and a change of elevation of -0.5. The returning segment to the centerline is a relative change of 7 feet with no change in elevation.

```
Starting point:
Change in offset [Grade/Slope/Close/Points/Undo/eXit]: p
Select point [Relative/Grade/Slope/Close/Undo/eXit]: (select the
   lower right point on surface 2)
```

```
Select point [Relative/Grade/Slope/Close/Undo/eXit]: (select the
   lower left point on surface 2)
Select point [Relative/Grade/Slope/Close/Undo/eXit]: (select the
   upper left point on surface 2)
Select point [Relative/Grade/Slope/Close/Undo/eXit]: s
Slope (3 for 3:1) [Relative/Grade/Points/Close/Undo/eXit]: -2
Change in offset: -1
Slope (3 for 3:1) [Relative/Grade/Points/Close/Undo/eXit]: r
Change in offset [Grade/Slope/Close/Points/Undo/eXit]: 0
Change in elev: -.5
Change in offset [Grade/Slope/Close/Points/Undo/eXit]: 7
Change in elev: 0
Change in offset [Grade/Slope/Close/Points/Undo/eXit]: x
```

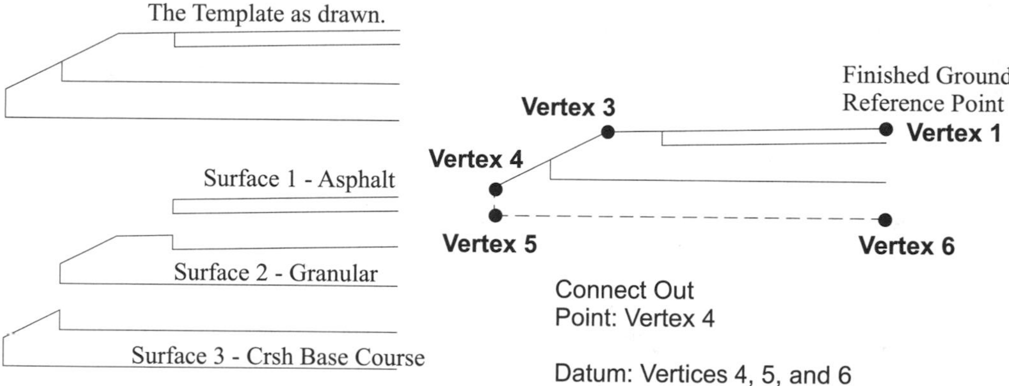

Figure 10–31

EXERCISE

Defining the CDT2 Template

The finished ground reference point is the upper right endpoint. All the surfaces within the template are symmetrical. The datum is the outer upper and lower left and lower right bottom endpoints of the Crushed Base surface (see Figure 10–31).

1. Define the template with the Define Template routine of the Templates cascade of the Cross Sections menu. Name the template CDT2. Use Figure 10–31 as a reference. The Finished Ground Reference point is Vertex 1 and the Connect-out point is Vertex 4. All of the surfaces are normal surfaces and there are no subassemblies attached to the template.

Editing the CDT2 Template

2. Edit the template to establish the top-of-surface and the transition zones (see Figure 10–32). The SRfcon option of the Edit Template routine defines the top-of-surface.

3. The transition points of a template are also in the Surface Control (sr) options list. All the selections are endpoint or intersection object snaps. Make sure you set a running object snap of endpoint and intersection for this exercise. The transition region and control points are the same point on the template (see Figure 10–32). The transition region points, L1 and R1, have the following properties:

   Dynamic

   Free

   Hold elevation

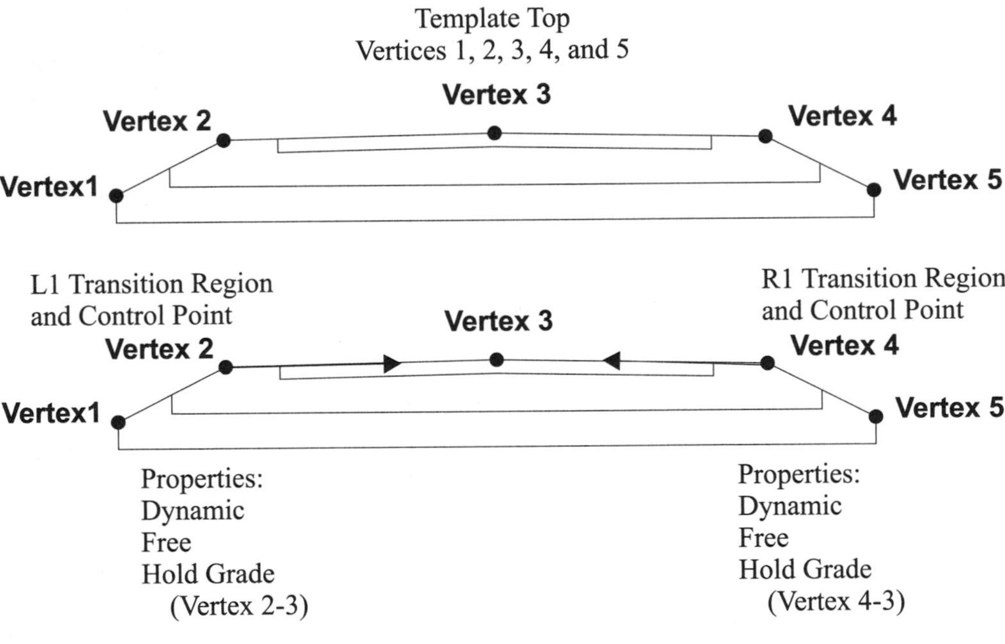

Figure 10–32

Processing the CDT2 Design

Since the new template uses the same slopes and transition alignments as the previous exercise, you need only change the name of the template in Template Control. Change the name of the template in the Template Control dialog box. When you exit the Design Control dialog box, the design processor calculates the new cross sections from the design data.

1. Open the Cdtrans drawing and save the changes in the CDTTPL drawing file.
2. Use the Set Current Alignment routine from the Cross Sections menu to set the Transcl alignment as the current alignment.

3. Assign the CDT2 template in the Design Control dialog box. The Edit Design Control routine is in the Design Control cascade of the Cross Sections menu. Choose the OK button to close both the Template Control and the Design Control dialog boxes. After you have exited the dialog boxes, the routine processes the data.

4. View the cross sections with the View/Edit Sections routine.

Reviewing the CDT2 Design

Notice that the Granular and the Crushed Base Course surfaces change with the transition region, but not the Asphalt surface. In fact, the Asphalt surface remains at the finished ground reference point while the remainder of the template is transitioning. The reason for this is that the Asphalt surface does not touch or intersect the vertical line passing through the transition region points. To remedy this situation, either the region point or the Asphalt surface needs to be modified so that the Asphalt surface touches or crosses the vertical line through the region point (see Figure 10–33).

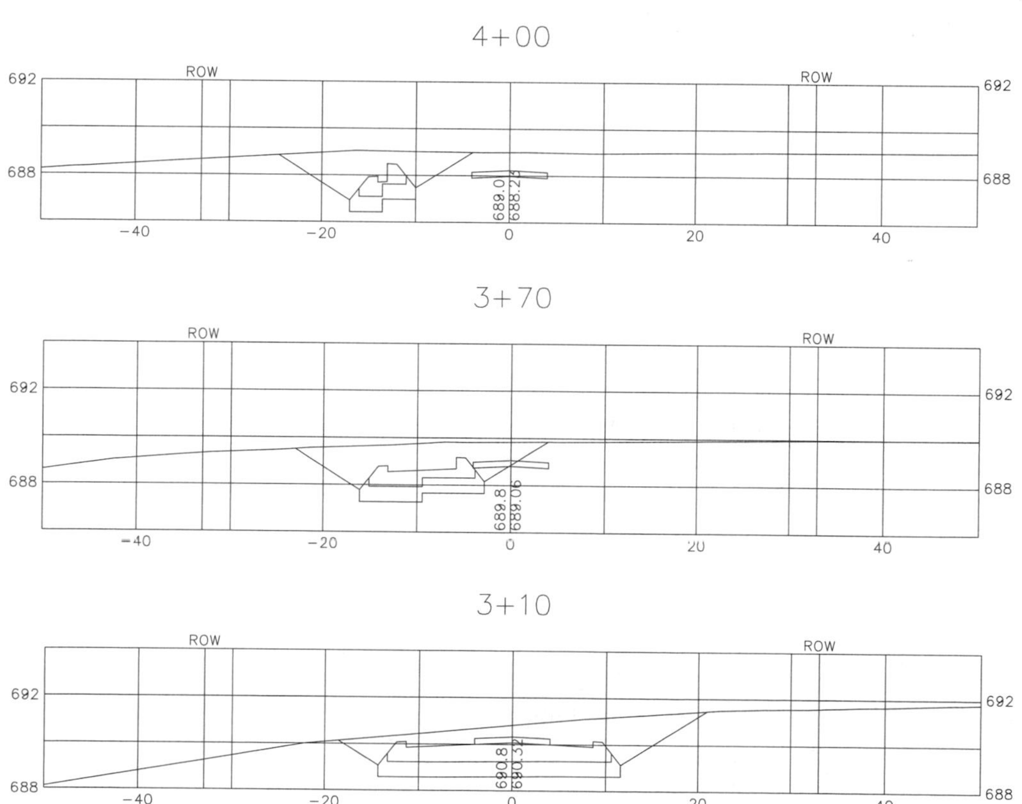

Figure 10–33

Editing the CDT2 Template

1. Open the CDTTPL drawing of the Tplate project and save the changes in the Cdtrans drawing.

2. Redefine the L1 and R1 transition region points to the outer vertex of the Asphalt surface. L1 region point is vertex 3 and R1 region point is vertex 5. Use the SRfcon option of the Edit Template routine to redefine the locations of L1 and R1. The hold grade for the template changes to vertex 3-4 on the left side and vertex 5-4 on the right side of the template. The control points remain the same points on the template. Each region has the properties of dynamic, free, and hold elevation (see Figure 10–34). Save the edited template as CDT2.

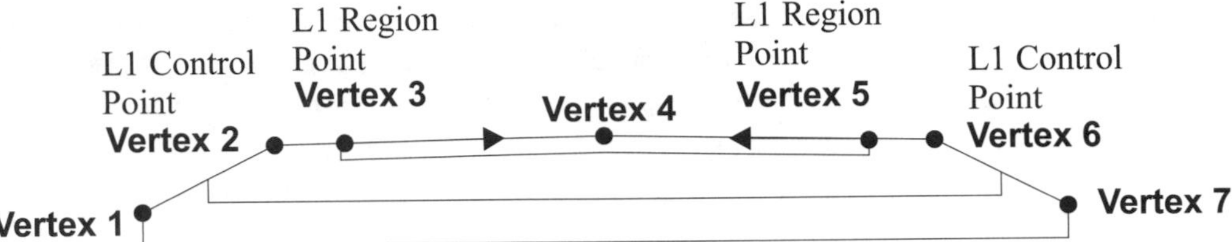

Figure 10–34

3. Exit and save the drawing.

4. Open the Cdtrans drawing.

5. Set the current alignment to Transcl with the Set Current Alignment routine of the Alignments menu.

6. The CDT2 template is already assigned to the design processor. Running the Edit Design Control routine will not reprocess the design with the new template definition. To process the design with the new template definition, select the Process Sections routine of the Design Control cascade of the Cross Sections menu. This routine will reprocess the design and create the new cross sections.

Reviewing the CDT2 Design

1. View the cross sections with the View/Edit Sections routine.

The transition of the roadway is successful and coherent. There is, however, a slight problem when the template collapses to the point of the disappearance of the region points (see Figure 10–35). It seems the template would behave better if you set the hold elevation property to hold grade.

2. There are some issues with the template as defined. Try defining the Transitional control point to be the same as the region point (the outer edge of the Asphalt surface).

EXERCISE

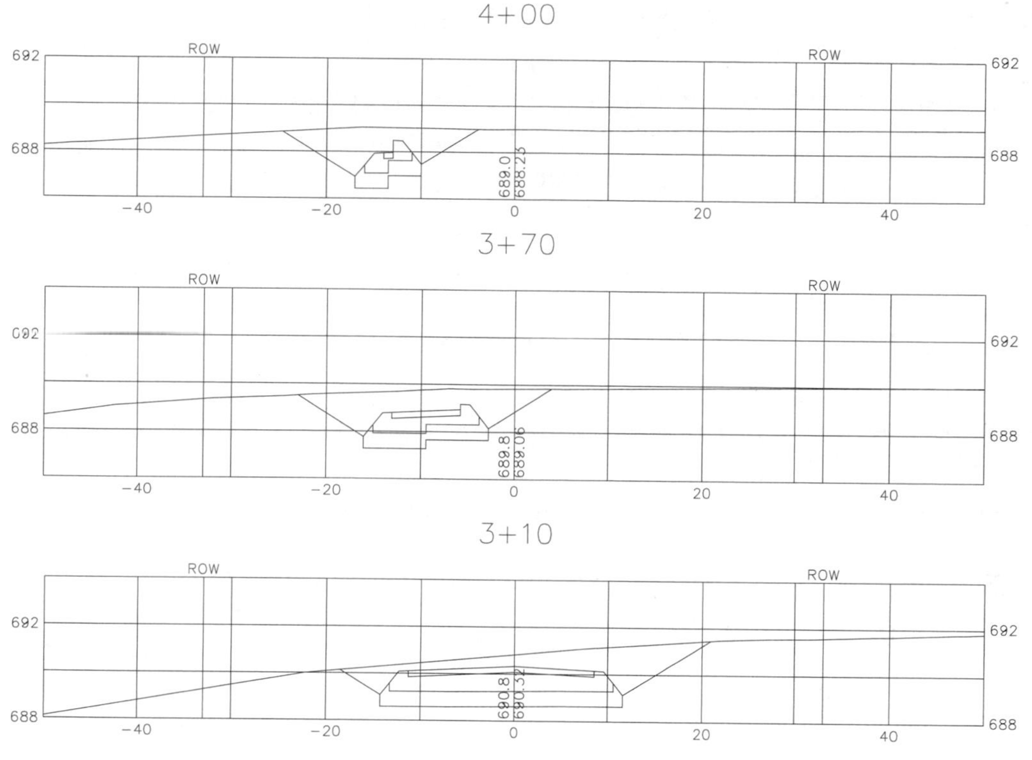

Figure 10–35

3. Make layer 0 (Zero) the current layer and freeze the layers TRANSCL, BSDK1L, and BSDK1R and the layers prefixed with TRANSCL.
4. Click the Save icon to save the drawing.

EXERCISE

UNIT 4: TRANSITIONING A REGION

The necessity of a separate control and region point occurs when there is a need to manipulate a zone within a template. The zone may include other template surfaces. A roadway with a median is an example of this type of problem.

A median template has to have the median as an open surface. The defining routine closes the surface when creating the right side of the template. The remaining surfaces are closed surfaces (see Figure 10–36). This assumes you are not using subassemblies for the outside curb and shoulder elements. If you are using subassemblies, then the outside curb and shoulder are subassemblies.

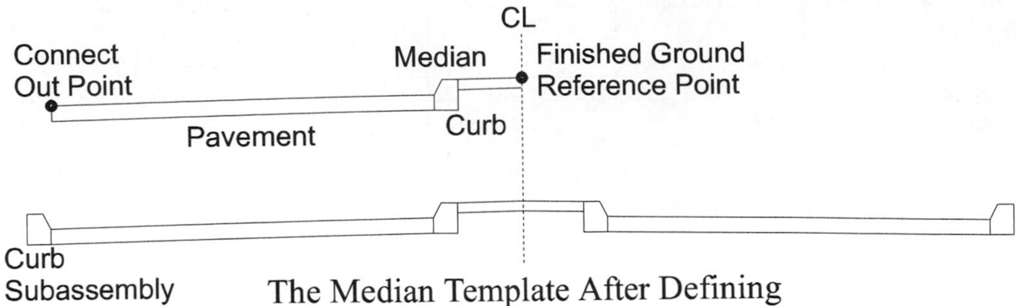

Figure 10–36

The transition control and region points control the motion of the template, the inner curb, and the median. The designer locates the transition control (L1 or R1) point at the inner edge-of-pavement and the region point at the back of curb (see Figure 10–37). If the property Hold Grade is set, the grade to hold is from the back of curb to the crown of the median. If Hold Elevation is set, the design processor looks at the difference in elevation between the Region point and the crown of the median.

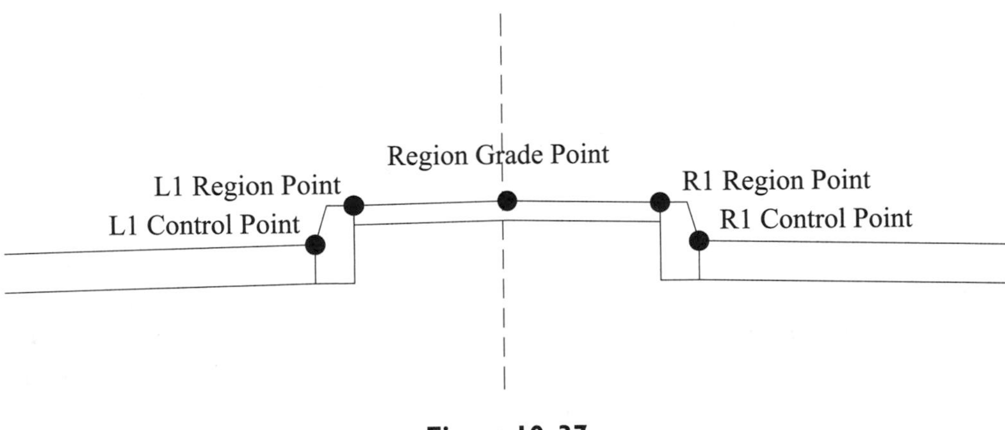

Figure 10–37

By defining such a template, the transition points have control over the central portion of the template. As the transition alignments move away from the centerline, the median widens. As the transition alignments converge, the median narrows or even disappears.

EXERCISE

Exercise

The new template is a two-lane roadway with a median between the pavement. The outer curb and shoulder are subassemblies. The inner curb and the pavement are closed surfaces. The median is an open surface. The template is symmetrical, so you need to draw the left portion only.

After you complete this exercise, you will be able to:

- Define a template with a median
- Process the design of the template
- View the behavior of the template surfaces within the cross sections

TRANSITIONING A REGION

1. Open the CDTTPL drawing of the tplate project.

The Median Template Subassemblies

2. Draw the curb subassembly with the Draw Template routine of the Cross Sections menu. When you define the subassembly, name it curb2. Start drawing the curb on the middle right side where the slant starts (vertex 1). Draw the subassembly counterclockwise (see Figure 10–38).

The following are the curb2 dimensions:

Width:

Top 0.35 feet

Bottom 0.5 feet

Height:

1 foot

The slant starts 0.5 feet up the curb.

The connect-in point is the starting point of the slant on the right side of the curb (vertex 1). The connect-out point is the upper left-corner of the curb (vertex 3). The datum is symmetrical and goes from the upper left side to the bottom left side and finally to the lower right side of the curb (vertices 3–5).

3. Define the curb subassembly.

The Define routine assumes that the top of any subassembly is a straight line from the connect-out to connect-in point. If you want to have the top-of-surface for the subassembly as a line traveling over the curb, you must run the Edit Template routine. The Top Surface option of the SRfcon group of Edit Subassembly defines the top of the subassembly. The vertices 3,2, and 1 define the top-of-surface of the subassembly (see Figure 10–38)

4. Create the shoulder subassembly linework. The subassembly is a -15:1 slope for 3 feet (see Figure 10–39). Name the subassembly medshdr. Draw the subassembly from right to left. The connect-in point is the right endpoint and the connect-out point is the left endpoint. The datum is symmetrical and goes from the left side to the right side.
5. Define the medshdr subassembly. The Define routine assumes that the top of the subassembly is a straight line from the connect-out to connect-in point. Since the subassembly is a straight line, there is nothing wrong with this assumption.

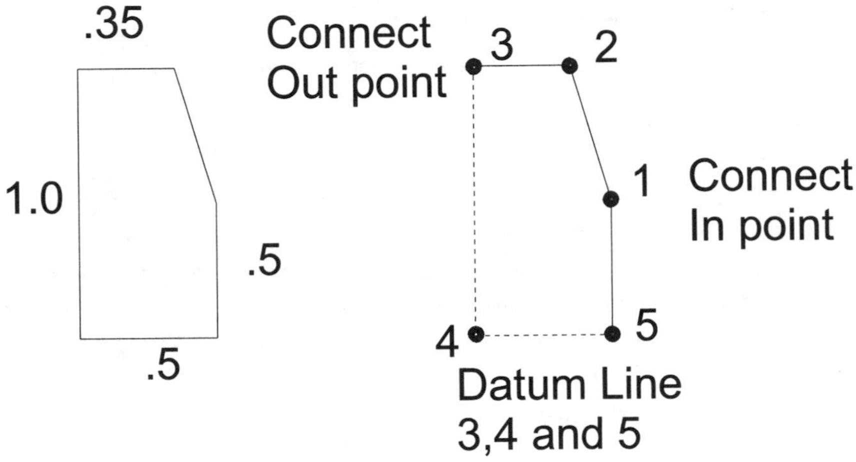

Figure 10–38

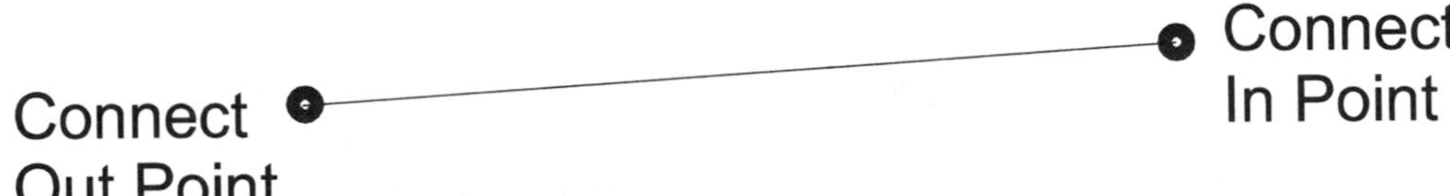

Figure 10–39

Drawing the Template Surfaces

6. Create the following surfaces: median, curb, and pavement (see Figure 10–36). After drawing the surfaces, define a template called median. Remember, the template is symmetrical so draw only the left side. Also, draw all the surfaces counterclockwise. Start by drawing the median surface, then the curb and pavement surfaces.

Median: An open surface with a -2% grade, a width of 2 feet, and a thickness of 3 inches (0.25). Use the material of Median for this surface.

Curb: A closed surface 1 foot high, with a base width of .5 feet, and a top width of .35 feet. The right side of the curb is straight and the left (pavement) side has a slope half way down. The curb surface is a mirror copy of the polyline that became the curb2 subassembly. Use the curb material for this surface

Pavement: A closed surface with a -2% grade, a width of 12.5 feet, and is 0.5 feet thick. Use the asphalt material for this surface.

Defining the Template

7. Define the template.

Each of the surfaces is symmetrical. When the Define routine prompts for the surfaces, identify the median, curb, and pavement surfaces. Name and select each of the surfaces as the routine prompts for their values.

The finished ground reference point is the crown of the median. The connect-out point is the upper outer endpoint of the pavement. The datum starts at the outer lower endpoint of the pavement and continues to the right outer edge of the curb. From the point on the curb, the datum goes up to the bottom of the median and ends at the right lower outside vertex of the centerline of the median (see Figure 10–36).

The Define routine presents the Subassembly Attachment dialog box. Use curb2 for the curb and the medshdr as the cut and fill shoulder (see Figure 10–40).

When the routine prompts for a name, assign the name of CDTMED.

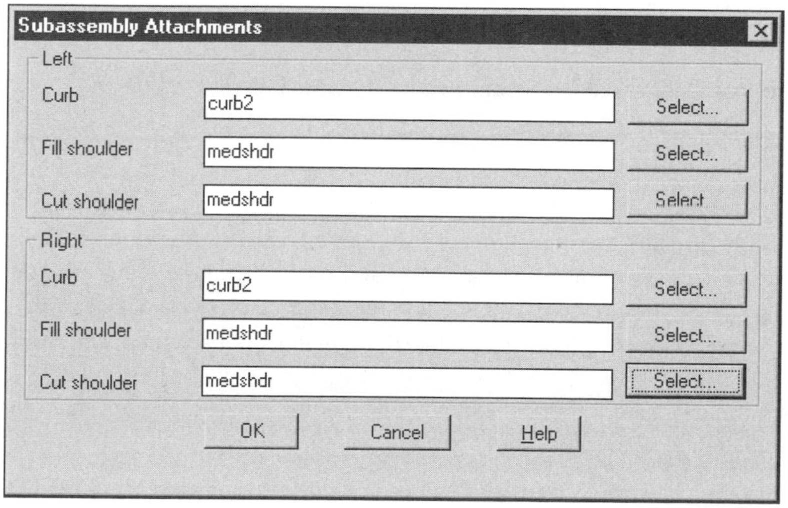

Subassembly Attachments

Left

Curb	curb2	Select...
Fill shoulder	medshdr	Select...
Cut shoulder	medshdr	Select...

Right

Curb	curb2	Select...
Fill shoulder	medshdr	Select...
Cut shoulder	medshdr	Select...

OK Cancel Help

Figure 10–40

EXERCISE

Editing the Median Template

After defining the template, you need to edit the template to define the transition and the top-of-surface points.

The location of the transition region point is to the right of the control point for L1 and the location of the transition region point is to the left of the control point for R1. The region point is the back of curb where the median intersects the curb on the right and left sides of the median.

8. Use the Edit Template routine to define the top-of-surface and the L1 and R1 control and region points.
9. First, define the top-of-surface line in the SRfcon option of the Edit Template routine (see Figure 10–41).

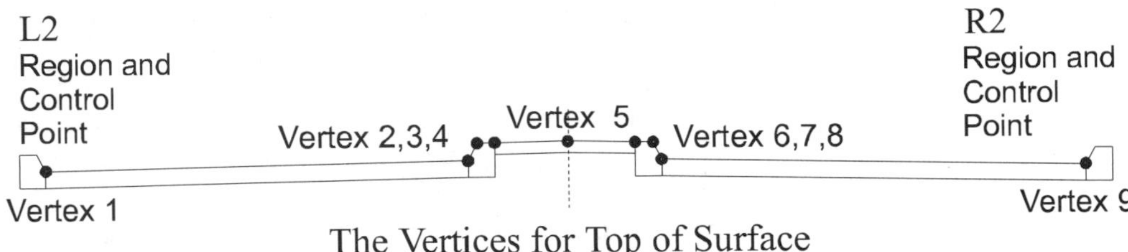

Figure 10–41

The L1 and R1 control points are the upper inner corner of the pavement surface. The region point is the intersection between the back-of-curb and the median. The grade is from the region point to the crown of the median. Assign each region the properties of dynamic, free, and hold grade (see Figure 10–37).

10. Also define L2 and R2 transition control points (see Figure 10–41).

    Vertex for L2 is vertex 1.

    Vertex for R2 is vertex 9.

The L2 and R2 region and control points are the same point on the template. Use the outer upper endpoint of the pavement as the region and control point. The grade for the point follows the grade of the upper pavement line. Use the nearest object snap to define the grade. The L2 and R2 regions have the properties dynamic, constrained, and hold elevation.

11. Save the modified template, exit the Edit Template routine, and erase the template line work from the display.
12. Open the drawing Cdtrans of the Cdtrans project and save the CDTTPL drawing.

EXERCISE

Defining the Median Centerline

Since the median will be traveling down a new alignment, this exercise needs three new horizontal alignments: MEDCL, BSDK2L, and BSDK2R. This means traveling the entire roadway design path from horizontal alignments, to profiles, to vertical alignments with sampling for profile and cross-section elevations.

The lines already exist in the drawing for the new centerline and transition alignments. There are several segments for each transitional alignment. Isolate the layer for the alignment and select the entities with a crossing window.

1. Define each transition alignment by starting with the BSDK2L, and end with MEDCL. Each alignment starts in the southwest and ends in the northeast corner of the drawing. When the Define Alignment From Objects routine prompts for a starting point, just press ENTER to accept the default value. All of the alignments start at station 0.00.
   - BSDK2L—Isolate the layer BSDK2L (use the Layer dialog box), and define the alignment. Describe the alignment as the Median Left Alignment.
   - BSDK2R—Isolate the layer BSDK2R (use the Layer dialog box), and define the alignment. Describe the alignment as the Median Right Alignment.
   - MEDCL—Isolate the layer MEDCL (use the Layer dialog box), and define the alignment. Describe the alignment as the Median Centerline Alignment.

If the MEDCL alignment is not the current alignment, use the Set Current Alignment routine in the Alignments menu to set MEDCL as the current alignment. Remember, all of the layers will have a MEDCL prefix because of the layer prefix set in the Prefix and Labels Defaults dialog box earlier in this section.

2. Sample along the MEDCL alignment using the Sample From Surface routine of Existing Ground of the Profiles menu. When the routine prompts for a surface, select Existing from the dialog box. Press ENTER to accept the beginning and ending station for sampling.
3. Pan the alignments to the left of the screen.
4. Run the Create Full Profile routine in the Create Profile cascade of the Profiles menu to create the MEDCL profile. Assign the profile layers different colors.
5. Save the drawing.
6. Draw the vertical centerline tangents with the Create Tangents routine of the FG Centerline Tangents cascade of the Profiles menu. Use the following information to draw the tangents:

The tangent lines at Station 0 and at the end of the alignment are to meet the existing ground.

EXERCISE

The first tangent is from 0+00 to 3+00 with an elevation of 691.00 at 3+00.

```
Command:
Alignment Name: MEDCL     Number: 10     Descr: Median Centerline
Starting Station: 0.00  Ending Station: 1206.07
Start Station: 0+00      Existing Elevation: 690.8371
Select point (or Station): s
Station <0>:  (press ENTER to accept)
Elevation <690.84>: (press ENTER to accept)
Station: 0+00   Elevation: 690.8371
Select point [Station/eXit/Undo/Length]: s
Enter station: 300
Select point [Grade/Elevation/Undo/eXit]: e
Elevation <690.84>: 691
Station: 3+00   Elevation: 691.0000  Last Grade: 0.0543
```

The second tangent is from 3+00 to 6+50 with an elevation of 682.00 at 6+50.

```
Select point [Station/eXit/Undo/Length]: s
Enter station: 650
Select point [Grade/Elevation/Undo/eXit]: e
Elevation <691>: 682
Station: 6+50   Elevation: 682.0000  Last Grade: -2.5714
```

The third tangent is from 6+50 to 9+50 with an elevation of 690.00 at 9+50.

```
Select point [Station/eXit/Undo/Length]: s
Enter station: 950
Select point [Grade/Elevation/Undo/eXit]: e
Elevation <682>: 690
Station: 9+50   Elevation: 690.0000  Last Grade: 2.6667
```

The last tangent is from 9+50 to the end of the existing ground. Use the Point option and an endpoint object snap to select the end of the alignment.

```
Select point [Station/eXit/Undo/Length]:
Select point [Station/eXit/Undo/Length]: _endp of
Station: 12+06.07   Elevation: 694.8013  Last Grade: 1.8750
Select point [Station/eXit/Undo/Length]: X
```

7. Define the vertical alignment for the alignment MEDCL with the Define FG Centerline of the FG Vertical Alignments cascade of the Profiles menu.

EXERCISE

8. In the Vertical Alignment Editor, create the following vertical curves (see Figure 10–42).

Station:

3+00 - 150 foot vertical curve

6+50 - 300 foot vertical curve

9+50 - 200 foot vertical curve

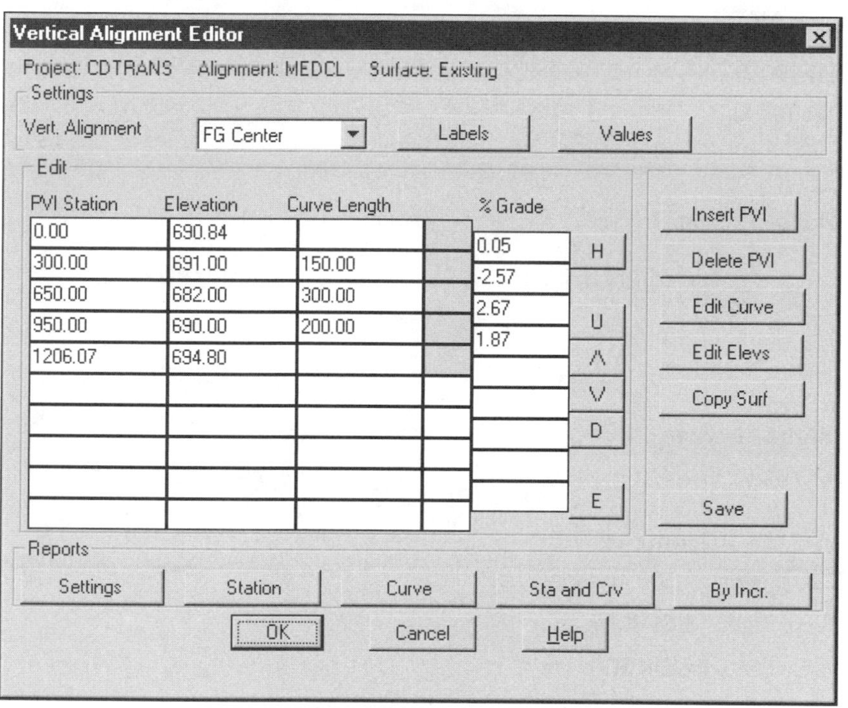

Figure 10–42

9. Select the OK button to exit the curve editor and the Vertical Editor, and save the changes to the vertical alignment. The routine does not import the new alignment until you manually import it into the drawing.

10. Use the Import routine of the FG Vertical Alignments cascade in the Profiles menu to replace and annotate the current alignment in the profile.

11. Next sample the cross-section data from the existing ground. Use the Sample From Surface routine of Existing Ground in the Cross Sections menu. Sample a swath of 75 feet at an interval of every 25 feet. Sample the beginning and end of the alignment and the PCs and PTs. Press ENTER to accept the beginning and ending stations.

Processing the Median Design

1. Run the Edit Design Control routine in the Design Control cascade of the Cross Sections menu. When the routine starts, select the OK button to accept the beginning and ending stations.

Set the following values.

Template Control:
CDTMED

Ditches:
None

Left			**Right**		
Slope:	Simple		Slope:	Simple	
	Typical	Maximum		Typical	Maximum
Fill:	4:1	2:1	Fill:	4:1	2:1
Cut:	5:1	3:1	Cut:	5:1	3:1

Right-of-ways		**Right-of-ways**	
Offset	40.00	Offset	40.00

Attach Alignment:
Set the First Right and First Left

First Right - **BSDK2R**

First Left - **BSDK2L**

2. Select the OK button to exit the Design Control dialog box and to process the design for the new cross sections.
3. View the cross sections for the median design with the View/Edit Sections routine of the Cross Sections menu (see Figure 10–43).

Notice that the median appears and disappears along the roadway. The template retains its composure as the control points cross over the centerline. This is due to the properties of the transition regions in the template.

EXERCISE

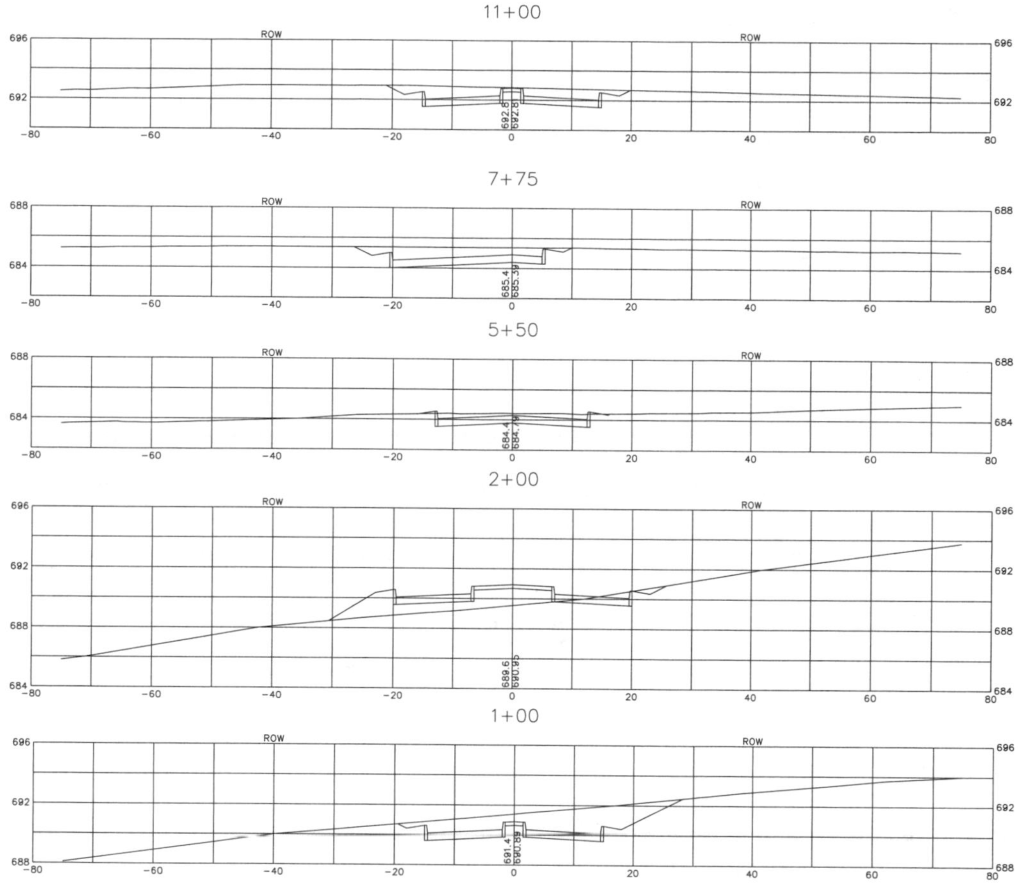

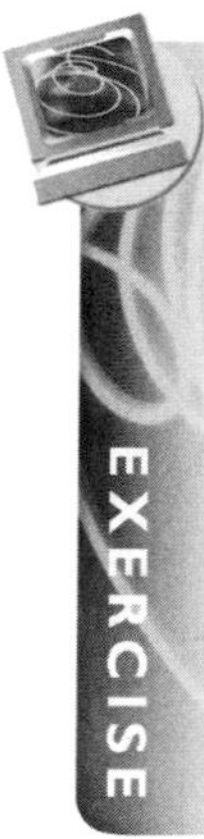

Figure 10–43

UNIT 5: IMPORTING THE TRANSITION CONTROL DATA

Even though the L2 and R2 transition control points are not being used, the Edit Design Control routine calculates their location along the roadway. The Import routines of the Ditch /Transition cascade place the control points as line work in either plan or profile view. After editing the lines, you may define the lines as horizontal and/or vertical transitional alignments. The Attach Alignment or Profiles dialog box of Edit Design Control assign the alignments to the template.

Exercise

After you complete this exercise, you will be able to:

- Import ditch and transition line work

IMPORTING TRANSITION CONTROL LINES

1. Zoom in to get a closer view of the MEDCL profile.
2. Import into the profile the L1, L2, R1, and R2 vertical transition lines. Use the Import Profile routine found in the Ditch/Transition cascade of the Cross Sections menu. Import one line at a time and undo the import when preparing to import the next line.
3. Zoom back and pan the drawing to get a better view of the MEDCL centerline.
4. Import L2 and R2 alignments into the plan view of the alignment. Edit the lines to represent a widening and narrowing transition of the roadway. If you create an alignment that attempts to move L2 or R2 towards L1 or R1, the transition will not occur. The reason for this is the constrained property assigned to L2 and R2. If you change the property to free, you will see L2 and R2 move towards L1 and R1.
5. Define the new alignments. Name the alignments, BSDK2R2 and BSDK2L2.

After defining the transitional alignments, make MEDCL the current alignment. Attach the BSDK2L2 and BSDK2R2 horizontal alignments to the template as the second left and right transition alignments. You can do this in the Attach Alignments dialog box of Edit Design Control. After attaching the alignments, select the OK button in both the Attach Alignments and Design Control dialog boxes. After you exit the dialog box, the design data will be reprocessed.

6. View the new cross sections with the View/Edit Sections routine.
7. Click the Save icon to save the drawing.

UNIT 6: DITCH DESIGN

When you apply ditch parameters to the template in the Design Control dialog box, the routine calculates the centerline and elevation of the ditch. You can import the lines into the profile or plan view using the Import routines of the Ditch/Transition cascade of the Profiles menu. After you modify the lines to better represent a ditch design, they can be defined as a new horizontal and vertical ditch control. The vertical alignments control the base point of the left or right ditch. The horizontal alignments control the offset location of the ditches.

DITCH SETTINGS

The Design Control dialog box sets conditional or preliminary ditch design values. When you import, edit, define, and apply the transitional alignments, they will improve and modify the prior ditch design.

When working with the ditch parameters, the Edit Design Control routine requires a minimum of two parameters to generate a ditch. When you apply a vertical alignment to the ditch, the Base elevation parameter toggle must be on and a ditch base width must be set to a value other than 0 (zero).

Exercise

After you complete this exercise, you will be able to:

- Use ditch settings in design control

DITCH DESIGN

1. If you are in the Cdtrans drawing, open the Design Control dialog box for the MEDCL centerline and set the following ditch parameters (see Figure 10–44). After setting the ditch parameters, select the OK button to exit both the Ditch and Design Control dialog boxes. After you exit the Design Control dialog box, the design processor evaluates the data.

Type:	Left and Right Cut/Fill
Foreslope:	ON 2:1 Slope
Ditch Depth:	1.5
Base Width:	0.5

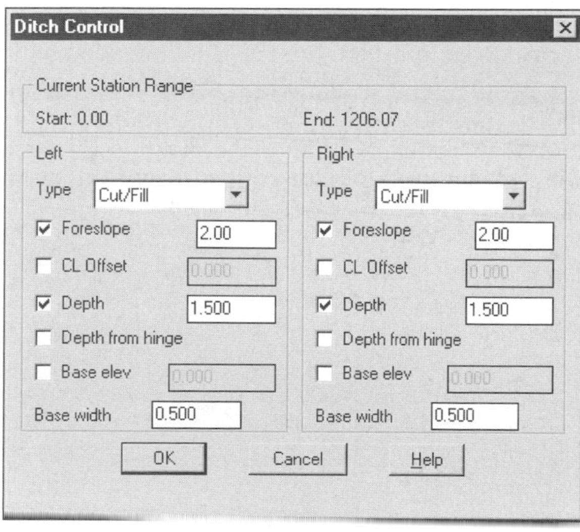

Figure 10–44

EXERCISE

2. View the new cross sections with the View/Edit Sections routine in the Cross Sections pull-down.
3. Import the right ditch line. Use the Import to Profile routine of the Ditch/Transition cascade of the Cross Sections menu.

```
Command:
Alignment Name: MEDCL     Number: 10     Descr: Median Centerline
Starting Station: 0.00  Ending Station: 1206.07
Profile to draw [Left/Right]: r
Right profile to draw [Ditch/Super/1/2/3/4/5/6/7/8]: d
Beginning station <0>: (press ENTER to accept)
Ending station <1206.07>: (press ENTER to accept)
Delete finished ground profile layer (Yes/No) <Yes>: (press ENTER
  to accept)
Erasing entities on layer <MEDCL-PDGR> ...
```

4. If you like, you can edit the right ditch profile line, define it as a profile alignment, and attach it to the design in the Attach Profiles dialog box of the Edit Design Control routine.
5. Click the Save icon and save the drawing.

The power of Civil Design is the manipulation of the roadway template. The design process, however, is not a single run down the centerline with all the parameters set. What you will find is that the design process is an iterative process. The initial values and first run through the design processor may only be the beginning of the process. After running the design through the processor, you will use routines that import the horizontal and/or vertical design results into the drawing. You will be able to modify them in cross-section, profile, and/or plan view. After you redefine the control, the design processor will calculate new cross sections for the modified design.

The next section looks at two more sets of controls over the design, daylight slope control and superelevation.

Slope Control and Superelevation

After you complete this section, you will be able to:

- Control slope and bench design
- Control depth and surface of sloping
- Control sloping with stepped parameters
- Superelevate a roadway design

SECTION OVERVIEW

Along with control over the template through transition alignment, you can control other aspects of the roadway design. These additional controls include slope variances, benching, and superelevation. Civil Design allows you to enter parameters that influence the type of cross-section slopes, benches, or templates the design processor uses while evaluating a design. Although slopes and benching are a part of the Edit Design Control routine, the default type of control is simplistic. The depth, stepped, and surface slope controls read external files to create their effect on the design.

UNIT 1

The first unit of this section reviews the simple method of daylight slope control. This unit also reviews the benching controls in the Edit Design Control routine.

UNIT 2

The depth control of the Edit Design Control routine reads an external file to set its conditional behavior. The depth control changes slopes depending on the depth of cut and fill. This design strategy is a part of the second unit of this section.

UNIT 3

The third unit of this section reviews the use of benching parameters to develop a roadway cross section. This process allows you to define repetitive benches in a roadway cross section.

UNIT 4

The surface control slope design process is the focus of the fourth unit of this section. Rather than using elevations, this method uses surfaces to control the slopes used to daylight the roadway design.

UNIT 5

The fifth unit of this section covers the use of superelevation parameters in a roadway design. The parameters can be the result of AASHTO curve tables assigned to the curves of an alignment or user-defined parameters assigned in the superelevation calculator.

The values of slopes and benches used in the exercises of this section are extreme numbers because these numbers produce changes that are easily recognizable. When you are working on a job, the design values would be less extreme, producing only subtle changes not easily seen.

UNIT 1: SIMPLE SLOPE CONTROL AND BENCHING

The design methodology, simple slope control, is in the Slope Control dialog box, accessed through the Edit Design Control routine. The slopes apply to the outer edges of the template. If the template does not have ditches, the slopes apply at the connect-out point of the template; if ditches are on, the slopes connect to the outside endpoint of the ditch base. If you are using subassemblies, the slopes connect to the connect-out point of the last subassembly. You can specify a typical and maximum slopes in the Slope Control dialog box of Edit Design Control (see Figure 11–1). The design processor uses the typical slope as much as possible to meet the conditions set in the dialog box, but if necessary, it will use the maximum slope to find a solution for a section. This is the simple method of slope control. If the simple method will not suffice, you can use the type settings in the Slope Control dialog box to set the type of external data to use to calculate slopes.

The design processor uses one of two conditions when searching for daylight. The first is the swath width, which is the offset distance of the sample set in the sampling of existing ground. The second method is the holding of the ROW (Right-of-Way) in the Edit Design Control routine. A design can specify a ROW in one of two ways.

The first is a constant offset distance from the centerline (see Figure 11–1) and the second is to define a ROW alignment. If you attach the ROW alignment to the template in the Attach Alignments dialog box of the Edit Design Control routine, the design is able to understand the fact that the width of the ROW is variable. If the daylight solution is to daylight at the ROW, the potential daylight point may be different in each cross section when using an alignment to define the ROW. In a way, the ROW and daylighting to the ROW is specifically controlling where a design ends.

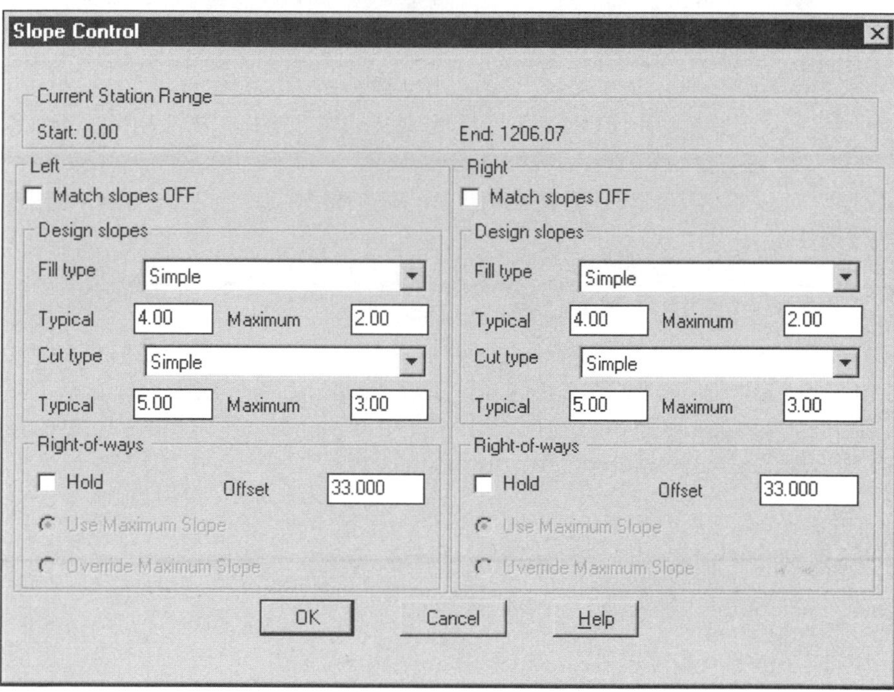

Figure 11–1

Benching allows you to place ledges that pool water and run them parallel to the road design. By adjusting the settings, you can have the bench act as a break in slope and allow the water to travel to a ditch in cut or away from the road in fill. Several conditions that affect the bench are out of the user's control. For further information, you can review the overview of benches in the online documentation. Search in the Help index under benches in roadway cross sections.

The bench settings are in the Bench Control dialog box of the Edit Design Control routine (see Figure 11–2). Here you can set the condition (cut or fill, right or left), the width of the bench, the height above the ditch base the bench occurs, and the grade of the bench. A positive grade slopes the bench ledge towards the centerline of the road. A negative grade slopes the bench ledge away from the centerline.

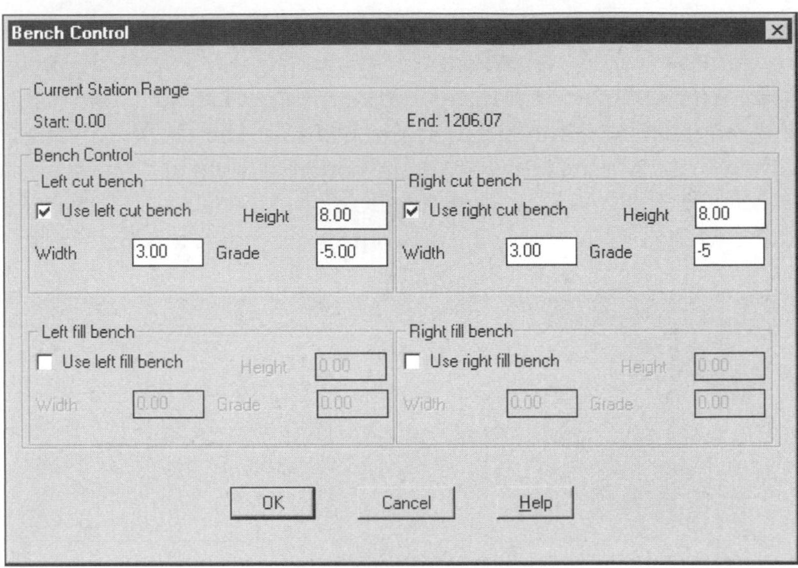

Figure 11–2

The bench parameters in this dialog box apply only to simple and depth type of slopes. The Stepped slope control is another type of benching. When you use surface control, all benching parameters are ignored.

Exercise

After you complete this exercise, you will be able to:

- Create an existing ground surface
- Define a horizontal centerline
- Create a profile
- Draft and define a vertical centerline
- Use slope and bench control in the Edit Design Control routine

SETTING UP THE DESIGN ENVIRONMENT

The following exercise reviews the slope control factors you specify while creating a roadway design.

Starting a New Drawing

1. Start a new drawing, Slopes, and assign the drawing to a new project, SlopeDesign (see Figures 11–3 and 11–4). Describe the project as working with slope settings in the Description area of the Project Defaults dialog box. Select the OK button to exit the Project Defaults and New Drawing dialog boxes.

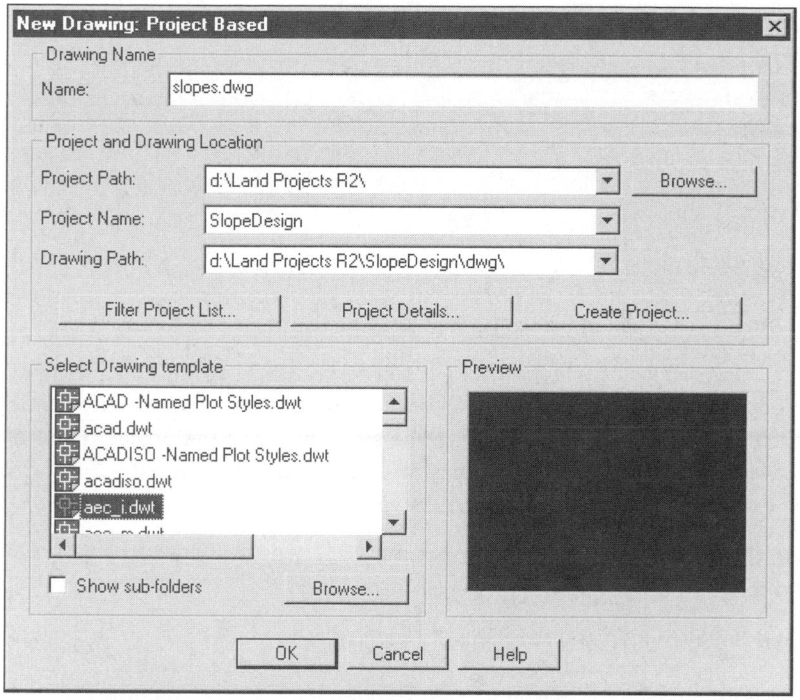

Figure 11–3

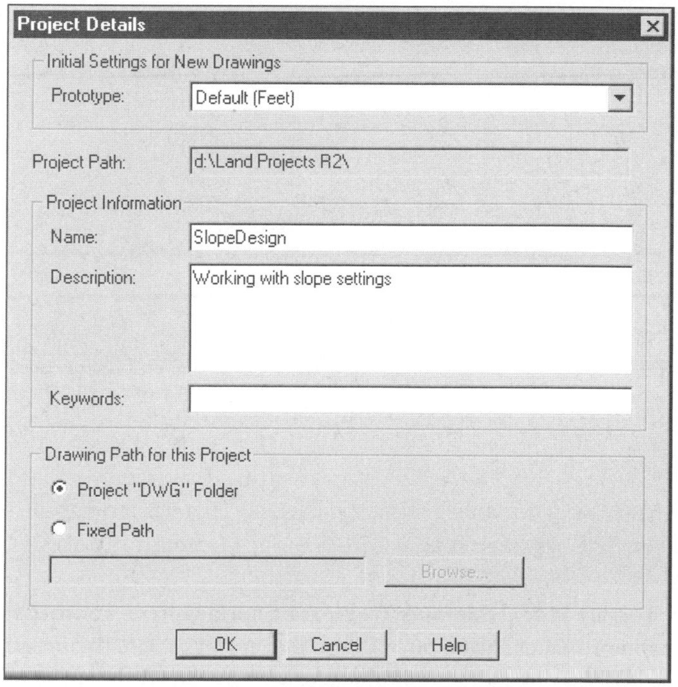

Figure 11–4

EXERCISE

2. Select the OK button to dismiss and accept the point database defaults.
3. In the Load Setup dialog box, select the Next button to set the following values.

Scale:
Horizontal Scale: 50.0
Vertical Scale: 5.0

Text Style:
Text Type: Leroy
Text Style: L100

4. Select the Next button at the bottom of the Load Setup dialog box until you reach the Save Settings panel. Save the settings under the profile name of LDD1-50. After saving the settings, select the Finish button. When the Finish dialog box appears, select the OK button to dismiss it.
5. Use the INSERT command and insert the *Slopesv* drawing. Insert the drawing at 0,0 and explode it upon insertion (see Figure 11–5).
6. Click the Save icon and save the drawing.

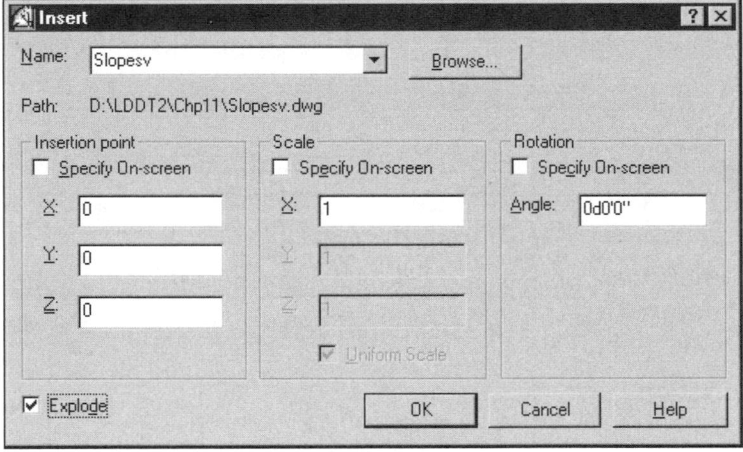

Figure 11–5

Creating Surfaces

There is data in the drawing for three surfaces. Each surface has a different set of contours on its own layer set. The first step is to create a surface for each set of data.

1. Use the Terrain Model Explorer to create a surface from contours for each of the following surfaces. Make sure that contour data is on, the weeding factor is 10.0 and 4.0, and the supplementing factor is 15.0 and 1.0 (see Figure 11–6). When creating each surface, make sure you toggle on Minimize Flat Triangles

Resulting from Contour Data (see Figure 11–7). After creating each surface, freeze the layers for that surface.

Existing Ground - Layers CONT-NML and CONT-HGH

Shale - Layer Shale

Limestone - Layer Lime

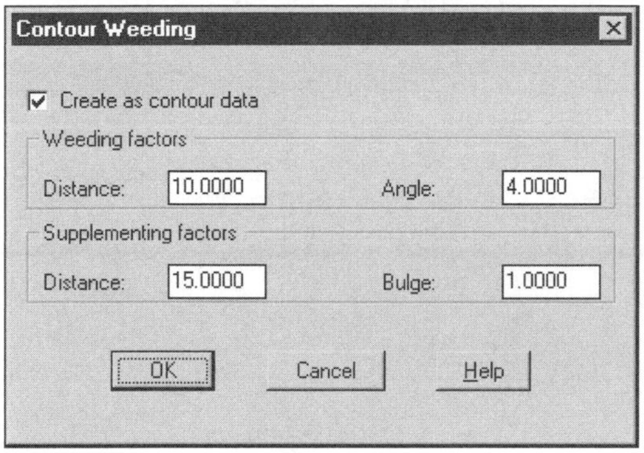

Contour Weeding
Create as contour data
Weeding factors
Distance: 10.0000 Angle: 4.0000
Supplementing factors
Distance: 15.0000 Bulge: 1.0000
OK Cancel Help

Figure 11–6

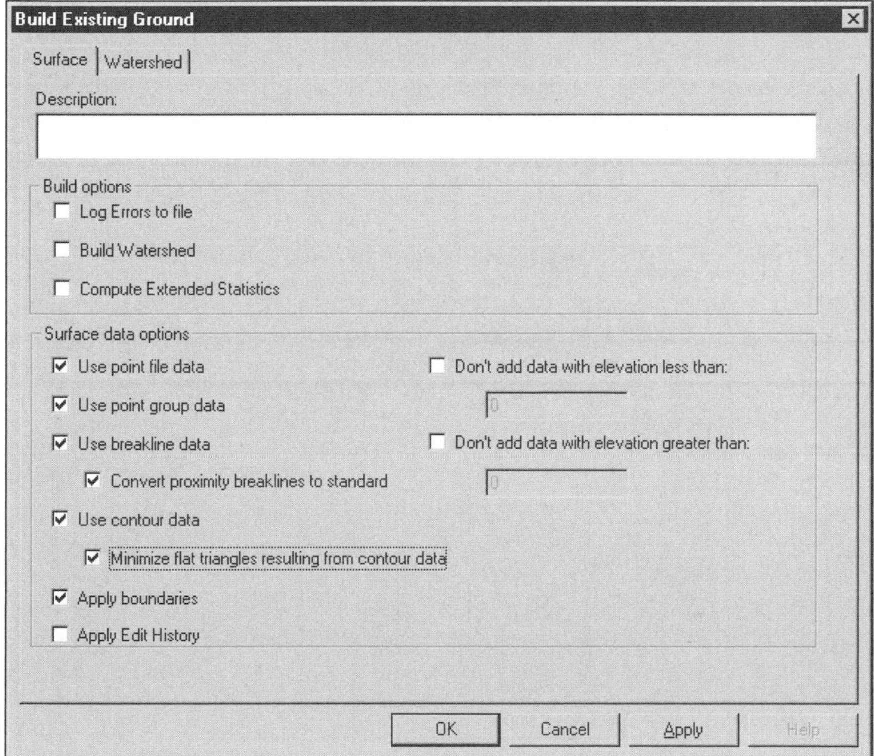

Figure 11–7

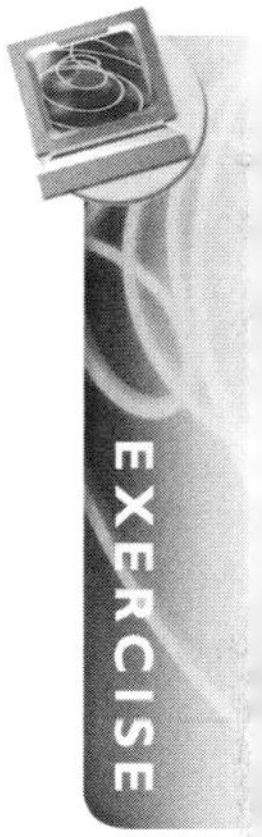

Defining the Centerline

1. Thaw or turn on the layer CL. Use the Define From Objects routine in the Alignments menu to define the Ascot centerline. The alignment starts at the lower left of the centerline line work. When the routine prompts for the starting point of the alignment, press ENTER to accept the default value. In the Define Alignment dialog box, set the starting station to 0.00 and set the description to Slope Exercise Alignment (see Figure 11–8).

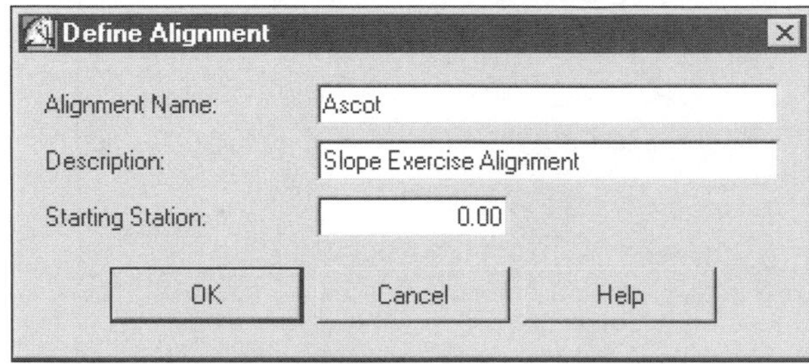

Figure 11–8

PROFILE SAMPLING AND CREATION

This exercise uses more than one surface when creating the design. Civil Design and Land Development Desktop refer to this as multiple surfaces. You need to complete two preliminary steps to indicate to the sampling routines, Profile and Cross Sections, that multiple surfaces are active. The first step is toggling on Multiple Surfaces. The second step is creating a list of the surfaces the routines are to sample. Both the Multiple Surfaces toggle and Multiple Surfaces list routines are in the Surfaces cascade of the Profiles (see Figure 11–9) and Cross Sections menus.

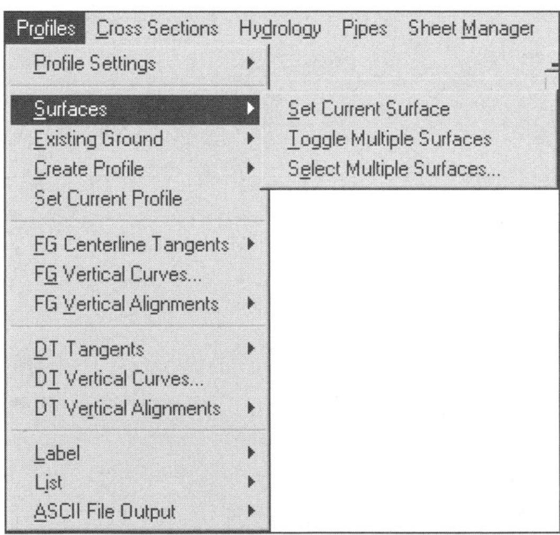

Figure 11–9

EXERCISE

1. Toggle on Multiple Surfaces by using the Toggle Multiple Surfaces routine in the Surfaces cascade of the Profiles menu. The routine prompts with a message on the command line that multiple surfaces are on.
2. Create the Surface list using the Select Multiple Surfaces routine in the Surfaces cascade of the Profiles menu (see Figure 11–10). First select Existing Ground and then hold down SHIFT and select the remaining surfaces. Holding down SHIFT is the only way to select more than one surface.

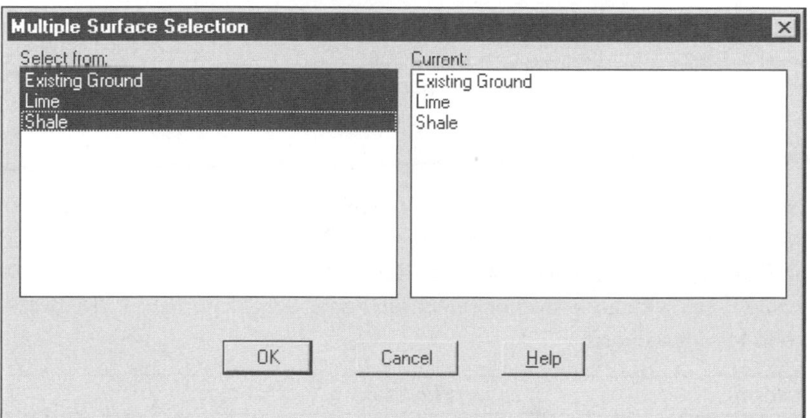

Figure 11–10

3. Sample the surfaces by using the Sample from Surface routine in the Existing Ground cascade of the Profiles menu. Select the OK button to accept the sampling defaults, and press ENTER twice to accept the beginning and ending sampling stations.

Since there are multiple surfaces, create one full profile using the Existing Ground surface. Within the context of the full profile, create one surface profile for each additional surface, lime and shale.

4. Create a full profile for the Existing Ground surface using the Full Profile routine of the Create Profile cascade of the Profiles menu. When the routine displays the Profile Surfaces dialog box, select Existing Ground (see Figure 11–11) and then select the OK button to review the Profile Generator defaults. Select the OK button to accept the values and build the profile. Select a point below and to the right of the centerline and surfaces.
5. Create one surface profile for each of the two remaining surfaces, lime and shale, within the full profile. Use the Surface Profile routine of the Create Profile cascade of the Profiles menu. The routine displays the Profile Surfaces dialog box. Select a surface name, then select the OK button, and the routine draws the surface profile in the Existing Ground full profile.

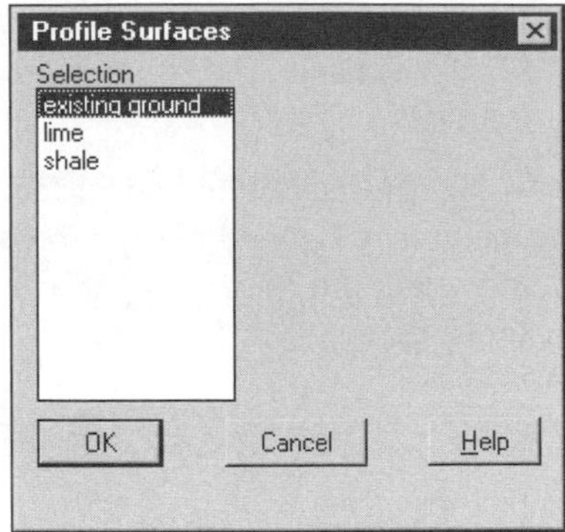

Figure 11–11

6. Create the tangents for a vertical alignment using the Create Tangents routine of the FG Centerline Tangents cascade of the Profiles menu. The first elevation is the Existing Ground elevation at station 0+00, and the elevation at the end of the profile is set with an endpoint selection. For the intermediate PVI elevations, use the following values:

Station	Elevation
0.00	362.55 (Existing Ground Elevation)
700.00	352.00
1350.00	355.00
1855.11	376.79 (Pick the end of Existing ground in the Profile)

7. Define the vertical alignment with the Define FG Centerline routine of the FG Vertical Centerline Alignments cascade of the Profiles menu. Enter the Vertical Alignment Editor. With multiple surfaces on, you will always need to select the surface to show in profile before entering the editor. The elevations in the editor represent the Existing Ground surface (see Figure 11–12). While in the editor add two vertical curves to the vertical alignment. Use the following values for the vertical curves. After entering in the vertical curve data, select OK to exit the dialog box, and select the Save button to save the changes to the vertical design.

Station	Elevation	Vertical Curve
0.00	362.55	(Existing Ground Elevation)
700.00	352.00	200.00
1350.00	355.00	200.00
1855.11	376.79	(Pick the end of Existing ground in the Profile)

8. Import and annotate the centerline when importing it into the drawing. Use the Import routine of the FG Vertical Alignments cascade of the Profiles menu.

EXERCISE

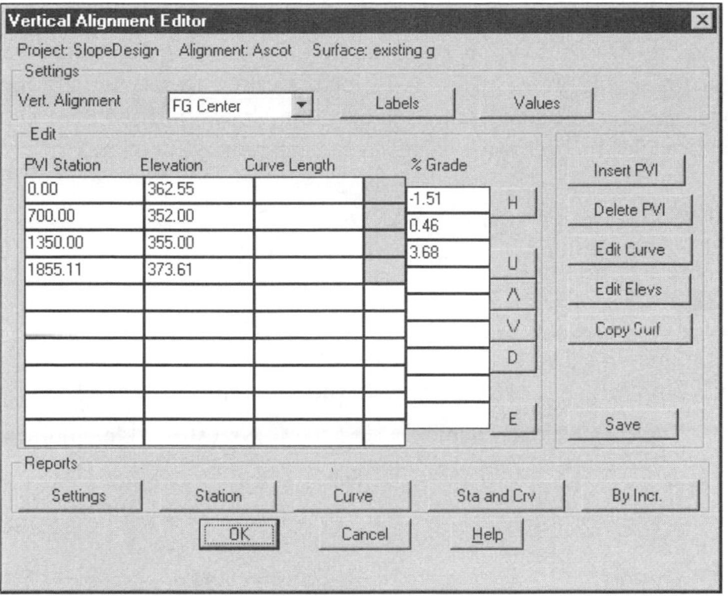

Figure 11–12

Cross Section Sampling

1. Sample for the cross sections along the roadway centerline. Use the Sample From Surface routine in the Existing Ground cascade of the Cross Sections menu. Make the swath width 85 feet and sample every 25 feet along tangents and curves. Also toggle on sampling at the beginning and end of the alignment and at the PCs and PTs (see Figure 11–13). Press ENTER twice when the routine prompts for the beginning and ending sampling stations.

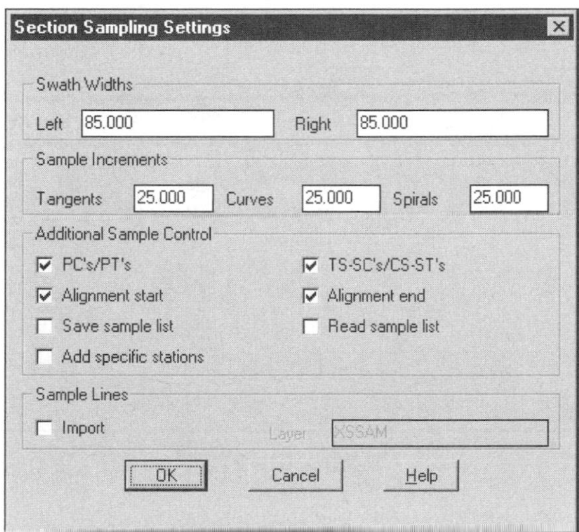

Figure 11–13

SLOPE CONTROL

1. Run the Edit Design Control routine of the Design Control cascade of the Cross Sections menu. Review Figure 11–14 for the values to set in the Slope Control dialog box, which you access by selecting the Slopes button in the Design Control dialog box. Set the following values in the appropriate section.

Template:

STD2L (Use datum 2)

Slopes:

Left

Type: Simple

	Typical	Maximum
Fill	4:1	2:1
Cut	5:1	3:1

Right-of-ways

Offset: 50.00

Right

Type: Simple

	Typical	Maximum
Fill	4:1	2:1
Cut	5:1	3:1

Right-of-ways

Offset: 50.00

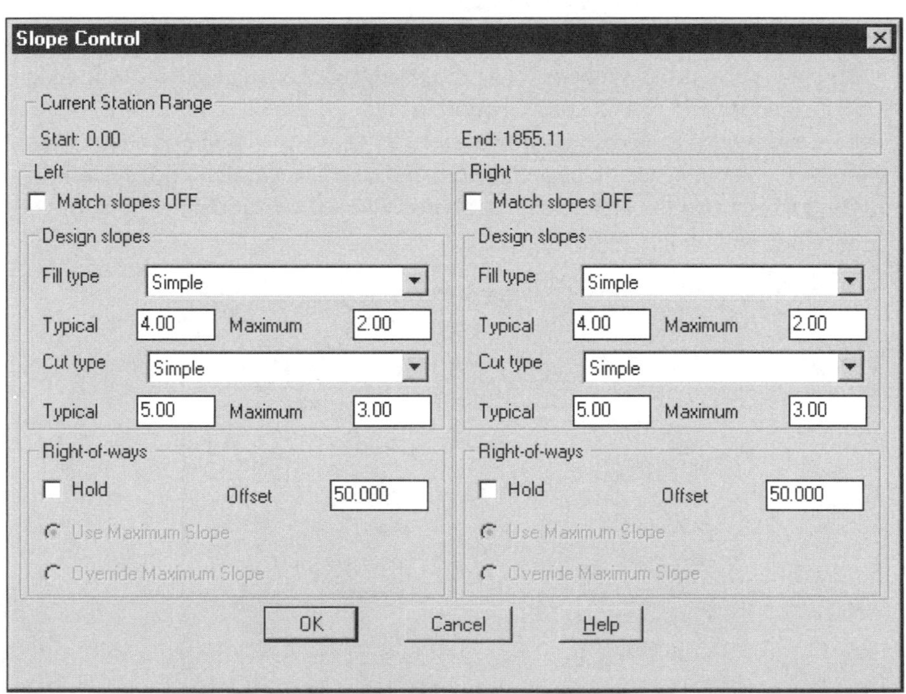

Figure 11–14

2. After setting the template and slope settings, select the OK button to process the design. There should be no errors.

3. Use the View/Edit Sections routine of the Sections menu to view the daylight slopes along the roadway. If you need to set up the view options, when you are in the View/Edit Sections routine, select the View option by entering the letter **v** and pressing ENTER to see the view options. Use Figure 11–15 as a guide for setting the values.

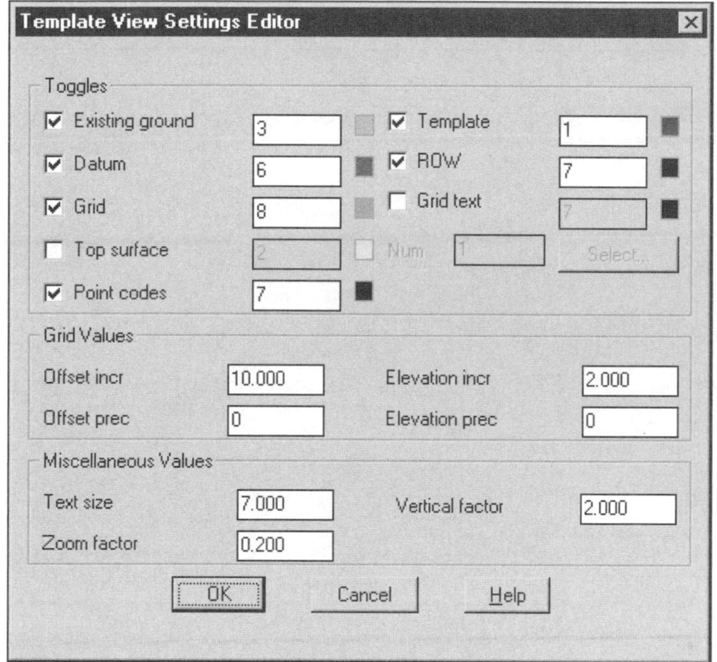

Figure 11–15

BENCHES

To add benches to the current design, set the benching information in the Bench Control dialog box of the Edit Design Control routine.

The upper half of the dialog box sets the cut bench parameters for the left and right side of the design. The bottom half of the dialog box sets the fill bench parameters for the left and right side of the design. In each section there is data for the height, width, and grade of the bench.

A bench occurs in one of the following situations: cut only, fill only, cut and fill, or None. The control also allows you to specify a right or left side bench. The height of a bench is the vertical distance before or between benches. The Simple bench routine measures the initial height from the ditch base elevation. The width of a bench is how far the bench recedes into the surface. The grade of a bench is the slope from the front to the back end of a bench. If the grade is negative, the bench slopes away from the centerline. If the grade is positive, the bench slopes towards the centerline.

EXERCISE

3. Run the Edit Design Control routine and set the following bench information for the roadway. Refer to Figure 11–16 for the settings.

Side	Type	Height	Width	Grade
Left	cut	5.0	5.0	-10
Right	cut	5.0	5.0	-10

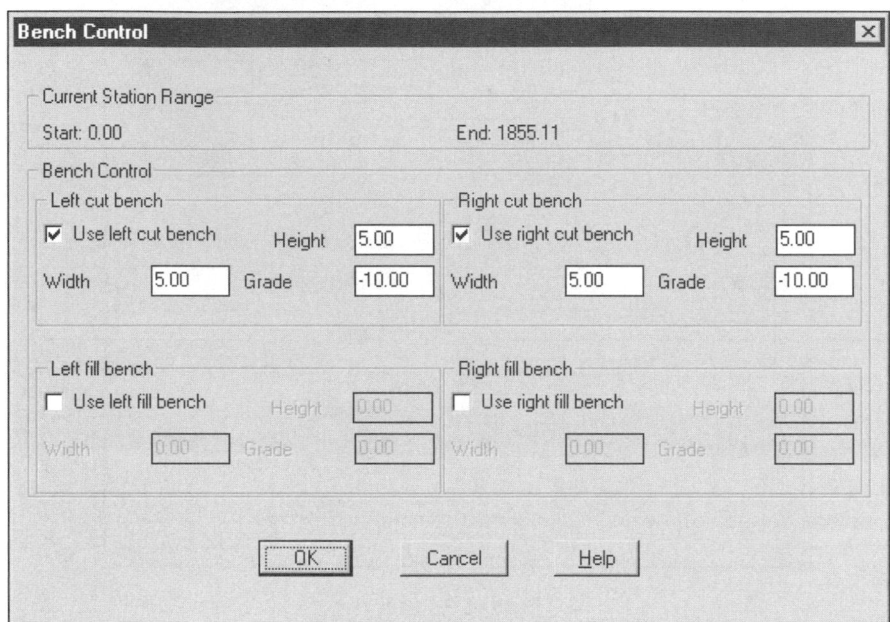

Figure 11–16

4. Select the OK button to close the Bench Control and Design Control dialog boxes. The Edit Design Control routine processes the new design data. There should be no errors. Select the OK button to close the Process Status dialog box.

5. View the new cross sections with the View/Edit Sections routine (see Figure 11–17). The major activity occurs around station 8+00 to 10+00.

6. Rerun the Edit Design Control routine and select the Bench button to display the Bench Control dialog box. Toggle off all benching parameters. Select the OK button until you exit the Design Control dialog box.

EXERCISE

Figure 11–17

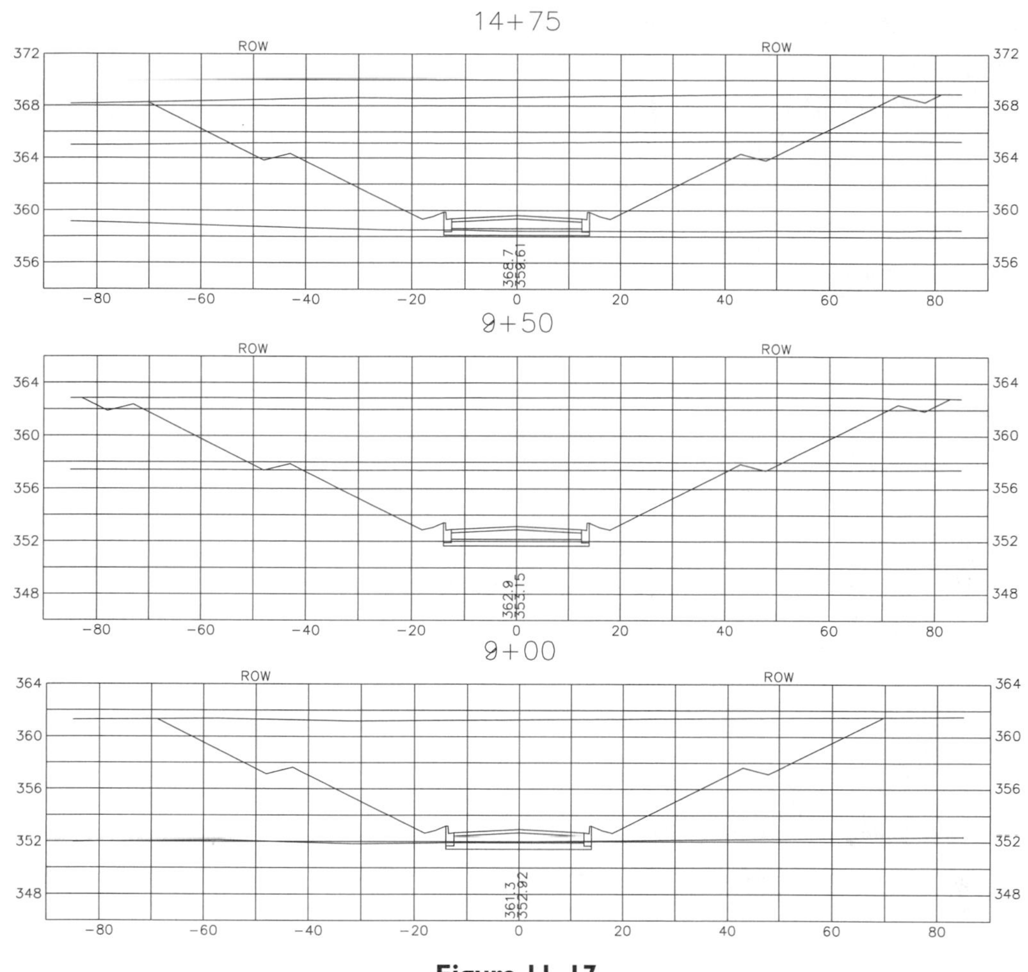

UNIT 2: DEPTH PARAMETERS

The Depth Parameters option of the Edit Design Control routine uses an external data file to influence a design solution. To use the depth parameters in evaluating a design, you set the slope type to Depth in the Design Slopes portion of the Slope Control dialog box of the Edit Design Control routine.

When you set the depth parameters, the external file specifies new typical and maximum slopes for a set of specific depths. These depths apply to fill or cut depths. As the design processor moves the roadway template along the centerline, the processor is aware of the depths listed in the parameter file. Depending upon the depth of the design and the entries in the file, the processor may change the typical and maximum daylight slopes.

You specify the depths that change the daylight slopes in the Depth Control Editor (see Figure 11–18). The first entry in the file is always 0.00. The list of depths is a sorted list, and as you enter new depths, the routine resorts the list. There are depth entries for cut and fill and typical and maximum acceptable slopes.

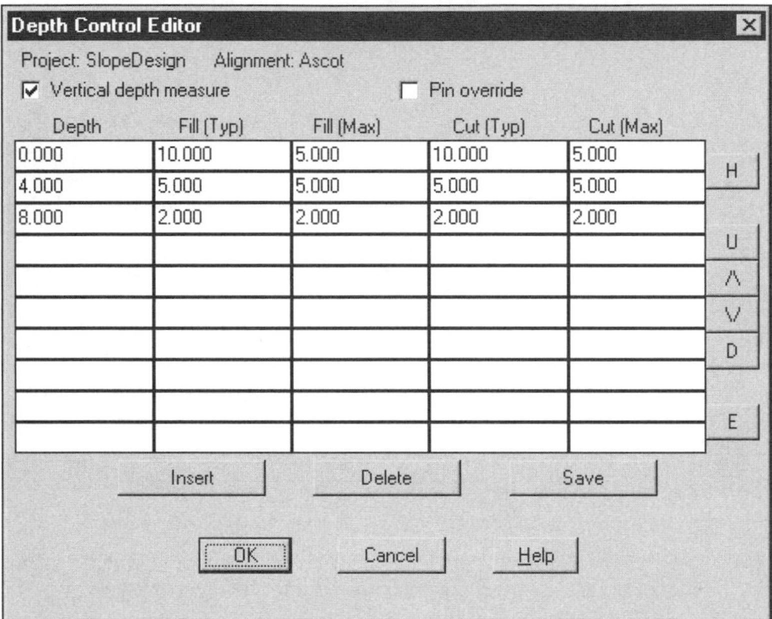

Figure 11–18

You can choose the method of measuring the depth. The first method is by calculating the distance from the connect-out point vertically to the existing ground. Civil Design uses this definition when you toggle on Vertical Depth Measure. The second method is measuring the depth of the template connect-out point to the elevation of the catch point. Civil Design uses this definition when you toggle off Vertical Depth Measure. The design processor uses the typical slope set in the Design Control dialog box to calculate the depth between the two points (the template connect-out point and the catch point). After determining the depth, the processor selects the appropriate slope value from the values set in the depth file.

You can also control the pinning of the slopes within the depth file. If you toggle off Pin override, the processor acts according to all previous rules. If you toggle on Pin override, the design processor will scan the depth file for a slope that produces a catch point before the swath width or ROW restriction is met. In this case the design processor selects a slope independent of the depths specified in the file. For example, the processor may select a slope value set for a depth of 6 feet, but apply the slope to a situation where the template is only 6 inches below the existing ground.

If the hold right-of-way and/or benching settings are on, they affect the depth slope control.

Exercise

After you complete this exercise, you will be able to:

- Use depth control of sloping

DEPTH SLOPE CONTROL

1. Enter the control values into the Depth Control Editor. Use the values found in Figure 11–18. The Depth Slopes routine is in the Design Control cascade of the Cross Sections menu. Save the values before exiting the dialog box.
2. Run the Edit Design Control routine to set the slope conditions. Select the Slopes button to display the Slope Control dialog box. The settings in the Slope Control dialog box control which external slope file to use. You set the influences by changing the type of slope to depth type. The settings independently control cut, fill, left, and right slopes.
3. Set the left and right cut slopes to depth (see Figure 11–19).

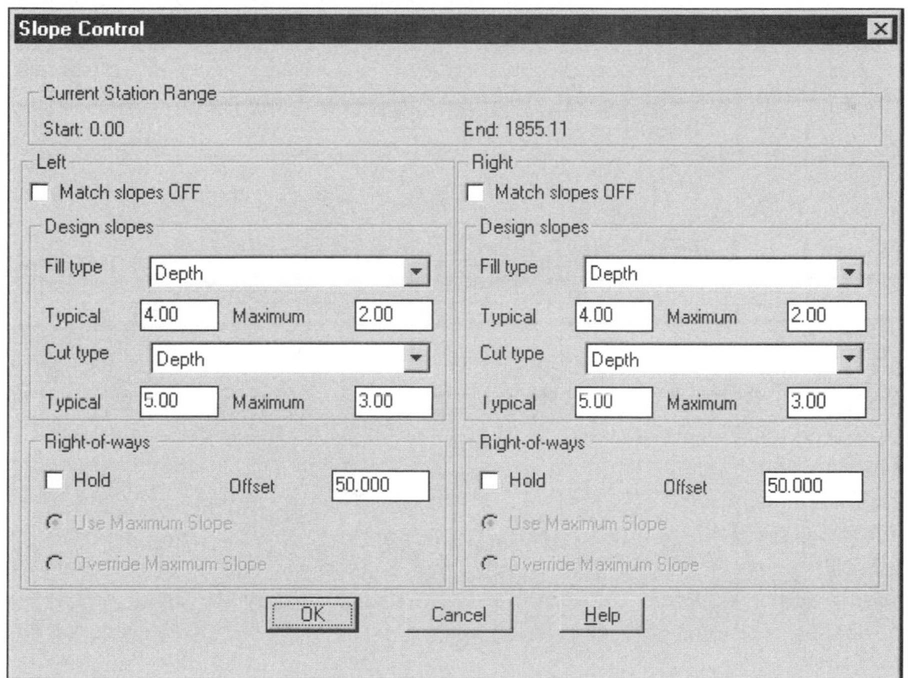

Figure 11–19

4. Exit the Slope Control and the Design Control dialog boxes by selecting the OK button until the processing of the design is complete.

5. Run the View/Edit Sections routine to see the cross sections. Go to station 7+00 to start viewing the depth parameters in action. The routine is in the Cross Sections menu. Figure 11–20 shows the results of processing the current slope data.

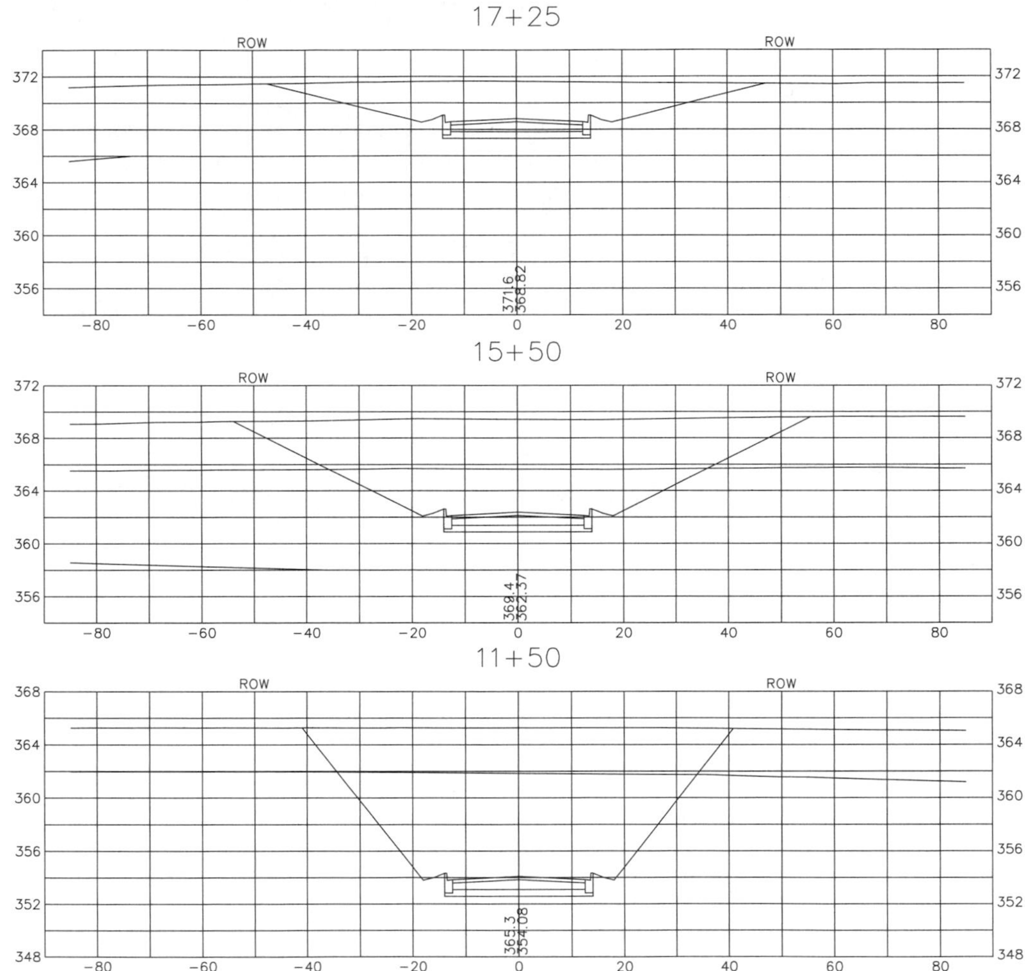

Figure 11–20

UNIT 3: STEPPED PARAMETERS

You create the stepped parameters similar to the way you create the depth slope parameters. The Slope Type settings of the Edit Design Control routine toggle on the stepped option. Once the option is on, the design processor reads an external data file containing the step (bench) parameters.

In the stepped file, the benches are either a distance above the hinge point or a depth below the match point elevation.

The format of the Stepped file is Depth at which the bench occurs, the slope to use until daylight or the next bench in Fill, and the slope to use until daylight or the next bench in Cut. The last two columns specify the width of the bench, and the grade of the bench (see Figure 11–21). In the bench settings, a negative grade for a bench slants it away from the centerline. In Stepped control, it is just the opposite.

There is a toggle at the top of the editor that determines the location of the first, and if necessary, subsequent benches. If you toggle on Hinge to Match, the design processor will use the elevation of the outside edge of the ditch to measure the depth for the bench. If you toggle on Match to Hinge, the design processor will use the elevation of the daylight surface to the outside of the ditch to calculate the location of the bench.

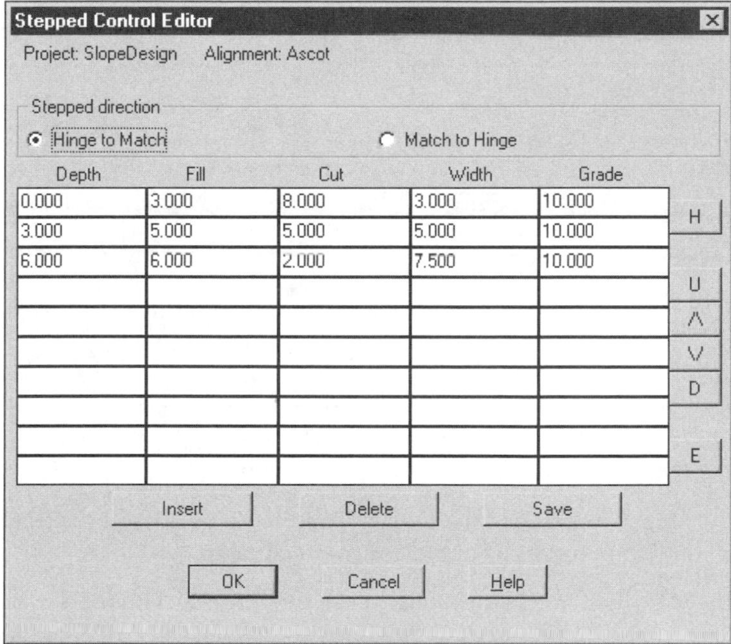

Figure 11–21

Exercise

After you complete this exercise, you will be able to:

- Use stepped control of sloping

STEPPED PARAMETERS

1. Enter the control values into the Stepped Control Editor. Use the values found in Figure 11–21. The Stepped Slopes routine is in the Design Control cascade of the Cross Sections menu. Save the values before exiting the dialog box. To start the data entry process, choose the Insert button and enter the values.
2. Run the Edit Design Control routine to set the slope conditions to Stepped. Select the Slopes button to display the Slope Control dialog box. Set the type of slope control to use by changing the type to stepped in the Slope Control dialog box (see Figure 11–22).

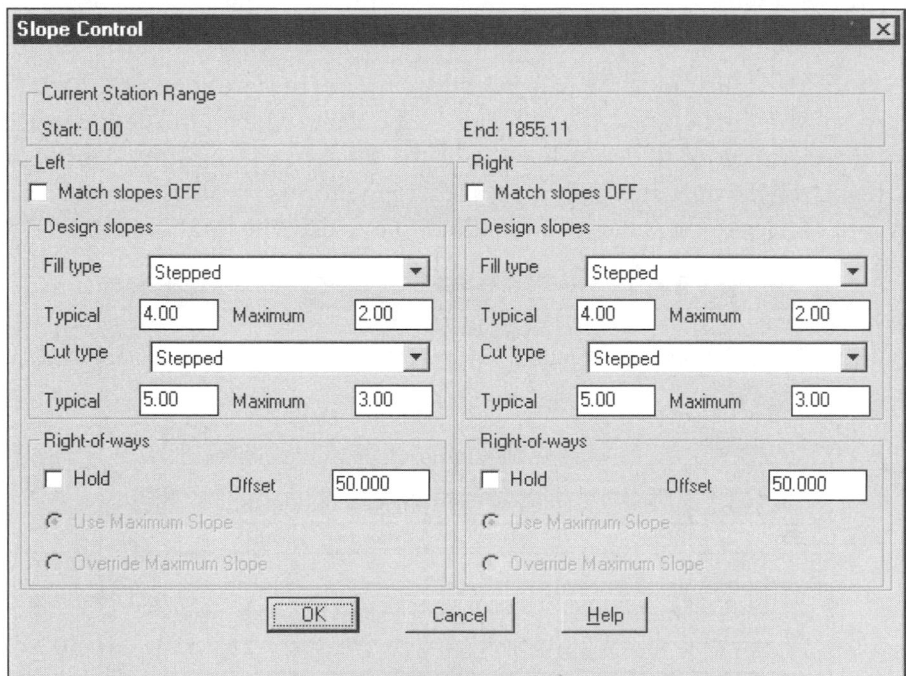

Figure 11–22

3. Exit the Slope Control and the Design Control dialog boxes by selecting the OK button until the processing of the design is complete.
4. Run the View/Edit Sections routine to see the cross sections. Go to station 8+00 to start viewing the Stepped parameters in action. Figure 11–23 contains some of the results from processing the current slope data.

11+00

ROW ROW

368 364 360 356 352 348

364.8 353.85

−80 −60 −40 −20 0 20 40 60 80

10+00

ROW ROW

364 360 356 352 348

363.9 353.38

−80 −60 −40 −20 0 20 40 60 80

9+00

ROW ROW

364 360 356 352 348

361.3 352.92

−80 −60 −40 −20 0 20 40 60 80

Figure 11–23

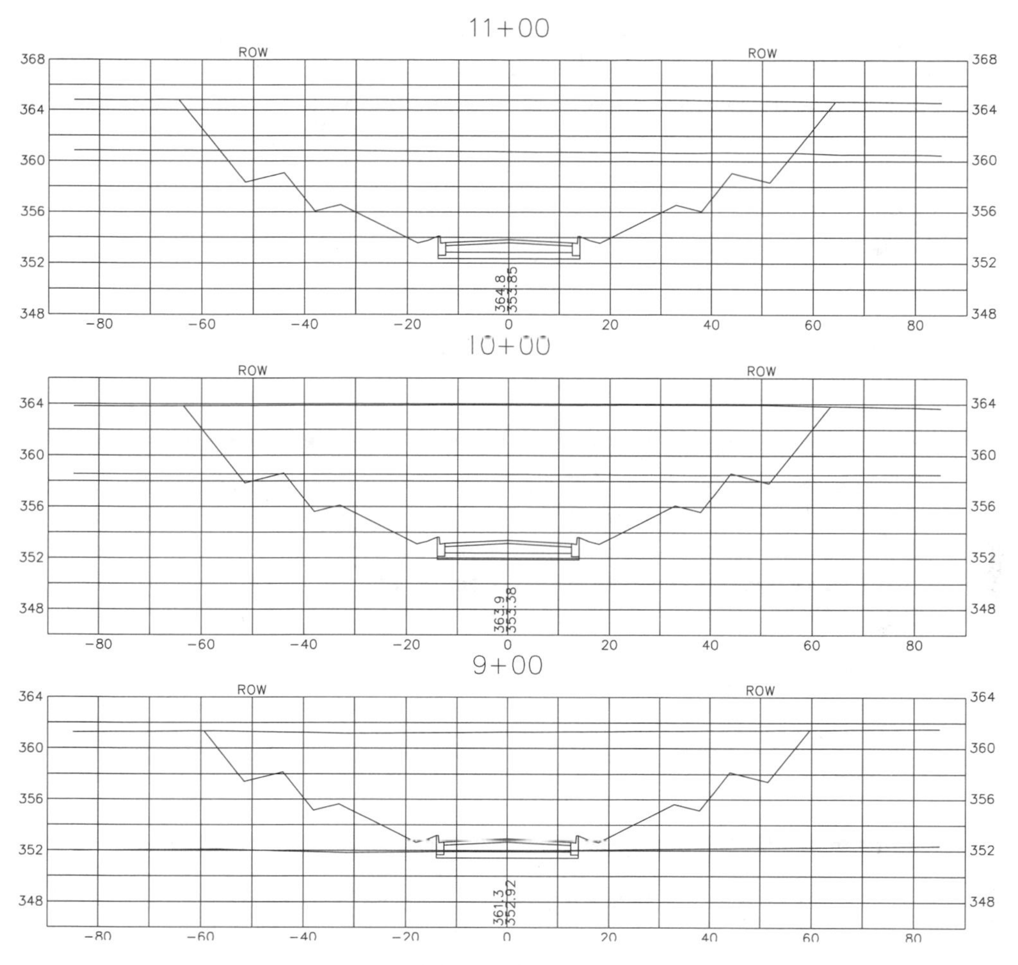

UNIT 4: SURFACE PARAMETERS

The surface option takes into account the fact that materials making up different surfaces have different cohesion values. A surface made from rock retains a steeper slope than a surface containing sand. The differing slopes are a part of the settings in the Slopes by Surface Type dialog box, which the design processor uses when evaluating the roadway design. When the template cuts through a surface with a low angle of repose, the surface parameters specify a shallow slope value. When the template cuts through a surface with a high angle of repose, the surface parameters specify a steep slope value.

In the Slopes by Surface Type dialog box, you state the name of the surface, its slope value, and, if necessary, a ledge width. The ledge does not need a grade because the ledge follows the surface slope. If you want a specific grade to occur on the bench, this is where you set the grade value.

EXERCISE

Exercise

After you complete this exercise, you will be able to:

- Use surface control of sloping

SURFACE PARAMETERS

1. Run the Surface Slopes routine in the Design Control cascade of the Cross Sections menu to set the surface slope data. The routine displays a New Surface Values dialog box (see Figure 11–24). When you select the OK button, the data appears in the Surface Control Editor dialog box (see Figure 11–25). Any editing of the surface data is done through this editor. Set the following values for Surface Slopes.

Surface Name	Slope
Existing Ground	5.0
Shale	10.0
Limestone	1.0

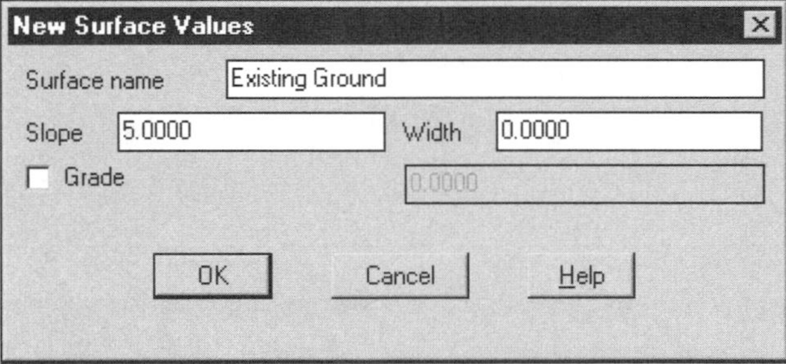

Figure 11–24

2. Run the Edit Design Control routine to set the slope type. Select the Slopes button to display the Slope Control dialog box. Change the type of slope control for Cut to Surface. When processing the design, the design processor will use the external control file (see Figure 11–26).
3. Select the OK button to exit the Slope Control and the Design Control dialog boxes and to process the new settings.

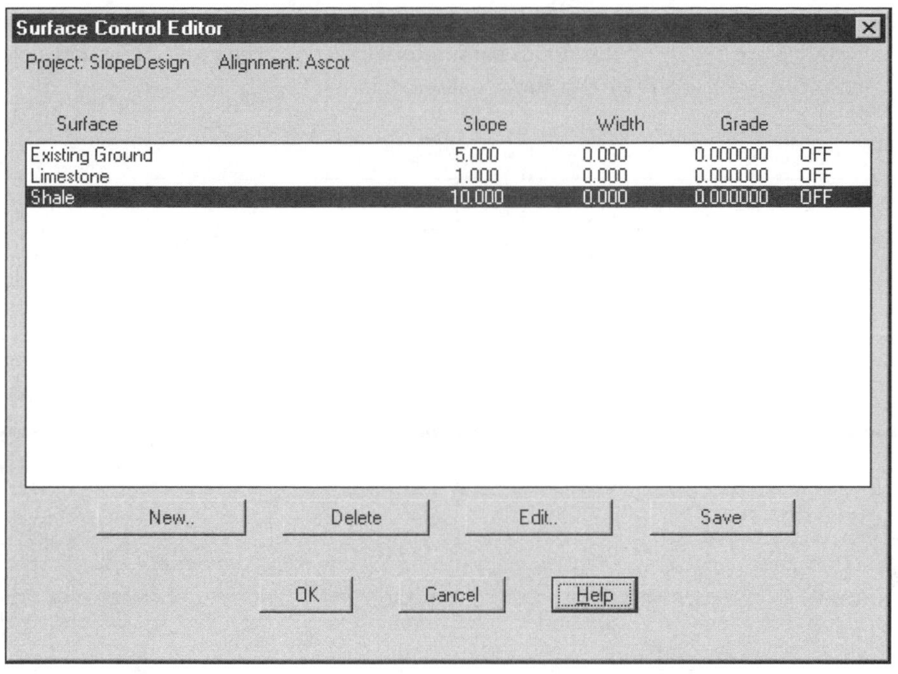

Figure 11–25

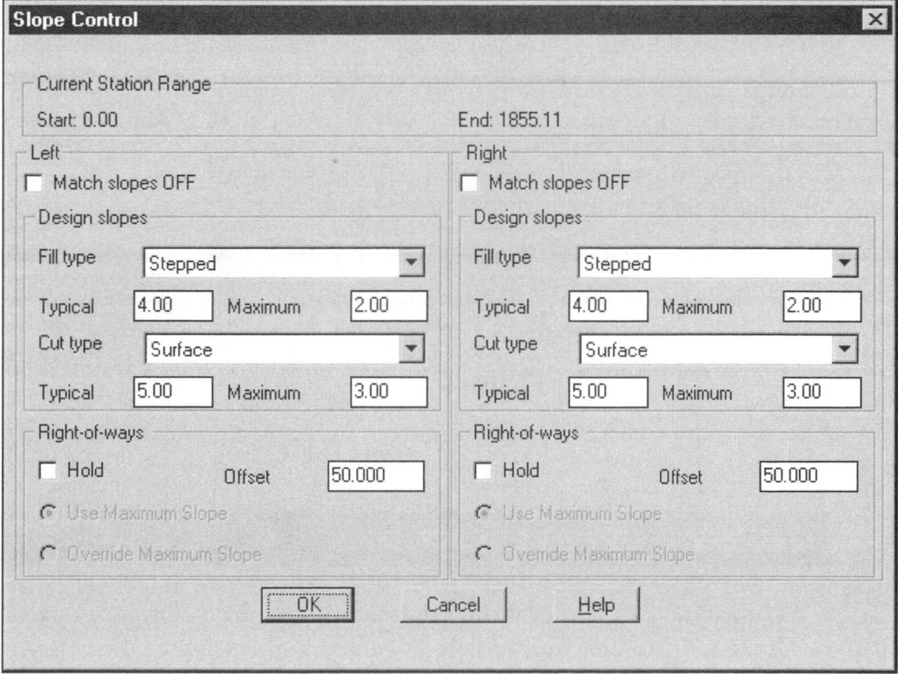

Figure 11–26

4. Run the View/Edit Sections routine to review the cross sections. Go to station 5+00 to start viewing the depth parameters in action. Figure 11–27 contains the results from processing the surface slope data.

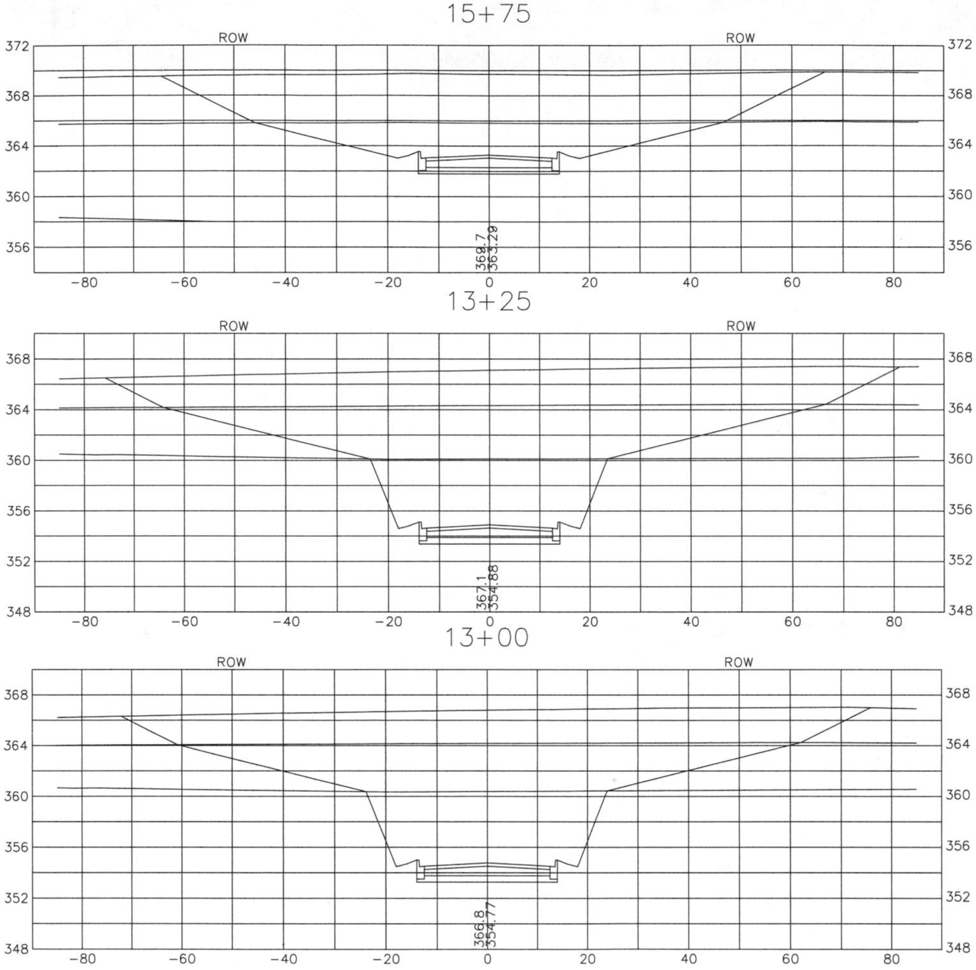

Figure 11–27

UNIT 5: SUPERELEVATIONS

Civil Design incorporates a superelevation processor into roadway design. For a simple template, the process of superelevating is relatively straightforward. The curve data can be custom set on a per-curve basis or on the basis of your own custom superelevation table, or the data can be placed on the curve using AASHTO superelevation table data.

The initial run of the design may produce most of the effects you want. However, you may need to control specific aspects of the design. The design process may be one of an initial run followed by the defining of alignments to control the specific aspects of the design.

As with transition alignments, Civil Design tools can import the superelevation results into plan and profile views. By manipulating the line work, you can create new alignments controlling specific behaviors of the template.

The catch to this process is that once you decide to use alignments to control specific aspects of the design, you have to use alignments to control all the aspects of the design. When using alignments, the designer must turn off the superelevation calculations. If you do not turn off superelevation, any attached alignments will not affect the template.

If you are designing a roadway with superelevation, the design process starts with superelevation on and may end with superelevations off because you are creating new alignments that reflect the superelevation calculations and user-defined modifications. If you apply the alignments to the template instead of continuing to use the superelevation calculations, the roadway design is a reflection of the alignments not a superelevation method.

The power of Civil Design is not necessarily in its ability to automate, but in its ability to process and allow for the review, editing, redefining, and applying of modified data. The Civil Design process is iterative and not necessarily a single run through.

When you use superelevation, the superelevation processor controls how the pavement rotates. Civil Design controls the method for crown removal and the values for the superelevation process.

When working with superelevations, you need to understand a few terms:

- *Runout*—The distance over which superelevation rotates the side of the pavement from a normal crown to no crown. Tangent runout distance is another name for runout.
- *Runoff*—The distance over which superelevation rotates the pavement from the removed crown state to superelevation.
- *Percentage of Runoff*—The amount of runoff that occurs before coming to the curve.
- *E value*—The maximum superelevation rate. The E rate is either ft/ft or m/m ratio. A 0.10 E value is a 10% grade.

If the roadway template contains a crown, you must decide how to handle the removal of the crown. The processor sets the removal of the crown with or without a runout distance. If you toggle on Crown Removal by Runout Distance, the routine removes the crown from one side of the template over a specified distance: the tangent runout distance. This distance, now occurring in the runoff distance, removes the crown from the remaining portion of the template. If you toggle off Remove Crown by Runout Distance, the routine rotates the pavement to remove the crown. After removing the crown, the routine continues to rotate the pavement at a constant rate to the beginning of the curve. As a result, the runout distance may not be the same distance for the removal of the entire pavement crown.

Civil Design supports five methods for calculating superelevations: A, B, C, D, and E. The following is a brief description of each method.

- *Method A*—Rotates a crowned pavement about the centerline alignment. Which edge-of-pavement the routine rotates depends upon the direction of the curve. A right-hand curve rotates the left side and a left-hand curve rotates the right side.
- *Method B*—Rotates a crowned pavement about the inside edge. The direction of the curve defines the inside edge; that is, right turn, right edge and left turn, left edge. The option forces the outside edge upward.
- *Method C*—Rotates a crowned pavement about the outside edge. The direction of the curve defines the outside edge; that is, right turn, left edge and left turn, right edge. The method forces the inside edge downward.
- *Method D*—Rotates a noncrowned pavement about the outside edge. The direction of the curve defines the outside edge; that is, right turn, left edge and left turn, right edge. The option forces the inside edge downward.
- *Method E*—Rotates a noncrowned pavement about the inside edge. The direction of the curve defines the inside edge; that is, right turn, right edge and left turn, left edge. The option forces the outside edge upward.

In the defaults portion of the Superelevation Control, set the default method, E value, the curve direction, the runout, runoff, and percentage of runoff values from transitioning in or out of curves.

After determining the method for superelevating, the type of crown removal, and the defaults, import the alignment data. Civil Design does not dynamically link the Superelevation Control to the alignment data files. If for any reason the alignment changes, the alignment must be reimported so the Superelevation Control can be aware of the changes.

After importing the alignment, set the curve values. Editing the curves sets the specific values for each curve along the alignment. You can insert and delete curves from within the editor.

Before starting superelevation, a template with a superelevation zone needs to be defined.

Exercise

After you complete this exercise, you will be able to:

- Edit a superelevation zone into a template
- Set superelevation parameters
- Superelevate a roadway design

SUPERELEVATION TEMPLATE

1. Open the CDTTPL drawing of the tplate project and save the changes to the current drawing.

2. Use the Draw Template routine of the Cross Sections menu to draw a new template, Super2L, for the superelevation exercise. The template consists of two surfaces, asphalt and crushed base course. The outside edge-of-pavement is the superelevation zone point, and the crown of the asphalt is the inner superelevation zone point. The outside edge of the shoulder is the outer superelevation zone point. This outside zone provides a roll over edge for the sloping pavement (see Figure 11–28).

   Asphalt Surface

   Offset 12.5 Feet, 2% grade, and a thickness of 4 inches (0.3333)

   Crushed Base Course

   Top of surface is the bottom of the Asphalt surface, shoulder width of 5 feet, and an end slope of 2:1 for 3 feet.

3. Use the Define Template routine of the Templates cascade of the Cross Sections menu to define the template, with both surfaces normal. Assign the appropriate materials, and identify the connect-out, final grade reference point, and datum points on the template (see Figure 11–28).

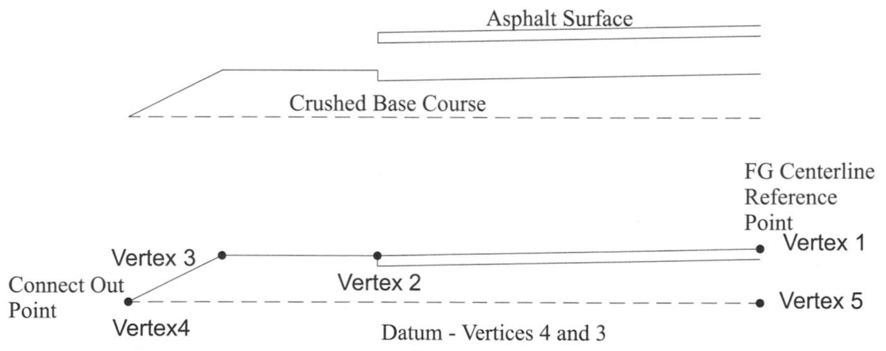

Figure 11–28

4. Use the Edit Template routine of the Templates cascade of the Cross Sections menu to edit the template and define the top-of-template with the Topsurf option of Surface Control. See Figure 11–29.

5. While still in the Edit Template routine use the Surface Control (sr) option and add the left and right superelevation definitions to the template. See Figure 11–29. The following is the command sequence for adding the superelevation points on the template. After adding the control points, exit and save the changes to the Super2L template.

```
Edsrf/SAve/eXit/ASsembly/Display/SRfcon/Redraw <eXit>: sr
Connect/Datum/Redraw/Super/Topsurf/TRansition/eXit <eXit>: s
Outer left superelevation point: (Vertex 3)
Inner superelevation reference point: (Vertex 4)
Outer rollover point: (Vertex 2)
Outer right superelevation point: (Vertex 5)
Inner superelevation reference point: (Vertex 4)
Outer rollover point: (Vertex 6)
Connect/Datum/Redraw/Super/Topsurf/TRansition/eXit <eXit>:
Edsrf/SAve/eXit/ASsembly/Display/SRfcon/Redraw <eXit>:
Save template [Yes/No] <Yes>:
Template name <super2l>:
Template exists. Overwrite [Yes/No]: Y
```

6. Reopen the Slopes drawing of the SlopeDesign project and save the changes to the current drawing.

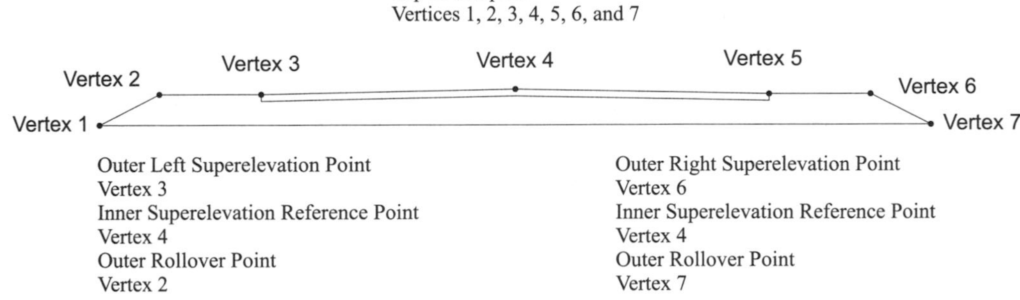

Figure 11–29

DEFINING THE CENTERLINE

1. Open the Layer dialog box and thaw the Supercl layer.
2. Use the Define From Objects routine to define the centerline. The centerline starts at the southwest end of the line work. When the routine prompts for the starting point, press ENTER to accept the default value. The routine displays the alignment dialog box. Enter the name of Superelevation and describe it as the Centerline for Superelevation.

3. Pan the centerline to the left side of the screen.
4. Place the surface prefix, *-, in the Layer Prefix box of the Labels and Prefix dialog box of the Settings cascade of the Profiles menu.
5. Toggle off multiple surfaces by running the Toggle Multiple Surfaces routine in the Surfaces cascade of the Profiles menu. Make sure the routine returns "multiple surfaces are off." If the routine prompts for a current surface, select Existing Ground.
6. Sample the elevations under the centerline with the Sample From Surface routine of the Existing Ground cascade in the Profiles menu. When the routine displays the sampling settings, select the OK button to dismiss the dialog box and press ENTER on the beginning and ending stations for sampling. If the routine should prompt for a surface, select Existing Ground.
7. Create a Full Profile using the Full Profile routine of the Create Profile cascade of the Profiles menu. Place the profile to the right of the roadway centerline.
8. Use the Create Tangents routine of the FG Centerline Tangents cascade of the Profiles menu and create the tangent lines in the profile. Use the existing elevation at station 0+00. Make a PVI at station 7+50 with an elevation of 353 and select the endpoint of the profile to set the last station and elevation of the vertical alignment (see Figure 11–30).

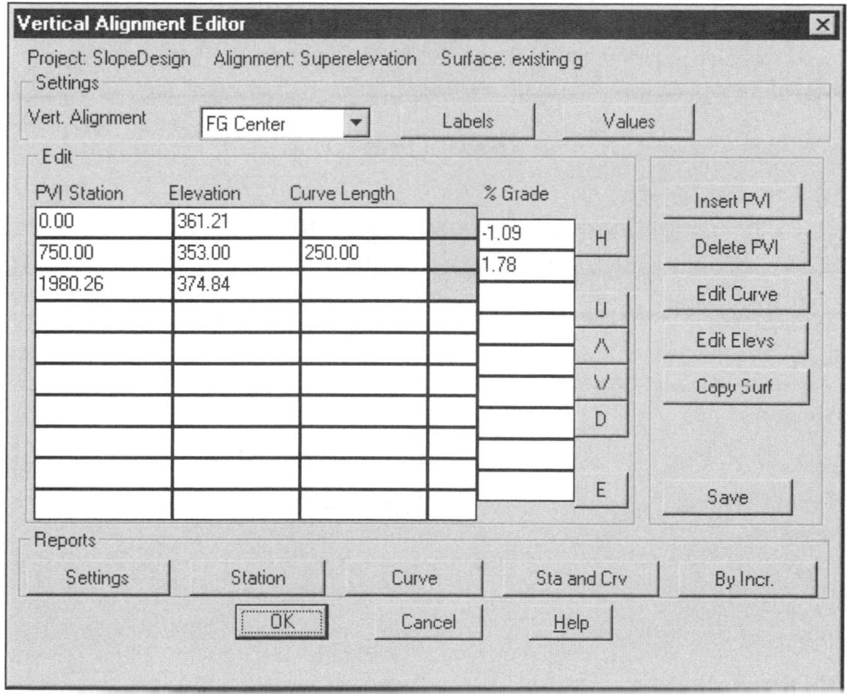

Figure 11–30

EXERCISE

9. Create a vertical curve of 250 feet for station 7+50.
10. Select the OK button to exit and save the changes.
11. After editing the vertical alignment, import and annotate the vertical alignment.
12. Sample the Existing Ground surface to develop the cross-section data. Use the settings found in Figure 11–31 to set your values.

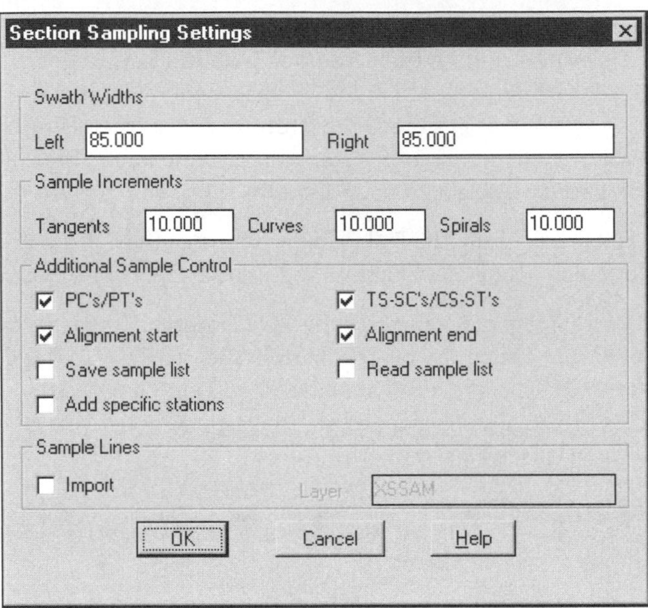

Figure 11–31

13. Open the Design Control dialog box, set the template to Super2L, and set the slopes to the following values.

Template:
Super2L

Slopes:

Left

Type:	Simple	
	Typical	Maximum
Fill	4:1	2:1
Cut	5:1	3:1

Right-of-ways
Offset: 50.00

Right

Type:	Simple	
	Typical	Maximum
Fill	4:1	2:1
Cut	5:1	3:1

Right-of-ways
Offset: 50.00

14. Open the Superelevation Control dialog box by selecting the Superelevation Parameters routine found in the Design Control cascade of the Cross Sections menu.

15. First, toggle on Superelevation Calculations and Crown Removal by Runout Distance (see Figure 11–32). The Superelevation Control dialog box uses the current alignment for its data.

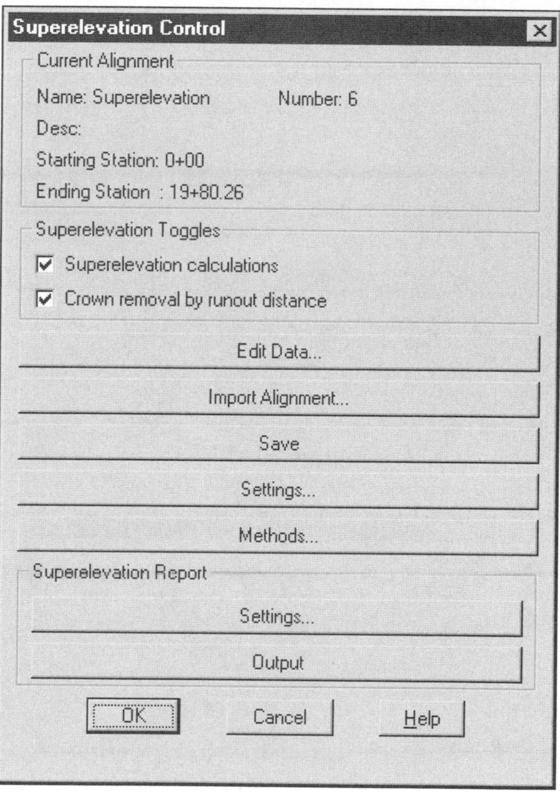

Figure 11–32

16. Select the Edit Data button to display the data for the first curve on the road. Set the parameters to match Figure 11–33.

17. Select the Next button to set the parameters for the second curve. Set the parameters to match Figure 11–34. Select the OK button to exit the Superelevation Curve Edit.

18. Select the Save button to save the superelevation data.

19. Select the OK button to process the design.

Superelevation Curve Edit

Curve Detail Information

Current Curve 1

PC sta: 4+03.79 PT sta: 8+22.01
Radius: 350.00 Length: 418.22
Spiral in: 0.00 TS sta: 4+03.79
Spiral out: 0.00 ST sta: 8+22.01

Curve Edit Information

Start sta 403.79 End sta 822.01 Method A
E value 0.080000 Direction Right Rollover 0.000000

Transition In
Runout 50.000
Runoff 100.000
% runoff 100.000

Transition Out
Runout 50.000
Runoff 100.000
% runoff 100.000

Next Previous Curve #... Station...
Info... Insert Curve... Delete Curve... Subgrades...
OK Cancel Help

Figure 11–33

Superelevation Curve Edit

Curve Detail Information

Current Curve 2

PC sta: 10+89.06 PT sta: 14+35.03
Radius: 350.00 Length: 345.97
Spiral in: 0.00 TS sta: 10+89.06
Spiral out: 0.00 ST sta: 14+35.03

Curve Edit Information

Start sta 1089.06 End sta 1435.03 Method A
E value 0.080000 Direction Left Rollover 0.000000

Transition In
Runout 50.000
Runoff 100.000
% runoff 100.000

Transition Out
Runout 50.000
Runoff 100.000
% runoff 100.000

Next Previous Curve #... Station...
Info... Insert Curve... Delete Curve... Subgrades...
OK Cancel Help

Figure 11–34

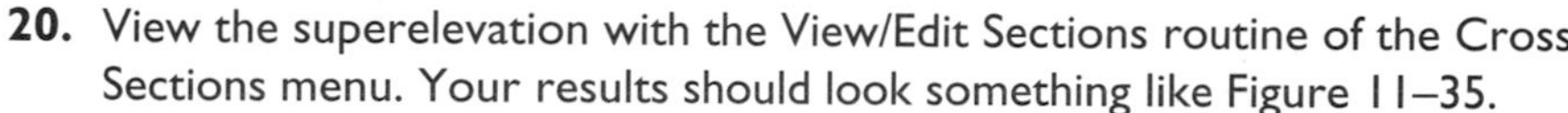

20. View the superelevation with the View/Edit Sections routine of the Cross Sections menu. Your results should look something like Figure 11–35.

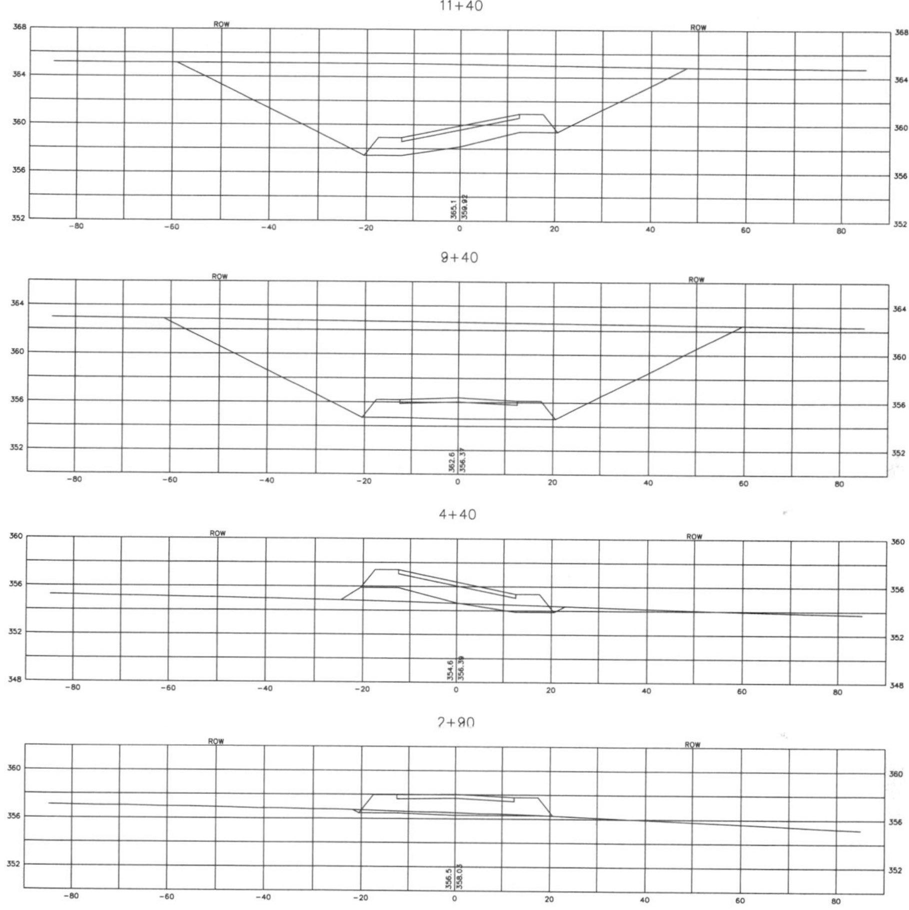

Figure 11–35

The roadway design capabilities in Civil Design are powerful and at times complex. Transition alignments, external slope control, and superelevations give you a strong set of design tools. The roadway design process is rigid, yet at every step along the way Civil Design supplies you with tools to create, analyze, edit, and annotate the design elements.

Hydrology

After you complete this section, you will be able to:

- Work with the structural calculators of Hydrology
- Calculate runoff with the TR-55 Rational calculator
- Calculate runoff with the TR-55 Graphical Peak Discharge calculator
- Calculate runoff with the TR-55 Tabular runoff calculator
- Use pond design tools to design a pond
- Evaluate a pond
- Create, view, and plot an HEC2 cross section

SECTION OVERVIEW

Civil Design implements the TR-55 and TR-20 methods of runoff calculations. The Hydrology menu consists of tools for the calculation and creation of structures, runoff amounts, pond routings, and HEC2 cross-section data. Even though Hydrology presents the tools as separate, many of the routines read data from related routines. For example, the Pond Routing routine uses a defined pond as the basis of a water detention system. The pond definition was created by routines in the Grading menu. The evaluation of the pond design references data from the structure calculators to define the inflow and outflow devices. The inflow data can be runoff or structure calculations, and the outflow can be structure calculations generated by the different calculators.

Each structure or runoff calculation is a part of the project data. The data is in the *HD* folder of the project. Hydrology uses and creates several support and data files. Many of the files are based upon files from the prototype, and depending upon the types of calculations you make, some files may not be used. Some of the calculation saved files have the name of *temp##.???*. The routines place numbers in the pound sign location (##), and the extension will vary depending upon which routine creates the file.

In addition to structures and runoff calculators, Hydrology uses several pond design tools from the Grading menu. These tools include templates and predefined ponds. After you create and define a pond, the Hydrology menu provides several pond performance calculation tools. These calculations include an editor for pond routing and outflow.

The HEC2 Output cascade contains routines that define, create data files for, and plot cross sections of HEC2 data sets. You can plot the cross sections in the drawing, export them to another program, or view their data.

UNIT I

The first unit of this section reviews the hydrology settings. The settings affect the units, precision, colors, plotting values, and the default text editor. These settings are available from the Settings routine of the Hydrology menu or the Drawing Settings routine of the Projects menu.

UNIT 2

The structure calculators are the focus of the second unit of this section. The Hydrology menu contains calculators that cover pipes, channels, weirs, orifices, risers, and culverts. Most of the calculators save their values to the *HD* folder of the project. Other routines in the Hydrology menu then reference the data file for their calculations.

UNIT 3

Civil Design supports three TR-55 calculators: rational, graphic peak discharge, and tabular. These calculators use external data files and multiple dialog boxes to calculate a final volume. The three calculators are the topic of Unit 3.

UNIT 4

The fourth unit covers the pond tools of the Grading menu. You can create a pond from contours, perimeters with volumes, and templates, analyze the characteristics, and annotate the values of the pond with the tools of the Grading menu. The unit also covers the creation and plotting of HEC2 cross sections.

UNIT 5

This unit covers the Routing Editor of the Hydrology menu.

UNIT 6

The last unit covers the HEC2 Output cascade of the Hydrology menu. The menu contains routines to define, import, export, and plot HEC2 cross sections.

UNIT 1: HYDROLOGY SETTINGS

The Hydrology Settings dialog box contains values that control and influence the calculation and reporting of hydrological values. You can call this dialog box from the Settings routine of the Hydrology menu (see Figure 12–1) or as individual dialog boxes from the Drawing Settings routine of the Projects menu (see Figure 12–2).

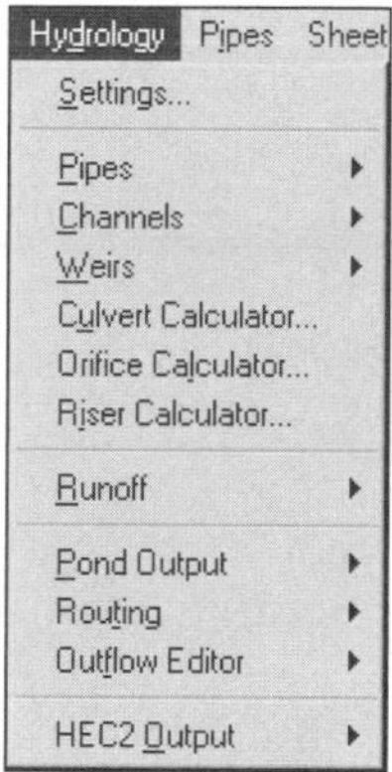

Figure 12–1

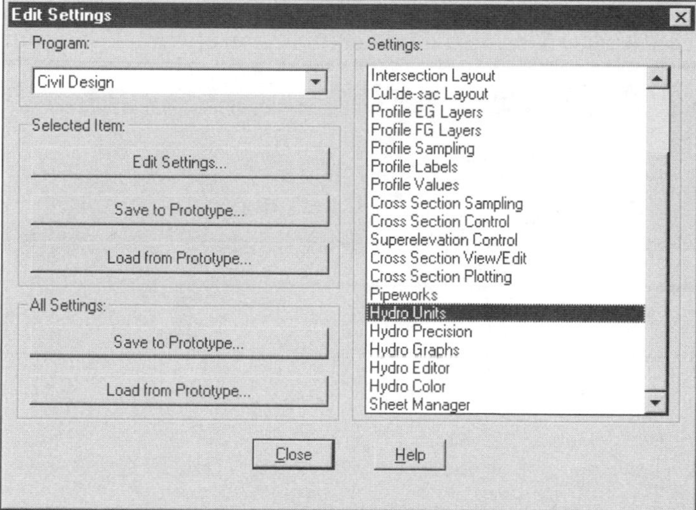

Figure 12–2

UNITS

The Unit Types dialog box sets the basic units for calculating and reporting runoff and structures (see Figure 12–3). These settings influence area, volumes, precision, velocities, and dimensions reporting.

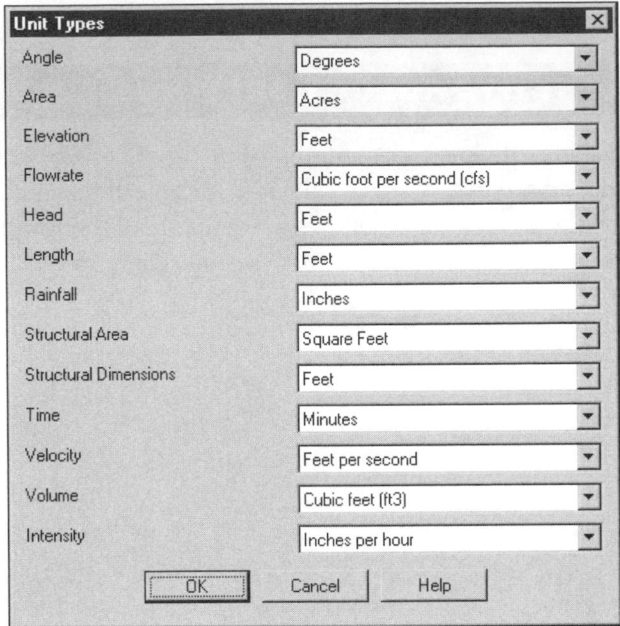

Figure 12–3

PRECISION

The Precision settings control the reporting of runoff and structure calculations (see Figure 12–4). All calculations are to the greatest precision of Land Development Desktop.

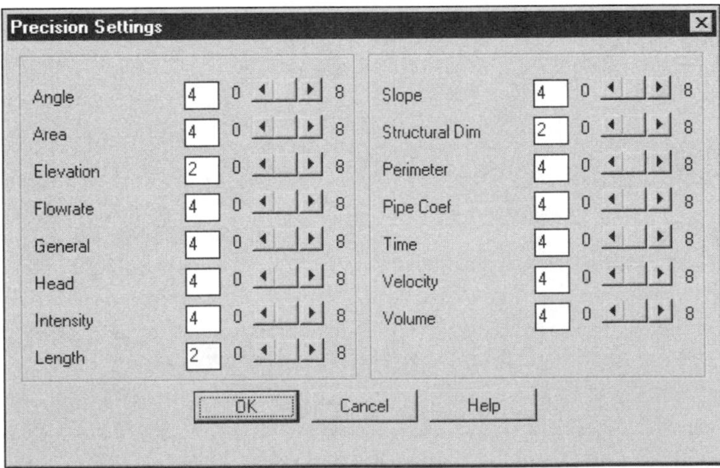

Figure 12–4

COLORS

The values in the Graph Settings dialog box influence the colors used by the performance and hydrographic charts in the program (see Figure 12–5).

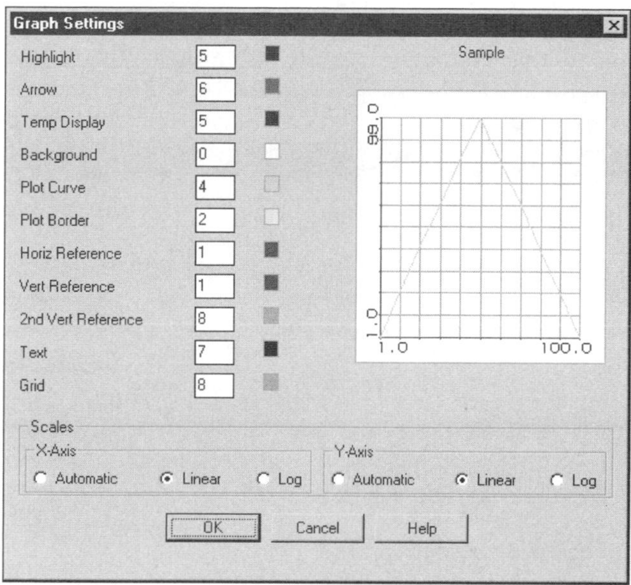

Figure 12–5

PLOT

The values in the Graphing Utility dialog box control the initial values of the HEC2 and performance chart plotting routine (see Figure 12–6). At the top of the dialog box are buttons that call subdialog boxes that control the name of the plot, the magnitude of the *X* and *Y*-axis, the grid spacing values, annotation, and ticks. The bottom portion of the dialog box controls loading, saving, creating, and plotting charts and sections into the drawing.

Graphing Utility
Preferences
Main | X Axis | Y Axis
Border | Grid | Ticks
Graphic Dimensions
Horizontal Dim ft 100.0000 Select
Vertical Dim ft 100.0000 Select
Curve Information
Preferences
New | Edit | Delete
New | Load | Save | Preview | Plot
OK | Cancel | Help

Figure 12–6

EDITOR

The Editor Settings dialog box allows you to call a text editor other than Notepad when editing files (see Figure 12–7).

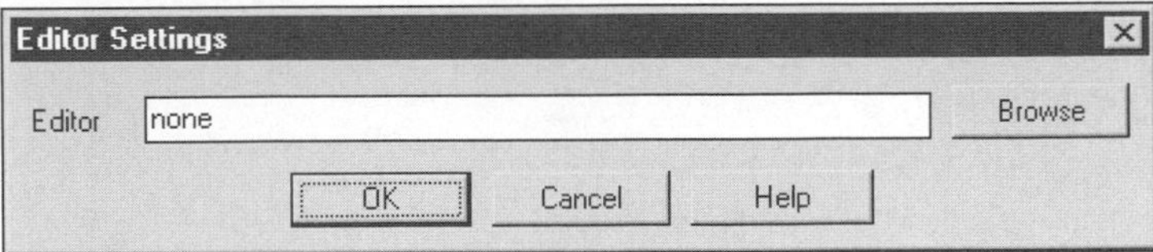

Figure 12–7

Exercise

After you complete this exercise, you will be able to:

- Edit and locate Hydrology settings

1. Start a new drawing *forpres*, and assign it to a new project. Use the LDDTR2 prototype and describe the project as Runoff Study (see Figures 12–8 and 12–9).

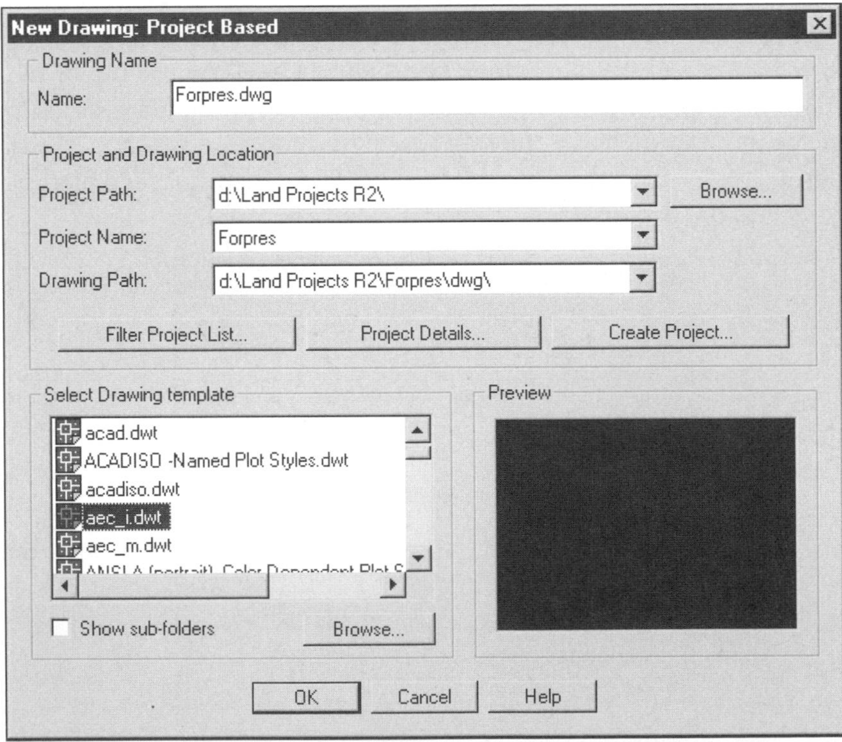

Figure 12–8

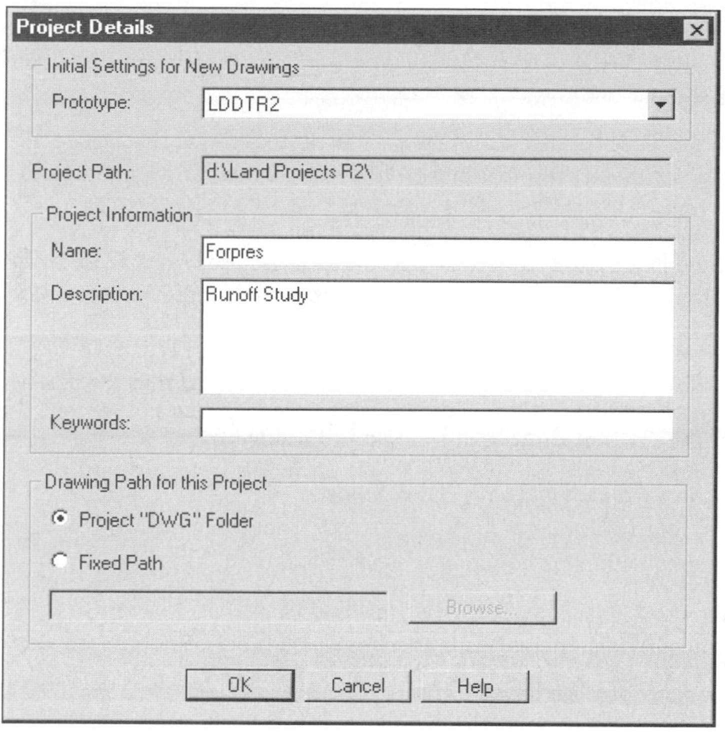

Figure 12–9

2. Select the OK button to accept default settings and to dismiss the Create Point Database dialog box.

3. When the Load Setup dialog box displays, select the LDD1-30 setup, then select the Load button, and then select the Finish button to dismiss the dialog box. When the Finish dialog box displays, select the OK button to dismiss it.

4. Use the INSERT command of the Desktop and insert the drawing *forpres* from the CD that comes with this book. Insert the drawing at 0,0 and explode it upon insertion.

5. Use the Zoom Extents routine to view the drawing.

6. Click the Save icon and save the drawing.

7. Select Settings from the Hydrology menu to display the Settings dialog box. The dialog box contains a set of buttons that call the same settings routines found in the Drawing Settings routine of Projects.

8. Select the Units button to view the current settings. Use Figure 12–3 to set your values. Select the OK button to set the values and to return to the Settings dialog box.

9. Select the Precision button to view the current settings. Use Figure 12–4 to set your values. Select the OK button to set the values and to return to the Settings dialog box.

10. Select the Colors button to view the current settings. Use Figure 12–5 to set your values. Select the OK button to set the values and to return to the Settings dialog box.

11. Select the Plot button to view the current settings. Use Figure 12–6 to set your values. Select the OK button to set the values and to return to the Settings dialog box.

12. Select the Editor button to view the current settings. Use Figure 12–7 to set your values. Select the OK button to set the values and to return to the Settings dialog box.

13. Select the OK button to exit the Settings dialog box.

14. Click the Save icon and save the drawing.

UNIT 2: STRUCTURE CALCULATORS

Hydrology provides several structure calculators including channels, culverts, pipes (pressure and gravity), orifices, and weirs. Each calculator takes user-entered and standard hydrographic data to compute its overall results. Examples of the standard data are the friction factors (Manning's n), rainfall amounts, or structural definitions (winged versus square entrance culverts).

The structural calculators will solve for an unknown value only if all of the required information is entered into the calculation dialog box. For example, calculating the diameter of an outlet orifice depends on flowrate, coefficients, headwater, and tailwater. After you enter the required known values, the program calculates the required orifice size to handle the numbers and displays the results in the dialog box.

You can save the calculation results of any calculator as project data. The calculators save the data in a text file located in the *HD* directory of the project. If appropriate, there is a rating curve of the structure and/or a hydrograph. You can use the tools within the calculator dialog box to place the curve or hydrograph into the drawing.

PIPE CALCULATORS

The Pipe calculators on the Pipes cascade of the Hydrology menu evaluate gravity and pressure systems. These calculators cover Manning's n, Darcy-Weisbach, and Hazen-Williams systems (see Figure 12–10).

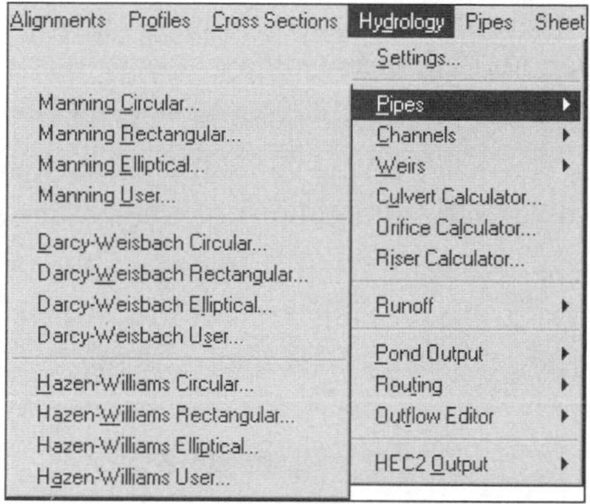

Figure 12–10

MANNING'S N

The Manning's n calculators are for circular, rectangular, elliptical, and user piping systems (see Figure 12–10). The pipe calculators are the same in form with additional entries for the specific types of pipe (see Figure 12–11). The elliptical calculator needs a major and semimajor axis. The rectangular calculator needs the width and height of the pipe. The user calculator requires a wetted area and perimeter to calculate its values.

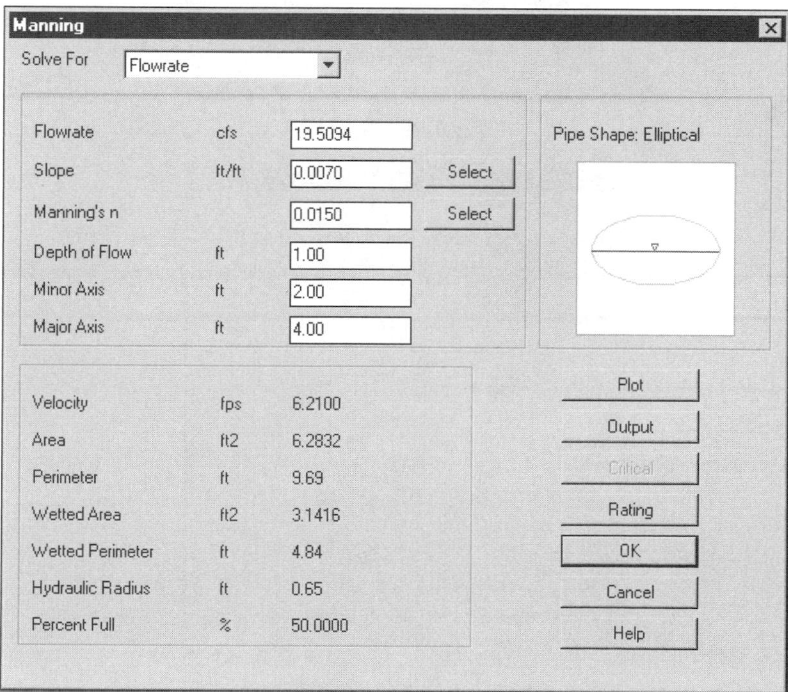

Figure 12–11

Each calculator solves for one of the following values: Flowrate, Slope, Manning's n, Depth of Flow, Diameter, and Diameter Full.

The data for the calculator includes flowrate, slope, Manning's n, depth of flow, and diameter. The output for the dialog box includes the velocity, area, perimeter, wetted area, wetted perimeter, hydraulic radius, and percentage full.

The Rating button creates a plot of the rating curve (see Figure 12–12). The Output button creates a text or WK1 file, or sends a report to a printer with the results of the calculations. The Plot button displays the Hydraulic Elements Curve of the calculation (see Figure 12–13).

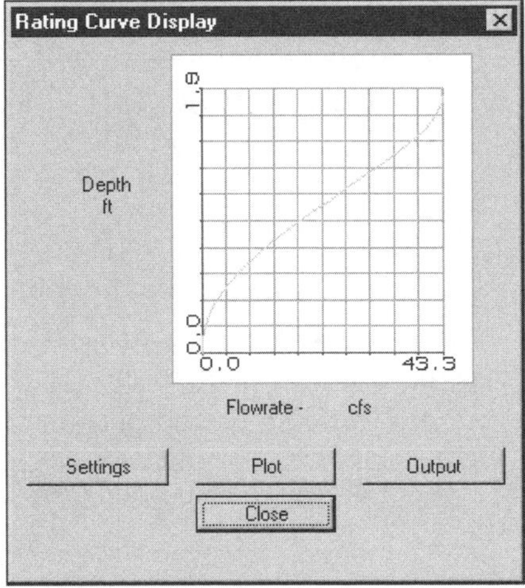

Figure 12–12

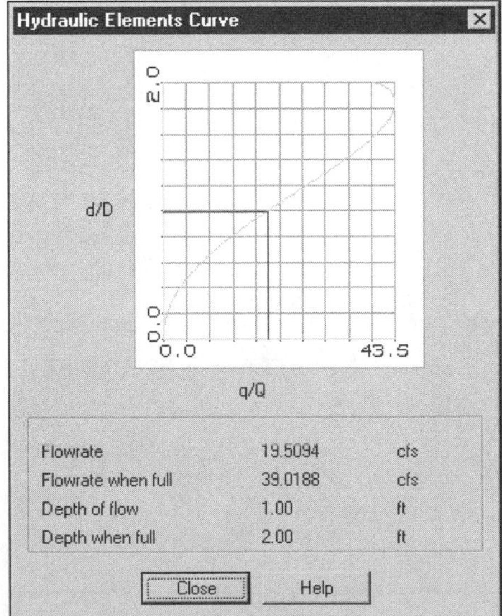

Figure 12–13

DARCY-WEISBACH CALCULATORS

The Darcy-Weisbach calculators solve for head loss in pressure piping systems (see Figure 12–14). Hydrology provides a calculator for circular, rectangular, elliptical, and user shapes.

Each calculator solves for flowrate, length, headloss, friction factor, and diameter. Enter the known values, flowrate, length, headloss, friction factor, and diameter, and the calculator returns the velocity, area, and perimeter of the pipe.

Each Output button prints a report about the calculations.

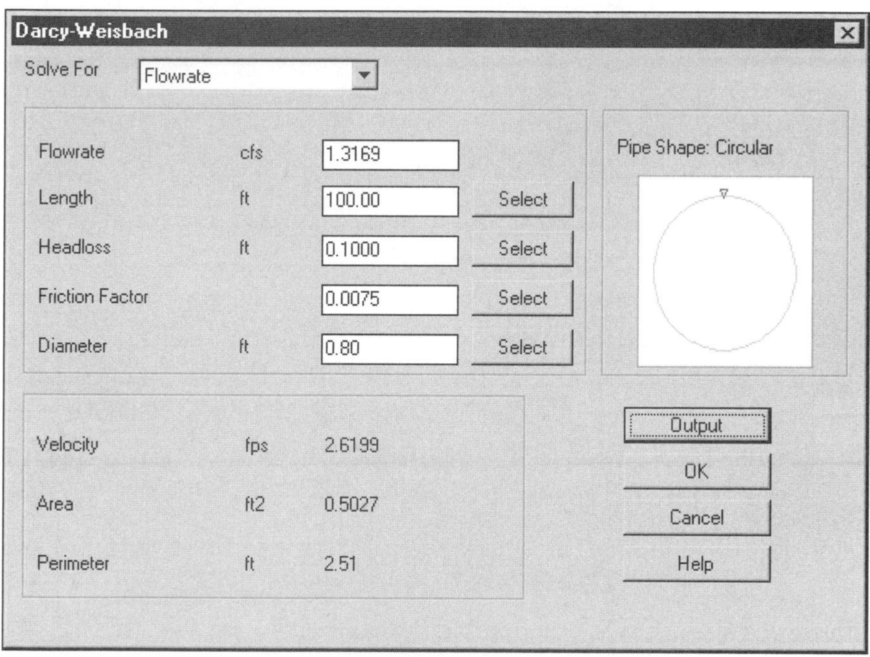

Figure 12–14

HAZEN-WILLIAMS CALCULATORS

The Hazen-Williams calculators solve for velocity in pressure piping systems (see Figure 12–15). Civil Design provides a calculator for circular, rectangular, elliptical, and user shapes.

Each calculator solves for flowrate, length, headloss, coefficient, and diameter. Enter the known values, flowrate, length, headloss, coefficient, and diameter, and the calculator returns the velocity, area, and perimeter of the pipe.

The Output button prints a report of the calculations.

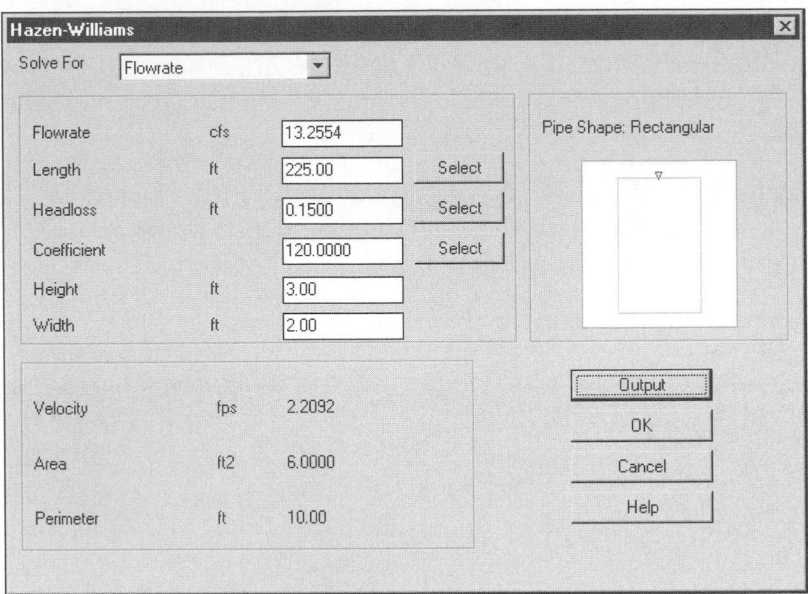

Figure 12–15

CHANNEL CALCULATORS

The Channel calculators consist of three types: Rectangular, Trapezoidal, and Advanced. Each calculator handles a different channel shape. Each calculator displays a dialog box similar to the Advanced Channel Calculator (see Figure 12–16). Each calculator solves for one of the following values: flowrate, slope, Manning's n, and depth of flow (see Figure 12–17).

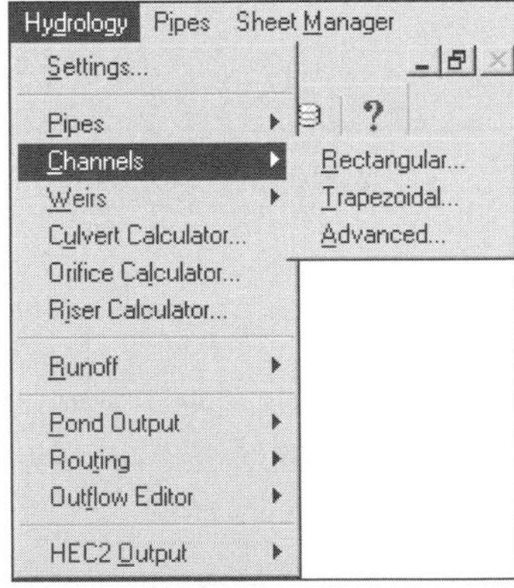

Figure 12–16

Each calculator contains:

- A Critical button that displays the critical values for the channel. The report includes specific energy, critical velocity, and froud number (see Figure 12–18).

- An Output button that outputs the calculated data to either a text or .WK1 file (spreadsheet file format).
- A Plot button that plots the performance rating of the current values to the screen (see Figure 12–19).
- A Rating button that plots the critical rating of the current values to the screen (see Figure 12–20).

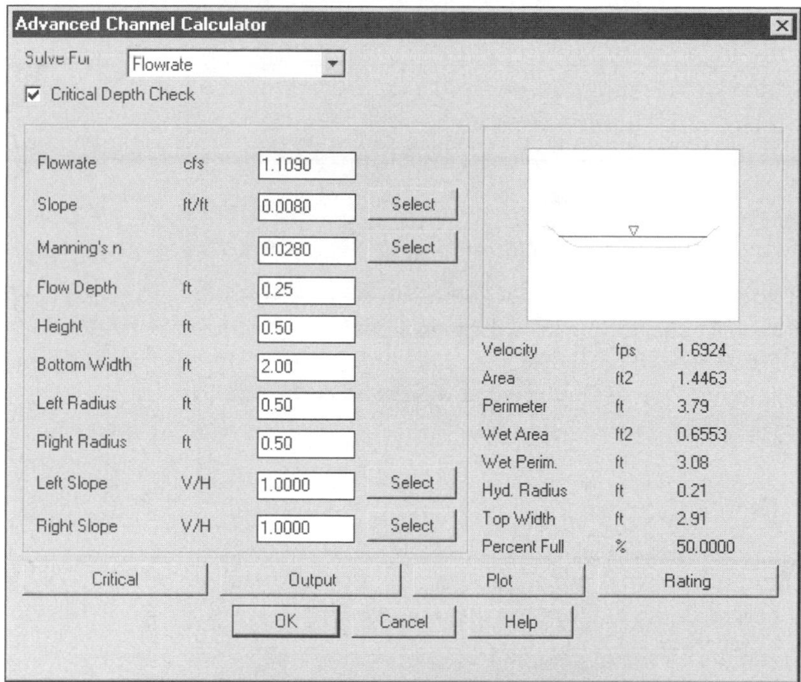

Figure 12–17

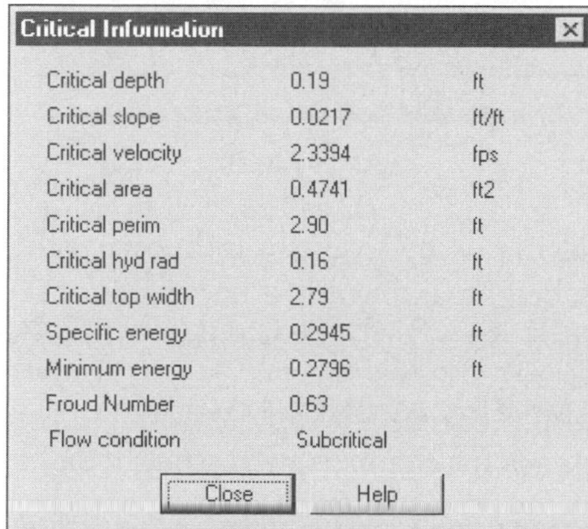

Figure 12–18

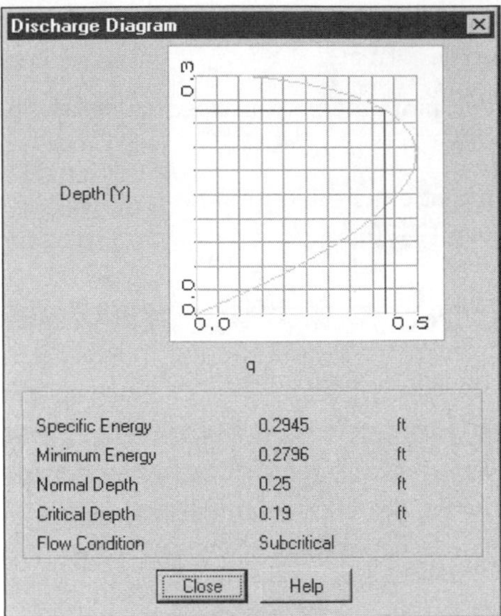

Figure 12–19

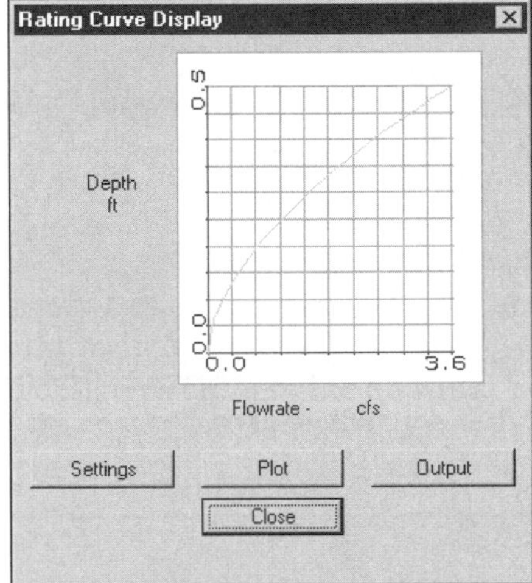

Figure 12–20

The results from the calculator include the unknown value and flowrate, slope, Manning's n, flow depth, height, and width. The calculator returns the velocity, area, perimeter, wet area, wet perimeter, hydrologic radius, top width, and percentage full.

RECTANGULAR CHANNEL CALCULATOR

The Rectangular calculator is for channels with straight sides.

TRAPEZOIDAL CHANNEL CALCULATOR

The Trapezoidal calculator is for channels with sloping sides.

ADVANCED CHANNEL CALCULATOR

The Advanced calculator is for channels with sloping sides and radii between the channel bottom and sides.

WEIR CALCULATORS

The Weir calculators cover Cipolleti, rectangular, and triangular weir shapes. Each calculator solves for flowrate, depth of flow, and coefficient, and also solves for additional values depending upon the type of weir (see Figures 12–21 and 12–22).

The Output button produces a report of the calculations. The Rating button produces a plot of the rating curve of the weir.

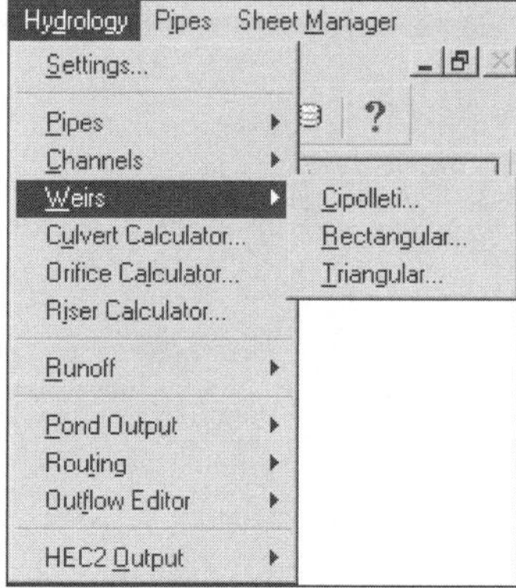

Figure 12–21

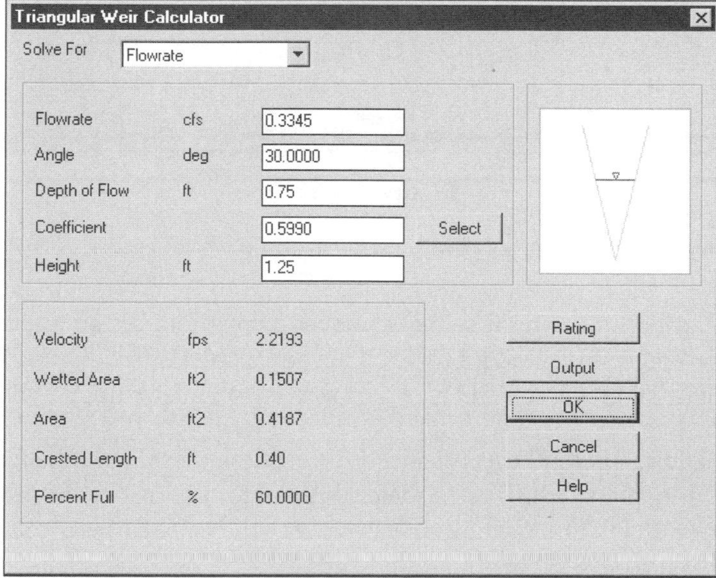

Figure 12–22

CULVERT CALCULATOR

The Culvert calculator calculates the numbers of a culvert design (see Figure 12–23). The culvert can be circular or square and contain multiple openings. The Culvert calculator is the most complex calculator in the structure set.

The Culvert Design dialog box, which is the main dialog box for the Culvert calculator, contains the final information for the evaluation of a culvert design. Some of these values come from calculations carried out in secondary dialog boxes.

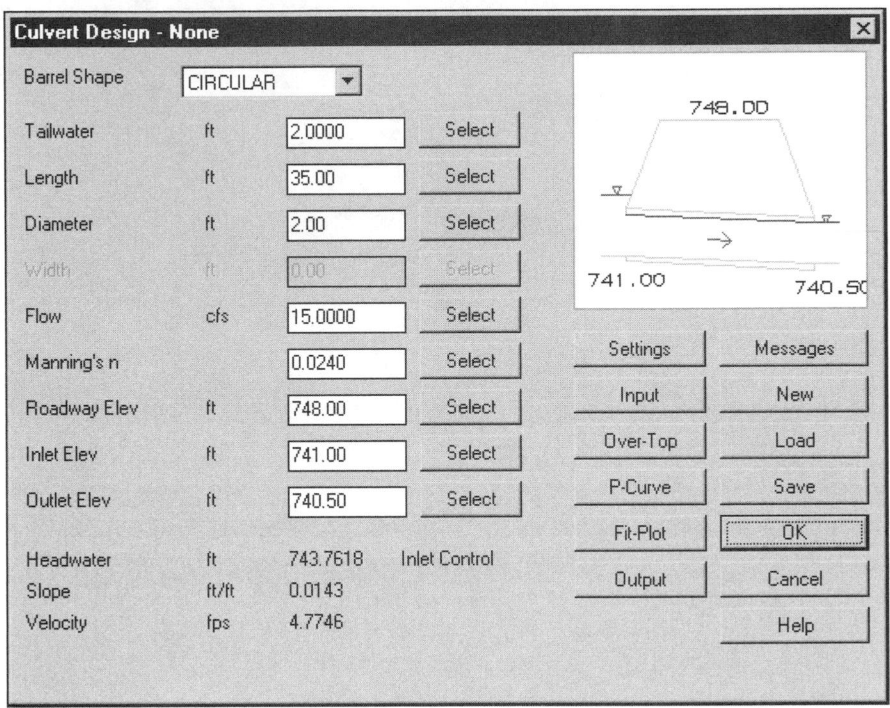

Figure 12–23

SETTINGS

The Settings button on the right side of the dialog box produces an additional dialog box that sets several essential values for the Culvert calculator (see Figure 12–24). The Culvert Settings dialog box sets the type of flow control: inlet, outlet, and optimum.

In the Control area, Inlet indicates that the culvert design evaluation is for headwater control at the inlet point. Outlet sets the evaluation for control at the outlet point of the culvert (tailwater). The Optimum button sets the calculator to solve for which control is the best for the culvert.

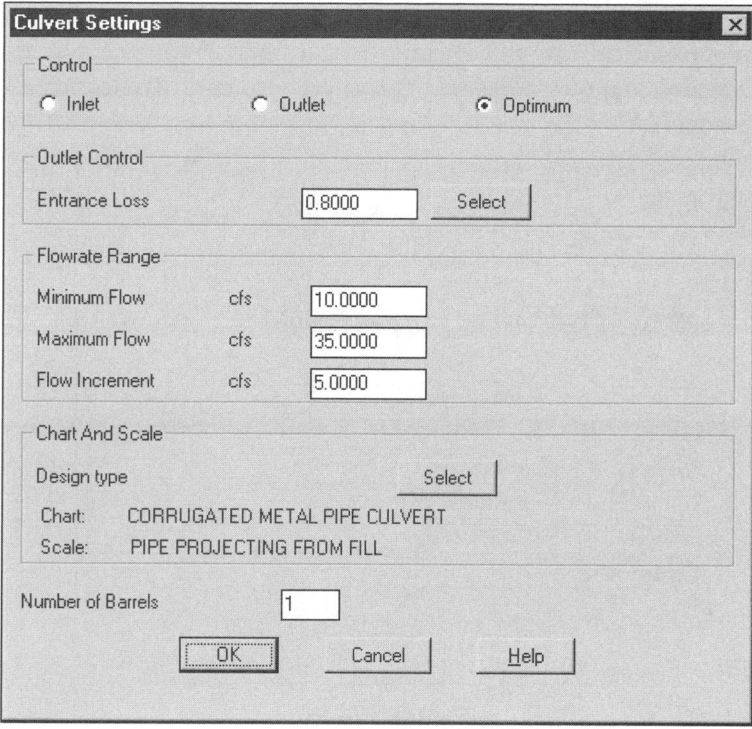

Figure 12–24

The Entrance Loss box contains a loss coefficient. You can enter a value, or if you choose the Select button, you can choose a value from a list of values in a data dialog box (see Figure 12–25).

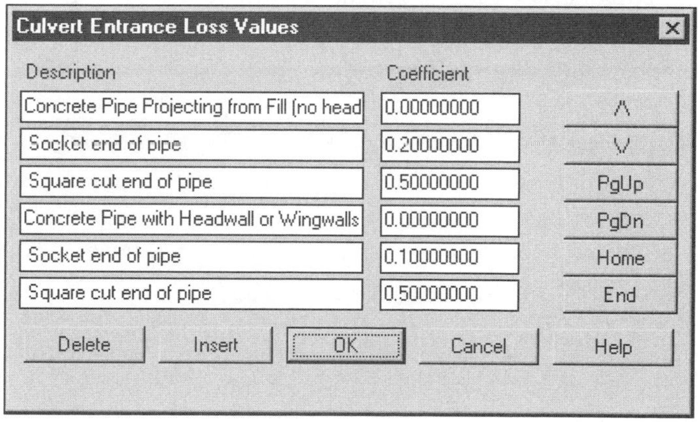

Culvert Entrance Loss Values

Description	Coefficient	
Concrete Pipe Projecting from Fill (no head	0.00000000	/\
Socket end of pipe	0.20000000	\/
Square cut end of pipe	0.50000000	PgUp
Concrete Pipe with Headwall or Wingwalls	0.00000000	PgDn
Socket end of pipe	0.10000000	Home
Square cut end of pipe	0.50000000	End

Delete | Insert | OK | Cancel | Help

Figure 12–25

The values in the Flowrate Range area control the *X* and *Y* axes and the increment for the P-curve plot of the culvert. The P-curve (performance) plot displays the headwater versus the flowrate data for the design. The values also control the increments in the report of the culvert.

The Design Type selection in the Chart and Scale area calls another dialog box that contains design types of culverts (see Figure 12–26).

The last part of the Culvert Settings dialog box controls the number of barrels in the culvert.

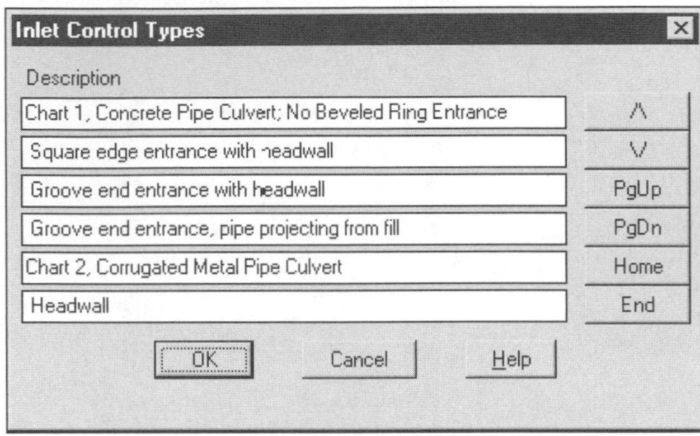

Figure 12–26

CULVERT TAILWATER

For the tailwater values in the Culvert Design dialog box, you can either enter a number, or choose the Select button. When you choose the Select button, the Tailwater Editor displays (see Figure 12–27). The Tailwater Only option sets a fixed height for the water level. By selecting Downstream Channel Flow, you encounter another dialog box, which is the Advanced Channel calculator. The dialog box can recall any prior calculations or calculate new values. The Tailwater Curve number can be from a file containing tailwater values or a hydrograph.

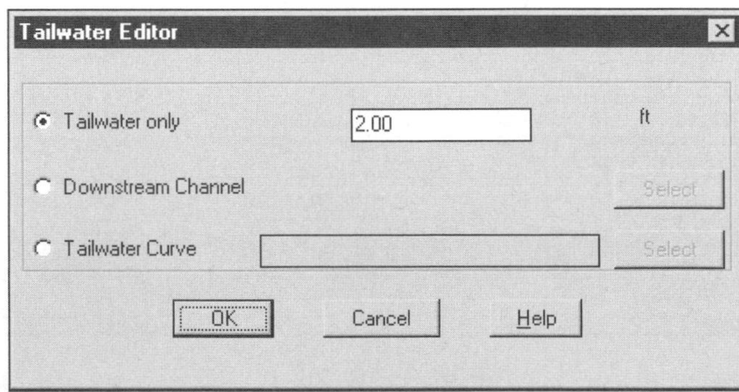

Figure 12–27

LENGTH

The length is the length of the pipes in the culvert. You can enter the length in the data area to the right of the length label or you can select two points from the drawing. You select the points in the drawing by clicking on the select button and selecting the points in the drawing.

DIAMETER

The diameter is the diameter of the culvert pipe(s). You set the diameter by entering the size in the area to the right of the Diameter label. You can set the size by clicking on the Select button and selecting the size from a diameter list.

FLOW

The flow values are either a fixed value or you can choose the Select button to select values from a hydrograph data file. The hydrograph data flow file is an output file either from the Rational, Tabular, Peak Discharge, or Inflow calculator.

MANNING'S N COEFFICIENTS

The Manning's n value is a fixed number, or if you choose the Select button, you can choose from a list of values (see Figure 12–28).

The remaining values in the main Culvert calculator dialog box are elevations for the road, inlet, and outlet.

The calculator computes the Over-Top condition, P-Curve values, and Fit-Plot. The Fit-Plot is the performance of the culvert. The Output button creates a summary report of the results from the calculator.

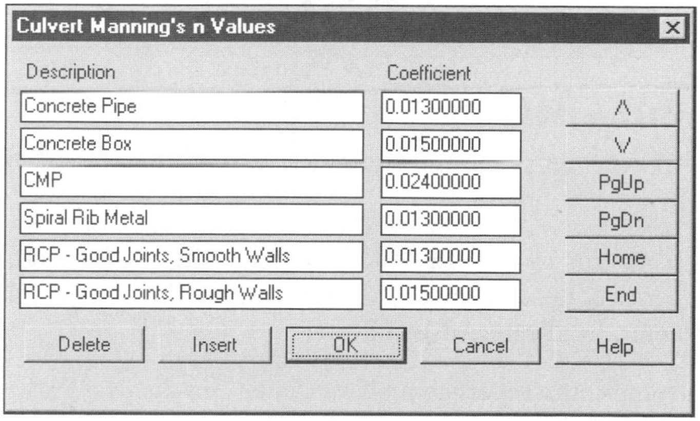

Culvert Manning's n Values

Description	Coefficient
Concrete Pipe	0.01300000
Concrete Box	0.01500000
CMP	0.02400000
Spiral Rib Metal	0.01300000
RCP - Good Joints, Smooth Walls	0.01300000
RCP - Good Joints, Rough Walls	0.01500000

/\ V PgUp PgDn Home End

Delete Insert OK Cancel Help

Figure 12–28

ORIFICE CALCULATOR

The Orifice calculator computes the hydraulic values for an orifice design (see Figure 12–29).

The calculator solves for one of the following values: flowrate, coefficient, area or diameter of the orifice, headwater, or tailwater. Enter the known values, flowrate, coefficient, diameter, headwater, tailwater, and velocity.

The Output button prints a report about the calculations, and the Rating button plots the calculations of the orifice.

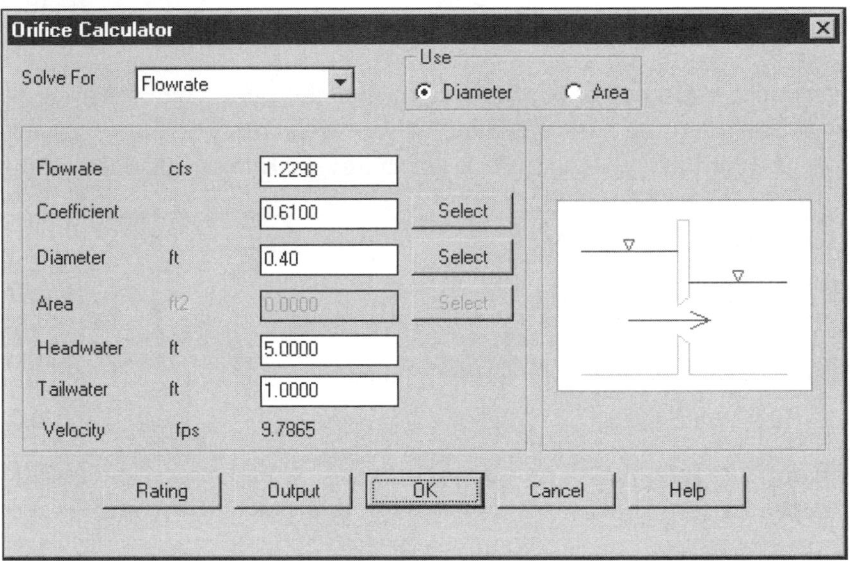

Figure 12–29

RISER CALCULATOR

The Riser calculator computes the hydraulic values for an orifice design (see Figure 12–30).

The calculator solves for one of the following values: flowrate, coefficient, area or diameter of the orifice, headwater, or tailwater. Enter the known values, flowrate, coefficient, diameter, headwater, and tailwater.

The Output button prints a report about the calculations and the Rating button plots the calculations of the orifice.

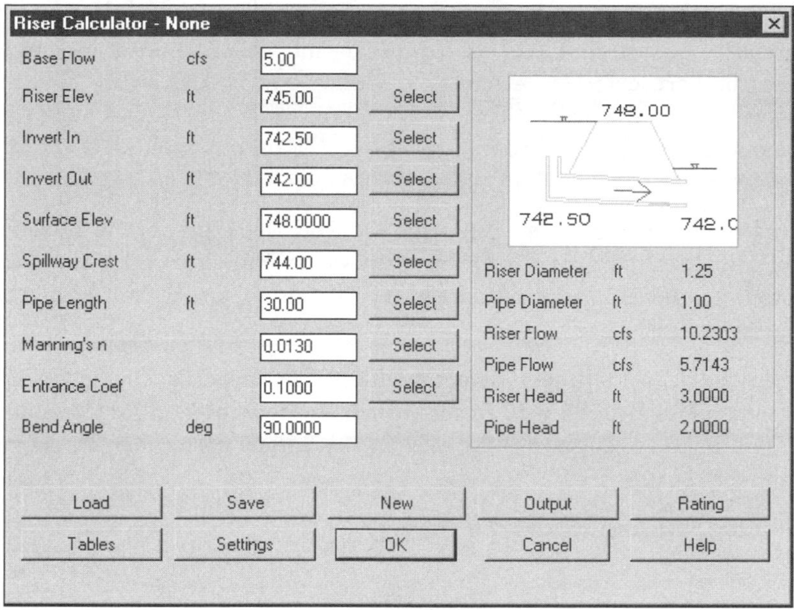

Figure 12–30

Exercise

After you complete this exercise, you will be able to:

- Use the structure calculators

1. If you are not in the *forpres.dwg* file, open the drawing.
2. From the Hydrology menu, select Manning Elliptical from the Pipes cascade. Set the Solve For value to Flowrate. Use Figure 12–11 to set the values. The Manning's n is the value for a concrete pipe. View the rating curve, the plot (Hydraulic elements), and the output for the calculation. After viewing the output, select the OK button to dismiss the calculator.
3. From the Hydrology menu, select the Advanced calculator from the Channels cascade. Set the Solve For value to Flowrate. Use Figure 12–17 to set the values. The Manning's n is the value for a coarse gravel channel. View the rating curve, the plot (discharge diagram), critical design values, and the output for the calculation. After viewing the output, select the OK button to dismiss the calculator.
4. From the Hydrology menu. select the Triangular calculator from the Weir cascade. Set the Solve For value to Flowrate. Use Figure 12–22 to set the values. View the rating curve and the output for the calculation. After viewing the output, select the OK button to dismiss the calculator.

EXERCISE

5. From the Hydrology menu, select the Culvert Calculator. Set the Barrel Shape to Circular. Use Figures 12–24, 12–25, 12–26, and 12–27 to set the values for the dialog box presented by selecting the Settings button. Use Figure 12–23 to enter the values into the main calculator dialog box. View the P-curve, the fit-plot, and the output for the calculator. After viewing the output, select the OK button to dismiss the calculator.
6. From the Hydrology menu, select the Orifice Calculator. Set the Solve For value to Flowrate. Use Figure 12–29 to set the values. View the rating curve and the output for the calculation. After viewing the output, select the OK button to dismiss the calculator.
7. From the Hydrology menu, select the Riser Calculator. Set the Solve For value to Flowrate. Use Figure 12–30 to set the values. View the rating curve and the output for the calculation. After viewing the output, select the OK button to dismiss the calculator.
8. Click the Save icon to save the drawing.

UNIT 3: RUNOFF CALCULATORS

The Soil Conservation Service (SCS) of the Department of Agriculture issued a document on urban watersheds called the TR-55 document. It covers the calculation of water runoff volumes, peak discharges, and pre- and post-development hydrographs.

The Civil Design implementation of TR-55 includes three calculators that handle increasing complexities of watersheds. The first is the Rational calculator. The Rational calculator handles simple homogeneous area calculations. The Graphical Peak Discharge calculator is for larger watersheds with a more complex structure. The Tabular calculator is for large watersheds having several different zones within its boundary. The Tabular method uses multiple graphical peak discharge calculations and times of travel through the watershed to calculate runoff values.

No matter which method calculates the runoff, each calculator uses a common set of data. These data sets include rainfall frequency, coefficients (curve numbers), and time of concentration.

RAINFALL FREQUENCY

The rainfall in a watershed is the foundation of runoff calculations (see Figure 12–31). Each county has rainfall values for storms of a preset duration. The typical duration of storms is 1, 2, 5, 10, 25, 50, and 100 years. Depending upon the requirements of the job, you need to select the correct amount of rainfall for the calculator.

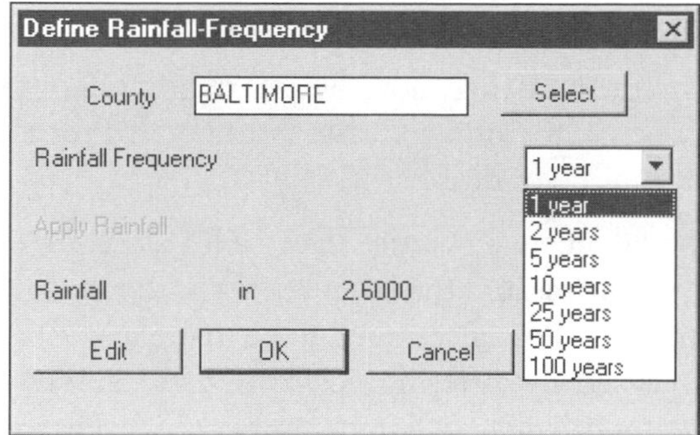

Figure 12–31

Civil Design does not supply a file for every county, so you must create a file for the locality or modify the installed rainfall file using the Rainfall Frequency Editor (see Figure 12–32). The file is the *county.rf* file in the *HD* folder of the prototype. When you start a new project, Land Development Desktop copies the file from the prototype into the *HD* directory of the project. The following is an excerpt of the file:

```
7               1     2     5     10    25    50    100
"ALLEGANY"      2.4   2.9   3.8   4.5   4.9   5.7   6.2
"ANNE ARUNDEL"  2.7   3.3   4.3   5.2   5.9   6.5   7.4
"BALTIMORE"     2.6   3.2   4.2   5.1   5.5   6.3   7.1
```

Rainfall-Frequency Editor

County	1	2	5	10	25	50	100	
ALLEGANY	2.4	2.9	3.8	4.5	4.9	5.7	6.2	/\
ANNE ARUN	2.7	3.3	4.3	5.2	5.9	6.5	7.4	\/
BALTIMORE	2.6	3.2	4.2	5.1	5.5	6.3	7.1	PgUp
CALVERT	2.8	3.4	4.4	5.3	6.1	6.7	7.6	PgDn
CAROLINE	2.8	3.4	4.6	5.3	6.0	6.8	7.6	Home
CARROLL	2.5	3.1	4.0	5.0	5.4	6.2	7.1	End

Frequency

Delete Insert OK Cancel Help

Figure 12–32

In the first line of the file, the number 7 indicates the number of columns for storm data and the numbers 1 through 100 are the column headings (year storm). The file is a space-delimited file. If you create a new file and replace the installed copy with the new file, the rainfall amounts for the counties are updated to represent your area.

SCS RUNOFF CURVE NUMBER EDITOR

The curve numbers with the hydrology module are straight out of the TR-55 document (see Figure 12–33). The SCS Runoff Curve Number Editor allows you to add or change the data water runoff conditions.

Many sites are not homogeneous in nature. Even a lot is a composite of pervious and impervious materials, the house and the lawn. Other watersheds have conditions that differ over their territory, so allowances need to be made in the calculations reflecting the different runoffs caused by differing absorption rates. The Composite Runoff Curve Number Calculator calculates a curve number that takes into consideration the area and runoff values of subareas of a watershed (see Figure 12–34).

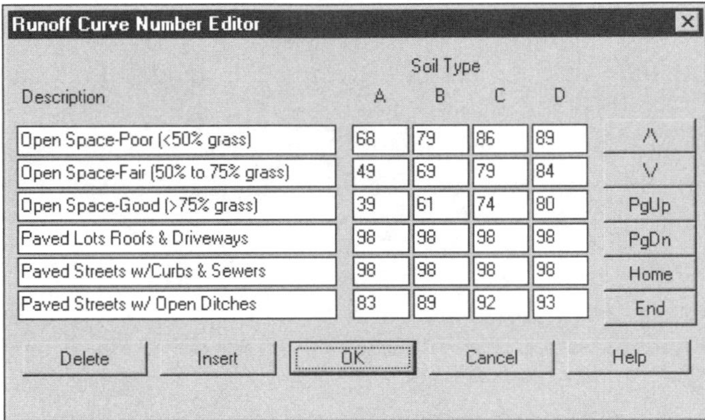

Figure 12–33

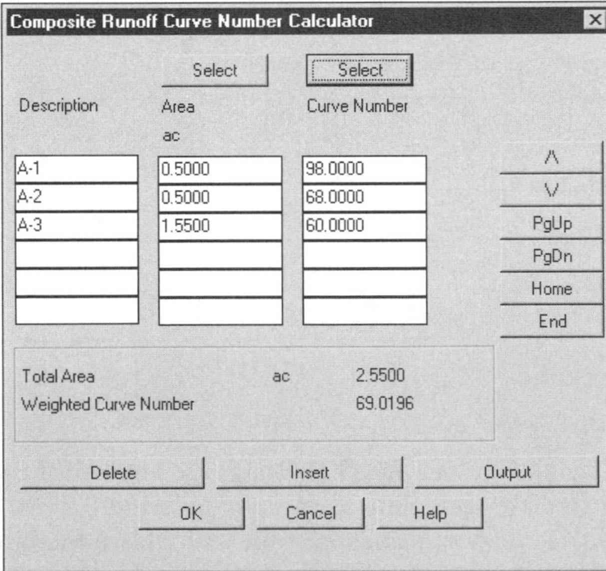

Figure 12–34

TIME OF CONCENTRATION

Time of concentration has three components: sheet, shallow, and channel flow (see Figure 12–35). Each watershed is different and may use only one component of the Time of Concentration calculations, whereas other watersheds may use a combination of components.

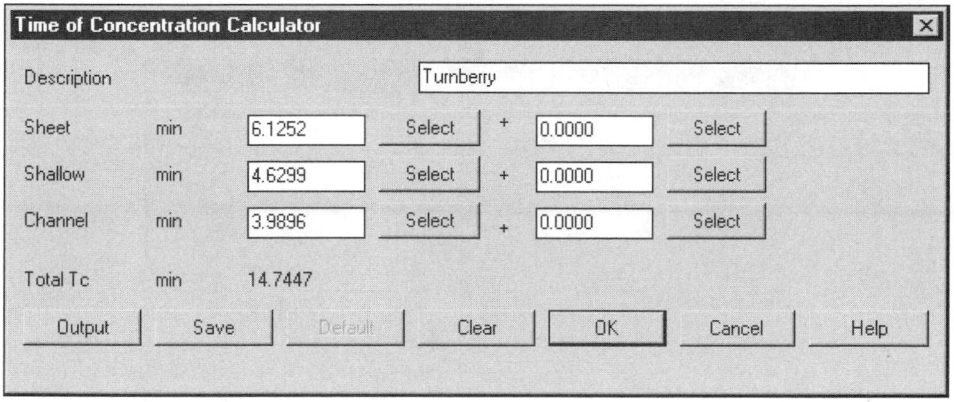

Figure 12–35

SHEET FLOW

Sheet flow is a flow of water over a plane surface. The flow represents the furthest hydraulic distance in the watershed. The maximum length of sheet flow is 300 feet.

The component elements for sheet flow are Manning's n, length of flow, two-year rainfall, and land slope (see Figure 12–36). The flow length and land slope can be selections from objects in the drawing.

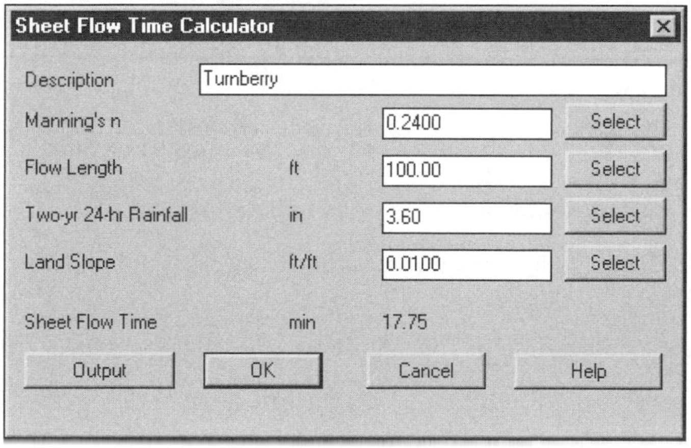

Figure 12–36

SHALLOW FLOW

Shallow flow is water that starts to concentrate into organized structures. This flow occurs between the sheet and channel flow of water.

The components for shallow flow are surface, length of flow, and watercourse slope (see Figure 12–37). The flow length and watercourse slope can be selections from objects in the drawing.

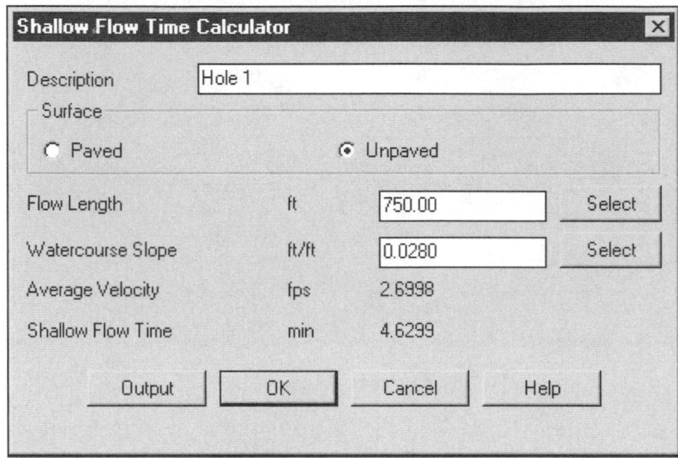

Figure 12–37

OPEN CHANNEL FLOW

Channel flow is a structure, natural or man-made, that carries water in a defined space.

The components of channel flow are cross-section flow area, wetter perimeter, channel slope, Manning's n, and flow length (see Figure 12–38). The flow length and channel slope can be selections from objects in the drawing.

Here are some rules about the values of time of concentration calculations:

- The minimum time of concentration value is 0.1 hour.
- The maximum time of concentration value for graphical peak discharge is 10 hours.
- The maximum time of concentration value for the Tabular method is 2 hours.

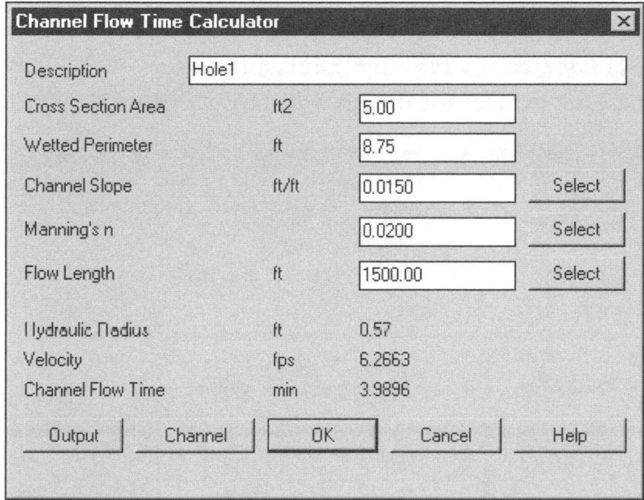

Figure 12–38

TIME OF TRAVEL

The time of travel calculator is a part of the Tabular method of TR-55 (see Figure 12–39). A watershed can be heterogeneous in nature and may have several natural divisions within its boundary. Water flowing from points in one subarea concentrate and then may travel through other subareas to discharge from the watershed. The time it takes the water to flow through other divisions within the watershed is the time of travel. The time of travel calculation has only two components because the water has already begun to concentrate. The two components are shallow and channel flow.

As mentioned previously, the runoff calculators use the components of rainfall, curve number, time of concentration, and time of travel. The complexity of the calculations increases when moving from the Rational to Graphical to Tabular calculator. Each calculator calls the aforementioned dialog boxes to generate the values for the method.

Time of Travel Calculator
Description Turnberry
Shallow min 9.0533 Select + 0.0000 Select
Channel min 5.2151 Select + 0.0000 Select
Total Tt min 14.2684
Output Save Default Clear OK Cancel Help

Figure 12–39

RATIONAL RUNOFF CALCULATOR

The Rational calculator computes runoff for small areas (see Figure 12–40). The calculations include allowances for intensity duration curves.

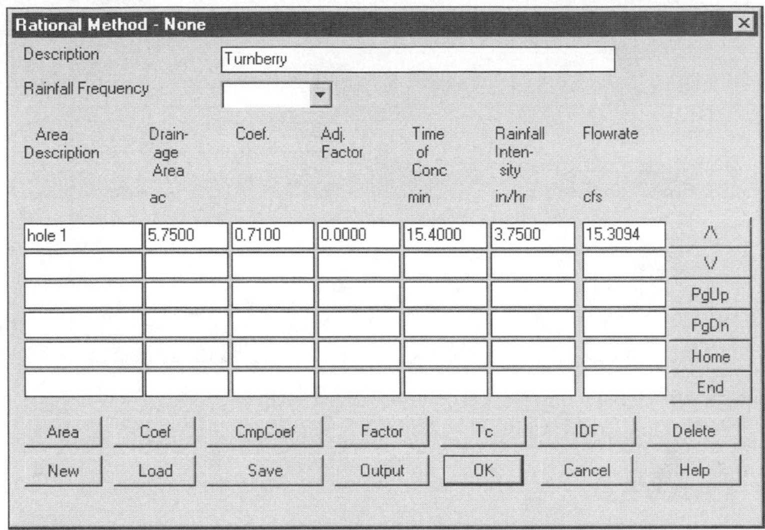

Figure 12–40

GRAPHICAL PEAK DISCHARGE CALCULATOR

The Graphical Peak Discharge calculator computes peak discharge from large, homogeneous watersheds using the TR-55 methodologies (see Figure 12–41).

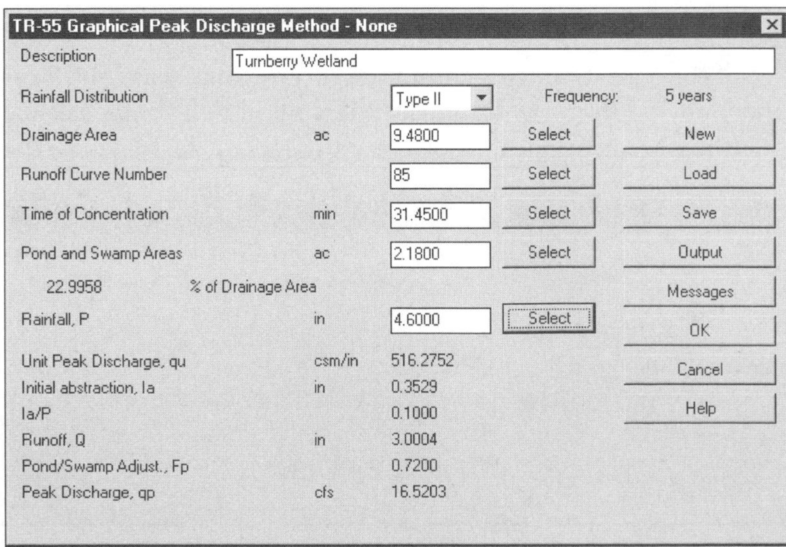

Figure 12–41

TABULAR RUNOFF CALCULATOR

The Tabular Runoff calculator is for a watershed boundary containing subdivisions and travel times for water moving from one zone to another zone within the boundary (see Figure 12–42).

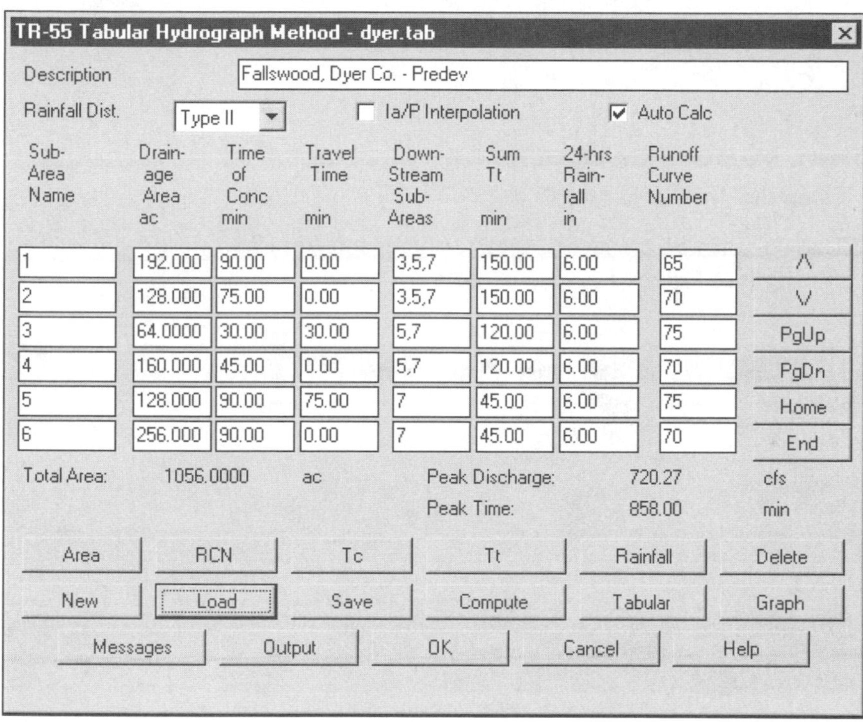

TR-55 Tabular Hydrograph Method - dyer.tab

Description: Fallswood, Dyer Co. - Predev

Rainfall Dist.: Type II — Ia/P Interpolation: ☐ — Auto Calc: ☑

Sub-Area Name	Drainage Area ac	Time of Conc min	Travel Time min	Down-Stream Sub-Areas	Sum Tt min	24-hrs Rainfall in	Runoff Curve Number
1	192.000	90.00	0.00	3,5,7	150.00	6.00	65
2	128.000	75.00	0.00	3,5,7	150.00	6.00	70
3	64.0000	30.00	30.00	5,7	120.00	6.00	75
4	160.000	45.00	0.00	5,7	120.00	6.00	70
5	128.000	90.00	75.00	7	45.00	6.00	75
6	256.000	90.00	0.00	7	45.00	6.00	70

Total Area: 1056.0000 ac — Peak Discharge: 720.27 cfs — Peak Time: 858.00 min

Figure 12–42

Exercise

The first step in hydrology is evaluating a site to understand and calculate its water runoff values. The Rational runoff calculator is for simple areas. As the sites become more complex, the Graphic Peak Discharge method handles the higher level of complexity. The most complex method assumes that the watershed contains subareas. The water from some subareas away from the discharge point must flow through subareas closer to the discharge point before adding their water to the outflow. The area represented by the *forpres* drawing is too small to be a part of the Graphical Peak Discharge or Tabular method of calculations. The Rational method is adequate in calculating the runoff for the area covered in the drawing. There are three areas supplying runoff to the detention pond. Each area is different in size and time of concentration and each empties into an inlet or ditch. Later these runoff values will become a part of the evaluation of the pond and its discharge.

The Graphical Peak Discharge and Tabular methods will calculate runoff amounts for other data sets.

After you complete this exercise, you will be able to:

- Calculate a Rational runoff amount
- Calculate a Graphical Peak Discharge
- Compute Tabular runoff values

RATIONAL RUNOFF CALCULATOR

1. Open the drawing *forpres.*
2. Thaw or turn on the layers AreaA, AreaB, and AreaC to view the runoff zones in the site (see Figures 12–43 and 12-44).

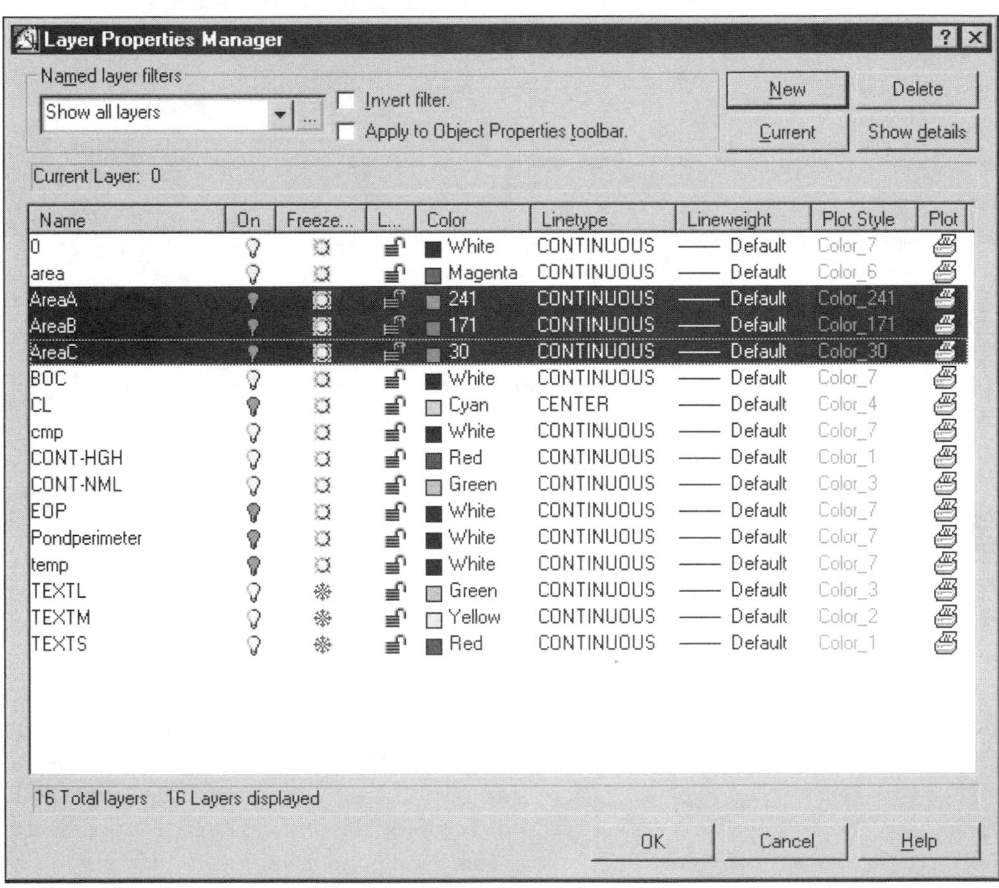

Figure 12–43

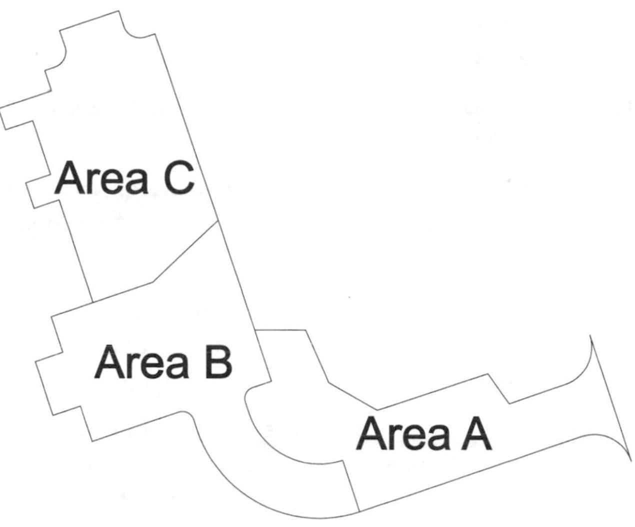

Figure 12–44

3. If you are not in the Civil Design package, change the menu palette to Civil Design.
4. Start the Rational Method routine of the Runoff cascade of the Hydrology menu. The routine presents the Rational Method dialog box (Figure 12–45). Enter the description as Mays Lake Detention Pond.

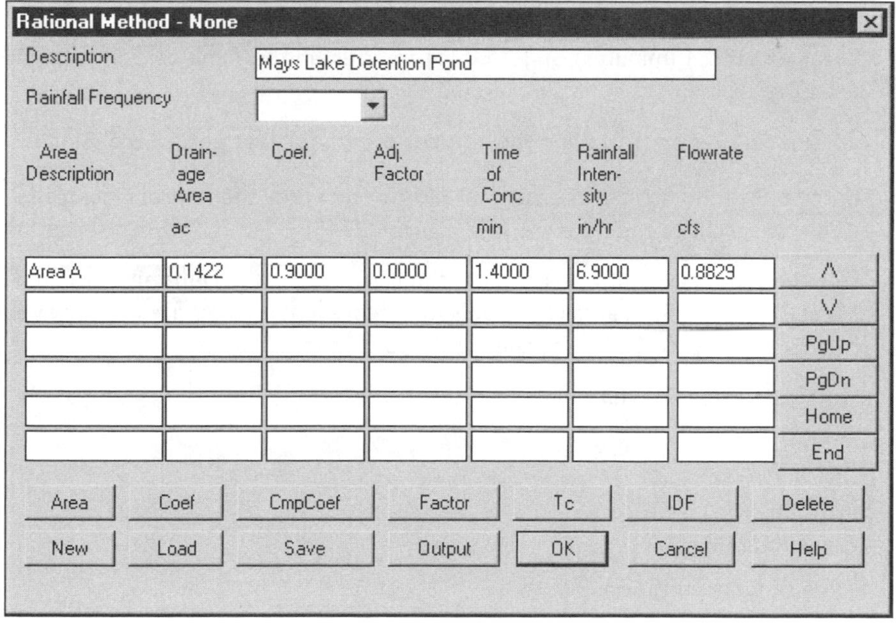

Figure 12–45

5. Select the box in the first row at the left side of the dialog box (Area Description column). Enter the description of Area A into the cell.
6. After entering the description in the first row, press ENTER. The routine moves on to the next box in the calculator.

The next box requires an area that comes from one of two places. You can manually calculate the area or, if you select the Area button at the bottom of the dialog box, the routine hides the dialog box and prompts you to select an object representing the area for this zone.

7. Select the polygon at the right of the project (driveway). After you select the polygon, the routine enters the area value from the polygon into the calculator and returns to the dialog box.

The next value needed for the zone is the Curve Number (the Coef. cell). If the zone is homogeneous, you enter the curve number based upon soil type or land use. If a zone is a combination of land use and soil types, a combined curve number needs to be calculated. The CmpCoef button displays a dialog box that calculates a composite number from several different curve numbers within a zone.

8. The current zone is a homogeneous material area, pavement, which has a curve number value of 0.90. Place the cursor into the Coef cell for the first area and enter a coefficient of 0.90.

The time it takes the water to concentrate is about 1.4 minutes. If the time in the Rational dialog box is in hours, enter **1.4/60**. The dialog box will calculate the percentage of one hour for you.

9. Enter either 1.4 (minutes) or 1.4/60 (of an hour) in the Time of Concentration cell (Tc).

The rainfall is for a 10-year storm and the amount of rain for that period is 6.9 inches.

10. Enter **6.9** in the Rainfall Intensity cell. After you enter the rainfall rate, press ENTER to calculate the flowrate (see Figure 12–45).
11. Use the same dialog box to calculate the flowrate for the remaining two zones (Area B and C). Both areas have polygons representing their area. Check your results with Figure 12–46 for the results for B and C.

The following data is for the remaining zones:

Area B

Area: 0.17

Curve Number: 90

Time of Concentration: 0.65

Rainfall Intensity (in/hr): 6.9

Flow Rate: 1.06

Area C

Area: 0.16

Curve Number: 90

Rainfall Intensity (in/hr): 6.9

Time of Concentration: 1.06

Flow Rate: 1.01

12. Save the rational calculations as *Mayslake.rat* and save the drawing.

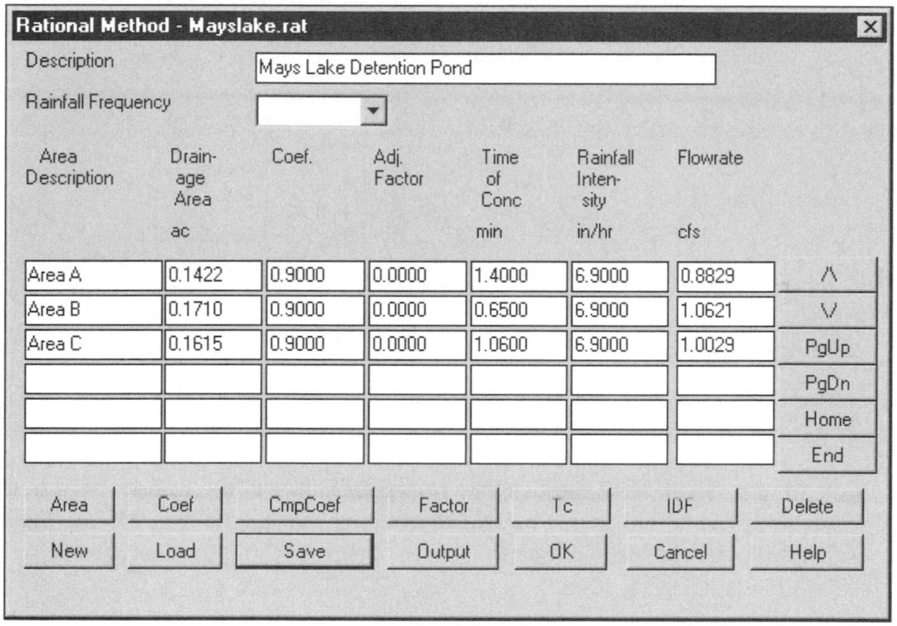

Rational Method - Mayslake.rat

Description: Mays Lake Detention Pond

Rainfall Frequency

Area Description	Drain-age Area (ac)	Coef.	Adj. Factor	Time of Conc (min)	Rainfall Inten-sity (in/hr)	Flowrate (cfs)
Area A	0.1422	0.9000	0.0000	1.4000	6.9000	0.8829
Area B	0.1710	0.9000	0.0000	0.6500	6.9000	1.0621
Area C	0.1615	0.9000	0.0000	1.0600	6.9000	1.0029

/\ | \/ | PgUp | PgDn | Home | End

Area | Coef | CmpCoef | Factor | Tc | IDF | Delete

New | Load | Save | Output | OK | Cancel | Help

Figure 12–46

GRAPHICAL PEAK DISCHARGE CALCULATOR

The Graphical Peak Discharge method calculates runoff for a single watershed. The watershed can have a composite curve number and more complex water flow characteristics. These complex flow characteristics show as sheet, shallow, and channel components of the Time of Concentration calculation (Tc) to move the water from the farthest point in the watershed to the discharge point.

With the increase in size, the curve numbers tend to be composite numbers rather than a single value because the larger areas tend to be more heterogeneous. When calculating the runoff amounts with the Graphical Peak Discharge method, there is an increase in the number of landuses and soils. The SCS defines a comprehensive list of landuse and soil conditions. The SCS divides the soil conditions list into four categories of soils: A, B, C, and D. The category A soil is the best drained and category D the worst.

The amount and intensity of rainfall are also more important. The routine requires a file containing the rainfall amounts for 1-, 2-, 5-, 10-, 25-, 50-, and 100-year storms. The file containing the amounts is the *county.rf* file. You need to modify or create a file containing the local county amounts.

Rainfall Frequency Editor

The first step in the process is to create two new entries in the *county.rf* file. The two counties are Dyer and DuPage. In addition to the amounts of rain, another factor is the type of storm. The Type I and IA storms are on the West Coast. These areas have wet winters and dry summers. Type III storms are along the Gulf of Mexico and the Eastern Seaboard. These areas have intense 24-hour rainfalls because of tropical storms. The remainder of the country is Type II.

The rain data for the counties is as follows:

Dyer Type II

1	2	5	10	25	50	100
2.94	3.60	4.36	5.37	6.00	7.29	8.73

DuPage Type II

1	2	5	10	25	50	100
2.51	3.04	3.80	4.47	5.51	6.46	7.5

1. Edit the rainfall file with the Rainfall Frequency Editor in the Runoff cascade of the Hydrology menu. The Insert button adds a new row to the file. Add the Dyer information to the file. Choose the Insert button and add the DuPage information to the file (see Figure 12–47).

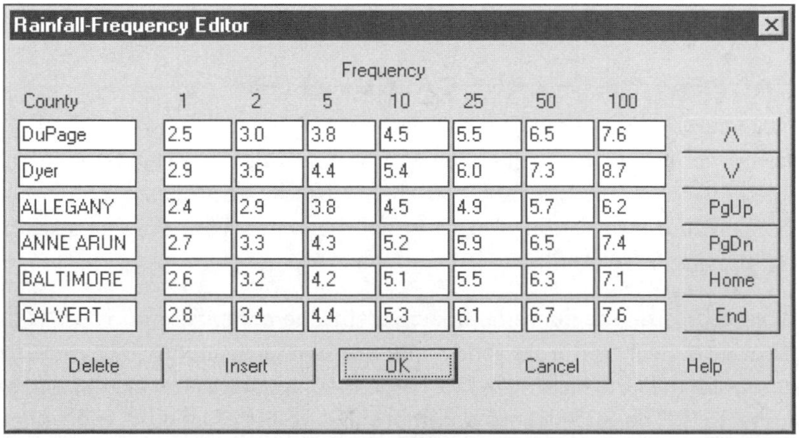

Rainfall-Frequency Editor

County	1	2	5	10	25	50	100	
DuPage	2.5	3.0	3.8	4.5	5.5	6.5	7.6	/\
Dyer	2.9	3.6	4.4	5.4	6.0	7.3	8.7	\/
ALLEGANY	2.4	2.9	3.8	4.5	4.9	5.7	6.2	PgUp
ANNE ARUN	2.7	3.3	4.3	5.2	5.9	6.5	7.4	PgDn
BALTIMORE	2.6	3.2	4.2	5.1	5.5	6.3	7.1	Home
CALVERT	2.8	3.4	4.4	5.3	6.1	6.7	7.6	End

Delete | Insert | OK | Cancel | Help

Figure 12–47

The data for the Dyer County Graphical Peak Discharge is:

Site Size: 250 acres

Soil	**Amount**	**Landuse**
Loring C	100 acres	acre Lots
Loring C	75 acres	Open Space good condition
Memphis B	75 acres	acre Lots

Time of Concentration:

Sheet Flow:

Condition: dense grass

Slope: 0.01 ft/ft

Length: 100 feet

Shallow Flow:

Condition: Unpaved

Slope: 0.01 ft/ft

Length: 1400 feet

Channel:

Condition: Manning's n 0.05

Flow Area: 27 square feet

Wetted Perimeter: 28.2 feet

Slope: 0.005 ft/ft

Length: 7300 feet

EXERCISE

The above data along with the rainfall amounts for Dyer County are enough to calculate a peak discharge for the site in Dyer County. See Figure 12–48 for a map of the area.

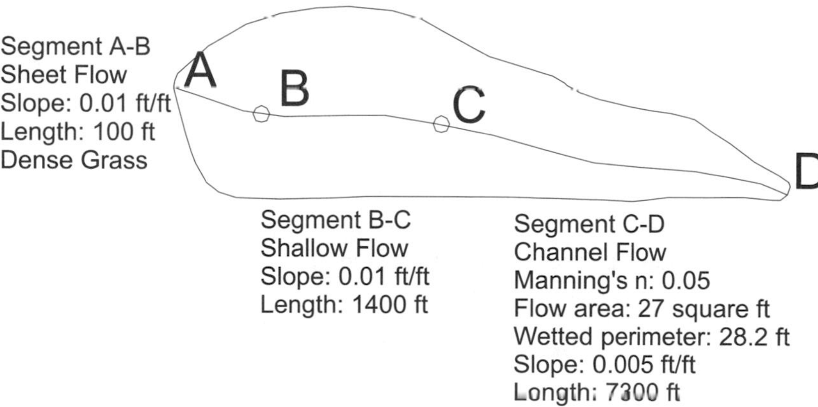

Figure 12–48

2. Run the Settings routine in the Hydrology menu and change the units and precision settings to the following values (see Figures 12–49 and 12–50):

```
Units:
Area:  Square Miles
       Precision:
       Area: 4
       Slope: 4
       Coefficients: 4
```

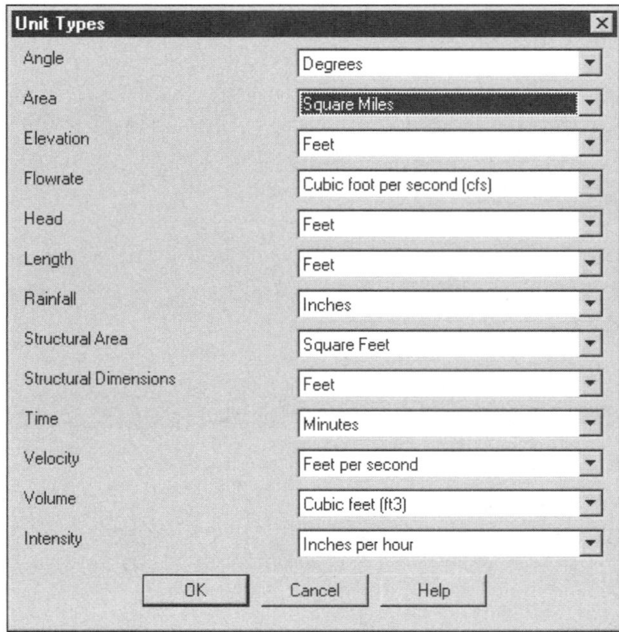

Figure 12–49

3. Run the TR-55 Graphical Method routine from the Runoff cascade of the Hydrology menu. The routine displays the Graphical Peak Discharge Method dialog box.

4. Enter the name of the site into the Description box: **Dyer Co**.

5. The type of rainfall for the area is a Type II rainstorm.

The area is in square miles, but the values for the current project are in acres. If you enter the number of acres divided by the number of acres in a square mile, the cell will calculate the correct number.

6. Enter **250/640** in the Drainage Area cell and press ENTER. The cell will respond with the value 0.3906, which is the percentage of one square mile represented by 250 acres.

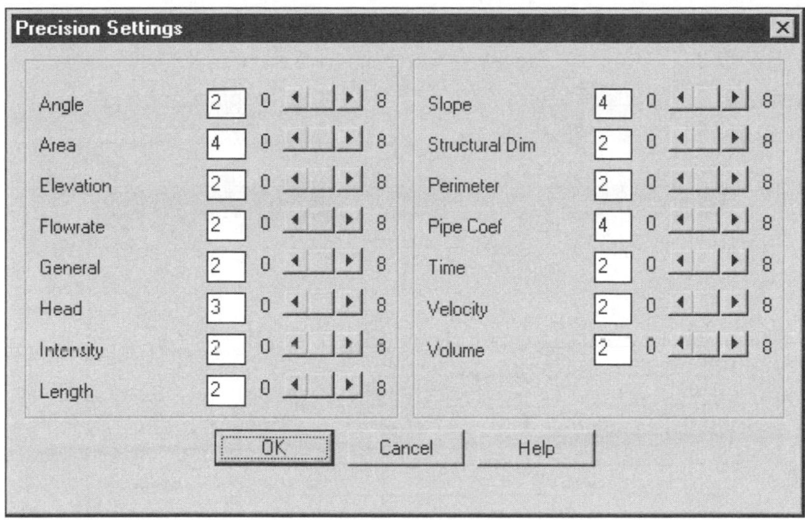

Figure 12–50

Composite Runoff Curve Numbers

The Runoff Curve Number is the next value for the site. The number is a composite value determined by the landuse and soil types of the watershed.

To establish the value of the curve number for the individual segments, choose the Select button above the heading Curve Number of the Composite Runoff Curve Number Calculator to view the SCS curve numbers (see Figure 12-51). This button displays the Curve numbers for Runoff Coefficients. Find the 1-acre lots with C type soil (Loring is a C soil type) (see the Runoff Coefficients dialog box in Figure 12–52). Select into the box containing 79 and then select the OK button to exit the dialog box and return to the Composite Runoff Curve Number Calculator. The curve number appears in the Curve Number box to the right of the area. Repeat this process for the remaining two sections of the site. The dialog box calculates a curve number at the bottom of the box. Select the OK button to exit the dialog box, and the Composite RCN number appears in the Run Curve Number cell of the Graphical Peak Discharge Dialog box.

7. Choose the Select button to the right of the Runoff Curve Number box. This displays a Runoff Coefficient dialog box that allows you to choose the method of calculating the curve number. Select the Composite Curve button to display the dialog box that develops the Composite Runoff Curve Number. Enter the values found below into the dialog box. Again the areas are in square miles. Enter the number of acres divided by 640 to get the correct area. Compare your results to Figure 12–51. After entering and checking the values, select the OK button to exit the current dialog box and return the main Graphical Peak Discharge dialog box.

Soil	Amount	Landuse
Loring C	100 acres	acre Lots
Loring C	75 acres	Open Space good condition
Memphis B	75 acres	acre Lots

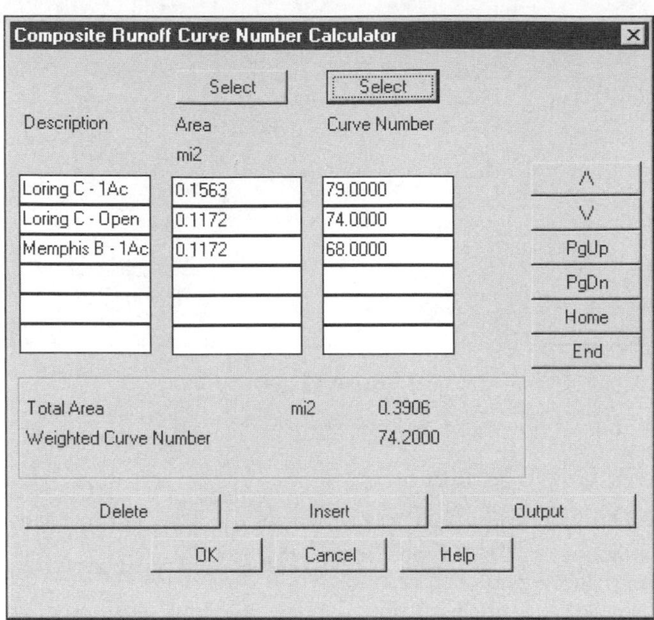

Figure 12–51

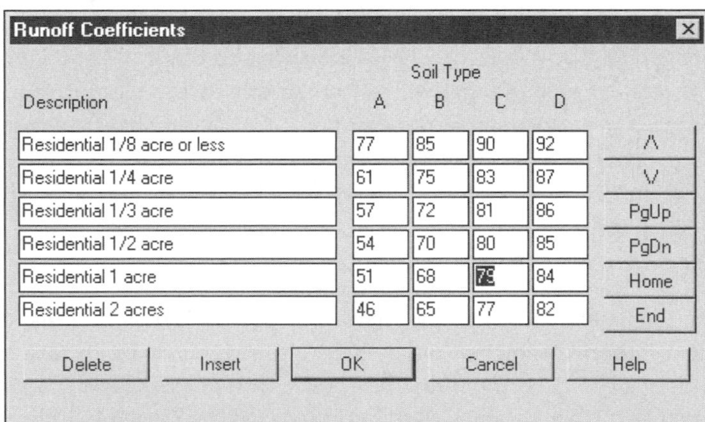

Figure 12–52

Time of Concentration

The next value in the Graphical Peak Discharge calculator is the time of concentration. The value for the current exercise is a combination of all three types of water flow.

8. Choose the Select button next to Time of Concentration. The button displays the main Time of Concentration dialog box. Enter the description of Dyer County in the Description area of the main Time of Concentration dialog box. The Select buttons to the right of the Sheet, Shallow, and Channel components of Tc call up their respective calculator dialog boxes.

Sheet Flow

9. Choose the Select button for sheet flow. Enter the data for the site. The calculation requires a curve number. Choose the Select button and select Grass, dense CN (70). Check your results with Figure 12–53. Select the OK button to exit the CN box.

Sheet Flow:

Condition: dense grass (Manning's n)

Length: 100 feet

Dyer County 2 year Rainfall: 3.60

Slope: 0.01 ft/ft

10. Enter the flow length and then choose the Select button to identify the two-year rainfall amount for Dyer County (see Figure 12–54). Select the OK button to exit the dialog box.

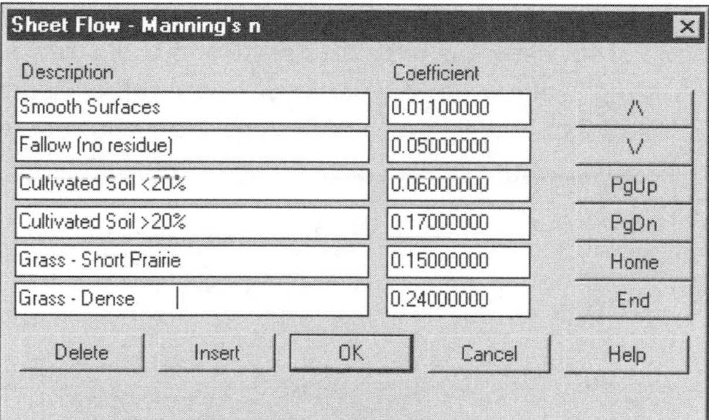

Figure 12–53

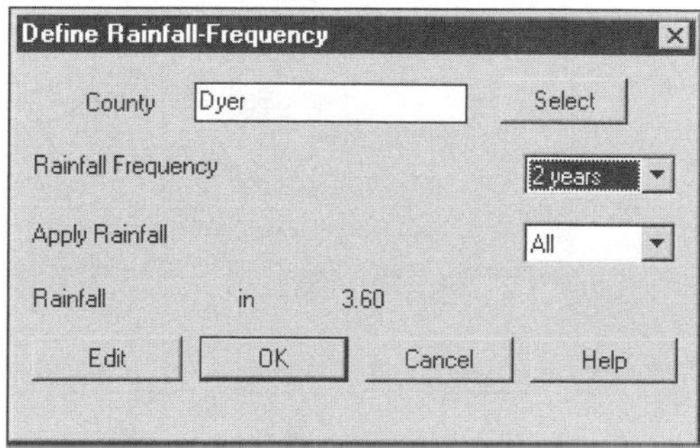

Figure 12–54

EXERCISE

11. Still in the Sheet Flow Time Calculator, enter the slope of the sheet flow.
12. Compare your values to Figure 12–55. Choose the OK button in the Sheet Flow Time Calculator to return to the main Time of Concentration dialog box.

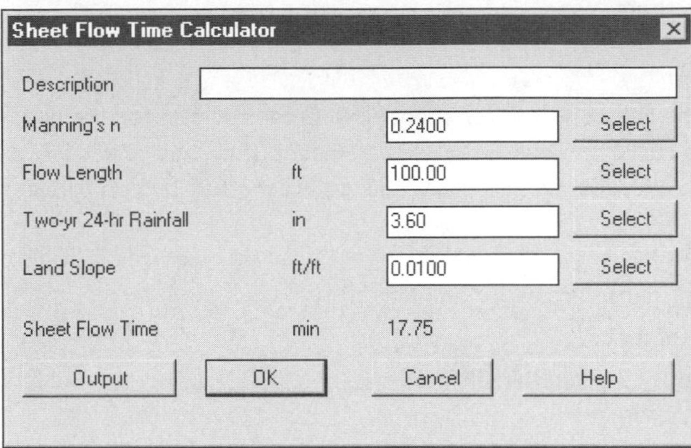

Figure 12–55

Shallow Flow

13. Choose the Select button to the right of Shallow Flow to display the calculator for this portion of Tc. Enter the data for the site and compare your dialog box to Figure 12–56. When your values are correct, choose the OK button in the Shallow Flow dialog box to exit back to the main Time of Concentration dialog box.

Shallow Flow:

Condition: Unpaved

Length: 1400 feet

Slope: 0.01 ft/ft

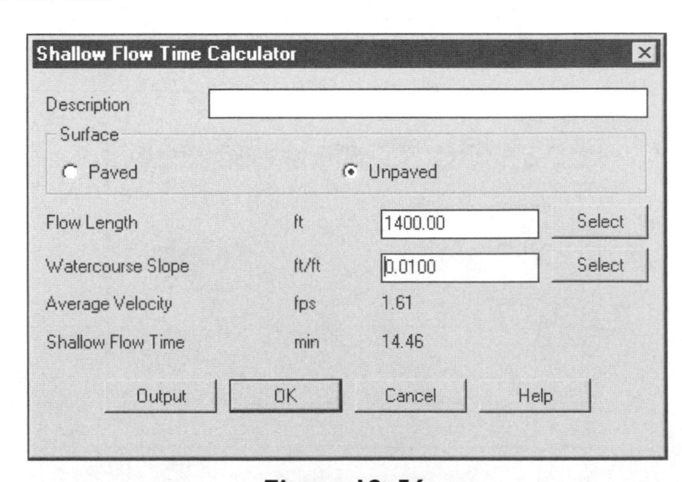

Shallow Flow Time Calculator

Description

Surface

Paved Unpaved

Flow Length ft 1400.00 Select

Watercourse Slope ft/ft 0.0100 Select

Average Velocity fps 1.61

Shallow Flow Time min 14.46

Output OK Cancel Help

Figure 12–56

Channel Flow

14. Choose the Select button to the right of Channel Flow to display the calculator. Enter the data for the site and compare your dialog box to Figure 12–57. When your values are correct, choose the OK button in the Channel Flow Time Calculator to return to the main Tc calculator.

Channel:

Flow Area: 27 square feet

Wetted Perimeter: 28.2 feet

Slope: 0.005 ft/ft

Condition: 0.05 (Manning's n)

Length: 7300 feet

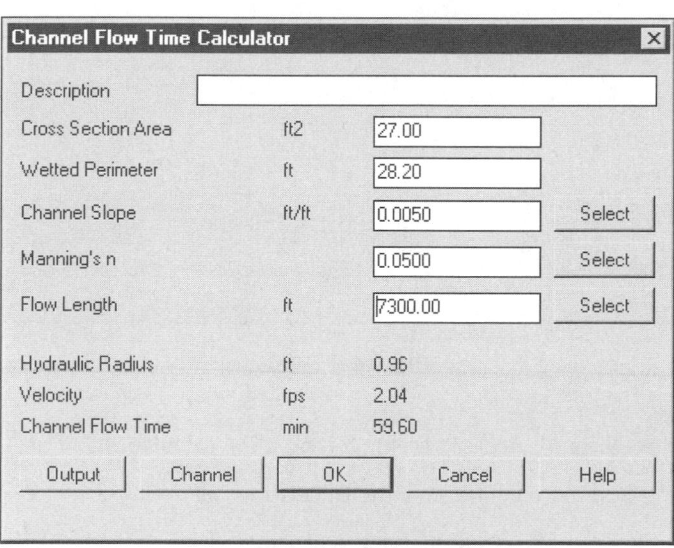

Figure 12–57

15. After you choose the OK button to exit the Time of Concentration dialog box, the time appears in the Time of Concentration cell of the Graphic Peak Discharge dialog box.

16. The next cell, Pond and Swamp Areas, contains the amount of area in swamp. This value modifies the discharge rate. The current site has no swamp area, so the number remains 0.00.

The last value is the amount of rain for the type of storm. The calculation is for a 25-year storm.

17. Choose the Select button and change the rainfall frequency to 25 for Dyer County. Exit the dialog box.

EXERCISE

The final results are values at the bottom of the dialog box (see Figure 12–58). The values are for Unit Peak Discharge, Initial abstraction, Ia/P, Runoff (Q), Pond and Swamp Adjustment values (Fp), and Peak Discharge (qp).

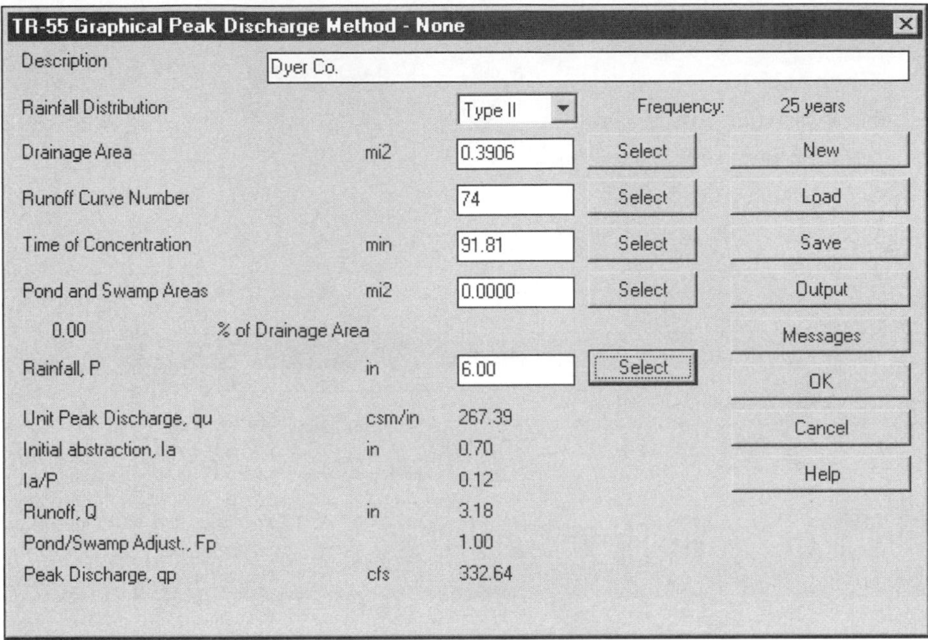

Figure 12–58

18. Select the Output button to create a report for calculations.

19. Select the Save button and save the calculations as *dyer.gpd*.

20. Exit the Graphical Peak Discharge calculator by selecting the OK button at the bottom of the dialog box.

TABULAR RUNOFF CALCULATOR

The Tabular method computes the characteristics of a single watershed that contains subareas with different characteristics and requires some of the collected water to travel from a subarea through other subareas to the discharge point. The results are both a hydrograph and runoff table. In essence, each subarea is its own graphic peak discharge calculation. The additional piece of information is the travel time through the other subareas to the discharge point.

The Tabular exercise has a seven-subarea watershed (see Figure 12–59). The discharge point is at the bottom of Area 7. The water concentrated in areas 1 and 2 travels through areas 3, 5, and 7 before reaching the discharge point. So, the calculation needs to consider collection and travel for the runoff of the site. In some ways the Tabular method is reminiscent of the Rational method. Each zone has its own line of data. The concentration and travel times produce output that has a time component. The output has an initial low volume discharge rising to a peak and then lowering. The time and amount of the peak are related to the time of concentration and travel times. A hydrograph is an *X* and *Y* axis plot representing the rise and fall of the runoff amounts.

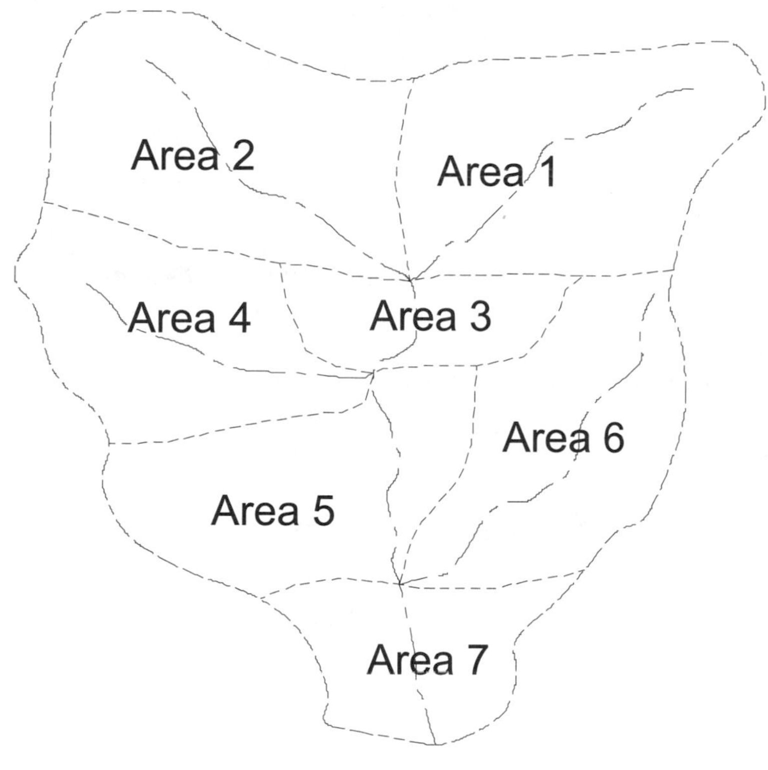

Figure 12–59

The data for the exercise is the following:

Fallswood, Dyer Co. — Predevelopment

Subarea	Area	T of C	Trvl T	Down Zones	24 Hr Rain	CN
1	0.3	1.50		3,5,7	6.0	65
2	0.2	1.25		3,5,7	6.0	70
3	0.1	0.50	0.50	5,7	6.0	75
4	0.25	0.75		5,7	6.0	70
5	0.2	1.50	1.25	7	6.0	75
6	0.4	1.50		7	6.0	70
7	0.2	1.25	0.75		6.0	75

1. Select Settings from the Hydrology menu and in the Units settings, set the time to hours. After setting the units, select the OK button to exit the units and settings dialog boxes.

2. Select TR-55 Tabular Method in the Runoff cascade of the Hydrology menu. The routine displays the Tabular calculator.

3. Enter the description for the job, Fallswood, Dyer Co. - Predevelopment.

4. Set the rainfall distribution to Type II.

5. Enter the data into the dialog box (see Figure 12–60). After entering the CN (curve number) for subarea 7, do not press ENTER, but select the Compute button to calculate the runoff numbers. Do NOT enter in the Sum Tt values. These values are calculated when you select the compute button in the next step. If you press the ENTER key after entering in the last row of information the Tabular button on the dialog box will be grayed out. If the Tabular button is grayed out you need to scroll down to row eight and delete this empty row. Once you delete the empty row, the Tabular key should be come active and you can view the tabular calculations.

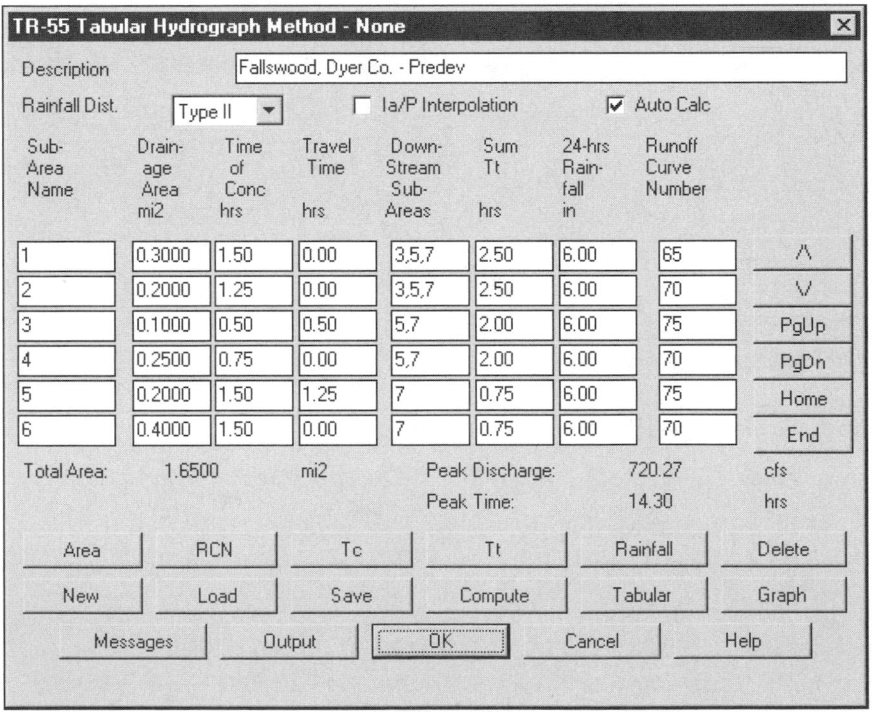

Figure 12–60

Note: If you press ENTER, the dialog box adds a new line of empty data to the calculations. The dialog box will not compute the values if there are any empty data line. You must delete the extra line of data to make the dialog box calculate the values.

Civil Design does not support the loading of a graphical peak discharge calculation into the Tabular calculator. If the subarea has a composite CN or complex time of concentration, you need to select the appropriate buttons at the bottom of the dialog box and enter the data. When you exit the dialog boxes, the values become a part of the cell.

6. Save the calculation as Fallpre.

7. Select the Tabular button to view the time/flow calculations. Exit the dialog box.

The results of the calculations display a table with the time and amount of flow passing through the system. The current runoff peaks at 14.3 hours (see Figure 12–61). Exit the dialog box.

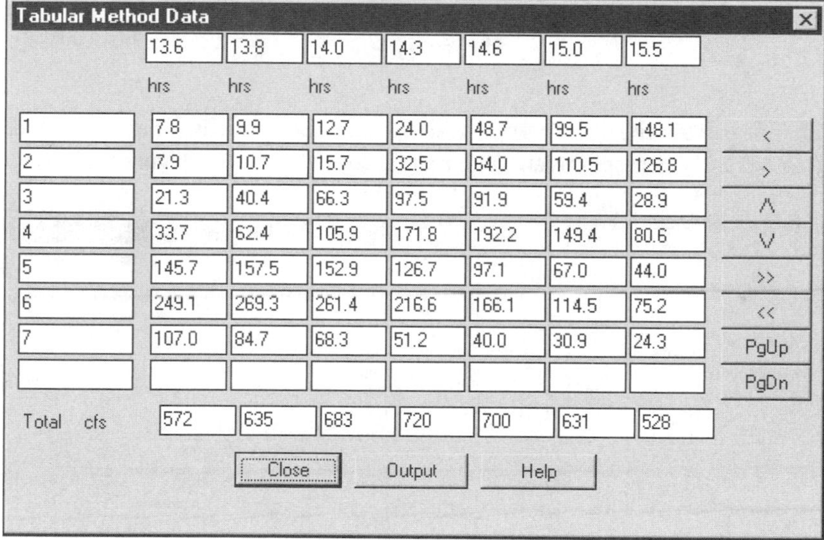

Tabular Method Data

	13.6 hrs	13.8 hrs	14.0 hrs	14.3 hrs	14.6 hrs	15.0 hrs	15.5 hrs
1	7.8	9.9	12.7	24.0	48.7	99.5	148.1
2	7.9	10.7	15.7	32.5	64.0	110.5	126.8
3	21.3	40.4	66.3	97.5	91.9	59.4	28.9
4	33.7	62.4	105.9	171.8	192.2	149.4	80.6
5	145.7	157.5	152.9	126.7	97.1	67.0	44.0
6	249.1	269.3	261.4	216.6	166.1	114.5	75.2
7	107.0	84.7	68.3	51.2	40.0	30.9	24.3
Total cfs	572	635	683	720	700	631	528

Figure 12–61

8. When you select the Graph button, the dialog box creates the hydrograph of the data (see Figure 12–62).

When developed, the site should produce a peak that is higher and sooner. The reason for this is the covering of absorbing surfaces and the restriction of the flow to specific areas within the site that concentrate and move the runoff faster through the system.

The post-development data is exactly the same as the predevelopment data except for the travel times. When you lower the travel times, the peak occurs sooner and at a higher level.

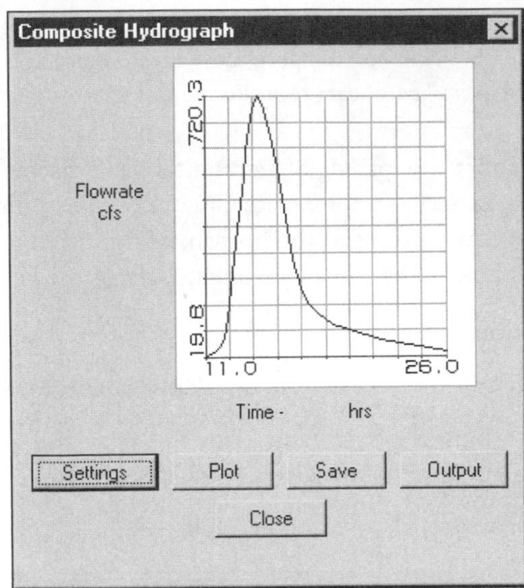

Figure 12–62

9. Change the description to post-development and the travel times for Areas 5 and 7 to 1.0 and 0.5 respectively. The results should look like Figure 12–63.

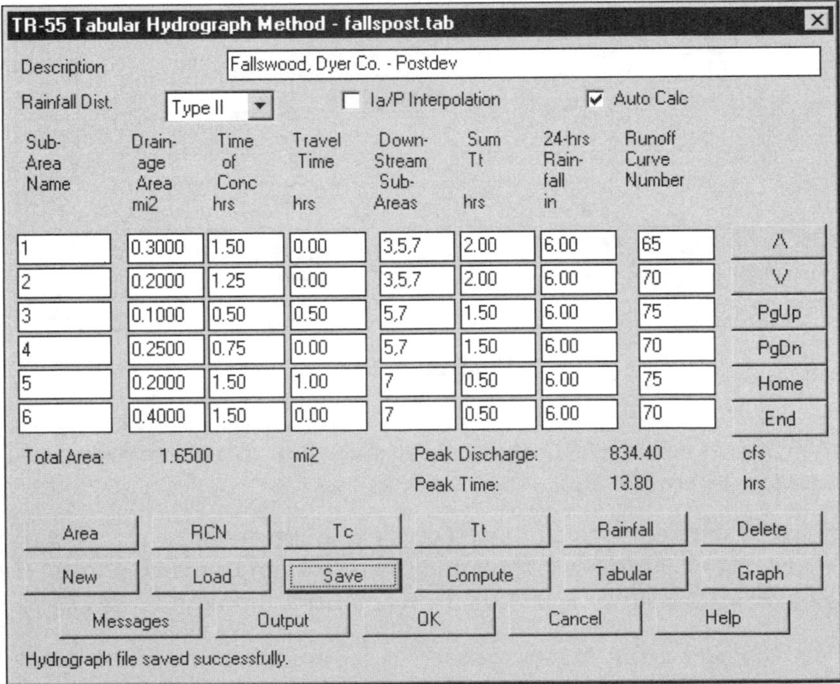

Figure 12–63

10. Save the data under the name Fallspost.

11. Select the Tabular button and view the results (see Figure 12–64). Exit the dialog box.

Tabular Method Data

	13.2 hrs	13.4 hrs	13.6 hrs	13.8 hrs	14.0 hrs	14.3 hrs	14.6 hrs
1	8.5	11.3	15.5	24.0	39.5	77.6	121.3
2	9.0	12.3	19.6	33.1	55.0	95.9	132.4
3	29.2	57.4	88.3	105.7	101.4	73.8	45.9
4	32.3	63.1	114.3	169.0	206.9	192.9	143.1
5	127.3	156.2	163.4	154.3	133.9	101.1	75.5
6	217.7	267.1	279.4	263.7	228.9	172.8	129.0
7	174.6	139.2	107.0	84.7	68.3	51.2	40.0
Total cfs	599	707	788	834	834	765	687

< > /\ \/ >> << PgUp PgDn

Close Output Help

Figure 12–64

12. Select the Graph button to view the hydrologic data (see Figure 12–65). Exit the dialog box.

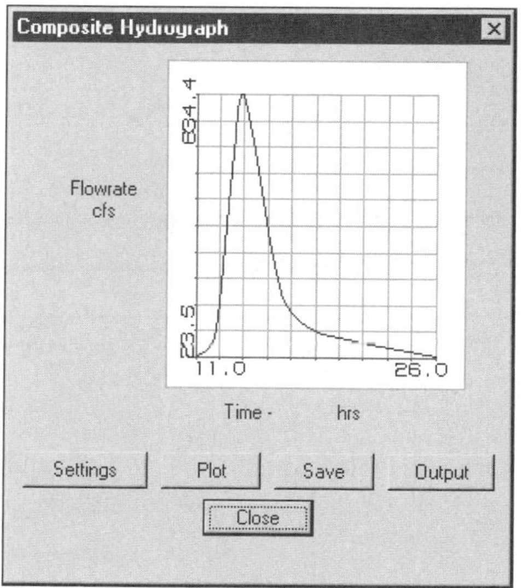

Figure 12–65

13. Select the OK button to exit the Plot and Tabular Calculation dialog boxes.

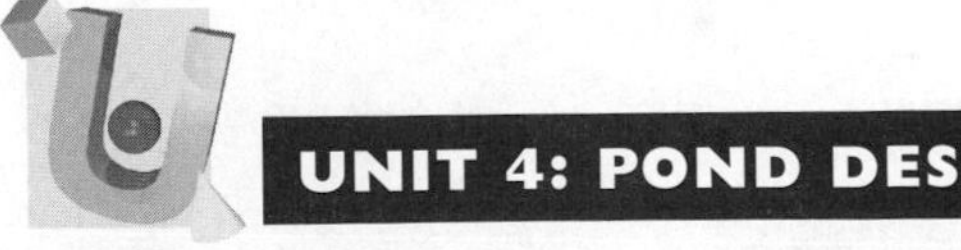

UNIT 4: POND DESIGN

POND SETTINGS

Ponds use the settings found in three dialog boxes. The settings of these boxes are found in the Drawing Settings routine for Civil Design or from the Pond Settings cascade of the Grading menu. Each routine displays a dialog box affecting specific pond elements, such as contours, breaklines, and benches.

CONTOURS

The top portion of the Pond Contours Settings dialog box sets the type of contours (see Figure 12–66). A pond contour can be relative to the elevation at the pond's parameter or an absolute elevation.

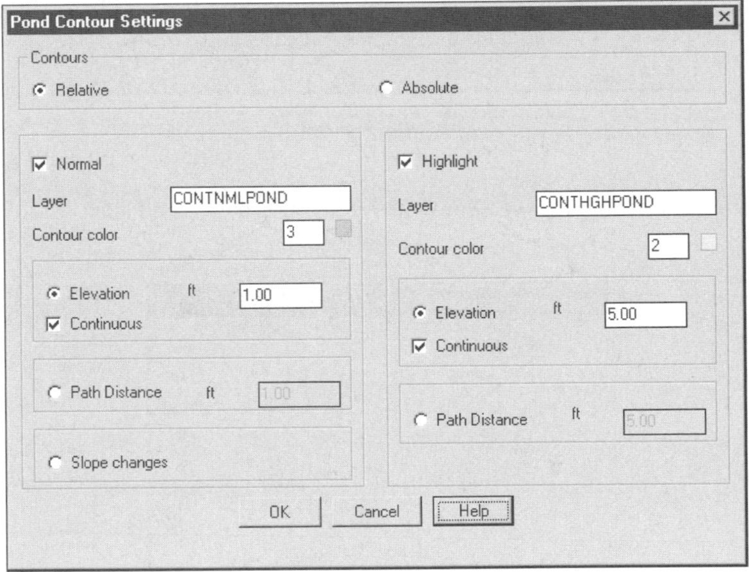

Figure 12–66

Relative

The Relative toggle indicates that the next contour is a distance from the elevation of the rim contour. If the elevation of the rim is 732.75 and the normal interval is 2 feet, the next contour is 730.75.

Absolute

The Absolute toggle indicates that the next contour is at the next interval below the rim contour. If the elevation of the rim is 732.75 and the normal interval is 2 feet, the next contour is 732.00.

Normal and Highlight

This setting is similar to the one in the contour routines in the Terrain menu. This dialog box group toggles the layers on and off and sets names, interval, and color.

Path Distance

This toggle creates contours at intervals along a slope template. The interval is independent of the normal and highlight contour settings. When this toggle is on, the normal and highlight contour settings are turned off.

Slope Changes

This toggle creates contours that represent changes in slopes of a pond template. When this toggle is on, it turns off the normal and highlight contour settings and the path distance.

SLOPE CONTROL LINES

The second pond settings dialog box, Pond Slope Control Settings, affects the creation of control (breaklines) for a pond (see Figure 12–67). The top portion of the Pond Slope Control Settings dialog box sets the layer name and color for pond slope control. The bottom portion of the dialog box controls how the routine creates the breakline. Normal breaklines create vertices along the breakline where they cross the normal contour lines of a pond. Highlighted breaklines create vertices along the breakline where they cross the major contour lines of a pond. The Slope Changes setting creates breakline data where the slope changes on the pond banks. The Draw to Bottom Polyline toggle forces the breakline to connect to the lowest polyline defining the pond.

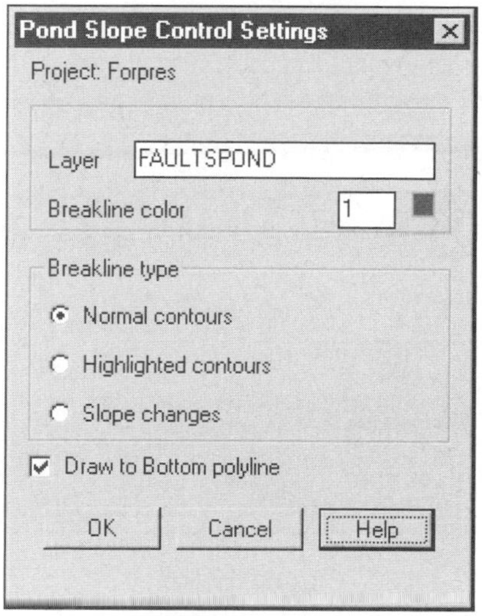

Figure 12–67

BENCHING SETTINGS

The Benching Settings dialog box sets a standard definition of a pond bench (see Figure 12–68). There are three controlling values to a bench. The first is the slope of the bench. The slope is always a grade value. The last two settings control the depth and width of the bench in the pond.

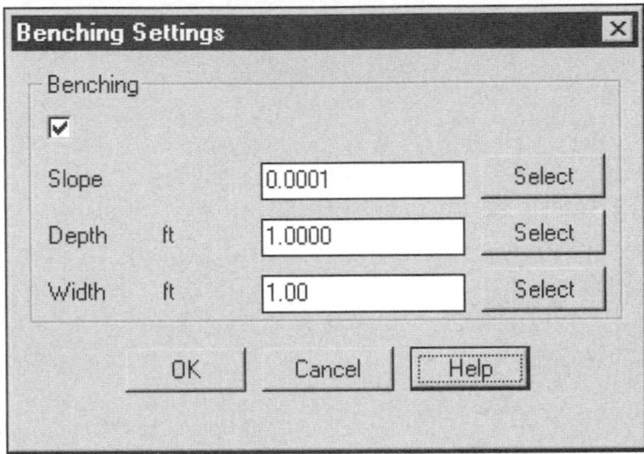

Figure 12–68

CREATING PONDS

Civil Design provides tools to design and define ponds in the Grading menu (see Figure 12–69). You can define a pond from existing contours or polylines in the drawing (the Define Pond cascade) or have a pond design be a result of a parametric pond routine (the Pond Perimeter cascade). There does not have to be a surface for the pond design process to work. All that is needed is a rim contour of a pond, to which design parameters are applied, or a selection of contours representing the elevations of the pond.

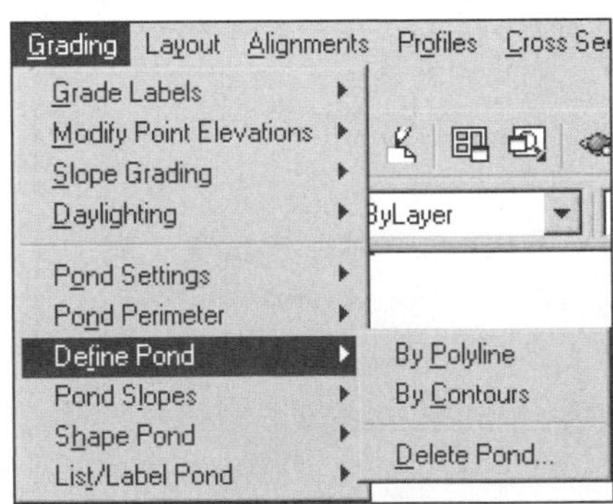

Figure 12–69

The parametric process includes using a predefined shape of pond or a rim contour (see Figure 12–70). The shape and depth of the pond are a result of applying a value to the shape or rim.

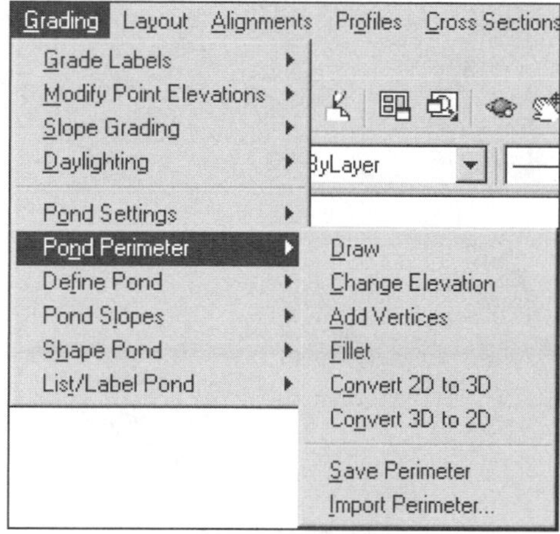

Figure 12–70

A pond definition may come from the application of one or more templates to the pond rim. Other tools, also in the Pond Slopes cascade, use a single slope or multiple slopes to determine the pond shape. The last method requires a volume to calculate the depth of the pond (see Figure 12–71).

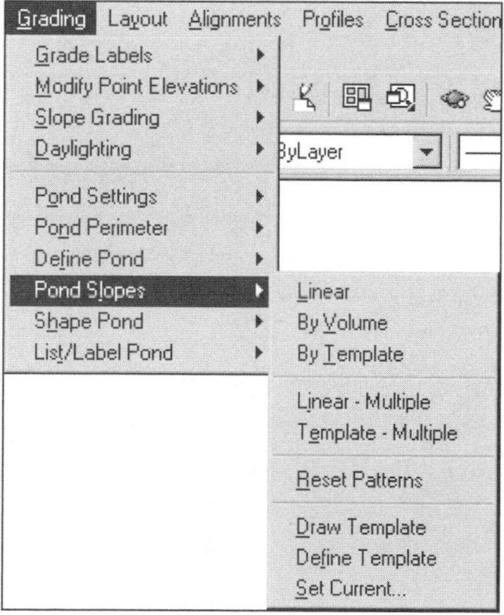

Figure 12–71

The Shape Pond cascade creates the contours representing the pond (see Figure 12–72). Routines in the cascade create contours, slope control lines (break lines), the bottom polyline, or all of the components.

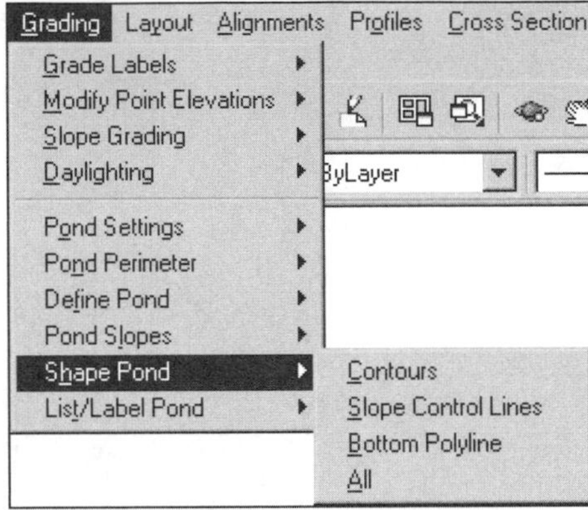

Figure 12–72

ANALYZING PONDS

Each pond design has characteristics that are important for the control and discharge of runoff. After you design a pond, these characteristics need to be known. The listing routines of the List/Label Pond cascade of the Grading menu report the properties, contour area, contour elevation, contour perimeter, and slope/grade between contours (see Figure 12–73).

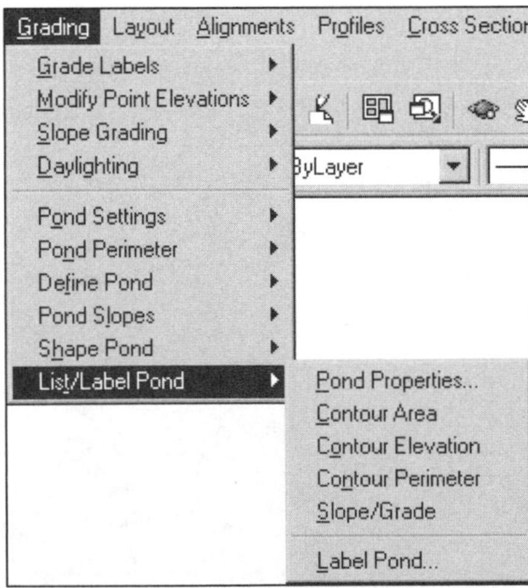

Figure 12–73

The Pond Output cascade of the Hydrology menu contains routines that export the characteristics of a pond to a file (see Figure 12–74). You have control over the order, delimiter, and precision of the numbers. The Output Editor by Pond routine reports the values from a defined pond. The Output Editor By Contours routine allows for the selection of a group of closed contours to define a pond for reporting. The final report is the Stage-Storage Curve report. This routine works with defined ponds and reports the average or conic volumes. After reporting on the pond, the routine displays a curve rating of the pond.

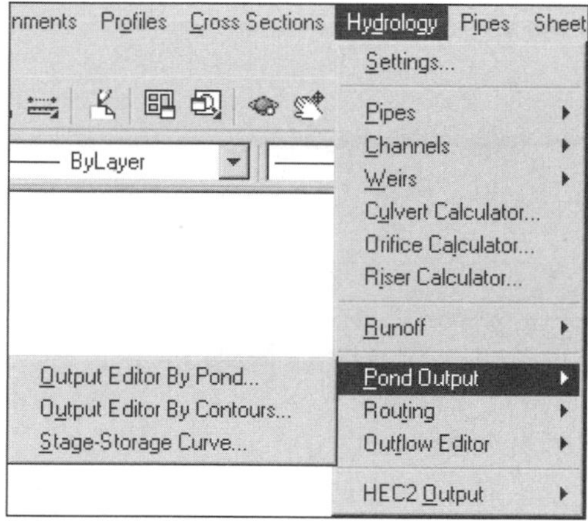

Figure 12–74

ROUTING EDITORS

The runoff calculations predict the flow into a discharge point. The discharge point can be a culvert, pipe, or pond. If the discharge point is a pond, the designer needs to know the behavior of the pond while accommodating the inflow. Generally, the pond is not a receptacle for collecting, but a way in which the designer can delay the discharge of collected water. Therefore, the final component of pond analysis is the discharging of the runoff. The runoff is discharged through a structure; for example, channel, culvert, pipe, orifice, or weir.

DETENTION BASIN STORAGE

The Detention Basin Storage routine calculates the TR-55 storage for a detention basin (see Figure 12–75). The routine is in the Routing cascade of the Hydrology menu. The calculator requires three main components: inflow, storage, and outflow values. The inflow values are results from a graphical peak discharge or tabular calculations, or a hydrograph. A defined pond supplies the storage capacity. The outflow data is an result a graphical peak discharge, tabular, or hydrograph output file, or an outflow structure. After you choose the Select button for Outflow, the routine presents the Outflow editor.

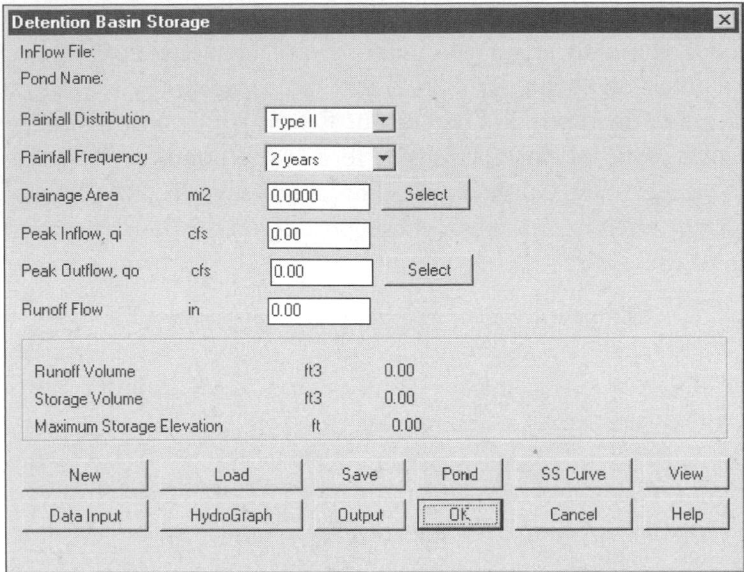

Figure 12–75

STORAGE INDICATION METHOD

The Storage Indication method loads four hydrographs: the pre- and post-development inflow, stage-storage, and stage discharge (see Figure 12–76). The routine is in the Routing cascade of the Hydrology menu. The calculator determines the amount and time of peak flow through the design. The purpose of the pond and outflow design is to delay and lessen the runoff.

Storage Indication Method - None

Description

Time Increment hrs 0.10

Time	Inflow	I1+I2/2	H1	S-O/2T	S+O/2T	H	OutFlow
hrs	cfs	ft3	ft	ft3	ft3	ft	cfs

^ V PgUp PgDn Home End

Peak Time: 0.00 hrs Peak Flow: 0.00 cfs

Input Curves | Storage Character | Routed Hydrograph | Multi-Hydrographs

New | Load | Save | Output | OK | Cancel | Help

Figure 12–76

THE OUTFLOW EDITORS

The Outflow Editor cascade of the Hydrology menu contains three routines that calculate the behavior of a pond with inflow and outflow data (See Figure 12-77).

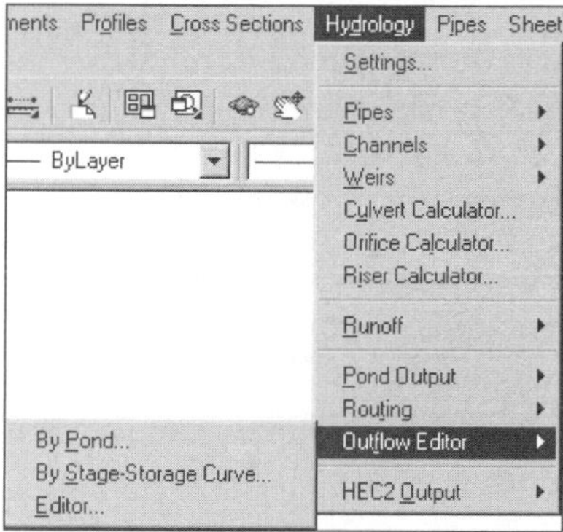

Figure 12–77

BY POND

The By Pond routine attaches runoff and structure calculations to a pond (see Figure 12–78). The top section of the dialog box displays the structures attached to the pond. The structures can be inflow or outflow devices. First you select the type of structure you want to add (inflow or outflow). Next, when you select the Add Structure button, the routine presents a list of structure types from which to choose. After you add all of the structures, the calculator reports the behavior of the pond and calculates the ratings for the pond. The ratings are composite and stage discharge.

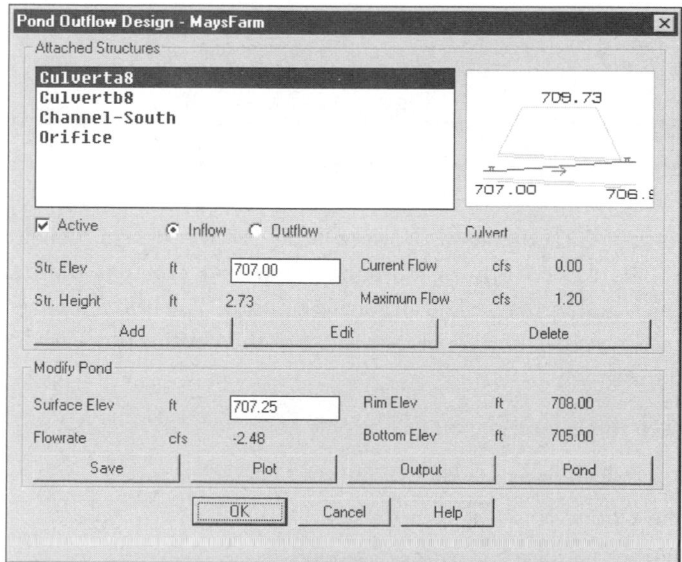

Figure 12–78

STAGE-STORAGE CURVE

The Stage-Storage Curve routine is similar to that of the Outflow editor; however, the routine works with Stage-Storage curves rather than defined ponds.

EDITOR

The Editor routine does not require either a defined pond or a stage-storage curve to function. The editor allows you to create a inflow and outflow for an amount of detention.

ANNOTATING PONDS

The Label Pond routine places text in the drawing representing the properties of a pond. The routine is in the List/Label Pond cascade of the Grading menu (see Figure 12–79). You can control the labeling of several pond properties. The routine labels the following properties: average area volume, conic volume, area, perimeter, pond elevation, and maximum depth.

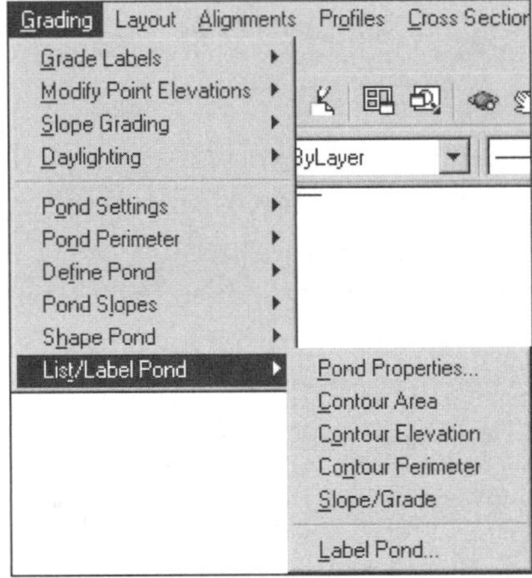

Figure 12–79

Exercise

The pond area of the Grading menu provides tools for the design, definition, and use of detention ponds. The top portion of the pond menu area contains the design tools. The middle portion has routines that define and analyze ponds. The bottom portion has routines to list and annotate ponds. The Hydrology menu has the routines to evaluate the effectiveness of the pond design.

After you complete this exercise, you will be able to:

- Use the pond design tools
- Define a pond from contours
- Evaluate the storage capacity of a pond
- Develop an inflow/outflow calculation

POND SETTINGS

1. Continuing in the *Forpres* drawing, select By Contours routine from the Pond Settings cascade of the Grading menu (See Figure 12–66). Set the following values and select the OK button to exit the dialog box.

   Relative: ON

Normal: ON	Highlight: ON
Elevations: 1.0	Elevations: 5.0

2. Select the Slope Control Lines routine in the Pond Settings cascade of the Grading menu to review its settings. After reviewing the settings, select the OK button to close the dialog box.
3. Select the Benching Settings routine in the Pond Settings cascade of the Grading menu. After reviewing the settings, select the OK button to close the dialog box.
4. Freeze the Cont-NML and Cont-HGH layers in the drawing.
5. While in the Layer dialog box, turn on the Pondperimeter layer. After turning on the Pondperimeter layer, select the OK button to close the layer dialog box.

Pond Creation

6. Use the By Volume routine in the Pond Slopes cascade of the Grading menu. This routine designs a pond by using a perimeter, slope, and volume to create a resulting pond. After you select the pond outline, set the volume to 15,000 cubic feet, and use a slope of 4:1. Name the pond MaysFarm Trial 1.
7. Use the UNDO command to remove the pond design.

EXERCISE

Pond by Template

1. Use the Draw Template routine from the Pond Slopes cascade of the Grading menu to draw a pond template. Draw the template to the right of the pond. A pond template is a cross section of the pond. After you select the starting point of the template, the first segment is relative to this point and travels 5 feet in X and –4 feet in Y. The second segment is 3 feet in X and –0.5 feet in Y from the last point. The last segment travels 6 feet in X and –4 feet in Y. The following is the routine sequence. The resulting polyline looks like the template in Figure 12–80.

```
Command: _pline
Specify start point:
Current line-width is 0.00
Specify next point or [Arc/Close/Halfwidth/Length/Undo/Width]: @5,-4
Specify next point or [Arc/Close/Halfwidth/Length/Undo/Width]: @3,-.05
Specify next point or [Arc/Close/Halfwidth/Length/Undo/Width]: @6,-4
Specify next point or [Arc/Close/Halfwidth/Length/Undo/Width]:
```

2. Run the Define Template routine of the Pond Slopes cascade of the Grading menu. Name the template Mays. Select the upper left end of the polyline when the routine prompts you to select the template. Answer No to defining another template. After defining the template, erase the polyline.

3. Apply a single template to the pond rim. Use the By Template routine in the Pond Slopes cascade in the Grading menu. First select the Mays template and select the OK button to close the dialog box. Next select the pond perimeter and when the routine asks for a pond name, enter MaysFarm Trial 1. Select the OK button to close the dialog box. Select the center of the pond as the offset side and apply the template to all vertices. Finally, answer Yes to the Shape Pond - All? prompt. The routine then creates the pond based upon the perimeter and the template.

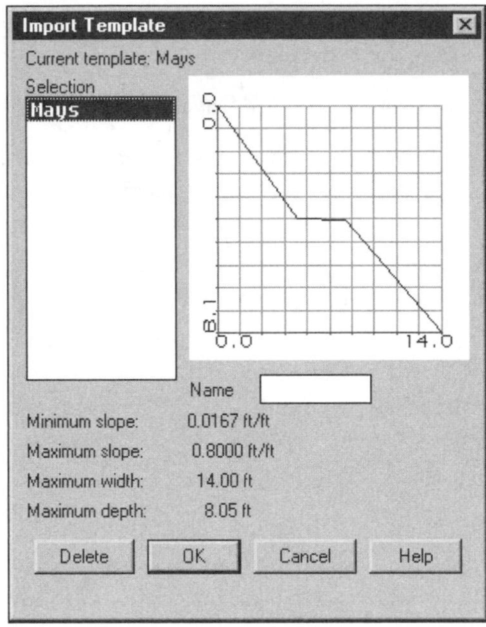

Figure 12–80

The routine proceeds as follows:

```
Command:
Current Template: Mays
Select polyline: (select the pond perimeter)
Select offset side: (select a point in the interior of the pond)
Attach template <Mays> to all vertices? (Yes/No) <Yes>:
Template <Mays> attached to all <52> vertices successfully.
Shape Pond - All? (Yes/No) <Yes>: y
```

```
Retrieving patterns...
Creating bottom polyline...
Processing slope control lines...
```

The routine creates the pond, faults, and contours.

4. Use the Pond Properties routine of List/Label Pond in the Grading menu to view the values of the pond.

Pond from Contour

Rather than creating a pond with a specific slope, volume, or with a template, Civil Design allows you to create pond definitions from existing contours. The contours 705 to 708, just east of the parking lot (in the center of the drawing on layers Cont-NML and Cont-HGH), constitute a pond. After you select the contours, the routine creates a pond definition.

1. Erase the current pond on the screen, including the pond perimeter line.
2. Open the Layer dialog box, turn off the Pondperimeter layer, and turn on the Cont-NML and Cont-HGH layers. Select the OK button to exit the Layer dialog box.
3. Run the By Contours routine in the Define Pond cascade of the Grading menu. Select the closed polylines in the center of the drawing. Use Figure 12–81 as a guide in selecting the contours. Name the pond MaysFarm.

Figure 12–81

4. Use the Pond Properties routine of the List/Label Pond cascade of the Grading menu to view the numbers of the pond (see Figure 12–82).
5. Save the drawing.

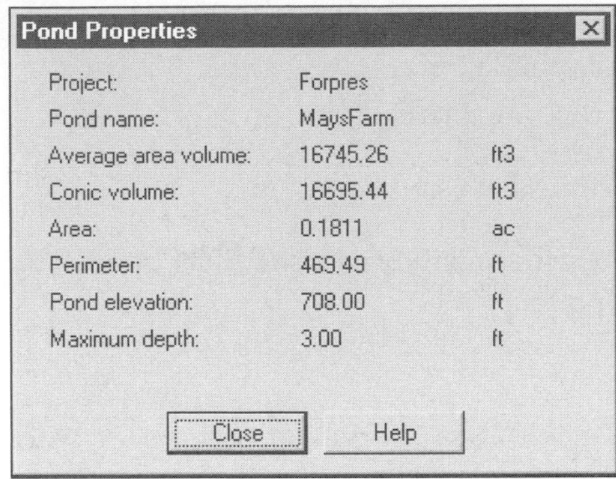

Figure 12–82

UNIT 5: POND OUTFLOW

The evaluation of a pond is a combination of structures, runoff calculations, and pond definitions. The By Pond routine of the Outflow Editor cascade of the Hydrology menu allows you to add existing or define new structure calculations to the inflow and outflow of a pond.

Exercise

The first part of this exercise uses the Culvert Calculator to calculate the characteristics of two inlets at the edge of the forest preserve parking lot and the channel calculator for the small channel draining the entranceway into the pond.

The second part of the exercise uses the Orifice Calculator to define the outflow device. Finally, in the Outflow Editor, you will join all of the elements into an evaluation of the design.

After you complete this exercise, you will be able to:

- Use the Culvert Calculator
- Evaluate the design
- View P-curve and over the top reports
- Use the Outflow Editor for pond evaluation

CULVERT CALCULATOR

Culvert A

1. Continuing in the *Forpres* drawing, open the Rational calculator of the Runoff cascade of the Hydrology menu and review the flowrate for Areas B and C. Note down the flowrates, because they will be a part of the culvert calculations.

2. Run the Culvert Calculator routine of the Hydrology menu.

3. First you need to set the values in the Culvert Settings dialog box of the Culvert Calculator. You call the dialog box by selecting the Settings button at the right side of the main dialog box. The following are the data to enter in the Settings dialog box (see Figure 12–83).

   Control: Optimum

   Outlet Control: 0.5 *(select Corrugated Metal Pipe with headwall)*

   Flow Range:

   Minimum: 0.90

   Maximum: 1.20

   Increment: 0.01

   Chart and Scale: Chart 2 – Under Corrugated Metal Pipe Culvert select

   Headwall

   Barrels: 1

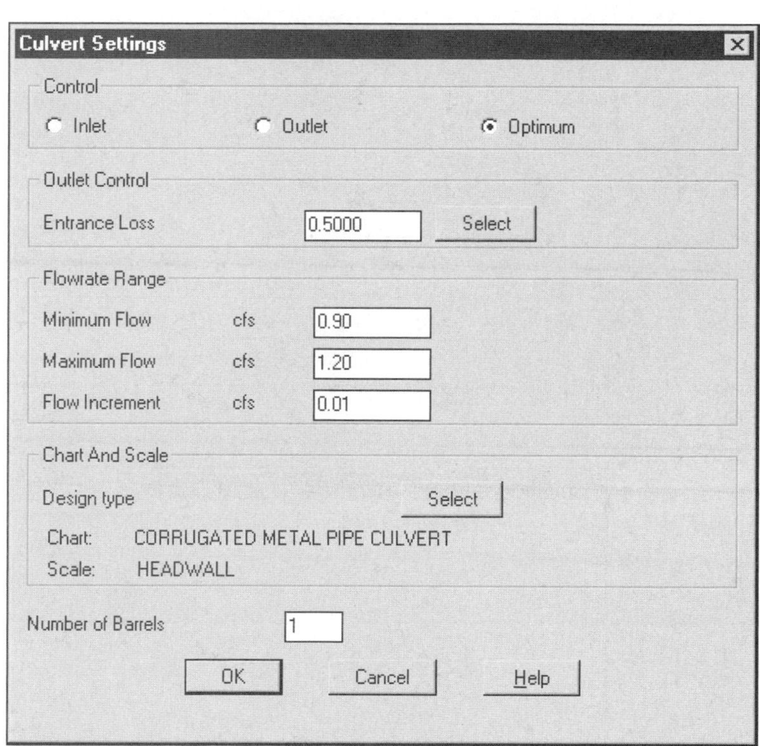

Figure 12–83

4. Select the OK button to exit the Culvert Settings dialog box and to return to the Culvert Design dialog box.

EXERCISE

5. Set the following values in the Culvert Design dialog box (see Figure 12–84). When entering in the length of the culvert, enter in 18 and do **not** include the foot mark. The tailwater and diameter of the barrel are 10 inches, but the units are in feet. When you enter in the values, you need to enter in 10/12 or 0.83 for the feet equivalent of 10 inches.

```
Tailwater: 10 Inches  (10/12 or 0.8333)
Length: 18 Feet
Diameter: 10 Inches   (10/12 or 0.8333)
Flow: 1.06
Manning's n: CMP (Corrugated Metal Pipe)
Elevation of Road: 709.33
Elevation of 1st Invert: 706.60
Elevation of 2nd Invert: 705.50
```

EXERCISE

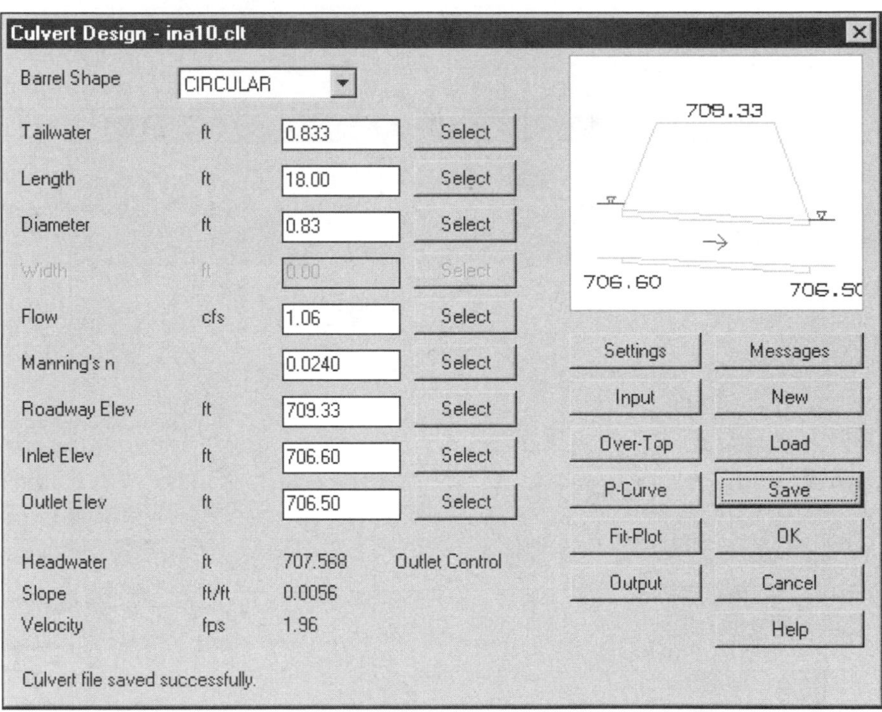

Figure 12–84

6. After entering in the values, select the P-Curve button to View the performance curve of the culvert. The Performance Curve Display dialog box displays.

7. From the Performance Curve Display dialog box select the Output button to view the report on the performance of the culvert. Close notepad without saving

the report. This returns you to the Performance Curve Display dialog box. Select the Close button to dismiss the dialog box and to return to the Culvert Design dialog box.

8. Select the Output button at the bottom right of the Culvert Design dialog box to create a report on the performance of the culvert. Before you view the report the routine displays a Culvert Output dialog box. In this dialog box you set the type of report (comprehensive or summary) and if you want to include any error messages (see Figure 12–85). Set the report to comprehensive and include error messages and click on the OK button to generate the report. Look for the flowrate of 1.06 and find its related outflow rate. The outflow rate should be around 2 fps.

9. Save the calculations as *ina10*.

The 10-inch pipe produced a 2 fps outflow. We would like to have an fps around 3 fps. Try using an 8-inch pipe and view the change in outflow fps.

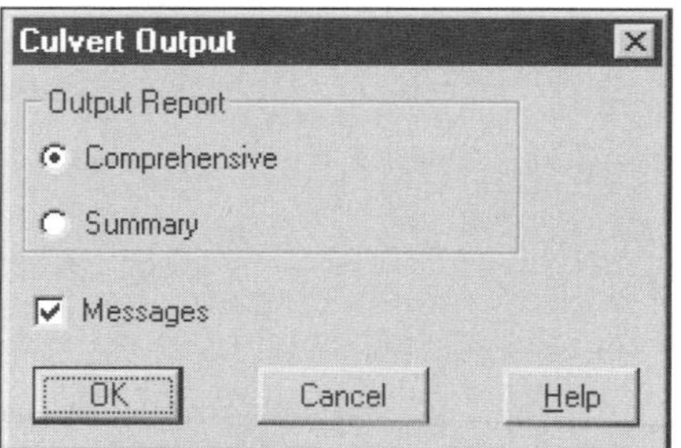

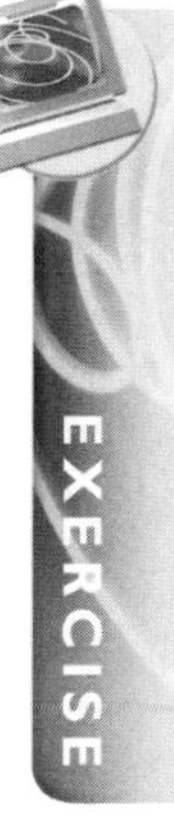

Figure 12–85

10. Change the Tailwater value to 8/12 (inches) and the pipe size to 8 inches. Make sure to press ENTER after you enter each value to force the recalculation of the parameters. View the results by selecting the Output button. Look for the flow rate of 1.06 and its associated outflow fps. The value is around 3 fps. Save the current calculations as *ina8*. Use Figure 12–86 as a reference.

```
Tailwater: 8 Inches  (8/12 or 0.6666)
Length: 18 Feet
Diameter: 8 Inches   (8/12 or 0.6666)
Flow: 1.06
Manning's n: CMP (Corrugated Metal Pipe)
Elevation of Road: 709.33
Elevation of 1st Invert: 706.60
Elevation of 2nd Invert: 705.50
```

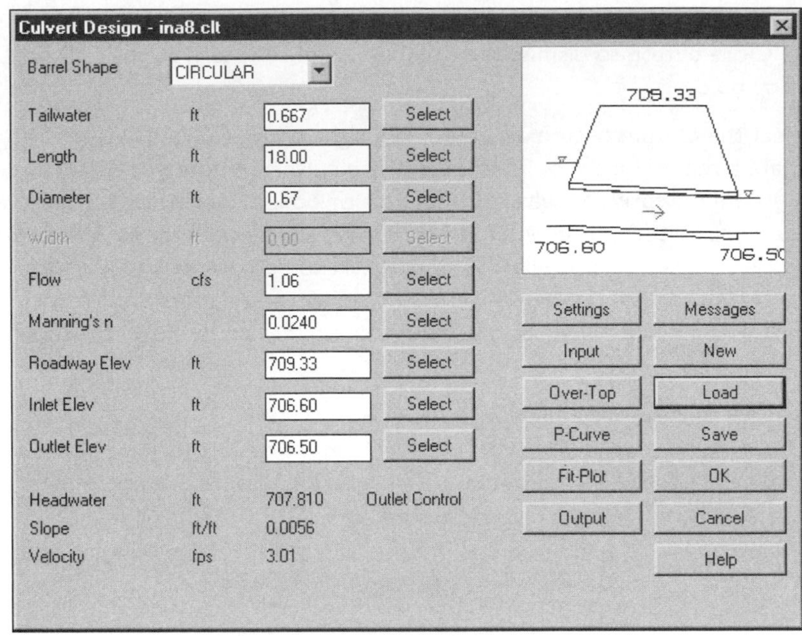

Figure 12–86

Culvert B

The values for the second inlet in the parking lot take the same values as currently present in the dialog box except for the flowrate and elevations.

1. In the Culvert Design dialog box, change the following values at the bottom center of the box:

   Flow: 1.01

   Manning's n: CMP (Corrugated Metal Pipe)

   Road Elevation: 708.26

   Elevation of 1st Invert: 706.10

   Elevation of 2nd Invert: 706.00

2. Change the Tailwater value to 10/12 (inches), the pipe size to 10/12 (inches), and leave the length of the pipe at 18 feet. Use Figure 12–87 as a reference. View the p-curve and the output results. In the output, look for the flowrate of 1.01 and its associated outflow fps. The rate is around 2 fps. The goal is to have a flow rate around 3 fps. Save the calculations as inb10.

```
Tailwater: 10 Inches  (10/12 or 0.8333)
Length: 18 Feet
Diameter: 10 Inches   (10/12 or 0.8333)
```

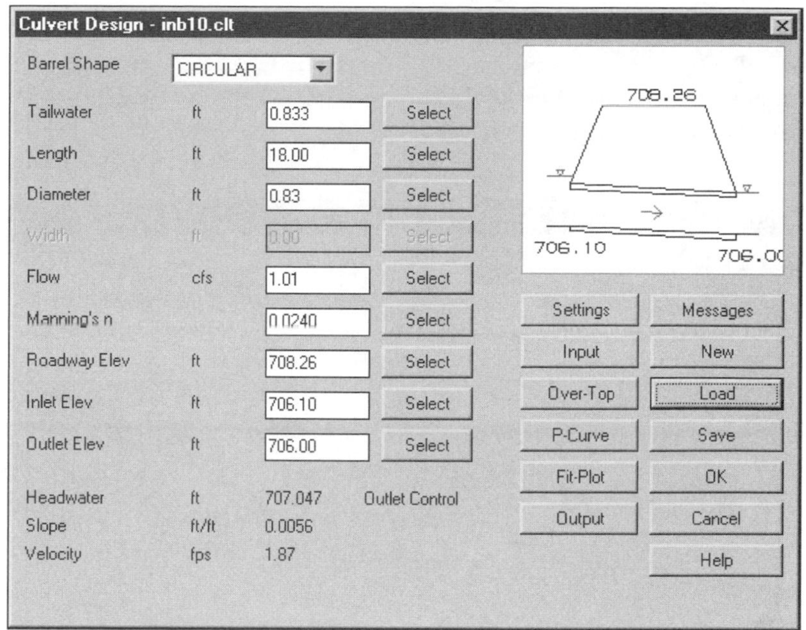

Figure 12–87

3. Change the Tailwater value to 8/12 (inches) and the pipe size to 8/10 (inches). Use Figure 12–88 as a reference. View the p-curve and the output results. In the output, look for the flowrate of 1.01 and its associated outflow fps. The rate is around 3 fps. The goal is to have a flow rate around 3 fps. View the results. Save the calculations as inb8.

```
Tailwater: 8 Inches  (8/12 or 0.6666)
Length: 18 Feet
Diameter: 8 Inches   (8/12 or 0.6666)
```

4. Select the OK button to exit the Culvert Design calculator.

The next structure is the channel carrying the water from the south to the north into the pond. The channel has a 2-feet base with 3:1 slopes. The channel is 1 foot deep.

5. Start the Trapezoidal calculator from the Channels cascade.

6. Enter the following values for the calculator (see Figure 12–89).

   Solve for: Depth of Flow
   Flow Rate: 0.4
   Slope: 0.03
   Manning's n: 0.0200 (Earthen channel)
   Depth of Flow: (Do not enter a value)

Height: 12"
Bottom Width: 24"
Left Slope: 1/3 (ft/ft) (0.3333)
Right Slope: 1/3 (ft/ft) (0.3333)

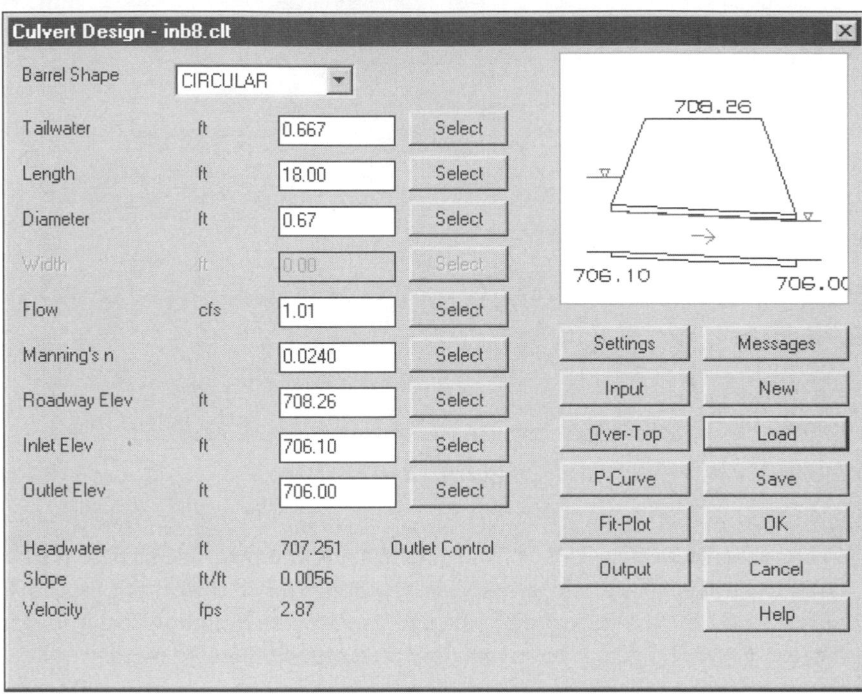

Figure 12–88

7. View the Discharge Diagram by selecting the plot button. Select the Close button to return to the Trapezoid Channel Calculator.
8. To View the rating of the channel first select the Rating button and then select the view button to view the rating curve. After viewing the rating curve, select the Close button on the Rating Curve Display and Rating Table Setup dialog boxes to return to the Trapezoid Channel Calculator.
9. Select Output button, next select the text format button, then select Yes to output critical computations to view the report. Select save from the File menu of Notepad to Saveas to save the information to a file. Name the file *schnl.txt*. After saving the file, select Exit from the File menu of Notepad to exit Notepad and to return to the Trapezoid calculator.
10. Select OK to exit the Trapezoid Channel Calculator.

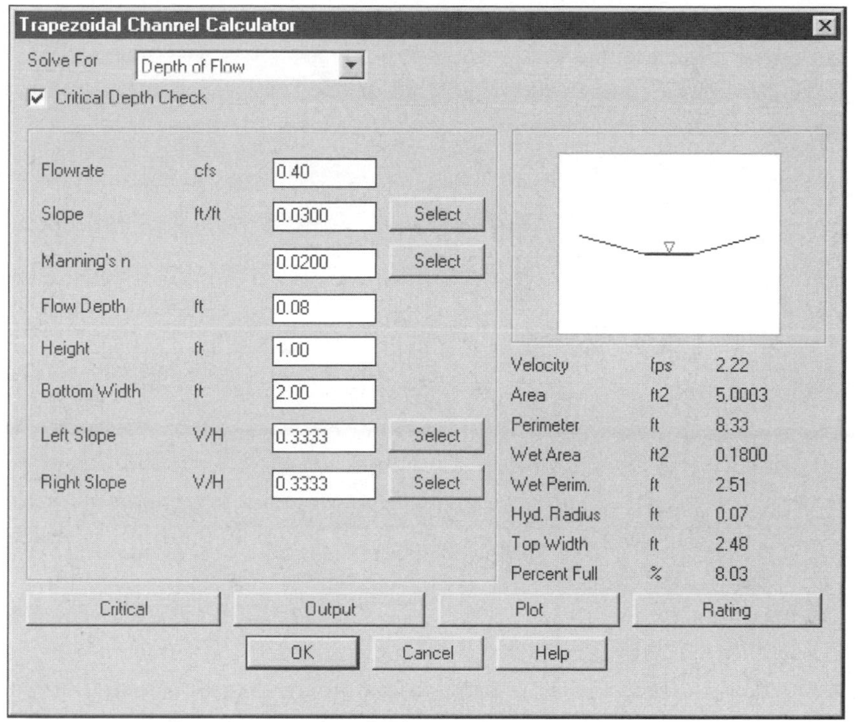

Figure 12–89

Orifice Calculations

The last structure to develop is the outlet orifice for the pond. The size of the pipe is arbitrarily set to 4 inches in diameter. A larger size would cause problems with debris clogging.

1. Select Orifice Calculator from the Hydrology menu (see Figure 12–90). Set the following values for the orifice.

   Solve for: Flowrate

   Flowrate: (Do not enter a value)

   Coefficient: 0.61 *(sharp edge)*

   Diameter: 4" (4/12)

   Headwater: 2.5

   Tailwater: 0.0

   The flowrate for 2.5 feet of headwater is 0.66.

2. Try the headwater values of 1.5 and 0.5. Note down the flow rates for all both headwater values.

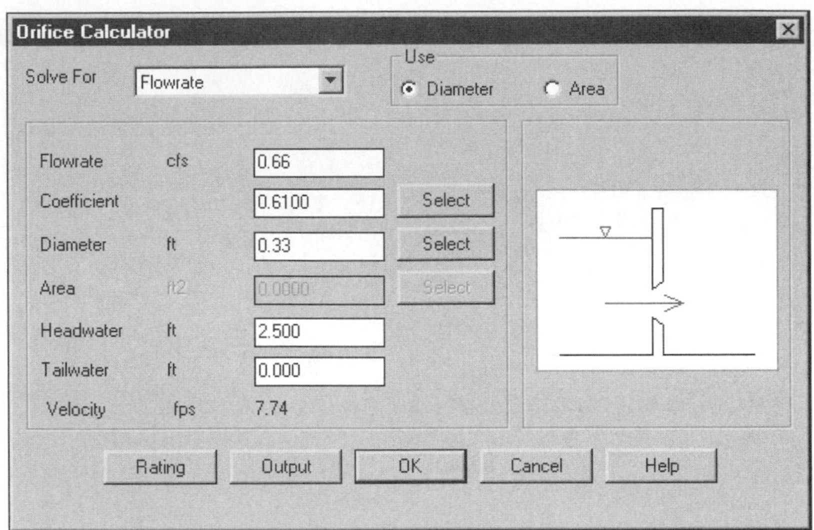

Figure 12–90

DISCHARGING FROM A POND

The intent is to draw the runoff from the parking areas into the detention pond. This redirection of water is by two culverts (inlets A and B) and a channel. The delay of discharging the water is the design of a pond. To control the amount of water outflowing into the ditch at the north end of the pond, the design uses an orifice. All of these elements have been an earlier part of this exercise. Now it is time to put the values together into an analysis of the design.

To put together the pieces and observe the results, run the By Pond routine of the Outflow Editor cascade of the Hydrology menu. The editor ties together the pond, inflow, and outflow structures to review the effects.

1. Select the By Pond routine of the Outflow Editor cascade of the Hydrology menu (see Figure 12–77). The routine presents a dialog box asking you to select a pond. Choose the MaysFarm pond. After you select the pond, the routine displays the Pond Outflow Design dialog box (see Figure 12–91).

The Pond Outflow Design dialog box contains the name of the Attached Structures (upper left), a schematic diagram (upper right), the out- or inflow values (upper center), and the pond elevations in the bottom portion of the dialog box. The radio buttons in the center of the dialog box identify the structure on the list as being an out- or inflow structure. Watch these buttons carefully because they have a tendency to change.

Note: First set the correct type of device (inflow/outflow, add a structure, check the button again to make sure it hasn't changed the type, and then save the values. It is not recommended to add several structures and then save.

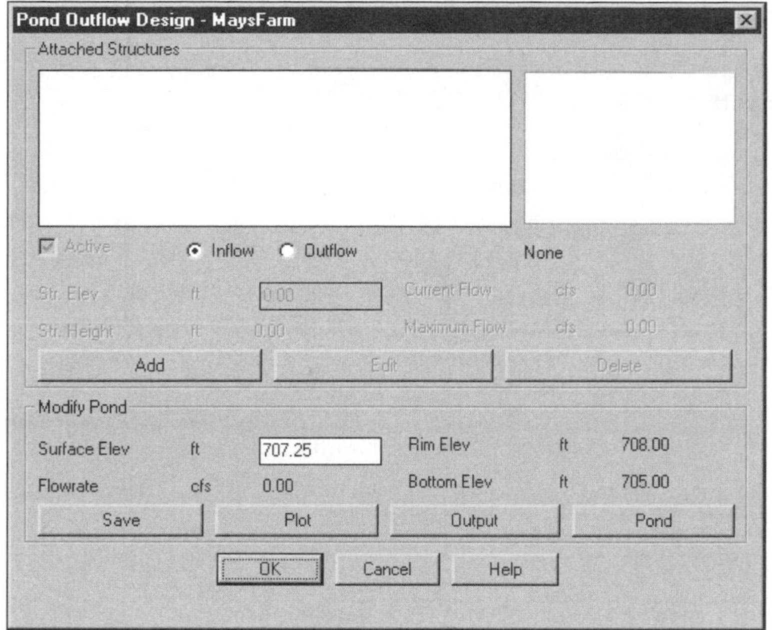

Figure 12–91

2. Select the Add button to add a structure. The routine displays a list, then choose Orifice. The orifice calculation from the previous exercise should be in the calculator (see Figure 12–90).
3. Choose the OK button for the Orifice calculator, name the structure Orifice, and select the OK button to exit the naming dialog box.
4. When in the Outflow calculator, make sure Orifice is highlighted, select the Outflow button, and then select the Save button.

The water elevation in the pond (surface elevation) is 708.00. If you change this elevation, the design evaluates the change in the outflow amounts of the orifice. When the pond is at 706.0, the outflow is around 0.30 (just above the Delete button). When the level is 707.5, the outflow is 0.68 cfs.

5. Change the pond level to different amounts.

The next step is to add the inlets. Remember, add the structures, one at a time. After adding each structure, be sure to set its type to inflow and then save.

6. Add inlet A to the Outflow calculator. Select Add, select Culvert from the list, and load the Culvert Calculator with the *ina8.clt* file.
7. Exit the Culvert Calculator and name the structure, Culverta8. Exit the naming box. When you're back in the Outflow calculator, make sure Culverta8 is highlighted, select the Inflow button, and then select the Save button.

The calculator shows the inflow of the inlet as a negative value.

8. Highlight Orifice and change the pond elevation (water level in the pond); view the outflow values.

The outflow values do not change because of the new structure. The outflow is the same amount, because only so much water can exit through the CMP. The additional water starts to collect in the detention pond before exiting from the pond. In this way the design delays the runoff and controls its speed into the surrounding territory. As with the pre- and post-development hydrographs of TR-55, the purpose of a detention pond is to modify the new peak charge. The modification to the discharge is to delay the peak and lessen its velocity. The pond and the orifice together accomplish these two tasks.

9. Add inlet B to the Outflow calculator. Select Add, select Culvert from the list, and load the Culvert Design Calculator with the *inb8.clt* file.
10. Exit the Culvert Calculator and name the structure Culvertb8. Exit the naming box. When you are back in the Outflow calculator, make sure Culvertb8 is highlighted, select the Inflow button, and then select the Save button.
11. Select Orifice and change the levels of the pond surface (see Figure 12–92).

When setting the pond level to 706.5 or higher, the calculator issues a warning. The warning is that the flowrate for inletb is exceeded. This is because the water now covers the tailwater side of the culvert. When the water is at 707.5 feet, it is ready to overtop the pond and almost fill the culvert (inlet).

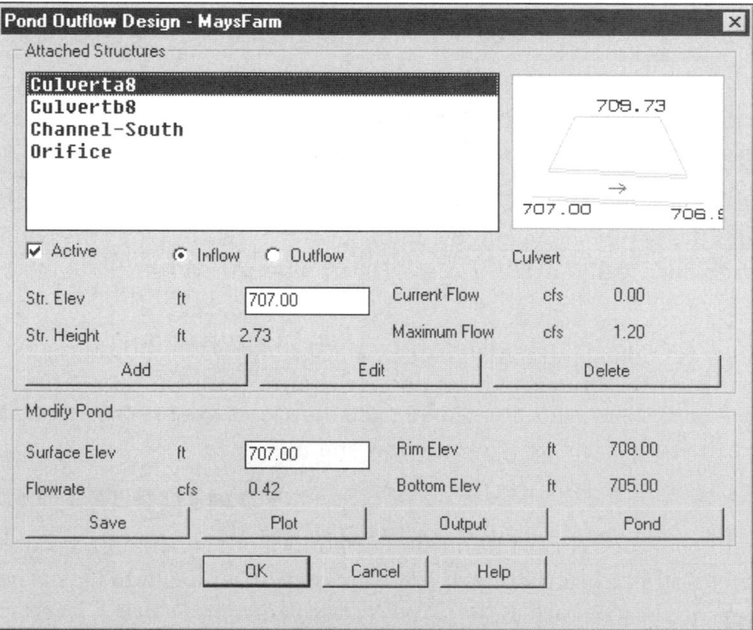

Figure 12–92

UNIT 6: HEC2 SECTIONS

The HEC2 Output cascade of the Hydrology menu contains routines to define, import, export, and plot HEC2 cross sections (see Figure 12–93). The HEC2 Single and Multiple routines create data sets for single or multiple cross sections. The data for the cross sections can be printed to a file. The Plot Single Section routine plots individual HEC2 cross sections in the current drawing. The Import HEC2-Sections routine uses the plotting utilities to import the cross sections into the drawing.

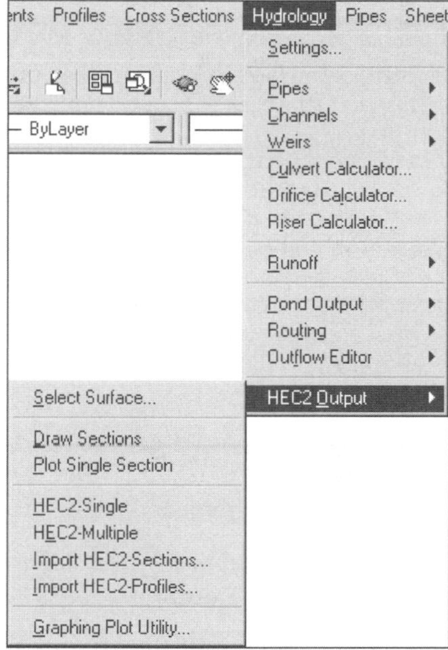

Figure 12–93

HEC2-SINGLE

The routine allows you to define single HEC2 cross sections, assign them their characteristics, and save their data. The cross sections are between two selected points containing surface data. After you select the cross-section points, the routine displays a dialog box that provides water flow data for the final cross section. When you select the Create File button in the HEC2 Output Settings dialog box, the routine displays the cross section data in Notepad. You save the data by selecting Save As from the File menu of Notepad.

HEC2-MULTIPLE

The routine allows you to define multiple HEC2 cross sections, assign them their characteristics, and save their data. The cross sections are along a centerline you define by selecting two points. After you define the centerline, the routine prompts for a cross-section interval. All sections must fall within a surface boundary. After you define the cross-section points, the routine displays a dialog box that provides water flow data for the final cross section. When you select the Create File button, the routine displays the cross-section data in Notebook. You save the data by selecting Save As from the File menu of Notebook.

IMPORT HEC2-SECTIONS

This routine imports any HEC2 data into the drawing as HEC2 cross sections. The plotting is done through the same dialog box that plots p-curve and ratings curves of hydrology structures.

Exercise

After you complete this exercise, you will be able to:

- Create, view, and plot HEC2 cross sections

HEC2 SECTIONS

Build a Terrain Model

1. Continuing in the *Forpres* drawing, if the CONT-NML and CONT-HGH layers are not visible, open the Layer dialog box and either thaw or turn them on.
2. Select the Terrain Model Explorer routine from the Terrain menu to create a new surface. Name the new surface Existing.
3. Create contour data from the contours. Select the contours by layer and select a contour on the CONT-NML and CONT-HGH layers. Use the following factors for weeding and supplementing the data:

   Weeding: 10.00 and 4.0

   Supplementing: 15.0 and 1.0

4. Build the surface with contour data only and toggle on Minimize Flat Triangles Resulting from Contour Data.

HEC2-Single

5. Run the HEC2-Single routine from the HEC2 Output cascade of the Hydrology menu. The routine prompts for two points to define a cross-section line. Select points that cross the ditch to the right of the pond and define about four cross

sections (south to north). After you define the series of cross sections, the routine displays a dialog box requesting the flow data for the cross sections (see Figure 12–94). Add the flow and Manning's n to your data. After you enter the flow data, select the Create File button. This displays the cross-section data that you can print or to save a file. Any program reading HEC2 cross sections can read the data from this file. From the File menu select Saveas and save the file as HEC2a.dat. Exit out of the program.

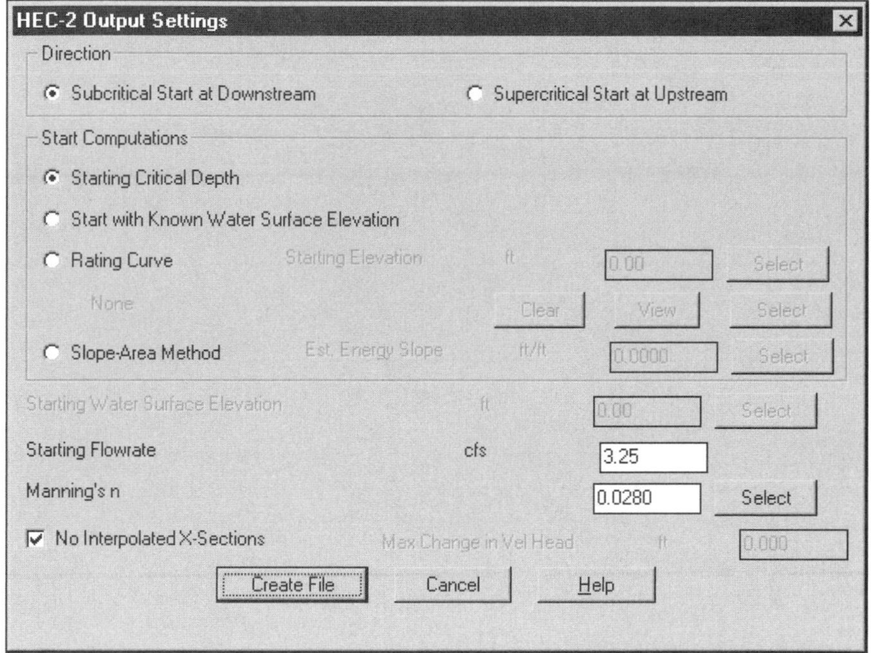

Figure 12–94

Import HEC2 Sections

6. Use the Import HEC2-Sections routine of the HEC2 Output cascade of the Hydrology menu to import the cross sections into the drawing. After selecting the HEC2a.dat file, the routine presents the plotting control dialog box (see Figure 12–95). This box allows for the plotting of almost any data set. The plotting control dialog box is the plotting utility for P-curves, rating curves, and hydrographs in the Hydrology menu.

7. Select the Preview button to preview a cross section. Then exit the preview by pressing any key. You return to the HEC2 Cross Section dialog box. Select the Next Section button to move to the next section, select the preview button to view the section, and press any key to return to the HEC2 Cross Section dialog box.

EXERCISE

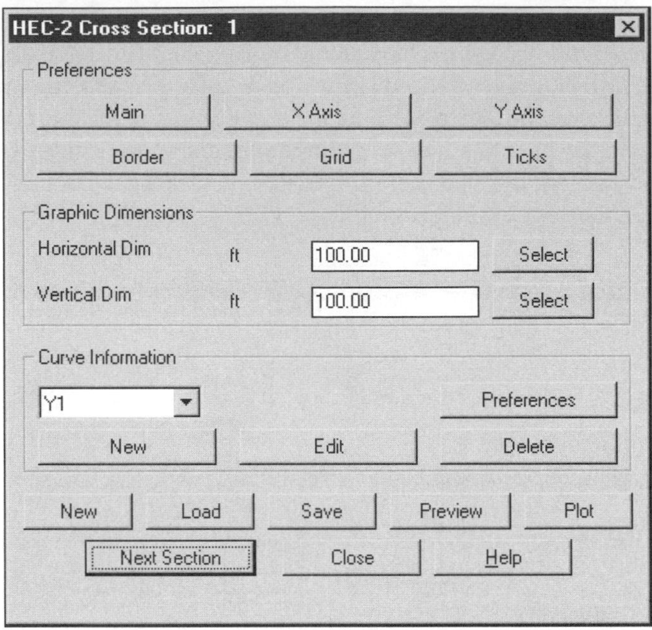

Figure 12–95

8. In the HEC2 Cross Section dialog box, choose the Plot button to place the current cross sections into the drawing. Place a few sections to the east of the site.

9. After placing the sections into the drawing, exit the HEC2 Cross Section routine.

HEC2-Multiple

The HEC2-Multiple routine creates a cross section from a centerline. First, define a centerline and the routine prompts for station and offset distances. After sampling the surface, the routine displays the HEC2 data in a dialog box. The Import HEC2-Multiple routine imports these cross sections into the drawing.

10. Use the ZOOM command to view the design area. There is a ditch running North and South just to the east of the pond. This is where you will develop the multiple cross sections.

11. Start the HEC2-Multiple routine from the HEC2 Output cascade of the Hydrology menu. The routine prompts for the first point of the centerline. Select a point just in the southern end of the channel. The routine prompts for a second point; select a point in the center of the channel at the northern end.

12. Next, the routine displays the flow characteristics dialog box. Use the same values found in Figure 12–94 to set your values.

13. Select the Create File button to save the cross-section data. First, the routine prompts you for the distance of the station increment (50 feet) and then the left and right offset (50 feet) distance. After setting the distances the routine displays the cross section data file.
14. Save the data by selecting Save As from the File menu of Notepad. Name the file HEC2b.dat. From the File menu of Notepad select Exit to exit Notepad.

You can use the Import HEC2 Sections routine to import the cross section into the current drawing.

15. Save the drawing.

The Hydrology menu of Civil Design contains tools to calculate, evaluate, and plot water-related questions that are a part of site development. These tools supply their data to other routines in Civil Design. For example, the runoff calculations are data for the flow into a catch basin or storm drainpipe run. As you saw in the examples of this section, the designing of a pond is related to the runoff and discharge properties of the site design.

The next section covers the tools in Land Development Desktop and Civil Design that create new surface data or site design. These new surfaces allow you to calculate volumes, post development runoff numbers, and drainpipe designs.

Grading and the Grading Object

After you complete this section, you will be able to:

- Use the three types of grading tools in Civil Design
- Create designs using point, breakline, and contour data
- Work with the Daylighting routines
- Design with the grading object

SECTION OVERVIEW

Section 6 of this book introduced some of the basic site design tools. The exercises for that section used point, contour, and 3D polyline routines to create second surface data. Some of the routines that are not a part of Section 6 are the focus of this section. The routines described in this section assign elevations to points from surfaces, 3D polylines, or from interpretation of elevations between controlling points. These routines are in the Points and Terrain menus of Land Development Desktop.

The grading routines of the Grading menu use two strategies. The Create Single and Create Multiple routines of the Daylight cascade of the Grading menu use the first strategy. This strategy has been a part of Civil Design since the Softdesk days. These routines work with 3D data in creating an intersection of a projected slope from a point to a design surface. The Slope Grading cascade of the Grading menu contains the second strategy and uses the grading object, which was new when the Desktop strategy was introduced. The grading object stores the conditions of its blending into a design surface. What is nice about this is the fact that the object does not even need to have a surface to grade to. It can grade for a distance or to an elevation, relative to or absolute from the object.

UNIT 1

The first unit of this section uses the point tools of the Points menu to create a stockpile. The elevations for the points creating the pile are from surface elevations and point grade/slope interpolations.

UNIT 2

The tools of the Daylight cascade of the Grading menu are the focus of the second unit. The Daylight tools allow you to use a single slope or differing slopes at each vertex of a 3D polyline to find the daylight intersection point.

UNIT 3

The third unit of the section works with the grading object found in the Slope Grading cascade of the Grading menu. This method is similar to the daylight process, but the grading object stores all of its grading conditions and simplifies the creation of a pile. Whether it is a pile, building pad, or grading off the back-of-curb of a parking lot, the grading object presents you with several opportunities to quickly evaluate different grading scenarios.

UNIT 1: GRADING WITH POINTS AND CONTOURS

The Points menu contains several tools that allow you to create a surface that represents your grading plan. The routines can assign elevations to points from existing surfaces in the project. Other routines can assign elevations to points by interpreting a grade or slope from a point or between controlling points. As with any task using points, when the time comes to create a surface, you may have to add breakline data to create the correct triangulation. To control the triangulation better, use contours and 3D polylines. When you create a surface from contours and 3D polylines, they are breakline data for the surface. However, contours and 3D polylines have problems with peninsulas and the loss of high and low point elevations. No matter which objects you employ, you will have to control problems with the surface data.

POINT DATA

The Create Points–Interpolate cascade of the Points menu has several point routines that would apply to creating points for a design surface. Many of the routines place points in a direction for a distance at a slope or grade from one point or between controlling points. The routines first establish a controlling elevation for the first and maybe a second point and then specify a series of values the routine needs to calculate a new point. The values include distance, elevation, grade, or slope. There is an Offset option for each new point placed into the drawing by the routines.

There are two routines in the Create Points–Slope cascade of the Points menu that are similar to and less complicated than the point interpolation routines of the Create Points-Interpolate cascade of the Points menu. The routines are the Slope/Grade–Distance and Slope/Grade–Elevation routines. The Distance and Elevation routines require only one point at which to start, and the second point does not need to be another existing point. When the routine creates the resulting points, the routine prompts about placing a point at the end of the distance from the starting point.

SLOPE/GRADE–DISTANCE

The Slope/Grade–Distance routine, which is on the Create Points-Slope cascade of the Points menu, prompts for a starting point, direction, a distance, and a slope or grade (see Figure 13–1). The routine asks you to select a distance and then gives you an opportunity to adjust the distance if necessary before proceeding. The routine then prompts for the number of intermediate points and whether the farthest distance is to have a point too. For example:

```
Beginning point:.g
Beginning point: Select point block:
x: 4859.7331 y: 5022.6703 Z: 721.00
Direction: Select point block: .g (toggles off Graphics mode)
Direction: (select a point)
Slope (Slope/Grade/<Infinite>: 5
Slope: 5:1, Grade: 20.00%
Distance <75.36>: 50
```

The routine produces a report about the elevations and slopes, then prompts for the number of intermediate points.

```
Enter number of intermediate points : 3
Offset <0.0>:
Add ending point (Yes/No): Y
```

SLOPE/GRADE–ELEVATION

The Slope/Create–Elevation routine of the Create Points-Slope cascade of the Points menu works much the same as the Distance routine, except that the primary values are elevations. If the routine does not attain the ending elevation over the default distance, it will continue placing points in the direction you show the routine. The routine prompts are as follows:

```
Beginning point:.g
Beginning point: Select point block:
x: 4859.7331 y: 5022.6703 Z: 721.00
Direction: .g
Direction: (select a point in the editor)
Slope (Slope/Grade/<Infinite>: 5
Slope: 5:1, Grade: 20.00%
Elevation <734.30>:
```

You will receive a report about the elevations and slopes. The routine then prompts for the number of intermediate points.

```
Enter number of intermediate points : 3
Offset <0.00>: (press ENTER)
Add ending point (Yes/No): Y
```

Again, if the routine does not attain the user elevation in the distance (the distance between points 1 and 2), the routine continues to place points in the vector direction until reaching the elevation.

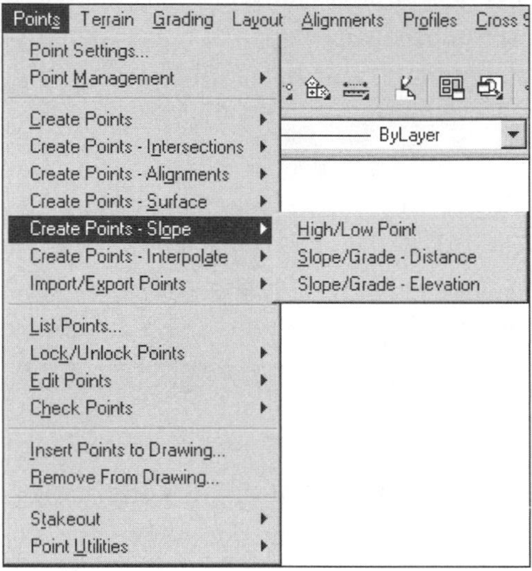

Figure 13–1

POLYLINE/CONTOUR VERTICES

The Polyline/Contour Vertices routine of the Create Points–Surface cascade of the Points menu places a point object at each vertex of the polyline (see Figure 13–2). The elevation of the point is the elevation of the surface at the vertex of the polyline.

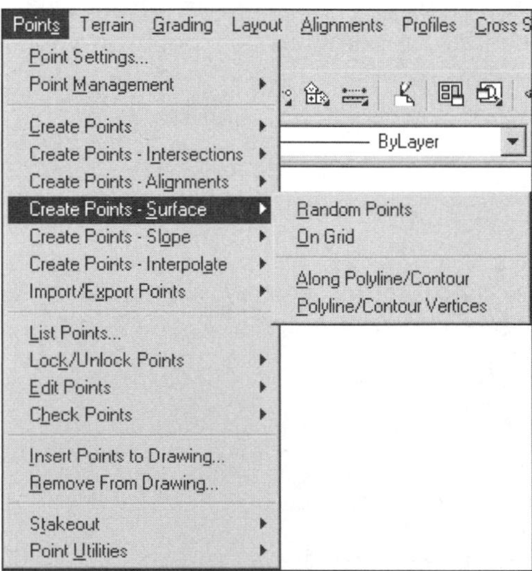

Figure 13–2

CONTOUR DATA

COPY BY SLOPE

The Copy by Slope routine of the Contour Utilities cascade of the Terrain menu creates new contours from existing contours (see Figure 13–3). After the routine prompts for the contour interval and the slope (negative slopes descend), you select the contour and select to the side of the contour where the slope occurs. The routine offsets once and prompts for the next contour to offset.

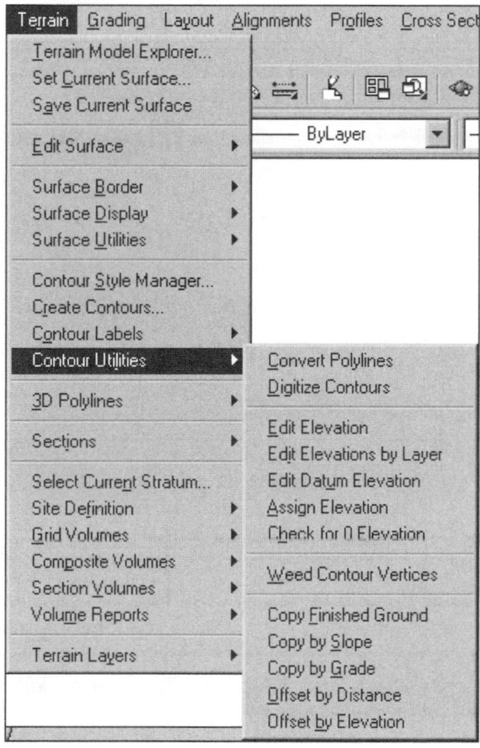

Figure 13–3

Exercise

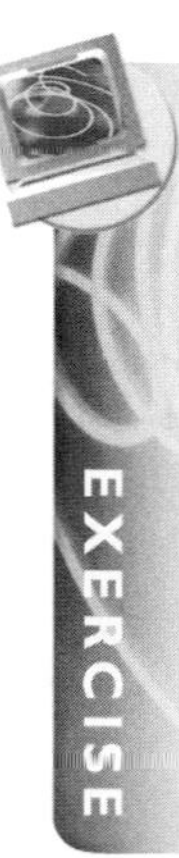

The task of this exercise is to develop a stockpile with contours, points, and breaklines.

After you complete this exercise, you will be able to:

- Create a stockpile design from the ground up
- Work with point, breakline, and contour data

STARTING A NEW DRAWING

1. Start a new drawing, *stockpile* (see Figures 13–4).

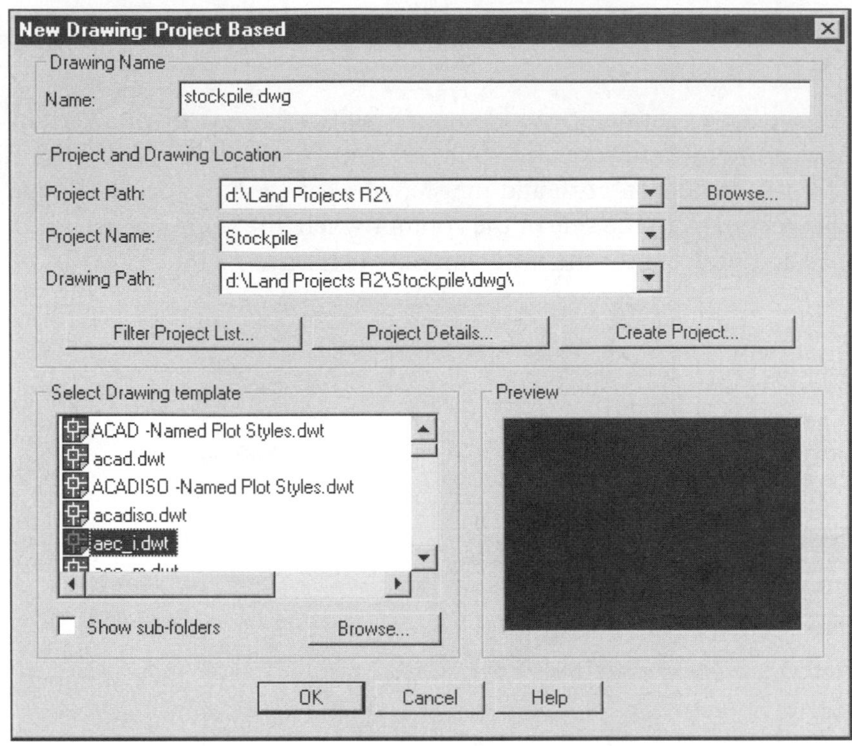

Figure 13–4

2. Assign the drawing to a new project, Stockpile. Use the LDDTR2 prototype see Figures 13–5). Describe the project as Grading Tools Exercise. Select the OK button to dismiss the Create Project and New Drawing dialog boxes.

3. Select the OK button to dismiss the point database defaults dialog box.

4. Select and load the LDD1-40 setup file to set the general drawing settings. After loading the setup file, select the OK button to dismiss the Load Settings dialog box.

5. Select the OK button to dismiss the Finish dialog box.

6. Insert the Stockpile drawing from the CD that accompanies this book. Insert the drawing at 0,0 and explode it upon insertion.

7. The inserted drawing has points that are not a part of the current project. Run the Modify Project routine of the Check Points cascade of the Points menu. In the Modify Project dialog box, toggle on Add Unregistered Points to Project Point Database (see Figure 13–6).

8. Freeze the CL, LOT, and RDS layers.

9. Run the Zoom Extents routine to center the drawing.
10. Click the Save Drawing icon and save the drawing.

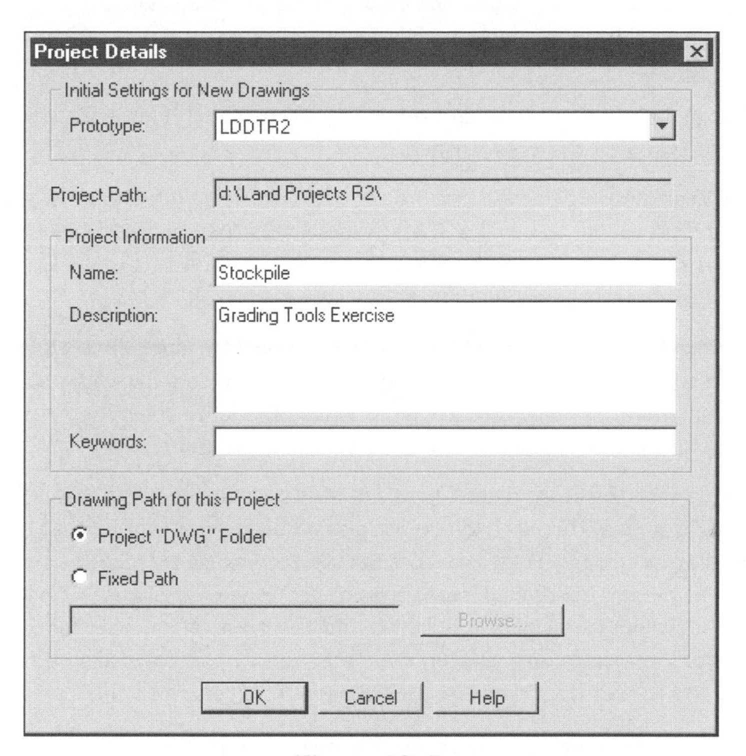

Project Details
Initial Settings for New Drawings
Prototype: LDDTR2
Project Path: d:\Land Projects R2\
Project Information
Name: Stockpile
Description: Grading Tools Exercise
Keywords:
Drawing Path for this Project
Project "DWG" Folder
Fixed Path
Browse...
OK Cancel Help

Figure 13–5

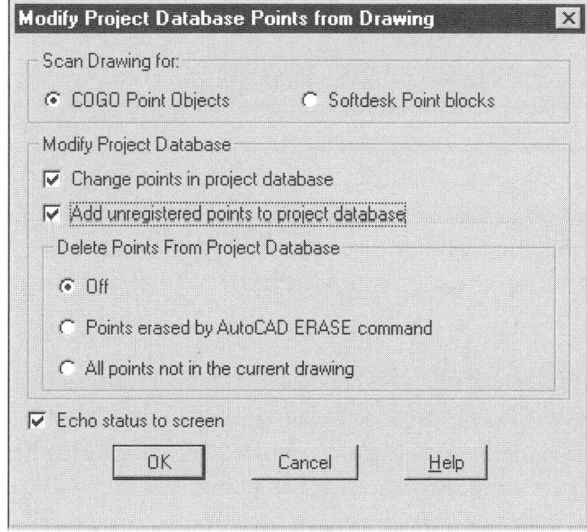

Figure 13–6

CREATE SURFACE BASE

1. Open the Terrain Model Explorer and select Terrain. After selecting Terrain, click the right mouse button and select Create Surface. Select the "plus" sign to expand the tree. Select Surface1, click the right mouse button, and rename the surface Base.
2. Select the "plus" sign to expand the list and select the "plus" sign again to view the data types for the base surface.
3. Select Breaklines, click the right mouse button, and select Define By Polyline from the shortcut menu. Select the 3D polyline that is the southern branch of the swale at the center of the site. Delete the existing object, and if the routine issues a warning about preexisting data, select Overwrite All.
4. Next select Contours, press the right mouse button, and select Add Contour data. In the Contour Weeding dialog box, set the weeding factors to 15.0 and 4.0 and supplementing factors to 20 and 0.5. Use the layer selection process and select a single contour. All the contours are on the same layer.
5. Finally, create the point data. There are no Point Groups. When you select Add Point Group, the Terrain Explorer displays a dialog box stating that there are no defined point groups. Select the Yes button to dismiss the dialog box and start the Point Group Manager. Create a point group with a Group Name of All using all of the points (see Figure 13–7). You do this by selecting the Build List button and setting the source to all points. Select the OK button to dismiss the dialog boxes and to exit the Point Group Manager. You return to the Terrain Explorer.

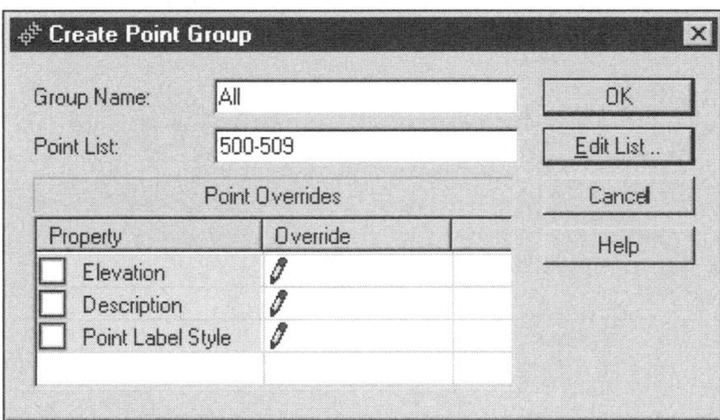

Figure 13–7

6. Again select Add Point Groups and select the All group.
7. Select the Base surface name, click the right mouse button, and select Build. The routine displays the Build Surface dialog box. Set your dialog box to match the settings in Figure 13–8. After setting the values, select the OK button to build the surface. After selecting the OK button to dismiss the Done Building Surface dialog box, close the Terrain Model Explorer.

8. Remove all of the points from the drawing and erase the swale breakline. Use the Remove From Drawing routine of the Points menu to remove the points.

9. Save the drawing by clicking the Save icon.

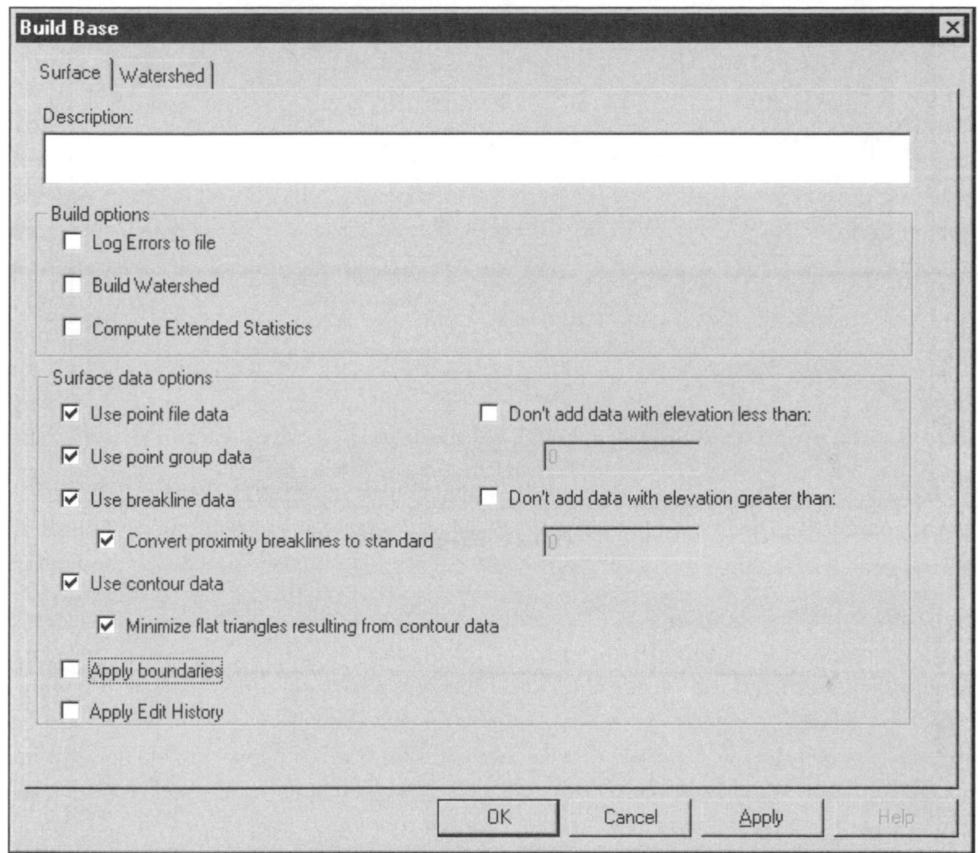

Figure 13–8

DESIGNING PILE1

1. In the Create tab of the Point Settings routine of the Points menu set the Descriptions to automatic with TOE as the description and the starting point number to 1000 (see Figure 13–9).

The stockpile has an elongated shape southwest-northeast, with a 3:1 slope on the northeast face, a 5:1 slope on the northwest and southeast flanks, and an 8:1 slope on the southwest side. Contours will make up the northeast face of the pile and the remainder of the pile will be from points (see Figure 13–10).

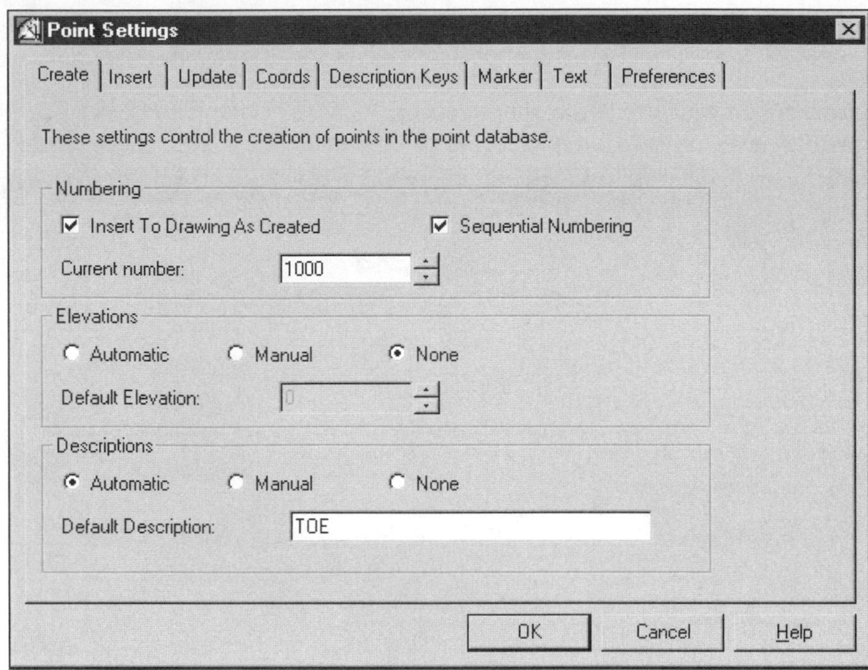

Figure 13–9

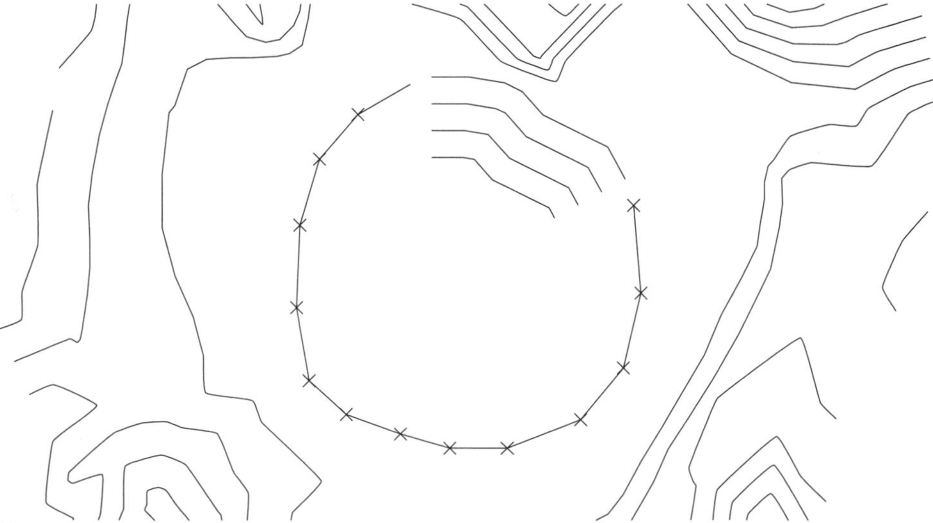

Figure 13–10

The first step will be the creation of the toe contour on the northeast part of the pile. The main concern here is what elevation to assign to the contour. If you assign the elevation of

732.00 feet to the contour, the elevation creates TIN triangles that generate cut volumes. If the elevation is 733.00 feet, the elevation creates areas of fill beyond the toe of the pile. Create a contour whose elevation is 732.50 feet. Use Figure 13–11 as a guide when drawing the toe contour. Make sure polar and object snap are off.

2. Draw the contour with the POLYLINE command. If you use the POLYLINE command and no object snaps, the resulting polyline has an elevation of 0 (zero). Use 13–11 as a guide to drawing the contour.

3. Use the Edit Elevation routine of the Contour Utilities cascade of the Terrain menu to edit the elevation of the polyline to 732.50 feet.

You can create a contour representation of the northeast face of the stockpile by using the Copy By Slope routine of the Contour Utilities cascade of the Terrain menu.

Figure 13–11

EXERCISE

Copy by Slope

In the current exercise, the Copy By Slope routine creates contours representing the face of the stockpile at a 3:1 slope. Create new contours from the existing toe contour. Set the slope value to a rise of 1 and a run of 3 and the contour interval of 5 feet. Offset the contour to the southwest. Offset three times, each time offsetting the new contour. List the elevation of the contours with the Contour Elevation routine of the Inquiry menu. Each contour should increase in elevation by 5 feet when listed from northeast to southwest (see Figure 13–10).

4. Create new contours with a 3:1 slope and a 5-foot interval. Use the Copy by Slope routine of the Contour Utilities cascade of the Terrain menu. Offset the new contours to the southwest. Do this three times.

5. List the contour elevations with the Contour Elevation routine in the Inquiry menu.

Polyline/Contour Vertices

The Polyline/Contour Vertices routine allows you to tie the toe of the slope onto a surface. The toe points need to be frequent enough to control the triangles that will connect the pile back into the base surface. The goal is to prevent points outside the toe line from connecting to points in the interior of the pile. If the space between toe points is too wide, the TIN Modeler will create triangles that connect a point that is outside the toe to a point inside the pile. It may be necessary later on to use the polyline as breakline. The toe breakline forces the Terrain Modeler to create triangles that go to and expand of the toe of the pile.

The routine places a point at each vertex on an entity. The elevation of each point is from the current surface (base).

1. Draw a polyline that represents the toe of the stockpile. The polyline should be a closed polyline with about 14 vertices for the placement of the polyline (see Figure 13–10).

Since the toe of the pile is point data, the slopes should be point data as well. To create the side slope points, you can use one of several point routines in the Points menu.

You may also want to consider that the slopes around the pile create the resulting pile elevations. To view the resulting elevations from traveling at a slope for a distance, then use the List Elevation @ Slope routine from the Inquiry menu.

2. Use the Polyline/Contour Vertices routine of the Create Points–Surface cascade of the Points menu to place point objects on the polyline. The routine places points at the vertices on the polyline and the point's elevation is the elevation of the current surface at the vertex.

3. Click the Save Drawing icon to save the drawing.

4. Use the List Elevation @ Slope routine to view the resulting elevations from the toe points and their respective slopes. Use a 5:1 slope for the northwest and southeast sides of the pile, and an 8:1 slope for the southwest side of the pile.

The routine prompts for a starting point. Since there are points on the polyline representing a starting point with the correct elevations, you should select one of the points.

Use either the *.p* toggle to identify a point by its number, or the *.g* toggle to select the point off the screen.

```
Command:
LDD
Elevation slope - Select point: .p
Elevation slope - Select point:
 >>Point number:
```

-or -

```
Command:
LDD
Elevation slope - Select point: .g
Elevation slope - Select point:
>>Select point object:
```

After you select the point block, the routine prompts for the slope and the distance from the starting point. After you set the variables, the routine returns the elevation of the second point. The direction of the slope does not matter, but what does is the resulting elevation.

```
>>Point number: 1003
 Slope (or Grade) <5.00:1>:
Slope: 5.00:1, Grade: 20.00
Distance <0>: 50
Elevation: 742.78
Elevation slope - Select point:
```

Slope/Grade–Distance

1. Run the Point Settings routine from the Points menu and set the following values in the Create tab.

```
Automatic descriptions: ON
Default description: SSIDE (for side slope)
```

The Slope/Grade-Distance routine prompts for a starting point. If there is no Select Point Object or Point Number prompt, toggle on either mode by typing **.g** or **.p**. Select a starting point object, one of the toe points on the northwest side of the pile. Next, the routine prompts for a direction from the beginning point. No existing points in the drawing represent the direction for the new points. Toggle off the point mode by typing **.g** or **.p**, and select a point toward the interior of the pile. The elevation of the second point is 0.0, so the first slope and grade is set to infinite. Set the slope to 5:1 and set the distance to 50.00. Next, the routine asks for the number of intermediate points and whether there is an offset to the point placement. Place three intermediate points with no offset and place a point object at the end, which in this case, is 50 feet away from the starting point.

The routine stays active and prompts for a new direction from the current starting point. You do not need to set any more points from this point. What needs to be done is to move to another point to continue developing the side slopes of the stockpile. By pressing ENTER at the direction request, you cause the routine to prompt for a new starting point. Before selecting the next point to start from, toggle on one of the point modes (.g or .p). Then the process starts over again for the next point. Again you need to toggle off the point mode to show the direction and toggle on a point mode to select the next beginning point.

This process repeats for every other point along the toe of the pile. Remember, the side slopes are 5:1 and the southwest side of the pile is at an 8:1 slope.

2. Go to the Create Points–Slope cascade of the Points menu and run the Slope/ Grade–Distance routine. When the routine prompts for the direction select a point near the center of the pile. Use Figure 13–12 as a guide.
3. Click the Save icon to save the drawing.

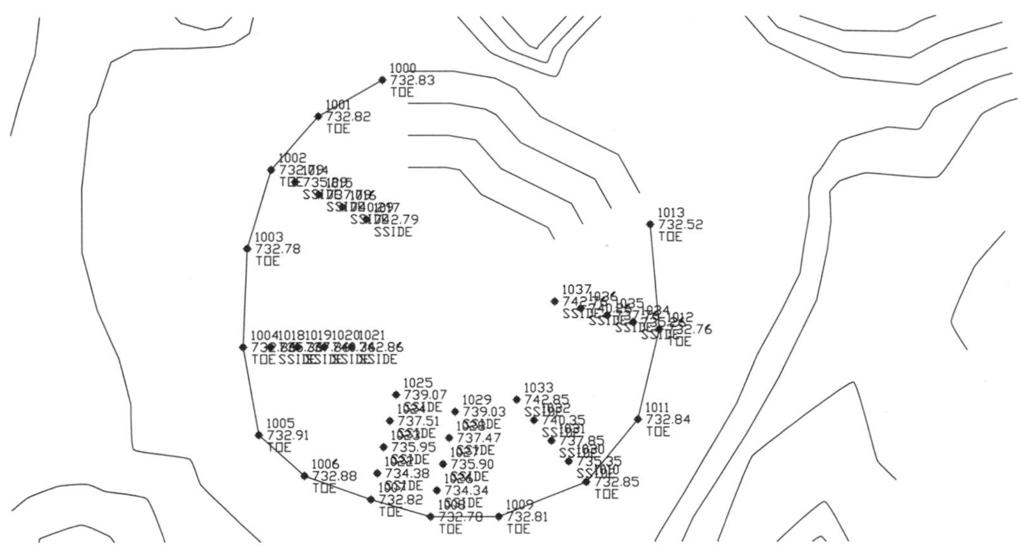

Figure 13–12

BUILDING SURFACE PILE1

1. There is enough data for another surface. The data for the surface is contour and point data.
2. Use the Point Group Manager and create a point group by selecting the points on the screen. Name the Group Pile1. After naming the group, select the Build List button to display the Point List dialog box. Toggle on Drawing Selection Set, select the Select button, and after the routine hides the dialog box, select the points. Select OK to exit the dialog boxes and Point Group Manager.

3. Go to the Terrain menu and select Terrain Explorer. Select Terrain and click the right mouse button, and select Create New Surface. Select the "plus" sign, click the right mouse button, and rename the surface to pile1. Then select the "plus" sign again to display the data list for the Pile1 surface.
4. Select Point Groups, click the right mouse button, and select Pile1 from the list.
5. Select Contour Data, press the right mouse button, and select Add Contour data. When the Contour Weeding dialog box displays, set the weeding factors to 15.0 and 4.0 and supplementing factors of 20.0 and 1.0. Instead of you selecting the contours by layer, use the entity option and select the contours from the display (***do not include the Toe polyline as contour data***).
6. Build the surface from point and contour data. Select the surface name, press the right mouse button, and select build surface. The routine displays the Build dialog box. Toggle on Minimize flat triangles resulting from contour data and then select the OK button to build the surface (see Figure 13–13). Select the OK button on the done dialog box to return to the Terrain Model Explorer.
7. Select the Pile1 surface name, click the right mouse button, and cascade out the Surface Display menu. Select the Quick View routine to view the triangles.

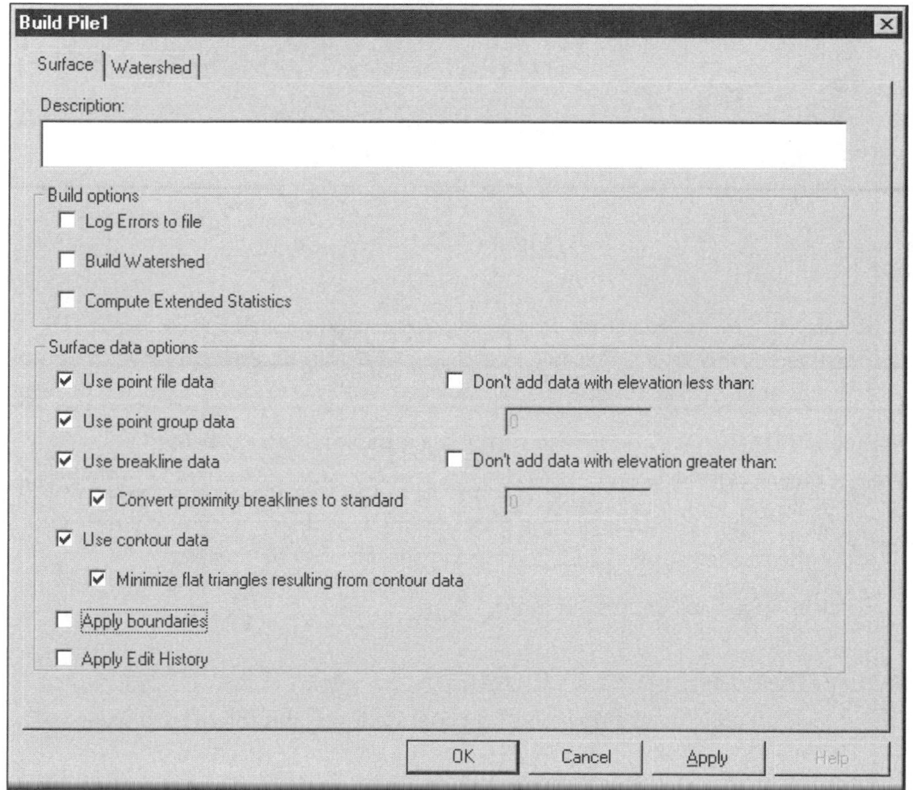

Figure 13–13

If there are triangles that connect a point from within the pile to the surrounding contours, the surface is not correct (see Figure 13–14). The triangles from the contours around the pile should connect with the toe line of the pile, not with points within the pile. If the surface is not correct, you need to establish the correct triangulation. The toe polyline of the pile needs to be a breakline representing daylight point of the pile slopes and the base surface.

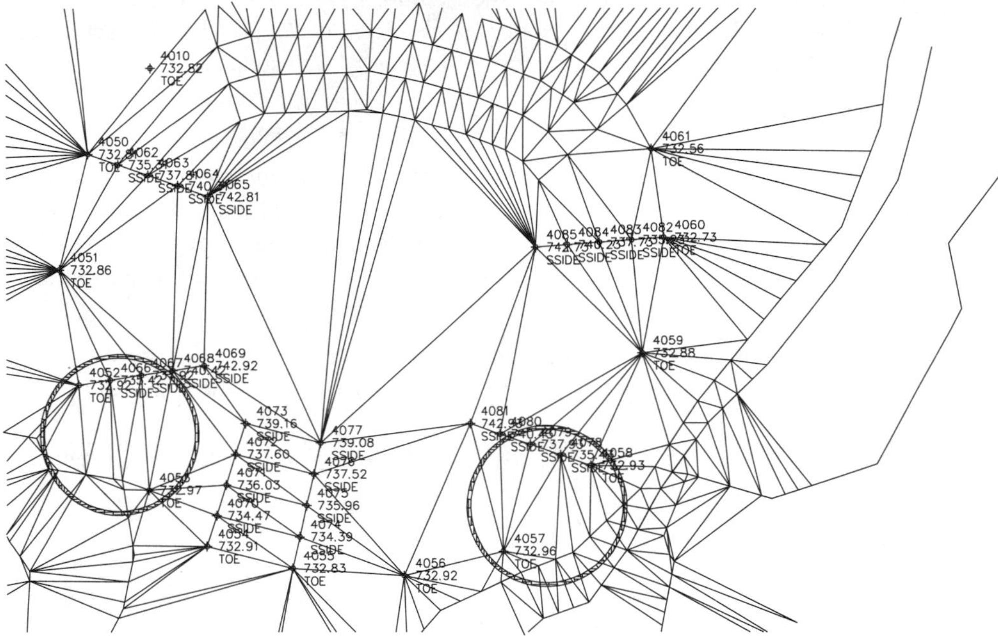

Figure 13–14

8. If you need to include the Toe polyline as a breakline, use the Proximity By Polyline routine of the Breakline cascade of Terrain Explorer. Name the breakline toe and answer Yes to deleting the object.
9. Rebuild the surface and return to the main Explorer dialog box.
10. Select the Pile1 surface name, click the right mouse button, and cascade out the Surface Display menu. From the shortcut menu, select the Quick View routine to view the triangulation. The triangulation should now be correct.
11. Click the Save icon and save the drawing.

CALCULATING THE PILE1 VOLUME

If the surface is correct, you are now able to generate a volume for the stockpile.

1. Use the ZOOM command to zoom out to view the entire project area and an area that is slightly larger that the site.

2. From the Volumes area (bottom) of the Terrain menu and run the Define Current Stratum. Name the stratum pile1, and set the surfaces to base (surface 1) and pile1 (surface 2). Select the OK button to exit the dialog box.
3. Use the Define Site routine of the Site Definition cascade of the Terrain menu to define a site that is larger than the surfaces. Set the rotation angle to 0 (zero) degrees and the M and N spacing to 5 (feet). Make sure you select an upper right point above and to the right of the contours. Press the ENTER key to accept the definition, name of the site, and to exit the routine.

Remember, when calculating the Grid and/or Composite volume, each method creates a new surface. Name each surface, pile1g, or pile1c—g for grid and c for composite.

If you have not done a volume calculation before, you should review section 6 of this book.

4. Use any or all of the volume methods to calculate a volume. If you use the Calculate Total Site Volume from either the Grid or Composite Volumes cascade of the Terrain menu, the routine will ask you to name a surface for the resulting volume calculations. Note down the resulting volumes.
5. Remove the points for pile1 with the Remove Points from Drawing routine found in the Points menu. Select the points by number range. Use the range of 1000–1100 to select the points.
6. Erase the contours that made the northeast face of the pile and the pline that represents the toe breakline of pile1.

EXERCISE

UNIT 2: GRADING WITH A 3D POLYLINE

One problem with the previous exercise is matching the elevations around the crown of the pile. If there had been large differences in elevation between the points around the toe of the pile, the elevations created for the top of pile would also vary greatly. The variations could make the pile mimic a volcano or have higher sides than the top contour on the northeast. It is difficult to match the elevations closely with this mixed bag of points and contours.

The next strategy is to use a top-down approach. This approach defines the top of the pile and projects slopes from each vertex of the crown down to the daylight surface. This strategy allows you to project a single slope (Create Single) from each vertex or varying the slope (Create Multiple) at each vertex (see Figure 13–15). The condition of the site or restrictions of the design may dictate that you use differing slopes a certain vertices around the design perimeter. In this case, you may have to vary the slopes to avoid creating a pile that crosses over the site boundary into land that should not be affect by the design. As you use the Create Single or Create Multiple routines, they will identify the daylight point of the vertex so you know the affect of the slope.

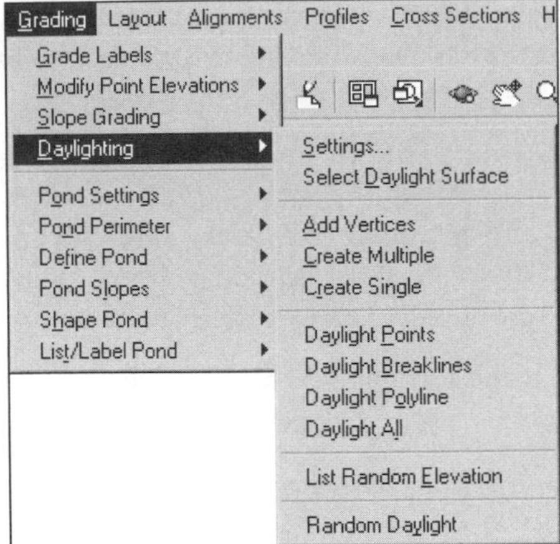

Figure 13–15

DAYLIGHT SETTINGS

The daylight settings consist of several conditions and values (see Figure 13–16). The first condition sets the type of daylight line; that is, the daylight object is a line segment or a polyline. The daylight line will eventually become a standard polyline (3D) breakline at the toe of the pile. The next group of toggles sets the search for cut, fill, or both. The next group of defaults specifies whether to prompt for the number of points between the starting point and the daylight line or just to use a set number of points.

You can set these values by calling the dialog box from the Settings routine of the Grade Label cascade of the Grading menu or from the Civil Design settings of the Drawing Settings routine of the Projects menu.

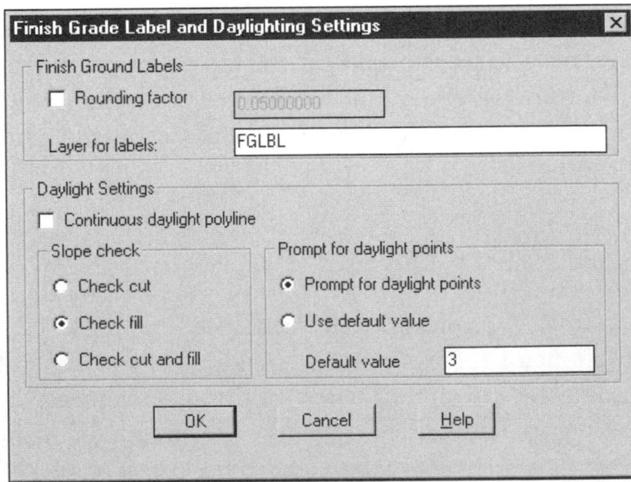

Figure 13–16

CREATE SINGLE

The Create Single routine in the Daylight cascade of the Grading menu calculates daylight from vertices on a 3D polyline (see Figure 13–16). Each vertex on the 3D polyline has the same slope value. The routine places the projected slope value as extended entity data at each vertex. The extended entity data represents the slope used while searching for daylight at the vertex. The daylight point is a blip (X) on the screen around the base of the pile. The X corresponds to the intersection of the daylight surface and a slope projected from a vertex on the polyline.

CREATE MULTIPLE

The Create Multiple routine in the Daylight cascade of the Grading menu calculates daylight from a vertices on a 3D polyline (see Figure 13–16). Each vertex on the polyline can have a different slope value. The routine places the projected slope value as extended entity data at each vertex. The extended entity data represents the slope used while searching for daylight at the vertex. The daylight point is a blip (X) on the screen around the base of the pile. The X corresponds to the intersection of the daylight surface and a slope projected from a vertex on the polyline.

The Transition option of Create Multiple allows you to vary the daylight slope over a series of vertices. When you use the Transition option, you identify the starting vertex and its slope, move past the intermediate vertices, and then select the ending vertex and set its slope. When you exit the Transition option, the routine applies a varying slope to each intermediate vertex on the polyline based upon the beginning and the ending slopes of the transition.

DAYLIGHT ALL

The Daylight All routine of the Daylighting cascade of the Grading menu creates points, breaklines, and 3D polylines representing the projected slopes, their resulting elevations, and the intersection of the projected slope with the daylight surface (see Figure 13–15). You can use all of the elements to create a new surface reflecting the elevations, slopes, and surface intersection.

The Daylight Points, Daylight Breaklines, and Daylight Polyline create the individual elements of the daylighting solution.

3D POLYLINES

The 3D polyline is a powerful surface design object. The object has an elevation at each vertex and when used as a surface breakline controls the triangulation. The routines are found in the 3D Polyline cascade of the Terrain menu (see Figure 13–17).

CREATE BY ELEVATION

The Create by Elevation routine from the 3D Polyline cascade of the Terrain menu draws a 3D polyline and assigns elevations to each vertex. The elevation can be a user entered value, an elevation from a surface, or elevations that vary over a series of vertices (Transition). If you draw an arc segment in the 3D polyline, the routine calculates vertices around an arc. There are not arc segments in a 3D Polyline.

The Transition option of Create by Elevation can affect the elevations of several vertices at one time. When you use the Transition option of Create by Elevation, you identify the starting vertex and its elevation, select the intermediate vertices, and then set the ending vertex with its elevation. When you exit the Transition option, the routine applies a calculated elevation to each intermediate vertex on the polyline based upon the beginning and the ending elevations of the transition.

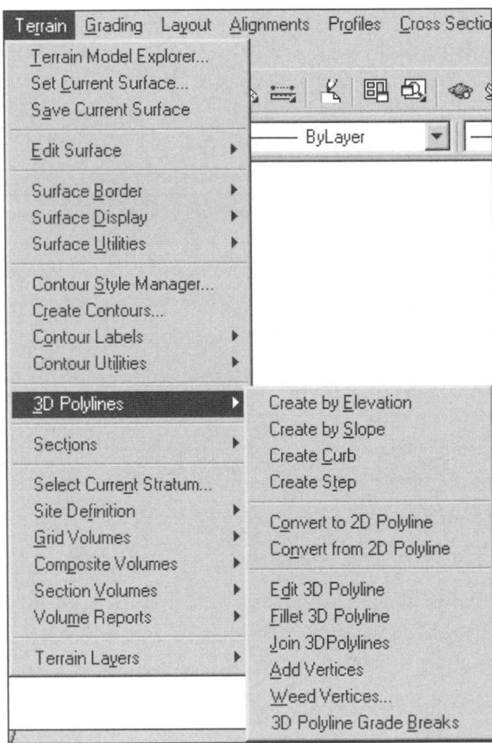

Figure 13–17

Exercise

It would seem better to draw the crown of the pile and then project slopes from the crown down to the base surface. This method allows the pile to have consistent elevations at the top and to maintain reasonable slopes from the crown to the toe.

After you complete this exercise, you will be able to:

- Create a stockpile design from the top down

- Work with 3D polylines
- Work with the daylight routines
- Work with point, breakline, and contour data
- Calculate the volume numbers

DESIGNING PILE2

1. Continue in the *Stockpile* drawing and use the Layer dialog box to create a new layer, PILE2, and make it the current layer.

The decision to create the top of the pile first leads to what type of 3D object to create. If the crown of the pile is to be the same elevation at each vertex, the line is a contour line. If each vertex has a different elevation, the object is a 3D polyline.

The 3D polyline is the most powerful tool when developing proposed surfaces. The reason for this power is twofold. First, the 3D polyline represents elevations without having to place points into the drawing. Second, a 3D polyline controls triangles when used as breakline data.

After drawing the 3D polyline, the Daylight routines (Create Single or Create Multiple) search for the intersection of a slope from the 3D-polyline vertex to a surface. The intersection point of the slope and the surface is the toe of the stockpile. In the current exercise, the toe must appear in the flat area in the center of the site, not in the surrounding areas. If the daylight intersection occurs outside the flat zone, you have to either move the crown of the pile or change the crown design. Once the design is correct, there are routines that create surface data from the Daylight points and breaklines.

EXERCISE

3D Polylines–Create by Elevation

1. Use the Create by Elevation routine in the 3D Polyline cascade of the Terrain menu. Draw a 3D polyline with approximately 14 vertices. Use Figure 13–18 as a guide for drawing the polyline. Close the polyline routine when you finish.

The elevations are 748 feet at the northeast and 743 feet at the southwest end. The elevations on the north and south sides gradually descend from 748 to 743 feet. Set the first two vertices on the northeast portion of the polyline to 748.00, and transition to the southwest side of the pile to the elevation of 743.00. You toggle on transition by typing the letter **T**. After toggling on transitioning, select four more vertices in the editor and end the transition by typing the letter **E**. The last vertex defaults to the elevation of 0 (zero), but set the ending elevation to 743.00. When drawing the polyline around the southwest side, set the vertices to 743.00. Finally, use the Transition option to return to the starting point in the northeast with four more vertices, going from 743.00 to 748.00. Again, type the letter **T** to start transition and the letter **E** to end transition. Place one more vertex at 748.00 and close.

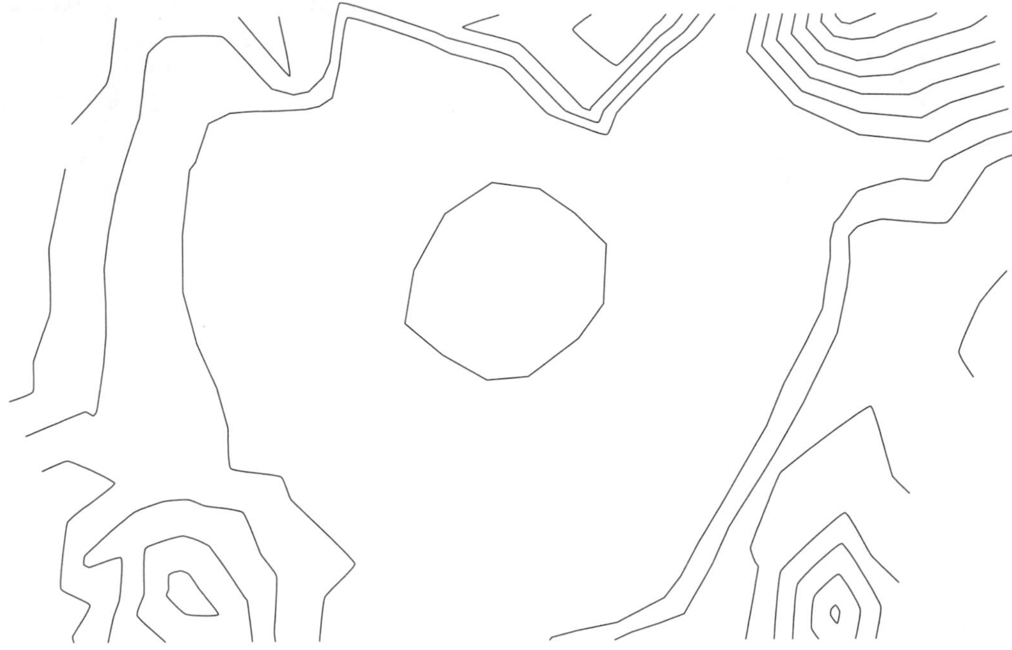

Figure 13–18

Daylight Slope Settings

You now have a 3D polyline representing the top of the pile. The design calls for different grades at varying locations around the pile. With elevations and slopes set, our interest turns to where these slopes intersect the base surface.

The daylight routines (Create Single and Create Multiple) identify a surface as the daylight surface. This surface is for the current editing session or until you identify another surface as the daylight surface. The daylight routines need to know with which surface to calculate an intersection.

1. Run the Points Settings routine of the Points menu and set the following defaults in the Create tab:

   `Current point number:` **`1100`**
   `Default description:` **`PILE2`**

2. In the Daylighting cascade of the Grading menu, use the Select Daylight Surface routine to set Base as the daylight surface.
3. Select the Settings routine of the Grade Labels cascade of the Grading menu and set the daylight defaults to the following (see Figure 13–16). After setting the values, select the OK button to exit the dialog box.

Continuous daylight line

Search for fill only

Prompt for points

Set the number of points to 3.

4. Save the drawing.

Grading with Create Single

1. Run the Create Single routine in the Daylighting cascade of the Grading menu. Set the slope to 3:1. The routine prompts for a side to look for daylight intersections. Choose a point just to the outside of the 3D polyline.

If you accidentally select a point in the center of the crown, the routine will create a volcano. The routine responds with blip crosses indicating the point of intersection around the pile.

2. Observe the location of the blips. If the blips occur outside the flat area, you need to adjust the location of the crown. Press ENTER to exit the routine. Compare your results to Figure 13–19. If you need to, move the location or edit the crown so the 3:1 slopes do not go outside of the flat area of the site.

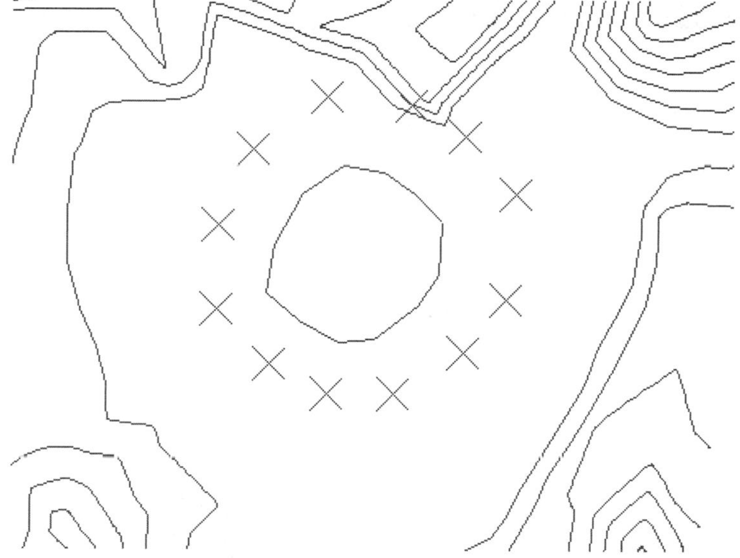

Figure 13–19

The problem with a single slope solution is that the stockpile needs to have different slopes around its perimeter. The design calls for a 3:1 slope on the northeast face, a 5:1 slope on the northwest and southeast sides, and an 8:1 slope on the southeast side. Currently, the slope values are correct only for the northeast side of the pile.

To be able to create different slopes for daylight, use the Create Multiple routine. The Create Multiple routine sets and/or changes the daylight slopes values for each vertex on the 3D polyline.

3. Redraw the screen to remove the daylight blips.

Grading with Create Multiple

1. Run the Create Multiple routine of the Daylighting cascade of the Grading menu. When the routine prompts you to select a polyline, select in the middle of the northeast portion crown. The routine asks you to select which side of the 3D polyline to search for daylight. You should select a point on the outside of the crown.

After you indicate the offset side, the routine reports the first vertex's values. These values include the elevation of the 3D polyline, elevation of the intersection point, and the difference in elevation. The routine then displays an option list that allows you to move to the next vertex, reset the data for this vertex or all vertices (set the slope value to 0), or specify a new slope for this vertex.

If the vertex has the correct slope, use the Next option to travel to the next vertex. Walk around the polyline vertices and change the fill slope to the appropriate values. You change the slope by typing the letter S and entering the correct slope. The values for the slopes are 3 for the northeast, 5 for the northwest and southeast, and 8 for the southwest. The routine places a blip X where the slope intersects the daylight surface. All daylight points should be within the flat central area of the site. If this is not the case, move or redraw the 3D polyline (see Figure 13–20).

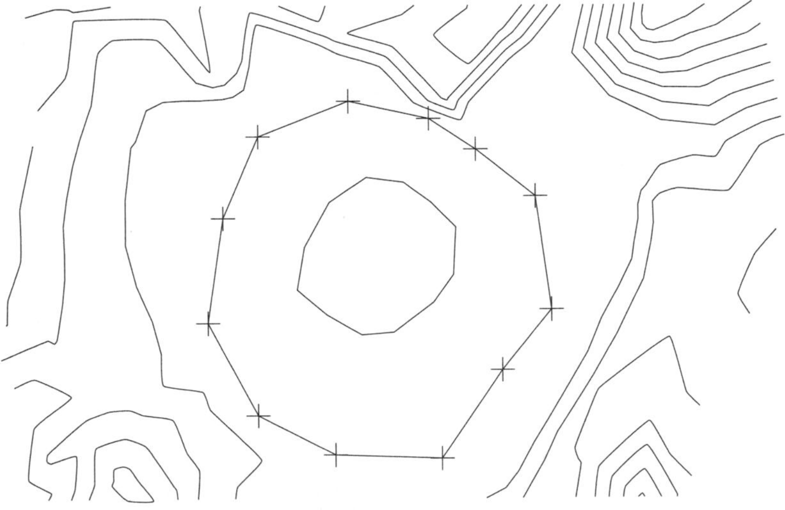

Figure 13–20

Transition Option

The Transition option creates a change of slope over a series of vertices. The Transition option works just like the elevation transition in the 3D polyline Create by Elevation routine.

EXERCISE

Set the starting slope, toggle on transition, advance to the vertex where the new slope occurs, end out of transition, and set the ending slope. The routine then calculates a transitional slope, between the starting and ending slope values, for each intervening vertex.

2. Rerun the Create Multiple routine to create new daylight data for the crown of the pile. The slopes around the pile vary when the slopes change from a 3:1 to a 5:1. The Transition option handles this change in slopes by varying the slope from 3:1 to 5:1 over a series of vertices. Use the transition option to vary the slopes between the various daylight slopes. The northeast face is a 3:1 slope, the southwest slope is 8:1, and the northwest and southeast sides are a 5:1 slope. Use the Transition option to change from one slope value to another.

BUILDING SURFACE PILE2

After you have set the daylight data for each vertex, if you are satisfied with the results, create the point and daylight line data by using the Daylight All routine. The Daylight Points, Daylight Breaklines, and Daylight Polyline routines create only their respective types of data. The Daylight All routine creates all the daylight points, breaklines, and polylines.

1. Use the Daylight All routine of the Daylighting cascade of the Grading menu. Select Base as the daylight surface. The routine creates the slope points and daylight polyline. Press ENTER at each vertex to create three intermediate slope points. Your drawing should look like Figure 13–21.

EXERCISE

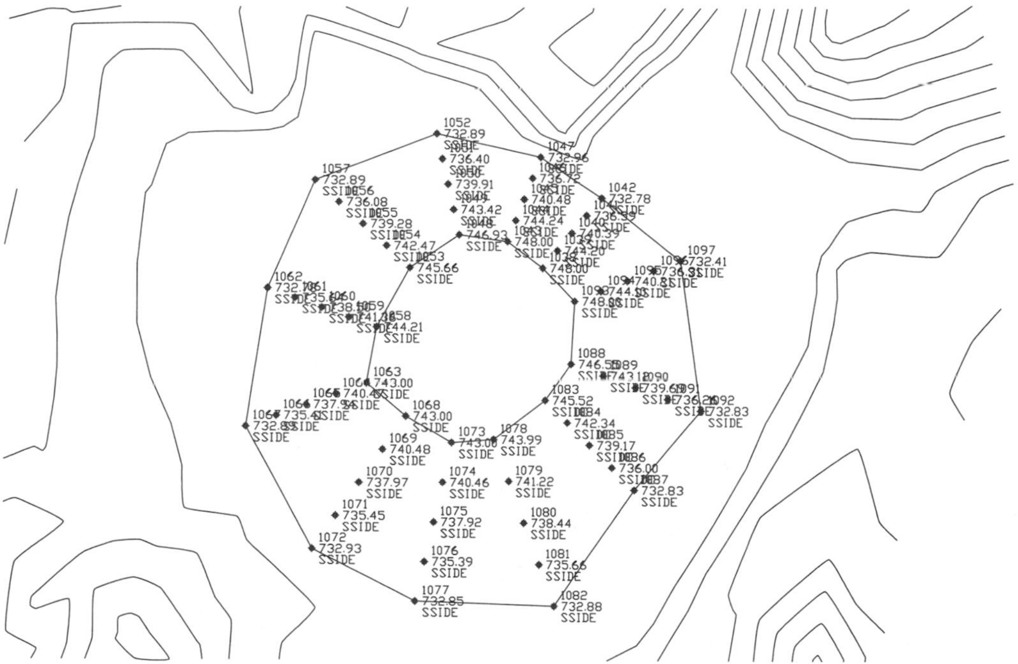

Figure 13–21

2. Run the Point Group Manager routine from the Point Management cascade of the Points menu. Create a point group, Pile2, by selecting the points on the screen.
3. Restore or open the Terrain Explorer and select Terrain. After selecting Terrain, click the right mouse button and select Create New Surface. Select Surface1, click the right mouse button, select Rename, and rename the surface to Pile2.
4. Select Point Groups and add the Pile2 point group.

The breakline data for the surface consists of the crown and toe daylight line. Both of the objects are 3D polylines, which means both are standard breaklines.

5. Use the Define by Polyline routine and name the breaklines pile2, and answer Yes to delete the objects.
6. Create the contour data and use the Layer option to select the contours from the excontour layer. Process the contours with a weeding factor of 15.0 and 4.0 and supplementing factors of 20.0 and 1.0. Select one of the contours to select all of the contours in the drawing.
7. Build the surface with the current settings. Make sure to toggle on Minimize Flat Spots from Contour Data.
8. View the surface with the Quick View routine of Surface Display. The surface should look similar to Figure 13–22.
9. Redraw the screen.
10. Click the Save icon to save the drawing.

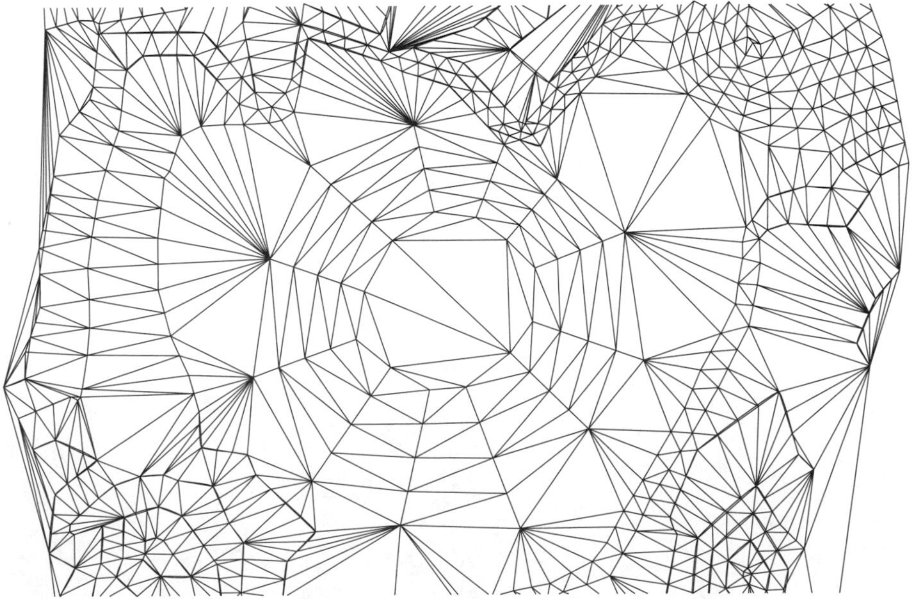

Figure 13–22

CALCULATING PILE2 VOLUME

If the surface is correct, you are now able to generate a volume for the stockpile.

Remember, when calculating the Grid and/or Composite volume, each method creates a new surface. Name each surface, pile1g, or pile1c—g for grid and c for composite.

If you have not done a volume calculation before, you should review section 6 of this book.

1. Use the ZOOM command to zoom out to view the entire project area.
2. From the Volumes area at the bottom of the Terrain menu, select the Define Current Stratum routine and define a new stratum. Name the stratum pile2, and set the surfaces to base (surface 1) and pile2 (surface 2).
3. Use any or all of the volume methods to calculate a volume. If you use the Calculate Total Site Volume from either the Grid or Composite Volumes cascade of the Terrain menu, the routine will ask you to name a surface for the resulting volume calculations. Note down the resulting volumes.

Remember, when calculating the Grid and/or Composite volume, each method creates a new surface. Name each surface pile1g, or pile1c—g for grid and c for composite.

4. Remove the points for pile2 with the Remove Points from Drawing routine found in the Points menu. Use the Select option to remove the points.
5. Erase the daylight polyline of pile2. Leave the crown polyline in the drawing.
6. Close the Terrain Explorer.
7. Click the Save icon to save the drawing.

UNIT 3: GRADING WITH A GRADING OBJECT

Land Development Desktop has a grading object. It is very similar to the multisloped 3D polyline of the Create Multiple routine. What is powerful about this object is its ability to always search for daylight even if you move or rotate the object. You can copy the object, and the new object is a clone of the first. There are many more choices for designing from the object than just setting a cut or fill slope. You can daylight to an elevation (relative to or absolute), a shear face (retaining wall), and daylight for a distance out from the object. All of these methods can occur on the same object. The object allows you to transition from one slope to another along its length.

The object can create a surface, contours, and a volume from its state. The state does not have to be the final condition of the polyline.

You can assign elevations to the object from a surface; each vertex is an average of all vertices' elevations from a surface, individually, and all the same elevation.

The routines affecting the grading object are in the Slope Grading cascade of the Grading menu (see Figure 13–23).

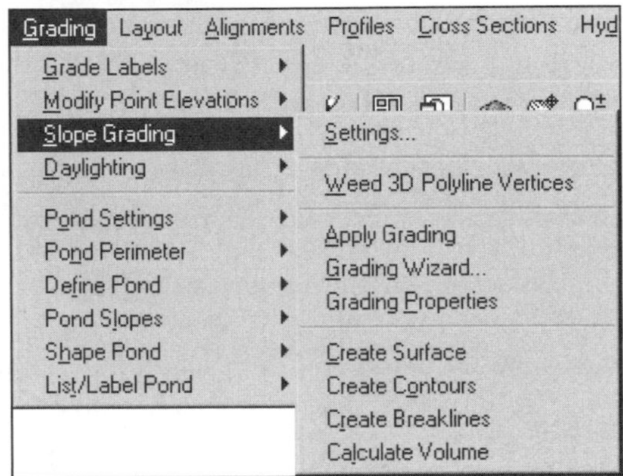

Figure 13–23

GRADING OBJECT SETTINGS

The Settings routine of the Slope Grading cascade of the Grading menu affects the behavior and how the grading object appears on the screen. You can call the Setting routine from the Slope Grading cascade or from the Civil Design settings of the Drawing Settings routine of the Projects menu. The dialog box contains several tabs each with values and controls you can set.

FOOTPRINT

When active, this allows you to set the elevations at the vertices along the grading object (see Figure 13–24). You can add or delete vertices as needed. The best strategy is to define the least complicated grading object and let the settings create the desired results.

TARGETS

The Target panel sets the daylight point for the vertex (see Figure 13–25). A daylight target can be a surface, an elevation (absolute or relative), and a distance from the grading object.

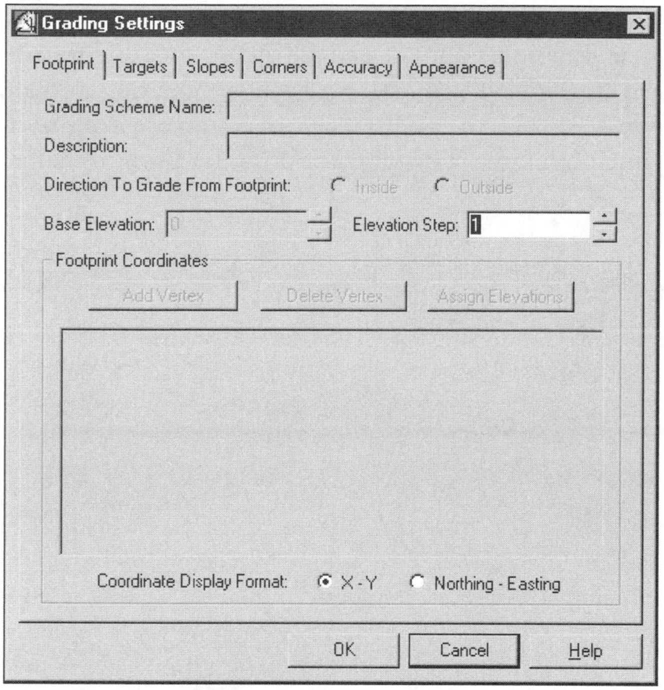

Grading Settings

Footprint | Targets | Slopes | Corners | Accuracy | Appearance

Grading Scheme Name:

Description:

Direction To Grade From Footprint: Inside Outside

Base Elevation: 0 Elevation Step: 1

Footprint Coordinates

Add Vertex | Delete Vertex | Assign Elevations

Coordinate Display Format: X - Y Northing - Easting

OK | Cancel | Help

Figure 13–24

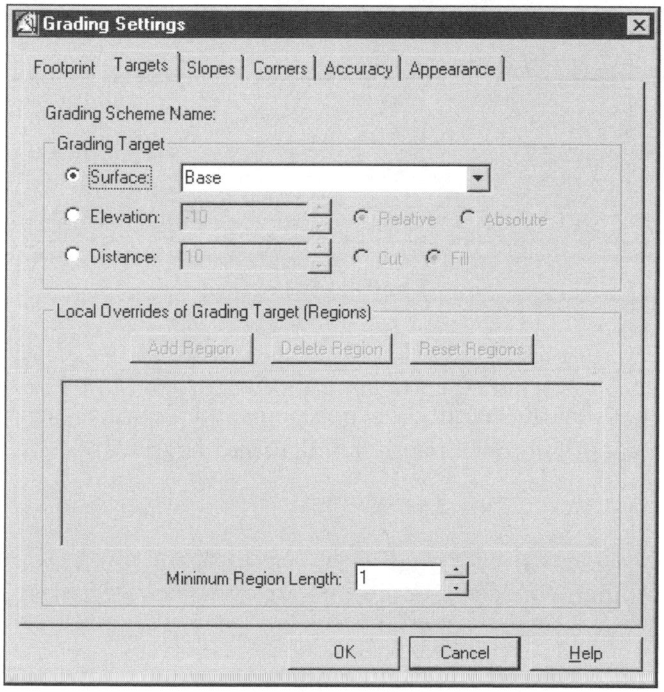

Grading Settings

Footprint | Targets | Slopes | Corners | Accuracy | Appearance

Grading Scheme Name:

Grading Target

Surface: Base

Elevation: -10 Relative Absolute

Distance: 10 Cut Fill

Local Overrides of Grading Target (Regions)

Add Region | Delete Region | Reset Regions

Minimum Region Length: 1

OK | Cancel | Help

Figure 13–25

SLOPES

The Slope panel sets the daylight cut and fill slopes (see Figure 13–26). You can use grades instead of slopes if you prefer. If you are creating a retaining wall, you can set the slope to vertical (cut or fill). You can assign any condition to a segment around the polyline perimeter. The segment does not have to match the segments on the polyline.

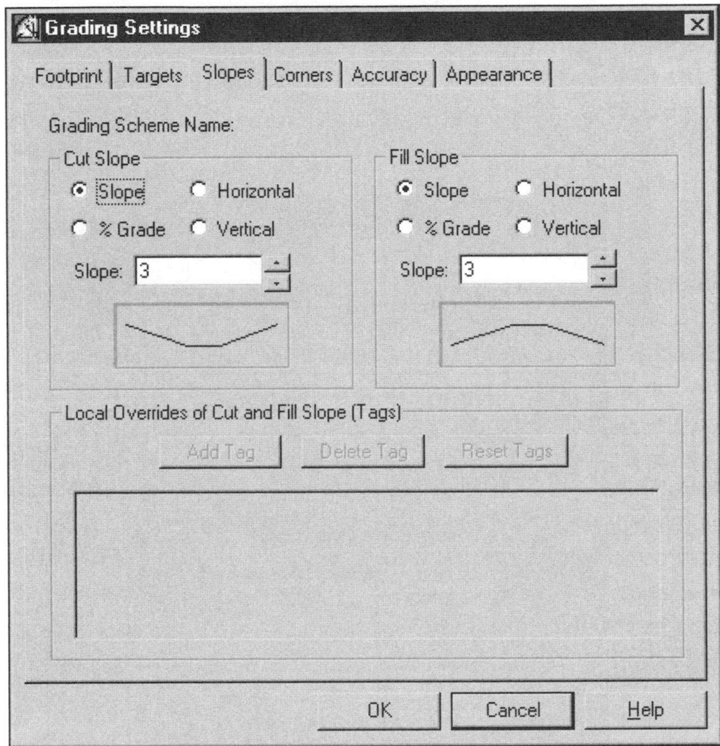

Figure 13–26

CORNERS

The Corners panel sets the method for going around a corner (see Figure 13–27). There are three types of corners: mitered, radial, and chamfered.

ACCURACY

The Accuracy panel sets the method of determining the daylight line (see Figure 13–28). The automatic spacing method creates slope projections at or with less spacing than the current setting.

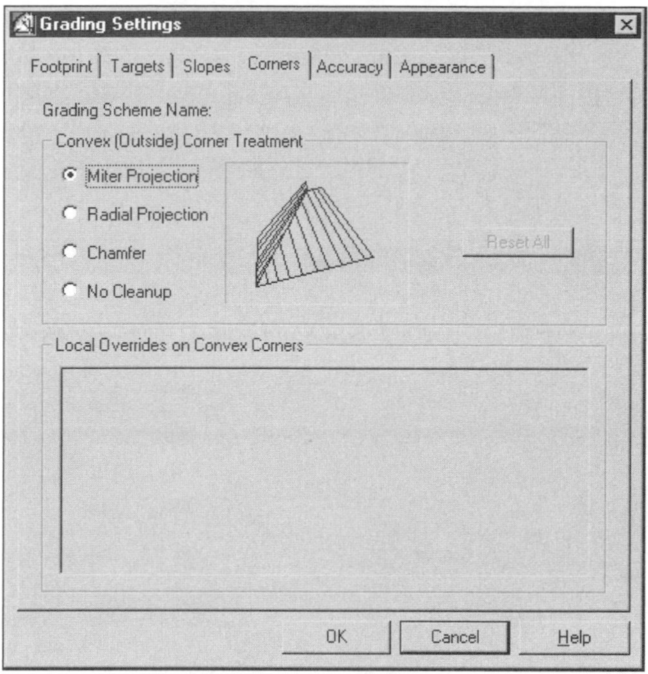

Figure 13–27

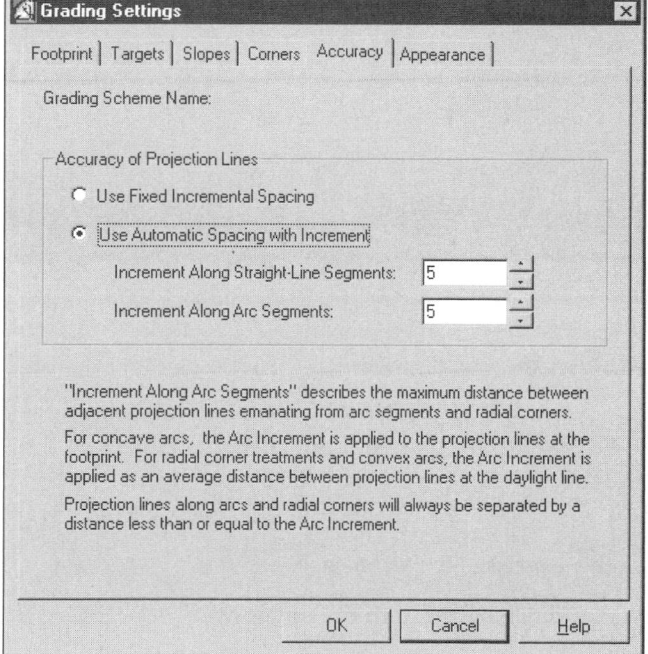

Figure 13–28

APPEARANCE

The Appearance panel sets the color for the solution of the grading object (see Figure 13–29). The colors of the solution change to reflect the solution currently found. If you should change the grading object's elevation or move it to interact with a hill or valley, the object will find the solution and if necessary change its colors to reflect the new situation.

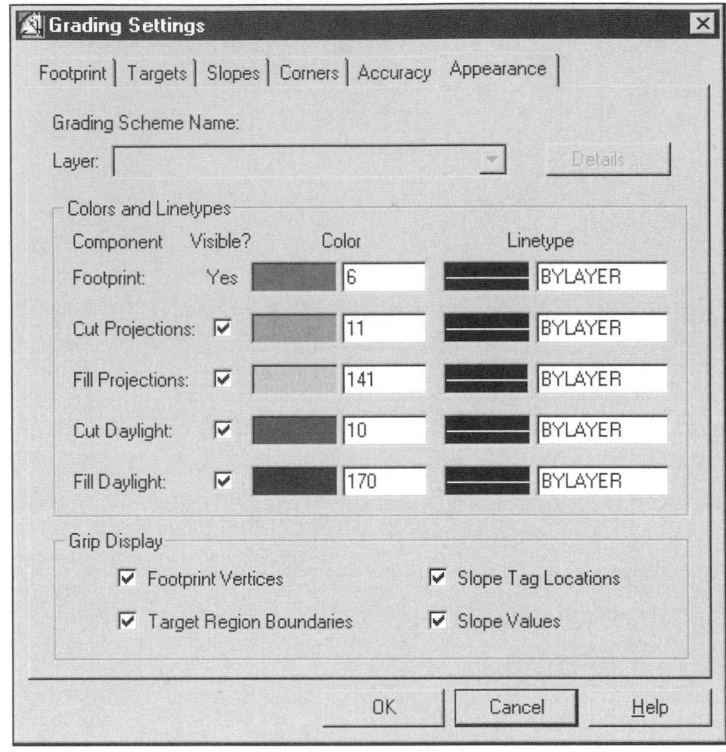

Figure 13–29

Exercise

As in the last exercise, it would seem better to draw the crown of the pile and then project slopes from the crown down to the base surface. This method allows the pile to have consistent elevations at the top and to maintain reasonable slopes from the crown to the toe. The grading object allows you to create a complex solution quickly from a grading object.

After you complete this exercise, you will be able to:

- Create a stockpile design from the top down
- Work with the grading object
- Calculate the volume numbers

DESIGNING PILE3

1. Continuing in the Stockpile drawing, you should see the crown and the contours. Use the PAN command and pan the crown to left side of the screen so you will be able can see both the crown and the Grading Wizard.

2. Run the Grading Wizard routine from the Slope Grading cascade of the Grading menu. Select the crown and then select a point on the outside to daylight to the outside of the crown. After you select the daylight direction, the routine displays the Grading Wizard.

3. In the Footprint panel, review the stations of the vertices along the perimeter of the crown. As you select each vertex in the lower portion of the panel, a marker moves along the crown polyline.

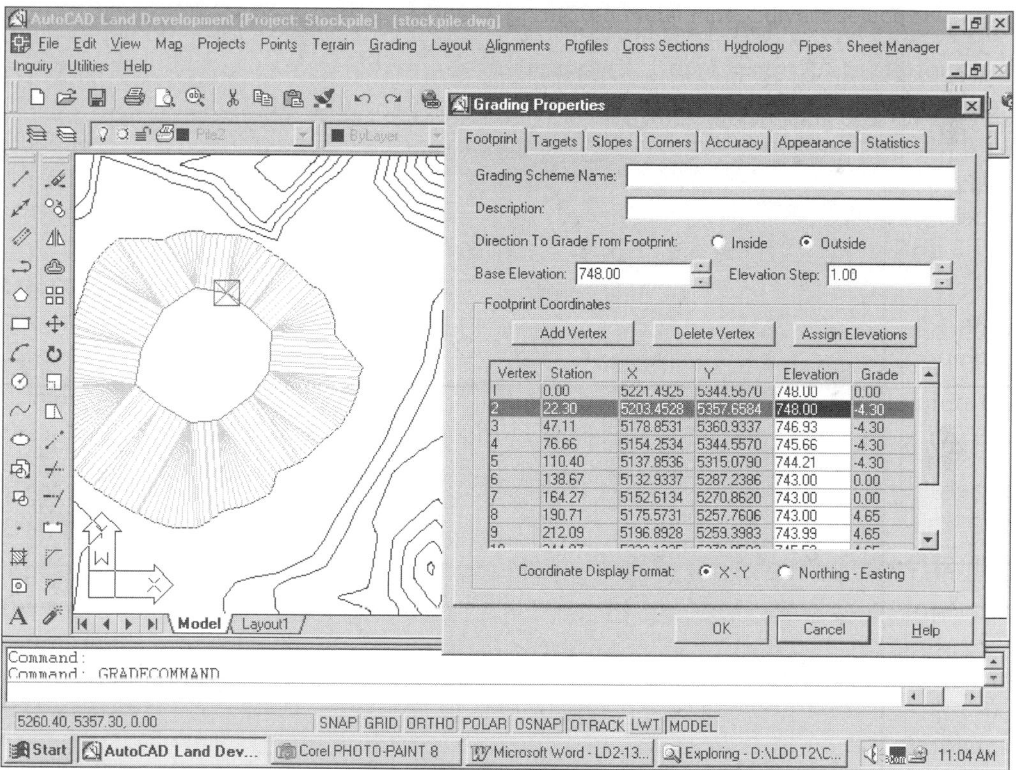

Figure 13–30

4. Review the settings in each panel. Set the values in the Target and Appearance tab as shown in Figures 13–25 and 29. After you set the values, select the Finish button to exit the Wizard. When the routine prompts you about deleting the polyline, answer No. The routine then draws the solution.

EXERCISE

5. Select the slope projection lines of the grading object, click the right mouse button, and then select Object Viewer from the shortcut menu. The grading object creates a 3D solution from the conditions you applied to it in the Grading Wizard (see Figure 13–31). To tilt the object up in the viewer, you click and hold the left mouse button down in the bottom circle of the green orbit icon and move the mouse to the center of the orbit circle. When your view of the object is from the side, you release the mouse button. To rotate the object, you click and hold the left mouse button down while selecting the right or left circle on the orbit icon and moving the cursor to the right or left. After you are done viewing the object close the Object Viewer by selecting the X in the upper right corner of the dialog box.

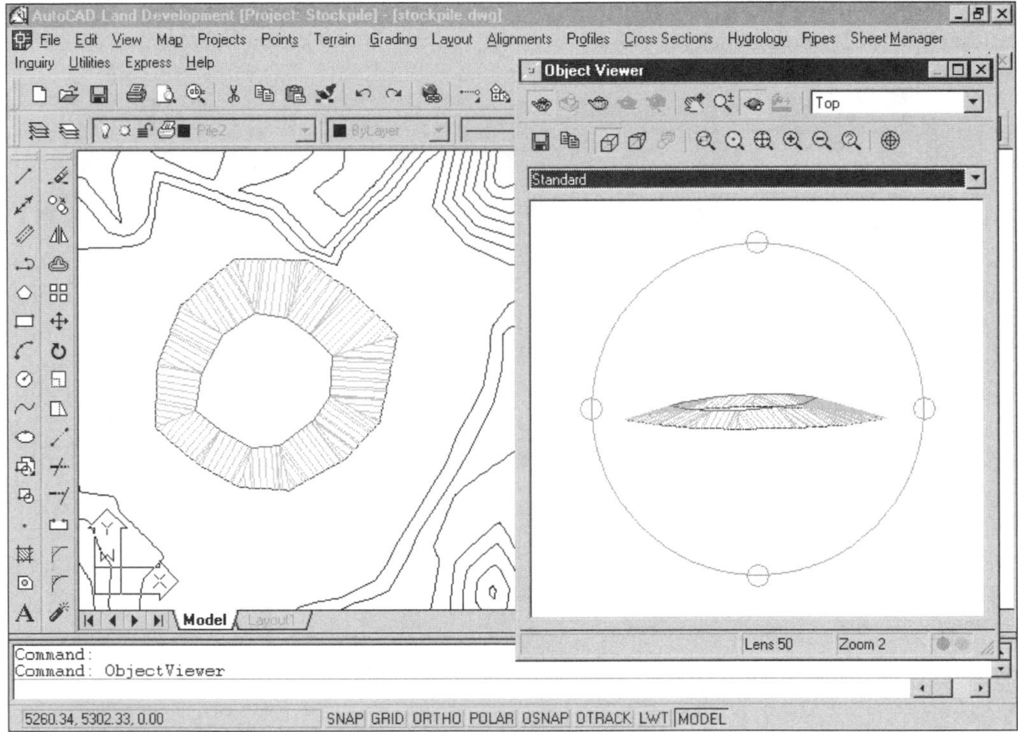

Figure 13–31

6. Again select the grading solution slope lines, click the right mouse button, and select grading properties. Change to the Footprint panel, set the base elevation to 720, then select the Finish button at the bottom of the dialog box to see the effect of the new elevation. The colors of the projection lines change because the solution is now cut rather than fill.

7. Select the slope projection lines of the grading object, click the right mouse button, and then select Object Viewer from the shortcut menu. The Grading

object creates a 3D solution from the conditions you applied to it in the grading Wizard (see Figure 13–32). Close the Object Viewer by selecting the X in the upper right corner of the dialog box.

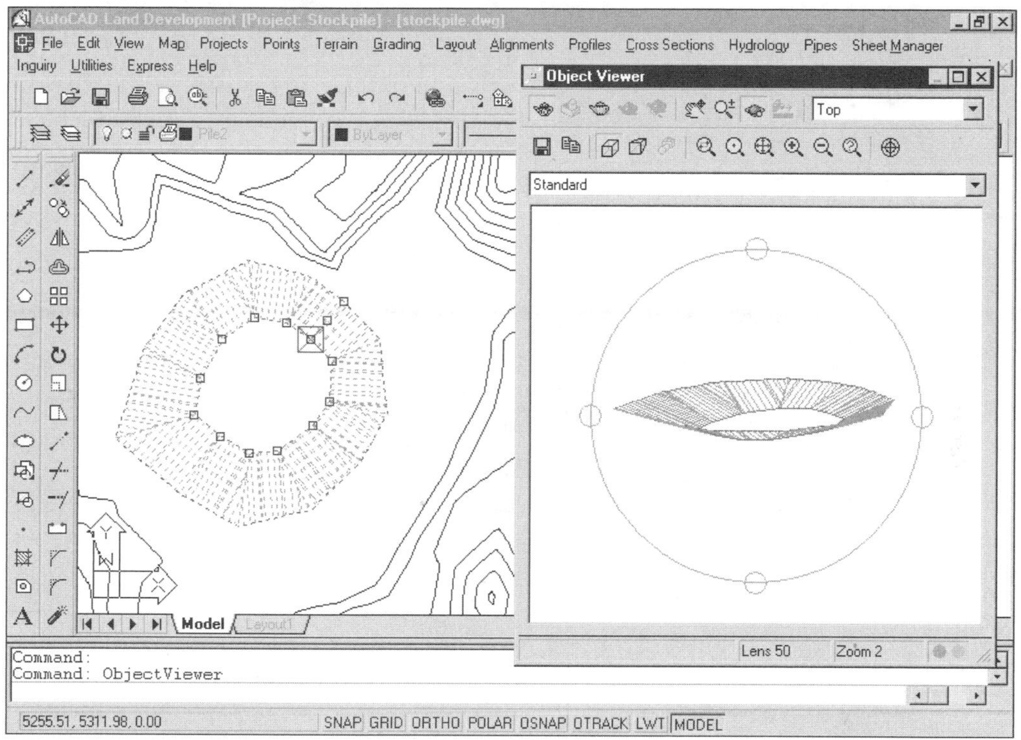

Figure 13–32

8. Select the grading slope lines, click the right mouse button, and select Grading Properties from the shortcut menu. Select the Footprint tab and reset the base elevation to 748. Select the Finish button to make the grading object a pile.

MULTIPLE SLOPE SOLUTION

1. Select the grading slope lines, click the right mouse button, and select Grading Properties from the shortcut menu. Select the Slopes tab to view the current slopes around the perimeter of the grading object. There should be two stations, 0.00 and the ending station.
2. Select the end station at the bottom of the dialog box and then select Add Tag. This adds an intermediate station for which you can set new slope values and conditions. Create new tags and edit their stations and elevations to match the following data. Your polyline crown may not be a long as the one in this exercise, however, the values should guide you in setting the values for your polyline. Use Figures 13–33 and 13–34 as a guide for entering in the slope data.

Station	Slope (Cut and Fill)
0.0	3:1
45.0	3:1
75.0	5:1
120.0	5:1
140.0	8:1
190.0	8:1
225.0	5:1
275.0	5:1
300.0	3:1

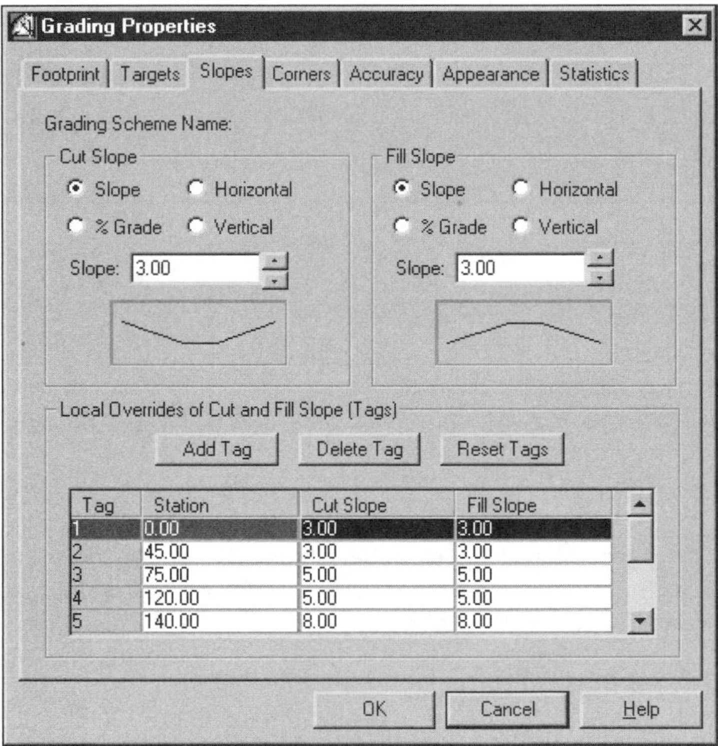

Figure 13–33

3. After entering the data, select the OK button to view the new solution.
4. Select the grading lines, click the right mouse button, and select Object Viewer from the shortcut menu. Rotate the grading object to view the variable slopes and the change of elevations from the front of the pile to the back. Exit the Object Viewer by selecting the X in the upper right corner.

EXERCISE

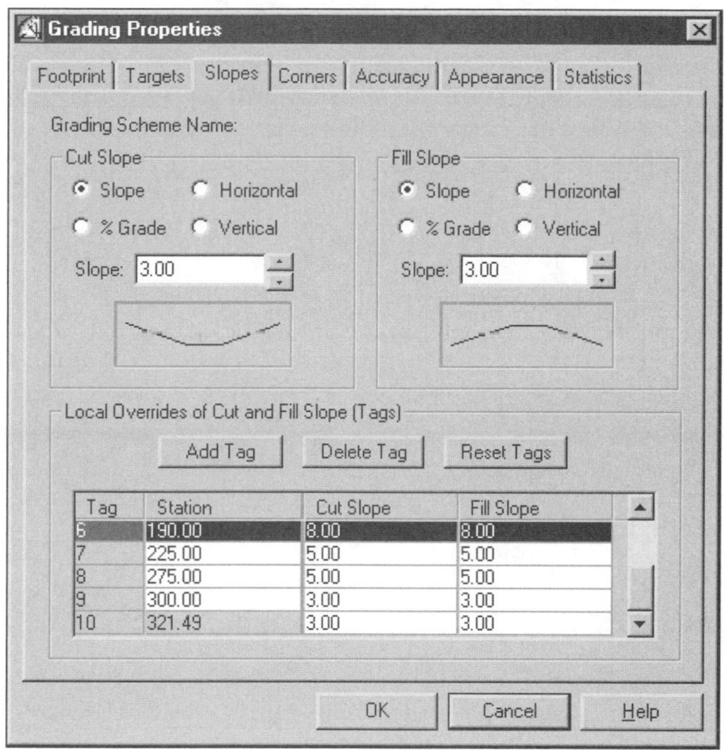

Figure 13–34

5. Select the grading slope lines, click the right mouse button, and select grading properties from the shortcut menu. Select the Statistics tab and then select the Calculate button. This calculates the cut and fill volume represented by the grading object. Select the Finish button to exit the grading Wizard.
6. Click the Save icon and save the drawing.

The grading tools of Civil Design include points, 3D polylines, and the grading object. The grading object is ideal for pad grading and simple pond design. The ability to move and copy the grading object allows you to place grading solutions quickly into a design. However, there are limitations to the grading object, especially when grading around curves where the slopes project towards another set of slopes. The grading object is not aware of situations when it encounters itself (see Figure 13–35).

The topic of the next section is the pipe design program of Civil Design. Once the grading design is done, the handling of storm and sanitary water is essential.

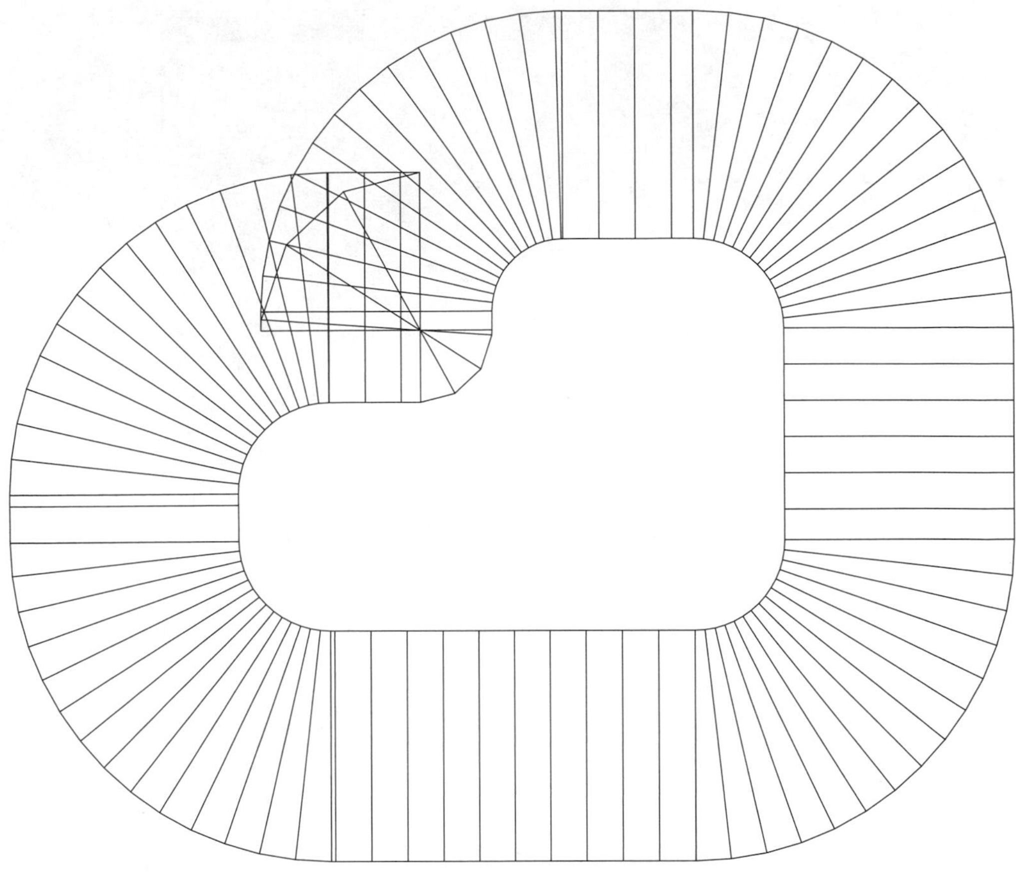

Figure 13–35

SECTION 14

Civil Design Pipes

After you complete this section, you will be able to:

- Define pipe-run-specific settings
- Define new pipe run structures
- Define pipe runs
- Edit and analyze pipe run data
- Anotate pipe runs in plan and profile view

SECTION OVERVIEW

The Civil Design program contains tools to create pipe runs or networks of piping (see Figure 14–1). The pipe design tool set is in the Pipes menu. The routines of the Pipes menu allow you to design storm-water and sanitary systems in a Desktop project. Even though you design a network, Civil Design stores the network as individual runs. When one run connects into another, the only link between them is a user-entered flow from one run to the next.

The pipe design routines and their resulting pipe runs are dependent upon a long series of default settings. These settings govern the type of structure, size of pipe, inverts, and a host of additional values.

Civil Design requires a defined horizontal alignment and probably a profile before editing the pipe routines. Pipe runs and alignments are closely related in Civil Design. You can associate a pipe run with an alignment or define the pipe run as its own alignment. If you define a pipe run as an alignment, you are able to create a profile based upon the definition of the pipe run.

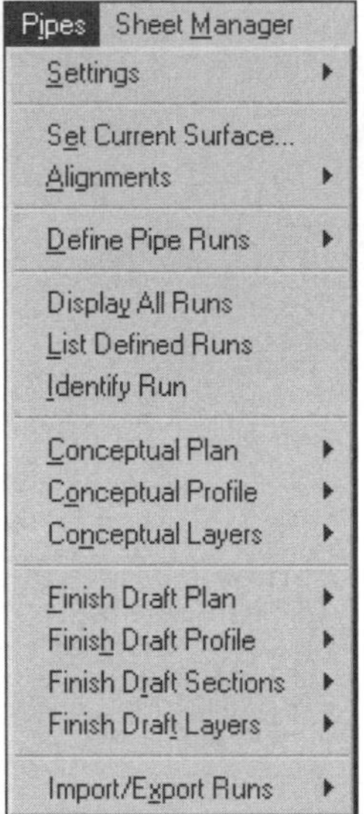

Figure 14–1

Civil Design ties all pipe runs to an alignment definition. The pipe run must be astride of a defined alignment; if not, the pipe run is the associated alignment. A pipe run cannot start or end beyond the associated alignment definition. If a pipe run is outside an alignment definition, the user runs the risk of a pipe routine not working or producing unpredictable results. If a pipe run is to start outside an alignment, the user will have to define dummy alignments to accommodate the pipe run. These dummy alignments maintain the correct roadway centerlines and still produce reliable, working pipe runs.

The development and editing of a pipe run design takes place in plan and profile view. When the design becomes final, the import routines place the line work and annotation into plan, profile, and cross sections.

You can edit a pipe run by three different methods: graphical, database, and spreadsheet/dialog box editor.

The routines on the Pipes menu run within the context of plan and profile views of the pipe run. There are graphical editing tools for plan or profile view. There is no tool that can edit all aspects of a pipe run. You can edit a majority of the pipe run in plan view with the Edit Graphical and Edit Data routines of the Conceptual Plan cascade.

The Edit Graphical routine of the Conceptual Plan cascade adds, deletes, moves, and edits most values of a pipe run. In the Edit Graphical routine of the Conceptual Profile cascade, you can edit the elevation of pipes and structures, but not their location.

The second method of editing pipe runs is directly within the pipe's database. In the database view, Civil Design presents the numbers behind the structure and pipe objects. The database dialog box allows for changes to many of the values but not all of them. The routine refuses to change a value it is not able to modify, but does not indicate which ones you cannot edit.

The spreadsheet/dialog box editor calculates and allows for changes and additions to the currently defined run. The additions include runoff amounts, flow values, headloss, and the editing of names and types of structures. The editor does not allow for the deleting or adding of structures. You can delete and add structures only with the plan editing tools.

Since Civil Design automates so many of the pipe tasks, the pipe run routines have an extensive set of defaults. These defaults affect the labeling of pipes, nodes (structures), and data values. Ideally the settings apply to all pipe runs, but this is not always the case. There is a difference between the piping and type of structures in sanitary and storm pipe networks. So setting the defaults to match the typical structure within the pipe run is a must.

One part of the settings consists of a library of structure definitions. These definitions contain descriptions, measurements, and symbology for plan and profile representations of the structures in a run. The user can develop new structures and add them to the pipe run library.

UNIT 1

Unit 1 focuses on the settings used by the Pipes program. As mentioned above, the list is extensive. The definition and annotation routines depend on and use the currently loaded set of settings. So, if you define different sets of settings, for example, storm, sewer, and catch basin, you must make sure the appropriate set of settings is loaded for the definition and annotation routines to work correctly.

UNIT 2

The definition and drawing of pipe runs is the topic of the second unit. You can import a run from an ASCII file or an Access database, define the run from an existing polyline, or draw the run by selecting points or entering stations and offsets to an associated alignment.

UNIT 3

The third unit reviews the process of editing a pipe run's data. The editing can be done in plan or profile view either graphically or in a spreadsheet format.

UNIT 4

Unit 4 covers the annotation of the pipe runs. Civil Design will annotate the runs in plan, profile, and cross section. All of the annotation that appears is a direct result of the currently loaded settings. If you have settings for different types of pipe runs, you must make sure you have the correct group of settings loaded.

UNIT 1: PIPES SETTINGS

The pipes portion of the Civil Design program uses an extensive set of settings to do its work (see Figure 14–2). The settings affect the typical structure and pipe of a newly defined pipe run, the annotation for plan, profile, and cross sections, hydrologic units, and precision. Since the structures in a storm sewer system are different from those in a sanitary system, it is important that you create a separate set of settings for storm and sanitary systems. You may also want to think about creating a set of settings for a storm-water system that is mainly catch basins rather than manhole structures.

If you are able to standardize on a set of defaults, it is important that you copy them into the prototype you use in the office. This way, every new job starts with the same settings and you are able to produce drawings with some consistency.

You access the Pipes Settings Editor dialog box from the Drawing Settings routine of the Projects menu or when you run the Edit routine of the Settings cascade of the Pipes menu of Civil Design (see Figure 14–3).

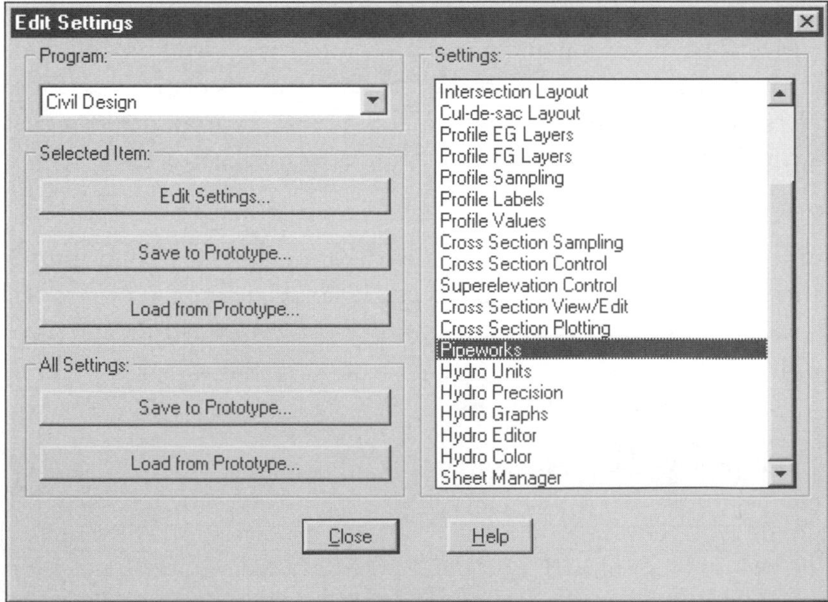

Figure 14–2

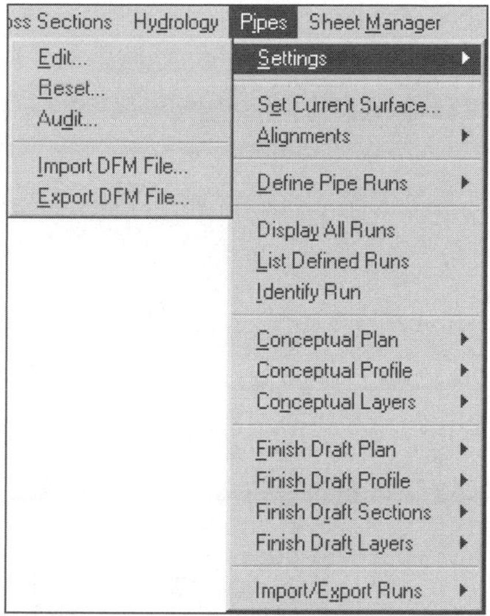

Figure 14–3

PIPE DRAFTING LABELS

PLAN PIPE DRAFTING SETTINGS

The Plan Pipe Drafting Settings dialog box controls the labeling of pipe segments of a pipe run in plan view (see Figure 14–4). The settings control the label position, type, prefixes, and suffixes.

Plan Pipe Drafting Settings
File: d:\land projects r2\abon\dwg\abon.dfm
Pipe Label Position Above
Line Type Single Line
Line Text
Label slope in %

Toggle	Prefix	Suffix	Precision
Size	DIA	in.	0 8 0
Slope	SLOPE	ft/ft	0 8 3
Material			
Label			
Arrow			
Length	LEN	'	0 8 2

OK Cancel Help

Figure 14–4

PROFILE PIPE DRAFTING SETTINGS

The Profile Pipe Drafting Settings dialog box controls the labeling of pipe segments of a pipe run in profile view (see Figure 14–5). The settings control the label position, type, prefixes, and suffixes.

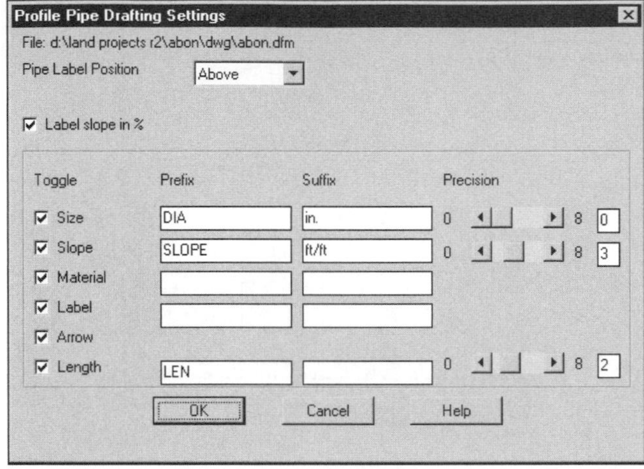

Figure 14–5

NODE DRAFTING LABELS

PLAN NODE LABEL SETTINGS

The Pipes-Plan Node Label Settings dialog box controls which elements are to be a part of a plan structure label set (see Figure 14–6). The elements include Station, Offset, Elevation, type of pipe run (Storm or Sewer), and Node Name.

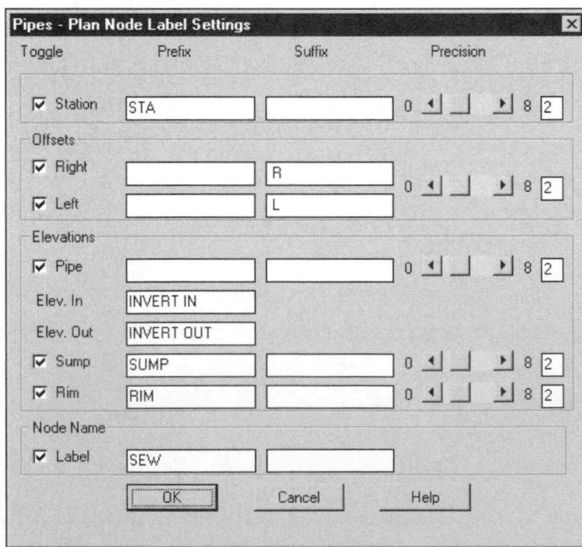

Figure 14–6

PROFILE NODE LABEL SETTINGS

The Pipes-Profile Node Label Settings dialog box controls which elements are to be a part of a profile structure label set (see Figure 14–7). The elements include Station, Offset, Elevation, type of pipe run (Storm or Sewer), and Node Name.

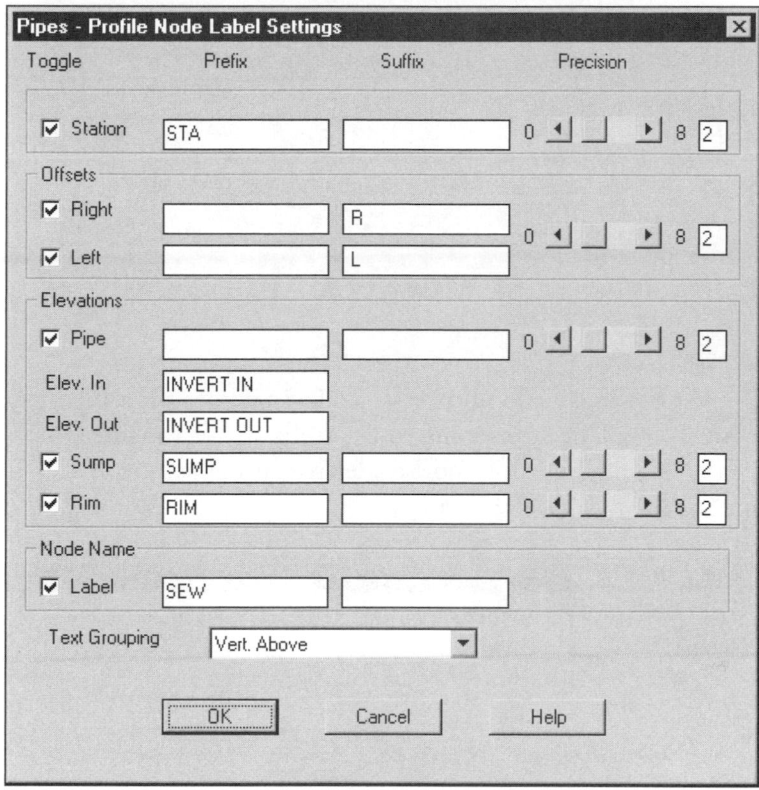

Figure 14–7

LAYER DATA

PLAN

The Plan Layer Settings dialog box sets the layer names for a pipe run in plan view (see Figure 14–8). The layer names use a prefix that is the name of the pipe run (indicated by an * asterisk), so an import or delete routine will erase only those layers used by a pipe run.

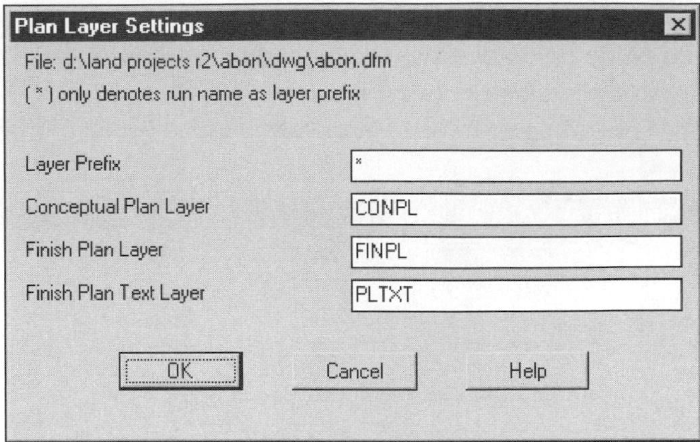

Figure 14–8

PROFILE

The Profile Layer Settings dialog box sets the layer names for a pipe run in profile view (see Figure 14–9). The layer names use a prefix that is the name of the pipe run (indicated by an * asterisk), so an import or delete routine will erase only those layers used by a pipe run.

Profile Layer Settings
File: d:\land projects r2\abon\dwg\abon.dfm
(*) only denotes run name as layer prefix
Layer Prefix: *
Conceptual Profile Layer: CONPR
Finish Profile Layer: FINPR
Finish Profile Text Layer: PRTXT
Hydraulic and Energy Gradeline Layer: HYD_GRD
OK | Cancel | Help

Figure 14–9

CROSS-SECTIONS

The Cross-Sections Layer Settings dialog box sets the layer names for a pipe run in cross sections (see Figure 14–10). The layer names use a prefix that is the name of the pipe run, so an import or delete routine will erase only those layers used by a pipe run. In this instance, the prefix is an * (asterisk).

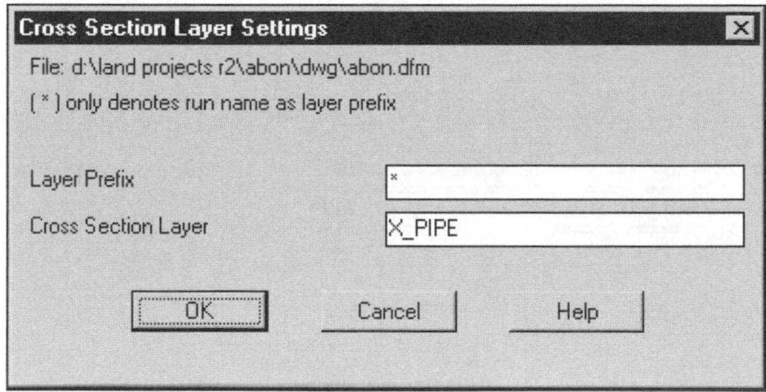

Figure 14–10

NODE AND PIPE DATA VALUES

PLAN SCALE EXAGGERATION

The scale exaggeration settings affect only the plan annotation of the pipe run (see the Nodes and Pipes-Plan Exaggeration Factor dialog box in Figure 14–11). The factors set in the dialog box scale the text and symbols by a simple multiplying factor.

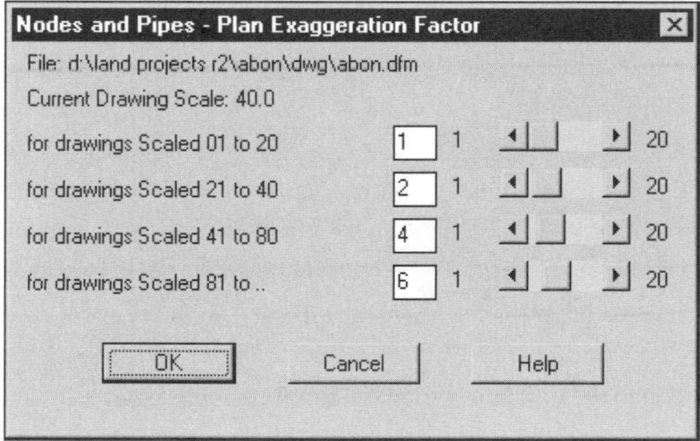

Figure 14–11

SLOPE CONTROL

The slope control settings provide the pipes calculator with the minimum and maximum slopes for a system (see the Pipe Slope Control dialog box in Figure 14–12). The calculator issues warnings when the slopes set for the run are exceeded. This dialog box is called from the Pipe Setting Editor dialog box of this group.

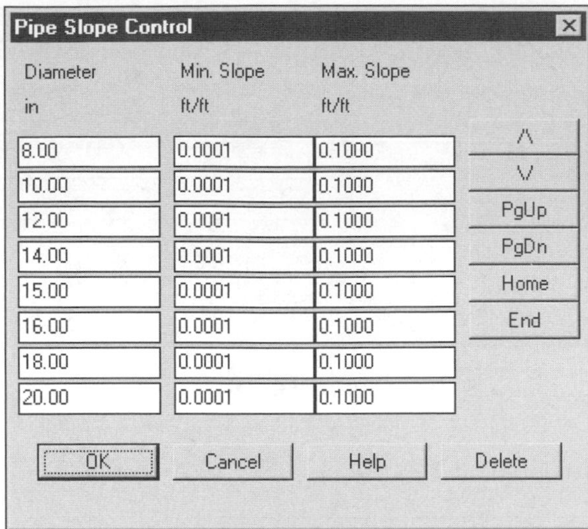

Figure 14–12

MATERIAL/COEFFICIENT

The Select Coefficient Table dialog box lists the material types and their associated coefficients (see Figure 14–13). The dialog box lists coefficients for Manning's n, Hazen-Williams Coefficient, and Darcy-Wiesbach Friction factor.

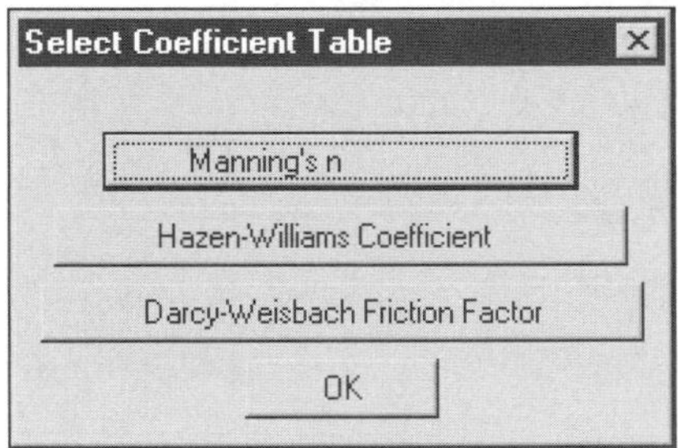

Figure 14–13

PIPE

The Pipe Data Settings dialog box contains the most important settings for pipes in a pipe run (see Figure 14–14). The settings in the dialog box are the default settings for pipes in the next defined pipe run. These settings become the initial data for pipes in a pipe run. The settings affect the pipe size, minimum and maximum slope (slope control), the formula to use in evaluating the pipe run, the material type, and the invert in and out drop from the rim.

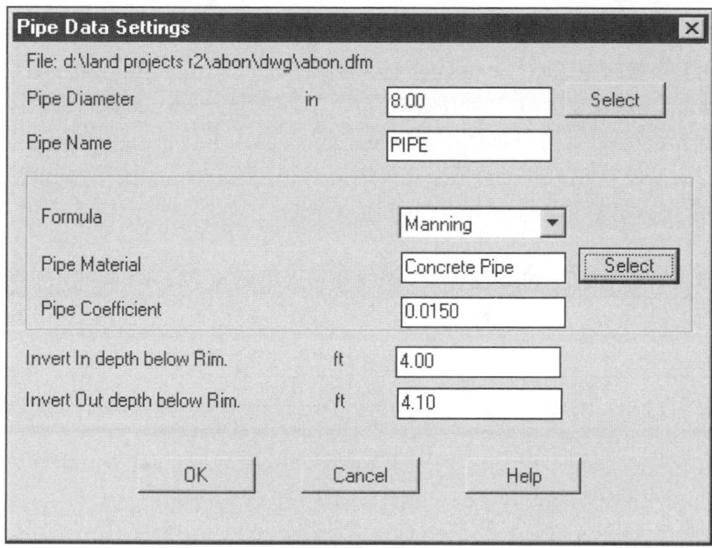

Figure 14–14

NODE

The Node Data Settings dialog box contains the most important settings for structures in a pipe run (see Figure 14–15). The settings in the dialog box are the default settings for structures in the next defined pipe run. These settings become the initial data for structures in a pipe run. The settings affect the structure label, structure library reference, and the in and out head loss values.

Node Data Settings
File: d:\land projects r2\abon\dwg\abon.dfm
Setting Node Data
Node Label Node
Structure Reference...Description
pmh Proposed Manhole
Node Head Losses
In 0.15 ft Out 0.15 ft
OK Cancel Help

Figure 14–15

STRUCTURE LIBRARY

The Structure Library Editor contains the structure name, the plan symbol, and the profile cross section of a structure (see Figure 14–16). The structure library comes with several predefined structures. However, you can define new structures and save them to the library. This library is important in that it defines what symbols the annotation routines use in plan view and how they draw the structure in profile view.

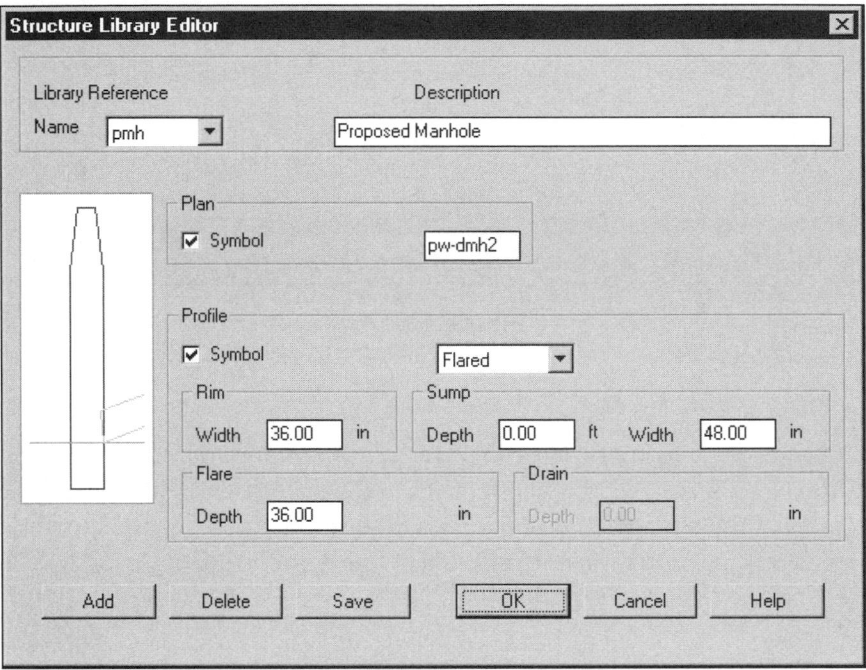

Figure 14–16

EDITORS

RUN EDITOR

The Run Editor is a group of settings affecting the calculations that the routines perform on a pipe run (see the Pipes-Run Editor Settings dialog box in Figure 14–17). The settings affect the method of calculation (upstream/downstream), whether the pipes are to be resized automatically, which formula to use in calculating pipe sizes, maximum pipe lengths, the calculation of a hydraulic grade line, and a peak adjustment factor.

You can access additional evaluation settings when you select the Range Set button at the bottom of the dialog box (see the Range Check Settings dialog box in Figure 14–18). These settings set the minimum and maximum values for water velocity in the pipes, another check for slope, and ground cover over the pipes (perpendicular or vertical).

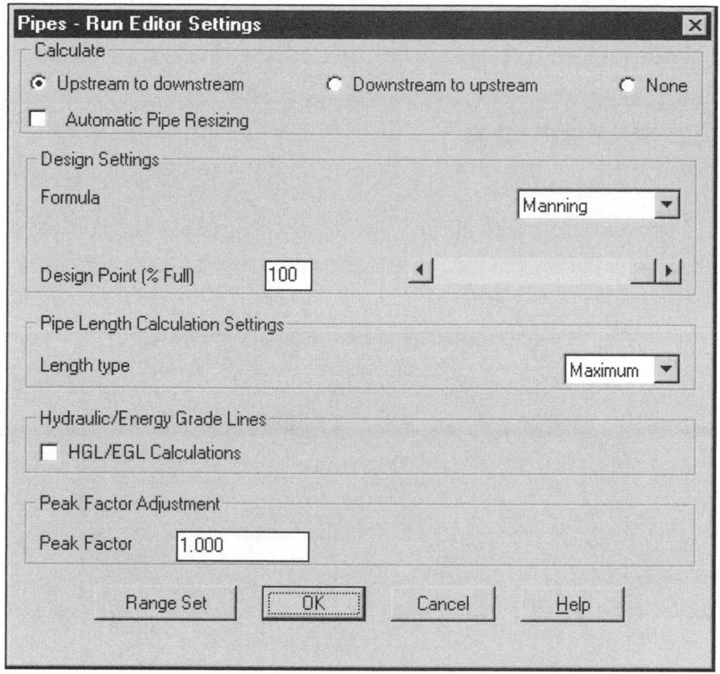

Figure 14–17

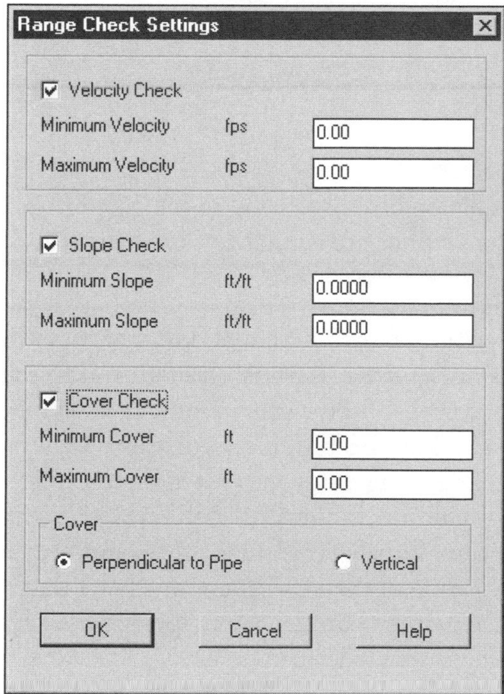

Figure 14–18

TEXT EDITOR

This dialog box allows you to use another text editor such as Notepad.

UNITS AND PRECISION

UNITS

The Hydraulic Units dialog box allows you to set the units for the pipe system (see Figure 14–19). The settings are similar to the settings in Hydrology.

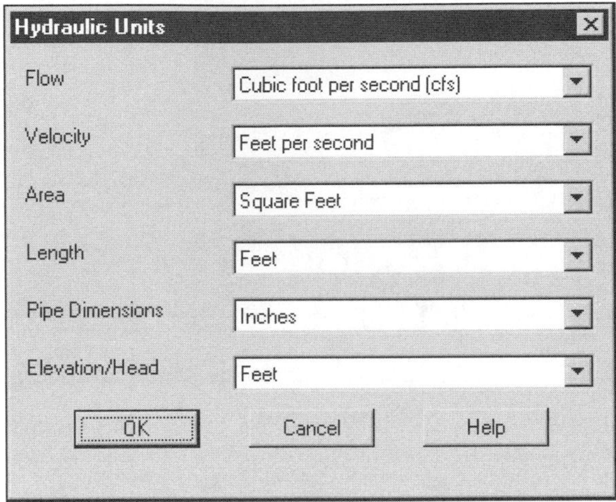

Figure 14–19

PRECISION

The Precision Settings dialog box allows you to set the precision for the pipe system (see Figure 14–20). The settings are similar to the settings in Hydrology.

All of these settings combine to create a run of a particular type. For example, the current settings are for a sanitary manhole run. The structures in the run do not have sumps, and the label identifies them as SEW structures. If we were designing a storm manhole run, there would probably be sumps in some or all of the structures, a different size pipe, and the label would be SSTM. Even the storm manhole run settings would not be correct if we were defining a run with catch basins. The structure types would be different, pipe sizes smaller, and the label different. The best strategy is to define as best you can a set of standards for each type of run and then use the Export DFM routine of the Setting cascade of the Pipes menu to create that set of defaults. When you get ready to define the pipe run, load the appropriate set of settings and then define the run using the just loaded settings.

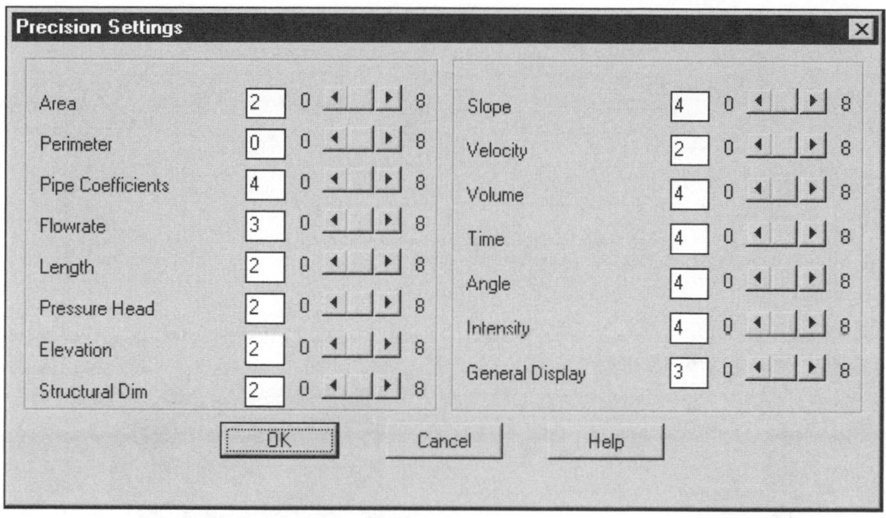

Figure 14–20

Exercise

This exercise reviews the settings for Pipes. The list is extensive, although most of the time you do not change many of the settings. What is necessary is to create three sets of settings. The first set of run data is for a sanitary sewer system, the second for storm-water manholes, and the last is for storm-water catch basins.

After you complete this exercise, you will be able to:

- Review the pipe settings
- Define pipe-run-specific settings
- Define new pipe run structures

EXERCISE SETUP

1. Start a new drawing, *Abon*, and assign it to a new project, Abon (see Figures 14–21 and 14–22). Use the LDDTR2 prototype and describe the project as the Pipeworks Exercise. Select the OK button to create the project and the drawing.

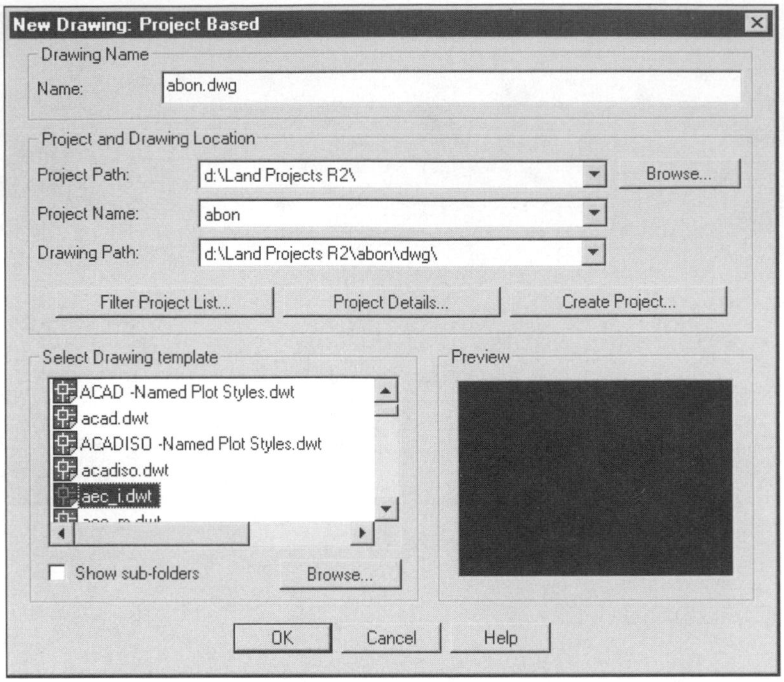

New Drawing: Project Based
Drawing Name
Name: abon.dwg
Project and Drawing Location
Project Path: d:\Land Projects R2\
Browse...
Project Name: abon
Drawing Path: d:\Land Projects R2\abon\dwg\
Filter Project List...
Project Details...
Create Project...
Select Drawing template
ACAD -Named Plot Styles.dwt
acad.dwt
ACADISO -Named Plot Styles.dwt
acadiso.dwt
aec_i.dwt
Show sub-folders
Browse...
Preview
OK
Cancel
Help

Figure 14–21

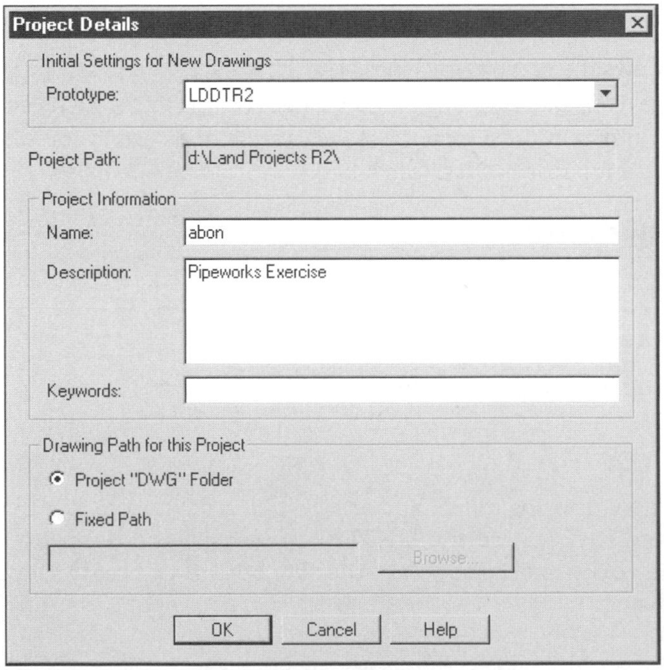

Project Details
Initial Settings for New Drawings
Prototype: LDDTR2
Project Path: d:\Land Projects R2\
Project Information
Name: abon
Description: Pipeworks Exercise
Keywords:
Drawing Path for this Project
Project "DWG" Folder
Fixed Path
Browse...
OK
Cancel
Help

Figure 14–22

2. When the Point Database settings dialog box appears, select the OK button to dismiss the dialog box.
3. The next dialog box is the Load Setup dialog box. Select the LDD1-40 setup, then the Load button at the top right, and finally select the Finish button at the bottom right to dismiss the dialog box. When the Finish dialog box displays, select the OK button to dismiss the dialog box.
4. Insert the *Abon.dwg* file from the CD that comes with this book. Insert the drawing at 0,0 and explode upon insertion.
5. Click the Save icon to save the drawing.

CREATING THE EXISTING SURFACE

1. Run the Terrain Model Explorer routine from the Terrain menu.
2. Create a new surface and name the surface Existing. Use contour data for the surface. The contours are on two layers, COMB-CONT-NML and COMB-CONT-HGH. Use the weeding factors of 15.0 and 4.0 and the supplementing factors of 20.0 and 1.0. Select the contours by layer.
3. Build the existing surface. Make sure Minimize Flat Areas and Use Boundary both are on. When the surface is made, select OK to exit the Done dialog box and return to the Terrain Modeler.
4. Review the resulting triangulation with the Quick View of Surface display either in Terrain Model Explorer or from the Terrain menu. There is false triangulation at the upper left of the project.

Use a boundary to control the false triangulation that occurs in the upper left of the contour data set.

5. Minimize the Terrain Model Explorer. Select the Surface Boundaries routine in the Surface Edit cascade of the Terrain menu. When the routine prompts to remove or add, press ENTER on add. The next question is to remove any existing boundaries. Answer yes to the question. Then select the polyline on the Boundary-Triangle layer, define the boundary as a show boundary, do not make the boundary a breakline, press ENTER on selecting another boundary, and answer yes to review the surface. The following is the command sequence.

```
Command:
RemoveAll/Add <Add>: (press ENTER)
Remove all existing boundary definitions (Yes/No) <No>: y
Select polyline for boundary:
Boundary definition [Show/Hide] <Show>: S
Make breaklines along edges (Yes/No) <No>: (press ENTER)
Select polyline for boundary:
View/Review surface (Yes/No) <Yes>:
Creating View ...13600
```

EXERCISE

6. Redraw the screen to remove the triangles.
7. Close down Terrain Explorer.
8. Use the Layer dialog box and freeze the contour and boundary-triangle layers.
9. Save the drawing by clicking the Save icon.

SANITARY SEWER SETUP

1. Select the Settings routine from the Pipes menu. This displays the Pipes Settings Editor.
2. In the Pipe Drafting Labels group (Plan and Profile), select the buttons to review their settings. In each dialog box (Plan and Profile) toggle on Label Slopes in percent (see Figures 14–4 and 14–5). If the Label prefix is set to EX, erase the prefix. After setting the values in each dialog box, select the OK button to return to the Pipes Settings Editor.
3. In the Node Drafting Labels group (Plan and Profile), select the buttons to review their settings (see Figure 14–6 and 14–7). In each dialog box (Plan and Profile), if the Node Name Label prefix is set to EX, erase the prefix. You can change SEW to Sewer or Sanitary if you like. After reviewing the settings in each dialog box, select the OK button to return to the Pipes Settings Editor.
4. In the Layer Data group, select each button (Plan, Profile, X-Section) to view the settings (see Figures 14–8, 14–9, and 14–10). After reviewing the settings, select the OK button to return to the Pipes Settings Editor.
5. In the Node and Pipe Data Values group, select the appropriate button and review the Scale Exaggeration, Slope Control, and Material/Coeffcient Settings dialog boxes (see Figures 14–11, 14–12, and 14–13). After reviewing each of the settings, select the OK button to return to the Pipes Settings Editor.
6. Select the Pipe button and set the pipe size to 12 inches. You have to choose the Select button to the right of the pipe size to display the Pipe Size and Slope dialog box. After selecting the pipe size, return to the Pipes Settings Editor. Set the formula to Manning and select Concrete Pipe as the material type. Again choose the Select button to the right of the Material Type, and in the new dialog box, select Concrete pipe. Next, set the Invert depth below Rim values to 15.0 for the invert In and to 15.1 for the Invert Out. Your dialog box should now look like Figure 14–23. After you set the values, select the OK button to exit to the Pipes Settings Editor.
7. Select the Structure Library Editor button to view the current structure list (see Figure 14–16). You view each structure by selecting its name from the drop-down list at the top of the dialog box. Review the structures, their size, flare, sump, and symbol names. View the pmh structure last.

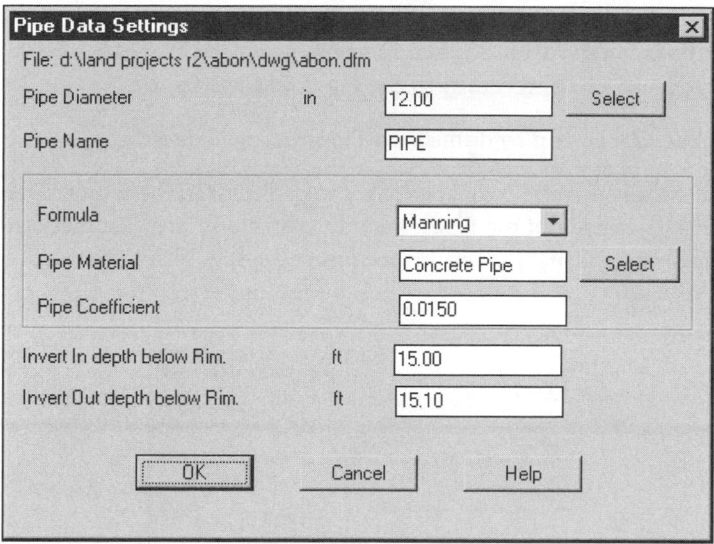

Figure 14–23

8. In Node and Pipe Data Values, select the Node button to display the Node Data Settings dialog box. Set the structure type to pmh, the node label to San #, and the headloss in and out to 0.15 (see Figure 14–24). After setting the values in the dialog box, select the OK button to exit and return to the main dialog box.

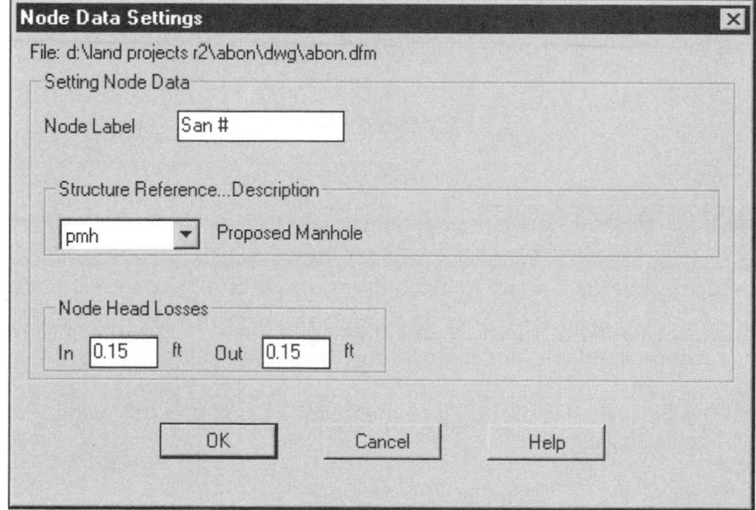

Figure 14–24

9. Select the Hydraulic Units button to view the current settings and change the Area to acres. After setting the values in the dialog box, select the OK button to exit and return to the main dialog box.

10. Select the Precision button to view the current settings and adjust them to match the settings in Figure 14–20. After setting the values in the dialog box, select the OK button to exit and return to the Pipes Settings Editor.

11. Select the OK button to dismiss the Pipe Settings Editor.

12. To make these settings permanent for the project, run the Export DFM routine of the Settings cascade of the Pipes menu to create a settings file that can be loaded when defining a sanitary pipe run (see Figure 14–25). Name the file *Sewer.dfm*.

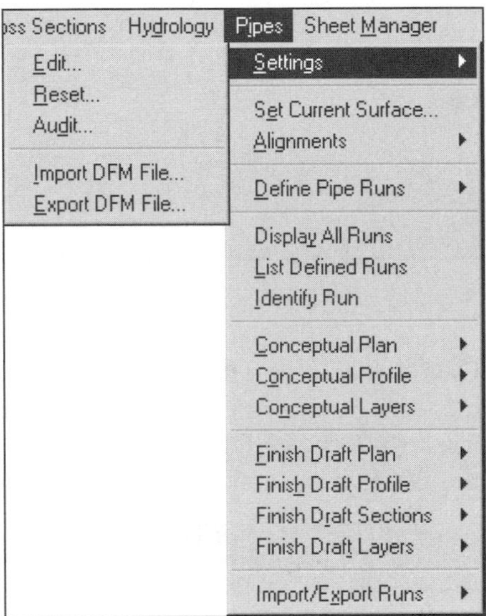

Figure 14–25

EXERCISE

STORM MANHOLE SETTINGS

The exported settings are good only for a sanitary pipe run. There should be files containing settings for a storm sewer and a storm catch basin run. In the next series of steps, you will adjust the type of structure, label settings, and pipe size. After making these changes, you will define current settings as settings for a storm manhole pipe run.

1. Rerun the Edit routine of the Pipes menu and change the following settings for a storm manhole run.

Pipes Drafting Labels:	Plan and Profile
Label Prefix:	Blank
Node Drafting Labels:	Plan and Profile
Node Name Prefix:	STM
Label Prefix:	Blank

2. Select the Structure Library Editor button to display the Structure Library Editor dialog box. At the top left of the dialog box, change the Library Reference to PMH. If you look at the values for PMH, you will notice that there is no sump associated with sanitary manholes. There needs to be a manhole structure with a sump to catch any debris that may enter the storm sewer system. This new structure will be the default structure for any storm manhole run.

3. Select the Add button at the lower left of the dialog box. The button displays a dialog box prompting for the name of the new structure, enter the name of **Psmh**, and select the OK button to dismiss the dialog box and return to the Structure Library Editor. The values from the previous structure are the defaults for the new structure. The new structure is similar to the pmh structure except for its description and sump (see Figure 14–26). Add or change the description to **Proposed Storm Manhole** and make the sump depth **4** feet. After making the changes, click on the Save button at the bottom of the Structure Library Editor to save the new structure to the Structure Library. Finally, select the OK button to exit the Structure Library Editor and return to the Pipe Settings Editor.

Description:	Proposed Storm Manhole
Sump:	4.0

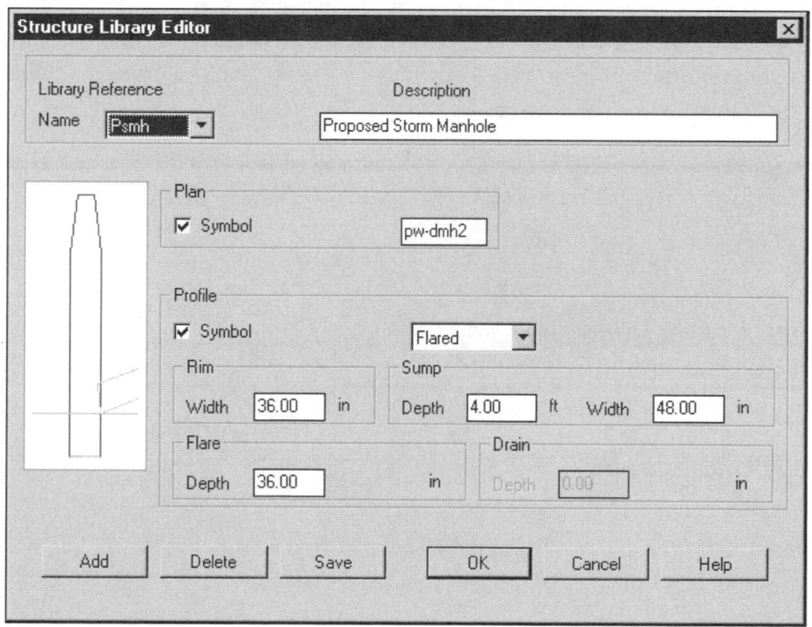

Figure 14–26

4. Next select the Pipe button to display the current settings for pipes. Change the default size of pipe to 10 inches. After you make the changes press the OK button to return to the Pipe Settings Editor.

EXERCISE

Pipe Data Values:

Pipe Size: 10 inches

5. Next select the Node button to display the current settings for node. Change the label, structure reference, and inverts. See the settings below. After you make the changes press the OK button to return to the Pipe Settings Editor.

Node Data Values:

Node Label: StmMH #

Structure Reference: Psmh

Invert In: 6.0

Invert Out: 6.1

6. After making the above changes, select the OK button to exit the Pipe Settings Editor dialog box. The settings represent a typical storm manhole run.
7. Select the Export DFM routine from the Settings cascade of the Pipes menu and export the current settings as *Storm.dfm*.

CATCH BASIN SETTINGS

The current settings are good only for a storm manhole run. There should be a file containing settings for a storm catch basin run. In the next series of steps, you will adjust the type of structure, label settings, and pipe size. After making these changes, you will define current settings as settings for a storm catch basin run.

1. Rerun the Edit routine of the Settings cascade of the Pipes menu and change the following for a catch basin run.

Pipes Drafting Labels: Plan and Profile

Label Prefix: Blank

Node Drafting Labels: Plan and Profile

Node Name Prefix: STMCB

Label Prefix: Blank

Pipe Data Values:

Pipe Size: 8 inches

Node Data Values: Node Label: StmCB #

Structure Reference: Pcb

Invert In: 4.0

Invert Out: 4.1

2. After making the above changes to the settings in the Pipe Settings Editor dialog box, select the OK button to exit the dialog box. The settings represent a typical catch basin run.

3. Select the Export DFM routine from the Settings cascade of the Pipes menu and export the current settings as *Catch.dfm*.

The next preliminary is to define the roadway centerline and create a profile for the segment we are going to use in the exercise. If you have not defined an alignment or worked with profiles before, you should review sections 7 and 8 of this book.

ROADWAY ALIGNMENT

1. Save the drawing by clicking the Save icon.
2. Turn off all of the layers except for the CL layer. The roadway alignment starts at the upper right and ends at the bottom left. Use the Define From Objects routine of the Alignments menu to define the centerline of the roadway. After you select the objects, the red X marks the reference point, so press ENTER to accept the default reference point. Name the alignment Rosewood and describe it as the Southwest Cul-de-Sac (see Figure 14–27). After entering the values, select the OK button to define and exit the dialog box.

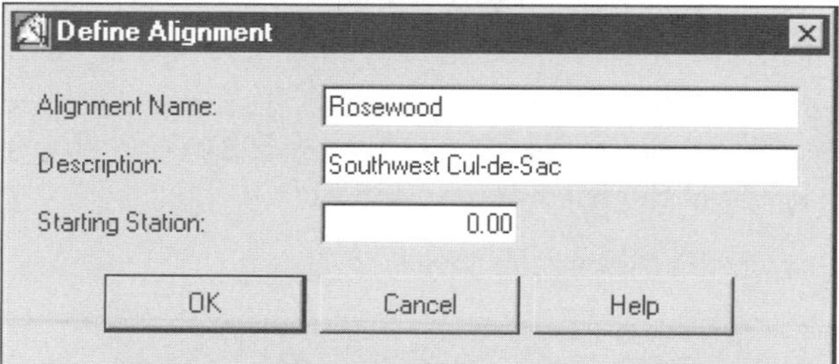

Figure 14–27

3. Turn on all of the layers except for the contours (COMB-CONT*) and surface boundary (Boundary-TRIANGLE) layers.
4. Set a layer prefix for the profiles by selecting the Label and Prefix routine from the Settings cascade of the Profiles menu. Set the prefix to *- (asterisk followed by a dash).
5. Run the Values routine of the Profile Settings cascade of the Profiles menu and make sure that the increments are all set to 50 in the Profile Value Setting dialog box (see Figure 14–28). Exit the dialog box.
6. Sample the elevations along the centerline using the Sample From Surface routine of the Existing Ground cascade of the Profiles menu. Sample from the beginning to the end of the roadway and select OK to dismiss the Profile Sampling Settings dialog box. If there is no current surface, you need to select the existing surface and then select OK to dismiss the dialog box.

EXERCISE

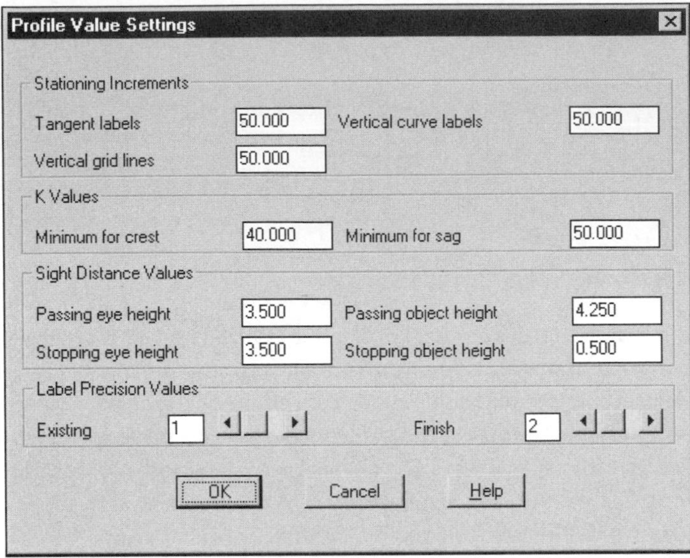

Figure 14–28

7. Create a Full profile for Rosewood; however, do not make it for its entire length. Start the profile at station 14+00 and a datum of 810 (see Figure 14–29). Select a point to the right of the roadway and contours.
8. Save the Drawing by clicking the Save icon.

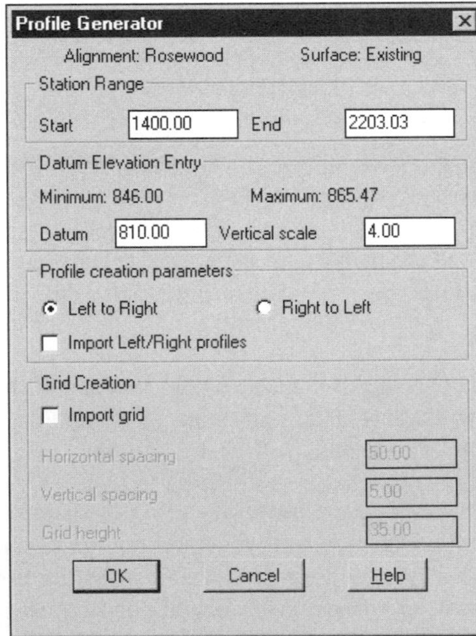

Figure 14–29

UNIT 2: CREATING PIPE RUNS

After you define the settings for the typical pipe runs in a design, the next step is defining the pipe runs. There are three methods for defining pipe runs. The first is drawing the pipe run in the drawing editor. The Draw Pipe Run routine prompts you for the location of the structure by either station/offset or by selecting a point in the editor as the location of the structure. The second method assigns the current settings to a selected existing polyline. The last method reads in an ASCII file, Lotus 1-2-3 spreadsheet, or Access database file to define a pipe run.

All of the pipe runs in a project are tied to an alignment (see Figure 14–30). Many times the sanitary and storm sewers are at specific stations and offsets to the roadway centerline. However, there are occasions when this is not true. A run of pipe and structures may cross an open area and not follow any defined alignment in the project. This pipe run will be its own alignment, and because it is its own alignment, it can have its own profile for the drawing documents. Each pipe run must have a unique name.

With the entire piping system as a set of connected runs, there is no real way to treat the runs as a network or system. It is up to you to manage the overall characteristics of the system. Where pipe runs join, it is essential for you to know how much flow exits from one run and enters into another. Also, when creating the connection, you need to define the connecting structure so it shows in the appropriate profiles. This is done with the Null and Null2 structure types. Null structures do not show in plan or profiles and Null2 structures show only in profile not plan views. If you encounter different situations, you will have to define your own null structure types to handle the situation.

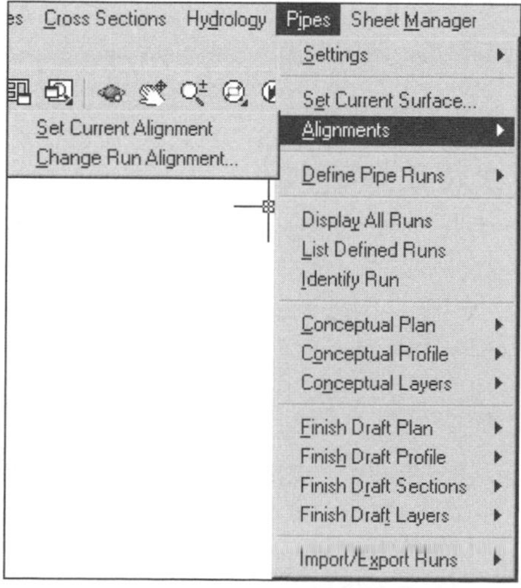

Figure 14–30

DRAW PIPE RUN

The Draw Pipe run routine of the Define Pipe Runs cascade of the Pipes menu draws a polyline on the screen representing the location of the pipe run (see Figure 14–31). The vertices of the polyline are the structures of the run. The vertices are the result of station and offsets entered by the user or from selections made in the graphics editor. The current settings for pipe runs are applied to the polyline. Before you draw the run, you must name the run.

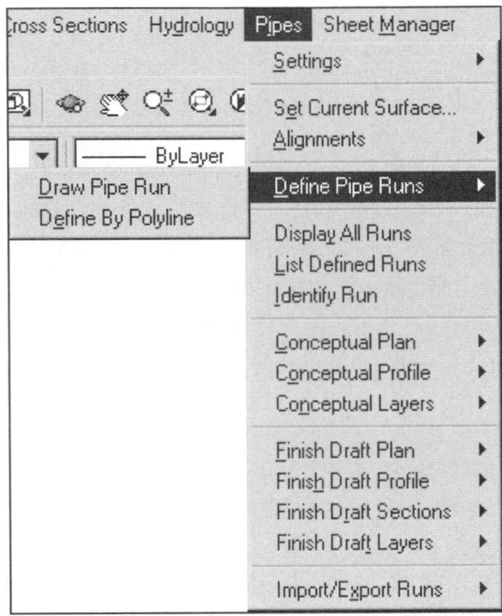

Figure 14–31

DEFINE BY POLYLINE

This routine applies the current pipe run settings to a selected polyline (see Figure 14–31). Before you select the run, you must name the run.

IMPORT/EXPORT PIPE RUNS

The routines of the Import/Export cascade create runs from external data and write pipe run definitions to a file (see Figure 14–32). Before you can import a run, you must define the format of the file. Before using the routine, read the online documentation covering the file format.

ACSII ASC FILE SETTINGS

This routine creates a format for the imported file (see the File Import Field Order dialog box in Figure 14–33). All of the data must be present in the file for the import to work. This means knowing the Northing/Easting of the structures. After importing the file, the routine creates the run on the screen and enters it into the pipe run database.

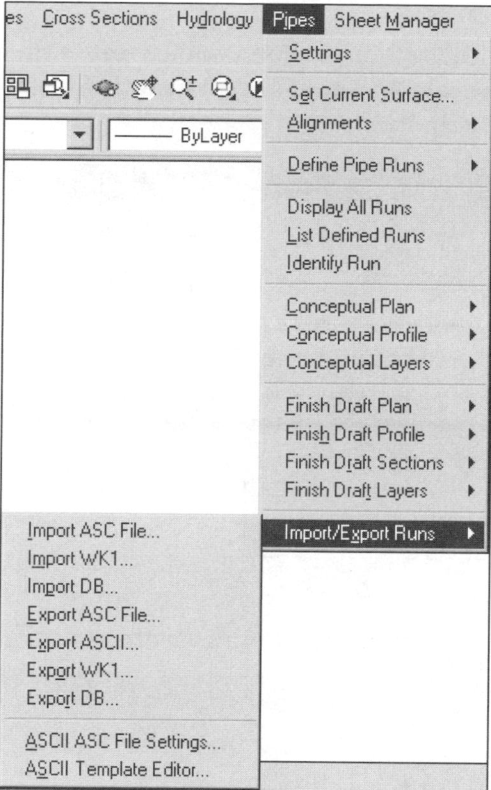

Figure 14–32

File Import Field Order

DEFAULT: d:\land projects r2\abon\pipewks\

Node Label	1	In Loss	8
Rim Elev	2	Out Loss	9
Invert In	3	Pipe Name	10
Invert Out	4	Node Size	11
Sump	5	Pipe Mater.	12
Northing	6	Pipe Size	13
Easting	7	Str Lib Ref	14

Clear Reset OK Cancel Help

Figure 14–33

ASCII TEMPLATE EDITOR

The ASCII Template Editor routine allows you to create a file format of your choice (see Figure 14–34). The column settings in the center of the dialog box are how you define the fields in each exported record.

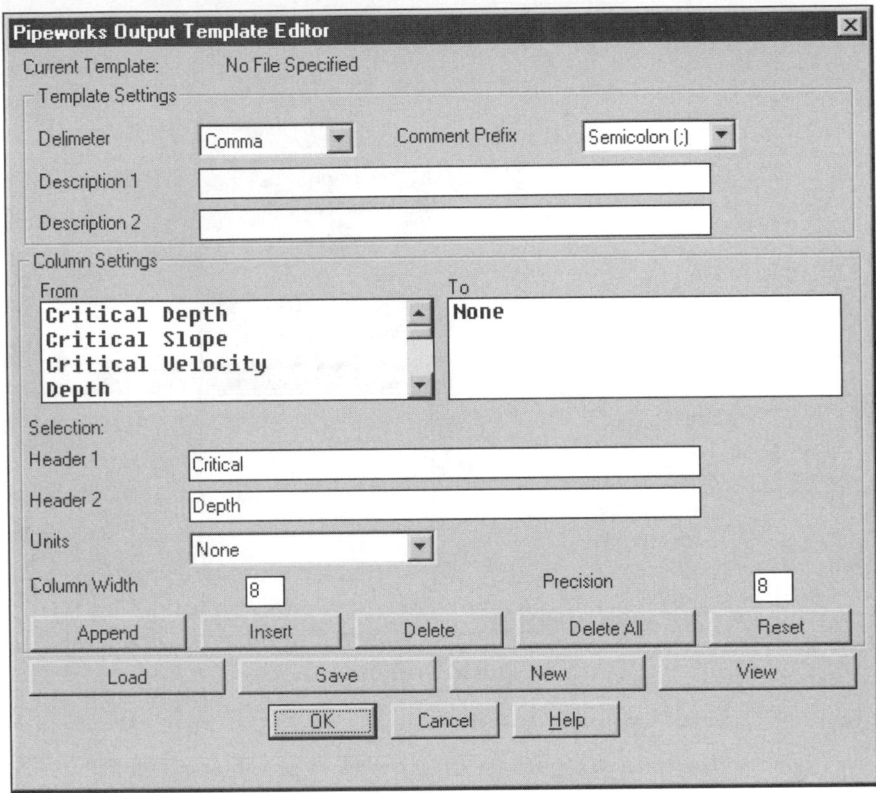

Figure 14–34

Exercise

After you complete this exercise, you will be able to:

- Define pipe runs
- Use the Draw Pipe Run routine
- Use the Define By Polyline routine

SANITARY RUN BY POLYLINE

The first run to define is the existing main branch of the sanitary sewer system. The line work for the run is already in the drawing and needs to have the settings applied to it with the Define By Polyline routine of the Define Pipe Runs cascade.

1. Turn on the layer SanStructure and zoom into the area of the line work for the existing sanitary pipe run by running the ZOOM command with the center option. Enter the coordinates 7450,8000 for the center of the screen and enter a height of 400. Use the Named View command and create the named view *Sanitary* from the current screen.
2. Set the current pipe run settings to sewer by loading the *sewer.dfm* file with the Import DFM routine of the Settings cascade of the Pipes menu.
3. Run the Define By Polyline routine from the Define Pipe Runs cascade of the Pipes menu and select the polyline between the two structures. The routine prompts you, asking whether it should use the existing surface to assign rim elevations. Select on to use the surface to set rim elevations. Finally, name the run ExSan. The final question is to reverse the flow of the run. The flow should be SE to NW. There should be an arrowhead at the top of the line; if there is not, answer Yes. You should not have to reverse the flow of the run. The next dialog box (see Figure 14–35) sets the relationship of the run to an alignment. This run is independent to the Rosewood alignment, so set the toggle to Create an Alignment from the pipe run. The routine then starts the process of defining the polyline segment as an alignment. The right endpoint is the starting or reference point; name the alignment ExSan, and describe it as Existing Sanitary Run (see Figure 14–36). You have now defined a pipe run and it is its own alignment. The ExSan alignment is now the current alignment.

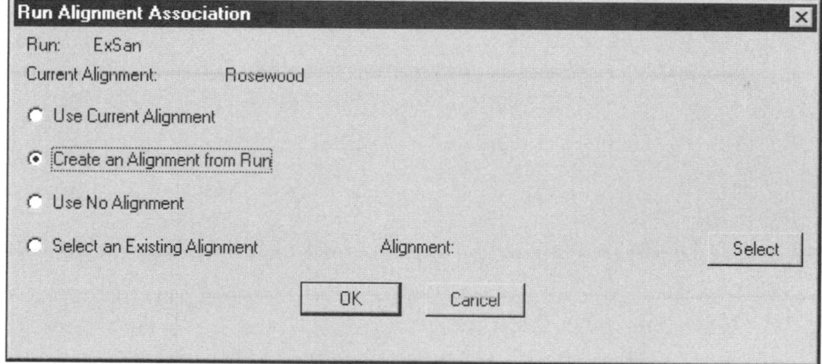

Figure 14–35

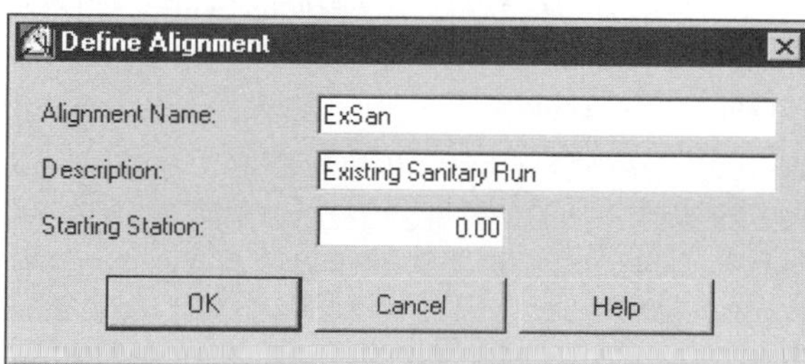

Figure 14–36

EXERCISE

EXERCISE

DRAW A SANITARY RUN

1. Make the Rosewood alignment the current alignment. Use the Set Current Alignment routine in the Alignment menu. If you can not select the centerline from the screen, press the right mouse button, and select Rosewood from the list of defined alignments.

2. Use the Draw Pipe Run routine of the Define Pipe Runs cascade of the Pipes menu to draw the new sanitary run down the middle of the road. Start the pipe run at station 14+50 and end the pipe run by connecting to the structure at station 21+00. The stationing of the pipe run is the same as the roadway. Name the pipe run Interceptor1. Use the Existing surface to assign rim elevations. Create a structure every 150 feet except for the last one—just connect it to the structure at station 21+00. You must add the structure to the run before specifying the next structure. When you get ready to place the last structure, zoom into the ExSan run structure in the middle of the street and use the insert object snap to place the point. After selecting the point, exit from the Add option and select the Save option to save the run. Finally, associate the pipe run with the Rosewood alignment and exit the routine. The following code is an example of the routine running and the ending sequence of options.

```
Command:
Enter new run name: Interceptor1
Select point: (eXit/Station): s (Station option)
Station: 1450
Offset: 0
eXit/Move/Add <Add>:  (press ENTER to accept)
Select point: (eXit/Station): s (Station option)
Station <1450.00>: 1600
 Offset <0.00>: (press ENTER to accept)
Run: Interceptor1     length: 150.00          nodes: 2
eXit/Next/Prev/Move/Delete/Undo/Save/Add <Add>: (press ENTER to accept)
Select point: (eXit/Station): s (Station option)
Station <1600.00>: 1750
 Offset <0.00>: (press ENTER to accept)
Run: Interceptor1     length: 299.94          nodes: 3
eXit/Next/Prev/Move/Delete/Undo/Save/Add <Add>: (press ENTER to accept)
Select point: (eXit/Station): s (Station option)
Station <1750.00>: 1900
 Offset <0.00>: (press ENTER to accept)
Run: Interceptor1     length: 449.94          nodes: 4
eXit/Next/Prev/Move/Delete/Undo/Save/Add <Add>: (press ENTER to accept)
Select point: (eXit/Station): '_zoom
>>Specify corner of window, enter a scale factor (nX or nXP), or
[All/Center/Dynamic/Extents/Previous/Scale/Window] <real time>: _w
```

```
>>Specify first corner: >>Specify opposite corner:
Select point: (eXit/Station): _ins of
Run: Interceptor1      length: 649.94         nodes: 5
eXit/Next/Prev/Move/Delete/Undo/Save/Add <Add>: (press ENTER to accept)
Select point: (eXit/Station): x (press ENTER to exit)
eXit/Next/Prev/Move/Delete/Undo/Save/Add <Add>: s  (press ENTER to save)
Run: Interceptor1      length: 649.94         nodes: 5
Add/eXit/Next/Prev/Move/DElete/DBase/Undo/Save <Save>: x
Run: Interceptor1      length: 649.94         nodes: 5
Command:
```

STORM RUN BY POLYLINE

1. Thaw or turn on the layer StmStructure and zoom the area by using the ZOOM command with the center option. Use the coordinates of 7400,8100 for the center point of the screen and as the height of 500.
2. Load the Storm.dfm file to set the settings to storm manholes. Use the Import DFM routine of the Settings cascade of the Pipes menu.
3. Start the Define By Polyline routine in the Define Pipe Runs cascade of the Pipes menu. Select the southerly polyline representing the pipe run that starts at the roadway edge and ends towards the northwest. When the routine prompts for a surface, turn on the Existing surface for rim elevations. Next name the run StmDetPond. After naming the pipe run, the next task is to set the direction of the run. This run takes the water collected in the cul-de-sac and places it into the detention pond in the northwest. The arrows should be pointing to the northwest. If they are not, tell the routine to reverse the flow of the run. Again create an alignment from the run while in the Run Association dialog box. This run is independent to the Rosewood alignment, so set the toggle to Create an Alignment from the pipe run. This is similar to what we did for the ExSan pipe run (see Figure 14–35). The routine then starts the process of defining the polyline segment as an alignment. The right endpoint is the starting or reference point of the alignment; name the alignment StmDetPond, and describe it as Detention Pond Run. You have now defined a pipe run and it is its own alignment. The StmDetPond alignment is now the current alignment.

EXERCISE

REVIEW SURFACE SLOPES

Before drawing the storm catch basin run, understand that the placement of the run is not a random process. Engineers would have calculated rainfall runoff and designed a surface that will shed water towards the structures you place into a project. It is hard to see the affect of the design on placing structures when you have no context or knowledge of the design. Before placing the catch basin run, look at the types and direction of slopes in the cul-de-sac area.

1. Zoom in on the cul-de-sac area until you can see the southern boundary of the surface as well as the storm and sanitary lines where they exit the current alignment.

2. Run the Slope Arrows routine from the Surface Display cascade of the Terrain menu. Set the routine for four user-defined slope ranges. Select the User Ranges button and set the values as seen in Figure 14–37.

Surface Range Definitions

Range Values

	Begin	End	Layer	Color
1 -	0.00	4.00	SRF-RNG1	1
2 -	4.00	8.00	SRF-RNG2	2
3 -	8.00	20.00	SRF-RNG3	3
4 -	20.00	100.00	SRF-RNG4	4
5 -				
6 -				
7 -				
8 -				

OK Cancel Next

Figure 14–37

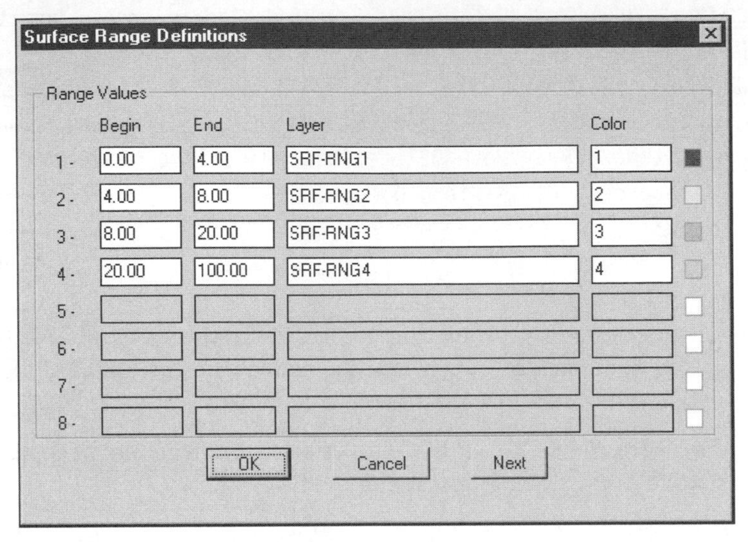

3. Use the scale factor of 15.0 when prompted; otherwise, select the OK buttons and press ENTER when the prompts and dialog boxes appear until you return to the command prompt. When you return to the command prompt, the routine draws slope arrows in the drawing representing the slope of the land. These arrows also show the paths the water takes as it drains over the surface. What we are looking for is: where do the arrows collectively point to so we can place the catchbasins in those locations? You should see something similar to Figure 14–38.

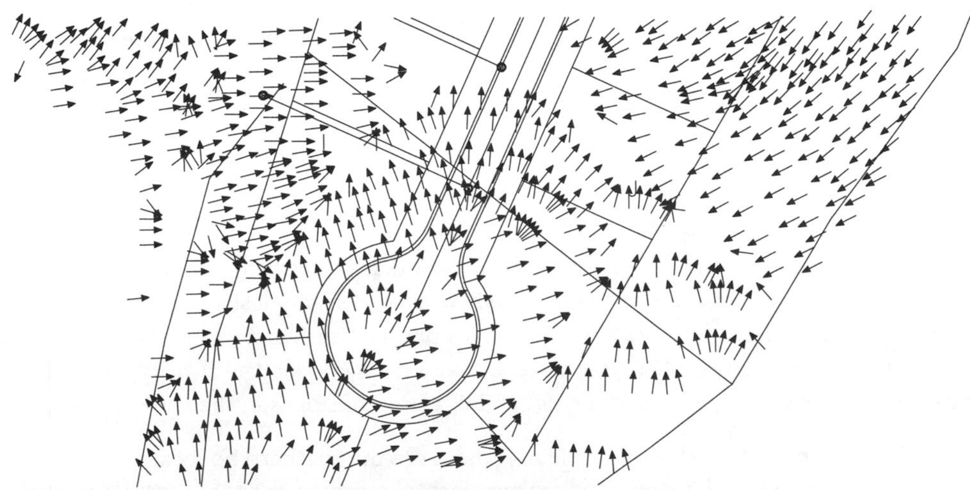

Figure 14–38

4. Erase the arrows with the Range Layers routine in the Terrain Layers cascade of the Terrain menu.

DRAW STORM CATCH BASIN RUNS

1. Change the current settings to be the catch basin settings (stmcb.dfm) by using the Import DFM File routine of the Settings cascade of the Pipes menu.
2. Use the Set Current Alignment routine of the Alignments menu to make Rosewood the current alignment.

The Catch basin run has two catch basins in the cul-de-sac and two in the street. One of the two catch basins in the street is opposite the storm manhole at station 20+20. The two catch basins in the cul-de-sac must have stations before the end of the Rosewood alignment. You will place them into the pipe run by selecting two points around the edge of the cul-de-sac. The routine will calculate the station and offset for the point relative to the current alignment (Rosewood).

3. Use the Draw Pipe Run routine of Define Pipe Runs cascade of the Pipes menu to create the catch basin run. Name the pipe run CBCuldesac and select the existing surface to supply rim elevations. Start by placing the two catch basins in the cul-de-sac; one on the northwest side and the other on the southeast side, both being inside of the stationing of the Rosewood alignment. The last three catch basins are at specific stations on the Rosewood alignment. The last catch basin sits inside of the storm manhole. We will have to edit the structure type later. Remember to add each catch basin to the run, to exit out of the point/station mode, and to save the run before exiting. When the Run Alignment Association dialog box appears, select Use Current Alignment.

Catch Basin	Station	Offset
CB #1	Selected Point (Northwest side of Cul-de-sac before the end of Rosewood)	
CB #2	Selected Point (Southeast side of Cul-de-sac before the end of Rosewood)	
CB #3	21+25	-12.50
CB #4	20+20	12.50
CB #5	20+20	12.50

```
Command:
Enter new run name: CBCuldesac
Select point: (eXit/Station): _nea to
eXit/Move/Add <Add>: (press ENTER to accept)
Select point: (eXit/Station): _nea to
Run: CBCuldesac        length: 80.81          nodes: 2
eXit/Next/Prev/Move/Delete/Undo/Save/Add <Add>: (press ENTER to accept)
Select point: (eXit/Station): s
```

EXERCISE

```
Select alignment:
Alignment Name: Rosewood  Number: 1   Descr: Southwest Cul-de-Sac
Starting Station: 0.00  Ending Station: 2203.03
Station <2175.71>: 2125
Offset <-39.47>: -12.5
Run: CBCuldesac        length: 138.25          nodes: 3
eXit/Next/Prev/Move/Delete/Undo/Save/Add <Add>: (press ENTER to accept)
Select point: (eXit/Station): s
Station <2125.00>: 2020
 Offset <-12.50>: (press ENTER to accept)
Run: CBCuldesac        length: 243.25          nodes: 4
eXit/Next/Prev/Move/Delete/Undo/Save/Add <Add>: (press ENTER to accept)
Select point: (eXit/Station): s
Station <2020.00>: (press ENTER to accept)
 Offset <-12.50>: 12.5
Run: CBCuldesac        length: 268.25          nodes: 5
eXit/Next/Prev/Move/Delete/Undo/Save/Add <Add>: (press ENTER to accept)
Select point: (eXit/Station): x
eXit/Next/Prev/Move/Delete/Undo/Save/Add <Add>: s
Run: CBCuldesac        length: 268.25          nodes: 5
Add/eXit/Next/Prev/Move/DElete/DBase/Undo/Save <Save>: x
Run: CBCuldesac        length: 268.25          nodes: 5
```

4. Click the Save icon to save the drawing.

UNIT 3: EDITING AND ANALYZING PIPE RUNS

You can edit and analyze pipe runs with the same routines. Civil Design has two types of editors to complete the necessary work. The first is a graphical editor, and the second is a spreadsheet format of the run. The editors are found in two cascade menus, Conceptual Plan and Conceptual Profile of the Pipes menu (see Figures 14–39 and 14–40).

The Conceptual Plan cascade has other editing routines. The Break, Join, and Delete routines create new runs from splitting up a single run, join two runs into one, or delete the information and/or the line work on the screen. The Reverse Flow routine starts the flow of the water in the run at the opposite end of the run. The Recalculate Rim routine is used when the surface that set the rim elevations changes or if the invert depth has changed and the inverts need updating. The

Check Plan routine updates the run when you edit the Access database or edit the run with Desktop editing tools.

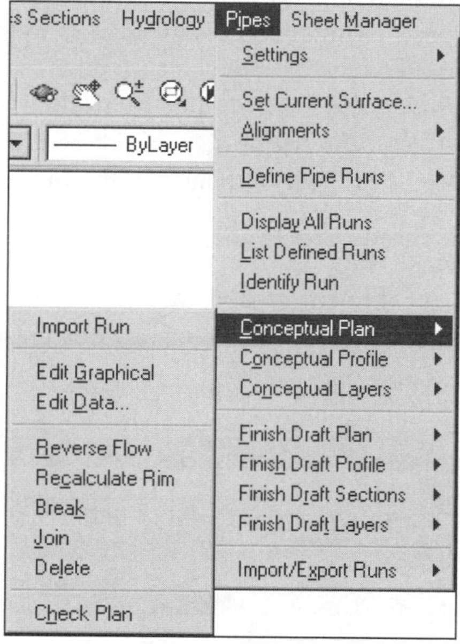

Figure 14–39

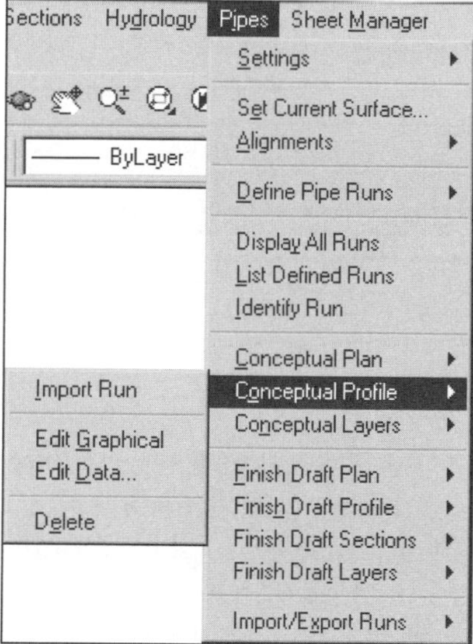

Figure 14–40

EDIT GRAPHICAL

The Edit Graphical routine is accessed from the Conceptual Plan and Conceptual Profile cascades. It works in plan or profile view and as its name implies, it graphically shows you where you are on the run and allows you to graphically rearrange the run. This editor edits and views all of the data in the run. The relocation of the structures of a run occurs only when using the Editor called from the Conceptual Plan cascade. The Database option (Dbase) of the editor displays all of the run data and it will gray out those items you are not allowed to edit in the dialog box (Figure 14–41). The Conceptual Plan version of this editor takes you from structure to structure. The Conceptual Profile version of this routine will take you from structure to pipe to structure in a profile. The profile editor does not show all of the run data, and you cannot relocate the structures from profile view. The emphasis of the profile editor is the inverts of the structures and pipes.

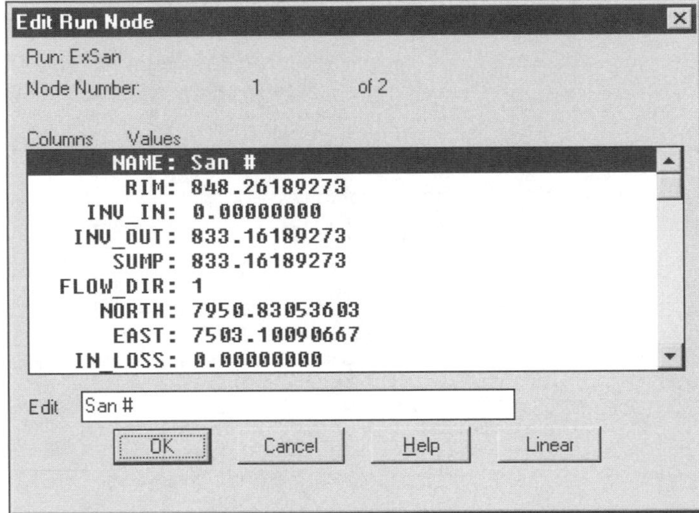

Figure 14–41

EDIT DATA

The Edit Data routine calls the Pipes Run Editor dialog box. The dialog box is a spreadsheet representation of the run (see Figure 14–42). You can use this editor to edit the run whether in plan or profile view. The editor displays all of the pertinent data and does not allow you to move or relocate any structures whether in plan or profile view. The editor can isolate the information in a set of preset views, for example, pipes, node, flow, and so on. When in Flow view, the routine can call calculations from the Hydrology area of the project.

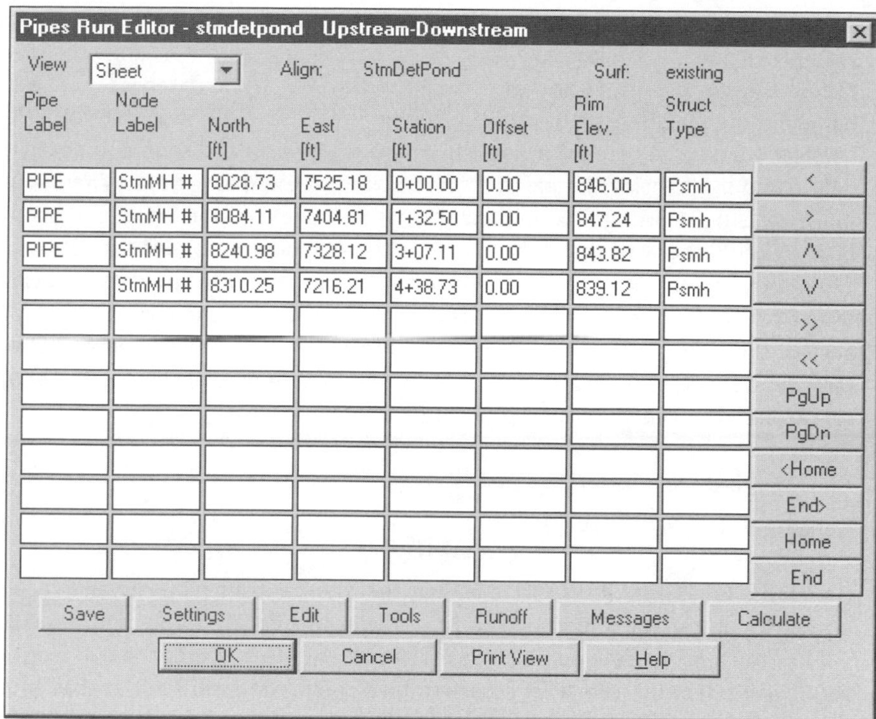

Figure 14–42

HAESTAD METHODS

All Civil Design pipe runs are readable into Haestad's Storm and Sewer CAD. The Haestad methods allow for a greater analysis and can treat the runs as a system. After you analyze and edit the runs in Haestad, you can return the results to the Pipes area and from that, update the drawing and project.

Exercise

After you complete this exercise, you will be able to:

- Edit and analyze pipe run data

GRAPHICAL PLAN EDITING OF A SANITARY PIPE RUN

1. Run the Edit Graphical routine from the Conceptual Plan cascade of the Pipes menu and select the ExSan run. If you cannot select the run from the screen, press ENTER and the routine will display a list of defined pipe runs. Then all you do is select the run from the list. There should be a blip "X" at the node nearest to where you select the run with an arrow pointing to the northwest at the second

node. The routine presents an option list that allows you to move, delete, add, view the database entries, save, exit, or go to the next node. Change the name of the first node to **San #1** and the second to **San #2** in the database. Edit the names in the Database (DBase) view. When in Dbase view, you first you select the name from the data list in the top of the dialog box and then you enter the new name by changing the name at the bottom of the dialog box. When you press ENTER, the dialog box updates the name in the top portion of the dialog box. When done with the changes, select the OK button to exit the dialog box. Then move to the next node and change its name. While in the database view, select other pieces of data. If you are not allowed to edit the data, the bottom area of the dialog box will be grayed out. Note down the Invert Out for San #1 (around 833.15 feet). Finally, save the changes before exiting the routine and you do not need to redefine the alignment before you exit.

Rename Node 1: **San #1**

Rename Node 2: **San#2**

2. Rerun the Edit Graphical routine to edit the Interceptor1 pipe run and rename its nodes. If you cannot select the run from the screen, press ENTER and the routine will display a list of defined pipe runs. Then all you do is select the run from the list. If you select the run at its southern end, the editing will start at that end. Edit the names in the Database (DBase) view. The top of the dialog box will indicate which structure you are at. The first structure is at 14+50 and the last is at 20+20. You will need to use the previous option to visit all of the nodes in the run and remember to save before you exit the routine. The fifth node is the same as the first node in the ExSan run. This means that its name should be **San #1**. This also means that the type of structure node 5 is needs to change, because the ExSan run also produces the plan symbol for San #1. In profile we need to have both ExSan and Interceptor1 show the same structure (San #1), because we will represent it on different profiles. To fix this problem, we need to have Node 5 of Interceptor1 (San #1) show in the profile annotation of Interceptor1, but not in the plan annotation of Interceptor1. So, you need to change the structure type of Node 5 to Null2 (no plan view, but a profile view). After you make and save the changes, exit the Edit Graphical routine.

Rename Node 1: **San #5**

Rename Node 2: **San#6**

Rename Node 3: **San #7**

Rename Node 4: **San#8**

Rename Node 5: **San #1**

Structure Type: **Null2**

EXERCISE

EDIT DATA OF SANITARY RUN

1. Run the Edit Data routine from the Conceptual Plan cascade of the Pipes menu and select the ExSan run. If you cannot select the run from the screen, press ENTER and the routine will display a list of defined pipe runs. Then all you do is select the run from the list. The dialog box issues a warning that there is a negative flow to the run; for now, select OK to dismiss the dialog box.
2. Switch to the Node view by clicking the drop arrow at the top left of the dialog box and select Node. This shows the structures and their pertinent data. Make sure the names are correct: San #1 and San #2.
3. Next switch to the Pipe view. The Finish Invert for the pipe is higher than the starting invert. Change the Finish Invert to 833.00. After you change the invert, the dialog box issues a warning that there is no flow in the system. Select the OK button to dismiss this dialog box.
4. Change the view to the Flow view and enter **3.75** as the Pipe Flow value.
5. Before you exit the editor, your Node, Pipe, and Flow views should look similar to Figures 14–43, 14–44, and 14–45.

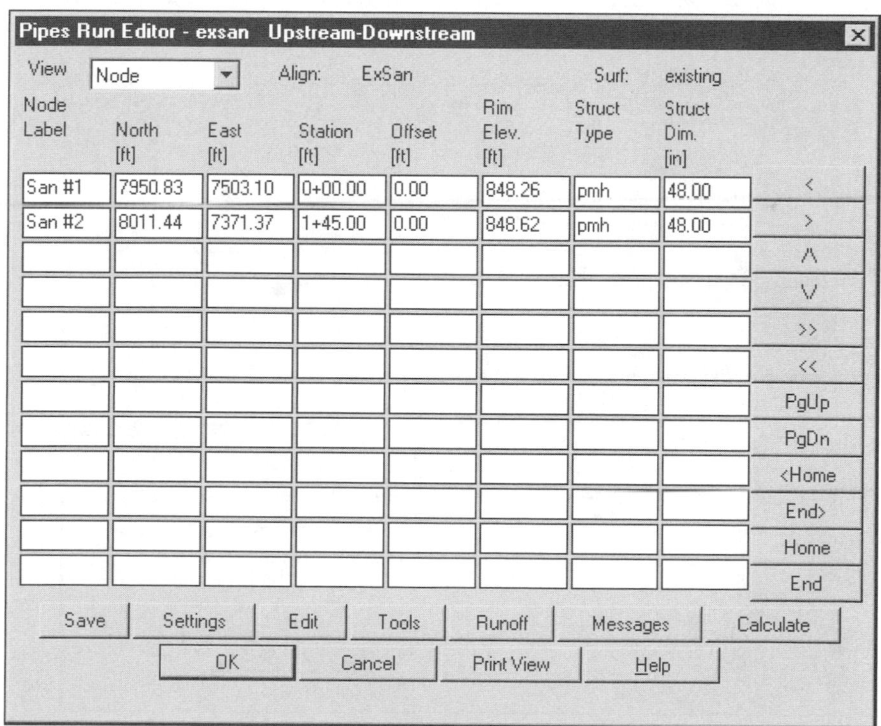

Figure 14–43

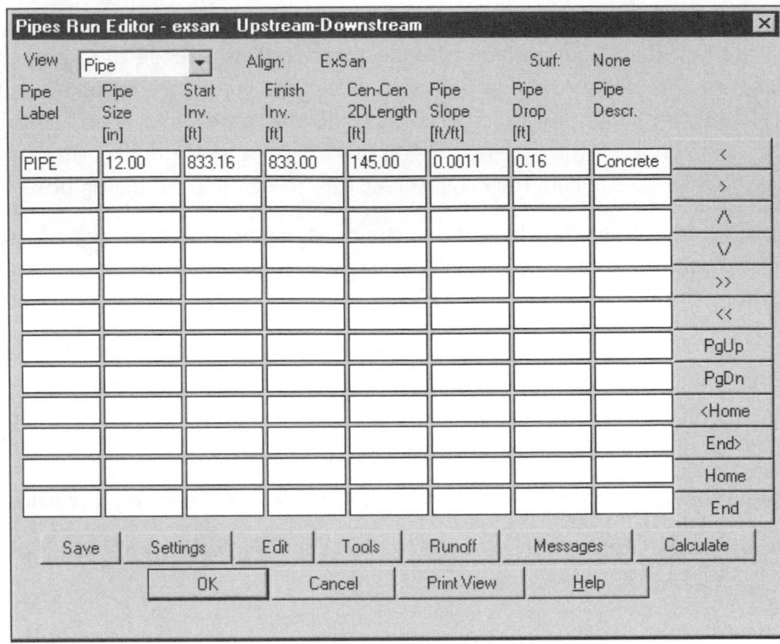

Figure 14–44

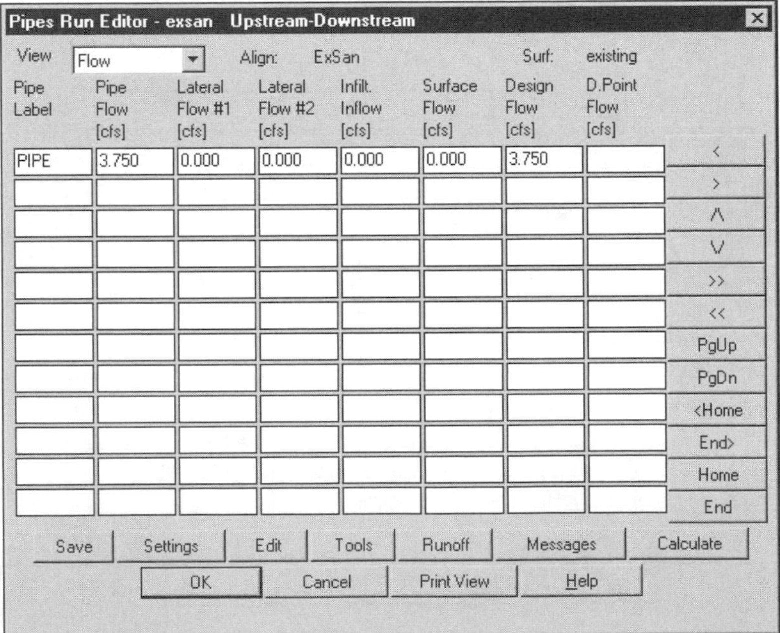

Figure 14–45

6. Select the Save button at the bottom of the dialog box and save the changes.
7. Exit the Editor for the ExSan run.

8. Rerun the Edit Data routine and select the Interceptor1 pipe run. If you cannot select the run from the screen, press ENTER and the routine will display a list of defined pipe runs. Then all you do is select the run from the list. The dialog box issues a warning that there is a negative flow to the run; for now, select OK to dismiss the dialog box.
9. Change to Node view and make sure the names are correct (see Figure 14–46).
10. Change to the Flow view and enter the following values. Each time you enter a flow value, the dialog box will issue a warning that there is a negative flow. Just select the OK button to proceed with the data entry. The reason for this is the inverts are not correct for the pipe run. You will edit their elevations in the profile view.

    Pipe Flow 1: **1.00** (CFS)

    Pipe Flow 2: **1.50** (CFS)

    Pipe Flow 3: **0.75** (CFS)

    Pipe Flow 4: **0.50** (CFS)

11. After reviewing the node names and entering the flow data, select the Save button to save the changes (see Figure 14–48).
12. Before you exit the editor, your Node, Pipe, and Flow views should look similar to Figures 14–46, 14–47, and 14–48.
13. Select the OK button to exit the Pipes Run Editor.

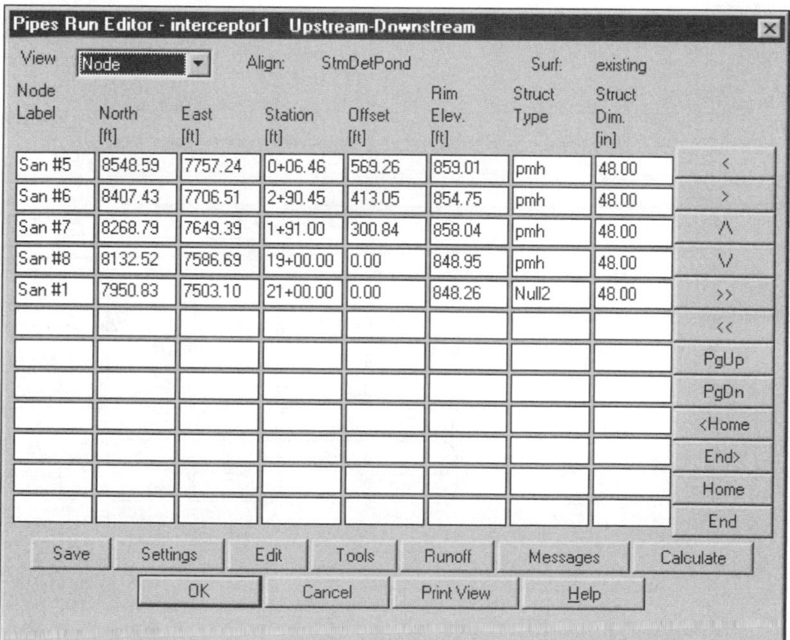

Pipes Run Editor - interceptor1 Upstream-Downstream

View: Node Align: StmDetPond Surf: existing

Node Label	North [ft]	East [ft]	Station [ft]	Offset [ft]	Rim Elev. [ft]	Struct Type	Struct Dim. [in]
San #5	8548.59	7757.24	0+06.46	569.26	859.01	pmh	48.00
San #6	8407.43	7706.51	2+90.45	413.05	854.75	pmh	48.00
San #7	8268.79	7649.39	1+91.00	300.84	858.04	pmh	48.00
San #8	8132.52	7586.69	19+00.00	0.00	848.95	pmh	48.00
San #1	7950.83	7503.10	21+00.00	0.00	848.26	Null2	48.00

Figure 14–46

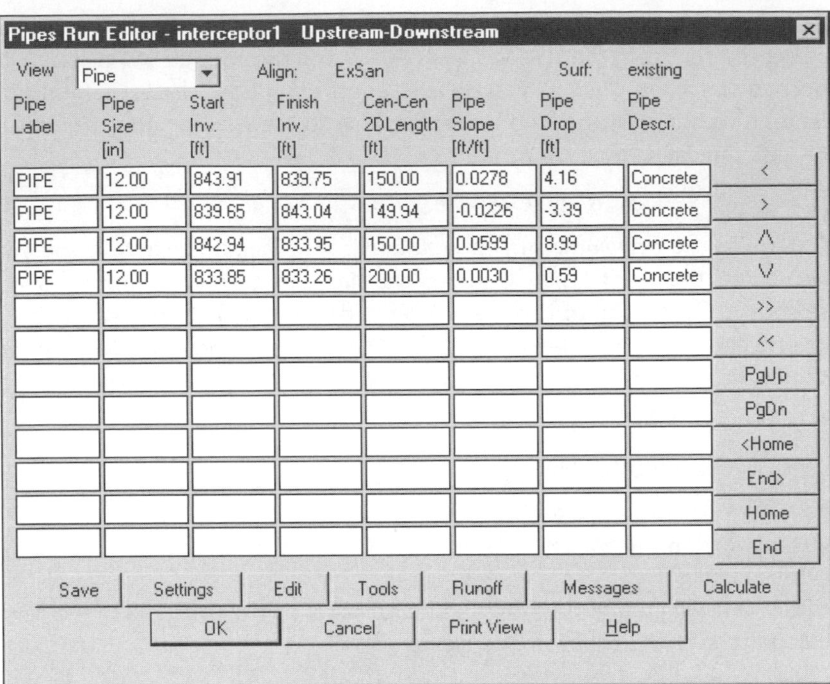

Pipe Label	Pipe Size [in]	Start Inv. [ft]	Finish Inv. [ft]	Cen-Cen 2DLength [ft]	Pipe Slope [ft/ft]	Pipe Drop [ft]	Pipe Descr.
PIPE	12.00	843.91	839.75	150.00	0.0278	4.16	Concrete
PIPE	12.00	839.65	843.04	149.94	-0.0226	-3.39	Concrete
PIPE	12.00	842.94	833.95	150.00	0.0599	8.99	Concrete
PIPE	12.00	833.85	833.26	200.00	0.0030	0.59	Concrete

Figure 14–47

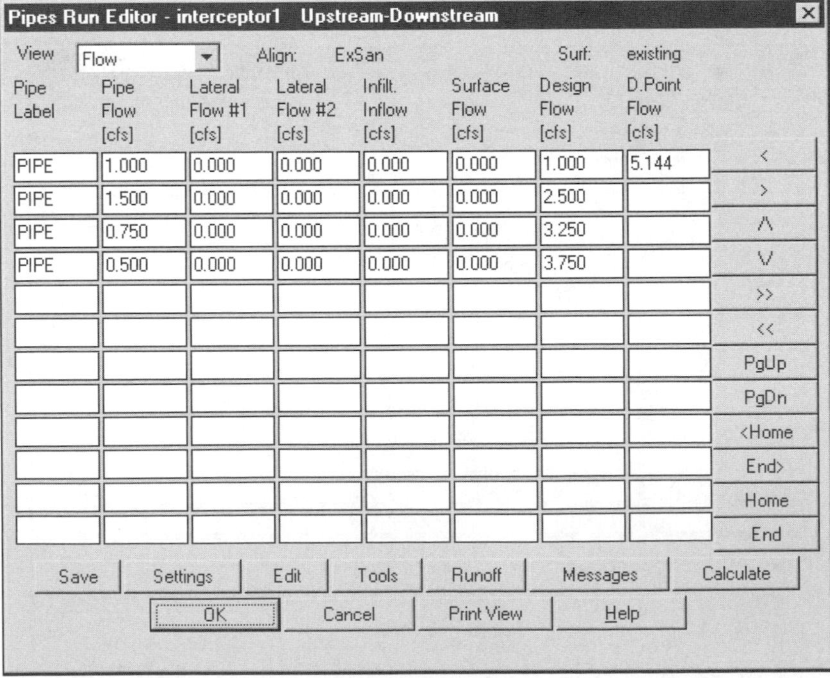

Pipe Label	Pipe Flow [cfs]	Lateral Flow #1 [cfs]	Lateral Flow #2 [cfs]	Infilt. Inflow [cfs]	Surface Flow [cfs]	Design Flow [cfs]	D.Point Flow [cfs]
PIPE	1.000	0.000	0.000	0.000	0.000	1.000	5.144
PIPE	1.500	0.000	0.000	0.000	0.000	2.500	
PIPE	0.750	0.000	0.000	0.000	0.000	3.250	
PIPE	0.500	0.000	0.000	0.000	0.000	3.750	

Figure 14–48

GRAPHICAL PLAN EDITING OF A STORM PIPE RUN

1. Run the Edit Graphical routine from the Conceptual Plan cascade of the Pipes menu and select the StmDetPond run. If you cannot select the run from the screen, press ENTER and the routine will display a list of defined pipe runs. Then all you do is select the run from the list. There should be a blip "X" at the node nearest to where you select the run with an arrow pointing to the northwest towards the pond. The routine presents an option list that allows you to move, delete, add, view the database entries, save, exit, or go to the next node.

2. Change the names of the nodes to Stm MH#1 - StmMH #4 in the database. The node numbers and the structure numbers are to be the opposite of the flow of the run. Change the name at the bottom of the dialog box and press ENTER to view the updating in the dialog box. When done with the change, select the OK button to exit the dialog box. Then move to the next node and change its name. While in the database view, select other pieces of data. If you are not allowed to edit the data, the bottom area of the dialog box will be grayed out. Note down the Invert Out for StmMH #4 (around 840.00 feet). Finally save the changes before exiting the routine, and you do not need to redefine the alignment before you exit.

   Rename Node 1: **StmMH #4**

   Rename Node 2: **StmMH #3**

   Rename Node 3: **StmMH #2**

   Rename Node 4: **StmMH #1**

EXERCISE

EDIT DATA OF STORM POND RUN

1. Run the Edit Data routine and select the StmDetPond run. If you cannot select the run from the screen, press ENTER and the routine will display a list of defined pipe runs. Then all you do is select the run from the list. Change to the Pipe view by selecting the view from the View drop down list at the top left of the dialog box and review the start and finish elevations of the pipes. The Finish Invert elevation for the first pipe is higher than the Start elevation. This is our negative flow. Change the Finish Invert elevation to 839.00 and press ENTER to change the value. The routine will issue a flow of 0.0 for the run. Select the OK button to dismiss the dialog box. After making the change, select the Save button to save the changes.

2. Review the values in the Node and Pipe views of the StmDetPond run. Your dialog boxes should look similar to Figures 14–49and 14–50.

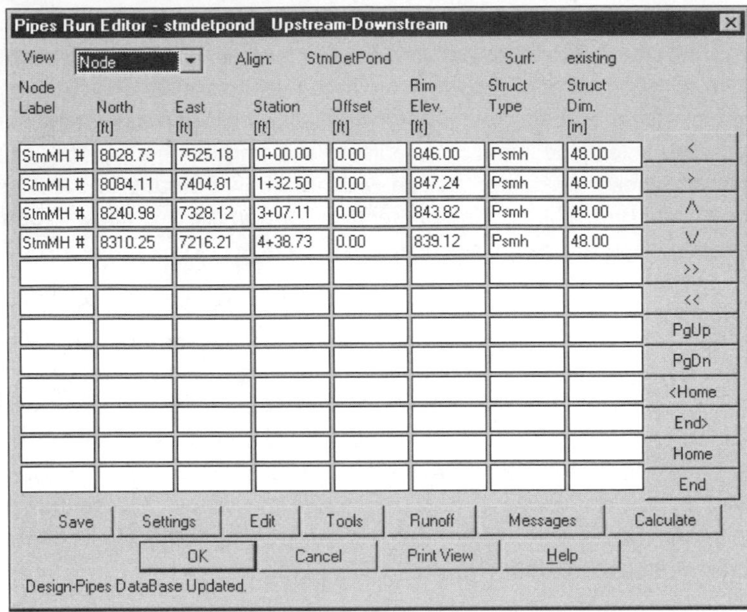

Figure 14–49

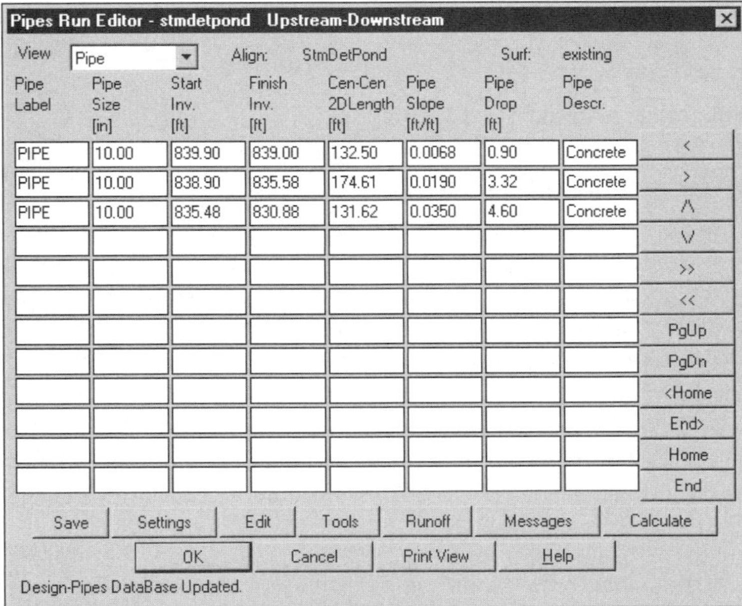

Figure 14–50

EDIT DATA OF CATCH BASIN RUN

1. Run the Edit Data routine and edit the CBCuldesac run. If you cannot select the run from the screen, press ENTER and the routine will display a list of defined pipe runs. Then all you do is select the run from the list. The routine displays the pipe

run in spreadsheet form. Select the OK button to dismiss the warning dialog box. Change to the Node view and review the structure types.

The last catch basin drawn in the run is really the storm manhole StmMH #4 a manhole structure with a sump. The first node (catch basin) is the next to last node drawn in the run. If you change the structure type to Null2, profile only, it will show a manhole with no sump (a sanitary manhole). There needs to be another Null structure that represents a storm manhole with a sump of 4', Null 3. The reason for a new null structure is that the StmDetPond run will provide the plan symbol, but we want to see the correct storm structure in the profiles of both runs.

2. Exit the Edit Data routine.
3. Run the Edit routine of the Settings cascade of the Pipes menu.
4. Select the Structure Library Editor button to display the structures and change the current structure to Null2. It shows a structure with no sump and for profile only. At the bottom of the dialog box, select the Add button to display the Name Assignment dialog box and enter the name **Null3**. After entering the name, select the OK button to dismiss the dialog box. Change the Sump Depth to 4.00 and select the Save button to save the new structure. Your dialog box should look like Figure 14–51. Select the OK button to exit the Structure Library Editor.

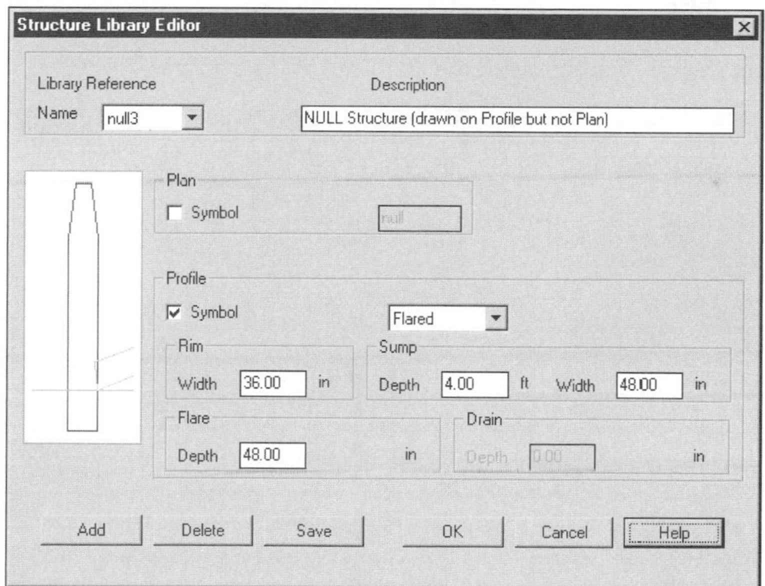

Figure 14–51

5. Rerun the Edit Data routine and select the CBCuldesac run. If you cannot select the run from the screen, press ENTER and the routine will display a list of defined pipe runs. Then all you do is select the run from the list. Change the names of the structures and change the structure type for the last node. This is done in the node view of the run.

EXERCISE

Rename Node 1: **StmCB #4**

Rename Node 2: **StmCB #3**

Rename Node 3: **StmCB #2**

Rename Node 4: **StmCB #1**

Rename Node 5: **StmMH #1**

Structure Type: **Null3**

6. Change to the Flow view and enter the flow values for the pipe flow. Again each time you enter a value, the routine will issue a negative flow. Just select the OK button to dismiss the dialog box. We will adjust the Invert to correct this problem in profile view. After making the changes, select the Save button to save the changes.

   Pipe 1 Flow: **1.350**

   Pipe 2 Flow: **0.850**

   Pipe 3 Flow: **1.250**

   Pipe 4 Flow: **1.500**

7. Before you exit the editor, your Node, Pipe, and Flow views should look similar to Figures 14–52, 14–53, and 14–54.
8. Select the OK button to exit the Edit Data dialog box.

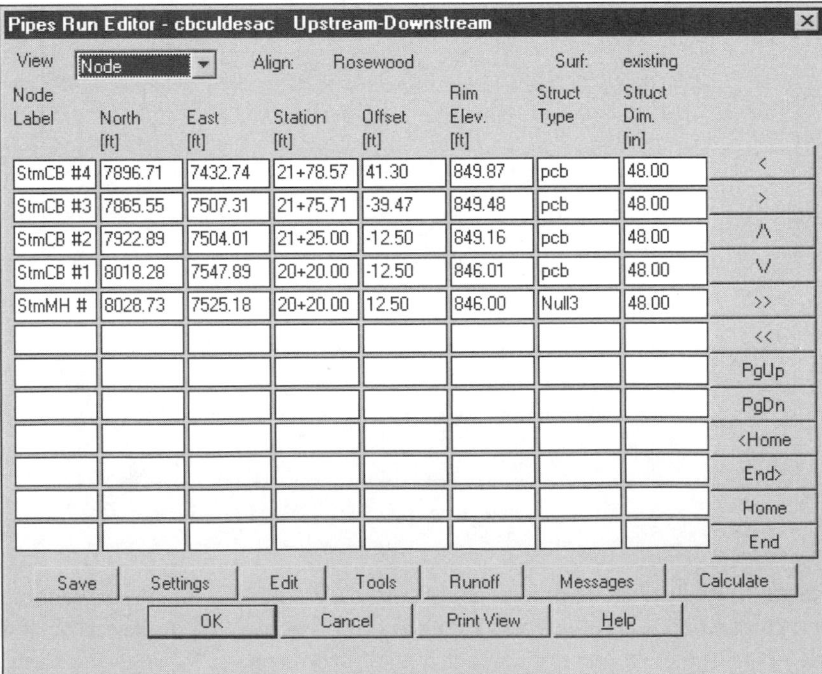

Node Label	North [ft]	East [ft]	Station [ft]	Offset [ft]	Rim Elev. [ft]	Struct Type	Struct Dim. [in]
StmCB #4	7896.71	7432.74	21+78.57	41.30	849.87	pcb	48.00
StmCB #3	7865.55	7507.31	21+75.71	-39.47	849.48	pcb	48.00
StmCB #2	7922.89	7504.01	21+25.00	-12.50	849.16	pcb	48.00
StmCB #1	8018.28	7547.89	20+20.00	-12.50	846.01	pcb	48.00
StmMH #	8028.73	7525.18	20+20.00	12.50	846.00	Null3	48.00

Figure 14–52

Pipes Run Editor - cbculdesac Upstream-Downstream

View: Pipe Align: Rosewood Surf: existing

Pipe Label	Pipe Size [in]	Start Inv. [ft]	Finish Inv. [ft]	Cen-Cen 2DLength [ft]	Pipe Slope [ft/ft]	Pipe Drop [ft]	Pipe Descr.
PIPE	8.00	845.77	845.48	80.81	0.0036	0.29	Concrete
PIPE	8.00	845.38	845.16	57.44	0.0037	0.21	Concrete
PIPE	8.00	845.06	842.01	105.00	0.0291	3.05	Concrete
PIPE	8.00	841.91	842.00	25.00	-0.0036	-0.09	Concrete

< > /\ \/ >> << PgUp PgDn <Home End> Home End

Save Settings Edit Tools Runoff Messages Calculate

OK Cancel Print View Help

Figure 14–53

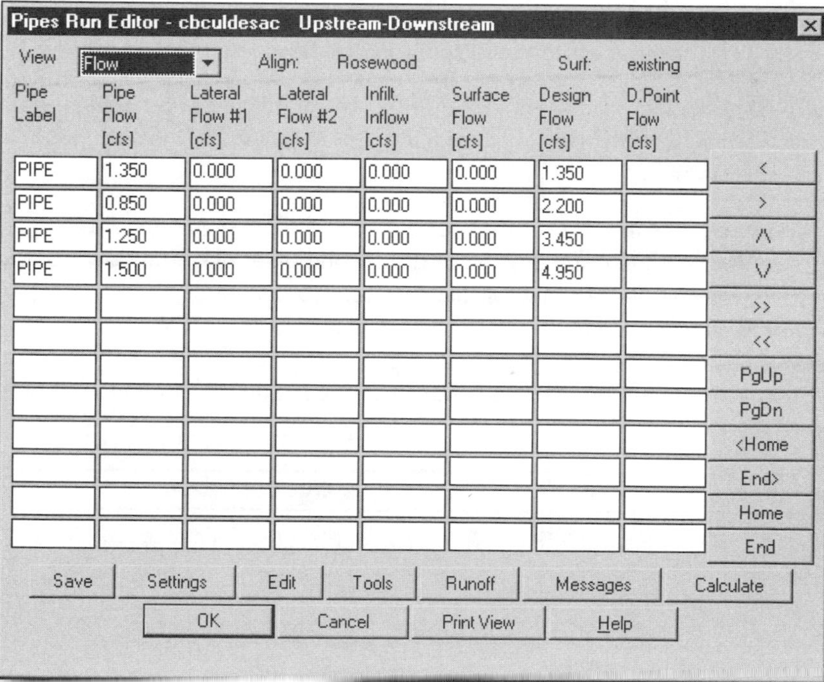

Pipes Run Editor - cbculdesac Upstream-Downstream

View: Flow Align: Rosewood Surf: existing

Pipe Label	Pipe Flow [cfs]	Lateral Flow #1 [cfs]	Lateral Flow #2 [cfs]	Infilt. Inflow [cfs]	Surface Flow [cfs]	Design Flow [cfs]	D.Point Flow [cfs]
PIPE	1.350	0.000	0.000	0.000	0.000	1.350	
PIPE	0.850	0.000	0.000	0.000	0.000	2.200	
PIPE	1.250	0.000	0.000	0.000	0.000	3.450	
PIPE	1.500	0.000	0.000	0.000	0.000	4.950	

< > /\ \/ >> << PgUp PgDn <Home End> Home End

Save Settings Edit Tools Runoff Messages Calculate

OK Cancel Print View Help

Figure 14–54

EDIT GRAPHICAL PROFILE–THE INTERCEPTOR1 RUN

1. Use the ZOOM and PAN commands to view the profile in the display area of the Desktop.
2. Use the Edit Data routine, Pipe view, to review the start elevation for the pipe of ExSan (833.16).
3. Exit out of the dialog box and rerun the Edit Data routine to review the pipe elevations for the Interceptor1 pipe run. In this case, the finish elevation of the last pipe is 833.26. So as long as the last finish elevation of Interceptor1 is higher that the start invert elevation of ExSan, things are OK. However, there are several elevations between the beginning of the pipe run and the end that produce a negative flow. Exit the Edit Data dialog box by selecting the OK button.
4. Use the Import Run routine of the Conceptual Profile cascade of the Pipes menu and import the Interceptor1 run. If you cannot select the run from the screen, press ENTER and the routine will display a list of defined pipe runs. Then all you do is select the run from the list. After you select the run and select the OK button, the routine places the run into the drawing. The pipe run is represented by a stick figure in the profile.
5. Run the Edit Graphical routine in the Conceptual Profile cascade of the Pipes menu and select a pipe representing the run in the profile. The routine marks the current editing location by placing a blip "X" on a pipe or structure nearest your selection point. The routine prompts you for the next action on the command line. The routine has two prompt lines one is for structures and the second is for pipes. Your screen should look like Figure 14–55.

```
NODE <San #6>-> (Next/Prev/In/Out/Rim/DBase/Undo/SAve/eXit) <eXit>: n
PIPE <PIPE>-> (Prev/SLope/In/Out/DBase/Undo/SAve/Graph/eXit/Next)
   <Next>:
```

6. Review the Pipe In and Out for the second and third segments of the pipe run. When you edit the out elevation of the second pipe, all of the downstream inverts will change. This means that you have to edit all of the downstream inverts to make the last invert match the elevation of 833.26. Use the following elevations to set the pipe inverts. The prompting can be confusing, so take care as you edit the values. When the editing is done, review the pipe ins and outs and save the changes.

Pipe Segment	In	Out
1	843.91	839.75
2	839.65	839.00
3	838.90	837.75
4	837.65	833.26

EXERCISE

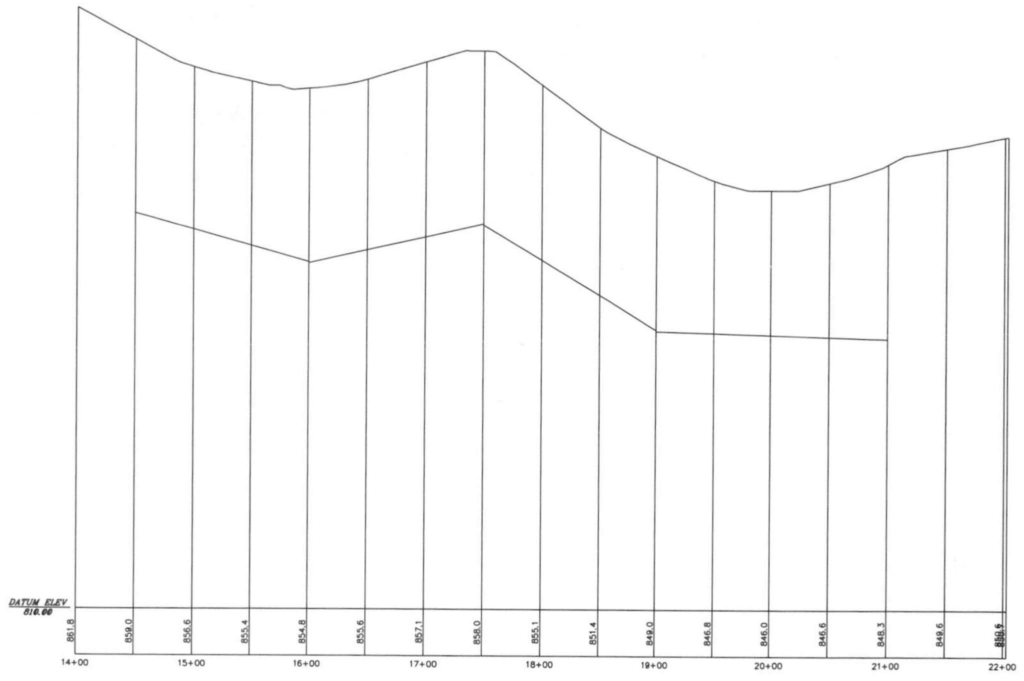

Figure 14–55

7. Save the changes, then exit the Edit Graphical routine and erase the polyline representing the Interceptor1 pipe run.

EDIT GRAPHICAL PROFILE–THE CBCULDESAC RUN

1. Use the Edit Data routine, Pipe view, to review the start elevation for the pipe between StmMH #4 and StmMH #3 of StmDetPond (839.90).

2. Exit out of the dialog box and rerun the Edit Data routine to review the pipe elevations for the CBCuldesac pipe run. In this case the finish elevation of the last pipe is 842.00. This elevation gives us our negative flow. So as long as the last finish elevation of CBCuldesac is higher than the start invert elevation of StmDetPond, things are OK. Exit the Edit Data dialog box by selecting the OK button. You will change the elevation of the finish pipe of the CBCuldesac run in the profile.

3. Use the Import Run routine of the Conceptual Profile cascade of the Pipes menu and import the CBCuldesac run. If you cannot select the run from the screen, press ENTER and the routine will display a list of defined pipe runs. Then all you do is select the run from the list. After you select the run and select the OK button, the routine places the run into the drawing. The pipe run is represented by a stick figure in the profile.

4. Run the Edit Graphical routine in the Conceptual Profile cascade of the Pipes menu and select a pipe representing the run in the profile. The routine marks the current editing location by placing a blip "X" on a pipe or structure nearest your selection point. The routine prompts you for the next action on the command line. You will probably have to zoom in to see the line work clearly. The routine has two prompt lines; one is for structures and the second is for pipes. Your screen should look like Figure 14–56.

```
NODE <San #6>-> (Next/Prev/In/Out/Rim/DBase/Undo/SAve/eXit) <eXit>: n
PIPE <PIPE>-> (Prev/SLope/In/Out/DBase/Undo/SAve/Graph/eXit/Next)
   <Next>:
```

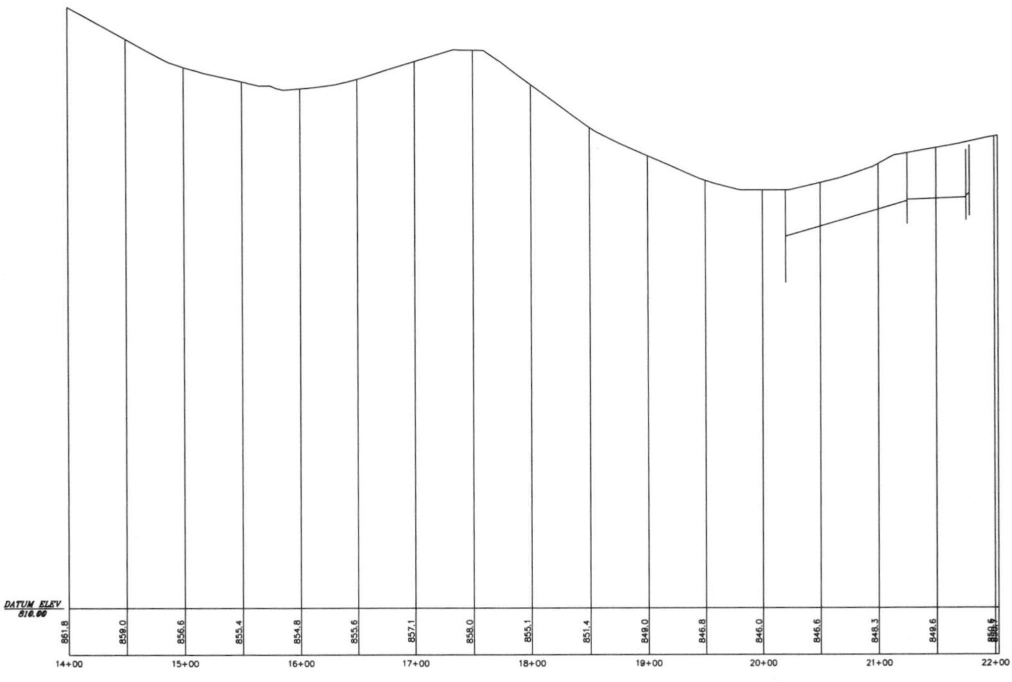

Figure 14–56

5. Use the Next option to move the blip "X" to the last pipe segment (left side of the run). Check the In elevation of the pipe (841.91) and check the Out elevation of the pipe (842.00). Change the Out elevation to 841.00. This makes it a down slope for the last pipe length and keeps it above the invert for the StmMH #4 structure (839.90).

6. Use the Save option to save the changes and then exit the Graphical Editor.

7. Erase the polyline representing the pipe run.

UNIT 4: ANNOTATING A PIPE RUN

The vast majority of the settings for pipes now come into play. The settings affect the labels that will appear in the drawing. If you attempt the profile labeling, undo it, and then relabel, you may see the labeling start to slowly move up and away from the profile. When you perform this sequence, you must run the Restore Text Areas routine of the Finish Draft Profile cascade of the Pipes menu. This resets the text positioning to the initial location in the profile.

PLAN ANNOTATION

The Finish Draft Plan cascade contains the routines annotating plan designs of pipe runs (see Figure 14–57). The Draw Pipes routine drafts the lines and annotates the current or selected run. The Special Lines routine displays a dialog box containing lettered lines that you can use between structures. The Delete Layer routine deletes any preexisting plan labeling.

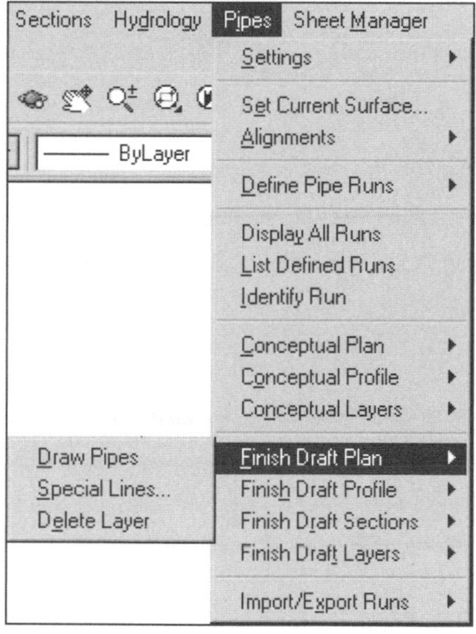

Figure 14–57

PROFILE ANNOTATION

The Finish Draft Profile cascade contains the routines that annotate in profile the pipe runs (see Figure 14–58). The Draw Pipes routine drafts the lines and annotates the current or selected run. The Delete Layer routine deletes any existing plan labeling.

The Hydraulic and Energy Gradeline routines label the grade lines of the pipe run. The hydraulic gradeline is a graphical representation of the sum of the static and dynamic heads versus the position along the pipeline. In a closed system, the gradeline is a line connecting the total head in the pipe from structure to structure. The energy gradeline is a graphical representation of the total energy of the flow through the system in terms of potential energy. The vertical difference between the hydraulic gradeline and the energy gradeline is the kinetic energy of the fluid per unit weight, V2/2g. The Align/Run Interferences routine labels the crossing of the alignment with the pipe run.

The user may make several attempts at drafting the final annotation. Each attempt at profile labeling moves the annotation above the previous set of annotation. The reason for this is that there may be more than one run in the profile. So, to separate the two groups of annotation, the routine places the next set of annotation above the last. As a result, the annotation climbs up the drawing. If you are attempting the labeling of the same run a second time, you need to restore the text area after deleting the incorrect text. The Restore Text Areas command forces the next annotation to occur in the correct position on the screen.

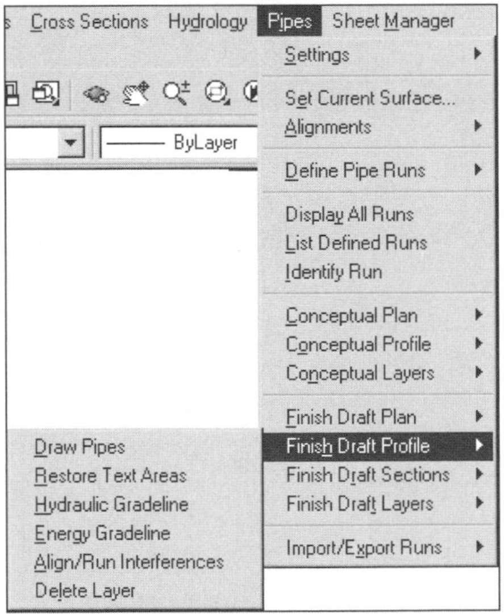

Figure 14–58

CROSS SECTION ANNOTATION

The Finish Draft Sections cascade contains routines to create and delete cross-section annotation (see Figure 14–59). The routines place the pipes and structure into the sections. The only condition with the annotation is that the cross sections be in a drawing containing a profile of the alignment.

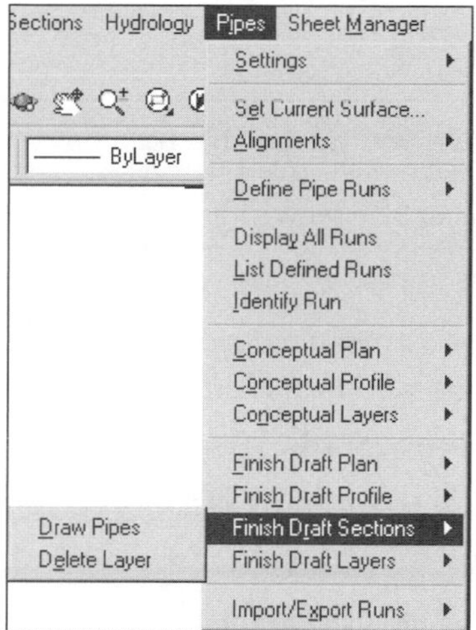

Figure 14–59

FINISH DRAFT LAYERS

The Finish Draft Layers cascade contains routines to manipulate the visibility of plan, profile, and section layers (see Figure 14–60).

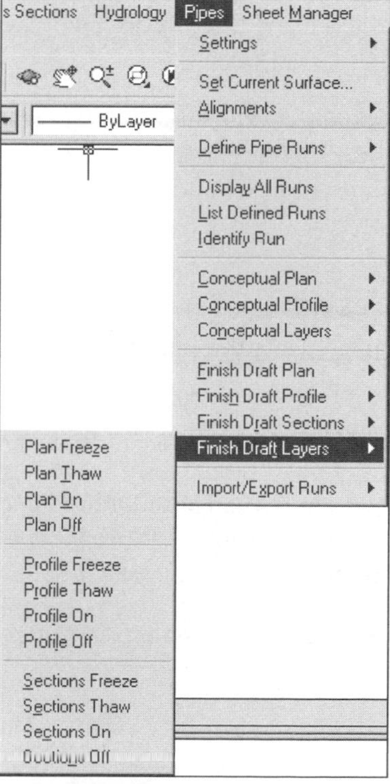

Figure 14–60

Exercise

After you complete this exercise, you will be able to:

- Annotate pipe runs in plan view
- Annotate pipe runs in profile view
- Annotate pipe runs in cross-section view

ANNOTATING CBCULDESAC

1. Your display still should show the profile for Rosewood. If not, zoom to see the Rosewood profile.

2. Import the StmCB.dfm file using the Import DFM file from the Settings cascade of Pipes.

3. Run the Draw Pipes routine of the Finish Draft Profile cascade of the Pipes menu. Annotate the CBCuldesac pipe run. After placing the annotation, you may have to shift some of the annotation around to make it fit.

ANNOTATING INTERCEPTOR1

1. Import the *Sewer.dfm* file using the Import DFM file from the Settings cascade of Pipes.

2. Run the Draw Pipes routine of the Finish Draft Profile cascade of the Pipes menu. Annotate the Interceptor1 pipe run. After placing the annotation, you may have to shift some of the annotation around to make it fit. Your drawing should now look like Figure 14–61.

ANNOTATING CBCULDESAC

1. Import the *StmCB.dfm* file using the Import DFM file from the Settings cascade of Pipes.

2. Use the ZOOM command to zoom to the cul-de-sac area.

3. Use the Draw Pipes routine of the Finish Draft Plan cascade of the Pipes menu. Annotate the CBCuldesac pipe run. After placing the annotation, you may have to shift some of the annotation around to make it fit.

4. Import the *Stm.dfm* file using the Import DFM file from the Settings cascade of Pipes.

5. Use the Draw Pipes routine of the Finish Draft Plan cascade of the Pipes menu. Annotate the StmDetPond pipe run. After placing the annotation, you may have to shift some of the annotation around to make it fit. Your drawing should now look like Figure 14–62.

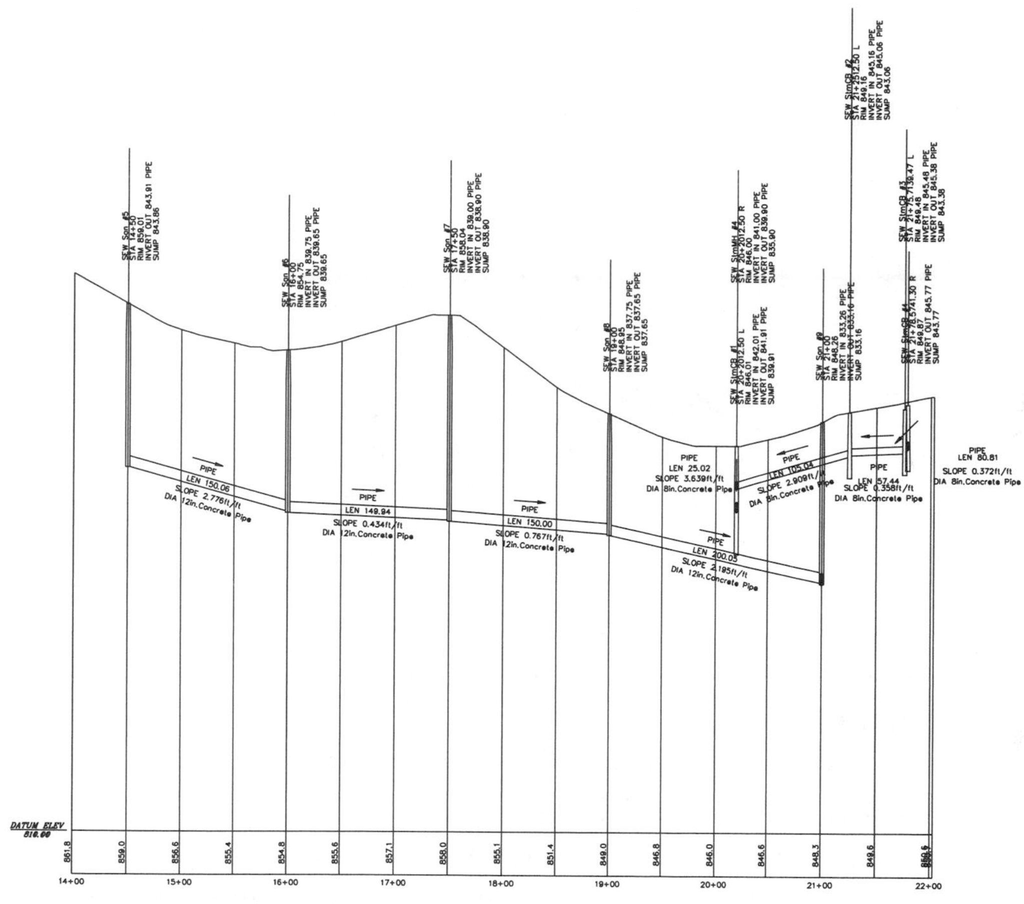

Figure 14–61

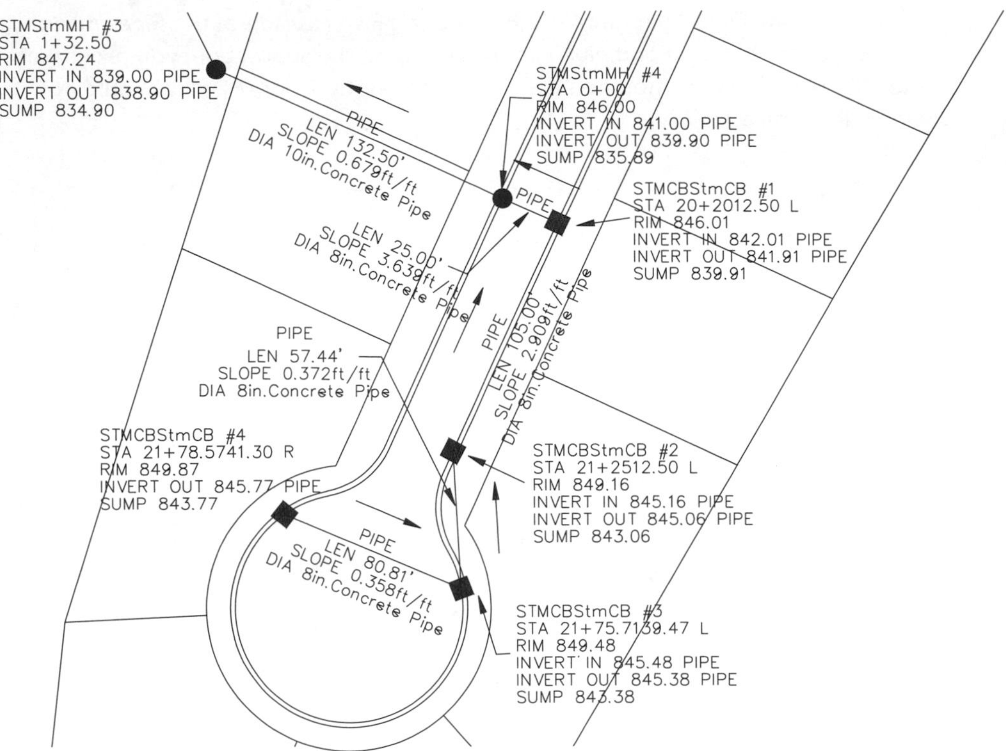

Figure 14–62

ANNOTATING PIPE CROSS SECTIONS

The last item to annotate is the pipes within the cross sections.

1. Zoom in and center on CBCuldesac.
2. Run the Sample From Surface routine of the Existing Ground cascade of the Cross Section menu to create cross sections for the Rosewood alignment. If the routine prompts you to select an alignment, click the right mouse button and select Rosewood from the list. Sample a swath of 75 feet, every 50 feet along tangent, curves, and spirals, at the beginning and end of the alignment, toggle on Save Sample List, and toggle on Add Specific Stations. Add the following stations:

   Stations to Add: 2020 and 2125.
3. Pan the drawing to the right so you have some room to the left of the subdivision.
4. Run the Settings routine from the Section Plot cascade of the Cross Sections menu. Select the Section Layout button to display the Layout Section dialog box (see Figure 14–63). Change the Rows Below Datum to 8. After setting the number, select the OK button to dismiss the Section Layout and Settings dialog box.

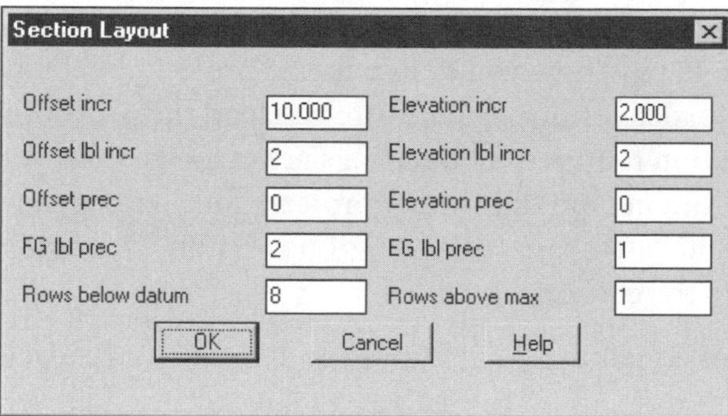

Figure 14–63

5. Use the Single routine of the Section Plot cascade of the Cross Sections menu to create cross sections for stations 2020, 2100, and 2125.
6. Use the Draw Pipes routine of the Finish Draft Sections cascade of the Pipes menu. Select Interceptor1 pipe run to place the pipe's cross sections into the sections. If you cannot select the run from the screen, press ENTER and the routine will display a list of defined pipe runs. Then all you do is select the run from the list. Answer Yes to erase the layers for Interceptor1.
7. Rerun the Draw Pipes routine of the Finish Draft Sections cascade of the Pipes menu and select the CBCuldesac pipe run. If you cannot select the run from the screen, press ENTER and the routine will display a list of defined pipe runs. Then all you do is select the run from the list. Answer Yes to erase the layers for CBCuldesac.

You will have to copy or draw the structure for Interceptor1 at station 21+00. The pipe for ExSan and the structure and pipe for CBCuldesac will have to be drawn in to the appropriate cross section (20+20) (see Figure 14–64).

8. Click the Save icon to save the drawing.

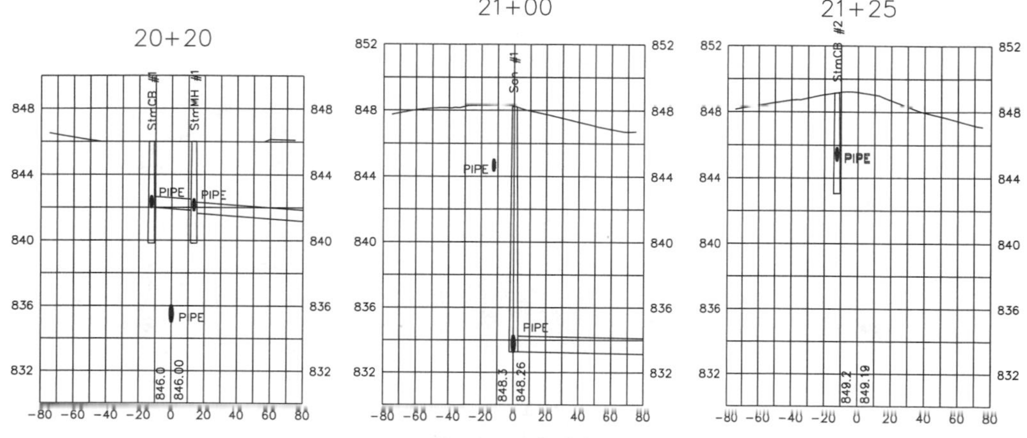

Figure 14–64

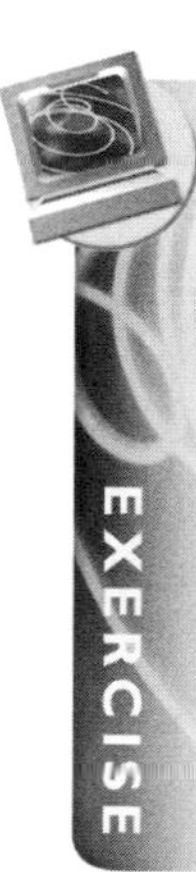

This ends the section on the Pipes routines of Civil Design. The routines create, edit, analyze, and annotate pipe runs you define in the project. The annotation is a result from the currently loaded settings. Make sure you load the appropriate settings before running the annotation routines. The settings are extensive and you need to define a set for the most common pipe runs in the project. This strategy of settings results in a sanitary, storm, and catch basin settings group. Before you define a run, you must make sure you have the correct set loaded.

This ends the discussion of the Civil Design add-in to Land Development Desktop. The next three sections focus on the Survey add-in for Land Development Desktop.

Autodesk Survey

After you complete this section, you will be able to:

- Use Survey module operating defaults
- The Survey command language
- Be familiar with the external files of Survey
- Use the point development tools of Survey
- Use the Survey Field Book file

SECTION OVERVIEW

The Survey add-in has its own special point routines and settings because of the focus on field survey methods. The routines in the Survey module reduce field data to coordinates and lines. With Land Development Desktop's description keys active, the reduction of field data also produces points with symbols. With the use of figure commands within the field data, Survey will produce line work. By combining figure commands and description keys with field data, Survey can draw a majority of the survey lines and annotate points with symbols.

UNIT 1

The first step in understanding the Survey module is to review its capabilities by reviewing its settings. The Survey settings are the focus of the first unit of this section.

UNIT 2

The focus of the second unit is to learn and understand the Survey Command or Field Book Language as it pertains to working with points. The Survey Command Language (SCL) contains commands to create, edit, and analyze points in a project. The online documentation reviews the commands, but it does not explain how to use the commands in the context of a survey.

UNIT 3

The third unit of this section looks at the process of taking written surveys and entering them into a field book. This allows you to create coordinates from surveys that do not use electronic data collection.

UNIT 4

The fourth unit of this section looks at the creation of line work through figures. The exercise uses existing points to draw a figure and reviews the analysis of this figure type.

UNIT 5

The last unit creates line work by adding figure commands to a field book.

The use of data collectors and traverse reduction is the focus of the next two sections of the book.

The Land Development Desktop convention for entering angles requires the letter *d* for degrees, ' (single) for minutes, and " (double quotes) for seconds. The surveying industry enters angles with a shorthand format: *dd.mmss*. The degrees represent the whole number, and the precision holds the degrees and minutes of the angle. The first two places after the decimal hold the minutes and the last two places hold the seconds of the angle. The values in the minutes and second position cannot be over 59.

Note: This format for entering angles is not the same as the decimal format of AutoCAD. Do not confuse them.

UNIT 1: SURVEY SETTINGS

The Survey module has several groups of settings. These values are in addition to the values of the drawing setup and the Land Development Desktop settings. The Survey settings are essential for the correct reduction of field information into points. Like Land Development Desktop and Civil Design, Survey defines most of its initial settings from a selected prototype of the project.

The Survey settings affect several basic surveying commands and methods. These settings include units of measure, the Survey graphics, the Survey Command Processor (SCP), the equipment library, and corrections for the equipment, points, figures, traverses, and sideshots.

You can adjust the default values in several places in the Survey menus. Some settings may be contrary to the drawing setup. For example, even if metric is set in the drawing setup, the change does not affect the measuring units for Survey. To work with metric units in Survey, you must set the Survey Units to metric.

The settings that affect the Survey add-in are found in the Drawing Settings routine of the Projects menu (see Figure 15–1) and the Survey menus.

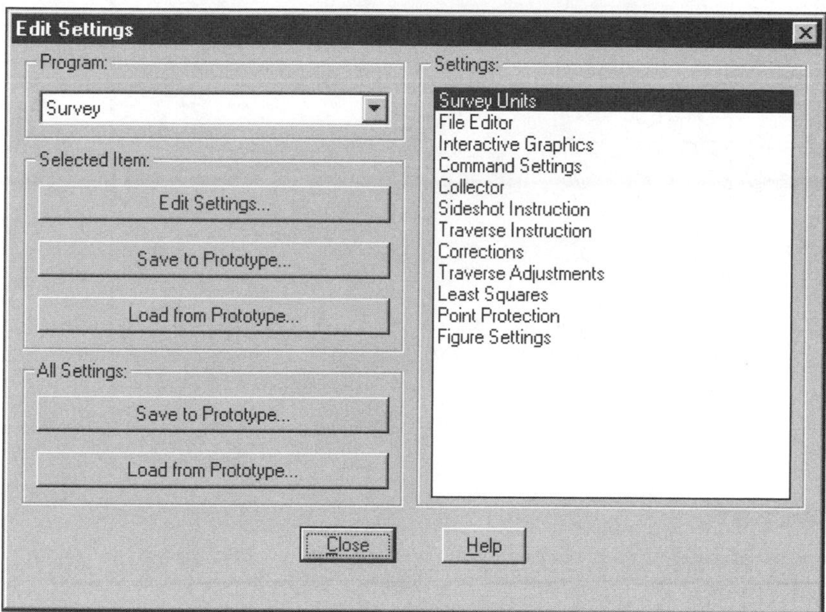

Figure 15–1

SURVEY UNITS

The Survey Measurement Units dialog box sets the units for the Survey Command Processor and survey routines (see Figure 15–2). The distance and angle units do not adjust themselves to the values in the drawing setup. You can set up a metric drawing and project, start the Survey Command Processor, and discover that the processor uses feet and degrees, minutes, and seconds as the default units. If metric is the default unit of measure for the drawing and project, you must make sure that the Survey Units are set correctly. This happens even if you select the metric prototype provided with Land Development Desktop.

The Survey Measurements Units settings are available only through the Drawing Settings routine of the Projects menu (see Figure 15–1).

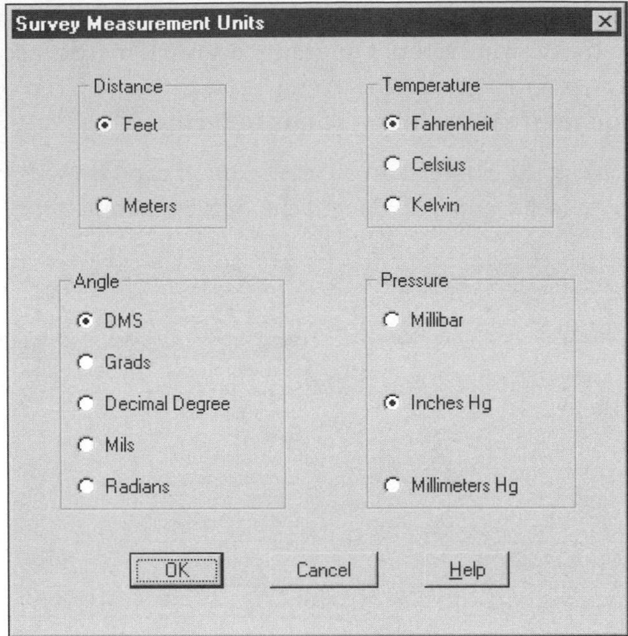

Figure 15–2

FILE EDITOR

The Survey External Editor Settings dialog box allows Survey to use another editor to view and edit Survey data (see Figure 15–3). The default editor is Windows Notepad.

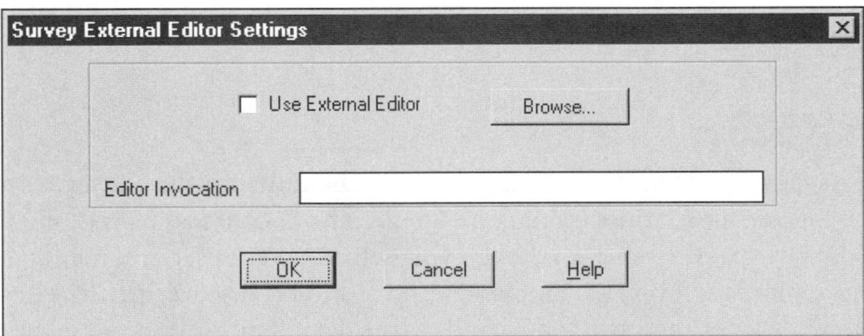

Figure 15–3

INTERACTIVE GRAPHICS

The Survey Interactive Graphics settings affect the display of information on the screen (see Figure 15–4). The main use of color is to display icons representing the surveying activity on the screen. When you are working in the Survey Command

Processor (SCP), with a field book, or the *batch.txt* file, the routines will display icons on the screen representing the activity in the file or the SCP. The colors of the icons represent specific activities occurring in the file.

The Interactive Graphics defaults are available only through the Drawing Settings routine from the Projects menu.

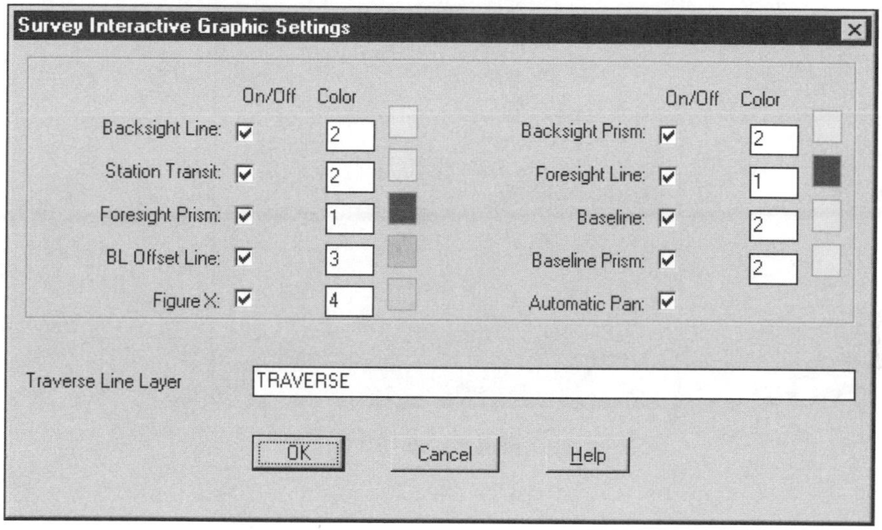

Figure 15–4

COMMAND SETTINGS

The Survey Command Settings pertain to the Survey Command Processor (SCP) and the operations it makes (see Figure 15–5). The Ditto Feature toggle repeats the last command by entering only the required data for the command. The SCP assumes that the new data is for the previous command type. The next four toggles, the Echo settings, allow the SCP to report to the command prompt the point course, figure course, point coordinates, and figure coordinates. Basically, these toggles report the direction from the last point and the new coordinates of the newly established point in the drawing. The next two toggles turn on the creation of the *batch.txt* and *output.txt* files. These two files are very important in the Survey module: The *batch.txt* file is a history file of commands entered from the keyboard or menu. The output file is a Survey-specific report file. These files are ASCII files and can be edited to correct data entry errors. The last two values in the dialog box set the names of the two files, *output.txt* and *batch.txt*.

The Survey Command settings are available only through the Drawing Settings routine from the Projects menu.

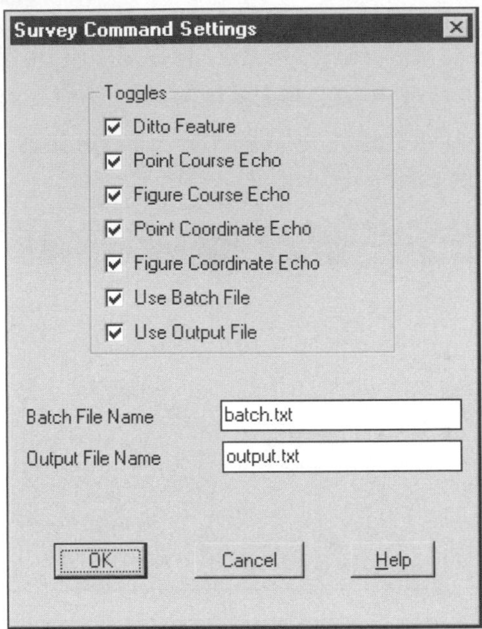

Figure 15–5

COLLECTOR

The settings for the Survey Data Collector define how Survey interprets and reports the processing of field data (see Figure 15–6). The top three settings in the dialog box affect how the routines report the lines read in from the field book. The last two toggles in this group control the appending of data to the observation database and the *batch.txt* file. Appending to the observation database is most important when creating traverses, traverse networks, and Least Square data sets for multiple field books.

CONVERT PRE-7.6 RAW FILES

The last group of settings in the dialog box, Convert Pre-7.6 Raw Files, affects how raw data is translated to the field book language using the Convert Pre 7.6 Raw Files routine of the Data Collection/Input menu. The Convert Coordinates to Raw Data setting concerns the downloading of data from a data collector. If the raw data is coordinates, when the data is converted, the field book will display angle and distance values.

The last two toggles in this group affect how the raw data draws line work when converted into a field book. The toggle Match Figure Name with Point Description creates figure commands related to the descriptions of points when the raw file is converted into the field book. The figure names are the descriptions of the points. When importing the Field Book file, the Import routine connects the related points to produce lines. As mentioned before, the name of the figures is the same as the description of the points. For example, all EOP1 and EOP2 points are connected to make the right and left edges-of-pavement for a survey. The toggle Remove Figure Name from Point Description allows the conversion routine to remove a figure name from a point's description. This allows you to create line work that includes points that do not have the same description. This second method requires the field crew to enter the figure name they are drawing and the description of the point.

DATA COLLECTION LINK

The Convert Pre-7.6 Raw Files settings do not affect the line generation method found in the Data Collection Link routine. The assumption in the Data Collection Link routine is that the description of the figure is the same as the name of the figure. Even though the Data Collection Link routine supports numerous data collectors, it supports only one method of creating lines. Field crews using data collectors may have to modify their field methods to take advantage of the power of drawing lines in the Data Collection Link routine. The online documentation of Survey covers specific collectors and the methods that are needed to successfully draw line work in the drawing from field observations.

The Data Collector settings are available through the Drawing Settings routine of the Projects menu and the Collector Settings routine of the Data Collection/Input menu.

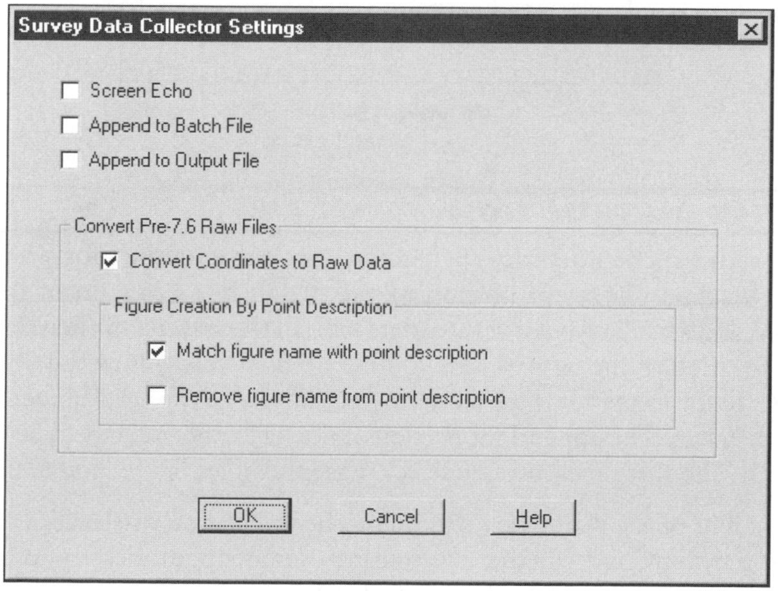

Figure 15–6

SIDESHOT INSTRUCTION

The Sideshot Instruction defaults apply to the Sideshot routine in the Data Collection/ Input menu or the SCP (see Figure 15–7). The dialog box affects the prompting of information by the data entry routines. The top portion of the dialog box contains three settings that toggle on or off the entry of elevations as Vertical Angles or Vertical Distances. The Command Toggles control calculating an elevation for a sideshot, the use of automatic point numbering, BS orientation (if not using a 0.00 angle), and whether to prompt for point descriptions.

The Sideshot Instruction settings are available through the Drawing Settings routine of the Projects menu and the Sideshot Settings routine of the Traverse/Sideshots cascade of the Data Collection/Input menu.

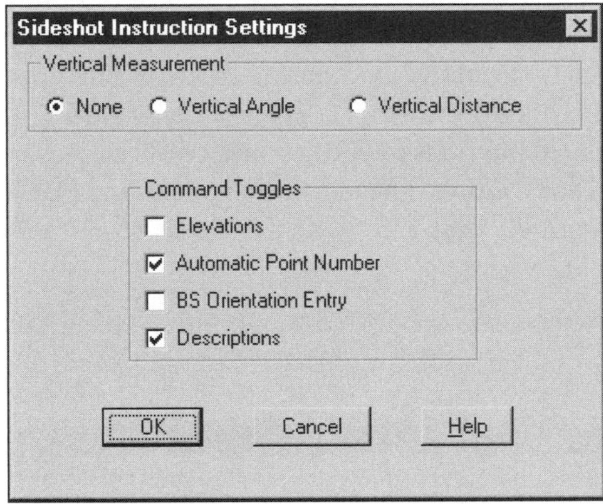

Figure 15–7

TRAVERSE INSTRUCTION

The Traverse Instruction Settings apply to the data entry routines for a traverse (see Figure 15–8). The upper left portion of the dialog box turns on or off vertical observations and sets the type of vertical data used in the traverse. When the surveyor enters an elevation of the traverse, the routine needs to know how the observations were made. The two methods for observing elevations are by vertical (zenith) angle or vertical distance. The upper right portion of the dialog box sets the type of angular observations. The three methods are Face1-Face2, Angles Right, and Deflections.

The bottom half of the dialog box toggles on the calculation of elevations as data is entered for a traverse and whether the routine prompts point descriptions. The F1 backsight , F2 backsight, and F2 foresight distance toggles set whether these values are prompted for when entering traverse data.

The Traverse Instruction settings are available through the Drawing Settings routine of the Projects menu and the Traverse Settings routine of the Traverse/Sideshots cascade of the Data Collection/Input menu.

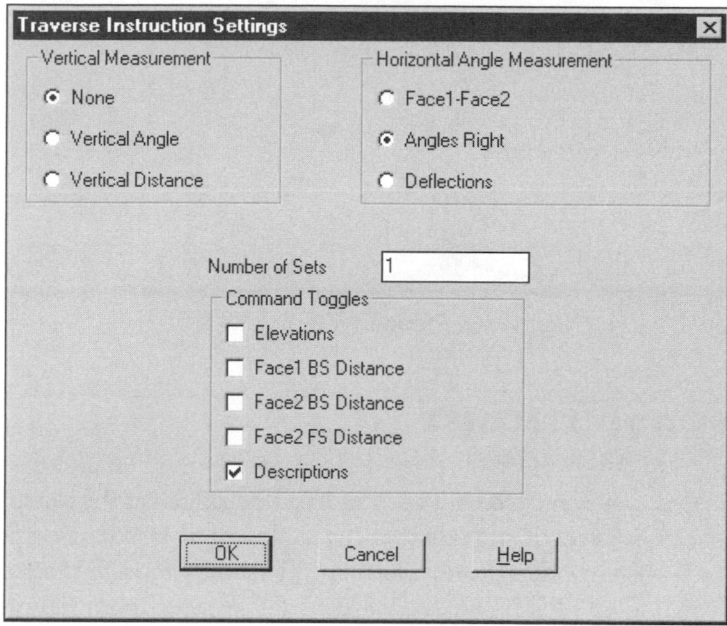

Figure 15–8

CORRECTIONS

The Survey Correction Settings dialog box toggles on the corrections applied to observation data as it is brought into a drawing (see Figure 15–9). Some of the corrections depend upon the settings found in the Equipment Library. Other corrections are straight mathematical calculations. The corrections you can apply to the data are Curvature & Refraction, Sea Level, Atmospheric Conditions, Horizontal and Vertical Collimation, Scale Factor, and EDM-Prism Eccentricity.

Generally, field crews with EDMs and data collectors make the data corrections in the field. As a result, you may not want Survey to do any corrections to the data. If you manually collect data and want to have corrections, this dialog box sets the corrections that Survey applies to the incoming data.

The Correction defaults are available in the Drawing Settings routine of the Projects menu.

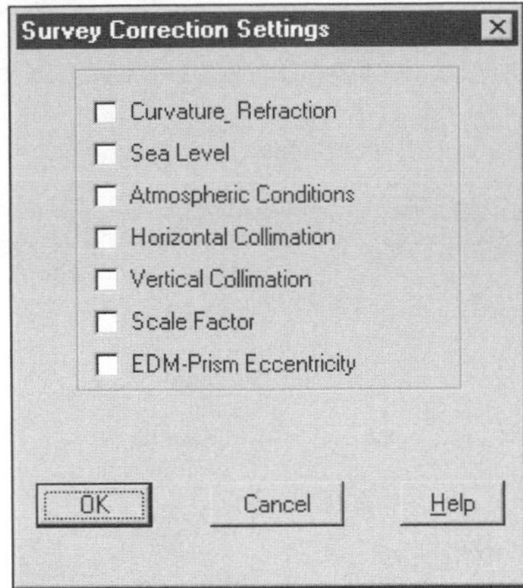

Figure 15–9

TRAVERSE ADJUSTMENTS

The Traverse Loop Adjustment Settings dialog box sets the default adjustment type for traverses analyzed in Survey (see Figure 15–10). The toggles at the top left determine whether an analysis is to occur and whether the angles are to be balanced. The Vertical Adjustment radio buttons determine whether to balance the elevations of a traverse and if so, by which method. The types of vertical adjustments are by length-weighted distribution or by equally distributing the error.

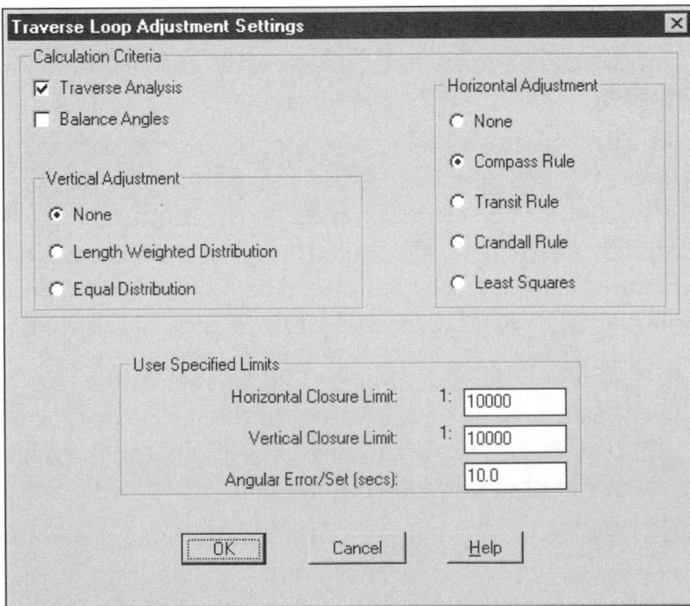

Figure 15–10

The horizontal adjustments include Compass Rule, Transit Rule, Crandall Rule, and Least Squares. All of the adjustment methods test for data blunders. The adjustment of the traverse is either as a 2D or 3D traverse.

The bottom portion of the dialog box sets the values for out-of-range error messages. Horizontal and vertical limits are in feet and angular limits are in seconds.

The Traverse defaults are available through the Drawing Settings routine of the Projects menu and the Adjustment Settings routine of the Traverse Loops cascade of the Analysis/Figures menu.

LEAST SQUARES

The Least Squares settings set the limits and error levels of the Least Squares analysis of field observations (see Figure 15–11). The upper left part of the dialog box determines the type of analysis, 2D or 3D. Directly below the type of analysis is a toggle for Blunder Detection. If this toggle is off, the analysis will distribute any error, no matter how severe, throughout the entire network. The upper right portion of the dialog box sets a value that determines whether a point has been solved for and how to represent its error as an ellipse. The lower portion of the dialog box sets the layers for the processing results and which elements to import into the drawing.

The Least Squares settings are available through the Drawing Settings routine in the Projects menu and the Least Squares Settings routine of the Analysis/Figures menu.

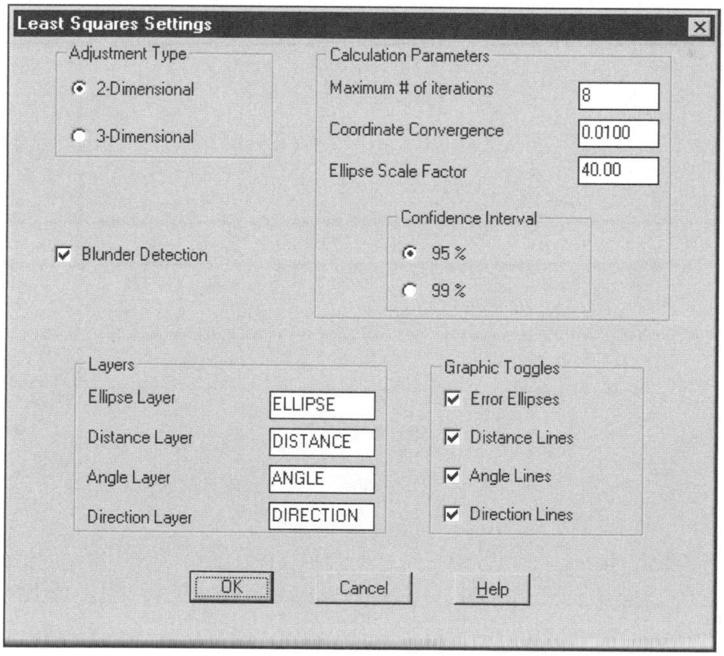

Figure 15–11

POINT PROTECTION

The Point Protection settings affect only the observations of a point (see the Survey Point Protection Settings dialog box in Figure 15–12). The settings take effect when importing a Field Book file or entering data in the SCP for a survey. A Field Book file may contain several observations of a single point; that is, a set of points for a traverse. The multiple observations of a point create an observation database that the Traverse routine evaluates. The Check Shots and Ignore All options do not do anything to any of the observations. The Average option uses all the observations to calculate the new coordinates of a point. The conditions you set with point protection allow you to realize potential problems in the field observations. If an observation exceeds the tolerance values set at the bottom of the dialog box, an error occurs and the Survey program prompts with an error message.

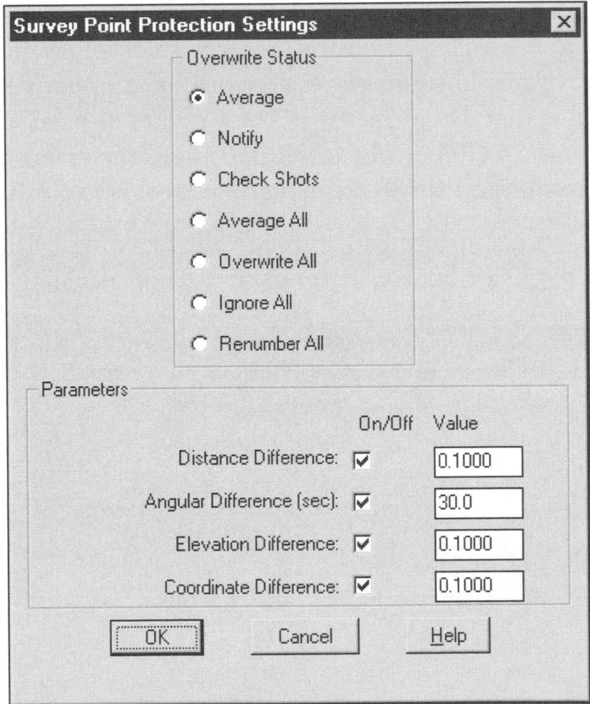

Figure 15–12

AVERAGE

The Average option processes all observations of a point to calculate a final coordinate for the point. Averaging occurs only when each observation falls within a tolerance level set at the bottom of the dialog box. If an observation is outside the tolerance level, the routine notifies you about the error.

NOTIFY

The Notify toggle forces a routine encountering a duplicate point to display an error message. The routine asks which point is correct and whether to delete or renumber the incorrect point. The routine then asks how any other duplicate points are to be handled.

This toggle makes F1/F2 (face 1 and face 2) shots duplicate point observations. If you are doing a traverse with F1/F2, this toggle should not be set.

CHECK SHOTS

If a point is out of the tolerance range, the routine notifies you of the condition. The routine does not modify the point.

AVERAGE ALL

The Average All option averages all multiple observations of a point. No warning is given. This is a global setting. The data becomes a part of the observation database and after a traverse analysis, the observations are reduced to a single point.

OVERWRITE ALL

If a routine comes across a duplicate point number, the routine discards the existing point and uses the new point coordinates. This is a global setting.

IGNORE ALL

The Ignore All option preserves the prior point and does not use the new definition of the point. This is a global setting. In essence, this toggle assures that the first observation is held and that all other observations of the point are disregarded.

RENUMBER ALL

When it finds a duplicate point number, Renumber All forces a routine to assign the duplicate point a new point number. This is a global setting.

At the bottom of the dialog box are settings that affect the type of errors to monitor and what you consider serious enough to know about. Toggle on one, all, or a combination of conditions. These conditions include a difference of distance, angle, elevation, and/or coordinate.

The Point Protection defaults are available only in the Drawing Settings routine from the Projects menu.

FIGURE SETTINGS

The top half of the Survey Figure Settings dialog box sets the default layer for figures (see Figure 15–13). If a figure name does not match a prefix, the Figure routine places the figure line work on this default layer.

The lower portion of the dialog box sets how the Survey routines create curve breakline data for a surface. Breakline data allows the Terrain Modeler to create a more accurate surface representation of linear features, for example, swales, river banks, edges-of-pavement, and so on. The mid-ordinate distance setting adds new vertices to the curve data. As the mid-ordinate distance becomes smaller, Survey creates more breakline vertices. The result of the additional data is more triangles along the arc segment, creating a better representation of the arc in the surface.

The Figure defaults are available through the Drawing Settings routine of the Projects menu and the Figure Settings routine of the Analysis/Figures menu.

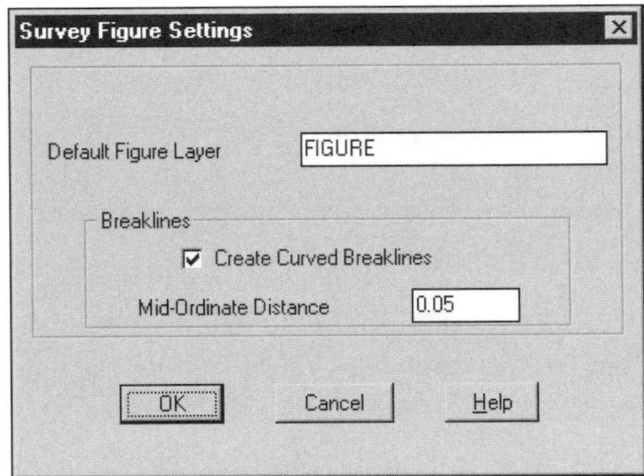

Figure 15–13

DATA FILES

The Survey add-in references several external files to process field data. These files supply the survey routines with command names and synonyms, equipment values, and prefixes from the figure prefix library.

COMMAND SYNONYMS

The Survey Command Processor has aliases for keywords in its vocabulary (see Figure 15–14). Instead of typing Angle-Distance, for example, you can use the SCP alias "ad" for the keywords. Aliases allow you to modify the Survey naming convention. If you are familiar with other coordinate geometry packages, the aliases allow the SCP to execute the correct command in terms familiar to you. See the Survey Command Language documentation for a list and explanation of the commands.

The Synonym defaults are available only in the Data Files routine of the Projects menu.

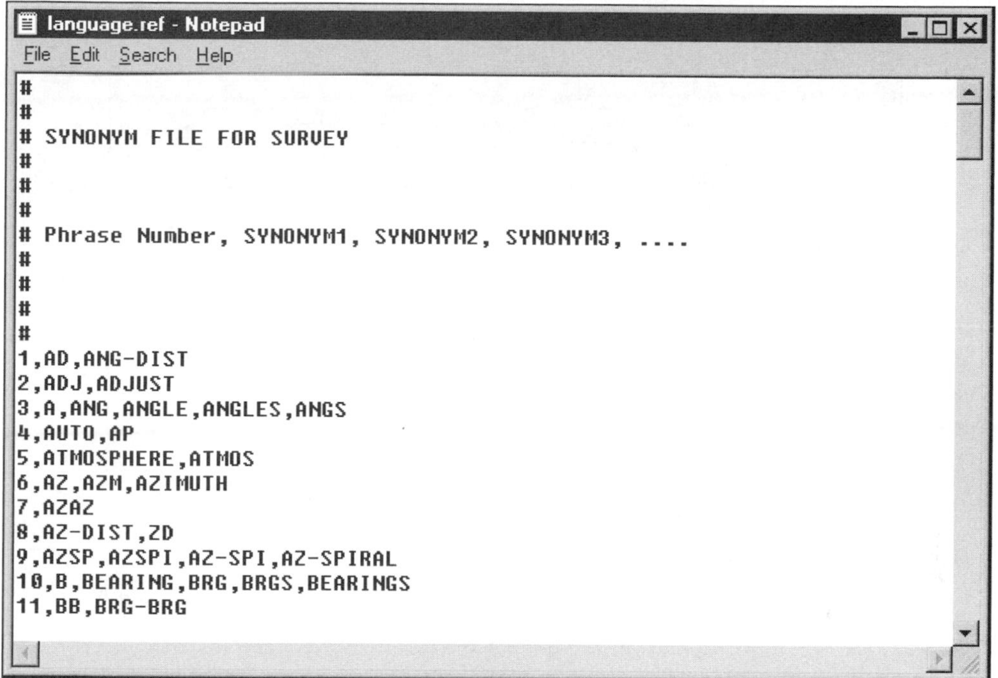
language.ref - Notepad

```
#
#
# SYNONYM FILE FOR SURVEY
#
#
#
# Phrase Number, SYNONYM1, SYNONYM2, SYNONYM3, ....
#
#
#
#
1,AD,ANG-DIST
2,ADJ,ADJUST
3,A,ANG,ANGLE,ANGLES,ANGS
4,AUTO,AP
5,ATMOSPHERE,ATMOS
6,AZ,AZM,AZIMUTH
7,AZAZ
8,AZ-DIST,ZD
9,AZSP,AZSPI,AZ-SPI,AZ-SPIRAL
10,B,BEARING,BRG,BRGS,BEARINGS
11,BB,BRG-BRG
```

Figure 15–14

EQUIPMENT

In the Survey Equipment Settings dialog box, you create an equipment library (see Figure 15–15). This library documents the error values for each instrument, which in turn are referenced by the corrections toggled on in the corrections dialog box. Again, if the instrument or data collector does the corrections, you do not need to develop an equipment library or toggle on any corrections.

These settings also provide data to the Least Squares routine for the calculation of standard deviation errors of distance and angle measurements. The Least Squares routine uses the current instrument values to interpolate the quality of observation data.

The instrument installed with Survey is a "typical" instrument that reflects the error values of most instruments.

The Equipment Settings are available through the Equipment Settings routine of the Analysis/Figures menu and the Data Files routine of the Projects menu.

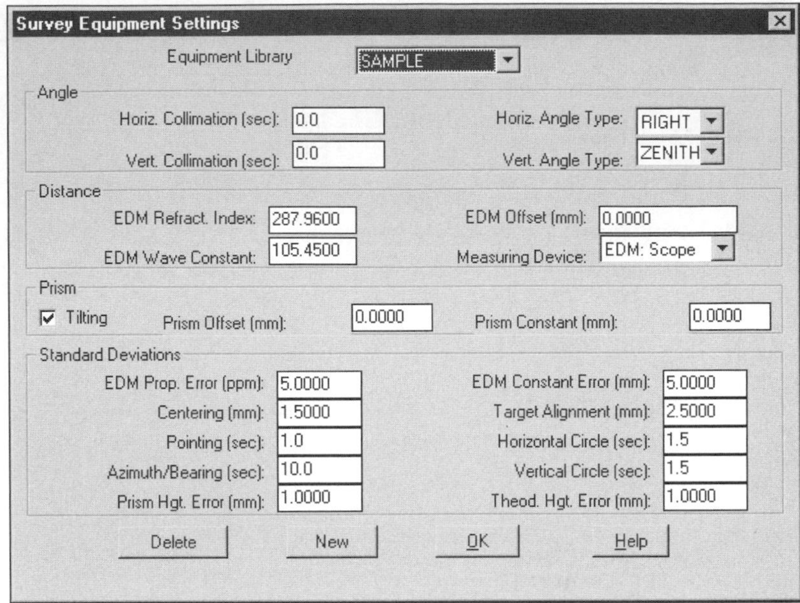

Figure 15–15

FIGURE PREFIX LIBRARY

The Figure commands draw lines directly from field observations, existing points, or by your transcription of field notes. The Figure Prefix library sorts the resulting figure line work onto specific layers in a drawing (see the Survey Figure Prefix Editor in Figure 15–16). Survey provides a simple starter library and allows you to modify and add to the prefix list at any time. The figure prefixes are available to all projects.

The Figure Prefix Library is available through the Figure Prefix Library routine of the Analysis/Figures menu and the Data File routine of the Projects menu.

Figure 15–16

EXERCISE

Exercise

After you complete this exercise, you will be able to:

- Use survey operating defaults

SURVEY SETTINGS

1. Start a new drawing, Lot35, and assign it to a new project, lot35 (see Figures 15–17 and 15–18). Describe the project as Survey Exercise 1 and use the LDDTR2 prototype. There is a copy of the LDDTR2 prototype on the CD that comes with the book.

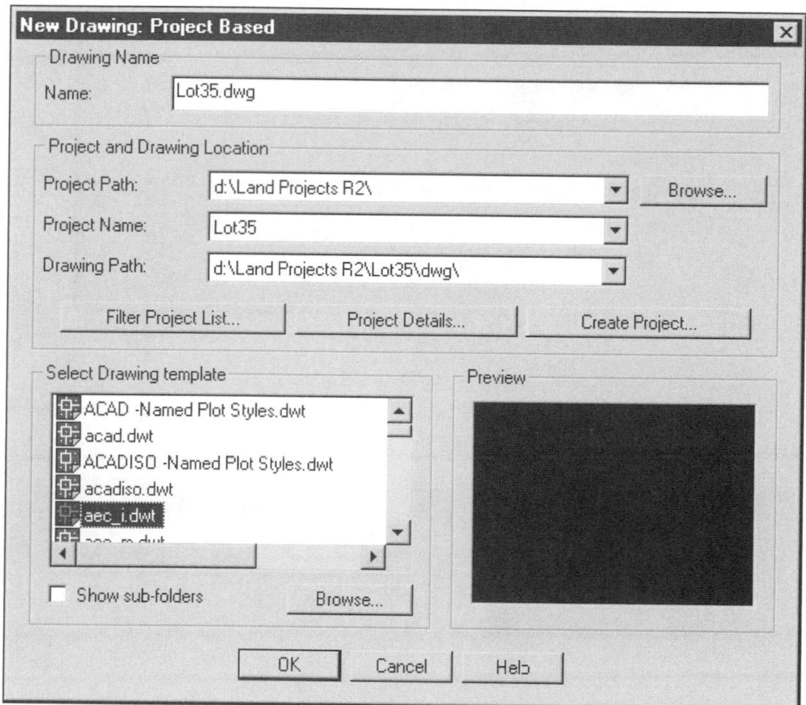

Figure 15–17

2. When Land Development Desktop displays the point database settings, select the OK button to dismiss the dialog box.
3. When the Load Settings dialog box displays, select and load the LDD1-20 setup. If you do not have the LDD1-20 setup, select the i20 setup. After you select the Load button, select the Finish button to exit the dialog box. The Desktop displays a Finish dialog box. Select the OK button to dismiss the Finish dialog box.
4. Select the Drawing Settings routine from the Projects menu to display the Edit Settings dialog box.

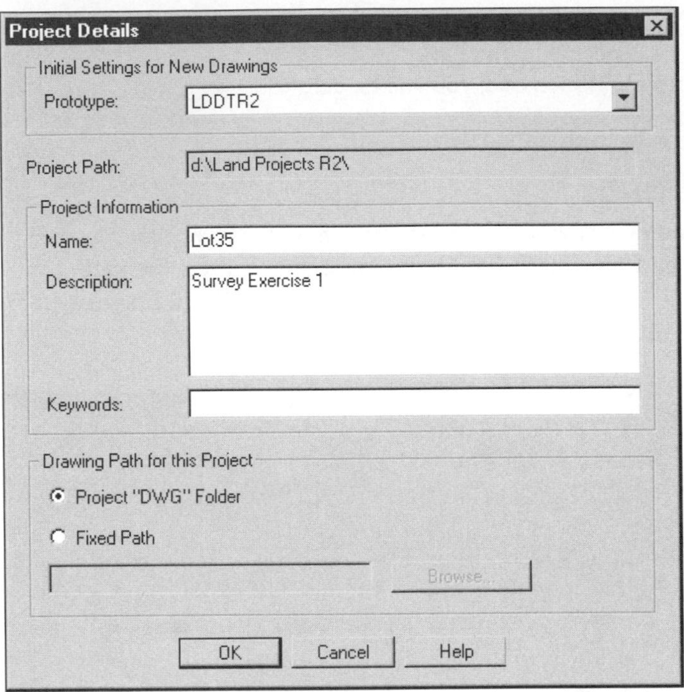

Figure 15–18

5. In the upper left of the dialog box, change the program to Survey (see Figure 15–19). This changes the Settings list to reflect the Survey program.

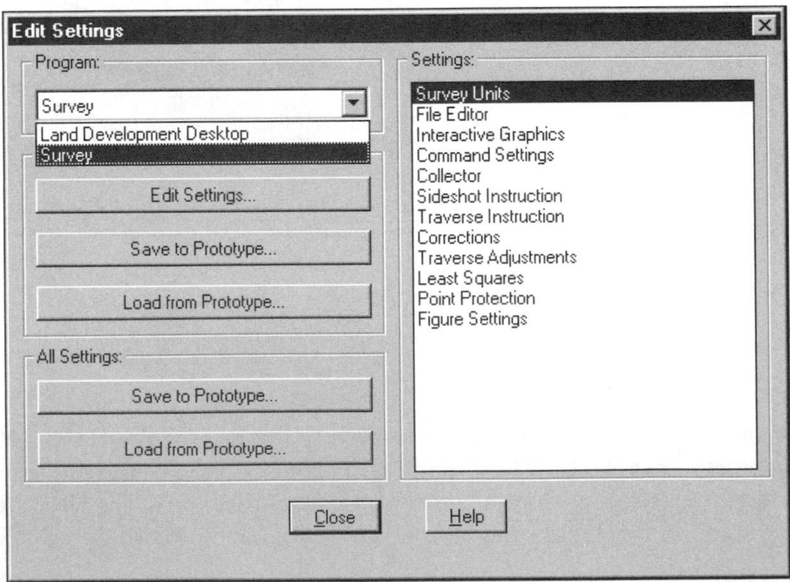

Figure 15–19

Survey Units

6. Select Survey Units from the Settings list and select the Edit Settings button to display the Measurement Units dialog box (see Figure 15–2). By default, all survey jobs are in imperial units even if you select the metric prototype.

7. Select the OK button to exit the Measurement Units dialog box and to return to the Edit Settings dialog box.

File Editor

8. Select File Editor from the Settings list and select the Edit Settings button. The Survey External Editor Settings dialog box (see Figure 15–3) allows you to specify a different text editor than Notepad.

9. Select the OK button to exit the Survey External Editor Settings dialog box and to return to the Edit Settings dialog box.

Interactive Graphics

10. Select Interactive Graphics from the Settings list and select the Edit Settings button. This dialog box allows you to specify different colors for icons displayed on the screen when working with the Survey Command Prompt, menu-selected commands, and when importing a field book (see Figure 15–4).

11. Select the OK button to exit the Survey Interactive Graphics Settings dialog box and to return to the Edit Settings dialog box.

EXERCISE

Command Settings

12. Select Command Settings from the Settings list and select the Edit Settings button. This dialog box allows you to modify the behavior of the SCP (see Figure 15–5).

13. Select the OK button to exit the Survey Command Settings dialog box and to return to the Edit Settings dialog box.

Collector

14. Select Collector Settings from the Settings list and select the Edit Settings button. This dialog box allows you to influence how the Convert Pre-7.6 Raw Files routine of the Data Collection/Input menu will convert pre-7.6 RAW files into field books (see Figure 15–6). Toggle on Convert Coordinates to Raw Data and Match Figure Name to Description. All other toggles should be off.

15. Select the OK button to exit the Survey Data Collector Settings dialog box and return to the Edit Settings dialog box.

Sideshot Instruction

16. Select Sideshot Instruction Settings from the Settings list and select the Edit Settings button. This dialog box controls the behavior of the Survey Command Prompt and menu-selected routines prompting for sideshot data (see Figure 15–7).

17. Select the OK button to exit the Sideshot Instruction Settings dialog box and return to the Edit Settings dialog box.

Traverse Instruction

18. Select Traverse Instruction Settings from the Settings list and select the Edit Settings button. This dialog box controls the behavior of the Survey Command Prompt and menu-selected routines prompting for Traverse data (see Figure 15–8).

19. Select the OK button to exit the Traverse Instruction Settings dialog box and return to the Edit Settings dialog box.

Corrections

20. Select Corrections from the Settings list and select the Edit Settings button. This dialog box allows you to select the type of corrections you want to apply to field observations (see Figure 15–9). Many data collectors and instruments apply corrections automatically, and you might not need to toggle on any of these settings.

21. Select the OK button to exit the Corrections dialog box and return to the Edit Settings dialog box.

Traverse Adjustments

22. Select Traverse Adjustments from the Settings list and select the Edit Settings button. This dialog box allows you to specify the type of balancing, the default adjustment, and the tolerances used in evaluating a traverse loop (see Figure 15–10).

23. Select the OK button to exit the Traverse Loop Adjustment Settings dialog box and return to the Edit Settings dialog box.

Least Squares

24. Select Least Squares from the Settings list and select the Edit Settings button. This dialog box allows you to specify the type of balancing, the default confidence interval, and the layers to use when evaluating a traverse loop by least squares (see Figure 15–11).

25. Select the OK button to exit the Least Squares Settings dialog box and to return to the Edit Settings dialog box.

Point Protection

26. Select Point Protection from the Settings list and select the Edit Settings button. This dialog box allows you to specify the default type of point protection and the point at which these errors occur (see Figure 15–12).

27. Select the OK button to exit the Point Protection Settings dialog box and return to the Edit Settings dialog box.

Figure Settings

28. Select Figure Settings from the Settings list and select the Edit Settings button. This dialog box allows you to specify the default name for figures that do not match a figure prefix and sets the mid-ordinate distance for breakline data around curves. Set the mid-ordinate distance to 0.5 (see Figure 15–13).
29. Select the OK button to exit the Figure Settings dialog box and return to the Edit Settings dialog box.
30. Select the Close button to close the Edit Settings dialog box.

DATA FILES

1. Select Data Files from the Projects menu to view the external data file that Land Development Desktop uses.
2. Change the program to Survey at the top, and the dialog box shows the Survey-specific data files (see Figure 15–20).

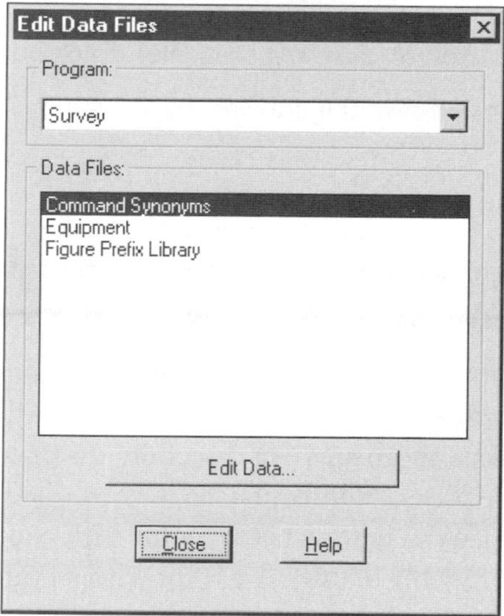

Figure 15–20

Command Synonyms

3. Select Command Synonyms in the central portion of the dialog box and select the Edit Data button. When you select the Edit Data button, the routine displays Notepad and allows you to edit or specify different aliases for commands in Survey (see Figure 15–14). You can add to the list to have additional synonyms available to you when using the SCP.
4. Close down Notepad to exit the Command Synonyms list and to return to the Edit Data Files dialog box.

EXERCISE

Equipment Error Settings

5. Select Equipment in the central portion of the dialog box and select the Edit Data button. This dialog box allows you to specify instrument characteristics for Survey (see Figure 15–15). Survey will use the values when determining error values for traverses and least square adjustments. You need to make a definition for each machine in your office.
6. Select the OK button to exit the Equipment dialog box and return to the Edit Data Files dialog box.

Figure Prefix Library

7. Select Figure Prefix Library in the central portion of the dialog box and select the Edit Data button. The Survey Figure Prefix Editor allows you to modify or add to a list of figure name prefixes (see Figure 15–16). The prefix of the figure's name also sets the layer for the resulting line work.
8. Select the OK button to exit the Equipment dialog box and to return to the Data Files dialog box.
9. Select the Close button to close the Data Files dialog box.
10. Click the Save icon and save the drawing.

UNIT 2: POINT TOOLS

The Survey module has its own set of point creation, editing, and analysis routines. These tools deal mainly with the reduction of field observation data into points. An unreduced observation has only an angle or direction, distance, and possibly a vertical measurement. All of these pieces of information, when added to an instrument setup, allow the surveyor to reduce the field data to a point with coordinates and possibly an elevation. The surveyor uniquely identifies a point with a point number and describes what the point represents with a description.

The Survey module has two types of routines for entering raw data into the drawing. The first method is direct data entry and the second is reading in a file. Each type of data entry routine, however, requires data entry to be done by different methods: a raw file from a data collector, a field book file, the instruction routines, the *batch.txt* file, or the Survey Command Prompt (SCP). The common element between the methods is the data. The data is written in the Survey Command Language.

By using the Survey Command Language and associated Survey tools, the surveyor produces points, point symbology, point sorting, traverse observations, and line work (figures). Point layer sorting and symbology are an interaction with the description keys of Land Development Desktop. Traverse observations are a result of the field measurement data and the use of figure commands within the field data to draw the lines. The mixing of field data and figure commands occurs either in the field, in the office as the field data is being processed, or as entries at the Survey Command Prompt.

MENU OR SURVEY COMMAND PROMPT

The data entry capabilities of the routines in the Data Collection/Input menu are limited. Both the menu routines and the SCP are unforgiving. If you are working on a manual survey, the best method for data entry is the field book. This unit reviews the routines of the menu system and the Survey Command Prompt. The use of the field book is covered in a later unit in this section.

THE DATA COLLECTION/INPUT MENU

The point routines in the Survey module are in the Control Points and Traverse/Sideshots cascades of the Data Collection/Input menu (see Figures 15–21 and 15–22). The point routines of these menus require direct data entry from the keyboard. After you enter the data, the routine executes a Survey Command Language command to calculate the point. While the point routines run, they display the entered data as Survey Command Language (SCL) input and echo on the command line the response from the Survey Command Processor. The primary purpose of the menu routines is to teach you the Survey Command Language and its data requirements by having the routine prompt you for the values and then viewing the resulting SCL format and response.

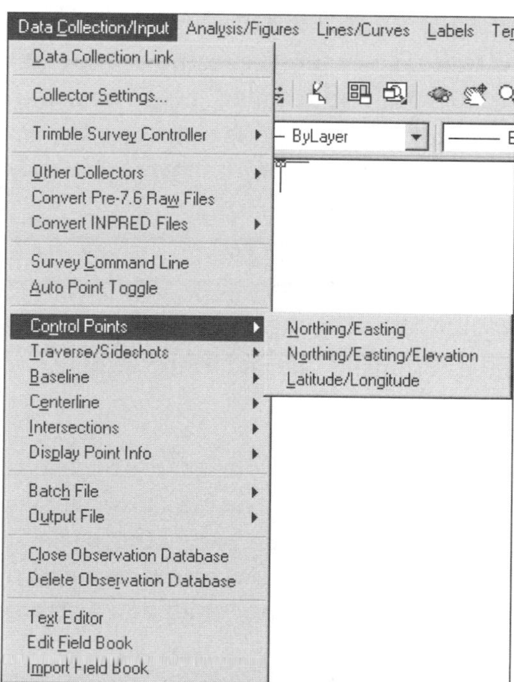

Figure 15–21

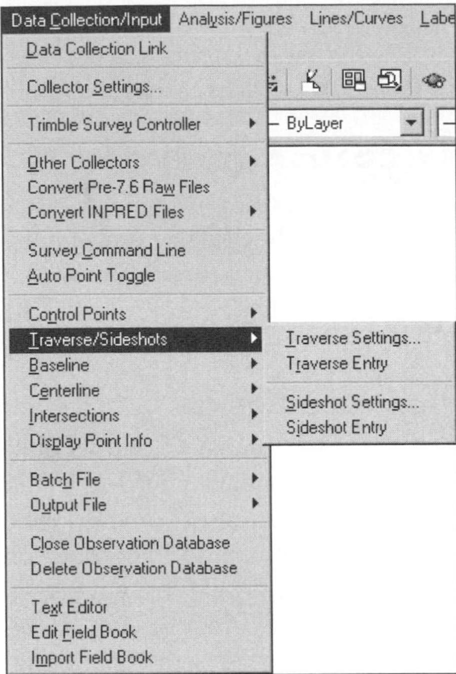

Figure 15–22

THE SURVEY COMMAND PROCESSOR

The Survey Command Processor is another method for point creation. The Survey Command Processor is a coordinate geometry program within the Survey module. The Survey Command Language is the language of the processor and is a terse language that assumes a knowledge and understanding of surveying methods, data requirements, and commands to calculate point data.

The Survey Command Processor replaces the Command: prompt of Land Development Desktop with a SURVEY: prompt. While you are in the SCP, you can enter the command name and its associated data, and after you press ENTER, the SCP displays the results on the command line and in the display.

Survey Command Format

When you look at the cmdhelp.ref file in Notepad, the file lists the various commands used by the Survey Command Processor. The file shows which portion of the format is required and what portion is optional. Any part of the format in parentheses denotes that part is optional and any part within brackets is required data. In the following example, the required data (in brackets) is the angle and distance values. The optional data (in parentheses) is the point number and description.

```
AD (point) [angle] [distance] (description)
```

SURVEY METHODS

A surveyor needs several pieces of information to start every survey: a known location and its elevation. If the surveyor does not have immediate access to these values, then he or she can make assumptions about their values. Generally, the initial point (benchmark) has an assumed coordinate of 5000, 5000 and an assumed elevation of 100. After finishing the work, the surveyor will research the point and assign actual coordinates and an elevation. If the coordinates change for the benchmark, all the points in the survey have to change as well. The same applies for the elevation. When the surveyor determines the correct elevation, all the points need to have their elevations adjusted using either in the SCP routine or the Datum routine of the Points menu. To adjust the location of points, the surveyor has many editing tools within the Desktop and Survey.

Before creating a point with the ANGLE-DISTANCE command (SCP), you must set other conditions. As mentioned before there is a required instrument location and elevation. In addition to the instrument location there needs to be either a backsight to another point or an azimuth measurement from the instrument's location. Both the coordinates for the instrument and the azimuth can be assumed. Generally, a backsight has known coordinates and the angle between the instrument to the backsight is 0.0000 degrees. This angle is set in the instrument. After you establish these values, the SCP is ready for a forward observation to a point (a foresight).

Working in the SCP is exactly the same as working in the field. The procedures for recording field data are the same for entering and processing data in the SCP. The Angle-Distance method is similar to the Setting Points By Turned Angle routine found on the Points menu.

As you enter the data, the SCP replicates the activity of the field crew in the drawing file. When you use the Survey Command Language to enter field data, the SCP is able to reduce the data into point coordinates and elevations. This requires you to understand field methods and terminology in order to successfully use field data to create points in the drawing.

ESTABLISHING A KNOWN POINT: THE BENCHMARK

The first step in creating a survey is to establish a benchmark. There may be only one or several benchmarks in the survey. If the coordinates of this point are known, the northing/easting point routine places the point into the drawing. If the northing, easting, and elevation of a point are known, the Northing/Easting Elevation routine places the point into the drawing. Both of these routines are in the Control Points cascade of the Data Collection/Input menu.

The initial point of a survey, the benchmark, is critical to the entire survey because the entire survey and all of its values are derived from the values of the benchmark. If the benchmark values are wrong, then the entire survey is in doubt.

The format for entering a point by northing and easting is as follows:

```
SURVEY: NE (point) [North] [East] (description)
```

-or-

```
SURVEY: NEZ (point) [North] [East] (elevation) (description)
```

The brackets and parentheses indicate the format of the routine as defined by the Survey language. The brackets indicate required values, and the parentheses indicate optional values.

Depending upon what is known or assumed, the surveyor enters the benchmark with one of the following two commands:

```
NE 1 5000.0000 5000.0000 Bench1
```

-or-

```
NEZ 1 5000.0000 5000.0000 721.23 Bench1
```

The first entry assumes a coordinate of 5000 north and 5000 east, a set of local coordinates; that is, assumed location. The entry does not have an elevation and a description of Bench1. Since the point does not have a known elevation, the benchmark is good only for horizontal work.

The second entry has an elevation along with assumed coordinates. The surveyor could have assigned the point an arbitrary elevation and adjusted the elevations later.

ESTABLISHING A SETUP: INSTRUMENT LOCATION OR STATION

The next step is to establish the instrument setup by locating an instrument on a benchmark. There may be one or several setups during the survey. After surveying a number of points, the field crew may establish one of the field points as the next setup point. Once a point is a part of the survey, the crew can station an instrument on that point.

If the survey does not include vertical measurements, the crew begins to survey the site from the setup because they only need to establish the coordinate location of points around the setup.

If the survey requires elevations, the crew needs to establish the vertical values of the station. This consists of the elevation of the benchmark and the crew measuring the height of the instrument above the benchmark, which then becomes a part of the station information, and finally, the crew measuring the height of the prism.

The format for setting up a station for a vertical survey is as follows:

```
SURVEY: STN [point] (instrument height) (description)
```

The entry for a station with no instrument height is:

```
SURVEY:STN 1 BM1
```

The entry for a station with an instrument height is:

```
SURVEY:STN 1 5.13 BM1
```

Setting a Backsight or Angle

The next value to establish is a reference or backsight angle. All the points that the field crew observes can be an angle from a known direction or angle. Basically, the crew records an angle (0) between the instrument and the backsight. A backsight is a point that establishes this initial reference angle.

The locations of all points from the benchmark depend upon the surveyor identifying a backsight. A point in the field is a turned right angle from the station and backsight reference angle (see Figure 15–23). Generally, the surveyor sets the angle between the instrument and the backsight as 0 degrees. The convention for a turned angle is positive as the instrument turns clockwise to locate a point. The surveyor refers to this method as a turned right angle.

The surveyor may also follow the azimuth convention. An azimuth assumes that north is 0 degrees and that angles increase when turning clockwise. If you use azimuths, you do not need to identify a point as a backsight (see Figure 15–24). If the field crew is not using an actual point for a backsight, they will sight a point in the field and arbitrarily set an azimuth to that point to establish the point as a backsight. The reason is that azimuth 0 (zero) is the instrument pointed to North. All of the observations made from this setup are in turned angles from north. To make this work, you need to establish the instrument setup point, define the azimuth to the backsight point, and then declare the dummy azimuth point as a backsight before making observations to any other points.

The format of the BACKSIGHT command is:

```
BS [point] (orientation)
```

To identify the backsight point number and the angle between the instrument and the backsight, use the BACKSIGHT command. In the above format the orientation is optional because if there is no value, the SCP assumes that the angle between the station and backsight is 0 degrees.

The following is a valid backsight entry:

```
SURVEY: BS 23
```

The entry sets point 23 as the backsight point with an angle of 0 degrees between the instrument and point 23.

After establishing a benchmark, an instrument setup, and a backsight or reference angle, the surveyor begins to record angles and distances (observations) to points at the site.

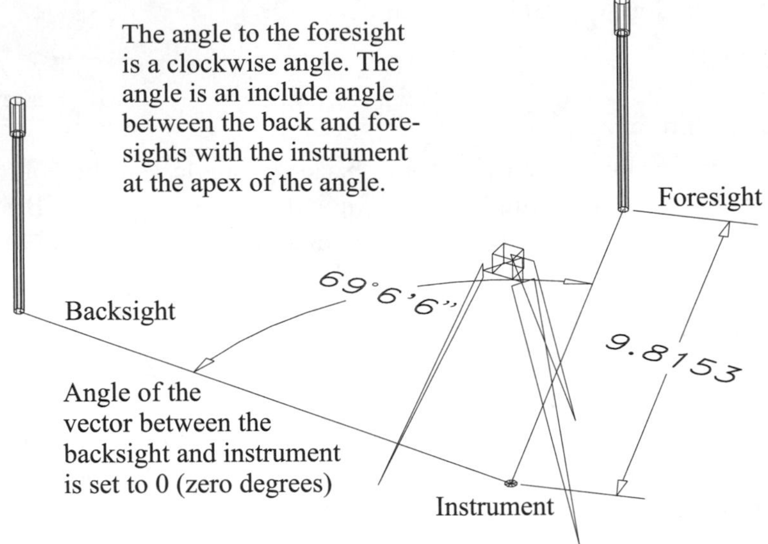

Figure 15–23

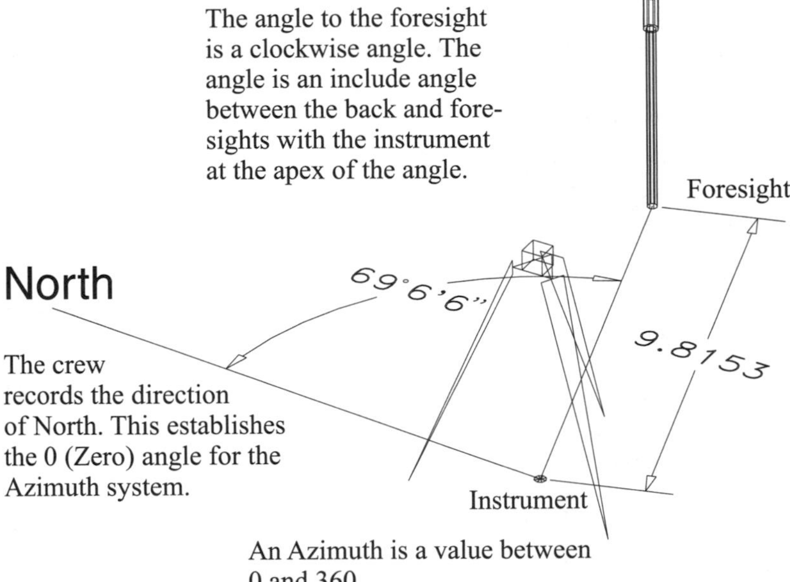

Figure 15–24

The second method for establishing a backsight is defining a dummy point representing an azimuth. This method does not require a second point with coordinates. It does require you to first define an azimuth between the station and backsight points and then to backsight to the point from the station. For example, you establish the coordinates for point 1 and then state an azimuth from point 1 to point 2. After establishing the azimuth, you then station on point 1, and then backsight to point 2. The following is a field book entry representing this method.

```
NE 1 5000.0000 5000.0000 100.00 IP
AZ 1 2 0.0000
STN 1
BS 2
```

ANALYZING POINTS

The Display Point Info cascade of the Data Collection/Input menu and the Survey Command Language have several routines to display information about points (see Figure 15–25). These routines display information on angles, bearings, distances, azimuths, and more.

The ditto option of the Survey Command Prompt (SCP) repeats the last command by typing only the data for the next query. Because there is only data on the command line, the SCP assumes that you are repeating the same command.

```
SURVEY>:A 1 4 3
Angle : 270-18-36
SURVEY:100 102 103
```

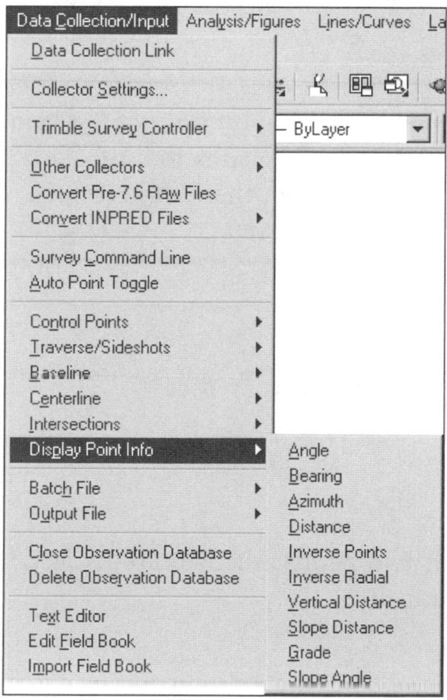

Figure 15–25

ANGLE

All angles are between three points: backsight, instrument, and foresight. This means that the middle point is the apex of the angle. The format for the Angle routine is:

A [Point 1] [Point 2] [Point 3]

```
SURVEY>: a 1 2 3
    Angle : 89-42-23
```

DISTANCE

The Distance routine returns the horizontal distance between two points. The format for distance information is:

D [Point 1] [Point 2]

```
SURVEY>: d 1 3
Distance:  130.07
```

AZIMUTH

To view the azimuth between points, invoke the Azimuth routine. The azimuth system measures angles clockwise with 0 degrees being north. The format for azimuth is:

AZ [point1] [point2]

```
SURVEY: AZ 1 2
SURVEY: 1 3
```

INVERSE POINTS

If you have a closed perimeter and you want to get the length of the perimeter and area of the closed box, use the Inverse Points routine. This routine prompts you for the points on the boundary, and when the routine returns to the starting point, it returns the area and length of the perimeter.

```
SURVEY: inv pts 1 2
SURVEY: 3
SURVEY: 4
SURVEY: 1 (the walk must end where it began)
Area: 4390.86
Perimeter: 320.91
```

GRADE

To find a grade between two points with elevations, use the Grade routine. The format for the Grade routine is:

Grade [point1] [point2]

```
SURVEY>: grade 3 4
    Grade: -0.62    Percent
```

SLOPE ANGLE

The Slope Angle routine returns the slope angle between two points. The format for the Slope Angle routine is:

Slope [point1] [point2]

```
SURVEY>: slope 1 2
    Slope angle: 0-02-44
```

SLOPE DISTANCE

The Slope Distance routine returns the 3D length of a line between two points. The format for the Slope Distance routine is:

SD [point1] [point2]

```
SURVEY>: sd 1 2
    Slope distance: 125.45
```

VERTICAL DISTANCE

The Vertical Distance routine returns the difference in elevation between two points. The format for the Vertical Distance routine is:

VD [point1] [point2]

```
SURVEY>: vd 2 4
Vertical distance -0.37
```

EDITING POINTS

The Survey Command Language provides a limited set of point editing tools. The tools edit only the elevation or description of a point. The only other editing tool deletes points from the point database. The Edit Points cascade of the Points menu holds the greatest variety and the easiest-to-use routines for point editing.

The following SCP commands edit points:

```
DEL PTS [point1] (point2) (entering point2 assumes a range of points)
MOD DESC [point1] [description]
MOD EL [point1] [elevation]
MOD ELS [point1] [point2] [elevation] (set a range of points to a single elevation)
MOD ELS BY [point1] [point2] [amount] (add an amount of change to a range
   of points)
```

ANNOTATING POINT DATA

The Survey module annotates points by using the description key matrix. You create and maintain the matrix using the Description Key Manager routine of the Point Management cascade of the Points menu.

EXERCISE

Exercise

The Survey module provides tools to query, enter data, and reduce field observations. The Data Collection/Input menu has routines that prompt you before calling the Survey Command Processor and returning results. Using the Survey Command Processor directly teams you up with a terse-language coordinate geometry program. To be able to use the SCP successfully, you need to know the command language and surveying principles.

This exercise uses both the prompt-driven menu routines and the Survey Command Processor (SCP). To invoke the SCP, type the letters **SV** at the Desktop command prompt while in the Survey menu.

You have to use more commands in the Survey module to set a point than you do when using a routine from the Points menu. The data entry in Survey that creates a new point includes specifying the point's number, coordinates, and description. All of these prompts occur no matter what the settings are for points in the Points menu. Review or print out the Survey commands before trying this exercise.

The next exercise is in conjunction with the drawing *Lot35* and Study Sheet A. Study Sheet A is a drawing on the CD that accompanies this book. The study sheets contain the layout and general information on Lot35. The assumed location of point #1 in the survey is a northing of 5000 and an easting of 5000.

After you complete this exercise, you will be able to:

- Use the point development tool menus
- Use the Survey Command Processor Language
- Use the field method of backsight, instrument, and foresight
- Use the Survey Language from the CMDLINE menu
- Enter commands at the Survey command prompt
- Query point data in the Survey module

TRANSCRIBING NOTES TO CREATE POINTS

1. If you are not in the *Lot35* drawing, open the drawing now. Change the menu to the Survey menu by using the Menu Palette routine of the Projects menu.

2. Review the survey plat for the details of Lot 35 (see Figure 15–26).

3. After you open the drawing, add the IP description key to the list. Use the Description Key Manager of the Point Management cascade of the Points menu. Use Figure 15–27 as a guide to entering the key.

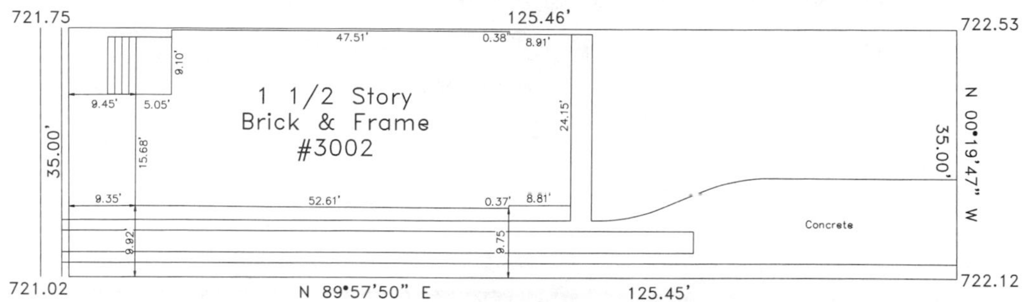

Figure 15–26

Create Description Key

DescKey Code: IP

OK

Cancel

Help

General | Scale/Rotate Symbol

Description Format: IRON

Point Layer: SET-MON-PNT

Symbol Insertion

Symbol Block Name: ip

Symbol Layer: SET-MON-SYM

Figure 15–27

4. Edit the Point Settings by using the Point Settings routine of the Points menu (see Figures 15–28 and 15–29). Make sure the current point number is set to 1. Set the following toggles:

Create Tab

Numbering:	Current Point Number: 1
Elevations:	Manual
Descriptions:	Manual

Insert Tab

Point Labeling Toggle:	ON
Current Point Label:	Active Description Keys Only

Point Settings

Create | Insert | Update | Coords | Description Keys | Marker | Text | Preferences

These settings control the creation of points in the point database.

Numbering

Insert To Drawing As Created   Sequential Numbering

Current number: 1

Elevations

Automatic   Manual   None

Default Elevation: 0

Descriptions

Automatic   Manual   None

Default Description: .

OK   Cancel   Help

Figure 15–28

Point Settings

Create | Insert | Update | Coords | Description Keys | Marker | Text | Preferences

These settings control the insertion of points in the AutoCAD drawing.

Search Path for Symbol Block Drawing Files

ogram files\land desktop r2\data\symbol manager\cogo\   Browse...

Insertion Elevation

Actual Elevation   If No Elevation Use: 0

Fixed Elevation   Fixed Elevation: 0

Point Labeling

Use Current Point Label Style When Inserting Points

Current Point Label Style:

active desckeys only

OK   Cancel   Help

Figure 15–29

5. From Study Sheet A or Figure 15–26, notice that Lot 35 consists of four lines of which only two have a bearing and a distance (the south and east boundary lines). The study sheet lists the two remaining lines as distances only (the north and west boundary lines).

6. Place point 1 into the drawing using the Northing/Easting/Elevation routine of the Control Points cascade of the Data Collection/Input menu. The location of point 1 is assumed to be 5000, 5000. Press ENTER twice to exit the routine after placing point 1.

```
Command:
Loading SURVEY Command Language . . . .
Enter new point number < 1 > :
Enter Northing : 5000
Enter Easting : 5000
Enter elevation <Null> : 721.02
SURVEY>  NEZ  1 5000 5000 721.02
 POINT 1       NORTH: 5000.0000    EAST: 5000.0000    EL: 721.0200
Enter new point number < 2 > :
Enter Northing :
```

7. Use the Zoom to Point routine in the Points menu and zoom to point 1 with a screen height of 50 feet.

8. You need to change the Description of point 1 to IP. Use the Mod Desc routine of the Survey Command Prompt. To start the Survey Command Processor, enter **sv** at the command prompt. The prompt changes from Command: to Survey: Enter the command with its data and press ENTER to execute the command. The description of point 1 should change to IRON and a symbol should appear at the point based upon the entry in the description key file. After you enter the data, type **exit** and press ENTER to exit the Survey Command Processor. The format of Modify Description is :

   Mod Desc [point] [desc]

```
Command: sv
SURVEY>: mod desc 1 IP
    Point 1   Desc: IP
SURVEY>: exit
```

9. Again use the Zoom to Point routine. Zoom to point 1 and use the height of 225.

10. Use the Pan routine to put point 1 in the lower left corner of the display. You may have to regenerate the drawing before panning point 1 to the lower left corner.

11. Click the Save icon to save the drawing.

SETTING POINTS 2 AND 3 BY BEARING/DISTANCE

Points 2 and 3 are at a bearing and distance from point 1. The Survey Command Language has a Bearing and Distance command. This command takes the entered bearing and distance

EXERCISE

and determines the coordinates of the point. Before entering the bearing and distance, you must establish a point from which the bearing and distance are calculated. In the Survey Command Language, you must station on a point (the STN command) to establish the point as the point the distance and angle is from. The sequence of commands is:

```
Station on point 1                          STN 1
Enter a Bearing and a Distance to point 2   BD 2 89.5750 1 125.45 IP
Station on point 2                          STN 2
Enter a Bearing and a Distance to point 3   BD 3 0.1947 4 35.00 IP
```

Survey follows the surveying convention for angle entries. To enter the angle of 34 degrees, 52 minutes, 18 seconds, you would type 34.5218 (*dd.mmss*) in Survey shorthand. Every Survey routine that uses an angle measurement uses this angular notation.

The Bearing method of angles divides a circle into four 90-degree segments. The top two quadrants, the northeast (quadrant 1) and northwest (quadrant 4), assume that a vector deflects from north to the east or west. So an angle of a vector varies from 0 degrees, a line traveling due north, to 90 degrees, a line traveling due east or west. The bottom two quadrants, the southeast (quadrant 2) and southwest (quadrant 3), assume that a vector deflects from south to east or west. So an angle of a vector varies from 0 degrees, a line traveling due south, to 90 degrees, a line traveling due east or west (see Figure 15–30).

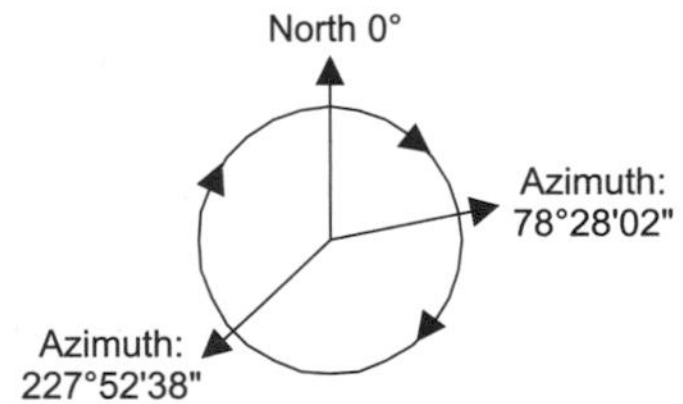

Azimuths:
Assumes North is 0°
Measures Angles Clockwise
Angles are between 0° and 360°

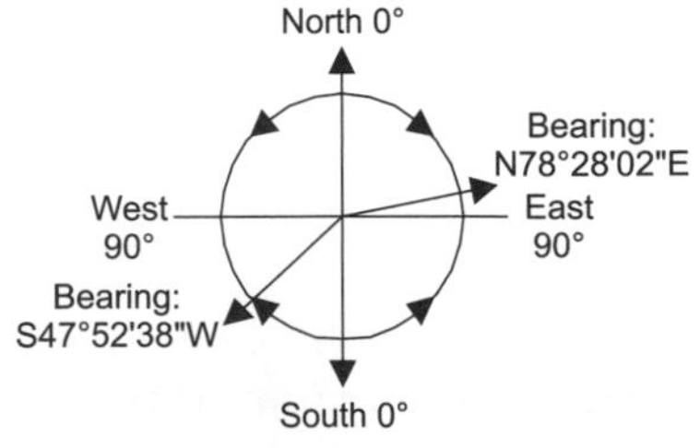

Bearing:
Assumes North or South is 0°
Angles are between 0° and 90°
East or West

Figure 15–30

1. Set points 2 and 3 using the BEARING-DISTANCE command of the Survey Command Prompt. To start the Survey Command Processor, enter **sv** at the command prompt. You must station yourself on point 1 before entering the bearing and distance to point 2. The BEARING-DISTANCE command allows you to enter the description of the point along with the point's data. The command does not allow for elevations. The format for the BEARING-DISTANCE command is:

   BD (point) [bearing] [quadrant] [distance] (description)

Point	Direction	Quadrant	Distance	Description
2	N89d57'50E"	1	125.45'	IP
3	N00d19'47W"	4	35.00'	IP

The data entry prompts are as follows:

```
Command: sv
SURVEY>: stn 1
    STATION       h.i.: <Null>
 POINT 1        NORTH: 5000.0000    EAST: 5000.0000      EL: 721.0200
SURVEY>: bd 2 89.5750 1 125.45 IP
     BEARING: N 89-57-50 E              DISTANCE: 125.4500
 POINT 2        NORTH: 5000.0791    EAST: 5125.4500      EL: <Null>
```

The first entry creates point 2.

```
SURVEY>: stn 2
    STATION       h.i.: <Null>
 POINT 2        NORTH: 5000.0791    EAST: 5125.4500      EL: <Null>
SURVEY>: bd 3 0.1947 4 35.00 IP
     BEARING: N 00-19-47 W              DISTANCE: 35.0000
 POINT 3        NORTH: 5035.0785    EAST: 5125.2486      EL: <Null>
```

This entry creates point 3.

2. After placing point 3 into the drawing, you need to add elevations to points 2 and 3. Use the MOD EL command to modify the elevations for the two points. After modifying the elevations of point 2 and 3, type **exit** at the command prompt and press ENTER to exit. The format for the MOD EL command is:

   MOD EL [point] [elevation]

Point	Elevation
2	722.12
3	722.33

EXERCISE

```
SURVEY>: MOD EL 2 722.12
    Point 2       Elevation: 722.12
SURVEY>: MOD EL 3 722.53
    Point 3       Elevation: 722.53
SURVEY>: EXIT
```

SETTING POINT 4 BY ARC/ARC INTERSECTION

The location of point 4 is more speculative. The distance from point 1 to point 4 is 35.00 feet and the distance from point 3 to point 4 is 125.46 feet. The best you can do is an intersection of two distances (Arcs) to establish point 4 (see Figure 15–31). The Arc/Arc routine of the Intersections cascade of the Data Collection/Input menu is the only method available for placing point 4. The routine operates by drawing two circles whose radii are the distance from each entered center point. If there are intersections, the routine places an X at each intersection. The routine then prompts you to select one or both intersections to place a point. The routine presents an option list that offers several methods of selecting the correct point.

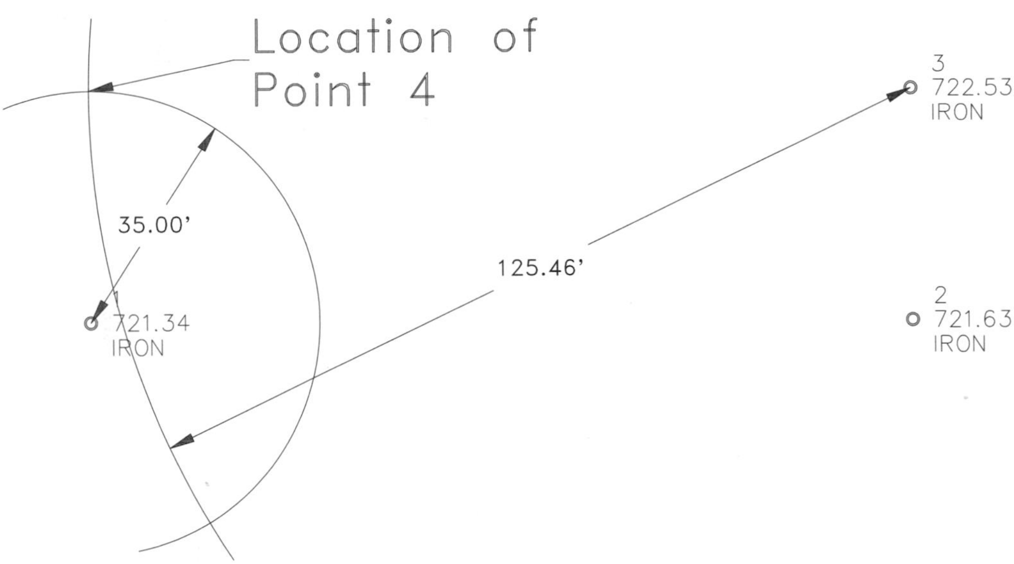

Figure 15–31

1. Place point 4 by using the Arc/Arc routine in the Intersections cascade of the Data Collection/Input menu (see Figure 15–32). Make sure to select the correct intersection point. The correct point is not always the first point on the list. The correct point is the most northerly of the two intersection points. After the routine creates the point, toggle to the text screen to view the SCP command that calculates the new point.

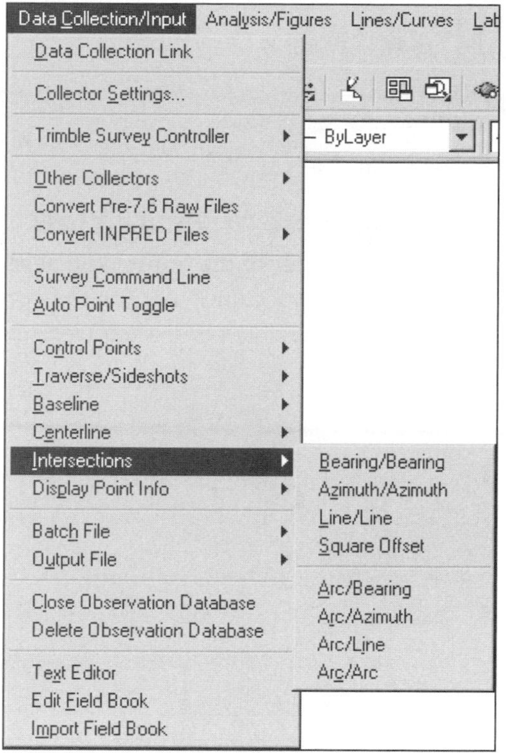

Figure 15–32

The prompts for the Arc/Arc routine are:

```
Command:
Enter arc center point number : 1
Enter radius of arc : 35
Enter arc center point number : 3
Enter radius of arc : 125.46
SURVEY>  RKRK  1 35 3 125.46
INTERSECTION # 1    NORTH:4969.982150    EAST:5017.998019
INTERSECTION # 2    NORTH:5034.999361    EAST:4999.788585
Select intersect [North/South/East/West/neaR/Far/1/2/All/Pick/eXit]
  <eXit>: 2
Enter new point number < 4 > :
SURVEY>  SAVE  2 4
POINT 4        NORTH: 5035.00      EAST: 4999.79      EL: <Null>
```

2. Use the Mod el and the Mod desc routines of the Survey Command Prompt to modify the elevation (721.75) and description (IP) of point 4.

```
Command: sv
SURVEY>: mod el 4 721.75
    Point 4       Elevation: 721.75
SURVEY>: mod desc 4 IP
    Point 4    Desc: IP
```

CREATING POINT 10 BY LINE/LINE INTERSECTION

To establish point 10, the southwest corner of the house, you need to offset the west and south property lines to establish the intersection point for the corner. In Survey this is a Line/Line intersection (see Figure 15–33).

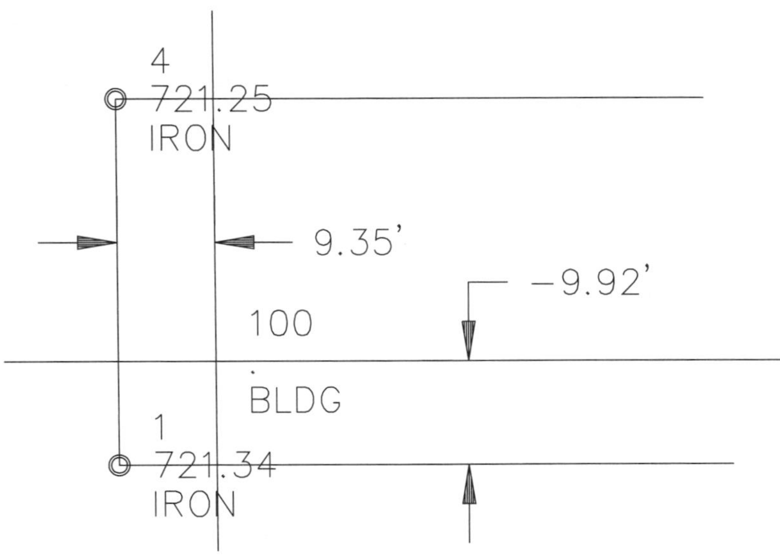

Figure 15–33

1. Use the Line/Line routine of the Intersections cascade of the Data Collection/ Input menu. Points 1 and 2 establish the first line and points 1 and 4 establish the second line. After establishing the lines, you enter the appropriate offset. An offset to the right is a positive value and an offset to the left is a negative value. You establish the right and left side of the line by standing at the beginning of the line and looking down towards the second endpoint. After the routine creates the point, toggle to the text screen to view the SCP command that calculates the new point.

   First Line: Points 1 & 2

   Offset from Line: -9.92 (Left Hand Offset)

   Second Line: Points 1 & 4

   Offset from Line: 9.35 (Right Hand Offset)

The following is the command sequence for establishing point 10.

```
Command:
Enter start point number : 1
Enter ahead point number : 2
Enter offset < 0.000000 > : -9.92
Enter start point number : 1
Enter ahead point number : 4
Enter offset < 0.000000 > : 9.35
SURVEY>  LINE-LINE  1 2 -9.92 1 4 9.35
INTERSECTION # 1   NORTH:5009.925857    EAST:5009.290213
Save intersect [Yes/No]? <No>: y
Enter new point number < 5 > : 10
SURVEY>  SAVE  1 10
POINT 10      NORTH: 5009.93      EAST: 5009.29      EL: <Null>
```

CREATING POINT 11 BY ARC/LINE INTERSECTION

Point 11 is a distance from point 10 and an offset from the line established by points 1 and 4.

1. Use the Arc/Line routine of the Intersections cascade of the Data Collection/ Input menu to place point 11. The routine first prompts for the arc center point, point 10. After you enter the center point, enter the radius of the arc, **15.68**. The routine prompts for the starting point of the line, point 1, and then prompts for the end point of the line, point 4. Finally, enter the offset, 9.45 feet, to establish the intersection to the right of the line. The routine will list two solutions; again, the solution is the most northerly of the two. The solution you want may not be the first solution on the list. Review the coordinates and blip Xs on the screen before selecting a solution. After the routine creates the point, toggle to the text screen to view the SCP command that calculates the new point.

The following is the command sequence to establish point 11.

```
Command:
Enter arc center point number : 10
Enter radius of arc : 15.68
Enter start point number : 1
Enter ahead point number : 4
Enter offset < 0.000000 > : 9.45
SURVEY>  RKLN  10 15.68 1 4 9.45
INTERSECTION # 1   NORTH:5025.605856    EAST:5009.295499
INTERSECTION # 2   NORTH:4994.247066    EAST:5009.484923
Select intersect [North/South/East/West/neaR/Far/1/2/All/Pick/eXit]
  <eXit>: 1
Enter new point number < 5 > : 11
SURVEY>  SAVE  1 11
 POINT 11      NORTH: 5025.61       EAST: 5009.30       EL: <Null>
```

EXERCISE

ANALYZING POINT DATA

1. Use some of the Display Point Info routines in the Data Collection/Input menu. Each of the routines prompt for a piece of information, and after you enter all of the required information, the routine invokes the SCP, which returns an answer. After the routine displays the information, toggle to the text screen to view the SCP command that calculates the new point.
2. Try using the SCP to display the point information. You start the SCP by typing the letters **sv** at the command prompt of Survey. At the Survey command prompt, type the word **help** and press ENTER. The response is a list of commands. Scroll down the list of commands to the Point Information area, and after exiting the listing, try a few of the commands.
3. After entering some commands in the SCP to exit from the SCP and return to the Desktop command prompt, type the word **exit** and press ENTER.

   Survey>: **exit** (*press* ENTER)
   Command:

VIEWING THE OUTPUT.TXT FILE

1. View the *output.txt* file by using the View File routine of the Output File cascade in the Data Collection/Input menu. The routines run from the Data Collection/Input menu record only their output and indent from the left margin. The output from commands entered in the Survey command prompt is at the left margin with both the command prompt entry and the result. The file may contain output from previous queries.
2. Exit the *Output.txt* file by closing down Notepad.
3. View the *batch.txt* file with the Edit Batch File routine of the Batch File cascade of the Data Collector/Input menu. The file contains all of the Survey Command Language entries. The routines run from the Data Collector/Input menu record their commands with an indent from the left margin. The commands from entries in the Survey command prompt are recorded at the left margin.
4. Exit the *Batch.txt* file by closing down Notepad.

SURVEY BASELINES

A baseline is a temporary linear construction between two points in the point database. Baselines exist only for the duration of the current editing session or until defining another baseline. After you define a baseline, inverse points to the baseline; that is, find the station and offset of the points relative to the baseline. A baseline routine sets points by station and offset from the baseline. The results of the baseline queries are in the *output.txt* file of Survey. There are no Survey defaults for baselines.

1. Create a baseline using points 1 and 4. The starting station is 0.00. Run the Define Baseline routine in the Baseline cascade of the Data Collection/Input menu.

2. Inverse the two house corner points (point 10 and 11) to the baseline. Enter the point numbers at the prompt, and the routine responds with the station and offset. Exit the routine by pressing ENTER until you return to the command prompt.
3. Click the Save icon and save the drawing.

EXERCISE

UNIT 3: FIGURE FROM POINTS

The Surveyor records field data as angles and distances from the reference angle and instrument setup. The surveyor can describe the angle in several different formats: turned angle (also F1 and F2), bearing, deflection, and azimuth (see Figure 15–34). The bearing is the same as the one used in the setting of points in the Lot 35 exercise. The new type of angle is a deflection angle, which is an include angle always less than 180 degrees. The surveyor turns the instrument 180 degrees before sighting a point. Even though there are several methods for observing points, most crews observe points using either the Angle-Distance or Face1 and Face2 methods.

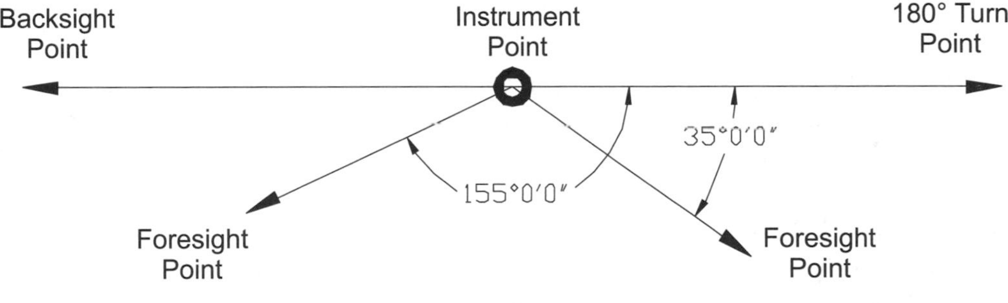

Figure 15–34

The formats of the different angle and distance observations are the following:

```
AD (point) [angle] [distance] (description)
F1 (point) [angle] [distance] (description)
F2 (point) [angle] [distance] (description)
BD (point) [angle] [distance] (description)
DD (point) [angle] [distance] (description)
AZ (point) [angle] [distance] (description)
```

The Survey indicates required data with a set of brackets. In the preceding command lines, you must enter an angle and a distance. Survey marks optional data with a set of parentheses. You have the option to enter a point number and description.

Accordingly, the following lines are valid entries:

```
AD 231.2451 35.2144
AZ 21 235.2332 53.8847
BD 23 45.3342 2 43.8835 PP
```

The first entry assumes that Auto point numbering is on, whereas the last two entries assume that Auto point numbering is off. The first entry reads: the angle to the next point is 231 degrees, 24 minutes, and 51 seconds at a distance of 35.2144 feet. The second entry adds the optional point number and reads: point 21 is an azimuth of 235 degrees, 23 minutes, and 32 seconds at a distance of 53.8847 feet from the instrument setup. The last entry adds the optional point number and description, and reads: point 23 is a bearing of 45 degrees, 33 minutes, and 42 seconds in quadrant 2. The distance to point 23 is 42.8835 feet and has a description of PP. If PP is a description key, upon importing the point, Survey will sort the point onto a layer, translate it, and if appropriate, place a symbol at the coordinates of the point.

Each of these entries locates a horizontal position for a point. The ANGLE-DISTANCE command does not allow the entry for a known elevation for the point, nor do the entries have any data to determine a point's elevation.

For the SCP to determine an elevation for a point, it needs the vertical information from the observation. This assumes that the vertical information about the instrument setup, a prism height, and method of recording vertical observation are known. You specify the type of vertical measurement within the command. Survey allows for two types of vertical measurements. The first, vertical distance (VD), also referred to as vertical difference, is a point's distance above or below a plane. The plane is parallel to the ground at the height of the instrument's scope (see Figure 15–35). The second, vertical angle (VA), is an angle between the sight line and a perpendicular passing through the vertical center of the instrument and scope (see Figure 15–36). It is from these vertical measurements that the SCP computes elevations.

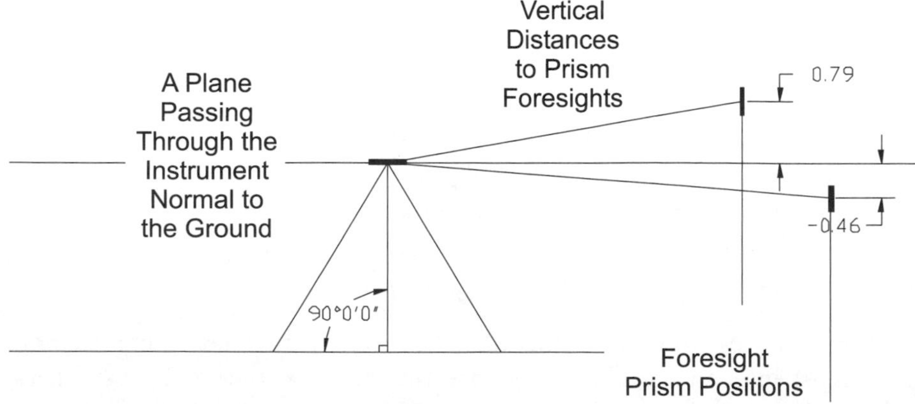

Figure 15–35

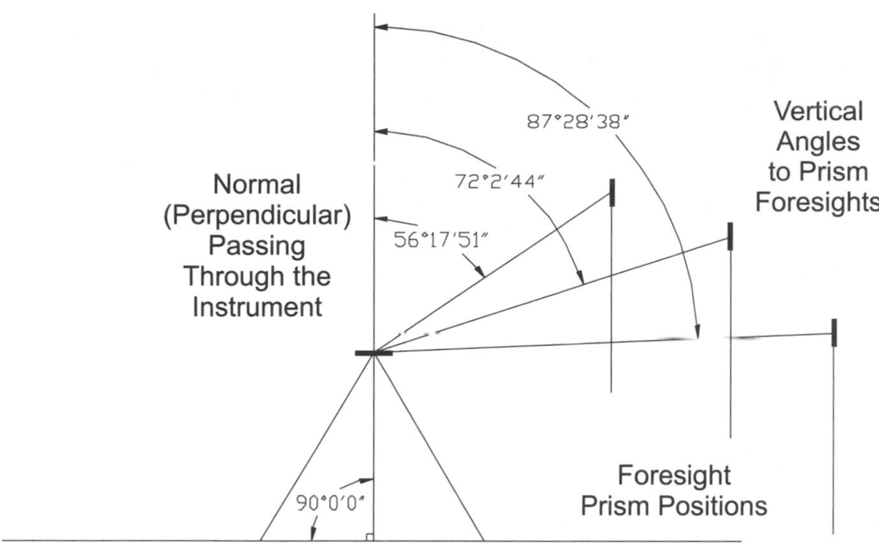

Figure 15–36

Each of the angle commands follows the same format for vertical measurements. The ANGLE-DISTANCE command structure with vertical measurements is as follows:

```
AD VA (point) [angle] [distance] [vertical angle] (description)
AD VD (point) [angle] [distance] [vertical distance] (description)
```

The format specifies the vertical measurement type after the angle-distance portion of the command. The format locates the measurement after the distance and before the description. Also you must specify whether the observation is a vertical angle (VA) or a vertical distance (VD).

Examples of the ANGLE-DISTANCE command formats with vertical observation data are:

```
AD VA 23 274.3544 124.54 89.5211 "IP"
AD VD 600 172.4435 56.48 -1.24 "CB"
```

The first entry translates as: point number 23 is an angle of 274 degrees, 35 minutes, and 44 seconds away from the backsight at a slope distance of 124.54 feet. The vertical angle is 89 degrees, 52 minutes, and 11 seconds, and is an iron pipe. The second entry reads: point 600 is an angle of 172 degrees, 44 minutes and 35 seconds away from the backsight at a distance of 56.48 feet. The vertical distance is –1.24 feet and is a catch basin.

If the surveyor entered a correct instrument setup and prism data, the preceding entries would produce coordinates and elevations.

If stationing an instrument on a point that has no elevation, the SCP will not calculate an elevation for any observations from that setup.

Any points you enter with fixed coordinates (NE or NEZ) will not change even if there is an error later in the survey of observing the location of the points. Sometimes the surveyor enters a point without an elevation and then later in the survey observes the point to assign it an elevation. When observing the point in the field to give the point an elevation, the Survey program ignores the results from the field observation and does not assign the existing point an observed elevation. Survey keeps the entered definition of the point and ignores all redefinitions. To assign the point an elevation, either compute the elevation and assign it to the first occurrence of the point or delete out all keyed-in point data that will be observed in the field. By not having the points already in the data set, Survey will assign the point its observed elevation.

BATCH.TXT AND OUTPUT.TXT FILES

While working in the SCP, the processor records the command sequence in an external file, the *batch.txt* file. The *batch.txt* file is a command history file for all Survey command activity. The *output.txt* file records the results of the command sequence whether in the Data Collector/Input menu or SCP.

Batch.txt File

The *batch.txt* file is a dynamic file. The Survey Command Prompt and menu-selected routines all append their commands to the file. Survey places the *batch.txt* file in the Survey subdirectory of the current project. You can edit the *batch.txt* file to correct typographical and data entry errors. When you rerun the *batch.txt* file, the corrections in the file correct any data entry errors. This may lead to duplicate points in the project. You may have to erase the points entered incorrectly before running the corrected batch.txt file.

Survey records SCP and menu-run commands differently in the *batch.txt* file. The command recording mechanism indents menu-called commands and records SCP commands at the left margin. This convention helps you identify where the command was called by looking at the justification of the command.

If you are correcting data in the *batch.txt* file that created erroneous point data, you must do more than just correct the information in the file. Before rerunning the *batch.txt* file, you must delete the points from the point database that the file recreates and corrects. If you do not delete the points before or during the rerunning of the *batch.txt* file, the Survey Command Processor will respond according to the point protection options set in the Survey defaults. Be careful when answering the questions about deleting points in the drawing during the Run *Batch.txt* routine. If you answer Yes to all the questions, the routine may delete all the points in the project.

The rerunning of the batch file also requires the deletion of the observation database. When rerunning the *batch.txt* file, the routine asks whether to delete the observation file. If the answer is no, the routine will stop and if the answer is yes, the Run *Batch.txt* routine continues. The observation database is the location of traverse data. If there is no traverse data in the file, deleting the file is unimportant. However, if traverse data is present in the file, deleting the observation database would be devastating.

Output.txt File

The next important file in the Survey module is the *output.txt* file. This file is not the *output.prn* file that other routines create when writing reports. The *output.txt* file, as with the *batch.txt,* is in the Survey subfolder of the current project. The *output.txt* file records the output from menu-called and Survey Command Prompt-entered commands. The file records the command and its results. As in the *batch.txt* file, the *output.txt* file records output from menu-called routines as indented and SCP commands and responses at the left margin of the file.

Field Book File

The most important file in the Survey module is the Field Book file. This file comes either from the conversion of raw field observations when using an electronic data collector or from someone typing data from a hand-written survey. The Field Book file is nothing more than a sequential list of observations from the field or field notes in the Survey Command Language. The order of data within a field book is the same as the procedures executed in the field. Because a field book reflects the rigid methodologies of surveying, the Field Book file is ideal for converting hand field notes into Survey data. Additionally, by using description keys and figure commands, the Field Book data will sort points, add symbols to points, and draw lines when importing the file.

Batch.txt versus Field Book File

The difference between a *batch.txt* file and a Field Book file is the extension of the file and the source of the data. You need only rename the *batch.txt* file to create a valid Field Book file. A Field Book file is a file with an extension of FBK and contains Survey Command Language commands. One advantage to a Field Book file is that any activity caused by routines called by the menu or Survey Command Processor does not append to or change the Field Book file. The source of information for a field book is field observations and the source for the *batch.txt* is the activity in the SCP or the menus.

CREATING A FIELD BOOK

To mimic field procedures in the SCP, you have to understand field methods. When a surveyor is in the field collecting topography (an existing ground survey), the surveyor follows a prescribed method of establishing and recording points. The two types of points the surveyor establishes, researches, or collects are benchmarks (reference points) and informational points (point observations). The surveyor locates his instrument at a point and identifies a second point as the backsight of the setup. The angle between the two points is 0 (zero). It is from this setup that the instrument turns an angle and records a distance to find the foresight (a head point). The surveyor records several points from each setup. There may be several setups within a single job. In Survey, the field book represents these procedures in file form.

Electronic File Conversions

One of the main functions of the Survey module is to translate raw field data into a field book. If you have a Sokkia data collector, TDS, Leica or other data collectors, Survey directly supports the downloading of their field data. The download routine (TDS DC Link) places the field file into the Survey subdirectory of the project. The Conversion routine of TDS DC Line converts the field file into a field book. Since the Survey Command Language represents field methods, the field book is a file made up of Survey Command Language commands representing the activities in the field.

Manual Field Books

A field book does not always start as a converted data collector file. You can create a field book from scratch by understanding field procedures, concepts, and the Survey Command Language. The written field survey is a paper copy of the field book. By determining the benchmark values and correcting any errors, you can transcribe paper notes into an electronic Field Book file.

A field crew rarely changes field collection methods during a survey. For example, the crew starts the survey by using angle and distances and then decides to start using azimuths and vertical differences. Every office and crew has a preferred method of surveying. The crew usually starts and finishes the job using the same mode.

As in the field, you must follow the same rules when writing a Field Book file. To observe points around an instrument, you need to establish a setup. A setup involves an instrument location and a backsight. The surveyor arbitrarily defines the angle between the instrument and the backsight as zero degrees. From this reference angle, the surveyor turns the instrument to view each foresight point. The crew records from the setup the foresight point angles and distances (observations). The surveyor records the angle as an angle from 0 degrees turned clockwise in the direction of the foresight point, that is, turn right angles. If vertical measurements are in use, there has to be an instrument and prism height and vertical distances or angles added to the data. The vertical data elements allow the SCP to calculate new elevations for the observed points. All of these elements are found in the Survey Command Language and the Field Book.

Exercise

This exercise covers the creation of a field book from a manual survey. The exercise converts the handwritten survey into a Field Book file using the procedures in the field and the Survey Command Language.

Before starting the next exercise, print out and review the Survey Command Language and field methods.

After you complete this exercise, you will be able to:

- Write and edit a field book
- Modify point data with a field book
- Create points by angles and distances
- Add description keys to the importing of a field book

CREATING A FIELD BOOK FROM A MANUAL SURVEY

This exercise consists of writing and editing a Field Book file. The exercise uses the Nichol's survey on the CD (Nichols.doc).

1. Start a new drawing, *Nichols*, using the *aec_i* drawing as the template drawing (see Figure 15–37). Assign the drawing to a new project, Nichols, and use the LDDTR2 Prototype for the project. Use the description of "A topo of 10/24/00" for the project (see Figure 15–38). Select the OK buttons to exit the Create Project and New Drawing dialog boxes. If you do not have the LDDTR2 prototype, there is a copy of it on the CD that comes with the book.

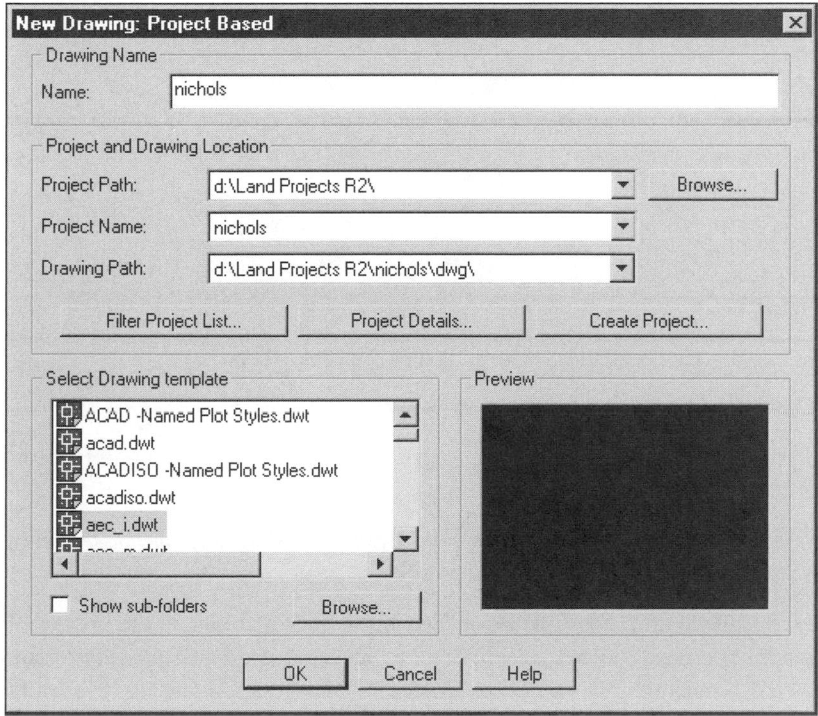

Figure 15–37

EXERCISE

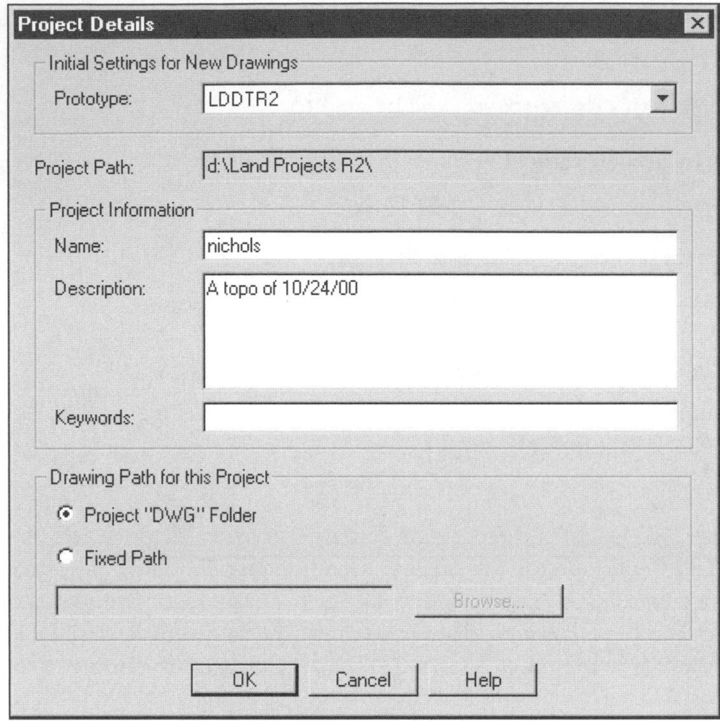

Figure 15–38

2. When the Point Database defaults dialog box appears, select the OK button to dismiss the dialog box.
3. Next, the Load Setup dialog box appears. Select the LDD1-40 setup and select the Load button to load the setup for the drawing. If you do not have the LDD1-40 setup, use the i40 setup instead. Select the Finish button to exit the Setup dialog box.
4. Select the OK button to dismiss the Finish dialog box.
5. Click the Save icon to save the drawing.
6. If you need to, change the menu palette to Survey. Use the Menu Palette routine of the Projects menu.
7. Print and review the field notes from the Nichols' Survey (*nichols.doc* file). The file is on the CD that comes with this book.

The Survey module locates the Field Book files in the Survey folder of the project. To create a file in the appropriate directory, all you need to do is use the Edit Field Book routine in the Data Collector/Input menu. If you enter the name of a nonexistent file, the routine asks if you want to create the file.

8. Choose the Edit Field Book routine from the Data Collection/Input menu. In the file name area, enter the file name of *nic1.fbk* and select the Open button. When the routine displays the Create File dialog box, answer Yes to creating the file.

The field notes of the file are the manually recorded field activities of the crew as they survey the Nichols' area. As you can see in the field notes, there are three main benchmarks at the site: points 1206, 1213, and 1226. Each point has known coordinates, a description, and only one does not have an elevation. There may be a time when point 1206 will need an elevation. For now you have enough information to start a field book.

9. Enter points 1213 and 1226 as NEZ survey commands and point 1206 as an NE survey command because the data contains coordinates for all three points and only the elevations for points 1213 and 1226. Use the coordinates and elevations in the field notes to create the data for the command. Remember to press the ENTER key after entering the line into the field book. This includes the last line of the field book.

The format for NEZ is:

```
NEZ (point) [North] [East] (elev) (desc)
```

The format for NE is:

```
NE (point) [North] [East] (desc)
```

The entry for point 1213 is:

```
NEZ 1213 2601.6437 1497.8024 634.52 "IP"
```

Enter the data for point 1226.

```
NEZ 1226 2623.0725 600.6777 636.43 "+"
```

The entry for point 1206 is:

```
NE 1206 2265.3919 1552.9250 "+"
```

10. Your field book should look like the following:

```
NEZ 1213 2601.6437 1497.8024 634.52 IP
NEZ 1226 2623.0725 660.6777 636.43 "+"
NE 1206 2265.3919 1552.9250 "+"
```

11. Save the file and exit the editor.

12. Go to the Data Collection/Input menu and import the *nic1.fbk* Field Book file.

The routine prompts for point and observations handling, which later prevents duplicate points from occurring in the point database.

EXERCISE

```
Erase all COGO points in the drawing? Y
Erase all existing observations? Y
Erase all figures? Y
```

13. Answer Yes to all the questions. Choose Zoom Extents to see the new points in the drawing.
14. Either use a routine from the Display Point Info cascade of the Data Collection/ Input menu or enter commands at the Survey Command Prompt and query the points for angles, directions, distances, grades, and so forth.

    For example:

    ```
    D 1206 1213
    D 1206 1226
    A 1213 1206 1226
    ```

15. If you are in the Survey Command Processor, you exit the processor by typing **EXIT** and pressing ENTER.
16. View the *batch.txt* file by selecting the Edit Batch File routine in the Batch File cascade of the Data Collector/Input menu. The file should be empty or nearly empty. Survey does not record any of the Field Book commands in the *batch.txt* file when you import a field book. The only commands in the file should be the query routines.
17. View the *output.txt* file by selecting the View File routine of the Output File cascade of the Data Collector/Input menu. The only commands in the file should be the query routines.
18. Exit the *output.txt* file viewer.

Establishing a Setup

The first set of points does not have field elevations. The crew already knows the elevations of the points, but they need to establish where the points are. The Northing/Easting and Angle-Distance routines establish this first set of points. The Angle-Distance routine does not allow data that has a known elevation added to the field observation. Traditionally, the observation assigns the elevation to the point. When you import the field book, the SCP calculates the point and elevation from observations. Here, however, we know the point's elevation, but we want to determine its location by field observations. To get around this problem, the ANG-DIST command will calculate the horizontal locations for the points. Later entries in the field book will modify the points' elevations with the MOD EL command. The SCP automatically assigns a NULL elevation to a point with no elevation.

Review the field survey notes to determine the field methods used by the crew. In the upper portion of the field book, they indicate that the angles they turn in the field are right angles, and the elevation observations are vertical distances. The crew notes these methods by marking the column as AngRt and Vdif. Since the crew wants to determine elevations, you need to be aware of any instrument and prism heights. The instrument height will probably change as the crew changes stations. The prism height will be constant, unless there are circumstances forcing a change to the prism's height. To miss these subtle changes is to render the survey in doubt.

This is a vertical survey and the initial benchmark is a point with known coordinates and an elevation. The crew establishes points 1213 and 1226 as benchmarks (stations).

Refer to the field notes for establishing the first setup. The first setup uses point 1213 as the location of the instrument. The backsight point is 1206, and there are no vertical observations from this point.

To establish point 1213 as a station, enter the STN command into the SCP. For now there is no vertical and the station command entry is quite simple. To establish point 1206 as the initial backsight point, enter the BS command into the field book. The BACKSIGHT command assumes 0 degrees unless you specify a value. The formats of the STATION and BACKSIGHT commands are as follows:

```
STN [point] (instrument height) (desc)
BS [point] (orientation) (no value for orientation assumes 0.0000)
```

After you establish an instrument setup, the next task is to establish a backsight. In reviewing the notes of the *Nichol.doc* file, the crew will use point 1213 as the initial station and point 1206 as a backsight. The new points are an angle and distance from the setup involving point 1213 as the instrument station. Point 1206 is the setup's backsight point and does not require an elevation. The purpose of point 1206 is to establish the 0-degree reference angle.

1. To establish a station (STN) and a backsight (BS), add the following commands to the Field Book file, *nich1.fbk*. Use the Edit Field Book routine found in the Data Collection/Input menu, and add the following two lines to the field book:

```
STN 1213
BS 1206
```

2. The Field Book file should now look like the following:

```
NEZ 1213 2601.6437 1497.8024 634.52 "IP"
NEZ 1226 2623.0725 660.6777 636.43 "+"
NE 1206 2265.3919 1552.9250 "+"
STN 1213
BS 1206
```

After establishing the first setup, review the *Nichols.doc* field notes. The crew records four observations from this setup. To reduce the observations into coordinate locations, enter the data in the field book. As the SCP reads the data, it calculates coordinates for each point. After calculating the coordinates, the SCP looks at the description. If the description matches a description key, the SCP follows the directions of the description key matrix.

EXERCISE

The next series of commands in your field book contain the data to calculate points 100–103. All the points are an angle and distance away from the current setup (instrument 1213 and backsight 1206). Remember, the angles are turned right angles from the backsight.

The catch to points 100–103 and 1248 is that the surveyor knows the elevations of the points but not their coordinate location. The Angle-Distance command of the Survey Command Language does not allow for the calculation of coordinates and the assignment of known elevation. So for now, enter the data to calculate only the coordinates for points 100–103 and 1248.

The command format and examples of the ANGLE-DISTANCE command are:

```
AD (point) [angle] [distance] (desc)
```

So the entry for point 1248 is:

```
AD 1248 200.2157 98.78
```

3. Add the ANGLE-DISTANCE commands to establish the following points in the Field Book file:

Point #	Angle	Distance	Description
1248	200.2157	98.78	
100	295.0930	372.86	"CPA"
101	278.4145	511.61	"CPD"
102	117.5913	495.78	"CPC"
103	99.0935	384.63	"CPB"

After you add the angle and distances to the field book, the file should look like the following:

```
NEZ 1213 2601.6437 1497.8024 634.52 "IP"
NEZ 1226 2623.0725 660.6777 636.43 "+"
NE 1206 2265.3919 1552.9250 "+"
STN 1213
BS 1206
AD 1248 200.2157 98.78
AD 100 295.0930 372.86 "CPA"
AD 101 278.4145 511.61 "CPD"
AD 102 117.5913 495.78 "CPC"
AD 103 99.0935 384.63 "CPB"
```

The SCP calculates the coordinates for the point with the ANGLE-DISTANCE command. If a point has a known elevation, later entries in the Field Book file need to modify the elevations of the located points. The only way to assign an elevation to a point that already exists is to use the SCP command, MOD EL (modify elevation).

The format and the example of the MOD EL command are:

```
Mod EL [point] [elevation]
Mod EL 100 633.96
```

Add the commands that modify the elevation of the following points:

Point #	Elevation
100	633.96
102	633.15
103	633.96

4. After you add the modify elevation information into the field book, the file should look like the following:

```
NEZ 1213 2601.6437 1497.8024 634.52 "IP"
NEZ 1226 2623.0725 660.6777 636.43 "+"
NE 1206 2265.3919 1552.9250 "+"
STN 1213
BS 1206
AD 1248 200.2157 98.78
AD 100 295.0930 372.86 "CPA"
AD 101 278.4145 511.61 "CPD"
AD 102 117.5913 495.78 "CPC"
AD 103 99.0935 384.63 "CPB"
MOD EL 100 633.96
MOD EL 102 633.15
MOD EL 103 633.96
```

5. Exit the Field Book Editor and save the file.
6. Import the field book using the Data Collection/Input menu.

The Import Field Book routine asks several questions before importing the file. The questions are:

```
Erase all COGO points in the drawing? Y
Erase all existing observations? Y
Erase all figures? Y
```

In this preliminary state, the answer to each of the above questions is yes.

EXERCISE

The new points enter the drawing and point database. The Import routine uses the Graphics Settings in the Survey defaults to display the activity in the Field Book file. In reality you are watching the field crew do the survey. You will see the instrument setup and the backsight and foresight lines appear on the screen as the survey is processed.

Survey uses the color yellow to indicate the station and backsight location. The routine places an instrument at the stations, a prism at the backsight, and marks the foresight points with a red prism and line.

Before going too far, you need to determine the description keys that apply to the points in the field book. By reviewing the field notes and the current description key list, you should be able to develop a complete description key list. The points 100–103 are CP (control points) points. Each description has a different letter after CP, the letters A–D. You want to match on CP, but you do not want to translate the key. The description key entry for CP is as follows:

Key	Translation	Layer	Symbol	Symbol Layer
CP@	$*	MON_PNT	BM	MON_SYM

This entry says the match is on the letters CP followed by a single alphabetical character (@). There is no translation of the key. The key is what the description will be in the drawing. The point goes to the layer MON_PNT (monument point) and the point receives a symbol BM. The symbol goes to a layer named MON_SYM (monument symbol).

Review the field notes, *Nichols.doc*, for points 200–229. The points represent a back-of-curb (BC) and front-of-walk (FW). In addition to the control points, the field crew observed other point objects: LP (light pole), PP (power pole), MH# (manhole), and an INB (inlet b). The crew also located points representing a bituminous slab, a gravel path, and a building. Some of these points are symbol locations and others are general descriptions. The descriptions LP, PP, MH#, and IN@ need to be description keys that place symbols into the drawing. The bituminous, gravel, and building descriptions need only to sort the points onto a specific layer.

Look through the field notes and create a description key matrix. You should first note down what description keys the crew uses and then fill in the remaining matrix values. There may be several descriptions using the same point and symbol layers.

7. Update the description key list with the new description keys. Add the keys with the Add Key routine of the Points menu.
8. Check the point settings in the Point Settings dialog box. Make sure the current point label style is active description keys only.

The current Field Book file established points 1213, 1206, 1248, and 100–103. The second portion of the field book will establish points 200–229. Please review the data for these points in the field notes.

EXERCISE

The points 200–229 are points with data to calculate elevations. This means that the next setup needs an instrument height with the STATION command and a prism height for the set of observations. Notice that near the end of the observations, the prism height changes from 4.65 to 6.0 even though the station does not change.

By using an exclamation point (!), you indicate a comment in the field book. The comments in the field book note that some of the points, although they are back-of-curb shots, are also points representing the front of a walk. Although the crew treats the points as only points, understand that some points create lines. The Survey figure commands name and draw line work in the drawing.

9. Add the following field notes to produce points 200–229. Enter the data into your Field Book file, *nic1.fbk*. Add the comments. A comment is always on the same line of code it describes. There is a completed field book file on the CD that comes with the book, *Nichols.fbk*. You can use this file if you should not have the time to enter in all of the data.

```
STN 1226 5.13                              !H.I. 5.13
BS 1213
PRISM 4.65
AD VD 200 92.1224 61.31 -0.89 "BC"    !BC/FW
AD VD 201 87.3738 61.13 -0.82 "BW"
AD VD 202 84.4515 17.12 -0.42 "BW"
AD VD 203 101.0837 17.49 -0.51 "BC"   !BC/FW
AD VD 204 252.2341 13.08 -0.32 "BC"   !BC/FW
AD VD 205 274.3455 12.68 -0.37 "BW"   !EDGE OF 10' GRAVEL WALK
AD VD 206 262.1906 33.27 -0.28 "BC"   !BC/FW
AD VD 207 259.3437 33.45 -0.74 "INB"
AD VD 208 265.3536 69.02 -0.15 "BC"   !BC/FW
AD VD 209 269.3947 69.87 0.08  "BW"   !CORNER OF BIT
AD VD 210 269.1905 123.72 0.28 "BW"   !EDGE OF BIT
AD VD 211 266.5835 123.95 0.12 "BC"   !BC/FW
AD VD 212 267.2954 168.00 0.23 "BC"   !BC/FW DEPRESSED WALK
AD VD 213 267.3912 187.20 0.31 "BC"   !BC/FW DEPRESSED WALK
AD VD 214 267.4530 204.83 0.32 "BC"   !BC/FW
AD VD 215 269.0846 205.23 0.47 "BW"   !EDGE OF BIT
AD VD 216 269.0922 188.70 0.41 "BW"   !NORTH END OF GRAVEL ALLEY
AD VD 217 269.1241 172.30 0.09 "BW"   !STH END OF GRAVEL ALLEY
AD VD 218 269.5533 118.54 0.47 "LP"
AD VD 219 275.5820 105.79 0.50 "BLDG"!NW BLDG CORNER
AD VD 220 275.3158 75.74 0.32 "BLDG" !SW BUILDING CORNER
AD VD 221 300.5235 77.02 -0.57 "BIT" !EDGE OF BIT
AD VD 222 295.2827 69.19 -0.58 "PP"
```

EXERCISE

```
AD VD 223 350.0500 36.62 -0.17 "MH1"  !4'DIA WATER MANHOLE
AD VD 224 346.2257 36.75 -0.08 "WV"  !WATER VALVE
AD VD 225 345.0415 38.24 -0.92 "GRAVEL" !GRAVEL PATH EDGE
AD VD 226 41.5340 50.00 -0.31 "GRAVEL"
AD VD 227 29.0440 57.77 -0.54 "PP"
PRISM 6.0
AD VD 228 275.1844 188.24 0.67 "PP"   "N END GRAVEL ALLEY E OF B
AD VD 229 274.0648 213.74 0.91 "BIT"   "N END GRAVEL ALLEY E 10
```

Exit the Field Book Editor and save the file.

11. Import the Field book, *nic1.fbk*, with the Import Field Book routine in the Data Collection/Input menu.

The Import Field Book routine asks several questions before importing the file. The questions are:

```
Erase all COGO points in the drawing? Y
Erase all existing observations? Y
Erase all figures? Y
```

In this preliminary state, the answer to each question is yes.

12. Zoom in on a point with a symbol to view the symbol.

13. Save the drawing file.

UNIT 4: SURVEY FIGURES

Figures are polylines representing lines and arcs that are the result of you manually transcribing written survey data or importing a field book. The routines that manually create figures are in the Analysis/Figures menu of Survey (see Figure 15–39). The routines in this menu are a subset of the Figure commands in the Survey Command Language. These routines create figures from existing points, bearings, and distances or when you transcribe manual field notes. The second type of figure comes from observation data found in the Field Book file. The figure commands in the field book come from field descriptions or from someone in the office entering the figure commands into a field book. The Import Field Book routine draws the line work from the figure commands found in the file.

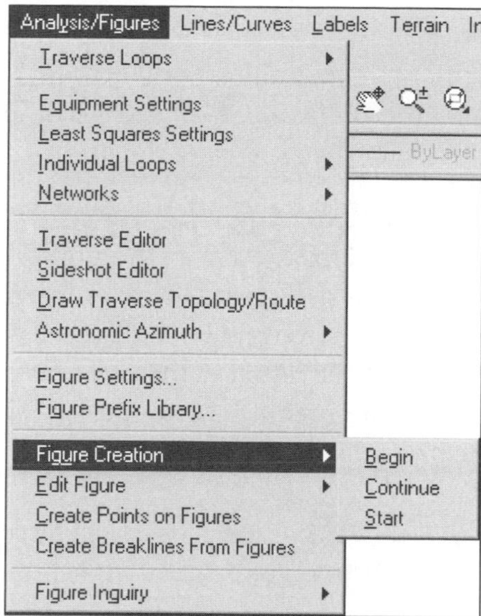

Figure 15–39

With a figure representing a linear feature like a roadway centerline or edge-of-pavement, a figure is an ideal Terrain breakline. A breakline controls the creation of surface triangles. You can convert a figure into a breakline with the Create Breaklines From Figures routine from the Analysis/Figures menu. A breakline preserves lines lost in the clutter of point data. If a figure has an arc segment, the mid-ordinate setting of Figure Settings controls the number of interpolated data points calculated by the Create Breaklines From Figures routine for a surface. The additional vertices reflect the mid-ordinate value and create additional surface data for the arc segment; that is, more chord segments along an arc. The greater number of the chord segments along the arc creates a better surface interpretation of the curve by triangles.

A figure and its points are bound together. The Inverse routine of the Figure Inquiry cascade of the Analysis/Figures menu lists the directions, distances, and point numbers of the figure. If a point moves, there is a routine to move the figure vertex to the new location of the point.

To convert an existing polyline into a figure, use the Change Name routine of the Edit Figure Name cascade of the Analysis/Figure menu and assign the name. The routine stores the name of the figure as extended entity data on the polyline. Even if it is a point at the vertices, Survey does not "attach" it to the figure. Either define a figure from existing points or add points to the figure after it is defined. The Create Points on Figures routine of the Analysis/Figures menu binds the figure and the points together after defining the figure.

FIGURE PREFIXES

Survey allows prefixes for figure names. The prefixes sort the resulting figures onto specific layers; for example, the prefixes, CL and CURB. The prefix CL sorts the figure name CLNichols to a centerline layer and the prefix curb sorts CURB-1 to the curb layer (see Figure 15–16).

Survey provides a short list of prefixes. Add to the prefix list by using the Figure Prefix Library routine in the Analysis/Figures menu or in the Figure Prefix Settings dialog box of the Drawing Settings routine of the Project menu. Survey stores the prefixes externally in the *figure.db* file found in the DATA\SURVEY folder of Land Development Desktop. The figure prefixes are not project-specific. You do not have the ability to print out the prefix list.

The initial Survey figure prefixes are:

```
BLDG - Building
CL - Centerline
CURB - Curb
DITCH - Dtich
EP - Edge_of_pavement
FENCE - Fence
SHORE - Shore
SW - Stone_wall
WALL - Wall
SWALK - Sidewalk
```

FIGURE CREATION ROUTINES

The figure generator creates a new figure with the Begin routine. The Continue routine implies that the data is continuing the figure from its last endpoint. The Start routine implies that the data is adding to the beginning endpoint of the figure. You cannot continue or start a closed figure because a closed figure is a closed Desktop polyline.

When you use the Figure Creation routine to begin a figure, the routine responds by asking for the name of the figure. After you name the figure, the routine prompts for the starting point of the figure, which can be a Desktop selection, a point number (the letter *.p* toggle), or a northing/easting (the letter *.n* toggle). From the start point, provide information on how to get to the next point on the figure. You can get to the next point by any of the following methods: a bearing or azimuth and a distance, a deflection and a distance, a right turn distance (left turns are a negative distance), a curve and associated data, a Desktop selection or object snap, a point number, a northing/easting, or close back to the beginning of the figure. If you type the appropriate toggle, the routine uses any of these direction methods.

If you are transcribing notes, you can start creating a figure by selecting the Begin routine of the Figure Creation cascade of the Analysis / Figures menu. The following lines represent the command sequence of the Begin routine.

```
Enter figure name: House
Figure Begun: House
Enter first point: (Desktop coordinates or object snap)
```

-or-

```
Enter first point: .p
Enter first point: Point Number: 202
```

-or-

```
Enter first point: .n
Enter first point: Northing: 5020.6547
Easting: 5348.2541
    NORTH: 5020.6547    EAST: 5348.2541
Enter point /BD/ZD/CUrve/<eXit> : (Desktop coordinates or object snap)
Enter point /BD/ZD/CUrve/<eXit> :
```

From the first point, you can only specify a bearing, azimuth, point, or curve. It is not until the second point is in place that the next point can be an angle-distance, deflection-distance, or a right turn. After the two initial points have been established, the first point represents the backsight and the second point represents an instrument setup. The angle-distance and deflection-distance assume that the measured angle between the second point and the first point is 0. The right turn (a negative value turns to the left) has to have a vector from which to turn 90 degrees. The routine understands the surveyor's convention of *ddd.mmss.*

The formats for the next vector are:

Bearing:

```
Enter bearing /AD/ZD/DD/RT/CUrve/PT/CLose/<eXit> : 45.1241
Enter quadrant: 1/2/3/4: 2
Enter distance: 34.33
```

Azimuth:

```
Enter azimuth /AD/BD/DD/RT/CUrve/PT/CLose/<eXit> : 145.1241
Enter distance: 321.83
```

Deflection Distance:

```
Enter deflection angle /AD/BD/ZD/RT/CUrve/PT/CLose/<eXit> : 15.2419
Enter distance : 82.40
```

Angle:

```
Enter angle /BD/ZD/DD/RT/CUrve/PT/CLose/<eXit> : 245.1154
Enter distance: 234.74
```

Right Turn:

```
Enter bearing /AD/ZD/DD/RT/CUrve/PT/CLose/<eXit> : RT
Enter distance: 73.98
```

When you are creating curves, the Curve routine prompts for the radius of the curve. You then must provide one additional curve value. The routine expects one of the following values: a tangent length, a chord distance, a delta angle, an external ordinate, a mid-ordinate, or a length of curve. A positive radius draws a clockwise arc and a negative radius draws a counter-clockwise curve.

When in the curve mode, the Figure routine prompts as follows:

```
Enter bearing /AD/ZD/DD/RT/CUrve/PT/CLose/<eXit> : CU
Enter radius /AD/BD/ZD/DD/RT/PT/CLose/<eXit> : 125.54
Select entry: Tan/Chord/Delta/Ext/Mid/<Length> : ch
Enter chord: 34.55
```

CLOSE FIGURE ANALYSIS

You can perform the analysis of a figure by using two routines: Mapcheck and Inverse (see Figure 15–40) on the Figure Inquiry cascade. Both routines produce a report about bearings, distances, closure, and other information about the figure. The precision set in the drawing setup, usually two decimal places, controls the depth of the analysis for Mapcheck. The Inverse routine provides the same information as Mapcheck, but the precision is to the depth of the Desktop precision (15 or 16 decimal places). In addition to the bearing and distances, the Inverse routine notes the points at each of the figure's vertices.

The Mapcheck and Inverse routines send their report of the figure to the text screen of Desktop. The Output Defaults routine does not redirect the reports to a file. The two routines do write their information to the *output.txt* file of Survey. Print the reports out of the *output.txt* file. Survey creates the *output.txt* file in the Survey subdirectory of the project.

A Survey figure is a polyline object and is convertible into a parcel. The transformation occurs in the Parcels menu of Land Development Desktop. Define the lot by using the Define From Polyline routine.

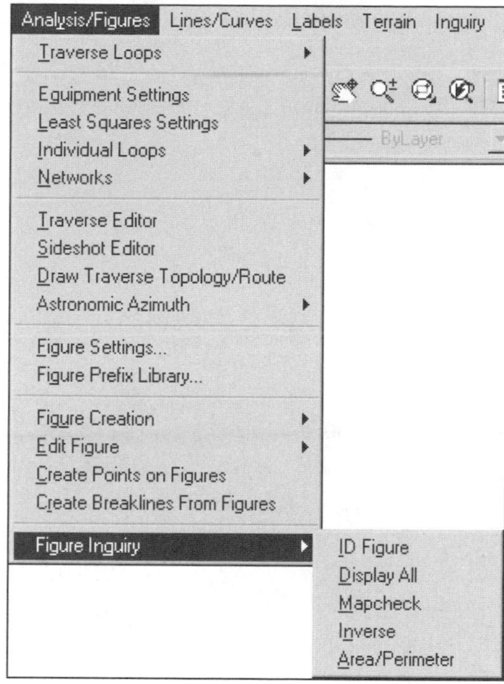

Figure 15–40

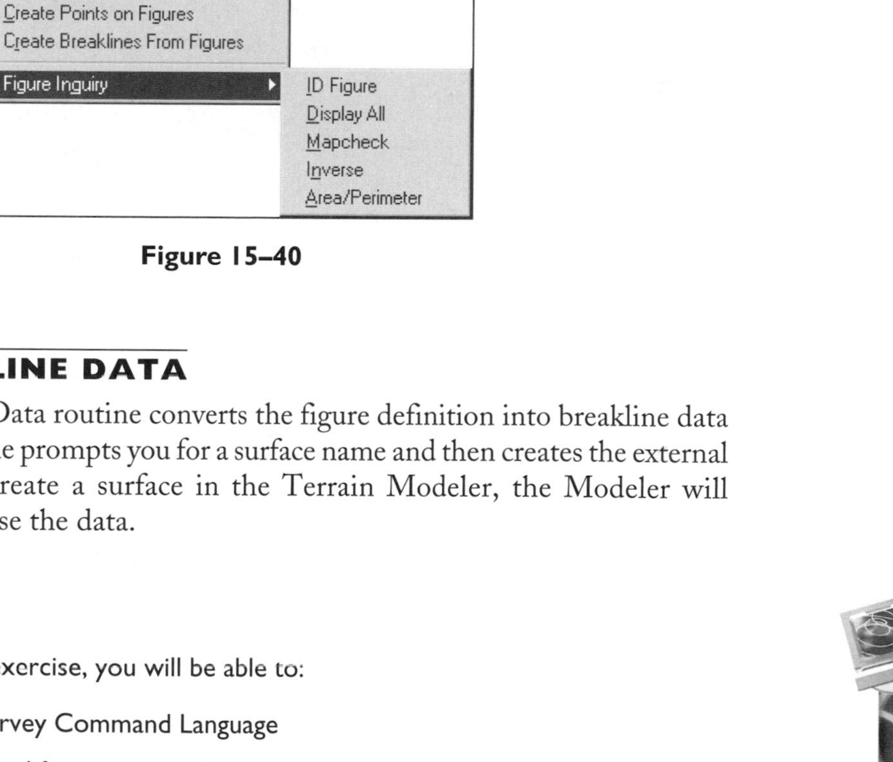

FIGURE BREAKLINE DATA

The Create Breakline Data routine converts the figure definition into breakline data for a surface. The routine prompts you for a surface name and then creates the external data file. When you create a surface in the Terrain Modeler, the Modeler will automatically see and use the data.

Exercise

After you complete this exercise, you will be able to:

- Revisit the Survey Command Language
- Analyze a closed figure

SURVEY FIGURES

1. Reopen the Lot35 drawing and save the changes to the Nichols drawing.
2. Use the Begin routine of the Figure Creation cascade of the Analysis/Figures menu to define the Lot35 figure. Start at point 1, select point numbers 2–4, and close on point 1.

The following defines the figure by point numbers (the *.p* option).

```
Command:
Loading SURVEY Command Language . . . .
Enter figure name : Lot35
Figure Begun: Lot35
Enter first point: .p
Enter first point:
 >>Point number: 1
NORTH: 5000.0000         EAST: 5000.0000
Enter point (BD/ZD/CUrve):
 >>Point number: 2
     BEARING: N 89-57-50 E              DISTANCE: 125.4500
    NORTH: 5000.0791         EAST: 5125.4500
Enter point (AD/BD/ZD/DD/RT/CUrve):
 >>Point number: 3
     BEARING: N 00-19-47 W              DISTANCE: 35.0000
    NORTH: 5035.0785         EAST: 5125.2486
Enter point (AD/BD/ZD/DD/RT/CUrve/CLose):
 >>Point number: 4
     BEARING: S 89-57-50 W              DISTANCE: 125.4600
    NORTH: 5034.9994         EAST: 4999.7886
Enter point (AD/BD/ZD/DD/RT/CUrve/CLose):
 >>Point number: .p
Enter point (AD/BD/ZD/DD/RT/CUrve/PT/CLose): cl
```

3. Use the Mapcheck and Inverse routines of the Figure Inquiry cascade of the Analysis/Figures menu to view the statistics of the figure. The output for the data is the *output.txt* file. Access the *output.txt* file from the CMDLINE menu in Survey. The listings do not show the point numbers (1-4).

FIGURE POINT DATA

The Create Points on Figures routine places points on the figure. The routine prompts you to select a figure from the screen and then prompts for a starting point number for the points. You cannot use an automatic elevation or description for the points. To give the points a description or elevation, they need to be edited.

1. Start the Create Points on Figures routine from the Analysis/Figures menu and select the Lot35 figure. Start the point numbers at point 100.
2. Use the Inverse routine of the Figure Inquiry cascade of the Analysis/Figures menu to view the statistics of the figure. This inverse now shows the point numbers 100-103.

UNIT 5: FIGURES FROM FIELD SURVEYS

If there is no access to an electronic data collector, or if you want to transcribe a written survey, you need to create a field book to create figures from field survey data. The field book contains the actions of the field crew in the Survey Command Language. Editing a field book from scratch is different from editing a field book or a raw file from a data collector.

You can control the creation of a figure by using the Begin command in the field book. The Begin command combines the name of the figure and the drawing action. In this situation, the figure name and the point descriptions do not need to match. To stop the drawing of a figure, you can issue the END command, with no name after the end, to begin a new figure, or to continue another figure. If you issue an END COMMAND, the following would be correct:

```
BEGIN house
AD VD 102 275.2241 157.33 -0.15 "BLDG CNR"
AD VD 103 273.2311 134.83 -0.05 "BLDG CNR"
END
```

The first line tells the Import Field Book routine that the next point is the start of a figure named housc. The next two points locate corners of the house. The Survey Command Processor draws a line connecting the two points. The last line tells the figure generator to stop drawing the house figure. The result of the four lines is a figure (polyline) between points 102 and 103 with a name of house.

You may be drawing several figures in a field book. The Import Field Book routine can draw only one figure at a time. Since the data may switch from one figure to another between points, Survey allows a Begin (new figure) or Continue command to stop the drawing of the current figure and to start a new or continue an existing. The routine then begins a new figure or continues drawing an existing figure.

In the Analysis/Figures menu, Survey provides a routine to create prefixes for figures. The figure name prefix sorts the resulting figure onto a user-specified layer. So, as with description keys, Survey can sort lines onto specific layers with figure prefixes. As a result, a centerline figure becomes a 2D polyline entity on the layer CL, a house becomes a 2D polyline entity on the building layer, and so forth. The default installation of Survey provides a short predefined list of prefixes. You can add to this list at any time to suit the needs of the office.

If you decide to use figure prefixes, this changes the BEGIN command to something like the following:

```
begin BLDG-house
AD VD 1002 275.2241 157.33 -0.15 "BLDG CNR"
AD VD 1003 273.2311 134.83 -0.05 "BLDG CNR"
END
```

If the field crew later takes additional shots of the building, to continue drawing the line work of the figure you would restart drawing a figure with one of two figure commands: CONTINUE or START.

The CONTINUE command tells the Import Field Book routine that the next point connects to the last polyline vertex in the figure; in the case of the building, it would be point 1003. If the next shot connects to point 1002 (the first polyline vertex of the figure), you are adding to the beginning of a figure. The START command is used when adding to the front of a figure. Be careful about the order of the new points. Each new point added to the figure with the START command must immediately precede the previous beginning point. For example, a building edge consists of points 1002–1003. Point 1002 is the beginning of the figure and 1003 is the current end (see Figure 15–41). If point 1004 continues the figure, add the points to the figure in their logical order. The order may not necessarily be the order in which the points appear in the field data. If points 1000 and 999 are points before the beginning point of the figure, you must add to the figure in their linear order. The linear order is the closest point to the beginning 1000, then the next point nearest to the new beginning point 999, and so on. This order may not be readily accessible to the office staff in the Field Book file, except for extensive editing of the file. In this situation, it may be better to draw the figure in the graphics editor.

The situation of the field crew recording cross sections also brings up the question of having to say "end figure" after only one shot on the figure's path (see Figure 15–42). For example, shots 101, 106, and 107 represent the right edge of pavement, shots 102, 105, and 108 represent the centerline, and shots 103, 104, and 109 represent the left edge-of-pavement. The sequence of data for the centerline in the data file would be the following:

```
AD VD 101 170.2421 125.45 "EP-R"
AD VD 102 172.3456 134.45 "CL-Smith"
AD VD 103 174.5421 145.65 "EP-L" !two consecutive edge-of-pavement shots
AD VD 104 178.5551 151.94 "EP-L"
AD VD 105 182.4712 154.01 "CL-Smith"
AD VD 106 184.3545 157.31 "EP-R" !two consecutive edge-of-pavement shots
AD VD 107 187.5347 161.58 "EP-R"
AD VD 108 189.2245 163.13 "CL-Smith"
AD VD 109 190.1154 165.94 "EP-L"
```

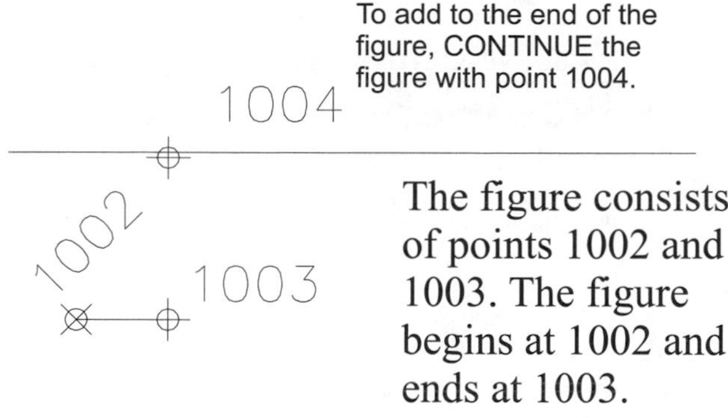

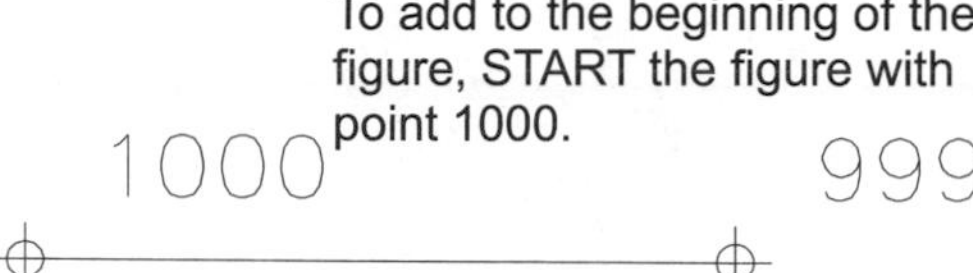

Figure 15–41

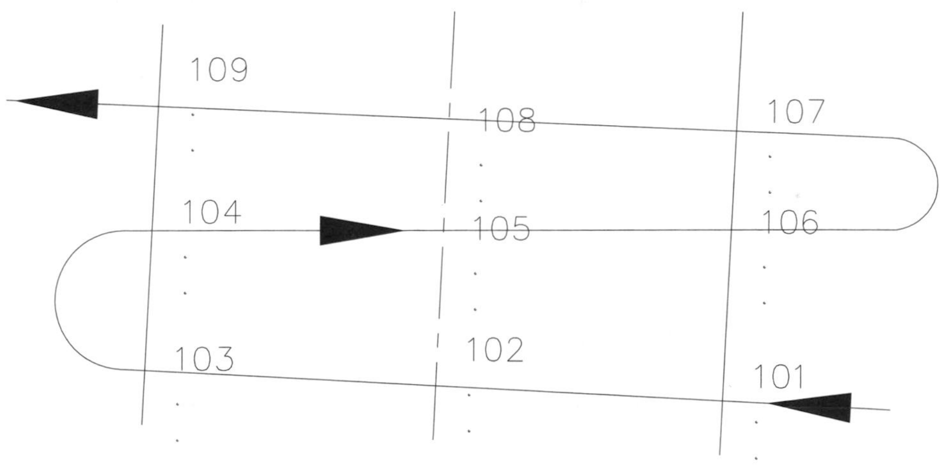

Figure 15–42

Edit the field book to create the figures as follows:

```
Begin EP-R
AD VD 101 170.2421 125.45 "EP-R"
End
Begin CL-Smith
AD VD 102 172.3456 134.45 "CL-Smith"
End
```

```
Begin EP-L
AD VD 103 174.5421 145.65 "EP-L" !two consecutive edge-of-pavement shots
AD VD 104 178.5551 151.94 "EP-L
End
CONTINUE CL-Smith
AD VD 105 182.4712 154.01 "CL-Smith"
End
CONTINUE EP-R
AD VD 106 184.3545 157.31 "EP-R" !two consecutive edge-of-pavement shots
AD VD 107 187.5347 161.58 "EP-R"
End
CONTINUE CL-Smith
AD VD 108 189.2245 163.13 "CL-Smith"
End
CONTINUE EP-L
AD VD 109 190.1154 165.94 "EP-L"
End
```

Too much work is involved to create the three figures by having to say "end" each time you want to begin or continue a figure. So the figure documentation says that a CONTINUE, BEGIN, or START command is the same as saying "end." When it encounters a CONTINUE, BEGIN, or START command, the SCP automatically stops drawing the current figure and continues or starts on the named figure so that you can revise the above portion of the Field Book file to look like the following:

```
Begin EP-R
AD VD 101 170.2421 125.45 "EP-R"
Begin CL-Smith
AD VD 102 172.3456 134.45 "CL-Smith"
Begin EP-L
AD VD 103 174.5421 145.65 "EP-L"
AD VD 104 178.5551 151.94 "EP-L"
CONTINUE CL-Smith
AD VD 105 182.4712 154.01 "CL-Smith"
CONTINUE EP-R
AD VD 106 184.3545 157.31 "EP-R"
AD VD 107 187.5347 161.58 "EP-R"
CONTINUE CL-Smith
AD VD 108 189.2245 163.13 "CL-Smith"
CONTINUE EP-L
AD VD 109 190.1154 165.94 "EP-L"
End
```

You should find the code cleaner, easier to read, and less work.

In the preceding example, it becomes obvious that the way in which a field crew records data directly affects the amount of work needed by the office to enter the figure commands in the data.

It is much simpler for the crew to record lines as a set of consecutive points because all the field crew has to do is enter the BEGIN command to start each line and one END command to finish the line drawing (see Figure 15–43). The figure commands for the preceding data set become:

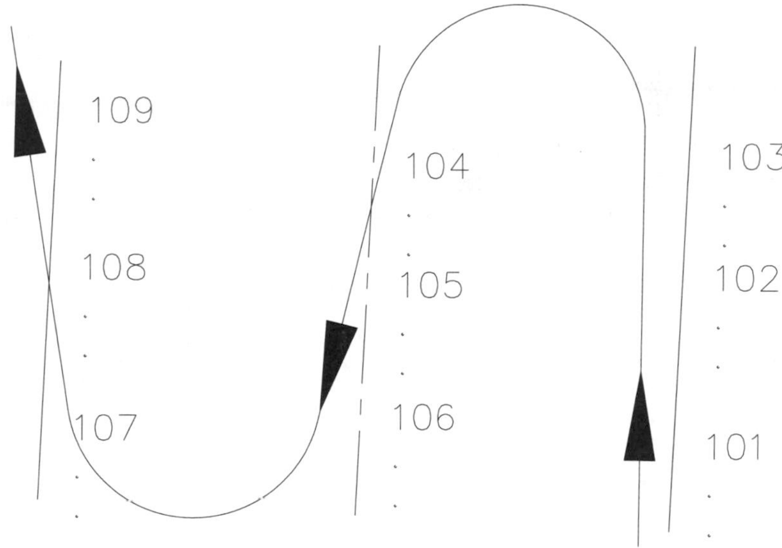

Figure 15–43

```
Begin EP-R
AD VD 101 170.2421 125.45 "EP-R"
AD VD 102 184.3545 157.31 "EP-R"
AD VD 103 187.5347 161.58 "EP-R"
Begin CL-Smith
AD VD 104 172.3456 134.45 "CL-Smith"
AD VD 105 182.4712 154.01 "CL-Smith"
AD VD 106 189.2245 163.13 "CL-Smith"
Begin EP-L
AD VD 107 174.5421 145.65 "EP-L"
AD VD 108 178.5551 151.94 "EP-L"
AD VD 110 190.1154 165.94 "EP-L"
End
```

Most field crews do not walk from one end of a line to the other. The crew surveys in a meandering, serpentine-like motion. This method makes it difficult for the office to edit the field book to add figure commands. If a figure references a point already in the data set, the crew adds a note to recall the point.

If the order of figures is so dispersed throughout the field book, you may find it more practical to solve the problem in a different way. You may want to create, name, and draw the figure at the end of the Field Book file, or if all else fails, draw the lines in the graphics editor.

The following sequence draws a figure by using the RECALL command.

```
AD VD 101 170.2421 125.45 "EP-R"
AD VD 102 172.3456 134.45 "CL-Smith"
AD VD 103 174.5421 145.65 "EP-L" !two consecutive edge-of-pavement shots
AD VD 104 178.5551 151.94 "EP-L
AD VD 105 182.4712 154.01 "CL-Smith"
AD VD 106 184.3545 157.31 "EP-R" !two consecutive edge-of-pavement shots
AD VD 107 187.5347 161.58 "EP-R"
AD VD 108 189.2245 163.13 "CL-Smith"
AD VD 109 190.1154 165.94 "EP-L"
Begin EP-R
Recall 101
Recall 106
Recall 107
Begin CL-Smith
Recall 102
Recall 105
Recall 108
Begin EP-L
Recall 103
Recall 104
Recall 109
End
```

The result of importing the preceding field book locates nine points and draws three 2D polylines. Each polyline has a figure name and each figure is on the layer specified by its prefix. If there are description keys of EP* and CL*, the points would be on layers you specify.

It may not be practical for the field crew to change their ways and enter figure commands. Therefore, it is a decision of whether to let the office enter the figure commands or not to implement the figure commands. The implementation of figure commands will be a compromise of field and office procedures.

FIGURE ARCS

What if the lineation has a curve? How does a curve become a part of a figure? You have limited choices for drawing arcs in the figure commands. The main arc figure command is C3 or CV3. The C3 command creates a curve from the next three points in the Field Book file. These commands are for point only curves. If you are transcribing survey notes, you may have more information about the arc than three points from a survey. The additional commands for drafting arcs are discussed below.

FIGURE RECTANGLES

The figure language contains commands to create rectangles by one of three methods. The first method closes a rectangle by connecting the last vertex of the figure to the first vertex. The second method closes a rectangle from the last vertex to the first by using two right angle segments. The last method creates a rectangle by establishing two points and specifying an offset to the parallel side.

The figure close commands are as follows:

```
CLOSE
CLOSE BLD ! The right turn command follows the command.
CLOSE RECT [offset]
```

See the documentation about these commands.

FIGURE OFFSETS

You can also offset a figure with the OFFSET FIGURE command. The command is as follows:

```
offset cl-Smith 12.5
```

The OFFSET command offsets the figure 12.5 feet to the right. The figure routines follow the same convention as stations and offsets to determine the side of offset. A negative offset is to the left and a positive offset is to the right.

TRANSCRIBING PLAT BOUNDARIES INTO FIGURES

So far the discussion has been about using field point data. What if the information about a figure is a direction and a distance from a plat of survey? The figure commands define figures by azimuths, angles, or bearings. You only need to add a direction and a distance to use the commands. There is no prohibition to mixing field points and angles and distances together to create a figure. In fact, you can create a field book without field data by using any reasonable measurements to produce line work on the screen. The re-creation of boundaries, lots, and so forth becomes a matter of writing a Field Book file. This is similar to what you did in the second exercise of this section.

The following is a list of the directional figure line commands:

```
FIG AD [angle] [distance]
FIG DD [angle] [distance]
FIG ZD [azimuth] [distance]
FIG BD [bearing] [quadrant] [distance]
FIG NE [Northing] [Easting]
```

If you have no point data to generate curves, how does the field book draw a curve? The figure commands include a CURVE command that requires you to know two values about the curve: the radius of the curve (positive for right hand turns, negative for left hand turns), and the delta, length, deflection angle, mid-ordinate, tangent, or chord value of the curve.

The following are the CURVE figure command combinations:

```
CRV DELTA [radius] [value]
CRV LENGTH [radius] [value]
CRV DEFL [radius] [value]
CRV MID [radius] [value]
CRV TAN [radius] [value]
CRV CHORD [radius] [value]
```

Lastly, you can put comments in the Field Book file by placing an "!" exclamation symbol in a line. The Import routine will ignore anything to the right of the "!."

EXERCISE

Exercise

With the understanding that the field crew records information about points and lines, you can add figure creation commands to the Field Book file. If you include the commands in the field book, the Import Field Book routine draws line work in the drawing.

In reviewing the printed field notes of the survey, you can identify two lines in the field notes, BC/FW and BW. The notes indicate that BC/FW is a back-of-curb that is also the front of a walk. The BW shots are the backside of the walk. You have to begin and continue the two figures in the file.

After you complete this exercise, you will be able to:

- Create lines and curves in the Survey module
- Revisit the Survey Command Language
- Use Figure commands in the Field Book file
- Edit and Import a Survey field book
- Use Survey figure prefixes

FIGURE COMMANDS IN A FIELD BOOK

1. Reopen the *Nichols* drawing and save the changes to the *Lot35* drawing.
2. If necessary, add two figure prefixes to the list. Use the values below to create the prefixes.

Prefix	Layer
CURB	CURB
SIDEWALK	SIDEWALK

3. Reedit the Nic1.fbk file with the Edit Field Book routine of the Data Collection/ Input menu or copy the completed Field Book file, *Nichols.fbk*, to the Survey directory of the Nichols project. The completed file, *Nichols.fbk*, is on the CD that comes with this book.

Remember, to create a figure, use the BEGIN command. To add to the end of an existing figure is to continue the figure. To add to the beginning of a figure is to start a figure. To stop the drawing of a figure, use the END command. If you want to stop one figure to work on another figure, use the BEGIN, START, or CONTINUE commands.

4. Add the following figure commands to the Field Book file at the lines for the observations of points 200-228.

```
STN 1226 5.13                          !H.I. 5.13
BS 1213
PRISM 4.65
BEGIN CURB-R
AD VD 200 92.1224 61.31 -0.89 "BC" !BC/FW
BEGIN SIDEWALK-R
AD VD 201 87.3738 61.13 -0.82 "BW"
AD VD 202 84.4515 17.12 -0.42 "BW"
CONT CURB-R
AD VD 203 101.0837 17.49 -0.51 "BC" !BC/FW
AD VD 204 252.2341 13.08 -0.32 "BC"  !BC/FW
CONT SIDEWALK-R
AD VD 205 274.3455 12.68 -0.37 "BW"  !EDGE OF 10' GRAVEL WALK
CONT CURB-R
AD VD 206 262.1906 33.27 -0.28 "BC"  !BC/FW
END
AD VD 207 259.3437 33.45 -0.74 "INB"
CONT CURB-R
AD VD 208 265.3536 69.02 -0.15 "BC"  !BC/FW
CONT SIDEWALK-R
AD VD 209 269.3947 69.87 0.08  "BW"  !CORNER OF BIT
```

```
AD VD 210 269.1905 123.72 0.28 "BW"  !EDGE OF BIT
CONT CURB-R
AD VD 211 266.5835 123.95 0.12 "BC"  !BC/FW
AD VD 212 267.2954 168.00 0.23 "BC"  !BC/FW DEPRESSED WALK
AD VD 213 267.3912 187.20 0.31 "BC"  !BC/FW DEPRESSED WALK
AD VD 214 267.4530 204.83 0.32 "BC"  !BC/FW
CONT SIDEWALK-R
AD VD 215 269.0846 205.23 0.47 "BW"  !EDGE OF BIT
AD VD 216 269.0922 188.70 0.41 "BW"  !NORTH END OF GRAVEL ALLEY
END
AD VD 217 269.1241 172.30 0.09 "BW"  !SOUTH END OF GRAVEL ALLEY
AD VD 218 269.5533 118.54 0.47 "LP"
BEGIN BLDG-1
AD VD 219 275.5820 105.79 0.50 "BLDG"!NW BUILDING CORNER
AD VD 220 275.3158 75.74 0.32 "BLDG" !SW BUILDING CORNER
END
AD VD 221 300.5235 77.02 -0.57 "BIT" !EDGE OF BIT
AD VD 222 295.2827 69.19 -0.58 "PP"
AD VD 223 350.0500 36.62 -0.17 "MH"  !4'DIA WATER MANHOLE
AD VD 224 346.2257 36.75 -0.08 "WV"
AD VD 225 345.0415 38.24 -0.92 "GRAVEL" !GRAVEL PATH EDGE
AD VD 226 41.5340 50.00 -0.31 "GRAVEL"
AD VD 227 29.0440 57.77 -0.54 "PP"
PRISM 6.0
AD VD 228 275.1844 188.24 0.67 "PP"   !N END GRAVEL ALLEY E OF B
```

5. Run the Import Field Book routine. Answer Yes to all of the questions and view the results. If, there are any errors in the Field Book file correct them and rerun the Import Field Book routine.

EXERCISE

The focus of the Survey add-in of Land Development Desktop is the reduction of field observations into points in the drawing. The tools are specific to this process and may not be something you come into contact with every day.

The Figure commands of the Survey Command Language create line work directly from the field observations. The creation of lines by manually editing the field book is a time-consuming and lengthy process. It would make more sense if the field crew entered the survey and figure commands into an electronic data collector and let another routine parse the data into observations and figure commands. This process is the focus of the next section.

SECTION 16

Survey Data Collection

After you complete this section, you will be able to:

- Use Figure commands as a part of the field data collection process
- Create lines and curves in the Survey module
- Use Figure commands in the Field Book file

SECTION OVERVIEW

One of the sources for Field Book data is data from electronic files collected in the field. The TDS software in Survey downloads these files into the Survey directory of the project. The TDS software, in addition to having downloading capabilities, has a routine that converts the raw data from the collector into a Survey field book. The format of the raw file is proprietary to the manufacturer who makes the instrument or data collector. The documentation that accompanies the instrument or data collector reviews the data format and describes the elements and sequence of the data.

The data in its raw format is difficult to read. You must research the format of the data, read through the file, and then interpret the data in the file. This is what the TDS software in Survey does. It reads the cryptic proprietary formats and converts them into the more readable format of the Survey Field Book. Three examples of the raw data in the formats Sokkia, TDS, and Geodimeter are shown below. Each file is of a different survey and each requires you to look at the format documentation to read its data.

The Sokkia file contains information about the initial setup, its elevation, instrument height, and prism height. From this setup, there are observations of points described as Ground shots and FLL (curb flow line).

Sokkia

```
08KI10001000.000001000.00000100.000000GROUND
08KI00021000.000001115.51000          GROUND
01NM]                000000                   00000031
  0.00000000
13PCP.C. mm Applied: 0.000
02TP00011000.000001000.00000100.0000005.45000000
07TP0001000290.0000000303.352222
03NM5.35000000
09F100010002115.52000091.1144444303.352222GROUND
09F100011001110.70000090.1394444290.872777B FLL
09F100011002126.32000090.2061111291.630000FLL
09F100011003125.21000090.4394444295.942222FLL
09F100011004119.40000090.8550000300.730000FLL
09F100011005105.88000091.1561111303.967777FLL
13NMC3
```

The TDS survey is also a initial setup and some observation data. The initial point is defined as STN 1. The data contains an error report on the initial backsight observation.

TDS (rw5)

```
JB,NMLOT,DT05-13-1998,TM12:32:19
MO,AD0,UN0,SF1.000000,EC0,E00.0000,AU0
SP,PN1,N 5000.00000,E 5000.00000,EL644.290,–BM MH RIM
SP,PN1,N 5000.00000,E 5000.00000,EL652.380,–STN 1
OC,OP1,N 5000.0000,E 5000.0000,EL652.380,–STN 1
SP,PN2,N 5000.00000,E 5000.00000,EL644.290,–STN 1
SP,PN2,N 5000.00000,E 5000.00000,EL644.290,–BM MH RIM
OC,OP1,N 5000.0000,E 5000.0000,EL652.380,–STN 1
BK,OP1,BP2,BS0.00000,BC0.00000
BK,OP1,BP2,BS0.00000,BC0.00000
–BS check 1-2: ZE92.0632,SD72.976,HD err=-72.927,VD err=2.495, Angular
  err=-99.3158
–BS Circle check : angular err= -99.3200
BK,OP1,BP2,BS0.00000,BC0.00000
–BS check 1-2: ZE92.0633,SD72.977,HD err=-72.928,VD err=2.496, Angular
  err=-0.0001
BK,OP1,BP2,BS0.00000,BC0.00010
OC,OP1,N 5000.0000,E 5000.0000,EL652.380,–STN 1
BK,OP1,BP2,BS0.00000,BC0.00010
LS,HI5.590,HR4.940
```

The Geodimeter data contains information about a CP (control point) and its location and elevation.

Geodimeter

```
54=199RC13
53=113-129
51=1999.0729
52=11.1101
56=90.0
74=29.00
2=1
94=CP
37=10000.000
38=50000.000
39=100.000
3=5.300
6=5.000
```

The problem you have with the various proprietary data formats is reading their data. The TDS software in Survey converts these various formats into the Survey Command Language and places the resulting translated data into a field book. What would be nice is if the data from the field could also draw line work. As discussed in the previous section, the field book can contain figure commands so that when the field book is imported into a drawing, the program can draw lines. Each collector requires a different encoding of figure commands. You need to read the collector and Survey documentation thoroughly to understand the data entry process.

UNIT 1

The focus of the first unit of this section is the settings and downloading routine of the TDS software within the Survey add-in. The downloading process is similar for each data collector. You need to read the documentation that comes with your data collector to see if there are any issues that need to be addressed before downloading the data.

UNIT 2

The second unit explains how to create figures directly from field observations. The exercise for this unit uses data from a Sokkia data collector. The process is similar for each data collector. You need to read the documentation that comes with your data collector and the Survey help file that covers your data collector to understand what must be done in the field to create line work when importing the Field Book.

UNIT 3

The third unit covers the process of uploading data to the data collector. The process is similar for each data collector. You need to read the documentation that comes with your data collector to see if there are any issues that need to be addressed before uploading the data. The uploading of coordinates from a project means you are able to use the field crew to collect initial data for a project and to set out in the field the final project design.

UNIT 1: DOWNLOADING DATA COLLECTOR RAW FILES

Survey provides a collector download program (TDS DC Link) that directly supports a large group of data collectors. The download program is the Data Collector Link routine of the Data Collection/Input menu (see Figure 16–1). When you select this routine, Survey loads the TDS DC Link program, where the downloading and file conversions occur (see Figure 16–2). The download routine creates a file with formatted raw data. The format of the file depends upon the data collector used.

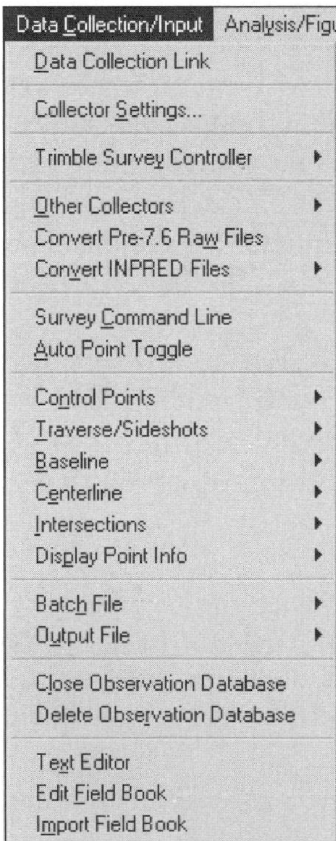

Figure 16–1

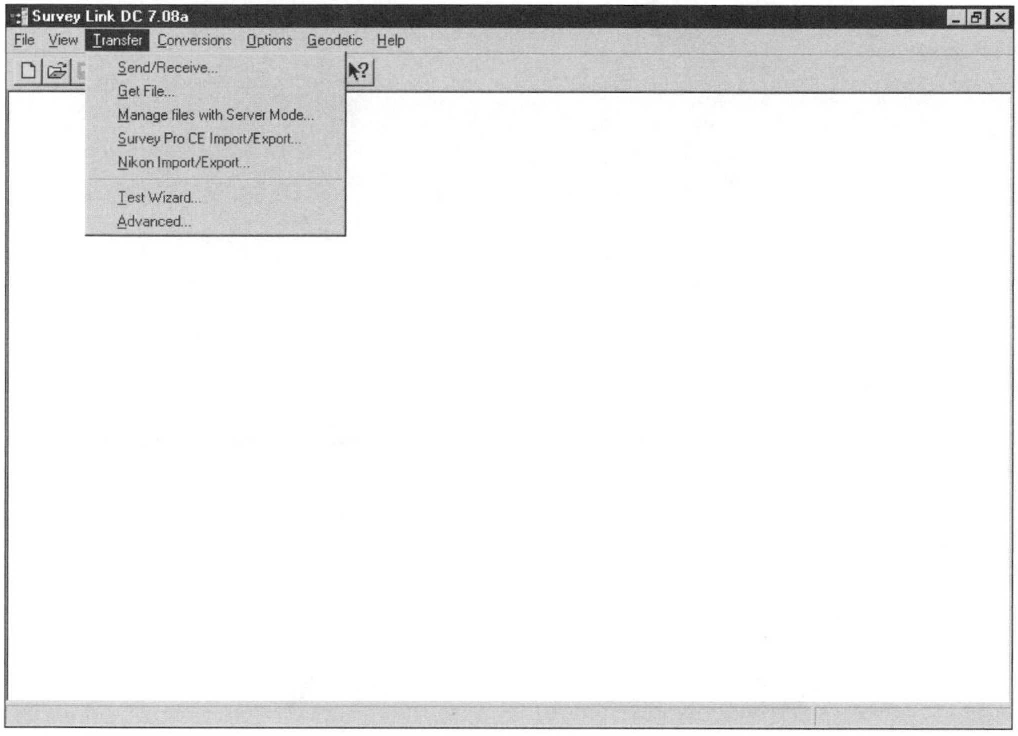

Figure 16–2

The Send/Receive routine of the Transfer menu communicates with the data collector (see the Setup dialog box in Figure 16–3). The communication takes place through the serial port on the computer and the data collector. Since the communication is serial, you must make sure that the settings of the Link program and the communication protocol of the collector are the same. The protocol of the Link program is set in the Setup dialog box. You can call this dialog box by selecting Send/ Receive from the Transfer menu, and then selecting the Setup button in the Transfer dialog box (see Figure 16–4). After setting up the communication protocol, you set the file name and location. Once all of these settings are complete, you start the downloading process.

After you create the raw data file, the next step is converting the file to either the Field Book or ASCII coordinate file format. This conversion is also done in the TDS software in Survey (see the Convert dialog box in Figure 16–5). You can call this dialog box by selecting Convert File Format from the Conversions menu. As mentioned in the previous section, all field book files have the extension of *fbk*.

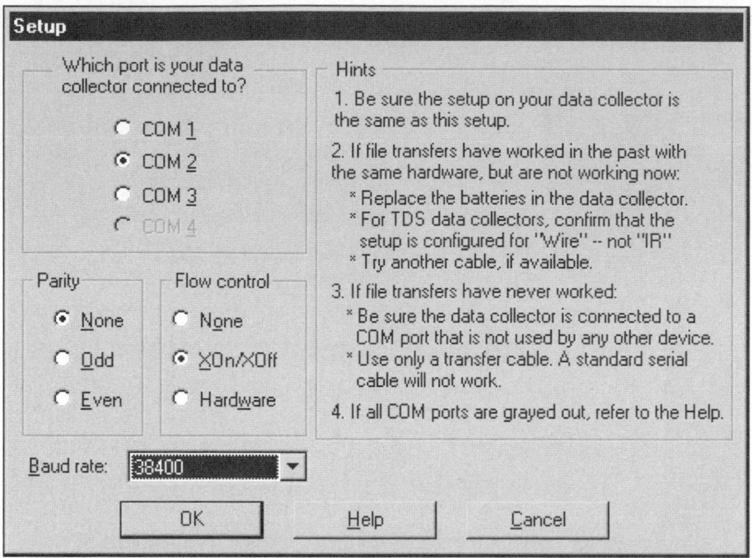

Figure 16–3

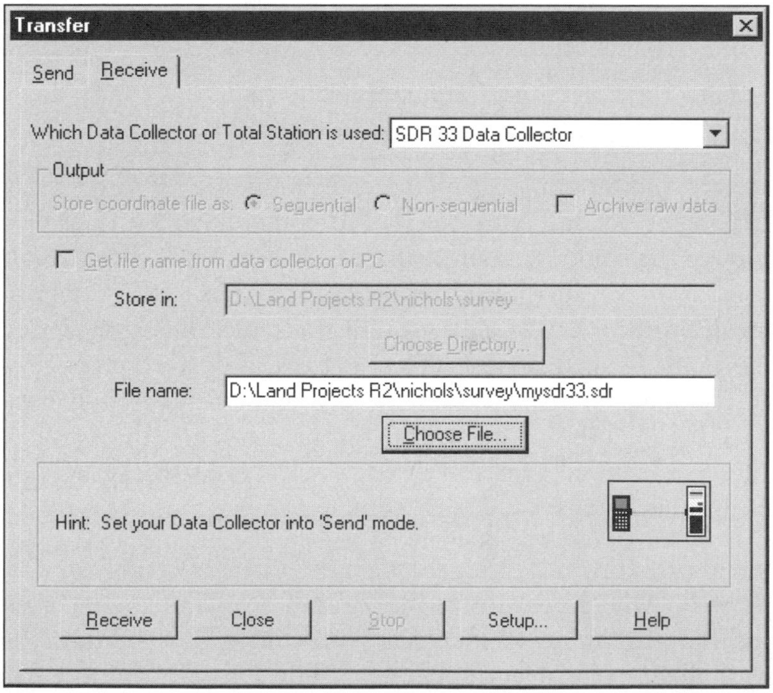

Figure 16–4

After successfully converting the raw file and reviewing the resulting Field Book file, you are ready to import the Field Book file. If you are using description keys, there must be a point style referencing the description set in the Point Settings dialog box of the Settings routine of the Points menu. When you import the Field Book file, the new points respond to the description key matrix by adding symbols to their coordinate locations.

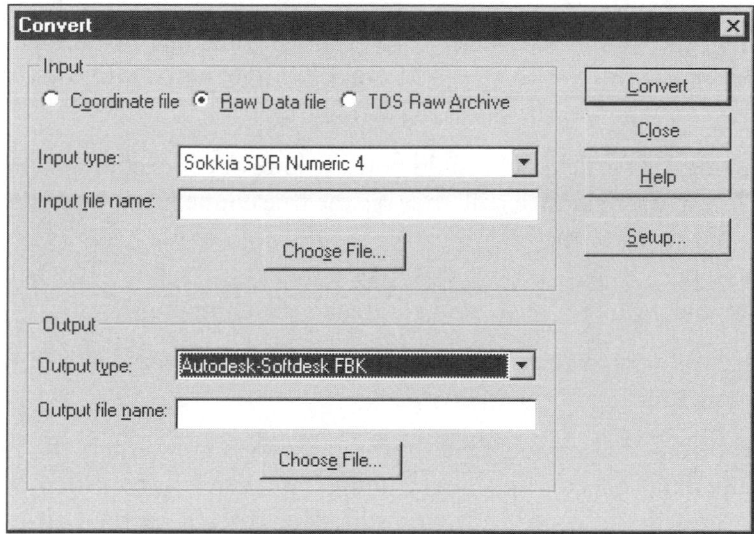

Figure 16–5

UNIT 2: CREATING FIGURES DIRECTLY FROM THE FIELD

The field crew members, as they survey a job, observe not only points but also lines. These lines represent walls, edges-of-pavements, centerlines, and so forth. The ability to create line work from field observations is the domain of the figure commands in the Survey Command Language.

In general, companies treat data collection, analysis, and drafting as discrete segments of labor. The field crew collects, the office surveyor analyzes, and drafters create the finished product. The problem is that the office rarely sees the site. When the office views the point data, each point carries the same weight when connecting points to create lines from the point data. This leads to incorrect line work and assumed point connections. Many times even the field notes do not help in solving ambiguities.

The field crew simply sees the site as it is. Each point they survey is an actual part of the world. To their eyes, the point data represents what is there logically, connected because it is not really a point. The data is a specific location within the context of the site. The field crew does not see a set of points without a frame of reference.

The office sees the site as only points. The office views the points without visual reality to tie the points into a cohesive frame of reference. A curb point 100 feet from another curb point does not necessarily mean there is a connection between them, especially when the drafter sees other curb points in the area. If there is ambiguous data, the drafter has to refer to the field notes to find the connections. Again, the field notes may not contain the necessary help.

It would seem ideal if the field crew could note the lines or connections in the electronic data they collect in the field. After all, they see the true connections. It may take a slight modification of their field methods to make the notation of lines easy. They may have to organize and preplan how and what is to be a part of the survey. The field crew will have to be familiar with the "logic" and command structure of figures.

If the field crew is a part of electronic data collection, they should also become a part of the electronic drafting of their surveys.

To achieve the goal of electronic field drafting, the field crew enters the figure commands into the data collector as codes or notes. The conversion of the raw file into a field book creates figure commands from the data entered by the field crew in the resulting field book file. Finally, when importing the field book into the drawing, the routine draws line work.

The main command placed into the data to begin drawing a line is B (B followed by a space). The field crew adds the figure command to the description (code) of the point, for example, B EOP1. The figure command is B for beginning new figures. The raw file conversion routine assumes that points with the same description are a part of the same figure. As a result of this coding, only those points with the same description become a part of the figure. The translator assumes that the next point of the survey connects to the end of the figure; that is, the survey moves down the figure in a logical connection order.

The command C3 can be placed in the description of a point or as a note indicating that the next three points are a curve definition. The code C3 cannot begin a new figure. If a curve point is the first point on a figure, you must use a note to begin the figure, and then you can embed the C3 into the description of the point or use another note.

The command S (Start) is to add to the beginning of existing figures. So, again you must begin the figure first and then use the Start code to add to the beginning of the figure. Start cannot begin a new figure.

The Import Field Book routine reads the Survey Command Language and Figure commands and draws the actual points and line work. The Import routine sorts, translates, and adds symbols to the points placed into the drawing if you toggle on description keys.

The figure command language of Survey is powerful. The only problem is the way you build each figure. Generally, the field crew does not survey figures from one end to the other, but walks portions of several lines (figures) at a time. For example, when surveying cross sections for a roadway, the crew may observe on each pass the left ditch, left edge-of-pavement, centerline, right edge-of-pavement, and the right ditch of a roadway. In the next pass of cross-section shots, the shots are in the opposite order. The only place where two consecutive points for the same figure may occur is when the crew wraps around to the next cross section. When the crew wraps around to start the next section, the first shot of the cross section is the right ditch that continues the last shot of the previous cross section. This, of course, assumes that the field crew is surveying just the roadway. What happens to the order when there are signs, trees, driveways, or walks for them to record?

With the raw file translator connecting only those points with the same description, it does not make any difference how the field crew surveys the area. The only assumption is that the field crew performs the survey in an order that will correctly draw the figure. It is important that the field crew be consistent with the descriptions because of the relationship of a description and a figure being processed.

Having the survey crew use figure commands while in the field requires time to organize and place commands into the survey data. If the field crew misses or incorrectly labels the codes, the correction of the data may not necessarily need to be done in the field. The data files are available to the office staff to edit and correct.

If the office needs to correct the data files, the process of placing and correcting figure commands is a series of iterative edits and imports of the field book. You need to view and review decisions made while editing the file. Because of confusing notes or relationships between points, you may not have any other option but to finish the line work in the drawing rather than completing the line work entirely in the field book. The field notes may just be too ambiguous to connect all the dots successfully.

If, for example, the field notes are inconclusive about the order of point connections along a roadway centerline, you may have to view the points in the graphics editor. It may be only after you see the data that a line is obvious. After viewing the point progression, edit the Field Book file to reflect the point connection order. When you reimport the field book, the new and correct connections are present in the drawing.

FIELD CODING

If you have a data collector, you can create figures directly from the data in the collector. You need to know three basic figure commands: B (B space), S (S space), and C3 (C3 space). These commands indicate the beginning of a new figure (B), the appending to the beginning of an existing figure (S), and the indication of the next three points as defining a curve (C3). The figure BEGIN (B) and START (S) commands are either a note in the data collector or a part of the description. If you are beginning a figure from a note, you press the note key on the data collector and enter BEGIN EOP1. The resulting entry in the data collector raw file (Sokkia format) is:

```
13NMBEGIN EOP1
```

This entry translates into the field book coding of:

```
BEGIN EOP1
```

If the field crew uses the BEGIN command as a part of the point's description, the BEGIN code is entered in as a part of the description of the point. In the following example, the field entry for the point's description is B EOP1. The resulting raw file (Sokkia format) entry would be as follows:

```
09F10001101534.140000090.1527777142.883888B EOP1
```

The above raw file entry, when converted into a field book entry by the Convert File Format routine from the Conversions menu, is the following:

```
BEGIN BEOP1
FC1 VA 1015 142.530200 34.1400000 90.091000 "EOP1"
```

Why is the edge-of-pavement EOP1? In most cases there are two edges-of-pavement, the right and left side. If you use only EOP, the translator assumes all EOP shots connect together. The resulting line work would travel from side to side across the centerline of the roadway. This is not the result you are looking for. The goal is to have line work representing the right and left sides of the edge-of-pavement. For the translator to be able to discern between the left and right sides, you need to indicate a difference between lines that are of the same generic type, that is edges-of-pavement (generic) for the right and left side (different). You would indicate this by coding one side as EOP1 and the other as EOP2. If the figure coding contains a basic root code indicating the generic figure type and a variable indicator for different lines of that type, the translator will automatically see the differences when creating the field book.

After the initial beginning shot, the surveyor enters only the description of the point (EOP1). The assumption of the TDS translator of a raw file into a field book is that it is to connect points with same description with figure line work. When the conversion routine translates the raw file to the field book, the routine adds the BEGIN, END, and CONTINUE commands into the file based upon the descriptions of each observation. If the field crew made errors entering the data, you may only have to add a few forgotten commands to the raw file to fix the resulting field book. If you are going to edit the raw file, be sure to edit a copy of the raw file, never the original file. After editing the raw file, you need to convert the edited raw file into a new field book file.

The following is an example of the BEGIN command embedded in a point's description in the raw file of a data collector (Sokkia):

```
09F10001101313.980000093.5022222118.279444FLL
09F10001101434.460000091.0016666141.967222FLL
09F10001101534.140000090.1527777142.883888B BCL
09F10001101616.680000091.0283333127.305000BCL
09F10001101710.880000094.0494444337.786111BCL
```

The field crew marks the beginning of a new figure with the addition of the B (B space) to the description field. When the translator reads this line, the translator takes out the B and places the BEGIN prior to the line in the Field Book file. The following is the resulting Field Book lines.

```
FC1 VA 1013 118.164600 13.9800000 93.300800 "FLL"
FC1 VA 1014 141.580200 34.4600000 91.000600 "FLL"
BEGIN BCL
FC1 VA 1015 142.530200 34.1400000 90.091000 "BCL"
FC1 VA 1016 127.181800 16.6800000 91.014200 "BCL"
FC1 VA 1017 337.471000 10.8800000 94.025800 "BCL"
```

The code above represents a crew walking down a line as they survey; however, most field crews walk a cross-section survey that takes them over several lines in a single pass. The following code (Sokkia) represents a cross-section walk over three lines. When creating figures from descriptions, all that is necessary is to give each line a unique name. If the curb line should end on the right hand side (BCR) and the figure starts again several feet later, you can begin the figure again with the same description.

Raw File Code

```
09F100011027110.23000089.8938888291.182222BCL
09F100011028150.68000090.9855555312.382222B FLR
09F100011029150.61000090.8255555312.679444B BCR
09F100011030151.93000090.9727777308.789444B CL
09F100011031134.52000091.0300000308.256111CL
09F100011032134.25000091.0300000312.400555FLR
09F100011033134.22000090.8422222312.699444BCR
09F100011034118.82000090.8633333313.568333BCR
09F100011035118.65000091.0783333313.257222FLR
09F100011036117.49000091.0783333308.633888CL
```

Field Book Code

```
FC1 VA 1027 291.105600 110.230000 89.533800 "BCL"
BEGIN FLR
FC1 VA 1028 312.225600 150.680000 90.590800 "FLR"
BEGIN BCR
FC1 VA 1029 312.404600 150.610000 90.493200 "BCR"
BEGIN CL
FC1 VA 1030 308.472200 151.930000 90.582200 "CL"
FC1 VA 1031 308.152200 134.520000 91.014800 "CL"
CONT FLR
FC1 VA 1032 312.240200 134.250000 91.014800 "FLR"
CONT BCR
FC1 VA 1033 312.415800 134.220000 90.503200 "BCR"
FC1 VA 1034 313.340600 118.820000 90.514800 "BCR"
CONT FLR
FC1 VA 1035 313.152600 118.650000 91.044200 "FLR"
CONT CL
FC1 VA 1036 308.380200 117.490000 91.044200 "CL"
```

All of the CONTINUE commands are a result of the assumption that the figure name matches the description. The parser identifies the beginning command of a figure and understands that if it encounters the same description again, the point is the continuation of the line that was begun.

If a point is a part of two lines, for example Back-of-Curb and Sidewalk, you must include one of the observations in one description group and recall the point for the other group. The figure command of recall tells the raw file translator that a previously observed point is to be reused. You enter the **RECALL** command as a note in the data collector. The entry for the note would be **RECALL 1005**. This is because you use one description for a point, but it participates in two description groups. The following code (Sokkia) shows the **RECALL 1005** note.

```
09F100011004119.40000090.8550000300.730000BCL
09F100011005105.88000091.1561111303.967777BCL
09F10001100640.310000092.1244444316.698333B SWK
13NMRECALL 1005
09F10001100729.400000092.6344444323.123333SWK
09F10001100817.150000094.1900000336.901111SWK
```

The above code recalls point 1005 to be a point in the SWK figure. Notice that the observation previously starts the sidewalk line. The following is the translation of the raw file into the field book file.

```
F1 VA 1004 300.43480 119.400 90.51180 "BCL"
F1 VA 1005 303.58040 105.880 91.09220 "BCL"
BEGIN SWK
F1 VA 1006 316.41540 40.310 92.07280 "SWK"
RECALL 1005
F1 VA 1007 323.07240 29.400 92.38040 "SWK"
F1 VA 1008 336.54040 17.150 94.11240 "SWK"
```

FIGURE LAYERS

When drafting line work, you assign the lines to layers. These layers represent generic information, for example, centerlines, back-of-curbs, boundary lines, etc. The Figure Prefix Library places line work created by the figure commands onto specific layers in the drawing. The Figure Prefix Library associates a layer name with the generic portion of the figure name. For example, the generic figure name of EOP assigns the resulting line work to the edge-of-pavement layer in the drawing. The Figure Prefix Library is universal to all projects.

For any of this strategy to work easily, standards need to be developed. Generally, the argument is about names of layers, their color, and linetype. However, the most important value is overlooked—getting the job done right. It is the content of the drawing and data files that is most important, not the layers. But once the standards are established, they need to be flexible and enforced.

BREAKLINES FROM FIGURES

You can create breaklines directly from figures in the drawing. These figures represent difficult areas on a surface for the Terrain Modeler to interpret; that is, linear features. The most difficult situation you will encounter in creating a surface is correctly representing the elevations around a curve. Since the field crew took only three shots around the curve, the triangles representing the surface model at the curve will have elevations that pull away from the arc of the curve. This is a result of the connection of the three points by lines representing chords of the observed curve. To help the TIN modeler create a better interpretation around the curve, Survey can create more data from the figure representing the curve. The Create Breaklines From Figures routine of the Analysis/Figures menu creates more data points around a curve from a figure. What the routine does is assume a constant slope between each of the points on the curve. Because of this constant slope, the routine can calculate an elevation at any point on the arc. By setting the mid-ordinate value to 0.1 (see Figure 16–6 for the Survey Figure Settings dialog box, which you can access through the Figure Settings routine of the Analysis/Figures menu), you create additional surface data about the arc. The mid-ordinate distance is the distance a chord segment is from an arc. If the mid-ordinate distance is 0.1, the routine starts at the first point on the arc and draws a chord segment whose distance is 0.1 of a unit from the arc. Where the end of the chord intersects the arc, the routine calculates the elevation of the intersection and creates an entry in a surface breakline data file. When the surface generator calculates the triangles along the arc, the triangles are much smaller and closer to the arc.

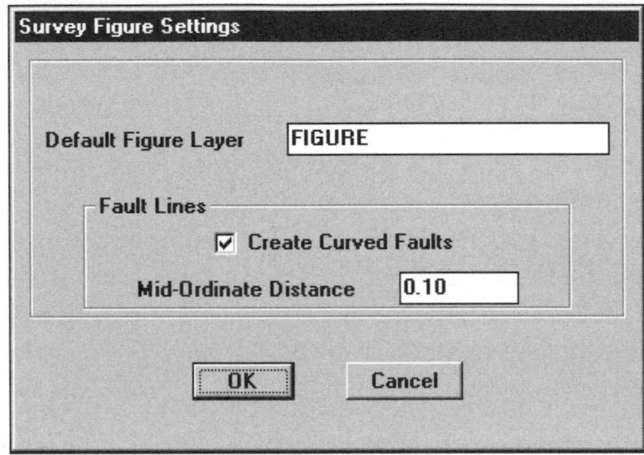

Figure 16–6

Exercise

After you complete this exercise, you will be able to:

- Use Figure commands as a part of the field data collection process
- Create lines and curves in the Survey module
- Revisit the Survey Command Language
- Use Figure commands in the Field Book file
- Edit and import a Survey field book
- Use Survey figure prefixes
- Create figure breakline data for the Terrain Modeler

CREATING FIGURES FROM FIELD OBSERVATIONS

1. Start a new drawing and name the drawing *Peoria* using the LDDTR2 prototype for a new project Peoria (see Figures 16–07 and 16–08). If you have not created the prototype, there is a copy of the LDDTR2 prototype on the CD that comes with this book.

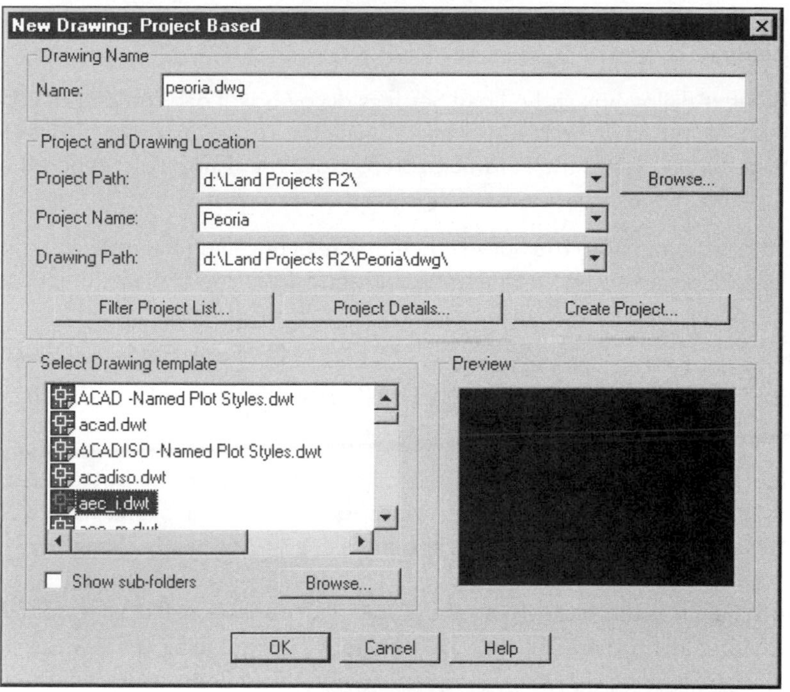

Figure 16–7

Figure 16–8

2. When the Point Database Settings dialog box appears, select the OK button to dismiss it.
3. The next dialog box is the Load Settings dialog box. Load the LDD1-20 setup file. If you do not have the LDD1-20 setup, use the i20 setup instead. After loading the setup file, select the Finish button to close the dialog box. When the Finish dialog box appears, select the OK button to dismiss it.
4. Change the menu to the Survey menu by selecting and loading the menu in the Menu Palette dialog box. Select Menu Palette from the Projects menu, and select the Survey menu.
5. Copy the file *Peoria.sdr* to the Survey folder of the project (Land Projects R2\PEORIA\SURVEY). The file is in the Survey directory of the CD that comes with this book.
6. View the SDR file by using the Text Editor routine from the bottom section of the Data Collection/Input menu. Print the file so you have a reference copy of it and after printing the file, exit without making or saving any changes.

The format of the file is the data format of the Sokkia Corporation. Each line contains specific information about the survey. The file has within it figure coding that, when parsed, will produce a field book that will draw lines representing flow lines, centerlines, and back-of-curbs. To make this all work, the first task is to code correctly the raw data. If there are some errors, you may have to edit the raw file (a copy of the raw file), or if the errors are not too bad, you can edit the field book.

The following is an excerpt from the *Peoria.sdr* file:

```
09F10001101313.980000093.5022222118.279444FLL
09F10001101434.460000091.0016666141.967222FLL
09F10001101534.140000090.1527777142.883888B BCL
09F10001101616.680000091.0283333127.305000BCL
09F10001101710.880000094.0494444337.786111BCL
```

The field crew marks the beginning of a new figure with the addition of the B (B space) to the description field. When the translator reads this line, the translator takes out the B and places the BEGIN prior to the line in the Field Book file. The following example shows the resulting field book lines.

```
FC1 VA 1013 118.164600 13.9800000 93.300800 "FLL"
FC1 VA 1014 141.580200 34.4600000 91.000600 "FLL"
BEGIN BCL
FC1 VA 1015 142.530200 34.1400000 90.091000 "BCL"
FC1 VA 1016 127.181800 16.6800000 91.014200 "BCL"
FC1 VA 1017 337.471000 10.8800000 94.025800 "BCL"
```

When the converter encounters the same description later in the raw file, it assumes that the point is a continuation of the figure. The translator then places a continue entry in the field book. The following is an example of the raw and field book file translation.

Raw File Code

```
09F100011027110.23000089.8938888291.182222BCL
09F100011028150.68000090.9855555312.382222B FLR
09F100011029150.61000090.8255555312.679444B BCR
09F100011030151.93000090.9727777308.789444B CL
09F100011031134.52000091.0300000308.256111CL
09F100011032134.25000091.0300000312.400555FLR
09F100011033134.22000090.8422222312.699444BCR
09F100011034118.82000090.8633333313.568333BCR
09F100011035118.65000091.0783333313.257222FLR
09F100011036117.49000091.0783333308.633888CL
```

Field Book Code

```
FC1 VA 1027 291.105600 110.230000 89.533800 "BCL"
BEGIN FLR
FC1 VA 1028 312.225600 150.680000 90.590800 "FLR"
BEGIN BCR
FC1 VA 1029 312.404600 150.610000 90.493200 "BCR"
BEGIN CL
FC1 VA 1030 308.472200 151.930000 90.582200 "CL"
FC1 VA 1031 308.152200 134.520000 91.014800 "CL"
CONT FLR
FC1 VA 1032 312.240200 134.250000 91.014800 "FLR"
CONT BCR
FC1 VA 1033 312.415800 134.220000 90.503200 "BCR"
FC1 VA 1034 313.340600 118.820000 90.514800 "BCR"
CONT FLR
FC1 VA 1035 313.152600 118.650000 91.044200 "FLR"
CONT CL
FC1 VA 1036 308.380200 117.490000 91.044200 "CL"
```

The Convert File Format routine of the Conversions menu assumes that similar described points are connect to each other. The file shows the field crew surveying lines in two ways. The first is by walking from the beginning to the end of a line (figure), and the second is by walking across several lines (figures). The *Peoria.sdr* file surveys the north side of the road as consecutive points and the south side of the road as cross section points.

EXERCISE

7. Start the TDS program by selecting Data Collection Link from the Data Collection/Input menu. This starts the TDS DC link program.

8. Use the Convert File Format routine of the Conversions menu to make a field book from the *Peoria.sdr* file. The routine displays a dialog box that prompts you for the name and type of file you are converting and the resulting file name and type. The *Peoria.sdr* file is a raw file and is a Sokkia SDR Numeric 4 format. The format of the resulting file is an Autodesk-Softdesk FBK. Use Figure 16-9 as a guide for setting up the file conversion.

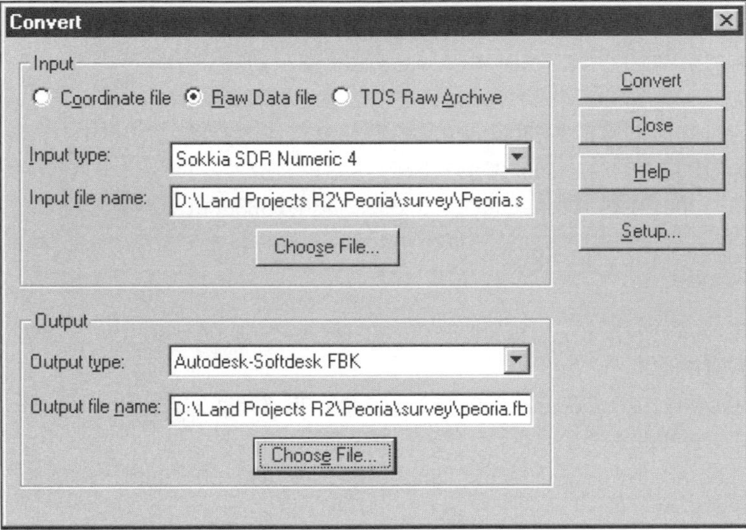

Figure 16–9

9. After you convert the file, the routine notifies you of the successful conversion. Select the OK buttons to exit the dialog boxes.

10. Exit from the TDS program and return to Survey.

11. Use the Import Field Book routine at the bottom of the Data Collection/Input menu to import the field book into the drawing. The routine prompts you with a number of questions. Answer Yes to all of the questions.

12. Zoom in to view the line work.

13. List the centerline figure and notice that the polyline is on the layer CL. CL is an entry in the Figure Prefix Library.

14. List the flowline and back-of-curb line to the north of the centerline. The two figures should be on the layer Figure. Neither FL nor BC is an entry in the prefix library.

PREFIX LIBRARY

1. Run the Figure Prefix Library routine from the Analysis/Figures menu. Add the prefixes FL, FN, and BC. The layer for FL is FLOW, the layer for FN is FENCE, and the layer for BC is CURB.
2. Run the layer command and create the FENCE. Assign the layer a color and the linetype of FENCELINE2. After you create the layer and assign the color and linetype, select OK to exit the Layer dialog box.
3. Reimport the field book into the drawing and answer Yes to all of the questions.
4. Zoom in and list the flowline and back-of-curb; they should now be on the FLOW and CURB layer. The fence line work should show the complex linetype of FENCELINE2.

EDIT A FIELD BOOK

The field crew omitted two C3 commands from the field data. You need to add the two figure curve commands to the field book to create the missing curves. Place the first C3 between points 1002 and 1003 for the flow line. The second C3 is between points 1022 and 1023.

1. Add the C3 commands to the following lines in the field book using the Edit Field Book routine at the bottom of the Data Collection/Input menu. Use the following excerpt as a guide in editing.

The following code indicates the location of the C3 for the curve between points 1003 and 1004.

```
BEGIN FLL
FC1 VA 1001 290.522200 110.700000 90.082200 "FLL"
FC1 VA 1002 291.374800 126.320000 90.122200 "FLL"
C3
FC1 VA 1003 295.563200 125.210000 90.262200 "FLL"
FC1 VA 1004 300.434800 119.400000 90.511800 "FLL"
FC1 VA 1005 303.580400 105.880000 91.092200 "FLL"
C3
FC1 VA 1006 316.415400 40.3100000 92.072800 "FLL"
FC1 VA 1007 323.072400 29.4000000 92.380400 "FLL"
FC1 VA 1008 336.540400 17.1500000 94.112400 "FLL"
FC1 VA 1009 336.134400 13.7100000 95.363000 "FLL"
END
```

The following code indicates the location of the C3 for the curve between points 1022 and 1023.

```
FC1 VA 1020 334.353600 16.8800000 92.460200 "BCL"
FC1 VA 1021 324.181400 25.7900000 91.512200 "BCL"
FC1 VA 1022 317.070400 37.1300000 91.281200 "BCL"
C3
```

```
FC1 VA 1023 303.422600 105.520000 90.533200 "BCL"
FC1 VA 1024 302.092200 113.880000 90.462800 "BCL"
FC1 VA 1025 296.122400 124.400000 90.155400 "BCL"
```

2. After making the changes, reimport the field book. Answer Yes to all of the questions, since all of the points are in the Field Book file.
3. Zoom to the northeast curb area (zoom to point 1023, height of 75) and view the new arc segment.

BREAKLINES FROM FIGURES

1. Set the mid-ordinate value to 0.05 for curve figure breaklines by setting the value in the Figure Settings dialog box in the Figure Settings routine of the Analysis/Figures menu (see Figure 16–10). After setting the value, select the OK button to dismiss the dialog box.

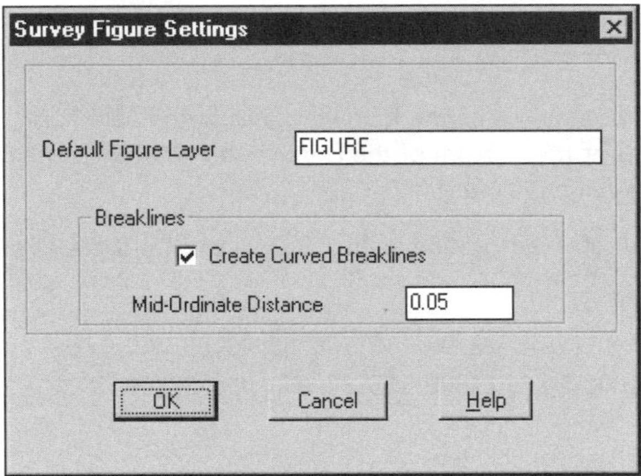

Figure 16–10

2. Select the Create Breakline From Figures routine found in the Analysis/Figures menu to create curve data for a surface. If there are no surfaces, you will see the New Surface dialog box. If there are preexisting surfaces, the routine will prompt you to select an existing surface or to create a new surface. Define a new surface and name the surface EG. After identifying the surface, the next step is to select which figures to use. You can select individual figures or all figures in the drawing. Use ALL only if you know that is the case; otherwise, select the figures you want in the surface. Most of the time you will be selecting all of the figures to create breaklines.
3. Open the Terrain Modeler from the Terrain menu. EG should be a surface on the list.
4. Select the "+" (plus sign) next to the surface name EG to expand the data list.

EXERCISE

5. Select Point Groups, click the right mouse button, and select Add Point Group. The routine responds with a dialog box indicating that there are no Point Groups defined and asking whether you want to define a point group. Select the Yes button. After you select the Yes button, the routine displays the Group Manager dialog box. Click the Create Group icon to display the Create Group dialog box. In the dialog box, name the group EG and select the Build List button. In the Build List dialog box, select all points. After selecting all the points, select the OK buttons until you exit the Point Group Manager. Close down the Point Group Manager and return to the Terrain Model Explorer.
6. Again select Point Groups, click the right mouse button, and select Add Point Group. This time, select the EG point group.
7. Build the surface by selecting the name of the surface, clicking the right mouse button, and selecting Build. Accept the default settings and select the OK button to process the data. Select the OK button to dismiss the Done Processing dialog box and close the Terrain Explorer.
8. Zoom into point 1025 with a height of 50 (Zoom to Point).
9. View the triangulation of the figure curve data by running the Quick View routine of the Surface Display cascade of the Terrain menu.
10. Save the drawing by clicking the Save icon.
11. Run the Create contours routine to create contours representing the surface. The range of elevations on the surface is quite small. You need to set the Minor interval to 0.25 and the Major to 1.0. Use Figure 16-11 as a reference to setting the values.

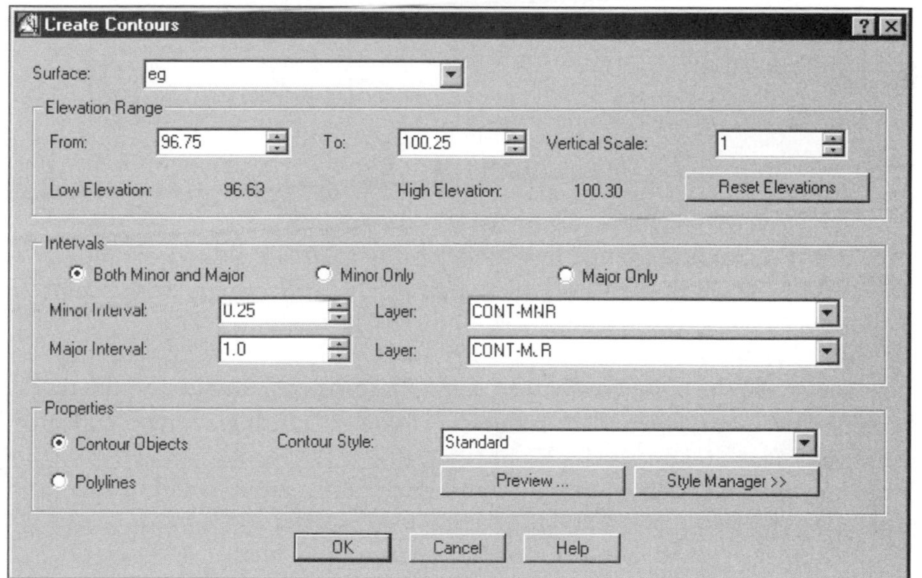

Figure 16–11

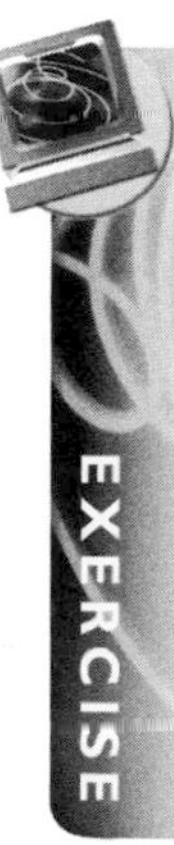

UNIT 3: UPLOADING TO THE DATA COLLECTOR

One of the advantages of having a data collector is the ability to upload coordinates of points into the data collector from points created in the project. These points then are set out in the field representing critical locations in the design.

The first step is to create the points. The second step is to create an Autodesk uploadable file for the TDS Link routine to send up to the data collector. The upload file has a specific extension, auf. The structure of the file is point number, northing, easting, elevation, and description. The creation of an uploadable file is the job of the Export routine of the Points menu (see Figure 16–12). One of the export formats is the Autodesk Uploadable file. The file can contain all of the points or the points of a point group. The last set is to send (upload) the file to the data collector. This is done in the Collector Link program of Survey (see Figure 16–13). In the Send tab of the Transfer dialog box, you identify the file and the destination collector. The file you send to the data collector can be renamed, and you can save a copy of the file on the hard drive in the collector's format.

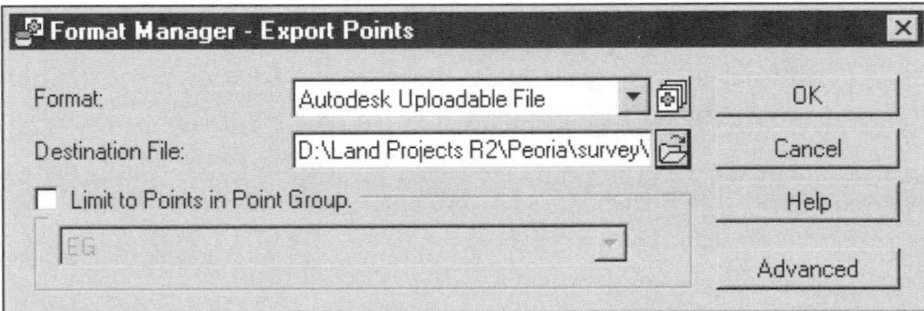

Figure 16–12

The power of the Survey module lies in the ability to create line work and breakline data for surfaces of a project. Using this capability greatly enhances the value of the field crew and shortens the time from field to design.

The next section of the book covers the traverse routines of the Survey add-in. Again, the most powerful tool is the data collector. If you do not have a data collector, the most powerful file you have is the Field Book file.

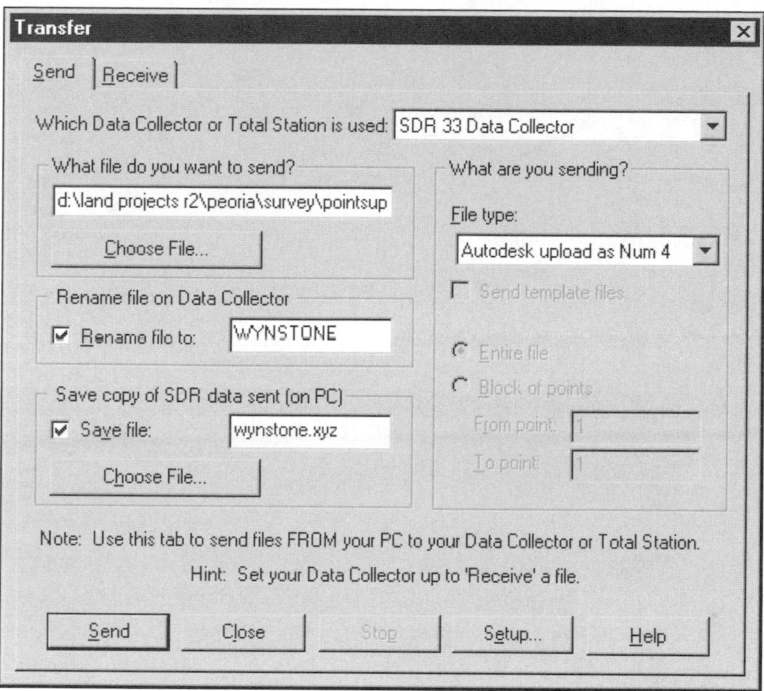
Transfer
Send
Receive
Which Data Collector or Total Station is used:
SDR 33 Data Collector
What file do you want to send?
d:\land projects r2\peoria\survey\pointsup
Choose File...
What are you sending?
File type:
Autodesk upload as Num 4
Send template files
Rename file on Data Collector
Rename file to:
WYNSTONE
Entire file
Block of points
Save copy of SDR data sent (on PC)
Save file:
wynstone.xyz
From point:
To point:
Choose File...
Note: Use this tab to send files FROM your PC to your Data Collector or Total Station.
Hint: Set your Data Collector up to 'Receive' a file.
Send
Close
Stop
Setup...
Help

Figure 16–13

SECTION 17

Survey Traverses

After you complete this section, you will be able to:

- Create observation data for a traverse
- Use data entry techniques for traverses
- Use traditional traverse reductions
- Use Least Squares traverse reductions
- Use traverse network reductions

SECTION OVERVIEW

The Survey add-in contains routines to complete an evaluation of traverse data collected in the field. The analysis can be of manually or electronically collected survey data. If the survey is manually collected, the field book is the easiest method of entering the data into the observation database. If the survey is collected electronically, you can evaluate data in the collector to reduce the field observations of the traverse. If you want to evaluate the traverse in Autodesk Survey, you must collect the data with no traverse reduction in the data collector, download it into a project, and analyze it with routines found in Survey. There are three other manual methods for entering the traverse data: the Instruction method (Traverse Entry of the Traverse/Sideshot cascade of the Data Collection/Input menu), the Survey Command Prompt traverse routine, and the Traverse Editor. The Survey Command Prompt and the Instruction method are the most unforgiving. The Traverse Editor is better but presents its own set of problems.

Many of the data collectors in use today have coordinate geometry programs within them. A traverse routine is usually among the onboard programs. The reduction of the field observations is done within the collector, and only the final coordinates are in the downloaded file. The coordinates represent a balanced traverse. When this occurs, the only function Survey provides is the downloading of the file and its conversion to a field book.

UNIT 1

The first unit of this section reviews the different methods of entering traverse data. Survey has five methods for entering traverse field data. The exercise covers entering data using the Traverse Entry routine of the Traverse/Sideshots cascade of the Data Collection/Input menu and using the Traverse Editor. The evaluation of the loop uses all of the reduction methods, Compass, Crandall, Transit, and Least Squares.

UNIT 2

A traverse may have more than one observation of a foresight and backsight and observations across the loop from other points in the traverse loop. The multiple observation points will become a part of a traditional traverse and its analysis. However, the cross loop observations are data only for Least Squares analysis of a loop. If there is more than one traverse that needs reduction, the Least Squares program is the only routine in Survey that accomplishes the task. This data is the focus of the second unit of this section.

UNIT 3

The last unit of this section reduces a traverse from a data collector raw file. The traverse has errors and procedural problems that you must correct before using the file.

UNIT 1: SURVEY TRAVERSE

The evaluation of field observations is necessary to prevent incorrect observations from creating bad points in the point database. The traverse is a method that evaluates the quality of the observations of a survey. Survey has five methods of traverse data entry. Each method is rigid in its methodology. Of all the methods, the easiest methods for data entry are the data collector and the Field Book. The five methods are the menu (Instruction) Traverse Entry routine, the Survey Command Prompt, the Traverse Editor, field observations (Field Book), and the data collector. The easiest method for entering traverse data is the Field Book method.

TRAVERSE DATA

The Traverse data entry and adjustment routines use and produce several external data files of which the observation database is the most important. If you make a mistake when entering data for a traverse while in the Traverse Entry routine or Survey Command Prompt traverse routine, the correction process is almost nonexistent. The best you can do is to remove the points in the point database and start over.

The routines do not allow the data entry process to go back and start over where the error occurred. If the data error occurs late in the entry process, you might be able to edit the *Batch.txt* file to correct the error and to complete the data entry in the file. By reading in the *Batch.txt* file containing the corrections, the file writes the correct data to the observation database (Traverse data) and you are ready to analyze the traverse. The routines to edit and rerun the batch file are in the Batch File cascade of the Data Collection/Input menu.

The Traverse adjustment routines reduce multiple observations of a point to coordinates and if there is vertical data, to points with elevations. No matter where the data comes from, the routines place the traverse data in the observation database. Survey locates the observation database in the Survey folder of the project.

The traditional traverse does not allow observations across the loop to other points in the traverse nor does it allow multiple observations of a point in the loop from different setups. The only multiple observations the traverse allows are Face1 and Face2. The Face1 and Face2 process consists of observing the point, turning the scope and turning the instrument 180 degrees, and reobserving the point.

A surveyor working on a project may have to survey a loop around the site of interest before surveying the interior. This loop is a traverse loop. The reason behind a traverse is to ensure that the measurements and locations of the site are as precise as possible. One problem with a traditional traverse loop is that it does not measure traverse points across the traverse loop. When attempting to cross from one side of the loop to the other, the surveyor may find that he or she is not where he or she ought to be. This occurs because the positions along the loop are only in reference to the preceding and next point along the loop. The traverse records only an angle at each occupied point and a distance along the loop. The traverse loop does not know anything about interior points and how they relate to the points in the external loop. As a result, crossing from one side of the loop to the other may not be as successful as you might think. Sideshots from the traverse loop do not help in understanding points in the interior of the loop because the adjustment routines adjust the location of the sideshots by the changes in the setup that recorded them.

The Least Squares method of analyzing point data does allow a loop around the site, and data across and through the site. This data ties interior points into the loop adjustment, resulting in network-wide adjustment with all points sharing the adjustment error. Thus, when crossing from one side of the loop to the other, you end up closer to the desired position.

ENTERING TRAVERSE DATA

The data for a traverse loop, a network (merged traverses), and a Least Squares analysis (a set of points) comes from the observation database of Survey. As you record field observations of points, the Survey module creates entries in an observation database. When defining a traverse loop or a network, the routine converts the observation database into a Traverse or Least Squares data file.

One or a combination of five methods can accomplish the recording of traverse and network data. The command sequences of the first two methods, the Traverse Entry routine of the Traverse/Sideshots cascade of the Data Collection/Input menu and the Survey Command Prompt, write survey commands to the *Batch.txt* file in the Survey subdirectory of the project. If any data entry errors occur, you can edit the *Batch.txt* file and replay it to generate the "corrected" observation database information. The third method is creating a traverse loop directly in the Traverse Editor. The next method is creating a traverse from field observations (a field book). The final method is storing the traverse data in a data collector.

METHOD 1: TRAVERSE ENTRY

The Traverse Entry routine found in the Traverse/Sideshots cascade of the Data Collection/Input menu prompts for the values that reflect the Traverse default settings. The exercise for this method uses a setup, backsight, a turned angle, and a distance to the foresight. After entering in the foresight, the Traverse routine moves the point of occupation (stationing) to the foresight and backsights the last occupied point. The routine then prompts for the new foresight data and continues until the first station is reoccupied for a closed traverse.

The Traverse Entry routine does not allow multiple observations at a station with the exception of Face1 and Face2 observations. The toggle for Face1 and Face2 observations is in the traverse settings. The Traverse Entry routine automatically writes the data to the observation database and the *Batch.txt* file. The initial backsight can be a false point. The direction to the backsight can be an azimuth, bearing, or northing/easting. The final foresight is an angle and distance to the original backsight creating a closed traverse.

You cannot correct any data entry errors in this method. The severity of the error determines where to go to correct the mistake(s). If the mistakes are severe enough, starting over from the beginning is the only choice. Before you can start over, the process requires erasing all points placed into the drawing from traverse observations and deleting the observation database. If the errors are not severe, editing the errors in the *batch.txt* file is sufficient. Once you decide to correct the errors in the *batch.txt* file, work only in the *batch.txt* file.

METHOD 2: THE SURVEY COMMAND PROMPT

Before using this method, you must set the Traverse settings and turn on Traversing in the SCP. If you turn on SCP Traversing, the SCP starts prompting for the observation data of each station while recording the information in the observation database and the *Batch.txt* file. The Survey Command Prompt method allows you to have multiple shots at a station only as Face1 and Face2 observations. The final foresight is an angle and distance to the original backsight, creating a closed traverse.

You cannot correct any data entry errors in this method. The severity of the error determines where to go to correct the mistake(s). If the mistakes are severe enough, starting over from the beginning is the only choice. Before you can start over, the process requires erasing all points placed into the drawing and deleting the observation database. If the errors are not severe, editing the errors in the *Batch.txt* file is sufficient. Once you decide to correct the errors in the *Batch.txt* file, work only in the *Batch.txt* file.

METHOD 3: THE TRAVERSE EDITOR

This method enters the traverse loop directly in the Traverse Editor. Set the defaults in the editor before you enter data. The defaults affect instrument station and foresight point data formats. The first line in the editor contains information about the instrument station and backsight. The second line contains the foresight point information. The data includes point number, angle, distance, prism, and description of the foresight point. When you exit the Traverse Editor, the routine generates the observation database and all associated files.

You must have all the data organized before entering information into the editor. The editor does not easily allow for the addition of new pieces of data because the editor is calculating values during data entry; therefore, correcting errors is difficult. It may be better to cancel out of the editor and reenter the data from the beginning rather than trying to edit the incorrect data.

METHOD 4: FIELD TRAVERSES—FIELD BOOK

The field book is a file containing the field observations of the traverse. The method creates traverse data from the data in a field book. The surveyor manually types the observations into a file using the Survey Command Language. When importing the Field Book file, Survey creates the observation database. The Define Loop routine of the Traverse Loops cascade of the Analysis/Figure menu creates a traverse loop definition reflecting the information found in the observation database. You may have to edit the Field Book file to correct errors made in the field. After you correct the Field Book data and reimport the Field Book, Survey creates a corrected observation database ready for the correct loop definition. This method is one of the easiest ways to create traverse data and correct errors.

METHOD 5: FIELD TRAVERSES—DATA COLLECTOR

In this method the observations for the traverse are observations entered into a data collector. After you download the data from the data collector, the file contains the observations in raw form. After converting the data and importing the field book into Survey, the process creates an observation database for the traverse. When you use the Define Loop routine of the Traverse Loops cascade of the Analysis/Figures menu, the routine will return the loop found in the observation database.

DEFINING A TRAVERSE

Once the observation database is present and correct, the next step is defining the traverse loop. A traverse loop contains data for each occupied point and backsight. After you define the loop, the last step is reviewing and adjusting the loop values.

TRAVERSE ADJUSTMENT

After you enter in a traverse the next step is to evaluate its quality. Survey has four methods of evaluating a traverse: Compass, Crandall, Transit, and Least Squares. The first three methods, Compass, Crandall, and Transit are traditional evaluation methods that evaluate a traverse loop by the loop's foresight, backsight, and angle data. Each method emphasizes a different data element in the traverse. If you have a traverse with multiple foresight/backsight, cross traverse, and/or external point observations, the only method you can use is Least Squares. You can use Least Squares to evaluate data that is a combination of traverse loops.

Exercise

After you complete this exercise, you will be able to:

- Create observation data for a traverse
- Use data entry techniques for traverses
- Use traditional traverse reductions
- Use Least Squares traverse reductions
- Use traverse network reductions

TRAVERSE DATA ENTRY

The first exercise creates a single observation traverse (see Figure 17–1). You should try one or two methods of entering the data. Be very careful when entering the data because many of the routines are unforgiving. If you make a mistake, you will probably end up modifying the *Batch.txt* file and rerunning the file to create a corrected observation database, entering the data again after some point cleanup, or creating a Field Book file to complete the data entry.

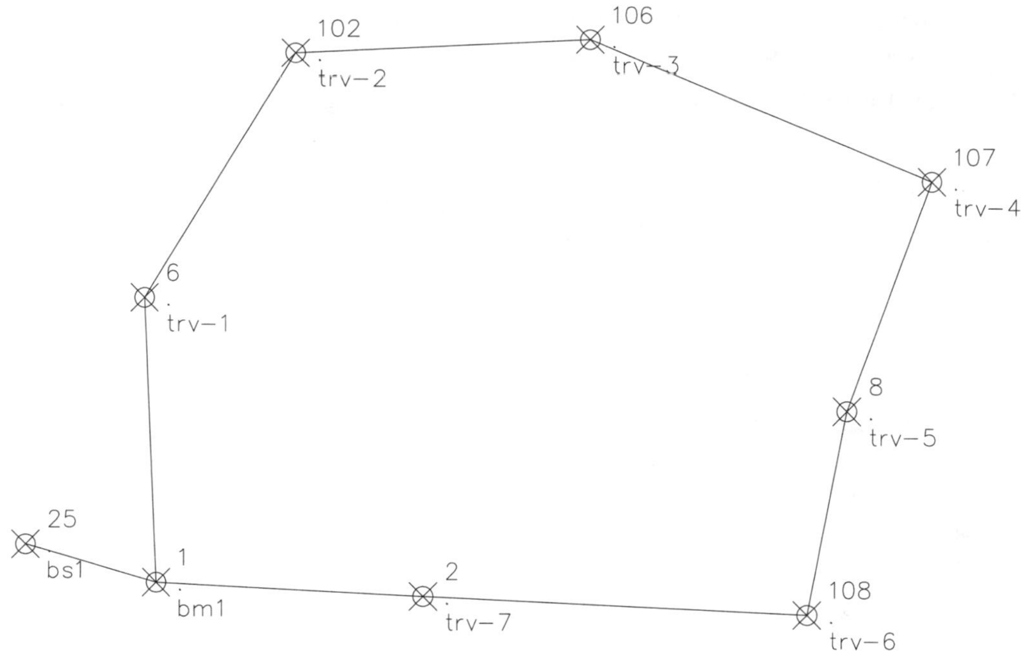

Figure 17–1

The format of the first line of traverse data is for the station:

```
1    25    0.0
```

This line states that the instrument is at point 1 and backsights to point 25. The angle between the two points is 0.0000 degrees.

The second line holds the data for the foresight observation.

```
6    71.3033 429.7884        Trv-1FS
```

This line of data indicates a foresight to point 6. The turned right angle is 71.3033 degrees *(ddd.mmss)* with a horizontal distance of 429.7884, and point 6 has the description of Trv-1FS.

If you make an error while entering data in the Traverse Entry routine or at the Survey Command Prompt, exit the routine. After exiting, you must do some housekeeping and start over from the beginning. This is because of the rigid handling of the observation database. You must remove any new points you have placed into the project during traverse data input except for points 1 and 25. You must also delete the observation database. If you make a mistake, the order of recovery is as follows:

1. Erase the new data entry points from the drawing and point database. Use the Erase routine of the Edit Points cascade of the Points menu to delete them. **Do not erase points 1 and 25**.

EXERCISE

2. Delete the observation database with the Delete Observation Database routine in the Data Collection/Input menu.

You should have only points 1 and 25 remaining in the drawing.

Traverse Data

STN	Pnt#	Angl	Hrz Dist	DescType
1	25	0.0		BS
	6	71.3033	429.7884	Trv-1FS
6	1	0.0		BS
	102	214.0533	438.5154	Trv-2FS
102	6	0.0		BS
	106	235.4233	451.0348	Trv-3FS
106	102	0.0		BS
	107	205.2035	564.9924	Trv-4FS
107	106	0.0		BS
	8	267.4337	369.7415	Trv-5FS
8	107	0.0		BS
	108	170.3710	313.1525	Trv-6FS
108	8	0.0		BS
	2	261.3654	590.6619	Trv-7FS
2	108	0.0		BS
	1	180.1813	405.0081	BM1FS
1	2	0.0		BS
	25	193.0433	209.9515	BSB1FS

Creating traverse data in the Traverse Editor or by creating a Field Book file is a little more forgiving. When you exit the Traverse Editor or import the Field Book file, the routine creates a new observation database.

TRAVERSE INPUT

1. Start a new drawing, *ManTrv*, and assign it to a new project, ManTrv. Use the LDDTR2 prototype for the project. Use Figure 17–2 and 17–3 as guides to defining the new drawing and project. Select the OK buttons to create the project and drawing.
2. When the Point Database Settings dialog box appears, select the OK button to dismiss the dialog box.

EXERCISE

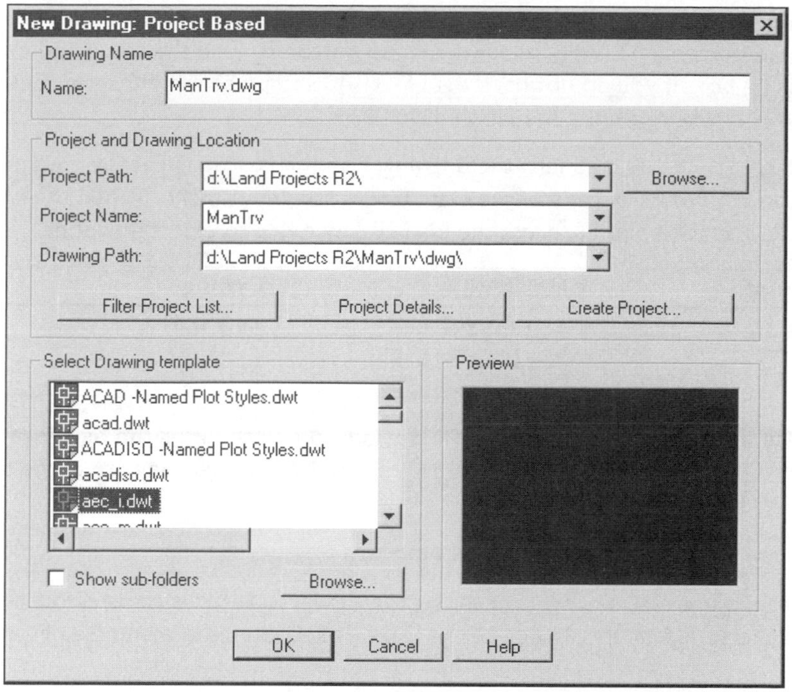

Figure 17–2

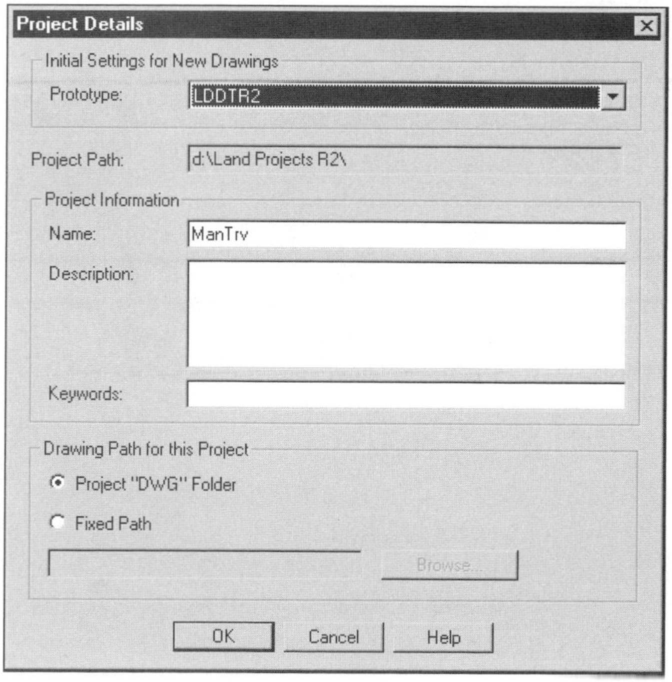

Figure 17–3

3. Next the Load Setup dialog box appears. Select the LDD1-40 setup, then the Load button, and finally select the Finish button to load the setup and exit the dialog box. If you do not have the LDD1-40 setup use the i40 setup. Select the OK button to dismiss the Finish dialog box.

4. Place points 1 and 25 into the drawing. Use the Northing/Easting routine of the Control Points cascade of the Data Collection/Input menu. When the routine asks you for point 26, press the ENTER key twice to exit the routine.

Point#	Northing	Easting
1	4715.0055	4464.1621
2	4772.4375	4262.2185

5. Use the Zoom to Point routine and zoom to point 1 using the height of 750. After zooming to the point, pan points 1 and 25 so that they are in the lower left of the screen.

6. Click the Save icon and save the drawing.

7. Set the traverse settings by running the Traverse Settings routine found in the Traverse/Sideshots cascade of the Data Collection/Input menu (see Figure 17–4).

The defaults are:

```
Vertical Measurements: None
Horizontal Angles: Angles Right
Descriptions: ON
```

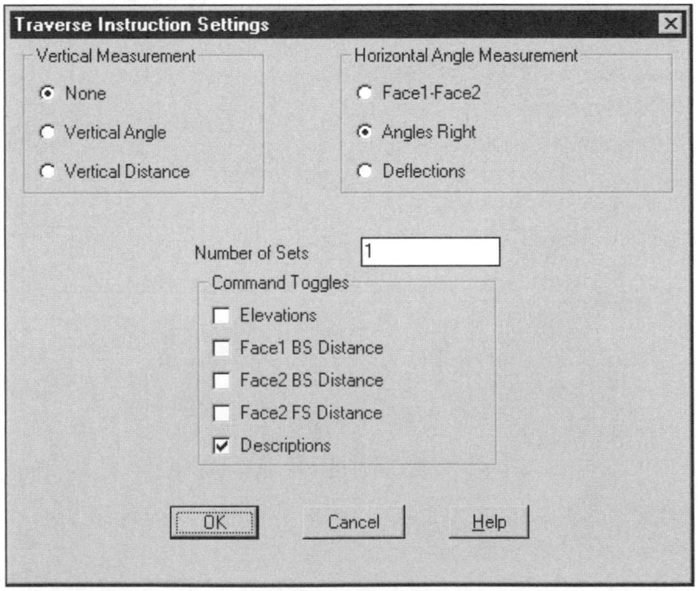

Traverse Instruction Settings

Vertical Measurement: (•) None, ( ) Vertical Angle, ( ) Vertical Distance

Horizontal Angle Measurement: ( ) Face1-Face2, (•) Angles Right, ( ) Deflections

Number of Sets: 1

Command Toggles: [ ] Elevations, [ ] Face1 BS Distance, [ ] Face2 BS Distance, [ ] Face2 FS Distance, [x] Descriptions

OK | Cancel | Help

Figure 17–4

METHOD 1: THE TRAVERSE ENTRY ROUTINE OF THE TRAVERSE/SIDESHOTS CASCADE

This Traverse Entry routine prompts for the required traverse values. After you enter the data, the routine calls the Survey Command Processor to calculate the northing and easting of the foresight point.

The Traverse Entry routine is rigid and unforgiving. Data entry mistakes require you to exit the entry routine and start over from the beginning. You must delete the new points from the database, retaining the original points 1 and 25. Also, the correction process requires the deletion of the observation database before starting the routine over to enter the data.

1. Select the Traverse Entry routine from the Traverse/Sideshots cascade of the Data Collection/Input menu. The following is a fragment of the routine's prompting for information. You need to complete the data entry to be able to move on to the next step, which is defining the traverse loop.

The routine prompts for the values.

```
Command:
Enter point number of setup  : 1
SURVEY>  STN 1
    STATION       h.i.: <Null>
  POINT 1        NORTH: 4715.0055   EAST: 4464.1621     EL: <Null>
Enter backsight point number  : 25
SURVEY>  BS 25
Enter point number < 2 > : 6
Enter foresight angle right : 71.3033
Enter foresight horizontal distance < 0.000000 > : 429.7884
Enter description  : TRV-1FS
SURVEY>  AD 6 71.303300 429.7884 "TRV-1FS"
     BEARING: N 02-36-55 W              DISTANCE: 429.7884
  POINT 6        NORTH: 5144.3462   EAST: 4444.5511     EL: <Null>
Advance to next setup (Yes/No)? <Yes>: (press ENTER to accept)
Enter point number of next setup  < 6 > : (press ENTER to accept)
```

The process repeats for each station and foresight. You occupy each new foresight point in the loop as the next new station, and each occupied point becomes the new backsight point. The routine knows the sequence. All you need to do is input the points and their observation data. From this new setup, the surveyor observes the next point along the loop until finally, the traverse reoccupies the starting point and records the observation of the initial backsight point as the final foresight. This method creates a traditional traverse loop.

If you make a mistake in entering the data, there are few options for recovery. You must exit the routine and return to the command prompt. Before start over, the recovery process has to delete any new points from the point database. The only points remaining in the drawing and database are points 1 and 25. Additionally, you need to delete the observation database.

Loop Definition

1. After entering the correct data, define the traverse loop. The routine is the Define Loop routine of the Traverse Loops cascade of the Analysis/Figures menu (see Figure 17–5). The routine responds with a dialog box listing the initial backsight, the occupied points, and final foresight (see Figure 17–6).

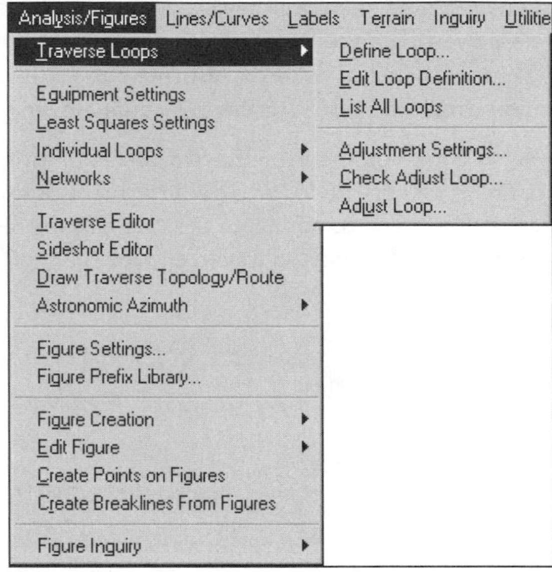

Figure 17–5

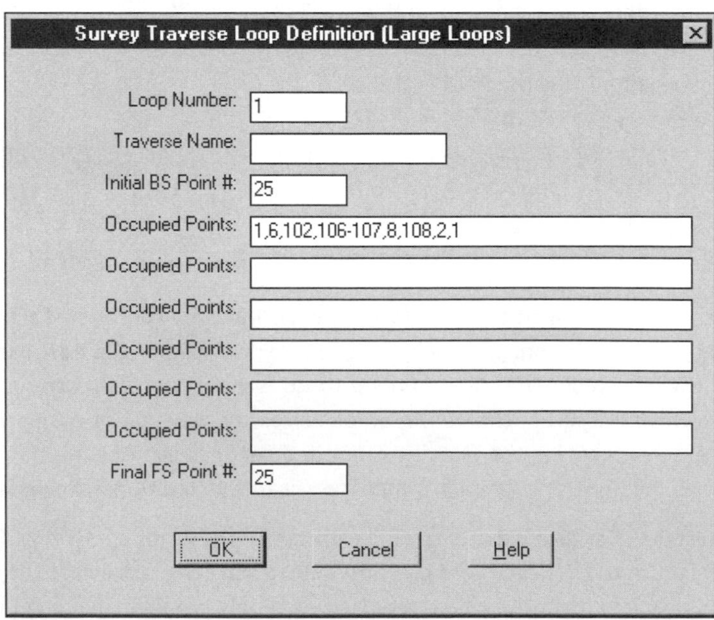

Figure 17–6

EXERCISE

Check the Loop Adjustments

1. You can review the loop data by using the Check Adjust Loop routine found in the Traverse Loops cascade of the Analysis/Figures menu (see Figure 17–5). First the routine displays a dialog box listing the defined loops in the project. Select loop number 1 and select the OK button to continue. The routine then presents a dialog box containing the methods of evaluating the traverse. Toggle on Balance Angles in the upper left of the dialog box. This toggle must be on to view the adjustment process (see Figure 17–7). Select the type of adjustment you want by selecting from the list at the upper right of the dialog box. After selecting the type of analysis, select the OK button to view the reduction numbers. Three dialog boxes will be presented to you in a sequential order: overall review of results, the raw traverse data, and the coordinate adjustment. Each report of the adjustment takes you into Notepad. After reviewing the values, close Notepad down to view the next screen of data. When done, you have the choice of viewing another adjustment method or exiting the routine.

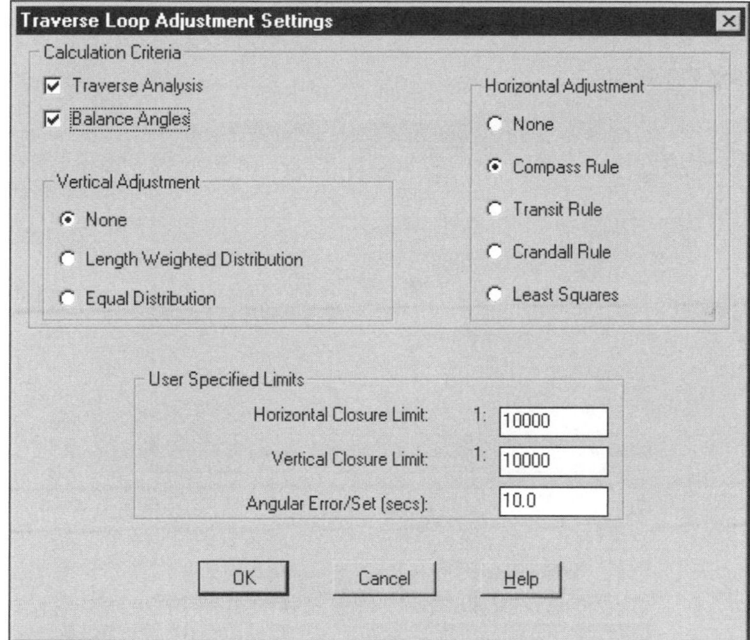

Figure 17–7

Adjust the Loop

1. After checking the data, adjust the traverse by the Compass, Transit, Crandall, or Least Squares method. Use the Adjust Loop routine of the Traverse Loops cascade of the Analysis/Figures menu. The routine calculates the adjustment values for each point in the traverse and displays Notepad with the same results as seen in the Check Adjust Loop routine. When you exit the Adjust Loop routine, it adjusts the coordinates for each point in the loop.

The define, check, and adjust steps are the same no matter which method you decide to use when working with traverse loops.

METHOD 3: ENTERING A TRAVERSE IN THE TRAVERSE EDITOR

1. Erase the new points from the traverse in the drawing and make sure to leave points 1 and 25.
2. Click the Save icon to save the drawing.
3. Start the Traverse Editor with the Traverse Editor routine in the Analysis/Figures menu. The routine first asks for the loop number to work with. Enter **2** (see Figure 17–8). After entering 2 as the loop number, select the OK button to exit the dialog box. The routine displays a new dialog box indicating that loop 2 does not exist, and asking whether you want to create it (see Figure 17–9). Select the Yes button to create the Loop 1 data set. The next dialog box prompts you for the initial station and backsight point. Enter point 1 and point 25 as the initial station and backsight points (see Figure 17–10). The routine finally displays the Traverse Editor (see Figure 17–11).
4. Review the defaults.

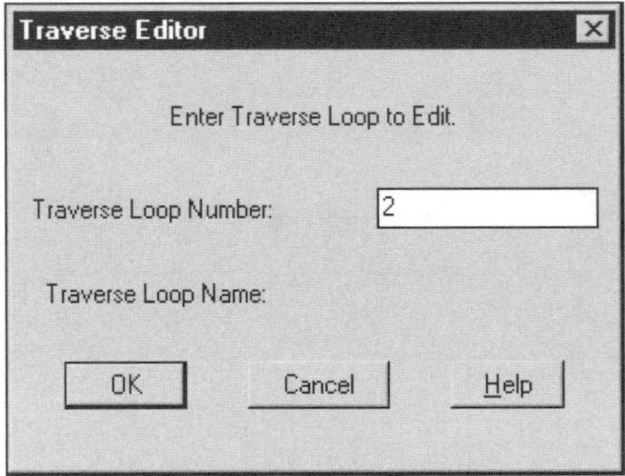

Figure 17–8

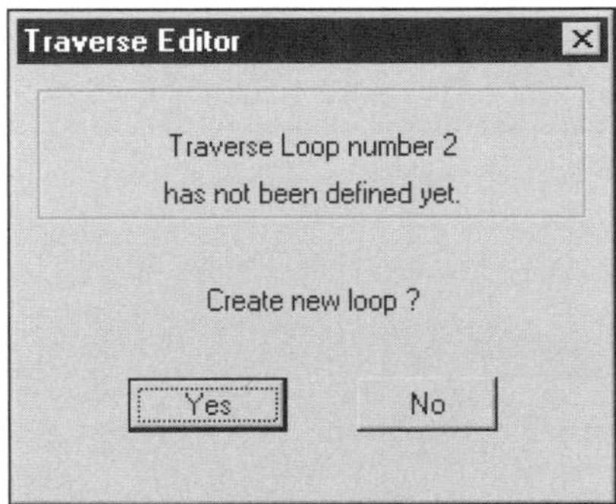

Figure 17–9

EXERCISE

Traverse Editor

Enter First Setup Point Numbers.

First Station Pt: 1

First Backsight Pt: 25

Traverse Loop Name: Exercise 1

OK | Cancel | Help

Figure 17–10

Traverse Editor

STN Point: 1 | BS Point: 25 | STN Coords

Theod Ht: <Null> | Bearing: N 74-07-28 W | North: 4715.0055

Instr Elv: <Null> | Distance: 209.95 | East: 4464.1621

Descript: | Descript: | Elev: <Null>

STN Pt	BS Pt		Angle	SlopeDist	VertAngle	Prism	DescKeys
1	25		0.0000				

Insert Stn | Insert Obs | Chg Theo Ht | OK | PgUp

Delete Stn | Delete Obs | Settings | Cancel | PgDn

Make FBK

Figure 17–11

The Traverse Editor does not recognize the defaults set in the Traverse Settings dialog box of the Data Collection/Input menu. There are two sets of defaults in the Traverse Editor; one set applies to the stations and the other to the foresight observations. You set the following defaults for the exercise first by reviewing the current settings. After you insert your first observation, set the observation settings before entering the first set of foresight data (see the Editor Settings dialog box in Figure 17–12, which you access by selecting the

Settings button in the Traverse Editor). Be sure to check the settings for the observations because sometimes the settings change without your intervention. After checking the settings, you can enter in the remaining data of the traverse.

Directions -	Angle (turned right)
Backsight -	Zero angle observation
Foresight -	Single observation
Distance -	Horizontal
Target -	None
Vertical -	None
Description -	Descriptions Only
Precision -	Angle 4, Linear 4 and elevation 2

The Traverse Editor has two lines of information. The first line is about the instrument and the backsight and the angle between them. This angle is 0.0000 for turned angles. The second line contains the information about the foresight point, its number, the turned angle, the slope distance, its description, and the height of the prism.

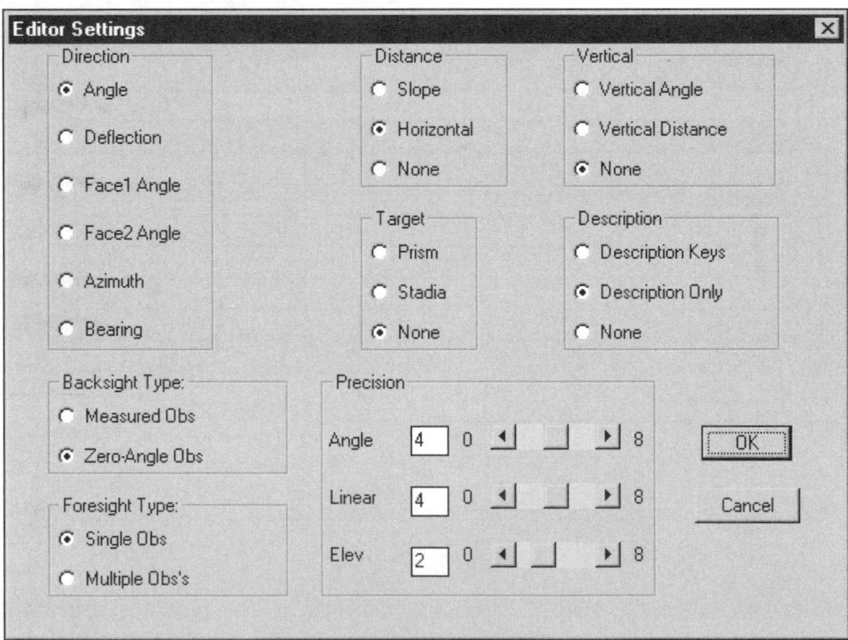

Figure 17–12

5. The Traverse Editor should look like Figure 17–13.
6. Exit the Traverse Editor by selecting the OK button.

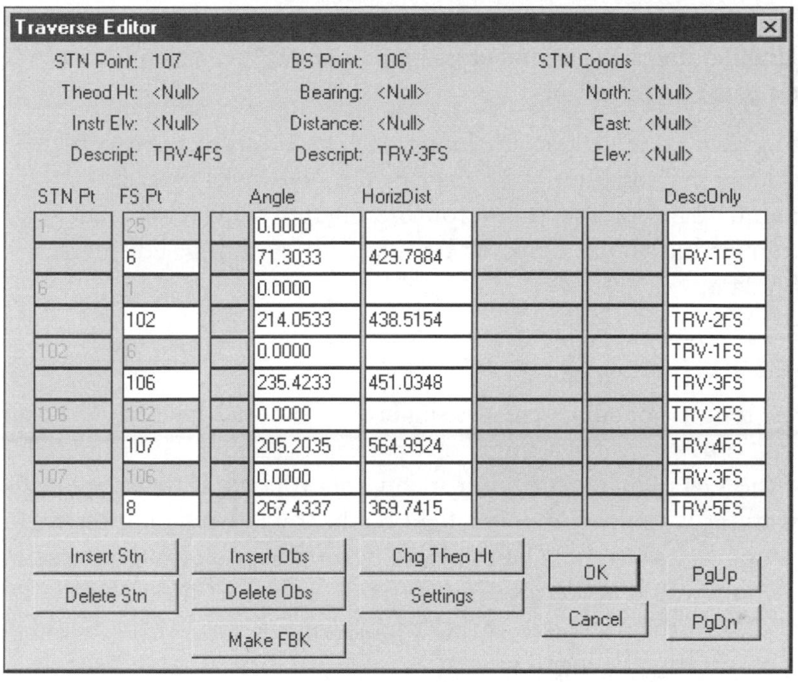

Figure 17–13

Check the Loop

7. Review the loop with the Check Adjust Loop routine found in the Traverse Loops cascade of the Analysis/Figures menu.

Adjust the Loop

8. Adjust the traverse by any of the following methods: Compass, Transit, Crandall, and Least Squares. The adjustments occur in the Adjust Loop routine. When you exit the Adjust Loop routine, it adjusts the coordinates for each point in the loop.
9. Click the Save icon to save the drawing file.

UNIT 2: MULTIPLE OBSERVATIONS/ CROSS LOOP OBSERVATIONS

The traverse routines only allow for the analysis of a single loop. No observations of angle and/or distance across the traverse can be a part of the analysis. In a traditional traverse, there is no allowance for this type of data.

If a data file contains multiple observations of traverse loop points, the Traverse routines will include them into the adjustment if they are Face1 or Face2 observations. The

Traverse routine excludes any cross loop observations. When you create the Least Squares data file, the multiple and cross loop observations will appear as a part of the observation data.

If evaluating two or more traverse loops, the surveyor creates a network from the individual loops by combining the data in the Networks cascade of the Analysis/ Figures menu. The Create Input routine creates a network data set from the combination of individual loops or by manually entering, editing, and merging observation data.

THE NETWORK

The traverse routines of Survey create a single observational database. When importing a field book, Survey creates a single observation database. The Import Field Book routine of the Data Collection/Input menu asks you about deleting existing points and observations when you import a field book. If you have multiple traverse loops and each loop has its own field book file, to delete the points and observations of the previous import would be devastating. When you import the multiple field books, you answer No to the questions of the Import routine and let the field book create additional observation data and points.

If you create observation data for two or more connected loops, the only way to analyze the two together is in the Networks cascade of the Analysis/Figures menu. The Create Input routine converts the data in the observation database into a network data file. The observation database must contain the observations for all the loops. The network data file is also a Least Squares data file of all loop observations. This routine creates the file *network.lsi* that holds the data. In the *network.lsi* file, the routine identifies the initial occupied points as known and the initial backsight, if a point, as known. Generally, the analysis holds two known points. Yet if the initial backsight is not a point, the routine holds only one point, the initial station point. The *network.lsi* file is the same format as the individual loop data file of Least Squares.

If there are multiple and cross loop observations, they appear in the data file, but not in the individual loop's data. In the network editor, you can manually add additional distance and angle measurements between or to points. The network editor is the Edit Input File routine of the Networks cascade of the Analysis/Figures menu.

To create the network from Field Book files, you can import as many Field Book files as needed to create the complete observation database. If you simply answer No to the Delete observation database prompt, the Field Book data appends to the observation database. Selecting the Create Input routine creates network data in the *network.lsi* file from the observations in the observation database. Again, you can add additional observations while in the network editor.

The Network data file consists of three elements: points known, points suspected, and observations between the points listed. The observations are either angle or distance measurements. Each distance and angle observation has a standard error that represents the accuracy of the measurement to the Least Squares routine. The higher the standard error value, the more freedom the routine has in evaluating the location of the point from the observation's data. This means that the surveyor can include in the analysis measurements from different sources; for example, tape measurements and paced distances. The surveyor only needs to place a reasonable standard error value with each measurement. The standard deviation of errors comes from the current instrument in the Equipment Settings dialog box of Survey settings.

The network Least Squares analysis distributes error throughout the network points, but does have blunder detecting capabilities. This distribution of error throughout the network would seem a better analysis of a site, rather than relying solely on perimeter traverse calculations. The traditional traverse calculations affect only the points on the perimeter of a site, and the corrections apply only to the same perimeter points. This leaves the measurements across the perimeter unadjusted for error.

The Least Squares and Network evaluation methods allow for the analysis of the interior as well as the exterior (the traverse) points of a site. If you use Least Squares analysis, all the points of the network adjust together. This network-wide error adjustment minimizes errors over the entire site.

Exercise

This exercise deals with a field survey that contains multiple observations (F1 and F2 observations), cross loop shots, and second loop sharing observations with the first traverse loop. The traditional traverse does not provide for cross loop shots and as a result does not list them in the traverse data. When you create the Least Squares data file, however, the cross loop observations become a part of the data set.

The focus of this exercise is the method of entering traverse data from a field book. The exercise contains two traverses that share points. You perform the analysis of the points by doing a Least Square Network Adjustment.

After you complete this exercise, you will be able to:

- Create observation data for a traverse
- Use field data entry techniques for traverses
- Use traditional traverse reductions
- Use Least Squares traverse reductions
- Use traverse network reductions

EXERCISE SETUP

1. Start a new drawing and name the drawing *FbkTrv* and create a new project, FbkTrv, using the LDDTR2 prototype (see Figures 17–14 and 17–15). Select the OK button in each dialog box to exit.

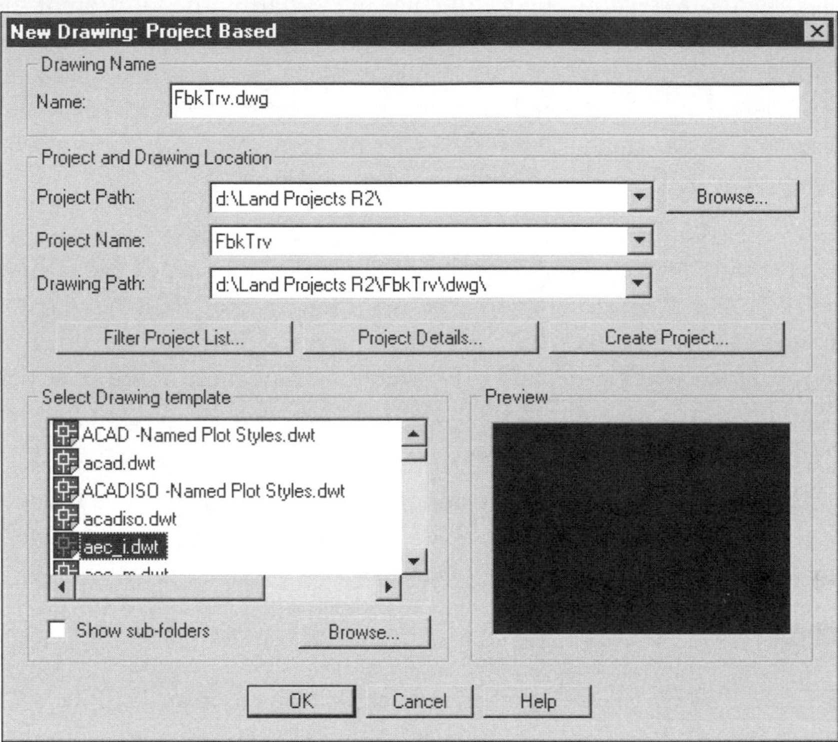

Figure 17–14

2. Select the OK button to dismiss the Point Database Settings dialog box.
3. The initialization process displays the Load setup dialog box. Select the LDD1-50 setup, then select the Load button to load the setup, and finally select the Finish button to exit the Load Setup dialog box. If you do not have the LDD1-50 setup, select the i50 setup. Select the OK button to dismiss the Finish dialog box.
4. Click the Save icon to save the drawing.
5. Change the menu to the Survey menu by using the Menu Palette routine of the Projects menu.
6. Copy the *trav1.fbk* and the *trav2.fbk* files from the CD to the Survey folder of the FbkTrv project.
7. Print a copy of each field book.

EXERCISE

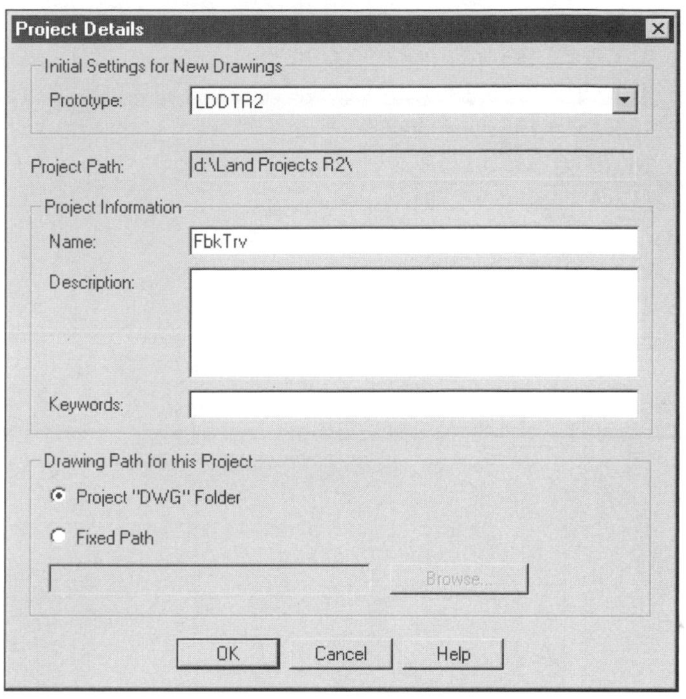

Figure 17–15

CREATING A LEAST SQUARES ANALYSIS DATA FILE

The traverse contains six legs with several cross loop observations (see Figure 17–16). The data for the loop is a Field Book file.

```
NE 6 5000.0000 5000.0000 IP
STN 6
AZ 6 5 0.0000
BS 5
F1 5 0.0000 1060.58 TRV-PT
F1 1 67.3735 619.85 TRV-1
F2 5 180.0002 1060.58 TRV-PT
F2 1 247.3745 619.85 TRV-1
F1 5 0.0000 1060.58 TRV-PT
F1 1 67.3736 619.85 TRV-1
F2 5 179.5953 1060.58 TRV-PT
F2 1 247.3731 619.85 TRV-1
STN 1
BS 6
F1 6 0.0000 619.86 IP
F1 2 133.4632 196.96 TRV-2
```

```
F2 6 179.5959 619.86 IP
F2 2 313.4623 196.96 TRV-2
F1 6 0.0000 619.86 IP
F1 2 133.4627 196.97 TRV-2
F2 6 180.0004 619.86 IP
F2 2 313.4628 196.97 TRV-2
F1 5 77.3424 1004.29 TRV-PT
F1 4 94.2740 941.78 TRV-4
F2 5 257.3420 1004.31 TRV-PT
F2 4 274.2741 941.77 TRV-4
```

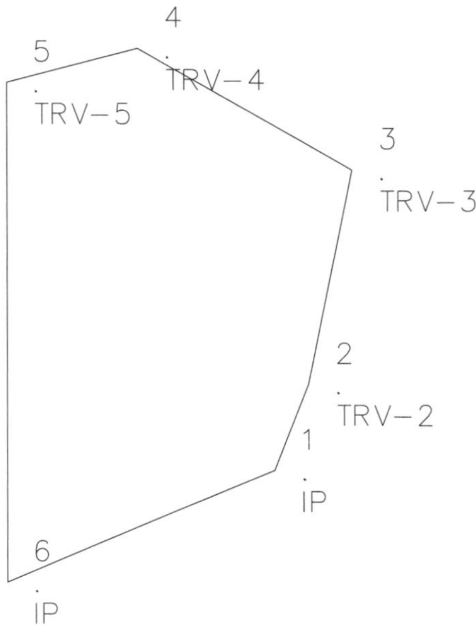

Figure 17–16

When the surveyor is at point number 1, he or she observes points across the traverse loop. The cross loop observations do not show in the traverse loop data; however, the observations do show when you create a data file for Least Squares or a network.

1. Use the Import Field Book routine of the Data Collection/Input menu and when prompted, answer Yes to the questions.

By answering Yes to the questions, you destroy any existing points and observations.

The Import Field Book routine will report some tolerance errors on some of the survey observations.

2. Press ENTER to average all of the errors during the importing of the file.

3. Define a traverse loop from the observation database. Use the Define Loop routine in the Traverse Loops cascade of the Analysis/Figures menu.
4. View the traverse data in the Traverse Editor. After reviewing the data, exit the Traverse Editor.
5. Create the Least Squares data for this traverse by running the Create Input File routine of the Individual Loops cascade of the Analysis/Figures menu.
6. View the Least Squares data with the Edit Input File routine of the Individual Loops cascade.

The only known point is point 6 and it is the only point without a question mark in front of it. The other points have question marks in front of them indicating that Least Squares will solve for their location. All of the angle-distance data in the field book is now distance and angle information in the data area of the file.

The data file now contains the cross loop observations. Least Squares is always a part of a traverse analysis, and as a result of this assumption, the data has to go through the traverse definition process to get to the Least Squares and Network routines.

If the surveyor ran an adjacent traverse to the current traverse, the surveyor has two options: to evaluate both loops individually or together as a network. The Process Input File of the Individual Loops cascade of the Analysis/Figures menu will process each loop. The Process Input File of the Networks cascade of the Analysis/Figures menu evaluates both traverses together.

EXERCISE

CREATING SECOND TRAVERSE DATA SET

The *trav2.fbk* file contains a second traverse that includes two points (5 and 6) from the first traverse.

When importing the second loop, the routine asks about removing Cogo and Survey points and deleting the observation database and figures. You do not need to delete anything, especially the observation database. The data set needs a combination of the observations from both loops to be able to create the data for the network.

The second traverse contains seven legs with several cross loop observations. The data for the loop is a Field Book file. The file, *trav2.fbk,* is in the Survey folder on the CD that accompanies this book.

1. Import the *trav2.fbk* field book. Answer No to the questions.

```
Command:
Erase all COGO points in Database [Yes/No]? <No>:
Erase all COGO points created by SURVEY [Yes/No]? <No>:
Erase all existing observations [Yes/No]? <No>:
Erase all figures (Yes/No)? <No>:
```

2. Use the Define Loop routine of the Traverse Loops cascade of Analysis/Figures menu. The Define Loop routine does not see the new loop; therefore, you need to manually define the new traverse loop.

3. Define loop number 2 (two). Change the loop values to those shown in Figure 17–17.

   The values are:
   Initial BS point: 8
   Occupied Points: 11,6,5,9-10,7-8,11
   Final FS Point: 6

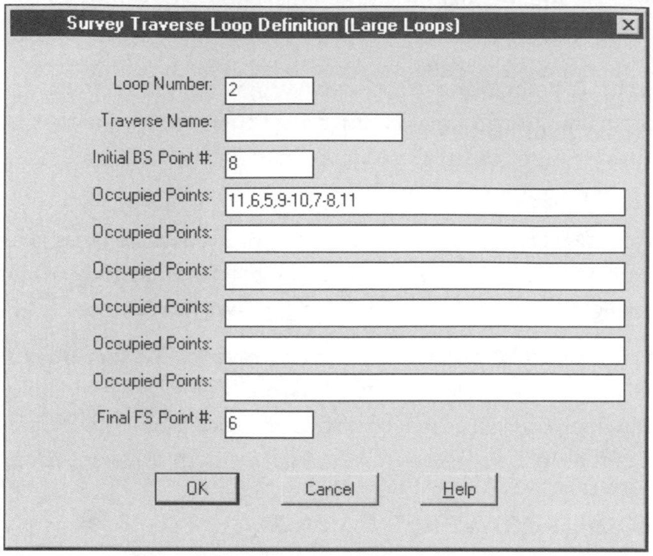

Survey Traverse Loop Definition (Large Loops)

Loop Number: 2
Traverse Name:
Initial BS Point #: 8
Occupied Points: 11,6,5,9-10,7-8,11
Occupied Points:
Occupied Points:
Occupied Points:
Occupied Points:
Occupied Points:
Final FS Point #: 6

OK | Cancel | Help

Figure 17–17

4. View the traverse data in the Traverse Editor. After viewing the data, exit the Traverse Editor.
5. Create Least Squares data by choosing the Create Input File routine of the Individual Loops cascade of the Analysis/Figures menu.
6. View the Least Squares data by selecting the Edit Input File of the Individual Loops cascade of the Analysis/Figures menu. The only known points are points 6 and 11, and they are the only points without a question mark in front of them. The other points have question marks in front of them indicating that Least Squares will solve for their location. All of the angle-distance data in the field book is now distance and angle information in the data area of the file.

CREATING A NETWORK DATA FILE

Now two traverse loops are a part of the project and share points. From this data the next step is to create a network data file. This file contains the data for both traverse loops.

1. Use the Create Input File routine of the Networks cascade of the Analysis/Figures menu to create a data set using traverse 1 and 2 data.

EXERCISE

2. Review the network data by choosing Edit Input File from the Networks cascade of the Analysis/Figures menu. The routine uses any defined traverses and creates a network data file.

You now view a data set that is a combination of the two individual loops. The file is a Least Squares data file.

You can control what points or lines are to be held by changing the locations of the question marks in the file. If you want to hold a point, remove the question mark in front of the NE line containing it to hold it. If you want to hold a line of the traverse, remove the question marks in front of the two points defining the line to hold the line.

EVALUATING A NETWORK DATA FILE

1. Run the Process Input File routine of the Networks cascade of the Analysis/Figures menu. Make sure you use the 2D Adjustment type. Review the output of the analysis. Some of the observations fail, but the overall traverse passes. You will need to decide if the failure of the individual observations is a result of blunders or bad data.
2. Import the adjusted coordinates and error ellipses using the Import Adjusted Coordinates routine of the Networks cascade of the Analysis/Figures menu.
3. Zoom in on some of the points to view its error ellipse.
4. Compare your results to Figure 17–18.
5. Click the Save icon and save the drawing.

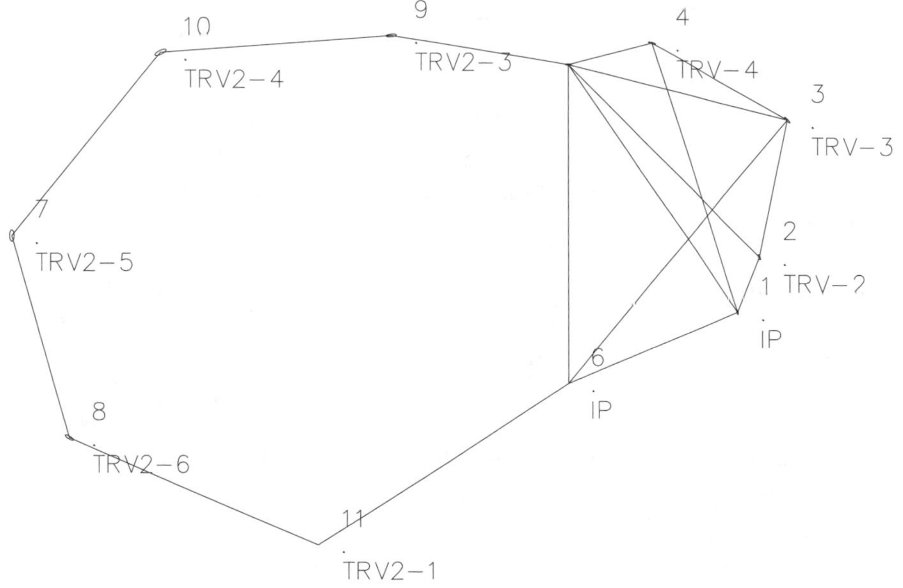

Figure 17–10

UNIT 3: ELECTRONIC DATA TRAVERSE

The last method, traverse data from an electronic field survey, is the focus of this unit. The exercise for this unit contains a traverse with sideshots. The survey contains errors and the loop definition needs editing before completing the evaluation of the data.

TRAVERSES AND SIDESHOTS

Many times a survey contains additional observations of points at a setup that are not a part of the traverse. Surveying calls these points sideshots. The adjustment of these points is relative to the adjustment of the setup point itself. Therefore, the location of a sideshot is totally dependent upon what happens with the traverse adjustment of the setup point.

The Traverse routines of Survey handle sideshot data. Sideshot data coexists with the data for any traverse and any method for traversing creates sideshots.

The Define Loop routine of the Traverse Loops cascade of the Analysis/Figures menu tries to follow the logical progression of a traverse. If the surveyor makes a mistake or creates dead ends in the field data, the traverse loop reflects these errors. If a loop contains errors, it will need editing to correct the progression of points. This assumes that the surveyor correctly observes the points in the field. If the data set does not contain the correct observations of the traverse points, there is no correction available.

The last traverse contains a survey with traverse and sideshot information. The surveyor, however, creates a dead end in the traverse data. The Define Loop routine is unable to correctly interpret the stations along the traverse loop. As a result, it is necessary to edit the correct progression of occupied points.

The surveyor indicates the points of the traverse in Figure 17–19. The points shown on this map are not the correct definition of the traverse. The surveyor either missed shooting the correct points or noted the wrong loop definition.

After you import the Field Book and run the Define Traverse loop routine, the loop definition is wrong. You need to view the Field Book file to solve the loop definition problem.

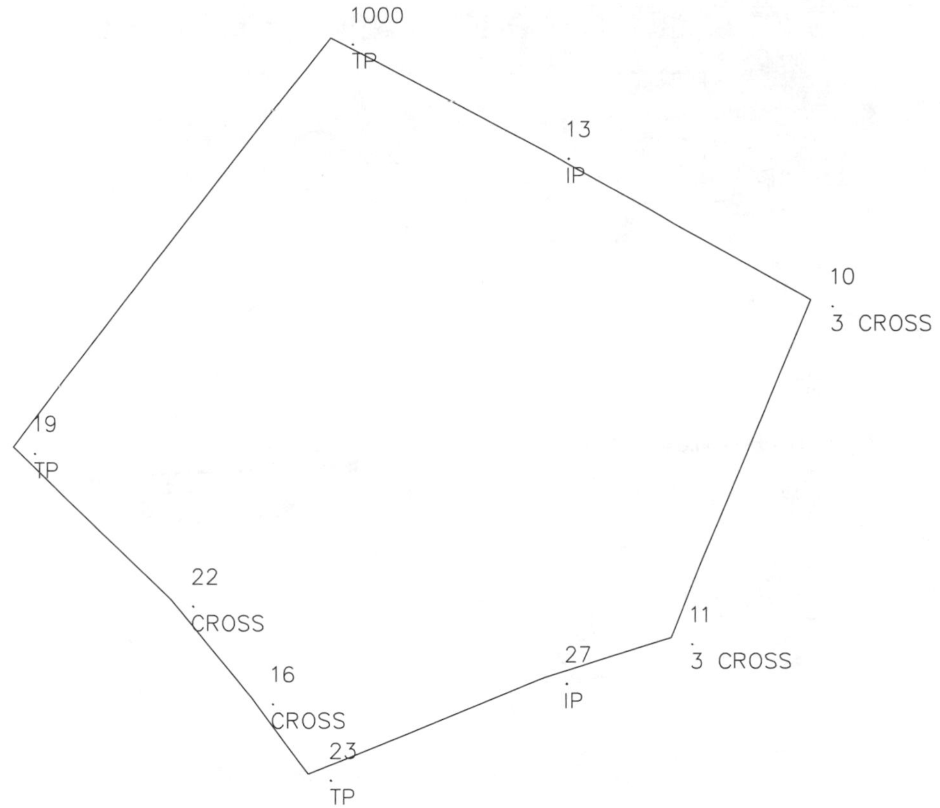

Figure 17–19

Exercise

EXERCISE

After you complete this exercise, you will be able to:

- Create observation data for a traverse
- Use field data entry techniques for traverses
- Use traditional traverse reductions
- Use Least Squares traverse reductions
- Edit a traverse loop definition to correct field errors

CREATING AND WORKING WITH TRAVERSES

1. Start a new drawing and name it *Oloop* and create a new project, Oloop, using the LDDTR2 prototype (see Figures 17–20 and 17–21).

New Drawing: Project Based
Drawing Name
Name: oloop.dwg
Project and Drawing Location
Project Path: d:\Land Projects R2\ Browse...
Project Name: oloop
Drawing Path: d:\Land Projects R2\oloop\dwg\
Filter Project List... Project Details... Create Project...
Select Drawing template
ACAD -Named Plot Styles.dwt
acad.dwt
ACADISO -Named Plot Styles.dwt
acadiso.dwt
aec_i.dwt
Show sub-folders Browse...
Preview
OK Cancel Help

Figure 17–20

Project Details
Initial Settings for New Drawings
Prototype: Default (Feet)
Project Path: d:\Land Projects R2\
Project Information
Name: oloop
Description:
Keywords:
Drawing Path for this Project
Project "DWG" Folder
Fixed Path
Browse...
OK Cancel Help

Figure 17–21

2. Select the OK button to dismiss the Point Database dialog box.

3. Load the LDD1-40 setup by selecting the Load button in the Load Setup dialog box. If you do not have the LDD1-40 setup, use the i40 setup. After loading the setup, select the Finish button to exit the Load Setup dialog box.

4. Select the OK button to dismiss the Finish dialog box.

5. If necessary, change the menu to the Survey menu by using the Menu Palette routine in the Projects menu.

6. Copy the *oloop.raw* file from the Survey directory found on the CD that comes with this book. Put the file into the Survey directory of the oloop project. Rename the copy to *oloop.sdr* and place another copy of the *oloop.raw* file into the folder. Print out either file to view the data.

The data contains Face1 and Face2 shots and an average coordinate for the point. The traverse was done in traverse mode, and the data collector actually reduced the observations into final coordinates. Lines 1 and 2 contain the observation and line 3 contains the reduced coordinates (09MC). The same is true for lines 5 and 7 and the mean coordinate (09MC) on line 8. The following is an excerpt from the file.

```
09F100130010603.15000089.4086111119.0155553 CROSS
09F200130010603.150000270.591111299.0183333 CROSS
09MC00130010603.11788690.0000000119.0156943 CROSS
13MDRpt dist: 250.060   <Null>
09F100130014250.06000089.7955555358.577777IP
13MDRpt dist: 250.050   <Null>
09F200130014250.050000270.206388178.579166IP
09MC00130014250.05339390.0000000358.577222IP
```

The TDS conversion routine will produce only the mean coordinates from the data. One way to ignore the mean coordinate (MC) entry in the file and to get the F1/F2 observations is to convert the file with the Convert Pre-7.6 Raw Files routine of the Data Collection/ Input menu. The only way to collect a traverse and use the TDS conversion routine is to collect the data as field observations and do not toggle on the traverse routine within the data collector.

6. Create a field book, olooptds, with the conversion routine in the Data Collection Link cascade of the Data/Collection/Input menu. Use Figure 17–22 to perform the conversion. Exit the converter and TDS link.

7. Use the Edit Field Book routine to view the resulting field book.

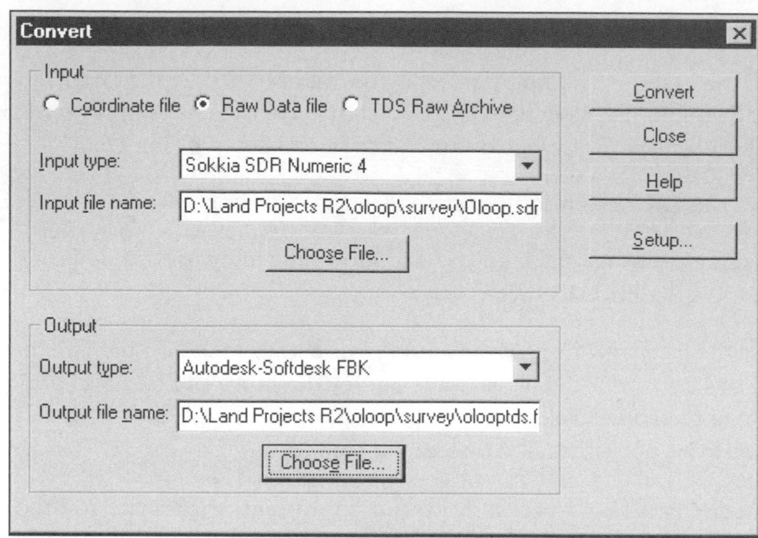

Figure 17–22

The excerpt below shows that the F1 and F2 observations are now just an F1 observation with the resolved distance.

```
STN  13 0.000 "IP"
PRISM  0.000
BS  10 119.01010
! ! Rpt dist: 603.150   <Null>
PRISM  0.000
F1 VA 10 119.01010 603.118 90.00000 "3 CROSS"
! ! Rpt dist: 250.060   <Null>
```

Later in the file, the points observed from the traverse loop are entered as NESS or sideshots with reduced coordinates. None of this data will create the observations we need to reduce the traverse.

```
STN  16 0.000 "CROSS"
PRISM  0.000
BS  22 320.27450
F1 VA 22 320.27450 250.480 89.31390 "CROSS"
NE SS  50 323.8662 750.4773  NULL  "BLGX"
NE SS  51 288.4465 706.2670  NULL  "BLGX+.08"
NE SS  52 275.5914 692.9580  NULL  "BLGX+.08"
NE SS  53 259.9325 705.5896  NULL  "BLGX"
```

8. Create a Field Book file from the *oloop.raw* file using the Convert Pre-7.6 Raw Files routine of the Data Collection/Input menu.
9. View the field book by selecting the Edit Field Book routine of the Data Collection/Input menu.
10. Import the field book using the Import Field Book routine of the Data Collection/Input menu to create the point and observation databases. Press ENTER to average any errors in the data.
11. Use the Define Loop routine of the Traverse Loops cascade of the Analysis/Figures menu.

Notice that the traverse loop does not match the diagram of the loop the surveyor drew (see Figure 17–19). The loop stops at point 16 and does not continue to points 23, 27, 11, and 10 (see Figure 17–23). For some reason, the Define Loop routine does not find a logical progression from point 16 to point 23.

Survey Traverse Loop Definition (Large Loops)

Loop Number: 1
Traverse Name:
Initial BS Point #: 11
Occupied Points: 10,13,1000,19,22
Occupied Points:
Occupied Points:
Occupied Points:
Occupied Points:
Occupied Points:
Final FS Point #: 16

OK | Cancel | Help

Figure 17–23

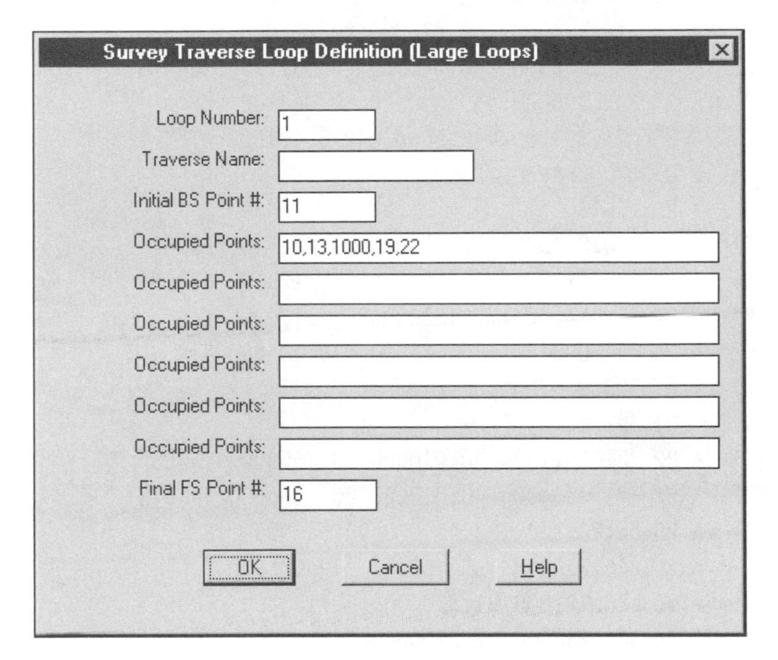

12. Review the Field Book file by using the Edit Field Book routine of the Data Collection/Figures menu. You may want to print out the file to better view the traverse data. Look for the stationing on point 22 and backsighting to point 19 in the file.

It is most important to view the activity between the occupation of point 16 and foresighting point 23. From point 22, the surveyor sights both points 16 and 23, and then stations the instrument on point 16. From point 16, however, the surveyor never foresights to point 23. As far as the traverse definition routine is concerned, the traverse comes to an end at point 16 because the surveyor never sights the next station on the loop from this point. It is only from point 22 that there are sights to point 23 (the next point on the traverse loop). In order for point 16 to be a part of the traverse, the crew needed to observe point 23 from the setup on point 16. To make the survey work, we need to drop point 16 from the traverse loop.

The following is that area in the Field Book.

```
STN 22 0.000000 "CROSS"
BS 19 313.284850
FC1 VA 19 313.284700 437.150000 90.322900 "TP"
FC2 VA 19 133.285000 437.160000 269.272600 "TP"
FC1 VA 16 140.274000 250.470000 90.285000 "CROSS"
FC2 VA 16 320.275300 250.470000 269.311900 "CROSS"
FC1 VA 23 141.390800 440.000000 89.465800 "TP"
FC2 VA 23 321.391700 440.000000 270.131500 "TP"
STN 16 0.000000 "CROSS"
BS 22 320.275150
FC1 VA 22 320.274500 250.460000 89.421400 "CROSS"
FC2 VA 22 140.275800 250.460000 270.173100 "CROSS"
FC1 VA 24 51.084600 170.120000 89.461500 "PK"
FC2 VA 24 231.090300 170.120000 270.134000 "PK"
STN 23 0.000000 "TP"
BS 22 321.391750
FC1 VA 22 321.391100 440.010000 90.260800 "CROSS"
FC2 VA 22 141.392400 440.010000 269.335500 "CROSS"
FC1 VA 25 66.245300 183.400000 89.515000 "IP"
FC2 VA 25 246.250200 183.400000 270.075600 "IP"
FC1 VA 26 128.251800 31.3700000 86.103200 "CROSS"
FC2 VA 26 308.253000 31.3700000 273.493600 "CROSS"
FC1 VA 27 68.035200 510.460000 89.281200 "IP"
FC2 VA 27 248.040000 510.470000 270.313300 "IP"
```

To correct the loop definition, edit it and correct the station order. To change the definition, edit the list of occupied stations and the final foresight in the dialog box. The Survey Traverse Loop Definition dialog box should look like Figure 17–24 after you have edited the stations and foresight point.

EXERCISE

13. Change the loop definition to match Figure 17–24. Use the Edit Loop Definition routine in the Traverse Loops cascade of the Analysis/Figures menu. Select the OK button to exit the editor.

Survey Traverse Loop Definition (Large Loops)

Loop Number: 1

Traverse Name:

Initial BS Point #: 11

Occupied Points: 10,13,1000,19,22-23,27,11

Occupied Points:

Occupied Points:

Occupied Points:

Occupied Points:

Occupied Points:

Final FS Point #: 10

OK | Cancel | Help

Figure 17–24

14. Check the loop adjustments to view the quality of the loop. Make sure to toggle on Balance Angles and 2D Adjustment Type to view the potential adjustments to the loop. After reviewing the adjustments, select the OK button to exit the dialog box.

15. Choose the adjustment you want to work with and if you like, print out the results.

16. Adjust the points, sideshots, and import the traverse figure into the drawing using the Process Input File routine of the Individual Loops cascade of the Analysis/ Figures menu. After you adjust the points, the routine prompts you to update the point database, draw the loop figure, and whether to adjust the sideshot points; answer Yes to these questions. Your drawing should now look like Figure 17–25.

If you adjusted the traverse with the Least Squares routines, the drawing includes error ellipses around the points of the traverse.

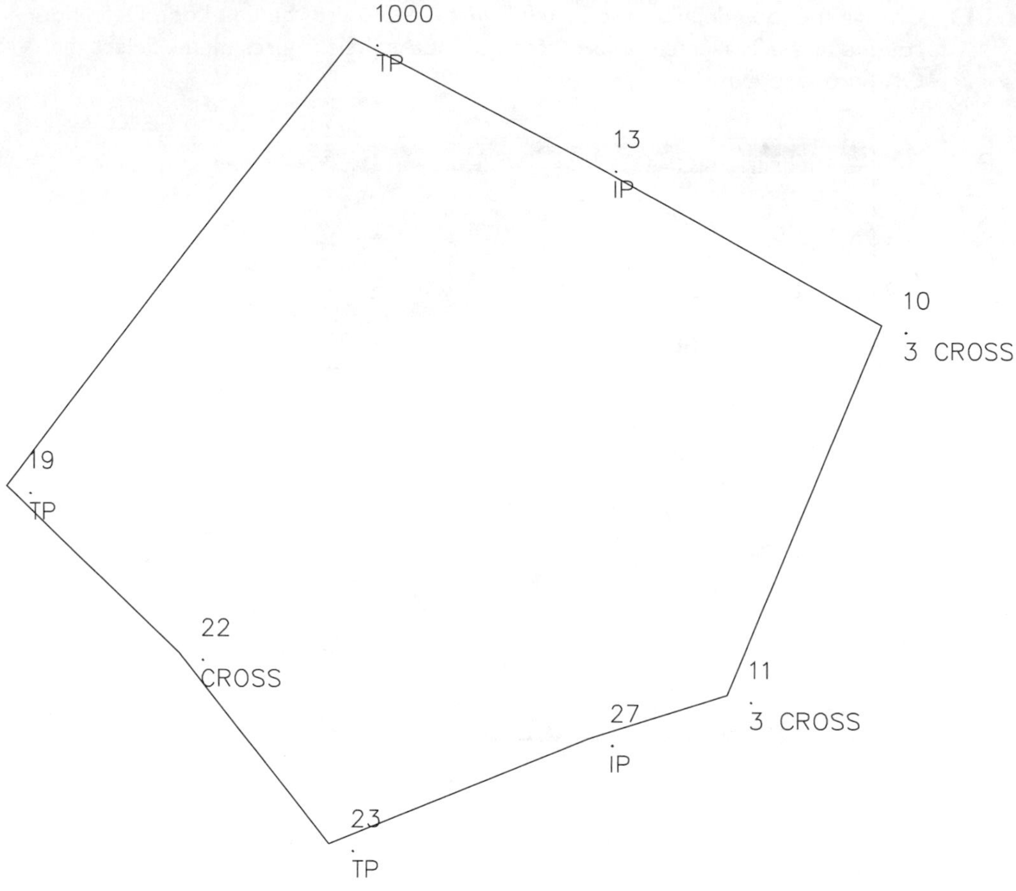

Figure 17–25

EXERCISE

This section covered the traverse routines of the Survey add-in. There are several ways to input the data: the Traverse Entry routine, the Survey Command Prompt, the Traverse Editor, the field book, and by converting a data collector file. The requirement for collecting a traverse in a data collector is not to turn on the traverse routine in the collector. If you do, the collector will reduce the observations and solve the traverse. To use the traverse routine in the Survey add-in, you must observe in topographic mode, or if possible, convert the file with the Convert Pre-7.6 Raw File conversion routine. You will need to experiment and decide where the traverse reduction is to take place.

The next section reviews the changes to Land Development Desktop, Civil Design, and Survey with the 2i release.

18

Land Development Desktop Extensions and 2i

After you complete this section, you will be able to:

- Use the iX extension
- Select and use a new file translator
- Use the slope annotation tool
- Evaluate the LDD2i release

Between the time of release of Land Development Desktop 2 and Land Development Desktop 2i, there were three feature releases. The three releases are Autodesk's Internet Extension (iX), the dtX (data) transformers, and the slope annotation utility. These releases reflect the Internet focus of Autodesk and the continued interest in developing and delivering simple but powerful tools to the user. All of these tools require Land Development Desktop 2 with patches. The three extensions became part of Land Development Desktop 2i. The 2i release features the ability to read DEM (Digital Elevation Models) as surface data; new set of standard translators for Mid/Mif, ArcView shapes, ArcInfo coverages, and Dgn design files; and a balancing routine for the cut and fill volume of a grading object.

UNIT 1

The focus of the first unit of this section is the Internet Extension (iX) of Autodesk. The iX displays AutoCAD Today, Autodesk Point A, or an HTML based CAD Manager's bulletin board and a new open drawing section.

UNIT 2

The second intermediate release consists of the dtX transformers. The new file translation engine is from Safe.com and includes two additional downloadable translators, one for SDTS and VML files. The file translators are the focus of the second unit of this section.

UNIT 3

The third unit focuses on the slope annotation tool. This tool allows you to annotate slopes between two points. The slope can be between two selected points on a surface or the slope from a single point on a surface.

UNIT 4

The Land Development Desktop 2i release is the focus of the last unit of the section. The 2i release is Land Develop Desktop 2 with the three intermediate tool sets plus some new routines. The new features in 2i are the ability to read in DEMs (Digital Elevation Models) as a surface, new basic file translators, and a balancing routine for the volume of a grading object.

UNIT 1: THE AUTODESK INTERNET EXTENSION

The need to communicate data, information, and drawings in a design project is a high priority. Whether the communication tool is the Internet or an intranet, the ability to publish (plot), view (dwf), and post messages was formerly something done outside of the Land Development Desktop environment. With the Internet Extensions, Autodesk introduces tools to ease the creation of vital Internet formats from within the Desktop. The installation of the Internet Extension creates a toolbar that reflects the new tools and abilities of the Desktop (see Figure 18–1). The icons in the toolbar are, from left to right, AutoCAD Today, Meet Now, Hyperlink, Publish to the WEB, and eTranmit.

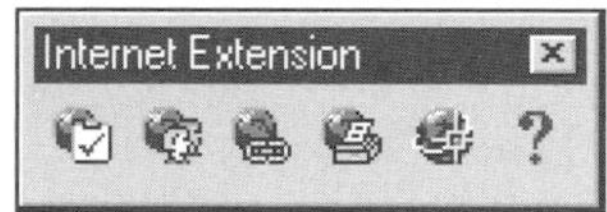

Figure 18–1

AUTOCAD TODAY

AutoCAD Land Development Desktop Today is an Internet/intranet information page (see Figure 18–2). The types of information shown in the dialog box depend on two things, the format of the dialog box contents and the type of Internet connection. The dialog box can be turned off or limited to the new open drawing section and CAD Manager's bulletin board. The All Today dialog box contains the open drawing and bulletin board areas as well as hyperlinks to Autodesk Point A, industry-specific information links, Buzzsaw.com, product catalogs, express tools request and purchase, and other features (see Figure 18–3).

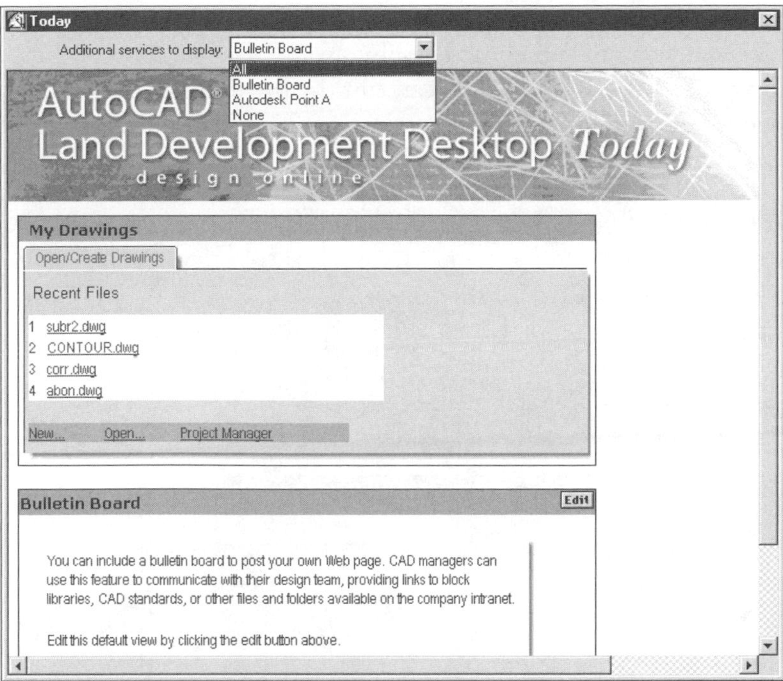

Figure 18–2

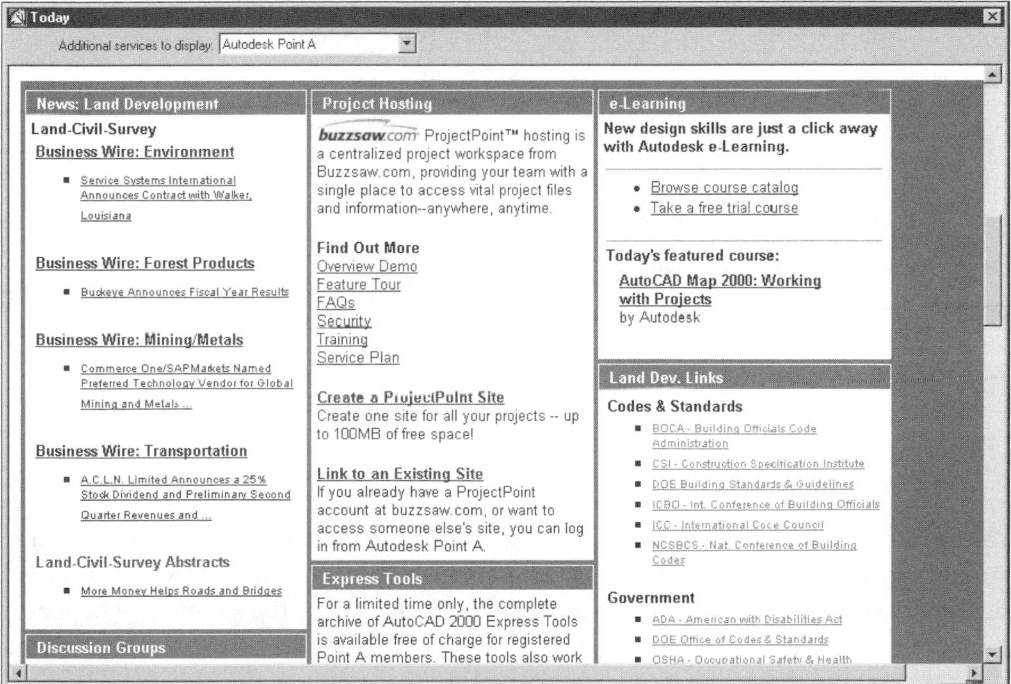

Figure 18–3

The Today view can be limited to a bulletin board that includes drawing use history and the CAD Manager's bulletin board (see Figure 18–4). The contents of the bulletin board are in a file, *editcadmgr.htm*. You can edit the file to change its contents so that everyone who is pointing to the server location is able to read the latest office-specific or project-specific information.

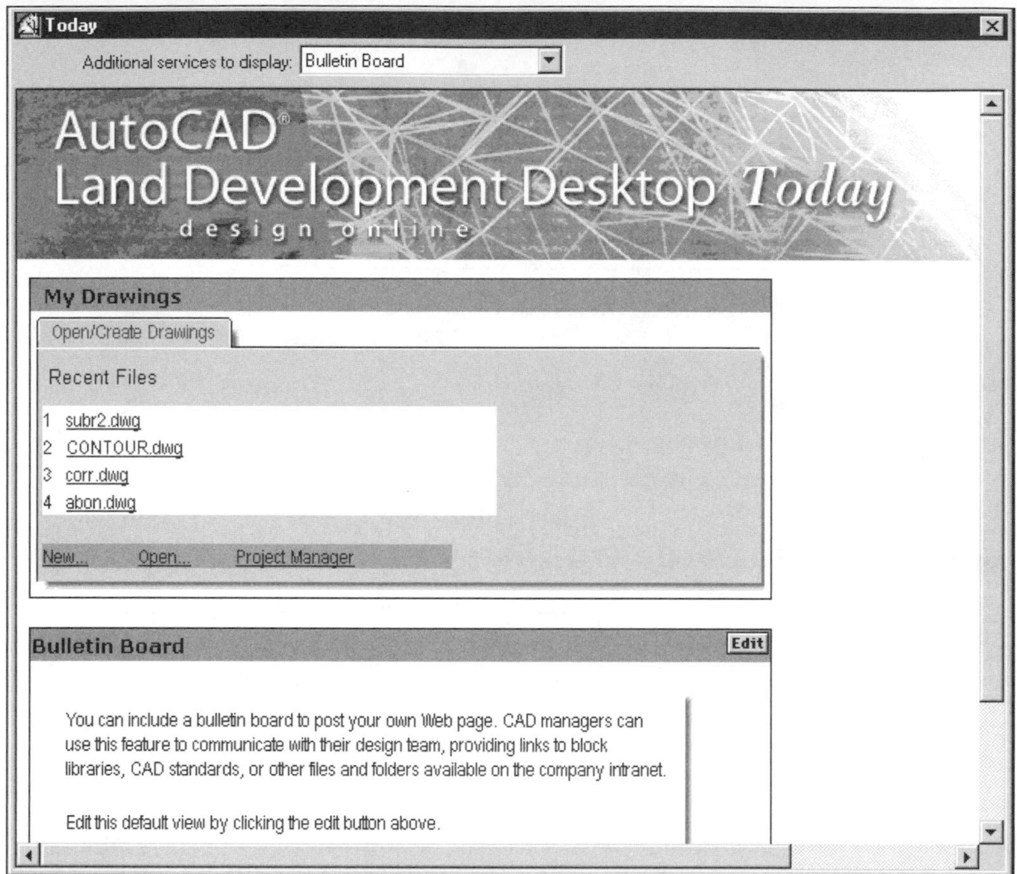

Figure 18–4

MEET NOW

The Meet Now program allows you to connect to other team members via a Net Meeting server provided by Autodesk. When you click the Meet Now icon, the routine launches into a wizard to set up your side of the meeting and then places you into a meeting that is currently taking place. Again, the quality of the experience depends upon the type of connection you have to the Internet. The Meet Now tool allows you to quickly set up and create a meeting with other team project members.

INSERT A HYPERLINK

The Insert a Hyperlink routine is an existing Desktop command. This routine allows you to place a hyperlink to any Desktop object. When you select an object and press the ENTER key, the routine displays the Insert Hyperlink dialog box (see Figure 18–5). After setting the link and the displayed text, you select the OK button to create the link and exit the dialog box.

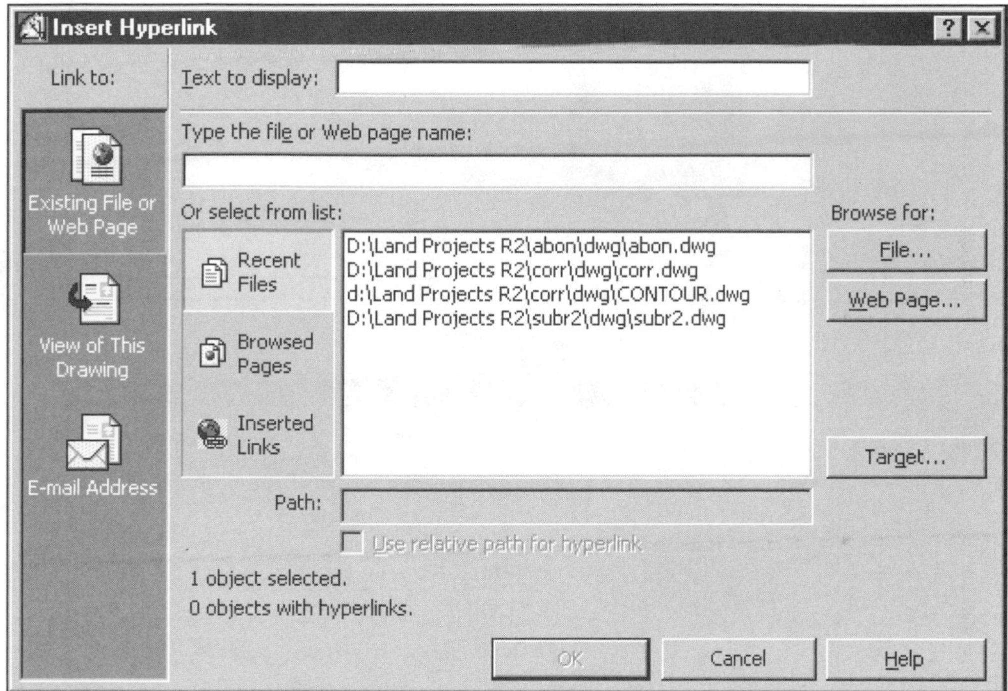

Figure 18–5

PUBLISH TO THE WEB

The Publish to the Web routine allows for the creation of an HTML page from selected Desktop drawings. When you click this icon, a wizard displays and walks you through the steps of publishing a Web page. The first panel asks you to identify whether this is a new Web page or an edit of an existing Web page (see Figure 18–6).

The second panel sets the name and description of the Web page (see Figure 18–7).

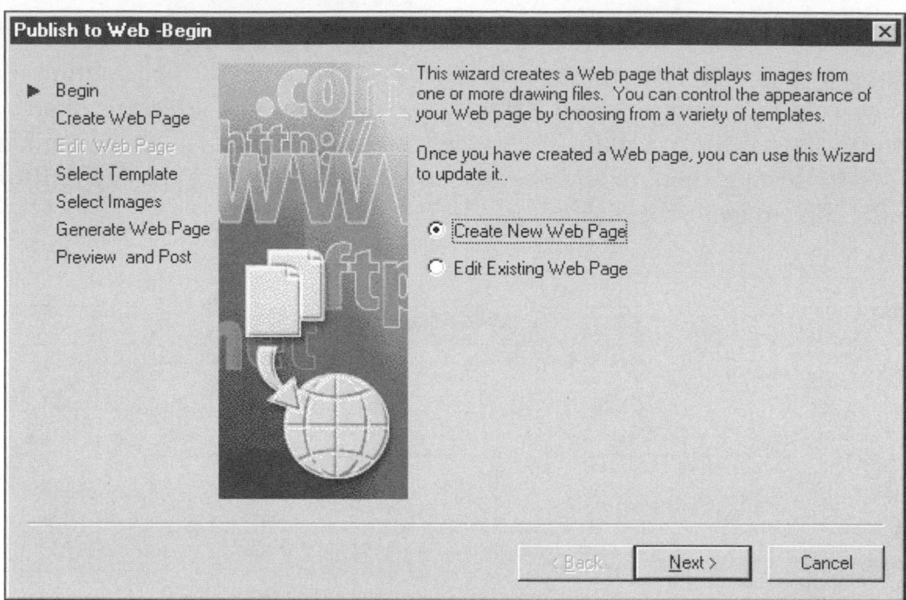

Figure 18–6

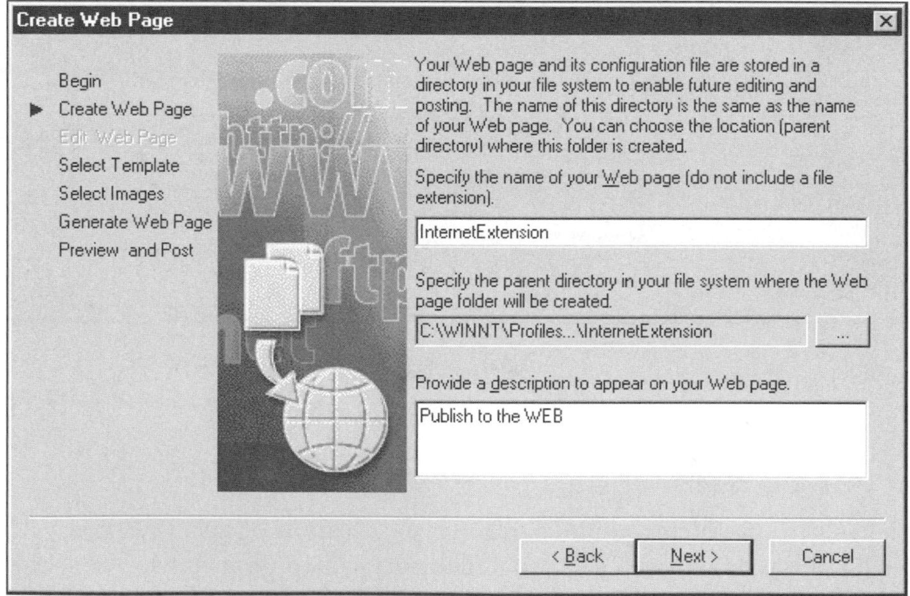

Figure 18–7

The third panel defines the template for the Web pages (see Figure 18–8).

The fourth page sets what drawing to use to create the Web page. The next panel controls how the images are produced (see Figure 18–9).

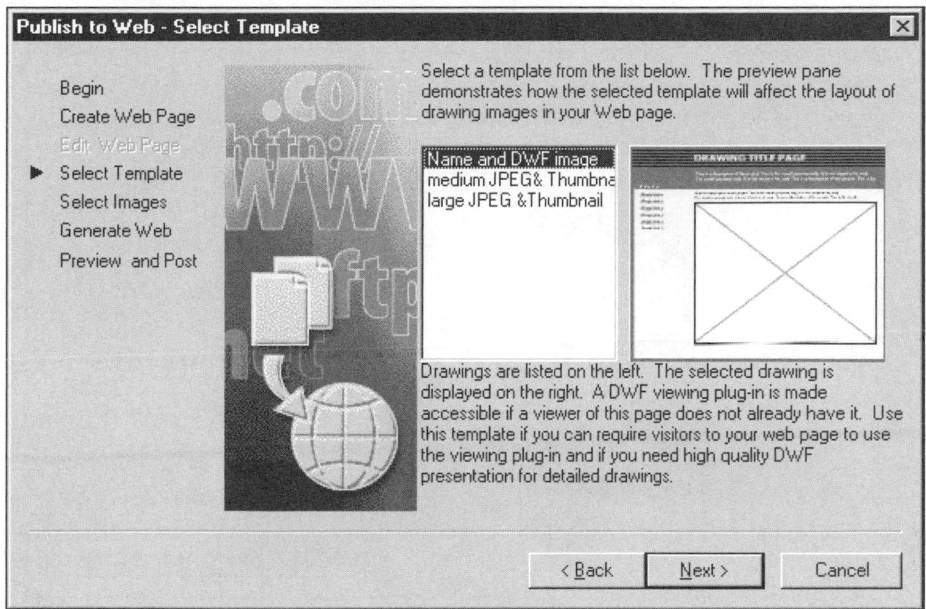

Figure 18–8

Publish to Web - Select Images

Begin
Create Web Page
Edit Web Page
Select Template
Select Images
Generate Web Page
Preview and Post

To add an image to your Web page, select a drawing and one of its layouts. Enter a label (the caption that appears under the drawing on your Web page) and optional description and then choose Add. To update an image, select it in List of Images, specify new settings (such as a different layout), and choose Update.

Image settings
Drawing:
D:\Land Pr...\subr2.dwg
Layout
Model
Label:
Site Topography
Description:

Add ->
Update ->
Remove
Move Up
Move Down

Image list
Model

< Back
Next >
Cancel

Figure 18–9

The final panel allows you to preview the results and post the results to the FTP site (see Figures 18–10 and 18–11).

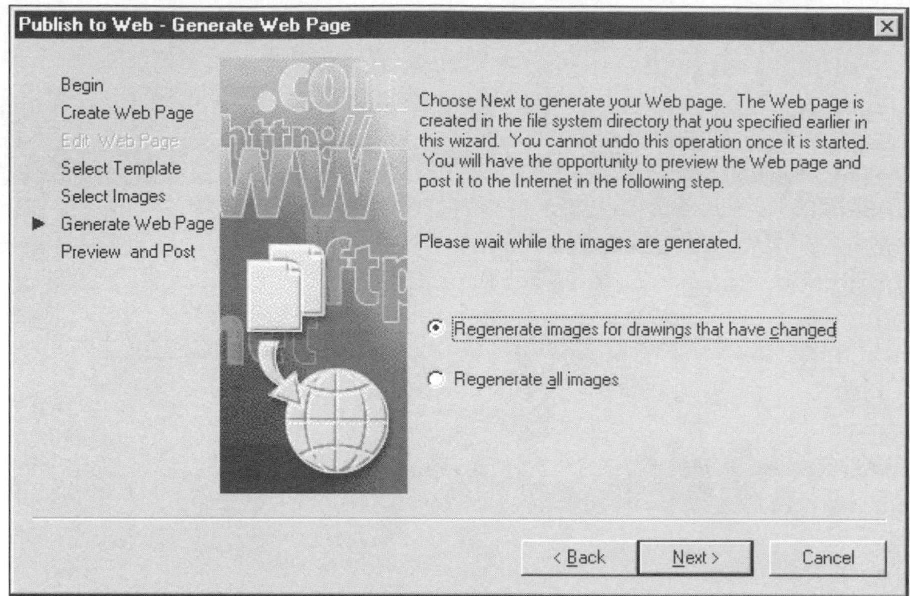

Figure 18–10

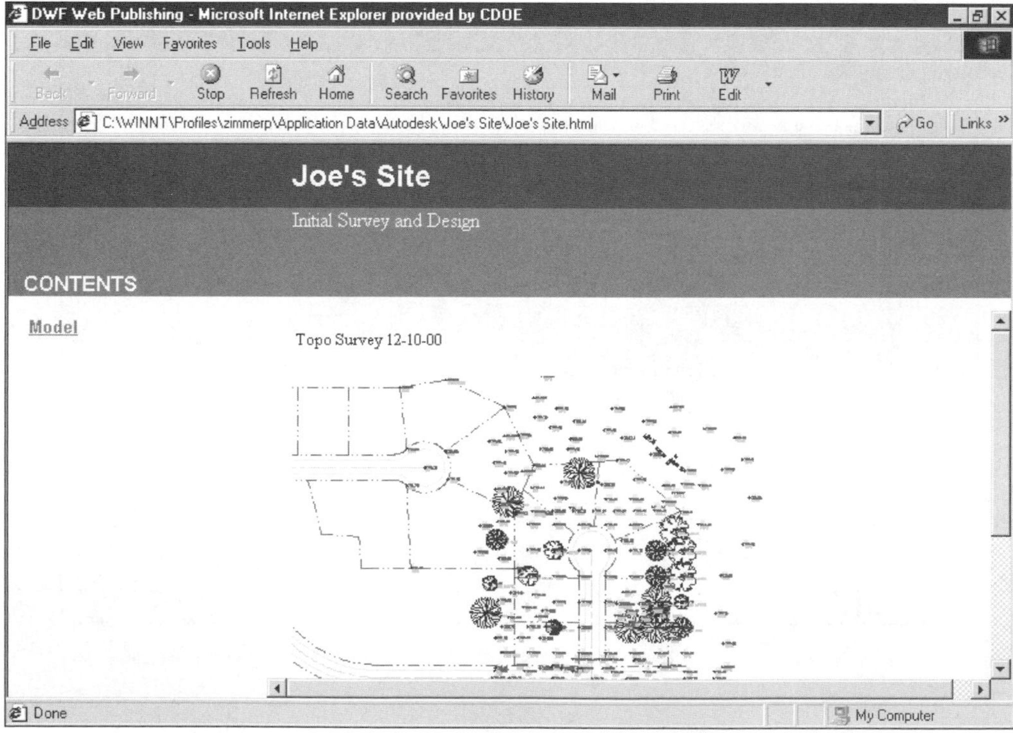

Figure 18–11

ETRANSMIT

The eTransmit routine allows for the creation of an email containing all of the files attached to the current drawing. When you transmit a drawing to a client, the client may not have the same fonts, xref directory structure, or images. To make sure you are sending to the client the correct files, eTransmit will create a zipped file containing all of the attached files. You can select which files to include or exclude from the zipped file. The eTransmit routine will create a transmittal report that contains the file names, locations, and other pertinent file information as a report. The report can be included in the email you send to the client.

UNIT 2: DTX TRANSLATORS

The dtX translators require you to download two parts to function. The first is the translator engine. This engine is the framework for the downloaded translators. The second part is the translators themselves. Additional translators are found on the Safe.com Web page or the Autodesk eStore.

You can purchase two translators to add to the capability of LDD. The two translators are SDTS and VML. The United States government publishes all spatial data in the SDTS standard. This standard includes raster, vector, and Digital Elevation Model (DEM) data. LDD can display the vectors for line work and use the DEM as surface data for a project. The VML translator allows you to take 3D vector data from the WEB and use it as data in LDD.

The largest suppliers of low cost or free data are the gisdatadepot.com and the USGS web sites. You can download files from these sites that have data on countries, states, and counties. For example if you go to the free data area of GisDataDepot, select United States, then statewide data, and then Illinois, you are able to download data collected on the state or county level. If you select the county level, you can select a county in the state and all of the information published on the site. For example Cook county Illinois has data on administrative boundaries, archaeology, Digital Elevation Models, geology, hydrology, land use, wet lands, public lands, survey control, transportation, utilities, and vegetation. All of these pieces of data are at the URL of www.gisdatadepot.com/catalog/ or you can go through PointA to the USGS site (Autodesk's portal). In Point A, select the GIS industry and in the Data & Imagery area select SDTS data from the USGS. When you click on the hyperlink you go to a new web page that contains headings for the types of data. The types of data are 1:250,000 and 1:24,000 DEMs and 1:2,000,000 1:100,000, and 1:24,000 DLG files. If you download a DEM from either the GisDataDepot or the USGS site the file is

a compressed file with an extension of .tar.gz. To extract the data from the file you must do two extractions. You can do each extraction with WinZip, however, you must make sure to toggle **OFF** tar file smart CR/LF conversion in the Miscellaneous tab of the Winzip configuration dialog box. The only additional piece of software is to have a copy of the public domain routine, SDTSEDEM to create the DEM file. Once you have this file you can use the DEM as a data source to a project in LDD.

UNIT 3: SLOPE ANNOTATION

The slope annotation routine of the Surface Utilities cascade of the Terrain menu labels the slope of a surface at a single point or between two points selected on the surface. The label is the result of the settings found in the Slope Display Settings dialog box (see Figure 18–12). One word of caution: when selecting two points, the routine does not evaluate any elevation changes between the two points. You can select two points and have a slope label appear that may ignore the fact that there is a swale between the two points (see Figure 18–13). The two points you select should represent a constant slope you want to annotate.

Figure 18–12

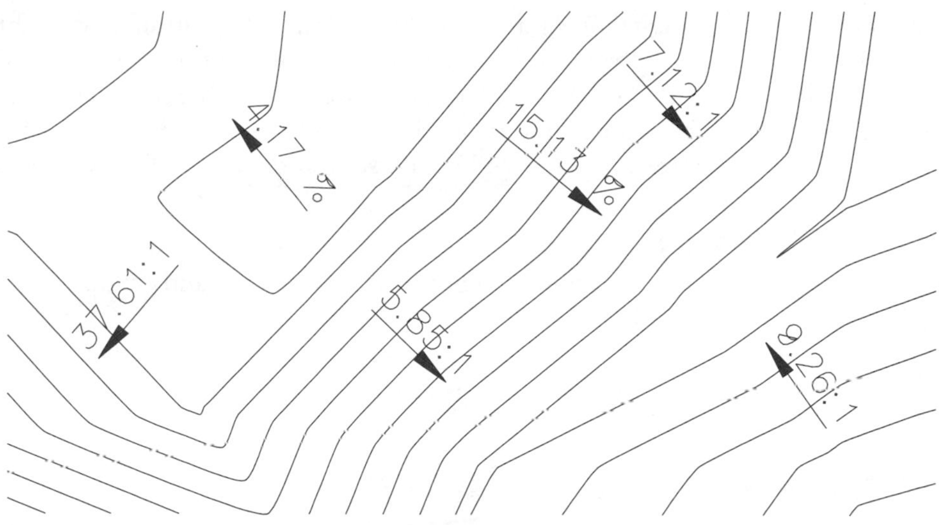

Figure 18–13

UNIT 4: LAND DEVELOPMENT DESKTOP 2i

Land Development Desktop 2i is the combination of the extension, the Autodesk Internet, dtX data translator, and the slope annotation with added feature enhancements. The enhancements include better plotter drivers, Internet-ready tools, and several performance enhancements. The major changes to Land Development Desktop include DEM data for surfaces and a balance volume grading routine.

FILE TRANSLATIONS

The original translators for ArcView Shapes, ArcInfo coverages, Mid/Mif, and DGN files are now by a set of more efficient routines. The translators are located in the Tools cascade of the Map menu. The translators are better, more efficient, and are able to handle data better than previous versions. The process or converting data files is the same. If your work is consistently translating data, you can define translator profiles that reflect the data you want to extract from the files. When you translate a file, all you need to do is to define or load a profile and the translator understands what you want to have extracted from the file. You are able to define layers and coordinate systems for the resulting data.

There are two translators available for Land Development Desktop 2i from the eStore. The first is a SDTS translator and the second is a VML translator. The SDTS translator reads SDTS files and creates spatial data that you can use in LDD. The vector data includes dlg (digital line graph) and DEM (digital elevation model) data. The dlg and DEM data can be used for preliminary studies or the basis for a GIS system. The USGS and the GIS Data Warehouse sites are excellent places to start you search for data. The data includes boundaries, transportation, flood plain, vegetation, and many more. The data can be for a state wide or county level study.

DEM SURFACES

The Terrain Model Explorer can now simply read in a DEM file and create a surface (see Figure 18–14) from the DEM data. The drawback to DEM files is their size. A DEM may be larger than the area of interest and create a very large surface file. To limit the scope and size of the DEM, use a surface boundary. These surfaces are ideal for preliminary site and hydrology studies.

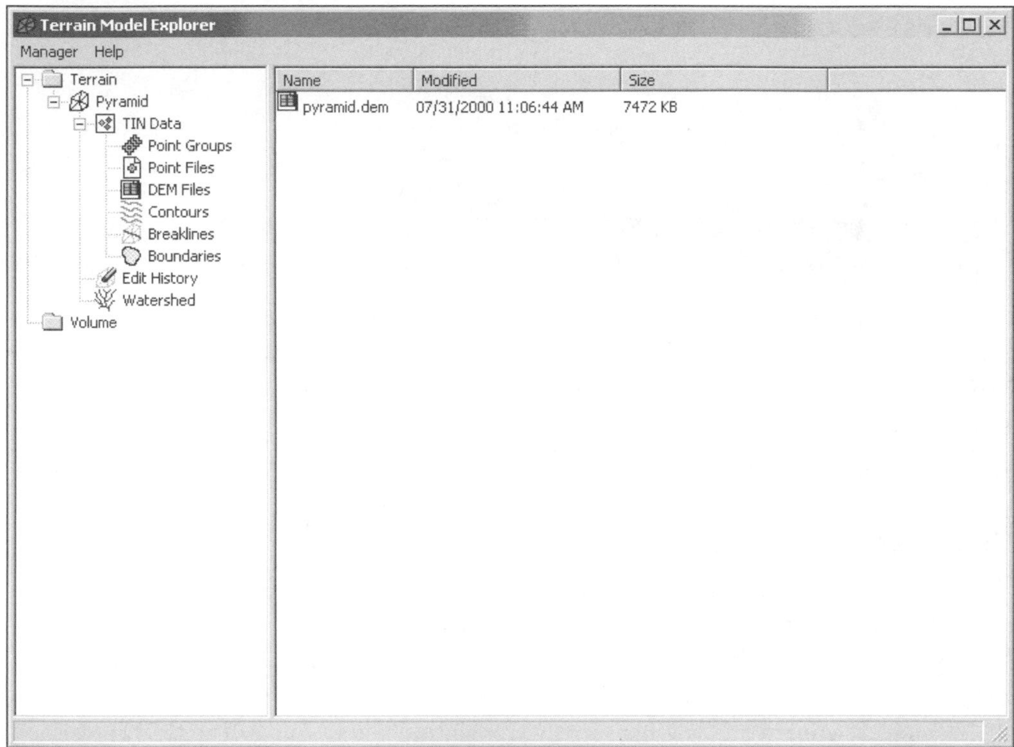

Figure 18–14

GRADING OBJECT VOLUME BALANCING

The ability to create grading objects and set their slopes, and elevations makes them a handy design tool (see Figure 18–15). However, in earlier releases it was difficult to balance the volumes created by the objects. The only method was to adjust the elevation of the object and ask for the resulting volume. Now the Balance

Grading Object routine allows you to set the target volume (cut or fill) and the variation you will accept. For example, you can target the balance to be 100 yards of fill, but the routine will accept a variation of 30 yards above or below the target (70–130 yards). The routine displays a dialog box that allows you to set the target and variance factor (see Figure 18–16).

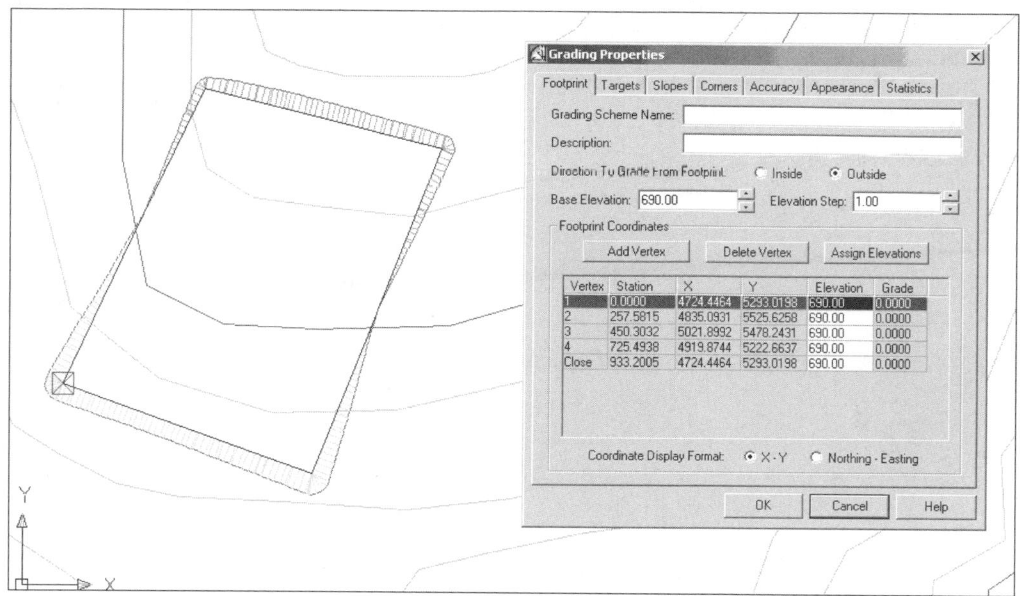

Figure 18–15

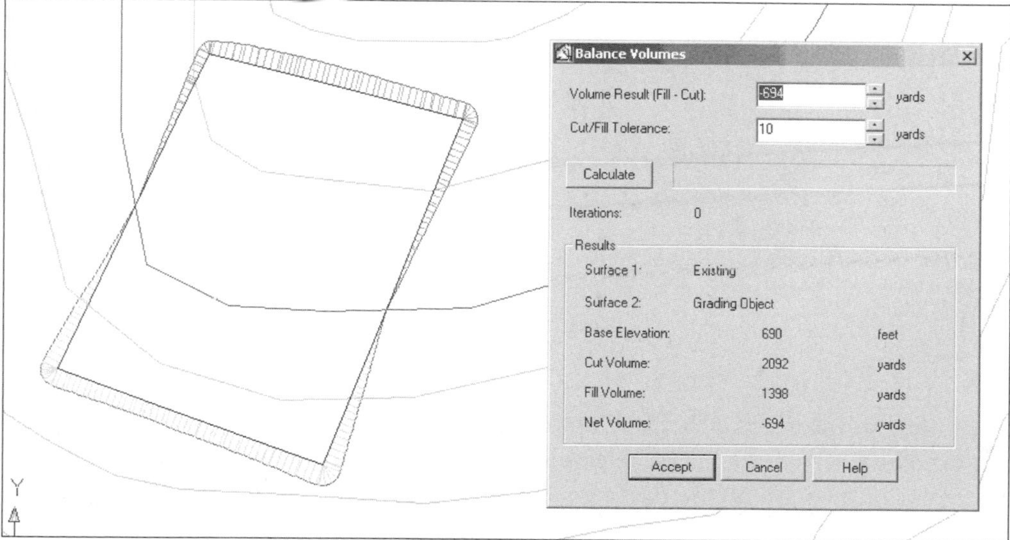

Figure 18–16

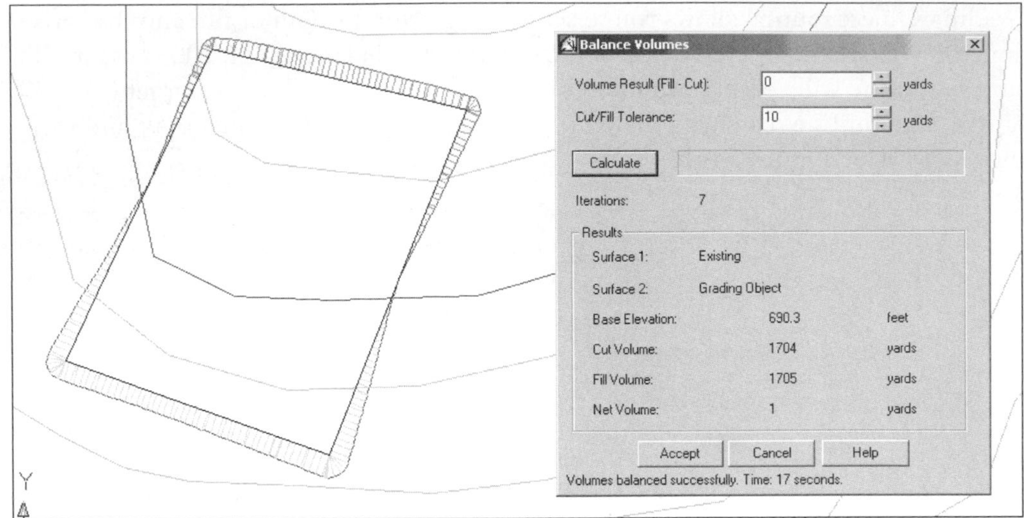

Figure 18–17

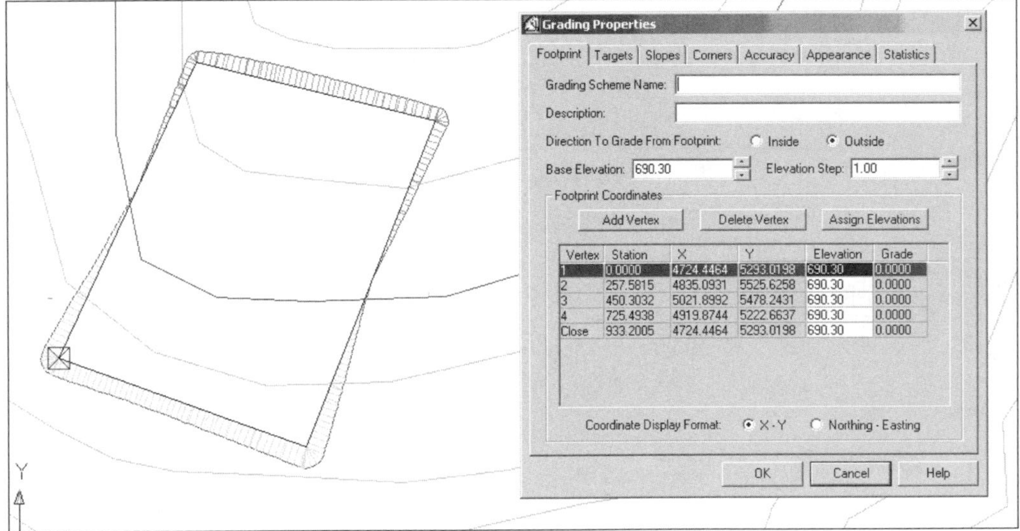

Vertex	Station	X	Y	Elevation	Grade
1	0.0000	4724.4464	5293.0198	690.30	0.0000
2	257.5815	4835.0931	5525.6258	690.30	0.0000
3	450.3032	5021.8992	5478.2431	690.30	0.0000
4	725.4938	4919.8744	5222.6637	690.30	0.0000
Close	933.2005	4724.4464	5293.0198	690.30	0.0000

Figure 18–18

The rolling releases for Land Development Desktop 2 provide the program with new and extended capabilities. The release of Land Development Desktop 2i continues the evolutionary path of the program. The enhancements to the program are a response to the requests of the users and the software designers of the Civil Design

Team. I strongly urge you to visit the Web page for Land Development Desktop at the Autodesk products site (www.autodesk.com) and enter your wish list items.

The Land Development Desk program and its associated products, Civil Design and Survey, provide you with a set of tools to tackle most any civil engineering design task. Land Development Desktop provides the base for creating and viewing data in the form of a drawing. The drawing views vectors, points, annotation, roadway centerlines, parcels, and surfaces that you create into a deliverable, a plotted drawing. Each of the data types is the basis for a design or a result of a survey.

The civil design program takes this basic data and allows you to create grading, site, and roadway designs. These designs use the grading, profile, cross section, hydrology, and piping tools.

The survey program allows you to create the points that define the lines and surfaces you encounter in a project when you start. Survey is also the program that allows you to place your design back into the ground after you create it in the drawing.

Land Development Desktop is a powerful program. It will evolve through the extension and major release cycles. Autodesk is mindful of the limitations of the program and will in the future address the needs of the user, you. You must understand that Land Development Desktop is a set of tools and as such is not a program that is aware of all possible situations.

INDEX

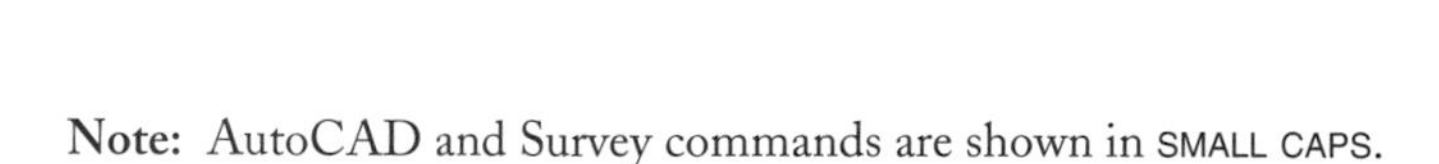

Note: AutoCAD and Survey commands are shown in SMALL CAPS.

B

C

E

F

G

J

L

M

S

U

V

W

Z

LICENSE AGREEMENT FOR AUTODESK PRESS

THOMSON LEARNING™

Educational Software/Data

You the customer, and Autodesk Press incur certain benefits, rights, and obligations to each other when you open this package and use the software/data it contains. BE SURE YOU READ THE LICENSE AGREEMENT CAREFULLY, SINCE BY USING THE SOFTWARE/DATA YOU INDICATE YOU HAVE READ, UNDERSTOOD, AND ACCEPTED THE TERMS OF THIS AGREEMENT.

Your rights:

1. You enjoy a non-exclusive license to use the enclosed software/data on a single microcomputer that is not part of a network or multi-machine system in consideration for payment of the required license fee, (which may be included in the purchase price of an accompanying print component), or receipt of this software/data, and your acceptance of the terms and conditions of this agreement.
2. You own the media on which the software/data is recorded, but you acknowledge that you do not own the software/data recorded on them. You also acknowledge that the software/data is furnished "as is," and contains copyrighted and/or proprietary and confidential information of Autodesk Press or its licensors.
3. If you do not accept the terms of this license agreement you may return the media within 30 days. However, you may not use the software during this period.

There are limitations on your rights:

1. You may not copy or print the software/data for any reason whatsoever, except to install it on a hard drive on a single microcomputer and to make one archival copy, unless copying or printing is expressly permitted in writing or statements recorded on the diskette(s).
2. You may not revise, translate, convert, disassemble or otherwise reverse engineer the software/data except that you may add to or rearrange any data recorded on the media as part of the normal use of the software/data.
3. You may not sell, license, lease, rent, loan, or otherwise distribute or network the software/data except that you may give the software/data to a student or and instructor for use at school or, temporarily at home.

Should you fail to abide by the Copyright Law of the United States as it applies to this software/data your license to use it will become invalid. You agree to erase or otherwise destroy the software/data immediately after receiving note of Autodesk Press' termination of this agreement for violation of its provisions.

Autodesk Press gives you a LIMITED WARRANTY covering the enclosed software/data. The LIMITED WARRANTY can be found in this product and/or the instructor's manual that accompanies it.

This license is the entire agreement between you and Autodesk Press interpreted and enforced under New York law.

Limited Warranty

Autodesk Press warrants to the original licensee/ purchaser of this copy of microcomputer software/ data and the media on which it is recorded that the media will be free from defects in material and workmanship for ninety (90) days from the date of original purchase. All implied warranties are limited in duration to this ninety (90) day period. THEREAFTER, ANY IMPLIED WARRANTIES, INCLUDING IMPLIED WARRANTIES OF MERCHANTABILITY AND FITNESS FOR A PARTICULAR PURPOSE ARE EXCLUDED. THIS WARRANTY IS IN LIEU OF ALL OTHER WARRANTIES, WHETHER ORAL OR WRITTEN, EXPRESSED OR IMPLIED.

If you believe the media is defective, please return it during the ninety day period to the address shown below. A defective diskette will be replaced without charge provided that it has not been subjected to misuse or damage.

This warranty does not extend to the software or information recorded on the media. The software and information are provided "AS IS." Any statements made about the utility of the software or information are not to be considered as express or implied warranties. Autodesk Press will not be liable for incidental or consequential damages of any kind incurred by you, the consumer, or any other user.

Some states do not allow the exclusion or limitation of incidental or consequential damages, or limitations on the duration of implied warranties, so the above limitation or exclusion may not apply to you. This warranty gives you specific legal rights, and you may also have other rights which vary from state to state. Address all correspondence to:

Autodesk Press
3 Columbia Circle
P. O. Box 15015
Albany, NY 12212-5015